Hodge Theory

Hodge Theory

Edited by

Eduardo Cattani
Fouad El Zein
Phillip A. Griffiths
Lê Dũng Tráng

Mathematical Notes 49

PRINCETON UNIVERSITY PRESS
PRINCETON AND OXFORD

Published by Princeton University Press
41 William Street, Princeton, New Jersey 08540

In the United Kingdom: Princeton University Press
6 Oxford Street, Woodstock, Oxfordshire, OX20 1TW

press.princeton.edu

Library of Congress Control Number 2013953395
ISBN 978-0-691-16134-1

British Library Cataloging-in-Publication Data is available

The publisher would like to acknowledge the authors of this volume for providing the camera-ready copy from which this book was printed.

This book has been composed in LaTeX

10 9 8 7 6 5 4 3 2 1

Contributors

Patrick Brosnan
Department of Mathematics
University of Maryland
College Park, MD 20742-4015, USA
pbrosnan@umd.edu

James Carlson
25 Murray Street, Apartment 7G
New York City, NY 10007-2361, USA
jxxcarlson@icloud.com

Eduardo Cattani
Department of Mathematics and Statistics
University of Massachusetts Amherst
Amherst, MA 01003-9305, USA
cattani@math.umass.edu

François Charles
Université Paris-Sud
Département de Mathématiques
Bâtiment 425
91405 Orsay Cedex, France
francois.charles@math.u-psud.fr

Mark Andrea de Cataldo
Department of Mathematics
Stony Brook University
Stony Brook, NY 11794-3651, USA

Fouad El Zein
Institut de Mathématiques du Jussieu
175 rue du Chevaleret
75013 Paris Cedex, France
elzein@math.jussieu.fr

Mark L. Green
Department of Mathematics
UCLA
Los Angeles, CA 90095-1555, USA
mlg@math.ucla.edu

Phillip A. Griffiths
Institute for Advanced Study
Einstein Drive
Princeton, NJ 08540-4907, USA
pg@ias.edu

Matt Kerr
Department of Mathematics, Box 1146
Washington University in St. Louis
St. Louis, MO 63130-4899, USA
matkerr@math.wustl.edu

Lê Dũng Tráng
Université de Aix Marseille
Centre de Mathématiques et Informatique
39, rue Joliot Curie
13453 Marseille Cedex 13, France
ledt@ictp.it

Luca Migliorini
Dipartimento di Matematica
Università di Bologna
Piazza di Porta S. Donato, 5
Bologna, Italy
luca.migliorini@unibo.it

Jacob P. Murre
Department of Mathematics
University of Leiden
P. O. Box 9512
2300 RA Leiden, The Netherlands
murre@math.leidenuniv.nl

Christian Schnell
Department of Mathematics
Stony Brook University
Stony Brook, NY 11794-3651, USA
christian.schnell@stonybrook.edu

Loring W. Tu
Department of Mathematics
Tufts University
Medford, MA 02155, USA
loring.tu@tufts.edu

Contents

Preface

Between 14 June and 2 July 2010, the *Summer School on Hodge Theory and Related Topics* and a related conference were hosted by the ICTP in Trieste, Italy. The organizers of the conference were E. Cattani, F. El Zein, P. Griffiths, Lê D. T., and L. Göttsche. Attending the summer school were a large and diverse group of young mathematicians, including students, many of whom were from countries in the developing world. The conference brought together leading active researchers in Hodge theory and related areas.

In the summer school, the lectures were intended to provide an introduction to Hodge theory, a subject with deep historical roots and one of the most vibrant areas in contemporary mathematics. Although there are a number of excellent texts treating various aspects of Hodge theory, the subject remains quite difficult to learn due in part to the breadth of background needed. An objective of the summer school was to give an exposition of basic aspects of the subject in an accessible framework and language, while presenting—in both the school and the conference—selected topics in Hodge theory that are at the forefront of current research. These dual goals are reflected in the contents of this volume, many of whose entries do in fact present both the fundamentals and the current state of the topics covered.

The lectures by Eduardo Cattani on Kähler manifolds provide a lucid and succinct account of the basic geometric objects that give rise to a Hodge structure in cohomology. Smooth projective complex algebraic varieties are Kähler manifolds, and the Hodge structures on their cohomology provide an extremely rich set of invariants of the variety. Moreover, they are the fundamental building blocks for the mixed Hodge structure in the cohomology of general complex algebraic varieties. The realization of cohomology by harmonic forms provides the connection between analysis and geometry, and the subtle implications of the Kähler condition on harmonic forms are explained in these lectures.

In modern Hodge theory, one of the most basic tools is the algebraic de Rham theorem as formulated by Grothendieck. This result, whose origins date from the work of Picard and Poincaré, relates the complex analytic and algebraic approaches to the topology of algebraic varieties via differential forms. The contribution to the volume by Fouad El Zein and Loring Tu presents a new treatment of this subject, from the basic sheaf-theoretic formulation of the classical de Rham theorem through the statement and proof of the final result.

Mixed Hodge structures are the subject of the lectures by El Zein and Lê Dung Trang. According to Deligne, the cohomology of a general complex algebraic variety carries a functorial mixed Hodge structure, one that is built up from the pure Hodge

structures on smooth projective varieties which arise when the original variety is com-
pactified and its singularities are resolved. This is a far-reaching, deep, and complex
topic, one whose essential aspects are addressed with clarity in their presentation.

The subject of period domains and period maps, covered in the lectures by Jim Carlson, involves Lie-theoretic and differential geometric aspects of Hodge theory. Among
the items discussed is the topic of infinitesimal period relations (transversality) and its
curvature implications, which are fundamental to the analysis of the limiting mixed
Hodge structures that reflect the behavior of the Hodge structures on the cohomology
of smooth varieties as those varieties acquire singularities.

The lectures by Luca Migliorini and Mark de Cataldo on the Hodge theory of maps
deal with the Hodge-theoretic aspects of arbitrary proper maps between general algebraic varieties. The fundamental result here, the decomposition theorem, shows how
variations of mixed Hodge structures combine with intersection cohomology to describe the deep Hodge-theoretic properties of the above maps. This subject is complementary to the lectures by Patrick Brosnan and El Zein on variations of mixed Hodge
structures; between the two, they provide the framework for Morihiko Saito's unifying
theory of mixed Hodge modules.

One of the very basic aspects of Hodge theory is the analysis of how the Hodge
structure associated to a smooth algebraic variety varies when the algebraic variety
degenerates to a singular one in an algebraic family. The fundamental concept here
is that of a limiting mixed Hodge structure. The analysis of how Hodge structures
degenerate is given in the lectures by Eduardo Cattani, a presentation that covers the
subject from its origins up through many of its most recent and deepest aspects.

The subject of the lectures by Brosnan and El Zein is variations of mixed Hodge
structures. This topic brings together the material from Cattani's lectures on variation
of Hodge structure and those by El Zein and Lê on mixed Hodge structures. As the
title suggests, it describes how the mixed Hodge structures vary in an arbitrary family
of algebraic varieties. The treatment given here presents a clear and efficient account
of this central aspect of Hodge theory.

One of the original purposes of Hodge theory was to understand the geometry of
algebraic varieties through the study of the algebraic subvarieties that lie in it. A central
theme is the interplay between Hodge theory and the Chow groups, and the conjectures
of Hodge and of Bloch–Beilinson serve to frame much of the current work in this area.
These topics are covered, from the basic definitions to the forefront of current research,
in the five lectures by Jacob Murre. The lectures by Mark Green are a continuation of
the topic of algebraic cycles as introduced in the lectures of Murre. Hodge theory again
appears because it provides—conjecturally—the basic invariants and resulting structure
of the group of cycles. In these lectures, the Hodge theory associated to a new algebraic
variety, whose function field is generated by the coefficients in the defining equation of
the original variety, is introduced as a method for studying algebraic cycles, one that
conjecturally reduces the basic open questions to a smaller and more fundamental set.
It also suggests how one may reduce many questions, such as the Hodge conjecture, to
the case of varieties defined over number fields.

A principal unsolved problem in Hodge theory is the Hodge conjecture, which
provides a Hodge-theoretic characterization of the fundamental cohomology classes

supported by algebraic cycles. For an algebraic variety defined over an abstract field, the Hodge conjecture implies that the Hodge-theoretic criterion for a cohomology class to support an algebraic cycle is independent of the embedding of that field into the complex numbers. Hodge classes with this property are called "absolute." The basic known result here is due to Deligne: it states that all Hodge classes on abelian varieties are absolute. This theorem relates the arithmetic and complex analytic aspects of these varieties, and a self-contained proof of it is presented in the article by François Charles and Christian Schnell.

Arithmetic automorphic representation theory is one of the most active areas in current mathematical research, centering around what is known as the Langlands program. The basic methods for studying the subject are the trace formula and the theory of Shimura varieties. The latter are algebraic varieties that arise from Hodge structures of weight 1; their study relates the analytic and arithmetic aspects of the algebraic varieties in question. The lectures by Matt Kerr provide a Hodge-theoretic approach to the study of Shimura varieties, one that is complementary to the standard, more algebraic, presentations of the subject.

The Summer School on Hodge Theory would not have been possible without the support from the Abdus Salam International Center of Theoretical Physics in Trieste, Italy. We are grateful for the hospitality of the entire staff and wish to thank, in particular, Lothar Göttsche, Ramadas Ramakrishnan, and Mabilo Koutou for their organizational help. The participation of graduate students and junior researchers from the USA was made possible by a grant from the National Science Foundation (NSF 1001125). We are also grateful to the Clay Mathematics Foundation for their generous support.

Chapter One

Introduction to Kähler Manifolds
by Eduardo Cattani

INTRODUCTION

This chapter is intended to provide an introduction to the basic results on the topology of compact Kähler manifolds that underlie and motivate Hodge theory. Although we have tried to define carefully the main objects of study, we often refer to the literature for proofs of the main results. We are fortunate in that there are several excellent books on this subject and we have freely drawn from them in the preparation of these notes, which make no claim of originality. The classical references remain the pioneering books by Weil [34], Chern [6], Morrow and Kodaira [17, 19], Wells [35], Kobayashi [15], Demailly [8], and Griffiths and Harris [10]. In these notes we refer most often to two superb recent additions to the literature: Voisin's two-volume work [30, 31] and Huybrechts' book [13].

We assume from the outset that the reader is familiar with the basic theory of smooth manifolds at the level of [1], [18], or [28]. The book by Bott and Tu [2] is an excellent introduction to the algebraic topology of smooth manifolds.

This chapter consists of five sections which correspond, roughly, to the five lectures in the course given during the Summer School at ICTP. There are also two appendices. The first collects some results on the linear algebra of complex vector spaces, Hodge structures, nilpotent linear transformations, and representations of $\mathfrak{sl}(2, \mathbb{C})$ and serves as an introduction to many other chapters in this volume. The second is due to Phillip Griffiths and contains a new proof of the Kähler identities by reduction to the symplectic case.

There are many exercises interspersed throughout the text, many of which ask the reader to prove or complete the proof of some result in the notes.

I am grateful to Loring Tu for his careful reading of this chapter.

1.1 COMPLEX MANIFOLDS

1.1.1 Definition and Examples

Let $U \subset \mathbb{C}^n$ be an open subset and $f \colon U \to \mathbb{C}$. We say that f is *holomorphic* if and only if it is holomorphic as a function of each variable separately; i.e., if we fix $z_\ell = a_\ell$, $\ell \neq j$, then the function $f(a_1, \ldots, z_j, \ldots, a_n)$ is a holomorphic function of z_j. A map $F = (f_1, \ldots, f_n) \colon U \to \mathbb{C}^n$ is said to be *holomorphic* if each component $f_k = f_k(z_1, \ldots, z_n)$ is holomorphic. If we identify $\mathbb{C}^n \cong \mathbb{R}^{2n}$, and set $z_j = x_j + iy_j$, $f_k = u_k + iv_k$, $j, k = 1, \ldots, n$, then the functions u_k, v_k are C^∞ functions of the variables $x_1, y_1, \ldots, x_n, y_n$ and satisfy the Cauchy–Riemann equations:

$$\frac{\partial u_k}{\partial x_j} = \frac{\partial v_k}{\partial y_j} \; ; \quad \frac{\partial u_k}{\partial y_j} = -\frac{\partial v_k}{\partial x_j}. \tag{1.1.1}$$

Conversely, if $(u_1, v_1, \ldots, u_n, v_n) \colon \mathbb{R}^{2n} \to \mathbb{R}^{2n}$ is a C^∞ map satisfying the Cauchy–Riemann equations (1.1.1), then the map $(u_1 + iv_1, \ldots, u_n + iv_n)$ is holomorphic. In other words, a C^∞ map $F \colon U \subset \mathbb{R}^{2n} \to \mathbb{R}^{2n}$ defines a holomorphic map $\mathbb{C}^n \to \mathbb{C}^n$ if and only if the differential DF of F, written in terms of the basis

$$\{\partial/\partial x_1, \ldots, \partial/\partial x_n, \partial/\partial y_1, \ldots, \partial/\partial y_n\} \tag{1.1.2}$$

of the tangent space $T_p(\mathbb{R}^{2n})$ and the basis $\{\partial/\partial u_1, \ldots, \partial/\partial u_n, \partial/\partial v_1, \ldots, \partial/\partial v_n\}$ of $T_{F(p)}(\mathbb{R}^{2n})$ is of the form

$$DF(p) = \begin{pmatrix} A & -B \\ B & A \end{pmatrix} \tag{1.1.3}$$

for all $p \in U$. Thus, it follows from Exercise A.1.4 in Appendix A that F is holomorphic if and only if $DF(p)$ defines a $\mathbb{C}$-linear map $\mathbb{C}^n \to \mathbb{C}^n$.

EXERCISE 1.1.1 Prove that a $2n \times 2n$ matrix is of the form (1.1.3) if and only if it commutes with the matrix J:

$$J := \begin{pmatrix} 0 & -I_n \\ I_n & 0 \end{pmatrix}, \tag{1.1.4}$$

where I_n denotes the $n \times n$ identity matrix.

DEFINITION 1.1.2 *A **complex structure** on a topological manifold M consists of a collection of coordinate charts (U_α, ϕ_α) satisfying the following conditions:*

(1) The sets U_α form an open covering of M.

(2) There is an integer n such that each $\phi_\alpha \colon U_\alpha \to \mathbb{C}^n$ is a homeomorphism of U_α onto an open subset of $\mathbb{C}^n$. We call n the complex dimension of M.

(3) If $U_\alpha \cap U_\beta \neq \emptyset$, the map

$$\phi_\beta \circ \phi_\alpha^{-1} \colon \phi_\alpha(U_\alpha \cap U_\beta) \to \phi_\beta(U_\alpha \cap U_\beta) \tag{1.1.5}$$

is holomorphic.

EXAMPLE 1.1.3 The simplest example of a complex manifold is $\mathbb{C}^n$ or any open subset of $\mathbb{C}^n$. For any $p \in \mathbb{C}^n$, the tangent space $T_p(\mathbb{C}^n) \cong \mathbb{R}^{2n}$ is identified, in the natural way, with $\mathbb{C}^n$ itself.

EXAMPLE 1.1.4 Since $\mathrm{GL}(n, \mathbb{C})$, the set of nonsingular $n \times n$ matrices with complex coefficients, is an open set in $\mathbb{C}^{n^2}$, we may view $\mathrm{GL}(n, \mathbb{C})$ as a complex manifold.

EXAMPLE 1.1.5 The basic example of a compact complex manifold is complex projective space which we will denote by $\mathbb{P}^n$. Recall that

$$\mathbb{P}^n := \left(\mathbb{C}^{n+1} \setminus \{0\} \right) / \mathbb{C}^*,$$

where $\mathbb{C}^*$ acts by componentwise multiplication. Given $z \in \mathbb{C}^{n+1} \setminus \{0\}$, let $[z]$ be its equivalence class in $\mathbb{P}^n$. The sets

$$U_i := \{[z] \in \mathbb{P}^n : z_i \neq 0\} \tag{1.1.6}$$

are open and the maps

$$\phi_i \colon U_i \to \mathbb{C}^n ; \quad \phi_i([z]) = \left(\frac{z_0}{z_i}, \ldots, \frac{z_{i-1}}{z_i}, \frac{z_{i+1}}{z_i}, \ldots, \frac{z_n}{z_i} \right) \tag{1.1.7}$$

define local coordinates such that the maps

$$\phi_i \circ \phi_j^{-1} \colon \phi_j(U_i \cap U_j) \to \phi_i(U_i \cap U_j) \tag{1.1.8}$$

are holomorphic.

In particular, if $n = 1$, $\mathbb{P}^1$ is covered by two coordinate neighborhoods (U_0, ϕ_0), (U_1, ϕ_1) with $\phi_i(U_i) = \mathbb{C}$. The coordinate change $\phi_1 \circ \phi_0^{-1} \colon \mathbb{C}^* \to \mathbb{C}^*$ is given by

$$\phi_1 \circ \phi_0^{-1}(z) = \phi_1([(1, z)]) = \frac{1}{z}.$$

Thus, this is the usual presentation of the sphere S^2 as the Riemann sphere, where we identify U_0 with $\mathbb{C}$ and denote the point $[(0, 1)]$ by ∞.

EXERCISE 1.1.6 Verify that the map (1.1.8) is holomorphic.

EXAMPLE 1.1.7 To each point $[z] \in \mathbb{P}^n$ we may associate the line spanned by z in $\mathbb{C}^{n+1}$; hence, we may regard $\mathbb{P}^n$ as the space of lines through the origin in $\mathbb{C}^{n+1}$. This construction may then be generalized by considering k-dimensional subspaces in $\mathbb{C}^n$. In this way one obtains the Grassmann manifold $\mathcal{G}(k, n)$. To define a complex manifold structure on $\mathcal{G}(k, n)$, we consider first of all the open set in $\mathbb{C}^{nk}$,

$$V(k, n) = \{W \in \mathcal{M}(n \times k, \mathbb{C}) : \mathrm{rank}(W) = k\}.$$

The Grassmann manifold $\mathcal{G}(k, n)$ may then be viewed as the quotient space

$$\mathcal{G}(k, n) := V(k, n) / \mathrm{GL}(k, \mathbb{C}),$$

where $GL(k, \mathbb{C})$ acts by right multiplication. Thus, $W, W' \in V(k, n)$ are in the same $GL(k, \mathbb{C})$-orbit if and only if the column vectors of W and W' span the same k-dimensional subspace $\Omega \subset \mathbb{C}^n$.

Given an index set $I = \{1 \le i_1 < \cdots < i_k \le n\}$ and $W \in V(k, n)$, we consider the $k \times k$ matrix W_I consisting of the I-rows of W and note that if $W \sim W'$ then, for every index set I, $\det(W_I) \ne 0$ if and only if $\det(W'_I) \ne 0$. We then define

$$U_I := \{[W] \in \mathcal{G}(k, n) : \det(W_I) \ne 0\}. \tag{1.1.9}$$

This is clearly an open set in $\mathcal{G}(k, n)$ and the map

$$\phi_I \colon U_I \to \mathbb{C}^{(n-k)k} ; \quad \phi_I([W]) = W_{I^c} \cdot W_I^{-1},$$

where I^c denotes the $(n-k)$-tuple of indices complementary to I. The map ϕ_I defines coordinates in U_I and one can easily verify that given index sets I and J, the maps

$$\phi_I \circ \phi_J^{-1} \colon \phi_J(U_I \cap U_J) \to \phi_I(U_I \cap U_J) \tag{1.1.10}$$

are holomorphic.

EXERCISE 1.1.8 Verify that the map (1.1.10) is holomorphic.

EXERCISE 1.1.9 Prove that both $\mathbb{P}^n$ and $\mathcal{G}(k, n)$ are compact.

The notion of a holomorphic map between complex manifolds is defined in a way completely analogous to that of a smooth map between C^∞ manifolds; i.e., if M and N are complex manifolds of dimension m and n respectively, a map $F \colon M \to N$ is said to be *holomorphic* if for each $p \in M$ there exist local coordinate systems (U, ϕ), (V, ψ) around p and $q = F(p)$, respectively, such that $F(U) \subset V$ and the map

$$\psi \circ F \circ \phi^{-1} \colon \phi(U) \subset \mathbb{C}^m \to \psi(V) \subset \mathbb{C}^n$$

is holomorphic. Given an open set $U \subset M$ we will denote by $\mathcal{O}(U)$ the ring of holomorphic functions $f \colon U \to \mathbb{C}$ and by $\mathcal{O}^*(U)$ the nowhere-zero holomorphic functions on U. A map between complex manifolds is said to be *biholomorphic* if it is holomorphic and has a holomorphic inverse.

The following result shows a striking difference between C^∞ and complex manifolds:

THEOREM 1.1.10 *If M is a compact, connected, complex manifold and $f \colon M \to \mathbb{C}$ is holomorphic, then f is constant.*

PROOF. The proof uses the fact that the maximum principle[1] holds for holomorphic functions of several complex variables (cf. [30, Theorem 1.21]) as well as the principle of analytic continuation[2] [30, Theorem 1.22]. □

---

[1] If $f \in \mathcal{O}(U)$, where $U \subset \mathbb{C}^n$ is open, has a local maximum at $p \in U$, then f is constant in a neighborhood of p.

[2] If $U \subset \mathbb{C}^n$ is a connected, open subset and $f \in \mathcal{O}(U)$ is constant on an open subset $V \subset U$, then f is constant on U.

Given a holomorphic map $F = (f_1, \ldots, f_n) \colon U \subset \mathbb{C}^n \to \mathbb{C}^n$ and $p \in U$, we can associate to F the $\mathbb{C}$-linear map

$$DF(p) \colon \mathbb{C}^n \to \mathbb{C}^n \, ; \quad DF(p)(v) = \left(\frac{\partial f_i}{\partial z_j}(p) \right) \cdot v,$$

where v is the column vector $(v_1, \ldots, v_n)^T \in \mathbb{C}^n$. The Cauchy–Riemann equations imply that if we regard F as a smooth map $\tilde{F} \colon U \subset \mathbb{R}^{2n} \to \mathbb{R}^{2n}$ then the matrix of the differential $D\tilde{F}(p)$ is of the form (1.1.3) and, clearly, $DF(p)$ is nonsingular if and only if $D\tilde{F}(p)$ is nonsingular. In that case, by the inverse function theorem, $\tilde{F}$ has a local inverse $\tilde{G}$ whose differential is given by $(D\tilde{F}(p))^{-1}$. By Exercise 1.1.1, the inverse of a nonsingular matrix of the form (1.1.3) is of the same form. Hence, it follows that $\tilde{G}$ is holomorphic and, consequently, F has a local holomorphic inverse. Thus we have:

THEOREM 1.1.11 (Holomorphic inverse function theorem) *Let $F \colon U \to V$ be a holomorphic map between open subsets $U, V \subset \mathbb{C}^n$. If $DF(p)$ is nonsingular for $p \in U$ then there exist open sets U', V' such that $p \in U' \subset U$ and $F(p) \in V' \subset V$ and such that $F \colon U' \to V'$ is a biholomorphic map.*

The fact that we have a holomorphic version of the inverse function theorem means that we may also extend the implicit function theorem or, more generally, the rank theorem:

THEOREM 1.1.12 (Rank theorem) *Let $F \colon U \to V$ be a holomorphic map between open subsets $U \subset \mathbb{C}^n$ and $V \subset \mathbb{C}^m$. If $DF(q)$ has rank k for all $q \in U$ then, given $p \in U$, there exist open sets U', V' such that $p \in U' \subset U$, $F(p) \in V' \subset V$, $F(U') \subset V'$, and biholomorphic maps $\phi \colon U' \to A$, $\psi \colon V' \to B$, where A and B are open sets of the origin in $\mathbb{C}^n$ and $\mathbb{C}^m$, respectively, so that the composition*

$$\psi \circ F \circ \phi^{-1} \colon A \to B$$

is the map $(z_1, \ldots, z_n) \in A \mapsto (z_1, \ldots, z_k, 0, \ldots, 0)$.

PROOF. We refer to [1, Theorem 7.1] or [28] for a proof in the C^∞ case which can easily be generalized to the holomorphic case.   □

Given a holomorphic map $F \colon M \to N$ between complex manifolds and $p \in M$, we may define the rank of F at p as

$$\operatorname{rank}_p(F) := \operatorname{rank}(D(\psi \circ F \circ \phi^{-1})(\phi(p))), \qquad (1.1.11)$$

for any local-coordinates expression of F around p.

EXERCISE 1.1.13 Prove that $\operatorname{rank}_p(F)$ is well defined by (1.1.11); i.e., it is independent of the choices of local coordinates.

We then have the following consequence of the rank theorem:

THEOREM 1.1.14   *Let $F: M \to N$ be a holomorphic map, let $q \in F(M)$ and let $X = F^{-1}(q)$. Suppose $\operatorname{rank}_x(F) = k$ for all x in an open set U containing X. Then, X is a complex manifold and*

$$\operatorname{codim}(X) := \dim M - \dim X = k.$$

PROOF.   The rank theorem implies that given $p \in X$ there exist local coordinates (U, ϕ) and (V, ψ) around p and q, respectively, such that $\psi(q) = 0$ and

$$\psi \circ F \circ \phi^{-1}(z_1, \dots, z_n) = (z_1, \dots, z_k, 0, \dots, 0).$$

Hence

$$\phi(U \cap X) = \{z \in \phi(U) : z_1 = \cdots = z_k = 0\}.$$

Hence $(U \cap X, p \circ \phi)$, where p denotes the projection onto the last $n - k$ coordinates in $\mathbb{C}^n$, defines local coordinates on X. It is easy to check that these coordinates are holomorphically compatible.   □

DEFINITION 1.1.15   *We will say that $N \subset M$ is a **complex submanifold** if we may cover M with coordinate patches (U_α, ϕ_α) such that*

$$\phi_\alpha(N \cap U_\alpha) = \{z \in \phi_\alpha(U) : z_1 = \cdots = z_k = 0\},$$

for some fixed k. In this case, as we saw above, N has the structure of an $(n - k)$-dimensional complex manifold.

PROPOSITION 1.1.16   *There are no compact complex submanifolds of $\mathbb{C}^n$ of dimension greater than 0.*

PROOF.   Suppose $M \subset \mathbb{C}^n$ is a submanifold. Then, each of the coordinate functions z_i restricts to a holomorphic function on M. But, if M is compact, it follows from Theorem 1.1.10 that z_i must be locally constant. Hence, $\dim M = 0$.   □

Remark.   The above result means that there is no chance for a Whitney embedding theorem in the holomorphic category. One of the major results of the theory of complex manifolds is the Kodaira embedding theorem (Theorem 1.3.14) which gives necessary and sufficient conditions for a compact complex manifold to embed in $\mathbb{P}^n$.

EXAMPLE 1.1.17   Let $f: \mathbb{C}^n \to \mathbb{C}$ be a holomorphic function and suppose $Z = f^{-1}(0) \neq \emptyset$. Then we say that 0 is a *regular value* for f if $\operatorname{rank}_p(f) = 1$ for all $p \in Z$; i.e., for each $p \in X$ there exists some $i = 1, \dots, n$, such that $\partial f / \partial z_i(p) \neq 0$. In this case, Z is a complex submanifold of $\mathbb{C}^n$ and $\operatorname{codim}(Z) = 1$. We call Z an *affine hypersurface*. More generally, given $F: \mathbb{C}^n \to \mathbb{C}^m$, we say that 0 is a *regular value* if $\operatorname{rank}_p(F) = m$ for all $p \in F^{-1}(0)$. In this case $F^{-1}(0)$ is either empty or is a submanifold of $\mathbb{C}^n$ of codimension m.

EXAMPLE 1.1.18 Let $P(z_0, \ldots, z_n)$ be a homogeneous polynomial of degree d. We set

$$X := \{[z] \in \mathbb{P}^n : P(z_0, \ldots, z_n) = 0\}.$$

We note that while P does not define a function on $\mathbb{P}^n$, the zero locus X is still well defined since P is a homogeneous polynomial. We assume now that the following regularity condition holds:

$$\left\{ z \in \mathbb{C}^{n+1} : \frac{\partial P}{\partial z_0}(z) = \cdots = \frac{\partial P}{\partial z_n}(z) = 0 \right\} = \{0\}; \qquad (1.1.12)$$

i.e., 0 is a regular value of the map $P|_{\mathbb{C}^{n+1}\setminus\{0\}}$. Then X is a hypersurface in $\mathbb{P}^n$.

To prove this we note that the requirements of Definition 1.1.15 are local. Hence, it is enough to check that $X \cap U_i$ is a submanifold of U_i for each i, in fact, that it is an affine hypersurface. Consider the case $i = 0$ and let $f: U_0 \cong \mathbb{C}^n \to \mathbb{C}$ be the function $f(u_1, \ldots, u_n) = P(1, u_1, \ldots, u_n)$. Set $u = (u_1, \ldots, u_n)$ and $\tilde{u} = (1, u_1, \ldots, u_n)$. Suppose $[\tilde{u}] \in U_0 \cap X$ and

$$\frac{\partial f}{\partial u_1}(u) = \cdots = \frac{\partial f}{\partial u_n}(u) = 0.$$

Then

$$\frac{\partial P}{\partial z_1}(\tilde{u}) = \cdots = \frac{\partial P}{\partial z_n}(\tilde{u}) = 0.$$

But, since P is a homogeneous polynomial of degree d, it follows from the Euler identity that

$$0 = d \cdot P(\tilde{u}) = \frac{\partial P}{\partial z_0}(\tilde{u}).$$

Hence, by (1.1.12), we would have $\tilde{u} = 0$, which is impossible. Hence 0 is a regular value of f and $X \cap U_0$ is an affine hypersurface.

EXERCISE 1.1.19 Let $P_1(z_0, \ldots, z_n), \ldots, P_m(z_0, \ldots, z_n)$ be homogeneous polynomials. Suppose that 0 is a regular value of the map

$$(P_1, \ldots, P_m): \mathbb{C}^{n+1}\setminus\{0\} \to \mathbb{C}^m.$$

Prove that

$$X = \{[z] \in \mathbb{P}^n : P_1([z]) = \cdots = P_m([z]) = 0\}$$

is a codimension-m submanifold of $\mathbb{P}^n$. X is called a complete intersection submanifold.

EXAMPLE 1.1.20 Consider the Grassmann manifold $\mathcal{G}(k, n)$ and let $I_1, \ldots, I_{\binom{n}{k}}$ denote all strictly increasing k-tuples $I \subset \{1, \ldots, n\}$. We then define

$$\mathfrak{p}: \mathcal{G}(k, n) \to \mathbb{P}^{N-1}; \quad \mathfrak{p}([W]) = [(\det(W_{I_1}), \ldots, \det(W_{I_N}))].$$

Note that the map $\mathfrak{p}$ is well defined since $W \sim W'$ implies that $W' = W \cdot M$ with $M \in \mathrm{GL}(k, \mathbb{C})$, and then for any index set I, $\det(W'_I) = \det(M) \det(W_I)$. We leave it to the reader to verify that the map $\mathfrak{p}$, which is usually called the *Plücker map*, is holomorphic.

EXERCISE 1.1.21 Consider the Plücker map $\mathfrak{p} \colon \mathcal{G}(2,4) \to \mathbb{P}^5$ and suppose that the index sets $I_1, \ldots, I_6$ are ordered lexicographically. Show that $\mathfrak{p}$ is a 1:1 holomorphic map from $\mathcal{G}(2,4)$ onto the subset

$$X = \{[z_0, \ldots, z_5] : z_0 z_5 - z_1 z_4 + z_2 z_3 = 0\}. \tag{1.1.13}$$

Prove that X is a hypersurface in $\mathbb{P}^5$. Compute $\mathrm{rank}_{[W]} \, \mathfrak{p}$ for $[W] \in \mathcal{G}(2,4)$.

EXAMPLE 1.1.22 We may define complex Lie groups in a manner completely analogous to the real, smooth case. A *complex Lie group* is a complex manifold G with a group structure such that the group operations are holomorphic. The basic example of a complex Lie group is $\mathrm{GL}(n, \mathbb{C})$. We have already observed that $\mathrm{GL}(n, \mathbb{C})$ is an open subset of $\mathbb{C}^{n^2}$ and the product of matrices is given by polynomial functions, while the inverse of a matrix is given by rational functions on the entries of the matrix. Other classical examples include the *special linear group* $\mathrm{SL}(n, \mathbb{C})$ and the *symplectic group* $\mathrm{Sp}(n, \mathbb{C})$. We recall the definition of the latter. Let Q be a symplectic form (cf. Definition B.1.1) on the $2n$-dimensional real vector space V, then

$$\mathrm{Sp}(V_{\mathbb{C}}, Q) := \{X \in \mathrm{End}(V_{\mathbb{C}}) : Q(Xu, Xv) = Q(u, v)\}. \tag{1.1.14}$$

We define $\mathrm{Sp}(V, Q)$ analogously. When $V = \mathbb{R}^{2n}$ and Q is defined by the matrix (1.1.4) we will denote these groups by $\mathrm{Sp}(n, \mathbb{C})$ and $\mathrm{Sp}(n, \mathbb{R})$. The choice of a symplectic basis for Q, as in (B.1.2), establishes isomorphisms $\mathrm{Sp}(V_{\mathbb{C}}, Q) \cong \mathrm{Sp}(n, \mathbb{C})$ and $\mathrm{Sp}(V, Q) \cong \mathrm{Sp}(n, \mathbb{R})$.

EXAMPLE 1.1.23 Let Q be a symplectic structure on a $2n$-dimensional, real vector space V. Consider the space

$$M = \{\Omega \in \mathcal{G}(n, V_{\mathbb{C}}) : Q(u, v) = 0 \text{ for all } u, v \in \Omega\}.$$

Let $\{e_1, \ldots, e_n, e_{n+1}, \ldots, e_{2n}\}$ be a basis of V in which the matrix of Q is as in (1.1.4). Then if

$$\Omega = [W] = \begin{bmatrix} W_1 \\ W_2 \end{bmatrix},$$

where W_1 and W_2 are $n \times n$ matrices, we have that $\Omega \in M$ if and only if

$$[W_1^T, W_2^T] \cdot \begin{pmatrix} 0 & -I_n \\ I_n & 0 \end{pmatrix} \cdot \begin{bmatrix} W_1 \\ W_2 \end{bmatrix} = W_2^T \cdot W_1 - W_1^T \cdot W_2 = 0.$$

Set $I_0 = \{1, \ldots, n\}$. Every element $\Omega \in M \cap U_{I_0}$, where U_{I_0} is as in (1.1.9), may be represented by a matrix of the form $\Omega = [I_n, Z]^T$ with $Z^T = Z$. It follows that $M \cap U_{I_0}$ is an $(n(n+1)/2)$-dimensional submanifold. Now, given an arbitrary $\Omega \in M$, there exists an element $X \in \mathrm{Sp}(V_{\mathbb{C}}, Q)$ such that $X \cdot \Omega = \Omega_0$, where $\Omega_0 = \mathrm{span}(e_1, \ldots, e_n)$. Since the elements of $\mathrm{Sp}(V_{\mathbb{C}}, Q)$ act by biholomorphisms on $\mathcal{G}(n, V_{\mathbb{C}})$, it follows that M is an $(n(n+1)/2)$-dimensional submanifold of $\mathcal{G}(n, V_{\mathbb{C}})$. Moreover, since M is a closed submanifold of the compact manifold $\mathcal{G}(k, n)$, M is also compact.

We will also be interested in considering the open set $D \subset M$ consisting of

$$D = D(V, Q) := \{\Omega \in M : i\, Q(w, \bar{w}) > 0 \text{ for all } 0 \neq w \in \Omega\}. \tag{1.1.15}$$

It follows that $\Omega \in D$ if and only if the Hermitian matrix

$$i \cdot [\bar{W}_1^T, \bar{W}_2^T] \cdot \begin{pmatrix} 0 & -I_n \\ I_n & 0 \end{pmatrix} \cdot \begin{bmatrix} W_1 \\ W_2 \end{bmatrix} = i(\bar{W}_2^T \cdot W_1 - \bar{W}_1^T \cdot W_2)$$

is positive definite. Note that, in particular, $D \subset U_{I_0}$ and that

$$D \cong \{Z \in \mathcal{M}(n, \mathbb{C}) : Z^T = Z \,;\, \mathrm{Im}(Z) = (1/2i)(Z - \bar{Z}) > 0\}, \tag{1.1.16}$$

where $\mathcal{M}(n, \mathbb{C})$ denotes the $n \times n$ complex matrices. If $n = 1$ then $M \cong \mathbb{C}$ and D is the upper half plane. We will call D the generalized *Siegel upper half-space*.

The elements of the complex lie group $\mathrm{Sp}(V_\mathbb{C}, Q) \cong \mathrm{Sp}(n, \mathbb{C})$ define biholomorphisms of $\mathcal{G}(n, V_\mathbb{C})$ preserving M. The subgroup

$$\mathrm{Sp}(V, Q) - \mathrm{Sp}(V_\mathbb{C}, Q) \cap \mathrm{GL}(V) \cong \mathrm{Sp}(n, \mathbb{R})$$

preserves D.

EXERCISE 1.1.24 Prove that relative to the description of D as in (1.1.16), the action of $\mathrm{Sp}(V, Q)$ is given by generalized fractional linear transformations

$$\begin{pmatrix} A & B \\ C & D \end{pmatrix} \cdot Z = (A \cdot Z + B) \cdot (C \cdot Z + D)^{-1}.$$

EXERCISE 1.1.25 Prove that the action of $\mathrm{Sp}(V, Q)$ on D is *transitive* in the sense that given any two points $\Omega, \Omega' \in D$, there exists $X \in \mathrm{Sp}(V, Q)$ such that $X \cdot \Omega = \Omega'$.

EXERCISE 1.1.26 Compute the *isotropy subgroup*

$$K := \{X \in \mathrm{Sp}(V, Q) : X \cdot \Omega_0 = \Omega_0\},$$

where $\Omega_0 = [I_n, i\, I_n]^T$.

EXAMPLE 1.1.27 Let $T_\Lambda := \mathbb{C}/\Lambda$, where $\Lambda \subset \mathbb{Z}^2$ is a rank-2 lattice in $\mathbb{C}$; i.e.,

$$\Lambda = \{m\,\omega_1 + n\,\omega_2 \,;\, m, n \in \mathbb{Z}\},$$

where ω_1, ω_2 are complex numbers linearly independent over $\mathbb{R}$. T_Λ is locally diffeomorphic to $\mathbb{C}$ and since the translations by elements in Λ are biholomorphisms of $\mathbb{C}$, T_Λ inherits a complex structure relative to which the natural projection

$$\pi_\Lambda \colon \mathbb{C} \to T_\Lambda$$

is a local biholomorphic map.

It is natural to ask whether, for different lattices Λ, Λ', the *complex tori* T_Λ, $T_{\Lambda'}$ are biholomorphic. Suppose $F\colon T_\Lambda \to T_{\Lambda'}$ is a biholomorphism. Then, since $\mathbb{C}$ is the universal covering of T_Λ, there exists a map $\tilde{F}\colon \mathbb{C} \to \mathbb{C}$ such that the diagram

$$
\begin{array}{ccc}
\mathbb{C} & \xrightarrow{\ \tilde{F}\ } & \mathbb{C} \\[2pt]
{\scriptstyle \pi_\Lambda}\big\downarrow & & \big\downarrow{\scriptstyle \pi_{\Lambda'}} \\[2pt]
\mathbb{C}/\Lambda & \xrightarrow{\ F\ } & \mathbb{C}/\Lambda'
\end{array}
$$

commutes. In particular, given $z \in \mathbb{C}$, $\lambda \in \Lambda$, there exists $\lambda' \in \Lambda'$ such that

$$\tilde{F}(z + \lambda) \;=\; \tilde{F}(z) + \lambda'.$$

This means that the derivative $\tilde{F}'$ must be Λ-periodic and, hence, it defines a holomorphic function on $\mathbb{C}/\Lambda$ which, by Theorem 1.1.10, must be constant. This implies that $\tilde{F}$ must be a linear map and, after translation if necessary, we may assume that $\tilde{F}(z) = \mu \cdot z$, $\mu = a + ib \in \mathbb{C}$. Conversely, any such linear map $\tilde{F}$ induces a biholomorphic map $\mathbb{C}/\Lambda \to \mathbb{C}/\tilde{F}(\Lambda)$. In particular, if $\{\omega_1, \omega_2\}$ is a $\mathbb{Z}$-basis of Λ then $\mathrm{Im}(\omega_2/\omega_1) \neq 0$ and we may assume without loss of generality that $\mathrm{Im}(\omega_2/\omega_1) > 0$. Setting $\tau = \omega_2/\omega_1$ we see that T_Λ is always biholomorphic to a torus T_τ associated with a lattice

$$\{m + n\tau \;;\; m, n \in \mathbb{Z}\}$$

with $\mathrm{Im}(\tau) > 0$.

Now, suppose the tori T_Λ, $T_{\Lambda'}$ are biholomorphic and let $\{\omega_1, \omega_2\}$ (resp. $\{\omega_1', \omega_2'\}$) be a $\mathbb{Z}$-basis of Λ (resp. Λ') as above. We have

$$\mu \cdot \omega_1 = m_{11}\omega_1' + m_{21}\omega_2' \;;\quad \mu \cdot \omega_2 = m_{12}\omega_1' + m_{22}\omega_2', \quad m_{ij} \in \mathbb{Z}.$$

Moreover, $m_{11}m_{22} - m_{12}m_{21} = 1$, since F is biholomorphic and therefore $\tilde{F}(\Lambda) = \Lambda'$. Hence

$$\tau \;=\; \frac{\omega_1}{\omega_2} \;=\; \frac{m_{11}\omega_1' + m_{21}\omega_2'}{m_{12}\omega_1' + m_{22}\omega_2'} \;=\; \frac{m_{11} + m_{21}\tau'}{m_{12} + m_{22}\tau'}.$$

Consequently, $T_\tau \cong T_{\tau'}$ if and only if τ and τ' are points in the upper half plane congruent under the action of the group $\mathrm{SL}(2, \mathbb{Z})$ by fractional linear transformations. We refer to Section 4.2 for a fuller discussion of this example.

Remark. Note that while all differentiable structures on the torus $S^1 \times S^1$ are equivalent, there is a continuous *moduli space* of different complex structures. This is one of the key differences between real and complex geometry and one which we will study using Hodge theory.

1.1.2 Holomorphic Vector Bundles

We may extend the notion of a smooth vector bundle to complex manifolds and holomorphic maps.

DEFINITION 1.1.28 *A **holomorphic vector bundle** E over a complex manifold M is a complex manifold E together with a holomorphic map $\pi\colon E \to M$ such that*

(1) *for each $x \in M$, the fiber $E_x = \pi^{-1}(x)$ is a complex vector space of dimension d (the rank of E);*

(2) *there exist an open covering $\{U_\alpha\}$ of M and biholomorphic maps*

$$\Phi_\alpha\colon \pi^{-1}(U_\alpha) \to U_\alpha \times \mathbb{C}^d$$

such that

(a) *$p_1(\Phi_\alpha(x)) = x$ for all $x \in U$, where $p_1\colon U_\alpha \times \mathbb{C}^d \to U_\alpha$ denotes projection on the first factor; and*

(b) *for every $x \in U_\alpha$, the map $p_2 \circ \Phi|_{E_x}\colon E_x \to \mathbb{C}^d$ is an isomorphism of complex vector spaces.*

We call E the *total space* of the bundle and M its *base*. The covering $\{U_\alpha\}$ is called a *trivializing cover* of M and the biholomorphisms $\{\Phi_\alpha\}$ *local trivializations*. When $d = 1$ we often refer to E as a *line bundle*.

We note that as in the case of smooth vector bundles, a holomorphic vector bundle may be described by *transition functions*, i.e., by a covering of M by open sets U_α together with holomorphic maps

$$g_{\alpha\beta}\colon U_\alpha \cap U_\beta \to \mathrm{GL}(d, \mathbb{C})$$

such that

$$g_{\alpha\beta} \cdot g_{\beta\gamma} = g_{\alpha\gamma} \tag{1.1.17}$$

on $U_\alpha \cap U_\beta \cap U_\gamma$. The maps $g_{\alpha\beta}$ are defined by the following commutative diagram:

$$\pi^{-1}(U_\alpha \cap U_\beta) \tag{1.1.18}$$

$$
\begin{array}{ccc}
 & \pi^{-1}(U_\alpha \cap U_\beta) & \\
\Phi_\beta \swarrow & & \searrow \Phi_\alpha \\
(U_\alpha \cap U_\beta) \times \mathbb{C}^d & \xrightarrow{(\mathrm{id}, g_{\alpha\beta})} & (U_\alpha \cap U_\beta) \times \mathbb{C}^d.
\end{array}
$$

In particular, a holomorphic line bundle over M is given by a collection $\{U_\alpha, g_{\alpha\beta}\}$, where U_α is an open cover of M and the $\{g_{\alpha\beta}\}$ are nowhere-zero holomorphic functions defined on $U_\alpha \cap U_\beta$, i.e., $g_{\alpha\beta} \in \mathcal{O}^*(U_\alpha \cap U_\beta)$ satisfying the *cocycle condition* (1.1.17).

EXAMPLE 1.1.29 The product $M \times \mathbb{C}^d$ with the natural projection may be viewed as vector bundle of rank d over the complex manifold M. It is called the *trivial bundle* over M.

EXAMPLE 1.1.30   We consider the *tautological* line bundle over $\mathbb{P}^n$. This is the bundle whose fiber over a point in $\mathbb{P}^n$ is the line in $\mathbb{C}^{n+1}$ defined by that point. More precisely, let

$$\mathcal{T} := \{([z], v) \in \mathbb{P}^n \times \mathbb{C}^{n+1} : v = \lambda z, \lambda \in \mathbb{C}\},$$

and let $\pi\colon \mathcal{T} \to \mathbb{P}^n$ be the projection to the first factor. Let U_i be as in (1.1.6). Then we can define

$$\Phi_i \colon \pi^{-1}(U_i) \to U_i \times \mathbb{C}$$

by

$$\Phi_i([z], v) = ([z], v_i).$$

The transition functions g_{ij} are defined by the diagram (1.1.18) and we have

$$\Phi_i \circ \Phi_j^{-1}([z], 1) = \Phi_i([z], (z_0/z_j, \ldots, 1, \ldots, z_n/z_j)) = ([z], z_i/z_j),$$

with the 1 in the jth position. Hence,

$$g_{ij} \colon U_i \cap U_j \to \mathrm{GL}(1, \mathbb{C}) \cong \mathbb{C}^*$$

is the map $[z] \mapsto z_i/z_j$. It is common to denote the tautological bundle as $\mathcal{O}(-1)$.

EXERCISE 1.1.31   Generalize the construction of the tautological bundle over projective space to obtain the universal rank-k bundle over the Grassmann manifold $\mathcal{G}(k, n)$. Consider the space

$$\mathcal{U} := \{(\Omega, v) \in \mathcal{G}(k, n) \times \mathbb{C}^n : v \in \Omega\}, \tag{1.1.19}$$

where we regard $\Omega \in \mathcal{G}(k, n)$ as a k-dimensional subspace of $\mathbb{C}^n$. Prove that $\mathcal{U}$ may be trivialized over the open sets U_I defined in Example 1.1.7 and compute the transition functions relative to these trivializations.

Let $\pi\colon E \to M$ be a holomorphic vector bundle and suppose $F\colon N \to M$ is a holomorphic map. Given a trivializing cover $\{(U_\alpha, \Phi_a)\}$ of E with transition functions $g_{\alpha\beta}\colon U_\alpha \cap U_\beta \to \mathrm{GL}(d, \mathbb{C})$, we define

$$h_{\alpha\beta}\colon F^{-1}(U_\alpha) \cap F^{-1}(U_\beta) \to \mathrm{GL}(d, \mathbb{C}) \,; \quad h_{\alpha\beta} := g_{\alpha\beta} \circ F. \tag{1.1.20}$$

It is easy to check that the functions $h_{\alpha\beta}$ satisfy the *cocycle condition* (1.1.17) and, therefore, define a holomorphic vector bundle over N denoted by $F^*(E)$, and called the *pull-back* bundle. Note that we have a commutative diagram:

$$
\begin{array}{ccc}
F^*(E) & \xrightarrow{\ \bar{F}\ } & E \\
\pi^* \downarrow & & \downarrow \pi \\
N & \xrightarrow{\ F\ } & M.
\end{array}
\tag{1.1.21}
$$

If L and L' are line bundles and $g^L_{\alpha\beta}$, $g^{L'}_{\alpha\beta}$ are their transition functions relative to a common trivializing cover, then the functions

$$h_{\alpha\beta} = g^L_{\alpha\beta} \cdot g^{L'}_{\alpha\beta}$$

satisfy (1.1.17) and define a new line bundle which we denote by $L \otimes L'$.

Similarly, the functions

$$h_{\alpha\beta} = (g^L_{\alpha\beta})^{-1}$$

also satisfy (1.1.17) and define a bundle, called the *dual bundle* of L and denoted by L^* or L^{-1}. Clearly $L \otimes L^*$ is the trivial line bundle over M. The dual bundle of the tautological bundle is called the *hyperplane bundle* over $\mathbb{P}^n$ and denoted by H or $\mathcal{O}(1)$. Note that the transition functions of H are $g^H_{ij} \in \mathcal{O}^*(U_i \cap U_j)$ defined by

$$g^H_{ij}([z]) := z_j/z_i. \qquad (1.1.22)$$

We may also extend the notion of sections to holomorphic vector bundles:

DEFINITION 1.1.32 *A **holomorphic section** of a holomorphic vector bundle $\pi \colon E \to M$ over an open set $U \subset M$ is a holomorphic map*

$$\sigma \colon U \to E$$

such that

$$\pi \circ \sigma = \mathrm{id}|_U . \qquad (1.1.23)$$

The sections of E over U form an $\mathcal{O}(U)$-module which will be denoted by $\mathcal{O}(U, E)$. The local sections over U of the trivial line bundle are precisely the ring $\mathcal{O}(U)$.

If $\pi \colon L \to M$ is a holomorphic line bundle and $g_{\alpha\beta}$ are the transition functions associated to a trivializing covering (U_α, Φ_α), then a section $\sigma \colon M \to L$ may be described by a collection of holomorphic functions $f_\alpha \in \mathcal{O}(U_\alpha)$ defined by

$$\sigma(x) = f_\alpha(x)\Phi_\alpha^{-1}(x, 1).$$

Hence, for $x \in U_\alpha \cap U_\beta$ we must have

$$f_\alpha(x) = g_{\alpha\beta}(x) \cdot f_\beta(x). \qquad (1.1.24)$$

EXAMPLE 1.1.33 Let $M = \mathbb{P}^n$ and let $U_i = \{[z] \in \mathbb{P}^n : z_i \neq 0\}$. Let $P \in \mathbb{C}[z_0, \ldots, z_n]$ be a homogeneous polynomial of degree d. For each $i = 0, \ldots, n$ define

$$f_i([z]) = \frac{P(z)}{z_i^d} \in \mathcal{O}(U_i).$$

In $U_i \cap U_j$, we then have

$$z_i^d \cdot f_i([z]) = P(z) = z_j^d \cdot f_j([z]),$$

and therefore

$$f_i([z]) = (z_j/z_i)^d \cdot f_j([z]).$$

This means that we can consider the polynomial $P(z)$ as defining a section of the line bundle over $\mathbb{P}^n$ with transition functions

$$g_{ij} = (z_j/z_i)^d ,$$

that is, of the bundle $H^d = \mathcal{O}(d)$. In fact, it is possible to prove that every global holomorphic section of the bundle $\mathcal{O}(d)$ is defined, as above, by a homogeneous polynomial of degree d. The proof of this fact requires Hartogs' theorem [13, Proposition 1.1.14] from the theory of holomorphic functions of several complex variables. We refer to [13, Proposition 2.4.1].

We note that, on the other hand, the tautological bundle has no nontrivial global holomorphic sections. Indeed, suppose $\sigma \in \mathcal{O}(\mathbb{P}^n, \mathcal{O}(-1))$ and let ℓ denote the global section of $\mathcal{O}(1)$ associated to a nonzero linear form ℓ. Then, the map

$$[z] \in \mathbb{P}^n \mapsto \ell([z])\sigma([z])$$

defines a global holomorphic function on the compact complex manifold $\mathbb{P}^n$, hence it must be constant. If that constant is nonzero then both σ and ℓ are nowhere zero which would imply that both $\mathcal{O}(-1)$ and $\mathcal{O}(1)$ are trivial bundles. Hence σ must be identically zero.

Note that given a section $\sigma \colon M \to E$ of a vector bundle E, the zero locus $\{x \in M : \sigma(x) = 0\}$ is a well-defined subset of M. Thus, we may view the projective hypersurface defined in Example 1.1.18 by a homogeneous polynomial of degree d as the zero locus of a section of $\mathcal{O}(d)$.

Remark. The discussion above means that one should think of sections of line bundles as locally defined holomorphic functions satisfying a suitable compatibility condition. Given a compact, connected, complex manifold, global sections of holomorphic line bundles (when they exist) often play the role that global smooth functions play in the study of smooth manifolds. In particular, one uses sections of line bundles to define embeddings of compact complex manifolds into projective space. This vague observation will be made precise later in the chapter.

Given a holomorphic vector bundle $\pi \colon E \to M$ and a local trivialization

$$\Phi \colon \pi^{-1}(U) \to U \times \mathbb{C}^d,$$

we may define a basis of local sections of E over U (a *local frame*) as follows. Let $e_1, \ldots, e_d$ denote the standard basis of $\mathbb{C}^d$ and for $x \in U$ set

$$\sigma_j(x) := \Phi^{-1}(x, e_j); \; j = 1, \ldots, d.$$

Then $\sigma_j(x) \in \mathcal{O}(U, E)$ and for each $x \in U$, the vectors $\sigma_1(x), \ldots, \sigma_d(x)$ are a basis of the d-dimensional vector space E_x (they are the image of the basis $e_1, \ldots, e_d$ by a linear isomorphism). In particular, if $\tau \colon U \to M$ is a map satisfying (1.1.23) we can write

$$\tau(x) = \sum_{j=1}^{d} f_j(x)\sigma_j(x)$$

and τ is holomorphic (resp. smooth) if and only if the functions $f_j \in \mathcal{O}(U)$ (resp. $f_j \in C^\infty(U)$).

Conversely, suppose $U \subset M$ is an open set and let $\sigma_1, \ldots, \sigma_d \in \mathcal{O}(U, E)$ be a *local frame*, i.e., holomorphic sections such that for each $x \in U$, we have $\sigma_1(x), \ldots, \sigma_d(x)$ is a basis of E_x. Then we may define a local trivialization

$$\Phi \colon \pi^{-1}(U) \to U \times \mathbb{C}^d$$

by

$$\Phi(v) := (\pi(v), (\lambda_1, \ldots, \lambda_d)),$$

where $v \in \pi^{-1}(U)$ and

$$v = \sum_{j=1}^{d} \lambda_j \, \sigma_j(\pi(v)).$$

1.2   DIFFERENTIAL FORMS ON COMPLEX MANIFOLDS

1.2.1   Almost Complex Manifolds

Let M be a complex manifold and (U_α, ϕ_α) coordinate charts covering M. Since the change-of-coordinate maps (1.1.5) are holomorphic, the matrix of the differential $D(\phi_\beta \circ \phi_\alpha^{-1})$ is of the form (1.1.3). This means that the map

$$J_p \colon T_p(M) \to T_p(M)$$

defined by

$$J\left(\frac{\partial}{\partial x_j}\right) := \frac{\partial}{\partial y_j} \, ; \quad J\left(\frac{\partial}{\partial y_j}\right) := -\frac{\partial}{\partial x_j} \tag{1.2.1}$$

is well defined. We note that J is a smooth $(1,1)$ tensor on M such that $J^2 = -I$ and therefore, for each $p \in M$, then J_p defines a complex structure on the real vector space $T_p(M)$ (cf. (A.1.4)).

DEFINITION 1.2.1   *An **almost complex structure** on a C^∞ (real) manifold M is a $(1,1)$ tensor J such that $J^2 = -I$. An almost complex manifold is a pair (M, J) where J is an almost complex structure on M. The almost complex structure J is said to be* integrable *if M has a complex structure inducing J.*

If (M, J) is an almost complex manifold then J_p is a complex structure on T_pM and therefore by Proposition A.1.2, M must be even-dimensional. Note also that (A.1.10) implies that if M has an almost complex structure then M is orientable.

EXERCISE 1.2.2   Let M be an orientable (and oriented) two-dimensional manifold and let $\langle \, , \, \rangle$ be a Riemannian metric on M. Given $p \in M$ let $v_1, v_2 \in T_p(M)$ be a positively oriented orthonormal basis. Prove that $J_p \colon T_p(M) \to T_p(M)$ defined by

$$J_p(v_1) = v_2 \, ; \quad J_p(v_2) = -v_1$$

defines an almost complex structure on M. Show, moreover, that if $\langle\!\langle \, , \, \rangle\!\rangle$ is a Riemannian metric conformally equivalent to $\langle \, , \, \rangle$, then the two metrics define the same almost complex structure.

The discussion above shows that if M is a complex manifold then the operator (1.2.1) defines an almost complex structure. Conversely, the Newlander–Nirenberg theorem gives a necessary and sufficient condition for an almost complex structure J to arise from a complex structure. This is given in terms of the *Nijenhuis* torsion of J:

EXERCISE 1.2.3 Let J be an almost complex structure on M. Prove that

$$N(X,Y) = [JX, JY] - [X,Y] - J[X, JY] - J[JX, Y] \qquad (1.2.2)$$

is a $(1,2)$ tensor satisfying $N(X,Y) = -N(Y, X)$. The tensor N is called the *torsion* of J.

EXERCISE 1.2.4 Let J be an almost complex structure on a two-dimensional manifold M. Prove that $N(X,Y) = 0$ for all vector fields X and Y on M.

THEOREM 1.2.5 (Newlander–Nirenberg [20]) *Let (M, J) be an almost complex manifold, then M has a complex structure inducing the almost complex structure J if and only if $N(X, Y) = 0$ for all vector fields X and Y on M.*

PROOF. We refer to [34, Proposition 2], [30, Section 2.2.3] for a proof in the special case when M is a real analytic manifold. □

Remark. Note that assuming the Newlander–Nirenberg theorem, it follows from Exercise 1.2.4 that the almost complex structure constructed in Exercise 1.2.2 is integrable. We may explicitly construct the complex structure on M by using local isothermal coordinates. Thus, a complex structure on an oriented, two-dimensional manifold M is equivalent to a Riemannian metric up to conformal equivalence.

In what follows we will be interested in studying complex manifolds; however, the notion of almost complex structures gives a very convenient way to distinguish those properties of complex manifolds that depend only on having a (smoothly varying) complex structure on each tangent space. Thus, we will not explore in depth the theory of almost complex manifolds except to note that there are many examples of almost complex structures which are not integrable, that is, do not come from a complex structure. One may also ask which even-dimensional orientable manifolds admit almost complex structures. For example, in the case of a sphere S^{2n}, it was shown by Borel and Serre that only S^2 and S^6 admit almost complex structures. This is related to the fact that S^1, S^3, and S^7 are the only parallelizable spheres. We point out that while it is easy to show that S^6 has a nonintegrable almost complex structure, it is still unknown whether S^6 has a complex structure.

1.2.2 Tangent and Cotangent Space

Let (M, J) be an almost complex manifold and $p \in M$. Let $T_p(M)$ denote the tangent space of M. Then J_p defines a complex structure on $T_p(M)$ and therefore, by Proposition A.1.2, the complexification $T_{p,\mathbb{C}}(M) := T_p(M) \otimes_{\mathbb{R}} \mathbb{C}$ decomposes as

$$T_{p,\mathbb{C}}(M) = T_p'(M) \oplus T_p''(M),$$

where $T_p''(M) = \overline{T_p'(M)}$ and $T_p'(M)$ is the i-eigenspace of J_p acting on $T_{p,\mathbb{C}}(M)$. Moreover, by Proposition A.1.3, the map $v \in T_p(M) \mapsto v - iJ_p(v)$ defines an isomorphism of complex vector spaces $(T_p(M), J_p) \cong T_p'(M)$.

If J is integrable, then given holomorphic local coordinates $\{z_1, \ldots, z_n\}$ around p, we may consider the local coordinate frame (1.1.2) and, given (1.2.1), we have that the above isomorphism maps

$$\partial/\partial x_j \mapsto \partial/\partial x_j - i\,\partial/\partial y_j.$$

We set

$$\frac{\partial}{\partial z_j} := \frac{1}{2}\left(\frac{\partial}{\partial x_j} - i\frac{\partial}{\partial y_j}\right) ; \quad \frac{\partial}{\partial \bar{z}_j} := \frac{1}{2}\left(\frac{\partial}{\partial x_j} + i\frac{\partial}{\partial y_j}\right) . \tag{1.2.3}$$

Then, the vectors $\partial/\partial z_j$ are a basis of the complex subspace $T_p'(M)$.

Remark. Given local coordinates $(U, \{z_1, \ldots, z_n\})$ on M, a function $f: U \to \mathbb{C}$ is holomorphic if the local coordinates expression $f(z_1, \ldots, z_n)$ satisfies the Cauchy–Riemann equations. This is equivalent to the condition

$$\frac{\partial}{\partial \bar{z}_j}(f) = 0$$

for all j. Moreover, in this case, $\frac{\partial}{\partial z_j}(f)$ coincides with the partial derivative of f with respect to z_j. This justifies the choice of notation. However, we point out that it makes sense to consider $\frac{\partial}{\partial \bar{z}_j}(f)$ even if f is only a C^∞ function.

We will refer to $T_p'(M)$ as the holomorphic tangent space[3] of M at p. We note that if $\{z_1, \ldots, z_n\}$ and $\{w_1, \ldots, w_n\}$ are local complex coordinates around p then the change of basis matrix from the basis $\{\partial/\partial z_j\}$ to the basis $\{\partial/\partial w_k\}$ is given by the matrix of holomorphic functions

$$\left(\frac{\partial w_k}{\partial z_j}\right).$$

Thus, the complex vector spaces $T_p'(M)$ define a holomorphic vector bundle $T^h(M)$ over M, the *holomorphic tangent bundle*.

EXAMPLE 1.2.6 Let M be an oriented real surface with a Riemannian metric. Let (U, x, y) be positively oriented, local isothermal coordinates on M; i.e., the coordinate vector fields $\partial/\partial x$, $\partial/\partial y$ are orthogonal and of the same length. Then $z = x + iy$ defines complex coordinates on M and the vector field $\partial/\partial z = \frac{1}{2}(\partial/\partial x - i\,\partial/\partial y)$ is a local holomorphic section of the holomorphic tangent bundle of M.

We can now characterize the tangent bundle and the holomorphic tangent bundle of $\mathbb{P}^n$:

---

[3]This construction makes sense even if J is not integrable. In that case, we may replace the coordinate frame (1.1.2) by a local frame $\{X_1, \ldots, X_n, Y_1, \ldots, Y_n\}$ such that $J(X_j) = Y_j$ and $J(Y_j) = -X_j$.

THEOREM 1.2.7   *The tangent bundle $T\mathbb{P}^n$ is equivalent to the bundle*

$$\mathrm{Hom}(\mathcal{T}, E/\mathcal{T}),$$

where $E = \mathbb{P}^n \times \mathbb{C}^{n+1}$ is the trivial bundle of rank $n+1$ on $\mathbb{P}^n$ and $\mathcal{T}$ is the tautological bundle defined in Example 1.1.30. The holomorphic tangent bundle may be identified with the subbundle

$$\mathrm{Hom}_{\mathbb{C}}(\mathcal{T}, E/\mathcal{T}).$$

PROOF.   We work in the holomorphic case; the statement about the smooth case follows identically. Consider the projection $\pi \colon \mathbb{C}^{n+1} \setminus \{0\} \to \mathbb{P}^n$. Given $\lambda \in \mathbb{C}^*$, let M_λ denote *multiplication by λ* in $\mathbb{C}^{n+1} \setminus \{0\}$. Then, for every $v \in \mathbb{C}^{n+1} \setminus \{0\}$ we may identify $T'(\mathbb{C}^{n+1} \setminus \{0\}) \cong \mathbb{C}^{n+1}$ and we have the following commutative diagram of $\mathbb{C}$-linear maps

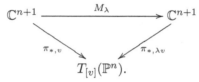

Now, the map $\pi_{*,v} \colon \mathbb{C}^{n+1} \to T'_{[v]}(\mathbb{P}^n)$ is surjective and its kernel is the line $L = \mathbb{C} \cdot v$. Hence we get a family of $\mathbb{C}$-linear isomorphisms

$$\mathfrak{p}_v \colon \mathbb{C}^{n+1}/L \to T'_{[v]}(\mathbb{P}^n); \quad v \in L,\ v \neq 0$$

with the relation

$$\mathfrak{p}_v = \lambda\,\mathfrak{p}_{\lambda v}.$$

We can now define a map

$$\Theta \colon \mathrm{Hom}_{\mathbb{C}}(\mathcal{T}, E/\mathcal{T}) \to T^h\mathbb{P}^n.$$

Let

$$\xi \in \mathrm{Hom}_{\mathbb{C}}(\mathcal{T}, E/\mathcal{T})_{[z]} = \mathrm{Hom}_{\mathbb{C}}(\mathcal{T}_{[z]}, (E/\mathcal{T})_{[z]}) \cong \mathrm{Hom}_{\mathbb{C}}(L, \mathbb{C}^{n+1}/L);$$

then we set

$$\Theta(\xi) := \mathfrak{p}_v(\xi(v)) \text{ for any } v \in L, v \neq 0.$$

Note that this is well defined since

$$\mathfrak{p}_{\lambda v}(\xi(v)) = \lambda^{-1}\mathfrak{p}_v(\lambda\,\xi(v)) = \mathfrak{p}_v(\xi(v)).$$

Alternatively one may define

$$\Theta(\xi) = \left.\frac{d}{dt}\right|_{t=0}(\gamma(t)),$$

where $\gamma(t)$ is the holomorphic curve through $[z]$ in $\mathbb{P}^n$ defined by

$$\gamma(t) := [v + t\xi(v)].$$

One then has to show that this map is well defined. It is straightforward, though tedious, to verify that Θ is an isomorphism of vector bundles. $\square$

EXERCISE 1.2.8   Prove that

$$T(\mathcal{G}(k,n)) \cong \operatorname{Hom}(\mathcal{U}, E/\mathcal{U}),$$

$$T^h(\mathcal{G}(k,n)) \cong \operatorname{Hom}_{\mathbb{C}}(\mathcal{U}, E/\mathcal{U}),$$

where $\mathcal{U}$ is the universal bundle over $\mathcal{G}(k,n)$ defined in Exercise 1.1.31 and E is the trivial bundle $E = \mathcal{G}(k,n) \times \mathbb{C}^n$.

As seen in Appendix A, a complex structure on a vector space induces a complex structure on the dual vector space. Thus, the complexification of the cotangent space $T^*_{p,\mathbb{C}}(M)$ decomposes as

$$T^*_{p,\mathbb{C}}(M) := T^{1,0}_p(M) \oplus T^{0,1}_p(M) \,; \quad T^{0,1}_p(M) = \overline{T^{1,0}_p(M)}.$$

Given local holomorphic coordinates $\{z_1, \ldots, z_n\}$, $z_j = x_j + i y_j$, the 1-forms $dz_j := dx_j + i\, dy_j$, $d\bar{z}_j = dx_j - i\, dy_j$ are the dual coframe to $\partial/\partial z_1, \ldots, \partial/\partial \bar{z}_n$ and consequently, $dz_1, \ldots, dz_n$ are a local holomorphic frame of the holomorphic bundle $T^{1,0}(M)$.

The complex structure on $T^*_p(M)$ induces a decomposition of the kth exterior product (cf. (A.1.12)):

$$\textstyle\bigwedge^k(T^*_{p,\mathbb{C}}(M)) = \bigoplus_{a+b=k} \bigwedge^{a,b}_p(M),$$

where

$$\textstyle\bigwedge^{a,b}_p(M) = \overbrace{T^{1,0}_p(M) \wedge \cdots \wedge T^{1,0}_p(M)}^{a \text{ times}} \wedge \overbrace{T^{0,1}_p(M) \wedge \cdots \wedge T^{0,1}_p(M)}^{b \text{ times}}. \qquad (1.2.4)$$

In this way, the smooth vector bundle $\bigwedge^k(T^*_{\mathbb{C}}(M))$ decomposes as a direct sum of C^∞ vector bundles

$$\textstyle\bigwedge^k(T^*_{\mathbb{C}}(M)) = \bigoplus_{a+b=k} \bigwedge^{a,b}(M). \qquad (1.2.5)$$

We will denote by $\mathcal{A}^k(U)$ (resp. $\mathcal{A}^{a,b}(U)$) the $C^\infty(U)$-module of local sections of the bundle $\bigwedge^k(T^*_{\mathbb{C}}(M))$ (resp. $\bigwedge^{a,b}(M)$) over U. We then have

$$\mathcal{A}^k(U) = \bigoplus_{a+b=k} \mathcal{A}^{a,b}(U). \qquad (1.2.6)$$

Note that given holomorphic coordinates $\{z_1, \ldots, z_n\}$, the local differential forms

$$dz_I \wedge d\bar{z}_J := dz_{i_1} \wedge \cdots \wedge dz_{i_a} \wedge d\bar{z}_{j_1} \wedge \cdots \wedge d\bar{z}_{j_b},$$

where I (resp. J) runs over all strictly increasing index sets $1 \le i_1 < \cdots < i_a \le n$ of length a (resp. $1 \le j_1 < \cdots < j_b \le n$ of length b) are a local frame for the bundle $\bigwedge^{a,b}(M)$.

We note that the bundles $\bigwedge^{k,0}(M)$ are holomorphic bundles of rank $\binom{n}{k}$. We denote them by Ω^k_M to emphasize that we are viewing them as holomorphic, rather than smooth, bundles. We denote the $\mathcal{O}(U)$-module of holomorphic sections by $\Omega^k(U)$. In particular, Ω^n_M is a holomorphic line bundle over M called the *canonical bundle* and usually denoted by K_M.

EXAMPLE 1.2.9 Let $M = \mathbb{P}^1$. Then as we saw in Example 1.1.5, M is covered by two coordinate neighborhoods (U_0, ϕ_0), (U_1, ϕ_1). The coordinate change is given by the map $\phi_1 \circ \phi_0^{-1} \colon \mathbb{C}^* \to \mathbb{C}^*$:

$$w = \phi_1 \circ \phi_0^{-1}(z) = \phi_1([(1, z)]) = 1/z.$$

This means that the local sections dz, dw of the holomorphic cotangent bundle are related by

$$dz = -(1/w)^2 \, dw.$$

It follows from (1.1.18) that $g_{01}[(z_0, z_1)] = -(z_0/z_1)^2$. Hence $K_{\mathbb{P}^1} \cong \mathcal{O}(-2) = T^2$.

EXERCISE 1.2.10 Find the transition functions for the holomorphic cotangent bundle of $\mathbb{P}^n$. Prove that $K_{\mathbb{P}^n} \cong \mathcal{O}(-n-1) = T^{n+1}$.

1.2.3 De Rham and Dolbeault Cohomologies

We recall that if $U \subset M$ is an open set in a smooth manifold M and $\mathcal{A}^k(U)$ denotes the space of $\mathbb{C}$-valued differential k-forms on U, then there exists a unique operator, the *exterior differential*,

$$d \colon \mathcal{A}^k(U) \to \mathcal{A}^{k+1}(U); \quad k \geq 0$$

satisfying the following properties:

(1) d is $\mathbb{C}$-linear.

(2) For $f \in \mathcal{A}^0(U) = C^\infty(U)$, df is the 1-form on U which acts on a vector field X by $df(X) := X(f)$.

(3) Given $\alpha \in \mathcal{A}^r(U)$, $\beta \in \mathcal{A}^s(U)$, the *Leibniz property* holds:

$$d(\alpha \wedge \beta) = d\alpha \wedge \beta + (-1)^r \alpha \wedge d\beta. \tag{1.2.7}$$

(4) $d \circ d = 0$.

It follows from (2) above that if $\{X_1, \ldots, X_m\}$ is a local frame on $U \subset M$ and $\{\xi_1, \ldots, \xi_m\}$ is the dual coframe, then given $f \in C^\infty(U)$ we have

$$df = \sum_{i=1}^m X_i(f) \, \xi_i \, .$$

In particular, if M is a complex manifold and $(U, \{z_1, \ldots, z_n\})$ are local coordinates, then for a function $f \in C^\infty(U)$ we have

$$df = \sum_{j=1}^n \frac{\partial f}{\partial x_j} dx_j + \sum_{j=1}^n \frac{\partial f}{\partial y_j} dy_j = \sum_{j=1}^n \frac{\partial f}{\partial z_j} dz_j + \sum_{j=1}^n \frac{\partial f}{\partial \bar{z}_j} d\bar{z}_j. \tag{1.2.8}$$

The properties of the operator d imply that for each open set U in M we have a complex:

$$\mathbb{C} \hookrightarrow C^\infty(U) \overset{d}{\to} \mathcal{A}^1(U) \overset{d}{\to} \cdots \overset{d}{\to} \mathcal{A}^{2n-1}(U) \overset{d}{\to} \mathcal{A}^{2n}(U). \qquad (1.2.9)$$

The quotients

$$H^k_{dR}(U, \mathbb{C}) := \frac{\ker\{d \colon \mathcal{A}^k(U) \to \mathcal{A}^{k+1}(U)\}}{d(\mathcal{A}^{k-1}(U))} \qquad (1.2.10)$$

are called the *de Rham cohomology groups* of U. The elements in

$$\mathcal{Z}^k(U) := \ker\{d \colon \mathcal{A}^k(U) \to \mathcal{A}^{k+1}(U)\}$$

are called *closed k-forms* and the elements in $\mathcal{B}^k(U) := d(\mathcal{A}^{k-1}(U))$ *exact k-forms*. We note that if U is connected then $H^0_{dR}(U, \mathbb{C}) \cong \mathbb{C}$. Unless there is the possibility of confusion we will drop the subscript since, in this chapter, we will only consider de Rham cohomology.

EXERCISE 1.2.11   Prove that the set of closed forms is a subring of the ring of differential forms and that the set of exact forms is an ideal in the ring of closed forms. Deduce that the de Rham cohomology

$$H^*(U, \mathbb{C}) := \bigoplus_{k \geq 0} H^k(U, \mathbb{C}) \qquad (1.2.11)$$

inherits a ring structure

$$[\alpha] \cup [\beta] := [\alpha \wedge \beta]. \qquad (1.2.12)$$

This is called the *cup product* on cohomology.

If $F \colon M \to N$ is a smooth map, then given an open set $V \subset N$, F induces maps

$$F^* \colon \mathcal{A}^k(V) \to \mathcal{A}^k(F^{-1}(V))$$

which commute with the exterior differential; i.e., F^* is a map of complexes. This implies that F^* defines a map between de Rham cohomology groups,

$$F^* \colon H^k(V, \mathbb{C}) \to H^k(F^{-1}(V), \mathbb{C}),$$

which satisfies the *chain rule* $(F \circ G)^* = G^* \circ F^*$. Since $(\mathrm{id})^* = \mathrm{id}$, it follows that if $F \colon M \to N$ is a diffeomorphism then $F^* \colon H^k(N, \mathbb{C}) \to H^k(M, \mathbb{C})$ is an isomorphism. In fact, the de Rham cohomology groups are a (smooth) homotopy invariant:

DEFINITION 1.2.12   *Let $f_0, f_1 \colon M \to N$ be smooth maps. Then f_0 is (smoothly) **homotopic** to f_1 if there exists a smooth map*

$$H \colon \mathbb{R} \times M \to N$$

such that $H(0, x) = f_0(x)$ and $H(1, x) = f_1(x)$ for all $x \in M$.

THEOREM 1.2.13   *Let $f_0, f_1 \colon M \to N$ be smoothly homotopic maps. Then*

$$f_0^* = f_1^* \colon H^k(N, \mathbb{C}) \to H^k(M, \mathbb{C}).$$

PROOF. We refer to [2, Section 4] for a proof of this important result.   □

COROLLARY 1.2.14 (Poincaré lemma)   *Let $U \subset M$ be a contractible open subset. Then $H^k(U, \mathbb{C})$ vanishes for all $k \geq 1$.*

PROOF. The result follows from Theorem 1.2.13 since in a contractible open set the identity map is homotopic to a constant map.   □

Hence, if U is contractible, the sequence

$$0 \to \mathbb{C} \hookrightarrow C^\infty(U) \xrightarrow{d} \mathcal{A}^1(U) \xrightarrow{d} \cdots \xrightarrow{d} \mathcal{A}^{2n-1}(U) \xrightarrow{d} \mathcal{A}^{2n}(U) \to 0 \qquad (1.2.13)$$

is exact.

The exterior differential operator is not of *pure bidegree* relative to the decomposition (1.2.6). Indeed, it follows from (1.2.8) that

$$d(\mathcal{A}^{a,b}(U)) \subset \mathcal{A}^{a+1,b}(U) \oplus \mathcal{A}^{a,b+1}(U). \qquad (1.2.14)$$

We remark that statement (1.2.14) makes sense for an almost complex manifold (M, J) and, indeed, its validity is equivalent to the integrability of the almost complex structure J; see [16, Theorem 2.8]. We write $d = \partial + \bar{\partial}$, where ∂ (resp. $\bar{\partial}$) is the component of d of bidegree $(1, 0)$ (resp. $(0, 1)$). From $d^2 = 0$ we obtain

$$\partial^2 = \bar{\partial}^2 = 0 \; ; \quad \partial \circ \bar{\partial} + \bar{\partial} \circ \partial = 0. \qquad (1.2.15)$$

EXERCISE 1.2.15   Generalize the Leibniz property to the operators ∂ and $\bar{\partial}$.

PROPOSITION 1.2.16   *Let M be a complex manifold and $U \subset M$ an open subset. Then*

$$\ker\{\bar{\partial} \colon \mathcal{A}^{p,0}(U) \to \mathcal{A}^{p,1}(U)\} = \Omega^p(U). \qquad (1.2.16)$$

PROOF. We may assume that $(U, \{z_1, \ldots, z_n\}$ is a coordinate neighborhood. Let $\alpha \in \mathcal{A}^{p,0}(U)$ and write $\alpha = \sum_I f_I \, dz_I$, where I runs over all increasing index sets $\{1 \leq i_1 < \cdots < i_p \leq n\}$. Then

$$\bar{\partial}\alpha = \sum_I \sum_{j=1}^n \frac{\partial f_I}{\partial \bar{z}_j} d\bar{z}_j \wedge dz_I = 0.$$

This implies that $\partial f_I / \partial \bar{z}_j = 0$ for all I and all j. Hence $f_I \in \mathcal{O}(U)$ for all I and α is a holomorphic p-form.   □

It follows then from (1.2.15) and (1.2.16) that, for each p, $0 \leq p \leq n$, we get a complex

$$0 \to \Omega^p(U) \hookrightarrow \mathcal{A}^{p,0}(U) \xrightarrow{\bar{\partial}} \mathcal{A}^{p,1}(U) \xrightarrow{\bar{\partial}} \cdots \xrightarrow{\bar{\partial}} \mathcal{A}^{p,n-1}(U) \xrightarrow{\bar{\partial}} \mathcal{A}^{p,n}(U) \to 0$$

called the *Dolbeault complex*. Its cohomology spaces are denoted by $H_{\bar{\partial}}^{p,q}(U)$ and called the *Dolbeault cohomology groups*.

EXERCISE 1.2.17 Let $\alpha \in \mathcal{A}^{p,q}(U)$. Prove that $\overline{\partial \alpha} = \bar{\partial} \bar{\alpha}$. Deduce that a form α is ∂-closed if and only if $\bar{\alpha}$ is $\bar{\partial}$-closed. Similarly for ∂-exact forms. Conclude that via conjugation, the study of ∂-cohomology reduces to the study of Dolbeault cohomology.

Given $a = (a_1, \ldots, a_n) \in \mathbb{C}^n$ and $\varepsilon = (\varepsilon_1, \ldots, \varepsilon_n) \in (\mathbb{R}_{>0} \cup \infty)^n$, we denote by

$$\Delta_\varepsilon(a) = \{z \in \mathbb{C}^n : |z_i - a_i| < \varepsilon_i\}$$

the n-dimensional *polydisk*. For $n = 1$, $a = 0$, $\varepsilon = 1$ we set $\Delta = \Delta_1(0)$, the *unit disk*, and $\Delta^* = \Delta \setminus \{0\}$, the *punctured unit disk*. The following result is known as the $\bar{\partial}$-Poincaré lemma:

THEOREM 1.2.18 *If $q \geq 1$ and α is a $\bar{\partial}$-closed (p, q)-form on a polydisk $\Delta_\varepsilon(a)$, then α is $\bar{\partial}$-exact; i.e.,*

$$H_{\bar{\partial}}^{p,q}(\Delta_\varepsilon(a)) = 0 \, ; \, q \geq 1 \, .$$

PROOF. We refer to [10, Chapter 0] or [13, Corollary 1.3.9] for a proof. $\square$

Hence, if $U = \Delta_\varepsilon(a)$ is a polydisk we have exact sequences:

$$0 \to \Omega^p(U) \hookrightarrow \mathcal{A}^{p,0}(U) \xrightarrow{\bar{\partial}} \mathcal{A}^{p,1}(U) \xrightarrow{\bar{\partial}} \cdots \xrightarrow{\bar{\partial}} \mathcal{A}^{p,n}(U) \to 0. \qquad (1.2.17)$$

Remark. Both the De Rham and Dolbeault cohomology groups may be realized as the sheaf cohomology groups of a constant sheaf. This is discussed in detail in Chapter 2. This will show, in particular, that even though our definition of the de Rham cohomology uses the differentiable structure, it is, in fact, a topological invariant. On the other hand, the Dolbeault cohomology groups depend essentially on the complex structure. This observation is at the core of Hodge theory.

1.3 SYMPLECTIC, HERMITIAN, AND KÄHLER STRUCTURES

In this section we will review the basic notions of Hermitian and Kähler metrics on complex manifolds. We begin by recalling the notion of a *symplectic* structure:

DEFINITION 1.3.1 *A **symplectic structure** on a 2d-dimensional manifold M is a closed 2-form $\omega \in \Lambda^2(M)$ such that $\Omega = \omega^d/d!$ is nowhere vanishing.*

Thus, if ω is a symplectic structure on M, at each $p \in M$, the form ω_p defines a symplectic structure Q_p on $T_p(M)$ (see Definition B.1.1). A *symplectic manifold* is a manifold M endowed with a symplectic structure ω.

The simplest example of a symplectic manifold is given by $\mathbb{R}^{2d}$ with coordinates denoted by $\{x_1, \ldots, x_d, y_1, \ldots, y_d\}$ and the 2-form

$$\omega_0 = \sum_{j=1}^{d} dx_j \wedge dy_j.$$

The classical Darboux theorem asserts that, locally, every symplectic manifold is *symplectomorphic* to $(\mathbb{R}^{2d}, \omega_0)$:

THEOREM 1.3.2 (Darboux theorem) *Let (M, ω) be a symplectic manifold. Then for each $p \in M$ there exists an open neighborhood U and local coordinates $\varphi \colon U \to \mathbb{R}^{2d}$ such that $\omega|_U = \varphi^*(\omega_0)$.*

PROOF. We refer to [23, Theorem 6.1] for a proof.   □

In what follows we will be particularly interested in symplectic structures on a complex manifold M compatible with the complex structure J:

DEFINITION 1.3.3 *Let M be a complex manifold and J its complex structure. A Riemannian metric g on M is said to be a **Hermitian metric** if and only if for each $p \in M$, the bilinear form g_p on the tangent space $T_p(M)$ is compatible with the complex structure J_p (cf. (B.2.1)).*

We recall from (B.2.2) in the second appendix to this chapter that given a symmetric bilinear form compatible with the complex structure we may define a J-invariant alternating form. Thus, given a Hermitian metric on M we may define a differential 2-form $\omega \in \mathcal{A}^2(M, \mathbb{C})$ by

$$\omega(X, Y) := g(JX, Y), \tag{1.3.1}$$

where we also denote by g the bilinear extension of g to the complexified tangent space. By Theorem B.2.1, we have

$$\omega \in \mathcal{A}^{1,1}(M) \quad \text{and} \quad \bar{\omega} = \omega. \tag{1.3.2}$$

We also recall that Theorem B.2.1 implies that every form ω as in (1.3.2) defines a symmetric $(1,1)$ tensor on M compatible with J and a Hermitian form H on the complex vector space $(T_p(M), J)$.

We express these objects in local coordinates: let $(U, \{z_1, \ldots, z_d\})$ be local complex coordinates on M. Then (1.3.2) implies that we may write

$$\omega := \frac{i}{2} \sum_{j,k=1}^{d} h_{jk} \, dz_j \wedge d\bar{z}_k \; ; \quad h_{kj} = \bar{h}_{jk}. \tag{1.3.3}$$

Hence $\omega(\partial/\partial z_j, \partial/\partial \bar{z}_k) = (i/2)\, h_{jk}$ from which it follows that

$$\omega(\partial/\partial x_j, \partial/\partial x_k) \;=\; -\operatorname{Im}(h_{jk}).$$

Moreover, we have

$$g(\partial/\partial x_j, \partial/\partial x_k) = \omega(\partial/\partial x_j, \partial/\partial y_k) \;=\; \operatorname{Re}(h_{jk}).$$

Hence g is positive definite if and only if the Hermitian matrix (h_{jk}) is positive definite. We may then restate Definition 1.3.3 by saying that a Hermitian structure is a $(1, 1)$ real form ω as in (1.3.3) such that the matrix (h_{jk}) is positive definite. By abuse of notation we will say that, in this case, the 2-form ω is positive.

1.3.1   Kähler Manifolds

DEFINITION 1.3.4   *A Hermitian metric on a manifold M is said to be a **Kähler metric** if and only if the 2-form ω is closed. We will say that a complex manifold is Kähler if and only if it admits a Kähler structure and refer to ω as a Kähler form.*

EXERCISE 1.3.5   Let (M, ω) be a Kähler manifold. Show that there are local coframes $\chi_1, \ldots, \chi_d$ in $\mathcal{A}^{1,0}(U)$ such that

$$\omega \;=\; \frac{i}{2}\sum_{j=1}^{d} \chi_j \wedge \bar{\chi}_j.$$

Clearly, every Kähler manifold M is symplectic. Moreover, if $\{z_1, \ldots, z_d\}$ are local coordinates on M and ω is a Kähler form on M then

$$
\begin{aligned}
\omega^d \;&=\; d!\,\left(\frac{i}{2}\right)^n \det((h_{ij})) \bigwedge_{j=1}^{d} (dz_j \wedge d\bar{z}_j) \\[2mm]
&=\; d!\,\det((h_{ij})) \bigwedge_{j=1}^{d} (dx_j \wedge dy_j),
\end{aligned}
$$

since $dz_j \wedge d\bar{z}_j = (2/i)dx_j \wedge dy_j$.

EXERCISE 1.3.6   Prove that $\omega^d/d!$ is the volume element of the Riemannian metric g defined by the Kähler form ω (see Exercise 1.4.3).

Thus we have a necessary condition for a compact complex manifold to be Kähler:

PROPOSITION 1.3.7   *If M is a compact Kähler manifold, then*

$$\dim H^{2k}(M, \mathbb{R}) > 0$$

for all $k = 0, \ldots, d$.

PROOF. Indeed, this is true of all compact symplectic manifolds as the forms ω^k, $k = 1, \ldots, d$ induce nonzero de Rham cohomology classes. Suppose, otherwise, that $\omega^k = d\alpha$. Then

$$\omega^d = d(\omega^{d-k} \wedge \alpha).$$

But then it would follow from the Stokes theorem that

$$\int_M \omega^d = 0,$$

which contradicts the fact that ω^d is a nonzero multiple of the volume element. □

Remark. As we will see below, the existence of a Kähler metric on a manifold imposes many other topological restrictions beyond those satisfied by symplectic manifolds. The earliest examples of compact symplectic manifolds with no Kähler structure are due to Thurston [25]. We refer to [32] for further details.

EXAMPLE 1.3.8 The affine space $\mathbb{C}^d$ with the form

$$\omega = \frac{i}{2} \sum_{j=1}^{d} dz_j \wedge d\bar{z}_j$$

is a Kähler manifold. The form ω gives the usual symplectic structure on $\mathbb{R}^{2d}$.

The following theorem may be seen as a generalization of Darboux's theorem to Kähler manifolds:

THEOREM 1.3.9 *Let M be a complex manifold and g a Kähler metric on M. Then, given $p \in M$, there exist local coordinates $(U, \{z_1, \ldots, z_d\})$ around p such that $z_j(p) = 0$ and*

$$\omega = \frac{i}{2} \sum_{j=1}^{d} h_{jk} \, dz_j \wedge d\bar{z}_k \,,$$

where the coefficients h_{jk} are of the form

$$h_{jk}(z) = \delta_{jk} + O(\|z\|^2). \tag{1.3.4}$$

PROOF. We refer to [30, Proposition 3.14] for a proof. □

EXAMPLE 1.3.10 We will construct a Kähler form on $\mathbb{P}^n$. We will do this by exhibiting a positive, real, closed $(1,1)$-form on $\mathbb{P}^n$. The resulting metric is called the Fubini–Study metric on $\mathbb{P}^n$.

Given $z \in \mathbb{C}^{n+1}$ we denote by

$$\|z\|^2 = |z_0|^2 + \cdots + |z_n|^2.$$

Let $U_j \subset \mathbb{P}^n$ be the open set (1.1.6) and let $\rho_j \in C^\infty(U_j)$ be the positive function

$$\rho_j([z]) := \frac{||z||^2}{|z_j|^2}, \tag{1.3.5}$$

and define $\omega_j \in \mathcal{A}^{1,1}(U_j)$ by

$$\omega_j := \frac{-1}{2\pi i} \partial\bar{\partial} \log(\rho_j). \tag{1.3.6}$$

Clearly, ω_j is a real, closed $(1,1)$-form. Moreover, on $U_j \cap U_k$ we have

$$\log(\rho_j) - \log(\rho_k) = \log|z_k|^2 - \log|z_j|^2 = \log(z_k\bar{z}_k) - \log(z_j\bar{z}_j).$$

Hence, since $\partial\bar{\partial}(\log(z_j\bar{z}_j)) = 0$, we have $\omega_j = \omega_k$ on $U_j \cap U_k$. Thus, the forms ω_j piece together to give a global, real, closed $(1,1)$-form ω on $\mathbb{P}^n$. We write

$$\omega = \frac{-1}{2\pi i} \partial\bar{\partial} \log(||z||^2). \tag{1.3.7}$$

It remains to show that ω is positive. We observe first of all that the expression (1.3.7) shows that if A is a unitary matrix and $\mu_A \colon \mathbb{P}^n \to \mathbb{P}^n$ is the biholomorphic map $\mu_A([z]) := [A \cdot z]$, then $\mu_A^*(\omega) = \omega$. Hence, since given any two points $[z], [z'] \in \mathbb{P}^n$ there exists a unitary matrix such that $\mu_a([z]) = [z']$, it suffices to prove that ω is positive definite at just one point, say $[1, 0, \ldots, 0] \in U_0$. In the coordinates $\{u_1, \ldots, u_n\}$ in U_0, we have $\rho_0(u) = 1 + ||u||^2$ and therefore

$$\bar{\partial}(\log\rho_0(u)) = \rho_0^{-1}(u) \sum_{k=1}^n u_k \, \bar{\partial}\bar{u}_k = \rho_0^{-1}(u) \sum_{k=1}^n u_k \, d\bar{u}_k,$$

$$\omega = \frac{i}{2\pi} \rho_0^{-2}(u) \left(\rho_0(u) \sum_{j=1}^n du_j \wedge d\bar{u}_j + \left(\sum_{j=1}^n \bar{u}_j du_j \right) \wedge \left(\sum_{j=1}^n u_k d\bar{u}_k \right) \right).$$

Hence, at the origin, we have

$$\omega = \frac{i}{2\pi} \sum_{j=1}^n du_j \wedge d\bar{u}_j,$$

which is a positive form.

The function $\log(\rho_j)$ in the above proof is called a Kähler *potential*. As the following result shows, every Kähler metric may be described by a (local) potential.

PROPOSITION 1.3.11 *Let M be a complex manifold and ω a Kähler form on M. Then for every $p \in M$ there exists an open set $U \subset M$ and a real function $v \in C^\infty(U)$ such that $\omega = i\partial\bar{\partial}(v)$.*

PROOF. Since $d\omega = 0$, it follows from the Poincaré lemma that in a neighborhood U' of p, we have $\omega = d\alpha$, where $\alpha \in \mathcal{A}^1(U', \mathbb{R})$. Hence, we may write $\alpha = \beta + \bar{\beta}$, where $\beta \in \mathcal{A}^{1,0}(U', \mathbb{R})$. Now, we can write

$$\omega = d\alpha = \partial\beta + \bar{\partial}\beta + \partial\bar{\beta} + \bar{\partial}\bar{\beta},$$

but, since ω is of type $(1,1)$ it follows that

$$\omega = \bar{\partial}\beta + \partial\bar{\beta} \quad \text{and} \quad \partial\beta = \bar{\partial}\bar{\beta} = 0.$$

We may now apply the $\bar{\partial}$-Poincaré lemma to conclude that there exists a neighborhood $U \subset U'$ around p where $\bar{\beta} = \bar{\partial}f$ for some ($\mathbb{C}$-valued) C^∞ function f on U. Hence

$$\omega = \bar{\partial}\partial\bar{f} + \partial\bar{\partial}f = \partial\bar{\partial}(f - \bar{f}) = 2i\,\partial\bar{\partial}(\mathrm{Im}(f)).$$

$\square$

THEOREM 1.3.12 *Let (M, ω) be a Kähler manifold and suppose $N \subset M$ is a complex submanifold. Then $(N, \omega|_N)$ is a Kähler manifold.*

PROOF. Let g denote the J-compatible Riemannian metric on M associated with ω. Then g restricts to a Riemannian metric on N, compatible with the complex structure on N, and whose associated 2-form is $\omega|_N$. Since $d(\omega|_N) = (d\omega)|_N = 0$, it follows that N is a Kähler manifold as well. $\square$

It follows from Theorem 1.3.12 that a necessary condition for a compact complex manifold M to have an embedding in $\mathbb{P}^n$ is that there exists a Kähler metric on M. Moreover, as we shall see below, for a submanifold of projective space, there exists a Kähler metric whose associated cohomology class satisfies a suitable integrality condition.

1.3.2 The Chern Class of a Holomorphic Line Bundle

The construction of the Kähler metric in $\mathbb{P}^n$ may be further understood in the context of Hermitian metrics on (line) bundles. We recall that a Hermitian metric on a $\mathbb{C}$-vector bundle $\pi \colon E \to M$ is given by a positive-definite Hermitian form

$$H_p \colon E_p \times E_p \to \mathbb{C}$$

on each fiber E_p, which is smooth in the sense that given sections $\sigma, \tau \in \Gamma(U, E)$, the function

$$H(\sigma, \tau)(p) := H_p(\sigma(p), \tau(p))$$

is C^∞ on U. Using partitions of unity, one can prove that every smooth vector bundle E has a Hermitian metric H.

In the case of a line bundle L, the Hermitian form H_p is completely determined by the value $H_p(v, v)$ on a nonzero element $v \in L_p$. In particular, if $\{(U_\alpha, \Phi_\alpha)\}$ is a cover of M by trivializing neighborhoods of L and $\sigma_\alpha \in \mathcal{O}(U_\alpha, L)$ is the local frame

$$\sigma_\alpha(x) = \Phi_\alpha^{-1}(x, 1); \quad x \in U_\alpha,$$

then a Hermitian metric H on L is determined by the collection of positive functions

$$\rho_\alpha := H(\sigma_\alpha, \sigma_\alpha) \in C^\infty(U_\alpha).$$

We note that if $U_\alpha \cap U_\beta \neq \emptyset$ then we have $\sigma_\beta = g_{\alpha\beta} \cdot \sigma_\alpha$ and, consequently, the functions ρ_α satisfy the compatibility condition

$$\rho_\beta = |g_{\alpha\beta}|^2 \rho_\alpha. \tag{1.3.8}$$

In particular, if L is a holomorphic line bundle then the transition functions $g_{\alpha\beta}$ are holomorphic and as in Example 1.3.10, we have

$$\partial\bar{\partial}\log(\rho_\alpha) = \partial\bar{\partial}\log(\rho_\beta)$$

on $U_\alpha \cap U_\beta$ and therefore the form

$$\frac{1}{2\pi i}\partial\bar{\partial}\log(\rho_\alpha) \tag{1.3.9}$$

is a global, real, closed $(1,1)$-form on M. The cohomology class

$$[(1/2\pi i)\,\partial\bar{\partial}\log(\rho_\alpha)] \in H^2(M, \mathbb{R}) \tag{1.3.10}$$

is called the *Chern class* of the vector bundle L and denoted by $c(L)$. The factor $1/2\pi$ is chosen so that the Chern class is actually an integral cohomology class:

$$c(L) \in H^2(M, \mathbb{Z}). \tag{1.3.11}$$

Recall that if $g_{\alpha\beta}$ are the transition functions for a bundle L then the functions $g_{\alpha\beta}^{-1}$ are the transition functions of the dual bundle L^*. In particular, if ρ_α are a collection of positive C^∞ functions defining a Hermitian metric on L then the functions ρ_α^{-1} define a Hermitian metric H^* on L^*. We call H^* the dual Hermitian metric. We then have

$$c(L^*) = -c(L). \tag{1.3.12}$$

DEFINITION 1.3.13 *A holomorphic line bundle* $L \to M$ *over a compact Kähler manifold is said to be* **positive** *if and only if there exists a Hermitian metric H on L for which the $(1,1)$-form (1.3.10) is positive. We say that L is* negative *if its dual bundle L^* is positive.*

We note that in Example 1.3.10 we have

$$|z_k|^2\,\rho_k([z]) = |z_j|^2\,\rho_j([z])$$

on $U_j \cap U_k$. Hence

$$\rho_k([z]) = \left|\frac{z_j}{z_k}\right|^2 \rho_j([z])$$

and, by (1.3.8), it follows that the functions ρ_j define a Hermitian metric on the tautological bundle $\mathcal{O}(-1)$. Hence, taking into account the sign change in (1.3.6), it follows that the Kähler class of the Fubini–Study metric agrees with the Chern class of the hyperplane bundle $\mathcal{O}(1)$. Thus,

$$c(\mathcal{O}(1)) \;=\; [\omega] \;=\; \left[\frac{i}{2\pi}\,\partial\bar\partial\log(\|z\|^2)\right] \tag{1.3.13}$$

and the hyperplane bundle $\mathcal{O}(1)$ is a positive line bundle. Moreover, if $M \subset \mathbb{P}^n$ is a complex submanifold then the restriction of $\mathcal{O}(1)$ to M is a positive line bundle over M. We can now state the following theorem:

THEOREM 1.3.14 (Kodaira embedding theorem)   *A compact complex manifold M may be embedded in $\mathbb{P}^n$ if and only if there exists a positive holomorphic line bundle $\pi\colon L \to M$.*

We refer to [30, Theorem 7.11], [19, Theorem 8.1], [10], [35, Theorem 4.1], and [13, Section 5.3] for various proofs of this theorem.

Remark.   The existence of a positive holomorphic line bundle $\pi\colon L \to M$ implies that M admits a Kähler metric whose Kähler class is integral. Conversely, any integral cohomology class represented by a closed $(1, 1)$-form is the Chern class of a line bundle (cf. [6, Section 6]), hence a compact complex manifold M may be embedded in $\mathbb{P}^n$ if and only if it admits a Kähler metric whose Kähler class is integral.

Recall (see [10, Section 1.3]) that Chow's theorem asserts that every analytic subvariety of $\mathbb{P}^n$ is algebraic. When this result is combined with the Kodaira embedding theorem, we obtain a characterization of complex projective varieties as those compact Kähler manifolds admitting a Kähler metric whose Kähler class is integral.

1.4   HARMONIC FORMS—HODGE THEOREM

1.4.1   Compact Real Manifolds

Unless otherwise specified, throughout Section 1.4.1 we will let M denote a compact, oriented, real, n-dimensional manifold with a Riemannian metric g. We recall that the metric on the tangent bundle TM induces a dual metric on the cotangent bundle T^*M such that the dual coframe of a local orthonormal frame $X_1, \ldots, X_n$ in $\Gamma(U, TM)$ is also orthonormal. We will denote the dual inner product by $\langle\,,\,\rangle$.

EXERCISE 1.4.1   Verify that this metric on T^*M is well defined; i.e., it is independent of the choice of local orthonormal frames.

We extend the inner product to the exterior bundles $\bigwedge^r(T^*M)$ by the specification that the local frame

$$\xi_I \;:=\; \xi_{i_1} \wedge \cdots \wedge \xi_{i_r},$$

where I runs over all strictly increasing index sets $\{1 \leq i_1 < \cdots < i_r \leq n\}$, is orthonormal.

EXERCISE 1.4.2 Verify that this metric on $\bigwedge^r T^* M$ is well defined; i.e., it is independent of the choice of local orthonormal frames, by proving that

$$\langle \alpha_1 \wedge \cdots \wedge \alpha_r, \beta_1 \wedge \cdots \wedge \beta_r \rangle = \det(\langle \alpha_i, \beta_j \rangle),$$

where $\alpha_i, \beta_j \in \mathcal{A}^1(U)$.
Hint: use the Cauchy–Binet formula for determinants.

Recall that given an oriented Riemannian manifold, the volume element is defined as the unique n-form $\Omega \in \mathcal{A}^n(M)$ such that

$$\Omega(p)(v_1, \ldots, v_n) = 1$$

for any positively oriented orthonormal basis $\{v_1, \ldots, v_n\}$ of $T_p(M)$. If $\xi_1, \ldots, \xi_n \in \mathcal{A}^1(U)$ is a positively oriented orthonormal coframe then

$$\Omega|_U = \xi_1 \wedge \cdots \wedge \xi_n.$$

EXERCISE 1.4.3 Prove that the volume element may be written as

$$\Omega = \sqrt{G} \, dx_1 \wedge \cdots \wedge dx_n \,,$$

where $\{x_1, \ldots, x_n\}$ are positively oriented local coordinates, $G = \det(g_{ij})$, and

$$g_{ij} := g(\partial/\partial x_i, \partial/\partial x_j).$$

We now define the *Hodge $*$-operator*. Let $\beta \in \mathcal{A}^r(M)$. Then $*\beta \in \mathcal{A}^{n-r}(M)$ is given by $(*\beta)(p) = *(\beta(p))$, where the $*$ operator on $\bigwedge^r T_p^*$ is defined as in (B.1.3). Therefore, for every $\alpha \in \mathcal{A}^r(M)$,

$$\alpha \wedge *\beta = \langle \alpha, \beta \rangle \, \Omega. \tag{1.4.1}$$

We extend the definition to $\mathcal{A}^r(M, \mathbb{C})$ by linearity.

EXERCISE 1.4.4 Suppose $\alpha_1, \ldots, \alpha_n \in T_p^*(M)$ is a positively oriented orthonormal basis. Let $I = \{1 \leq i_1 < \cdots < i_r \leq n\}$ be an index set and I^c its complement. Prove that

$$*(\alpha_I) = \operatorname{sign}(I, I^c) \, \alpha_{I^c} \,, \tag{1.4.2}$$

where $\operatorname{sign}(I, I^c)$ is the sign of the permutation $\{I, I^c\}$.

EXERCISE 1.4.5 Prove that $*$ is an isometry and that $*^2$ acting on $\mathcal{A}^r(M)$ equals $(-1)^{r(n-r)} I$.

Suppose now that M is compact. We can then define an L^2 inner product on the space of r-forms on M by

$$(\alpha, \beta) := \int_M \alpha \wedge *\beta = \int_M \langle \alpha(p), \beta(p) \rangle \, \Omega \,; \quad \alpha, \beta \in \mathcal{A}^r(M). \tag{1.4.3}$$

PROPOSITION 1.4.6   *The bilinear form* $(\bullet, \bullet)$ *is a positive-definite inner product on* $\mathcal{A}^r(M)$.

PROOF.   First of all we check that $(\bullet, \bullet)$ is symmetric:

$$(\beta, \alpha) = \int_M \beta \wedge *\alpha = (-1)^{r(n-r)} \int_M *(*\beta) \wedge *\alpha = \int_M *\alpha \wedge *(*\beta) = (\alpha, \beta).$$

Now, given $0 \neq \alpha \in \mathcal{A}^r(M)$, we have

$$(\alpha, \alpha) = \int_M \alpha \wedge *\alpha = \int_M \langle \alpha, \alpha \rangle \, \Omega > 0$$

since $\langle \alpha, \alpha \rangle$ is a nonnegative function which is not identically zero.   □

PROPOSITION 1.4.7   *The operator* $\delta \colon \mathcal{A}^{r+1}(M) \to \mathcal{A}^r(M)$, *defined by*

$$\delta := (-1)^{nr+1} * d *, \tag{1.4.4}$$

is the formal adjoint of d; that is,

$$(d\alpha, \beta) = (\alpha, \delta\beta) \quad \textit{for all } \alpha \in \mathcal{A}^r(M), \beta \in \mathcal{A}^{r+1}(M). \tag{1.4.5}$$

PROOF.

$$
\begin{aligned}
(d\alpha, \beta) &= \int_M d\alpha \wedge *\beta = \int_M d(\alpha \wedge *\beta) - (-1)^r \int_M \alpha \wedge d * \beta \\
&= -(-1)^r (-1)^{r(n-r)} \int_M \alpha \wedge *(* d * \beta) = \int_M \alpha \wedge *\delta\beta \\
&= (\alpha, \delta\beta).
\end{aligned}
$$

□

Remark.   Note that if $\dim M$ is even then $\delta = - * d *$ independently of the degree of the form. Since we will be interested in applying these results in the case of complex manifolds which, as real manifolds, are even-dimensional, we will make that assumption from now on.

We now define the *Laplace–Beltrami operator* of (M, g) by

$$\Delta \colon \mathcal{A}^r(M) \to \mathcal{A}^r(M); \quad \Delta\alpha := d\delta\alpha + \delta d\alpha.$$

PROPOSITION 1.4.8   *The operators* d, δ, $*$ *and* Δ *satisfy the following properties:*

(1) Δ *is self-adjoint; i.e.,* $(\Delta\alpha, \beta) = (\alpha, \Delta\beta)$.

(2) $[\Delta, d] = [\Delta, \delta] = [\Delta, *] = 0$.

(3) $\Delta\alpha = 0$ *if and only if* $d\alpha = \delta\alpha = 0$.

PROOF. We leave the first two items as exercises. Note that given operators D_1, D_2, the bracket $[D_1, D_2] = D_1 \circ D_2 - D_2 \circ D_1$. Thus, (2) states that the Laplacian Δ commutes with d, δ, and $*$.

Clearly, if $d\alpha = \delta\alpha = 0$ then we have $\Delta\alpha = 0$. Conversely, suppose $\alpha \in \mathcal{A}^r(M)$ and $\Delta\alpha = 0$. Then

$$0 = (\Delta\alpha, \alpha) = (d\delta\alpha + \delta d\alpha, \alpha) = (\delta\alpha, \delta\alpha) + (d\alpha, d\alpha).$$

Hence $d\alpha = \delta\alpha = 0$. □

DEFINITION 1.4.9 *A form $\alpha \in \mathcal{A}^r(M)$ is said to be* **harmonic** *if $\Delta\alpha = 0$ or, equivalently, if α is closed and coclosed, i.e., $d\alpha = \delta\alpha = 0$.*

EXERCISE 1.4.10 Let M be a compact, connected, oriented, Riemannian manifold. Show that the only harmonic functions on M are the constant functions.

EXERCISE 1.4.11 Let $\alpha \in \mathcal{A}^r(M)$ be closed. Show that $*\alpha$ is closed if and only if α is harmonic.

The following result shows that harmonic forms are very special within a given de Rham cohomology class:

PROPOSITION 1.4.12 *A closed r-form α is harmonic if and only if $||\alpha||^2$ is a local minimum within the de Rham cohomology class of α. Moreover, in any given de Rham cohomology class there is at most one harmonic form.*

PROOF. Let $\alpha \in \mathcal{A}^r(M)$ be such that $||\alpha||^2$ is a local minimum within the de Rham cohomology class of α. Then, for every $\beta \in \mathcal{A}^{r-1}(M)$, the function $\nu(t) := ||\alpha + t\,d\beta||^2$ has a local minimum at $t = 0$. In particular,

$$\nu'(0) = 2(\alpha, d\beta) = 2(\delta\alpha, \beta) = 0 \quad \text{for all} \quad \beta \in \mathcal{A}^{r-1}(M).$$

Hence, $\delta\alpha = 0$ and α is harmonic. Now, if α is harmonic, then

$$||\alpha + d\beta||^2 = ||\alpha||^2 + ||d\beta||^2 + 2(\alpha, d\beta) = ||\alpha||^2 + ||d\beta||^2 \geq ||\alpha||^2$$

and equality holds only if $d\beta = 0$. This proves the uniqueness statement. □

Hodge's theorem asserts that, in fact, every de Rham cohomology class contains a (unique) harmonic form. More precisely, we have the following theorem:

THEOREM 1.4.13 (Hodge theorem) *Let M be a compact Riemannian manifold and let $\mathcal{H}^r(M)$ denote the vector space of harmonic r-forms on M. Then*

(1) $\mathcal{H}^r(M)$ is finite-dimensional for all r;

(2) we have the following decomposition of the space of r-forms:

$$\begin{aligned} \mathcal{A}^r(M) &= \Delta(\mathcal{A}^r(M)) \oplus \mathcal{H}^r(M) \\ &= d\delta(\mathcal{A}^r(M)) \oplus \delta d(\mathcal{A}^r(M)) \oplus \mathcal{H}^r(M) \\ &= d(\mathcal{A}^{r-1}(M)) \oplus \delta(\mathcal{A}^{r+1}(M)) \oplus \mathcal{H}^r(M). \end{aligned}$$

The proof of this fundamental result involves the theory of elliptic differential operators on a manifold. We refer to [10, Chapter 0], [33, Chapter 6], and [35, Chapter 4].

Since d and δ are formal adjoints of each other it follows that

$$(\ker(d), \mathrm{Im}(\delta)) = (\ker(\delta), \mathrm{Im}(d)) = 0$$

and, consequently, if $\alpha \in \mathcal{Z}^r(M)$ and we write

$$\alpha = d\beta + \delta\gamma + \mu\,; \quad \beta \in \mathcal{A}^{r-1}(M), \gamma \in \mathcal{A}^{r+1}(M), \mu \in \mathcal{H}^r(M),$$

then

$$0 = (\alpha, \delta\gamma) = (\delta\gamma, \delta\gamma)$$

and therefore $\delta\gamma = 0$. Hence, $[\alpha] = [\mu]$. By the uniqueness statement in Proposition 1.4.12 we get

$$H^r(M, \mathbb{R}) \cong \mathcal{H}^r(M). \tag{1.4.6}$$

COROLLARY 1.4.14 *Let M be a compact, oriented, n-dimensional manifold. Then $H^r(M, \mathbb{R})$ is finite-dimensional for all r.*

COROLLARY 1.4.15 (Poincaré duality) *Let M be a compact, oriented, n-dimensional manifold. Then the bilinear pairing*

$$\int_M : H^r(M, \mathbb{R}) \times H^{n-r}(M, \mathbb{R}) \to \mathbb{R} \tag{1.4.7}$$

that maps $(\alpha, \beta) \mapsto \int_M \alpha \wedge \beta$ is nondegenerate. Hence

$$\left(H^{n-r}(M, \mathbb{R})\right)^* \cong H^r(M, \mathbb{R}).$$

PROOF. We may assume without loss of generality that M is a Riemannian manifold. Then, the Hodge star operator commutes with the Laplacian and defines an isomorphism:

$$\mathcal{H}^r(M) \cong \mathcal{H}^{n-r}(M).$$

Hence if $0 \neq \alpha \in \mathcal{H}^r(M)$ we have $*\alpha \in \mathcal{H}^{n-r}(M)$ and

$$\int_M \alpha \wedge *\alpha = (\alpha, \alpha) \neq 0.$$

$\square$

EXERCISE 1.4.16 Prove that the pairing (1.4.7) is well defined.

1.4.2 The $\bar{\partial}$-Laplacian

Let (M, J, ω) be a compact Kähler manifold and, as before, let g denote the associated Riemannian metric. Consider the L^2 inner product $(\bullet, \bullet)$ on $\mathcal{A}^*(M)$ defined in (1.4.3). Let $*$ be the corresponding star operator and $\delta = -*d*$ the adjoint of d. We extend these operators linearly to $\mathcal{A}^*(M, \mathbb{C})$. It follows from (B.2.4) that

$$*(\mathcal{A}^{p,q}(M)) \subset \mathcal{A}^{n-q,n-p}(M). \tag{1.4.8}$$

We write

$$\delta = -*d* = -*\bar{\partial}* - *\partial*,$$

and set

$$\partial^* := -*\bar{\partial}*; \quad \bar{\partial}^* := -*\partial*. \tag{1.4.9}$$

Note that $\bar{\partial}^*$ is indeed the conjugate of ∂^* and that ∂^* is pure of type $(-1, 0)$ and that $\bar{\partial}^*$ is pure of type $(0, -1)$ (see Exercise B.2.3).

EXERCISE 1.4.17 Let M be a compact, complex, n-dimensional manifold and $\alpha \in \mathcal{A}^{2n-1}(M, \mathbb{C})$. Prove that

$$\int_M \partial\alpha = \int_M \bar{\partial}\alpha = 0.$$

PROPOSITION 1.4.18 *The operator $\partial^* := -*\bar{\partial}*$ (resp. $\bar{\partial}^* := -*\partial*$) is the formal adjoint of ∂ (resp. $\bar{\partial}$) relative to the Hermitian extension $(\bullet, \bullet)^h$ of $(\bullet, \bullet)$ to $\mathcal{A}^*(M, \mathbb{C})$.*

PROOF. Given Exercise 1.4.17 and the Leibniz property for the operators $\partial, \bar{\partial}$, the proof of the first statement is analogous to that of Proposition 1.4.7. The details are left as an exercise. □

We can now define Laplace–Beltrami operators:

$$\Delta_\partial = \partial\partial^* + \partial^*\partial; \quad \Delta_{\bar{\partial}} = \bar{\partial}\bar{\partial}^* + \bar{\partial}^*\bar{\partial}. \tag{1.4.10}$$

The operators Δ_∂ and $\Delta_{\bar{\partial}}$ are of bidegree $(0, 0)$; i.e., they map forms of bidegree (p, q) to forms of the same bidegree. In particular, if $\alpha \in \mathcal{A}^k(U)$ is decomposed according to (1.2.6) as

$$\alpha = \alpha^{k,0} + \alpha^{k-1,1} + \cdots + \alpha^{0,k},$$

then $\Delta_{\bar{\partial}}(\alpha) = 0$ if and only if $\Delta_{\bar{\partial}}(\alpha^{p,q}) = 0$ for all p, q. The operators Δ_∂ and $\Delta_{\bar{\partial}}$ are elliptic and, consequently, the Hodge theorem remains valid for them. Thus if we set

$$\mathcal{H}_{\bar{\partial}}^{p,q}(M) := \{\alpha \in \mathcal{A}^{p,q}(M) : \Delta_{\bar{\partial}}(\alpha) = 0\}, \tag{1.4.11}$$

we have

$$H_{\bar{\partial}}^{p,q}(M) \cong \mathcal{H}_{\bar{\partial}}^{p,q}(M). \tag{1.4.12}$$

1.5   COHOMOLOGY OF COMPACT KÄHLER MANIFOLDS

1.5.1   The Kähler Identities

DEFINITION 1.5.1   *Let (M, ω) be an n-dimensional, compact, Kähler manifold. We define*

$$L_\omega \colon \mathcal{A}^k(M) \to \mathcal{A}^{k+2}(M) \, ; \quad L_\omega(\alpha) = \omega \wedge \alpha. \tag{1.5.1}$$

Let Λ_ω be the adjoint of L_ω relative to the inner product on $(\bullet, \bullet)$.

EXERCISE 1.5.2   Prove that for $\alpha \in \mathcal{A}^k(M)$, then $\Lambda_\omega \alpha = (-1)^k * L_\omega * \alpha$.

If there is no chance of confusion we will drop the subscript ω. It is clear, however, that the *Lefschetz operators* L and Λ depend on the choice of a Kähler form ω. We extend these operators linearly to $\mathcal{A}^k(M, \mathbb{C})$. It is easy to check that Λ is then the adjoint of L relative to the Hermitian extension of $(\bullet, \bullet)$ to $\mathcal{A}^k(M, \mathbb{C})$.

The following result describes the Kähler identities which describe the commutation relations among the differential operators $d, \partial, \bar{\partial}$ and the Lefschetz operators.

THEOREM 1.5.3 (Kähler identities)   *Let (M, ω) be a compact, Kähler manifold. Then the following identities hold:*

(1) $[\partial, L] \ = \ [\bar{\partial}, L] \ = \ [\partial^*, \Lambda] \ = \ [\bar{\partial}^*, \Lambda] \ = \ 0.$

(2) $[\bar{\partial}^*, L] = i\partial \, ; \quad [\partial^*, L] = -i\bar{\partial} \, ; \quad [\bar{\partial}, \Lambda] = i\partial^* \, ; \quad [\partial, \Lambda] = -i\bar{\partial}^*$

One of the standard ways to prove these identities makes use of the fact that they are of a local nature and only involve the coefficients of the Kähler metric up to first order. On the other hand, Theorem 1.3.9 asserts that a Kähler metric agrees with the standard Hermitian metric on $\mathbb{C}^n$ up to order two. Thus, it suffices to verify the identities in that case. This is done by a direct computation. This is the approach in [10] and [30, Proposition 6.5]. In Appendix B we describe a conceptually simpler proof due to Phillip Griffiths that reduces Theorem 1.5.3 to similar statements in the symplectic case. Since, by Darboux's theorem, a symplectic manifold is locally symplectomorphic to $\mathbb{R}^{2n}$ with the standard symplectic structure, the proof reduces to that case.

A remarkable consequence of the Kähler identities is the fact that on a compact Kähler manifold, the Laplacians Δ and $\Delta_{\bar{\partial}}$ are multiples of each other:

THEOREM 1.5.4   *Let M be a compact Kähler manifold. Then*

$$\Delta \ = \ 2\Delta_{\bar{\partial}}. \tag{1.5.2}$$

PROOF.   Note first of all that Theorem 1.5.3(2) yields

$$i(\partial\bar{\partial}^* + \bar{\partial}^*\partial) \ = \ \partial[\Lambda, \partial] + [\Lambda, \partial]\partial \ = \ \partial\Lambda\partial - \partial\Lambda\partial \ = \ 0.$$

Therefore,

$$
\begin{aligned}
\Delta_\partial &= \partial\partial^* + \partial^*\partial = i\,\partial[\Lambda,\bar\partial] + i\,[\Lambda,\bar\partial]\partial \\
&= i\,(\partial\Lambda\bar\partial - \partial\bar\partial\Lambda + \Lambda\bar\partial\partial - \bar\partial\Lambda\partial) \\
&= i\,\left(([\partial,\Lambda]\bar\partial + \Lambda\partial\bar\partial) - \partial\bar\partial\Lambda + \Lambda\bar\partial\partial - (\bar\partial[\Lambda,\partial] + \bar\partial\partial\Lambda)\right) \\
&= i\,\left(\Lambda(\partial\bar\partial + \bar\partial\partial) + (\partial\bar\partial + \bar\partial\partial)\Lambda - i\,(\bar\partial\bar\partial^* + \bar\partial^*\bar\partial)\right) \\
&= \Delta_{\bar\partial}.
\end{aligned}
$$

These two identities together yield (1.5.2). □

1.5.2 The Hodge Decomposition Theorem

Theorem 1.5.4 has a remarkable consequence: suppose $\alpha \in \mathcal{H}^k(M,\mathbb{C})$ is decomposed according to (1.2.6) as

$$
\alpha = \alpha^{k,0} + \alpha^{k-1,1} + \cdots + \alpha^{0,k};
$$

then since $\Delta = 2\Delta_{\bar\partial}$, the form α is $\Delta_{\bar\partial}$-harmonic and consequently, the components $\alpha^{p,q}$ are $\Lambda_{\bar\partial}$-harmonic and hence, Δ-harmonic as well. Therefore, if we set for $p+q = k$,

$$
\mathcal{H}^{p,q}(M) := \mathcal{H}^k(M,\mathbb{C}) \cap \mathcal{A}^{p,q}(M), \tag{1.5.3}
$$

we get

$$
\mathcal{H}^k(M,\mathbb{C}) \cong \bigoplus_{p+q=k} \mathcal{H}^{p,q}(M). \tag{1.5.4}
$$

Moreover, since Δ is a real operator, it follows that

$$
\mathcal{H}^{q,p}(M) = \overline{\mathcal{H}^{p,q}(M)}. \tag{1.5.5}
$$

If we combine these results with the Hodge theorem we get the following result:

THEOREM 1.5.5 (Hodge decomposition theorem)   *Let M be a compact Kähler manifold and let $H^{p,q}(M)$ be the space of de Rham cohomology classes in $H^{p+q}(M,\mathbb{C})$ that have a representative of bidegree (p,q). Then,*

$$
H^{p,q}(M) \cong H^{p,q}_{\bar\partial}(M) \cong \mathcal{H}^{p,q}(M) \tag{1.5.6}
$$

and

$$
H^k(M,\mathbb{C}) \cong \bigoplus_{p+q=k} H^{p,q}(M). \tag{1.5.7}
$$

Moreover, $H^{q,p}(M) = \overline{H^{p,q}(M)}$.

Remark. In view of Definition A.4.1, Theorem 1.5.5 may be restated as follows: the subspaces $(H(M,\mathbb{C}))^{p,q} \cong H^{p,q}_{\bar\partial}(M)$ define a Hodge structure of weight k on de Rham cohomology groups $H^k(M,\mathbb{R})$.

We will define $h^{p,q} = \dim_{\mathbb{C}} H^{p,q}(M)$. These are the so-called *Hodge numbers* of M. Note that the *Betti numbers* b^k, that is, the dimensions of the kth cohomology space, are given by

$$b^k = \sum_{p+q=k} h^{p,q}. \tag{1.5.8}$$

In particular, the Hodge decomposition theorem implies a new restriction on the cohomology of a compact Kähler manifold:

COROLLARY 1.5.6   *The odd Betti numbers of a compact Kähler manifold are even.*

PROOF. This assertion follows from (1.5.8) together with the fact that $h^{p,q} = h^{q,p}$. □

Remark.   The examples constructed by Thurston in [25] of complex symplectic manifolds with no Kähler structure are manifolds which do not satisfy Corollary 1.5.6.

Remark.   As pointed out in Exercise 1.2.11, the de Rham cohomology $H^*(M, \mathbb{C})$ is an algebra under the cup product. We note that the Hodge decomposition (1.5.7) is compatible with the algebra structure in the sense that

$$H^{p,q} \cup H^{p',q'} \subset H^{p+p',q+q'}. \tag{1.5.9}$$

This additional topological restriction for a compact, complex, symplectic manifold to have a Kähler metric has been successfully exploited by Voisin [32] to obtain remarkable examples of non-Kähler, symplectic manifolds.

Let M be a compact, n-dimensional Kähler manifold and $X \subset M$ a complex submanifold of codimension k. We may define a linear map:

$$\int_X : H^{2(n-k)}(M, \mathbb{C}) \to \mathbb{C}; \quad [\alpha] \mapsto \int_X \alpha|_X. \tag{1.5.10}$$

This map defines an element in $(H^{2(n-k)}(M, \mathbb{C}))^*$ and, therefore, by Corollary 1.4.15, a cohomology class $\eta_X \in H^{2k}(M, \mathbb{C})$ defined by the property that for all $[\alpha] \in H^{2(n-k)}(M, \mathbb{C})$,

$$\int_M \alpha \wedge \eta_X = \int_X \alpha|_X. \tag{1.5.11}$$

The class η_X is called the *Poincaré dual* of X and one can show that

$$\eta_X \in H^{k,k}(M) \cap H^{2k}(M, \mathbb{Z}). \tag{1.5.12}$$

One can also prove that the construction of the Poincaré dual may be extended to singular analytic subvarieties (cf. [10, 13]).

The following establishes a deep connection between the algebraic and analytic aspects of a smooth projective variety[12].

HODGE CONJECTURE *Let M be a smooth, projective manifold. Then*

$$H^{k,k}(M, \mathbb{Q}) := H^{k,k}(M) \cap H^{2k}(M, \mathbb{Q})$$

is generated, as a $\mathbb{Q}$-vector space, by the Poincaré duals of analytic subvarieties of M.

The Hodge conjecture is one of the remaining six Clay millennium problems [7]. It should be pointed out that all natural generalizations of the Hodge conjecture to compact Kähler manifolds fail; see [36, 29].

1.5.3 Lefschetz Theorems and Hodge–Riemann Bilinear Relations

Let (M, ω) be a compact Kähler manifold and let

$$\mathcal{A}^*(M, \mathbb{C}) := \bigoplus_{k=0}^{2n} \mathcal{A}^k(M, \mathbb{C}).$$

We can consider the operators L and Λ acting on $\mathcal{A}^*(M, \mathbb{C})$ and define a semisimple linear map $Y \colon \mathcal{A}^*(M, \mathbb{C}) \to \mathcal{A}^*(M, \mathbb{C})$ by

$$Y := \sum_{k=0}^{2n} (k - n)\pi_k,$$

where $\pi_k \colon \mathcal{A}^*(M, \mathbb{C}) \to \mathcal{A}^k(M, \mathbb{C})$ is the natural projection. Clearly L and Y are defined pointwise and, because of Exercise 1.5.2, so is Λ. Thus, it follows from Corollary B.2.5 that the operators $\{L, \Lambda, Y\}$ define an $\mathfrak{sl}_2$-triple.

We will now show how the Kähler identities imply that the Laplace–Beltrami operator Δ commutes with these operators and consequently, we get a (finite-dimensional) $\mathfrak{sl}_2$-representation on the space of harmonic forms $\mathcal{H}^*(M)$.

THEOREM 1.5.7 *Let (M, ω) be a Kähler manifold. Then, Δ commutes with L, Λ, and Y.*

PROOF. Clearly $[\Delta, L] = 0$ if and only if $[\Delta_\partial, L] = 0$. We have

$$
\begin{aligned}
[\Delta_\partial, L] &= [\partial\partial^* + \partial^*\partial, L] \\
&= \partial\partial^* L - L\partial\partial^* + \partial^*\partial L - L\partial^*\partial \\
&= \partial([\partial^*, L] + L\partial^*) - L\partial\partial^* + ([\partial^*, L] + L\partial^*)\partial - L\partial^*\partial \\
&= -i\partial\bar\partial - i\bar\partial\partial \\
&= 0.
\end{aligned}
$$

The identity $[\Delta, \Lambda] = 0$ follows by taking adjoints and $[\Delta, Y] = 0$ since Δ preserves the degree of a form. $\qquad\square$

We can now define an $\mathfrak{sl}_2$-representation on the de Rham cohomology of a compact Kähler manifold:

THEOREM 1.5.8 *The operators L, Y, and Λ define a real representation of $\mathfrak{sl}(2,\mathbb{C})$ on the de Rham cohomology $H^*(M,\mathbb{C})$. Moreover, these operators commute with the Weil operators of the Hodge structures on the subspaces $H^k(M,\mathbb{R})$.*

PROOF. This is a direct consequence of Theorem 1.5.7. The last statement follows from the fact that L, Y, and Λ are of bidegree $(1,1)$, $(0,0)$ and $(-1,-1)$, respectively.

$\square$

COROLLARY 1.5.9 (Hard Lefschetz theorem) *Let (M,ω) be an n-dimensional, compact Kähler manifold. For each $k \leq n$ the map*

$$L_\omega^k \colon H^{n-k}(M,\mathbb{C}) \to H^{n+k}(M,\mathbb{C}) \tag{1.5.13}$$

is an isomorphism.

PROOF. This follows from the results in Section A.3, in particular, Exercise A.3.7.

$\square$

We note, in particular, that for $j \leq k \leq n$, the maps

$$L^j \colon H^{n-k}(M,\mathbb{C}) \to H^{n-k+2j}(M,\mathbb{C})$$

are injective. This observation together with the hard Lefschetz theorem imply further cohomological restrictions on a compact Kähler manifold:

THEOREM 1.5.10 *The Betti and Hodge numbers of a compact Kähler manifold satisfy*

(1) $b^{n-k} = b^{n+k}$; $h^{p,q} = h^{q,p} = h^{n-q,n-p} = h^{n-p,n-q}$;

(2) $b^0 \leq b^2 \leq b^4 \leq \cdots$;

(3) $b^1 \leq b^3 \leq b^5 \leq \cdots$.

In both cases the inequalities continue up to, at most, the middle degree.

DEFINITION 1.5.11 *Let (M,ω) be a compact, n-dimensional Kähler manifold. For each index $k = p+q \leq n$, we define the **primitive cohomology** spaces*

$$H_0^{p,q}(M) \ := \ \ker\{L_\omega^{n-k+1} \colon H^{p,q}(M) \to H^{n-q+1,n-p+1}(M)\}, \tag{1.5.14}$$

$$H_0^k(M,\mathbb{C}) \ := \ \bigoplus_{p+q=k} H_0^{p,q}(M). \tag{1.5.15}$$

From Proposition A.3.9, we now have the following theorem:

THEOREM 1.5.12 (Lefschetz decomposition) *Let (M,ω) be an n-dimensional, compact Kähler manifold. For each $k = p+q \leq n$, we have*

$$H^k(M,\mathbb{C}) \ = \ H_0^k(M,\mathbb{C}) \oplus L_\omega(H^{k-2}(M,\mathbb{C})), \tag{1.5.16}$$

$$H^{p,q}(M) \ = \ H_0^{p,q}(M) \oplus L_\omega(H^{p-1,q-1}(M)). \tag{1.5.17}$$

The following result, whose proof may be found in [13, Proposition 1.2.31], relates the Hodge star operator with the $\mathfrak{sl}_2$-action.

PROPOSITION 1.5.13 *Let $\alpha \in \mathcal{P}^k(M, \mathbb{C})$. Then*

$$*L^j(\alpha) \;=\; (-1)^{k(k+1)/2} \frac{j!}{(n-k-j)!} \cdot L^{n-k-j}(C(\alpha)), \qquad (1.5.18)$$

where C is the Weil operator in $\mathcal{A}^k(M, \mathbb{C})$.

DEFINITION 1.5.14 *Let (M, ω) be an n-dimensional, compact, Kähler manifold. Let k be such that $0 \leq k \leq n$. We define a bilinear form*

$$Q_k = Q \colon H^k(M, \mathbb{C}) \times H^k(M, \mathbb{C}) \to \mathbb{C},$$

$$Q_k(\alpha, \beta) \;:=\; (-1)^{k(k-1)/2} \int_M \alpha \wedge \beta \wedge \omega^{n-k}. \qquad (1.5.19)$$

EXERCISE 1.5.15 Prove that Q is well defined; i.e., it is independent of our choice of representative in the cohomology class.

THEOREM 1.5.16 (Hodge–Riemann bilinear relations) *The bilinear form Q satisfies the following properties:*

(1) Q_k is symmetric if k is even and skew symmetric if k is odd.

(2) $Q(L_\omega \alpha, \beta) + Q(\alpha, L_\omega \beta) = 0$; we say that L_ω is an infinitesimal isomorphism of Q.

(3) $Q(H^{p,q}(M), H^{p',q'}(M)) = 0$ unless $p' = q$ and $q' = p$.

(4) If $0 \neq \alpha \in H_0^{p,q}(M)$ then
$$Q(C\alpha, \bar{\alpha}) > 0. \qquad (1.5.20)$$

PROOF. The first statement is clear. For the second note that the difference between the two terms is the preceding sign which changes as we switch from $k + 2$ to k. The third assertion follows from the fact that the integral vanishes unless the bidegree of the integrand is (n, n) and, for that to happen, we must have $p' = q$ and $q' = p$.

Therefore, we only need to show the positivity condition (4). Let $\alpha \in H_0^{p,q}(M)$. It follows from Proposition 1.5.13 that

$$(-1)^{k(k+1)/2} \omega^{n-k} \wedge \bar{\alpha} = *^{-1}(n-k)! \, C(\bar{\alpha}).$$

On the other hand, on $H^k(M)$, we have $C^2 = (-1)^k I = *^2$ and therefore

$$Q(C\alpha, \bar{\alpha}) \;=\; \int_M \alpha \wedge *\bar{\alpha} \;=\; (\alpha, \alpha)^h > 0.$$

$\square$

Properties (3) and (4) in Theorem 1.5.16 are called the first and second *Hodge–Riemann bilinear relations*. In view of Definition A.4.7 we may say that the Hodge–Riemann bilinear relations amount to the statement that the Hodge structure in the primitive cohomology $H_0^k(M, \mathbb{R})$ is polarized by the *intersection form* Q defined by (1.5.19).

EXAMPLE 1.5.17   Let $X = X_g$ denote a compact Riemann surface of genus g. Then we know that $H^1(X, \mathbb{Z}) \cong \mathbb{Z}^{2g}$. The Hodge decomposition in degree 1 is of the form

$$H^1(X, \mathbb{C}) = H^{1,0}(X) \oplus \overline{H^{1,0}(X)},$$

where $H^{1,0}(X)$ consists of the 1-forms on X which, locally, are of the form $f(z)\, dz$, with $f(z)$ holomorphic. The form Q on $H^1(X, \mathbb{C})$ is alternating and given by

$$Q(\alpha, \beta) = \int_X \alpha \wedge \beta.$$

The Hodge–Riemann bilinear relations then take the form $Q(H^{1,0}(X), H^{1,0}(X)) = 0$ and, since $H_0^{1,0}(X) = H^{1,0}(X)$,

$$iQ(\alpha, \bar{\alpha}) = i \int_X \alpha \wedge \bar{\alpha} > 0$$

if α is a nonzero form in $H^{1,0}(X)$. Note that, locally,

$$i\alpha \wedge \bar{\alpha} = i|f(z)|^2 dz \wedge d\bar{z} = 2|f(z)|^2 dx \wedge dy,$$

so both bilinear relations are clear in this case. We note that it follows that $H^{1,0}(X)$ defines a point in the complex manifold $D = D(H^1(X, \mathbb{R}), Q)$ defined in Example 1.1.23.

EXAMPLE 1.5.18   Suppose now that (M, ω) is a compact, connected, Kähler surface and let us consider the Hodge structure in the middle cohomology $H^2(X, \mathbb{R})$. We have the Hodge decomposition

$$H^2(X, \mathbb{C}) = H^{2,0}(X) \oplus H^{1,1}(X) \oplus H^{0,2}(X) ; \quad H^{0,2}(X) = \overline{H^{2,0}(X)}.$$

Moreover, $H_0^{2,0}(X) = H^{2,0}(X)$, while

$$H^{1,1}(X) = H_0^{1,1}(X) \oplus L_\omega H^{0,0}(X) = H_0^{1,1}(X) \oplus \mathbb{C} \cdot \omega$$

and

$$H_0^{1,1}(X) = \{\alpha \in H^{1,1}(X) : [\omega \wedge \alpha] = 0\}.$$

The polarization form on $H^2(X, \mathbb{R})$ is given by

$$Q(\alpha, \beta) = -\int_X \alpha \wedge \beta.$$

and the second Hodge–Riemann bilinear relation is equivalent to the following statements

$$\int_X \alpha \wedge \bar{\alpha} > 0, \quad \text{if} \quad 0 \neq \alpha \in H^{2,0}(X),$$

$$\int_X \omega^2 > 0,$$

$$\int_X \beta \wedge \bar{\beta} < 0, \quad \text{if} \quad 0 \neq \beta \in H^{1,1}_0(X).$$

We note that the first two statements are easy to verify, but that is not the case with the last one. We point out that the integration form $\mathcal{I}(\alpha, \beta) = -Q(\alpha, \beta)$ has index $(+, \cdots, +, -)$ in $H^{1,1}(X) \cap H^2(X, \mathbb{R})$; i.e., $\mathcal{I}$ is a hyperbolic symmetric bilinear form. Such forms satisfy the reverse Cauchy–Schwarz inequality: if $\mathcal{I}(\alpha, \alpha) \geq 0$, then

$$\mathcal{I}(\alpha, \beta)^2 \geq \mathcal{I}(\alpha, \alpha) \cdot \mathcal{I}(\beta, \beta) \qquad (1.5.21)$$

for all $\beta \in H^{1,1}(X) \cap H^2(X, \mathbb{R})$.

The inequality (1.5.21) is called Hodge's inequality and plays a central role in the study of algebraic surfaces. Via Poincaré duals it may be interpreted as an inequality between intersection indexes of curves in an algebraic surface or, in other words, about the number of points where two curves intersect. If the ambient surface is an *algebraic torus*, $X = \mathbb{C}^* \times \mathbb{C}^*$, then a curve is the zero locus of a Laurent polynomial in two variables and a classical result of Bernstein–Kushnirenko–Khovanskii says that, generically on the coefficients of the polynomials, the intersection indexes may be computed combinatorially from the Newton polytope of the defining polynomials (see Khovanskii's appendix in [4] for a full account of this circle of ideas). This relationship between the Hodge inequality, and combinatorics led Khovanskii and Teissier [24] to give (independent) proofs of the classical Alexandrov–Fenchel inequality for mixed volumes of polytopes using the Hodge inequality, and set the basis for a fruitful interaction between algebraic geometry and combinatorics. In particular, motivated by problems in convex geometry, Gromov [11] stated a generalization of the hard Lefschetz theorem, Lefschetz decomposition and Hodge–Riemann bilinear relations to the case of *mixed* Kähler forms. We give a precise statement in the case of the hard Lefschetz theorem and refer to [26, 27, 9, 5] for further details.

Kähler classes are real, $(1, 1)$ cohomology classes satisfying a positivity condition and define a cone $\mathcal{K} \subset H^{1,1}(M) \cap H^2(M, \mathbb{R})$. We have the following theorem:

THEOREM 1.5.19 (Mixed hard Lefschetz theorem)   *Let M be a compact Kähler manifold of dimension n. Let $\omega_1, \ldots, \omega_k \in \mathcal{K}$, $1 \leq k \leq n$. Then the map*

$$L_{\omega_1} \cdots L_{\omega_k} \colon H^{n-k}(M, \mathbb{C}) \to H^{n+k}(M, \mathbb{C})$$

is an isomorphism.

As mentioned above, this result was originally formulated by Gromov who proved it in the $(1, 1)$ case (note that the operators involved preserve the Hodge decomposition). Later, Timorin [26, 27] proved it in the linear algebra case and in the case of simplicial toric varieties. Dinh and Nguyên [9] proved it in the form stated above. In [5] the author gave a proof in the context of variations of Hodge structure which unifies those previous results as well as similar results in other contexts [14, 3].

Appendix A

Linear Algebra

A.1 REAL AND COMPLEX VECTOR SPACES

Here we will review some basic facts about finite-dimensional real and complex vector spaces that are used throughout this volume.

We begin by recalling the notion of *complexification* of a real vector space. Given a vector space V over $\mathbb{R}$ we define

$$V_{\mathbb{C}} := V \otimes_{\mathbb{R}} \mathbb{C}. \tag{A.1.1}$$

We can formally write $v \otimes (a + ib) = av + b(v \otimes i)$, $a, b \in \mathbb{R}$, and setting $iv := v \otimes i$ we may write $V_{\mathbb{C}} = V \oplus iV$. Scalar multiplication by complex numbers is then given by

$$(a + ib)(v_1 + iv_2) = (av_1 - bv_2) + i(av_2 + bv_1) ; \quad v_1, v_2 \in V; \ a, b \in \mathbb{R}.$$

Note that $\dim_{\mathbb{R}} V = \dim_{\mathbb{C}} V_{\mathbb{C}}$ and, in fact, if $\{e_1, \ldots, e_n\}$ is a basis of V (over $\mathbb{R}$), then $\{e_1, \ldots, e_n\}$ is also a basis of $V_{\mathbb{C}}$ (over $\mathbb{C}$).

The usual conjugation of complex numbers induces a conjugation operator on $V_{\mathbb{C}}$:

$$\sigma(v \otimes \alpha) := v \otimes \bar{\alpha} ; \quad v \in V, \ \alpha \in \mathbb{C},$$

or, formally, $\sigma(v_1 + iv_2) = v_1 - iv_2$, $v_1, v_2 \in V$. Clearly for $w \in V_{\mathbb{C}}$, we have that $w \in V$ if and only if $\sigma(w) = w$. If there is no possibility of confusion we will write $\sigma(w) = \bar{w}$, $w \in V_{\mathbb{C}}$.

Conversely, if W is a complex vector space over $\mathbb{C}$, then $W = V_{\mathbb{C}}$ for a real vector space V if and only if W has a conjugation σ; i.e., a map $\sigma \colon W \to W$ such that σ^2 is the identity, σ is additive linear, and

$$\sigma(\alpha w) = \bar{\alpha} \sigma(w) ; \quad w \in W; \ \alpha \in \mathbb{C}.$$

The set of fixed points $V := \{w \in W : \sigma(w) = w\}$ is a real vector space and $W = V_{\mathbb{C}}$. We call V a *real form* of W.

If V, V' are real vector spaces we denote by $\mathrm{Hom}_{\mathbb{R}}(V, V')$ the vector space of $\mathbb{R}$-linear maps from V to V'. It is easy to check that

$$(\mathrm{Hom}_{\mathbb{R}}(V, V'))_{\mathbb{C}} \cong \mathrm{Hom}_{\mathbb{C}}(V_{\mathbb{C}}, V'_{\mathbb{C}}) \tag{A.1.2}$$

and that if σ, σ' are the conjugation operators on $V_{\mathbb{C}}$ and $V'_{\mathbb{C}}$, respectively, then the conjugation operator on $\mathrm{Hom}_{\mathbb{C}}(V_{\mathbb{C}}, V'_{\mathbb{C}})$ is given by

$$\sigma_{\mathrm{Hom}}(T) = \sigma' \circ T \circ \sigma ; \quad T \in (\mathrm{Hom}_{\mathbb{R}}(V, V'))_{\mathbb{C}} ,$$

or in more traditional notation,

$$\bar{T}(w) = \overline{T(\bar{w})}; \quad w \in V_{\mathbb{C}}.$$

Thus, the group of real automorphisms of V may be viewed as the subgroup

$$\mathrm{GL}(V) = \{T \in \mathrm{GL}(V_{\mathbb{C}}) : \sigma \circ T = T \circ \sigma\} \subset \mathrm{GL}(V_{\mathbb{C}}).$$

If we choose $V' = \mathbb{R}$, then (A.1.2) becomes

$$(V^*)_{\mathbb{C}} \cong (V_{\mathbb{C}})^*,$$

where, as always, $V^* = \mathrm{Hom}_{\mathbb{R}}(V, \mathbb{R})$ and $(V_{\mathbb{C}})^* = \mathrm{Hom}_{\mathbb{C}}(V_{\mathbb{C}}, \mathbb{C})$ are the dual vector spaces. Thus, we may drop the parentheses and write simply $V_{\mathbb{C}}^*$. Note that for $\alpha \in V_{\mathbb{C}}^*$, its conjugate $\bar{\alpha}$ is

$$\bar{\alpha}(w) = \overline{\alpha(\bar{w})}; \quad w \in V_{\mathbb{C}}.$$

We may similarly extend the notion of complexification to the tensor products

$$T^{a,b}(V) := \overbrace{V \otimes \cdots \otimes V}^{a \text{ times}} \otimes \overbrace{V^* \otimes \cdots \otimes V^*}^{b \text{ times}}, \tag{A.1.3}$$

and to the exterior algebra $\bigwedge^r(V^*)$ and we have

$$\left(T^{a,b}(V)\right)_{\mathbb{C}} \cong T^{a,b}(V_{\mathbb{C}}); \quad \left(\bigwedge^r(V^*)\right)_{\mathbb{C}} \cong \bigwedge^r(V_{\mathbb{C}}^*).$$

In particular, a tensor $B \in T^{0,2}(V)$, which defines a bilinear form

$$B \colon V \times V \to \mathbb{R},$$

may be viewed as an element in $T^{0,2}(V_{\mathbb{C}})$ and defines a bilinear form

$$B \colon V_{\mathbb{C}} \times V_{\mathbb{C}} \to \mathbb{C}$$

satisfying $\bar{B} = B$. Explicitly, given $v_1, v_2 \in V$ we set

$$B(iv_1, v_2) = B(v_1, iv_2) = iB(v_1, v_2)$$

and extend linearly. A bilinear form $B \colon V_{\mathbb{C}} \times V_{\mathbb{C}} \to \mathbb{C}$ is real if and only if

$$B(w, w') = \overline{B(\bar{w}, \bar{w}')}$$

for all $w, w' \in V_{\mathbb{C}}$. Similarly, thinking of elements $\alpha \in \bigwedge^r(V^*)$ as alternating multilinear maps

$$\alpha \colon \overbrace{V \times \cdots \times V}^{r \text{ times}} \to \mathbb{R},$$

we may view them as alternating multilinear maps

$$\alpha \colon V_{\mathbb{C}} \times \cdots \times V_{\mathbb{C}} \to \mathbb{R}$$

satisfying

$$\alpha(w_1, \ldots, w_r) = \overline{\alpha(\bar{w}_1, \ldots, \bar{w}_r)}$$

for all $w_1, \ldots, w_r \in V_{\mathbb{C}}$.

On the other hand, given a $\mathbb{C}$-vector space W we may think of it as a real vector space simply by *forgetting* that we are allowed to multiply by complex numbers and restricting ourselves to multiplication by real numbers (this procedure is called *restriction of scalars*). To remind ourselves that we are only able to multiply by real numbers we write $W^{\mathbb{R}}$ when we are thinking of W as a real vector space. Note that

$$\dim_{\mathbb{R}}(W^{\mathbb{R}}) = 2 \dim_{\mathbb{C}}(W),$$

and that if $\{e_1, \ldots, e_n\}$ is a $\mathbb{C}$-basis of W then $\{e_1, \ldots, e_n, ie_1, \ldots, ie_n\}$ is a basis of $W^{\mathbb{R}}$.

It is now natural to ask when a real vector space V is obtained from a complex vector space W by restriction of scalars. Clearly, a necessary condition is that $\dim_{\mathbb{R}}(V)$ be even. But, there is additional structure on $V = W^{\mathbb{R}}$ coming from the fact that W is a $\mathbb{C}$-vector space. Indeed, multiplication by i in W induces an $\mathbb{R}$-linear map

$$J \colon W^{\mathbb{R}} \to W^{\mathbb{R}} ; \quad J(w) := i\,w$$

satisfying $J^2 = -I$, where I denotes the identity map.

Conversely, let V be a $2n$-dimensional real vector space and $J \colon V \to V$ a linear map such that $J^2 = -I$. Then we may define a $\mathbb{C}$-vector space structure on V by

$$(a + ib) * v := a\,v + b\,J(v). \tag{A.1.4}$$

We say that J is a *complex structure* on V and we will often denote by (V, J) the complex vector space consisting of the points in V endowed with the scalar multiplication[1] (A.1.4).

EXERCISE A.1.1 Let V be a real vector space and $J \colon V \to V$ a linear map such that $J^2 = -I$. Prove that there exists a basis $\{e_1, \ldots, e_n, f_1, \ldots, f_n\}$ of V such that the matrix of J in this basis is of the form

$$J = \begin{pmatrix} 0 & -I_n \\ I_n & 0 \end{pmatrix}, \tag{A.1.5}$$

where I_n denotes the $(n \times n)$ identity matrix.

PROPOSITION A.1.2 *Let V be a real vector space. Then the following are equivalent:*

(1) V has a complex structure J.

---

[1] We will use $*$ to denote complex multiplication in (V, J) to distinguish from the notation λv, $\lambda \in \mathbb{C}$ which is traditionally used to represent the point $(v \otimes \lambda) \in V_{\mathbb{C}}$. We will most often identify (V, J) with a complex subspace of $V_{\mathbb{C}}$ as in Proposition A.1.2 and therefore there will be no chance of confusion.

(2) The complexification $V_{\mathbb{C}}$ admits a decomposition

$$V_{\mathbb{C}} = W_+ \oplus W_- \tag{A.1.6}$$

where $W_\pm \subset V_{\mathbb{C}}$ are complex subspaces such that $\overline{W_\pm} = W_\mp$.

PROOF. Suppose $J : V \to V$ is a linear map such that $J^2 = -I$. Then we may extend J to a map $J : V_{\mathbb{C}} \to V_{\mathbb{C}}$. Since $J^2 = -I$, the only possible eigenvalues for J are $\pm i$. Let $W_\pm$ denote the $\pm i$-eigenspace of J. Clearly $W_+ \cap W_- = \{0\}$. Moreover, since any $v \in V_{\mathbb{C}}$ may be written as

$$v = \frac{1}{2}(v - iJv) + \frac{1}{2}(v + iJv),$$

and $v \mp iJv \in W_\pm$, it follows that

$$V_{\mathbb{C}} = W_+ \oplus W_-.$$

Suppose now that $w \in W_\pm$. Since J is a real map we have

$$J(\bar{w}) = \overline{J(w)} = \overline{\pm i\, w} = \mp i\, w.$$

Hence, $W_\pm = \overline{W_\mp}$ and we obtain the decomposition (A.1.6).

Conversely, given the decomposition (A.1.6), we define a linear map $J : V_{\mathbb{C}} \to V_{\mathbb{C}}$ by the requirement that $J(w) = \pm iw$ if $w \in W_\pm$. It is easy to check that $J^2 = -I$ and that the assumption that $\overline{W_\pm} = W_\mp$ implies that J is a real map; i.e., $\bar{J} = J$. $\square$

PROPOSITION A.1.3 *Let V be a real vector space with a complex structure J. Then, the map $\phi : (V, J) \to W_+$ defined by $\phi(v) = v - iJ(v)$ is an isomorphism of complex vector spaces.*

PROOF. We verify first of all that $\phi(v) \in W_+$; that is, $J(\phi(v)) = i\phi(v)$:

$$J(\phi(v)) = J(v - iJ(v)) = J(v) - iJ^2(v) = J(v) + iv = i(v - iJ(v)) = i\phi(v).$$

Next we check that the map is $\mathbb{C}$-linear. Let $a, b \in \mathbb{R}$, $v \in V$:

$$
\begin{aligned}
\phi((a + ib) * v) &= \phi(av + bJ(v)) = (av + bJ(v)) - iJ(av + bJ(v)) \\
&= (av + bJ(v)) + i(bv - aJ(v)) = (a + ib)(v - iJ(v)) \\
&= (a + ib)\phi(v).
\end{aligned}
$$

We leave it to the reader to verify that if $w \in W_+$ then $w = \phi(\frac{1}{2}(w + \bar{w}))$ and, therefore, ϕ is an isomorphism. $\square$

Suppose now that (V, J) is a $2n$-dimensional real vector space with a complex structure J and let $T \in \mathrm{GL}(V)$. Then T is a complex linear map if and only if $T(iv) = iT(v)$, i.e., if and only if

$$T \circ J = J \circ T. \tag{A.1.7}$$

EXERCISE A.1.4 Let V be a real vector space and $J\colon V \to V$ a complex structure on V. Prove an $\mathbb{R}$-linear map $T\colon V \to V$ is $\mathbb{C}$-linear if and only if the matrix of T, written in terms of a basis as in Exercise A.1.1, is of the form

$$\begin{pmatrix} A & -B \\ B & A \end{pmatrix},\qquad\qquad (A.1.8)$$

where A, B are $(n \times n)$ real matrices.

If $T \in \mathrm{GL}(V)$ satisfies (A.1.7), then the extension of T to the complexification $V_{\mathbb{C}}$ continues to satisfy the commutation relation (A.1.7). In particular, such a map T must preserve the eigenspaces of $J\colon V_{\mathbb{C}} \to V_{\mathbb{C}}$. Now, if $\{e_1, \ldots, e_n, f_1, \ldots, f_n\}$ is a basis of V as in Exercise A.1.1, then $w_i = \frac{1}{2}(e_i - iJe_i) = \frac{1}{2}(e_i - if_i)$, $i = 1, \ldots, n$ are a basis of W_+ and the conjugate vectors $\bar{w}_i = \frac{1}{2}(e_i + iJe_i) = \frac{1}{2}(e_i - if_i)$ $i = 1, \ldots, n$ are a basis of W_-. In this basis, the extension of T to $V_{\mathbb{C}}$ is written as

$$\begin{pmatrix} A + iB & 0 \\ 0 & A - iB \end{pmatrix}.\qquad\qquad (A.1.9)$$

We note, in particular, that if $T \in \mathrm{GL}(V)$ satisfies (A.1.7) then $\det(T) > 0$. Indeed, the determinant is unchanged after complexification and in terms of the basis $\{w_1, \ldots, w_n, w_1, \ldots, \bar{w}_n\}$, the matrix of T is as in (A.1.9) and we have

$$\det(T) = |\det(A + iB)|^2.\qquad\qquad (A.1.10)$$

If J is a complex structure on the real vector space W then the dual map $J^*\colon W^* \to W^*$ is a complex structure on the dual space W^*. The corresponding decomposition (A.1.6) on the complexification $W_{\mathbb{C}}^*$ is given by

$$W_{\mathbb{C}}^* = W_+^* \oplus W_-^*,\qquad\qquad (A.1.11)$$

where $W_{\pm}^* := \{\alpha \in W_{\mathbb{C}}^* : \alpha|_{W_{\mp}} = 0\}$. Indeed, if $\alpha \in W_{\mathbb{C}}^*$ is such that $J^*(\alpha) = i\alpha$ and $w \in W_-$ then we have

$$i\alpha(w) = (J^*(\alpha))(w) = \alpha(J(w)) = \alpha(-iw) = -i\alpha(w),$$

which implies that $\alpha(w) = 0$, and the statement follows for dimensional reasons. The decomposition (A.1.11) now induces a bigrading of the exterior product

$$V_{\mathbb{C}} := \bigwedge\nolimits^* W_{\mathbb{C}}^* = \bigoplus\nolimits_{p,q} V^{p,q},\qquad\qquad (A.1.12)$$

where

$$V^{p,q} := \bigwedge\nolimits^p W_+^* \wedge \bigwedge\nolimits^q W_-^*.$$

The complex structure J extends to an operator on V satisfying $J\alpha = i^{p-q}\alpha$ if $\alpha \in V^{p,q}$ and consequently,

$$J^2(\alpha) = (-1)^{\deg(\alpha)}\alpha.$$

A.2   THE WEIGHT FILTRATION OF A NILPOTENT TRANSFORMATION

In this section we will construct a filtration canonically attached to a nilpotent linear transformation and study its relationship with representations of the Lie algebra $\mathfrak{sl}(2, \mathbb{R})$. This filtration plays a key role in the study of degenerations of variations of Hodge structure.

Throughout this section, $N \colon V \to V$ will be a nilpotent linear transformation of *nilpotency index* k; i.e., k, the first positive integer such that $N^{k+1} = 0$. Given an integer $m \le k$, we will say that a subspace $A \subset V$ is a *Jordan block* of weight m if A has a basis $\{f_0^A, f_1^A, \ldots, f_m^A\}$ such that $N(f_i^A) = f_{i+1}^A$, where we set $f_{m+1}^A = 0$. It is convenient to re-index the basis as

$$e_{m-2j}^A := f_j^A.$$

Note that if m is even then the index of e^A takes even values from m to $-m$, while if m is odd then it takes odd values from m to $-m$.

Given a nilpotent transformation $N \colon V \to V$ of nilpotency index k we may decompose V as a direct sum

$$V = \bigoplus_{m=0}^{k} U_m ,$$

where U_m is the direct sum of all Jordan blocks of weight m. In fact, these subspaces U_m are unique. Clearly, each U_m decomposes further as

$$U_m = \bigoplus_{j=0}^{m} U_{m,m-2j} ,$$

where $U_{m,m-2j}$ is the subspace spanned by all basis vectors e_{m-2j}^A as A runs over all Jordan blocks of weight m. We now define

$$E_\ell = E_\ell(N) = \bigoplus_{m=0}^{k} U_{m,\ell}. \tag{A.2.1}$$

THEOREM A.2.1   *The decomposition* (A.2.1) *satisfies*

(1) $N(E_\ell) \subset E_{\ell-2}$;

(2) *for* $\ell \ge 0$, $N^\ell \colon E_\ell \to E_{-\ell}$ *is an isomorphism.*

PROOF.   The first statement is clear since for any Jordan block A, then $N(f_j^A) = f_{j+1}^A$ which implies the assertion. Suppose now that ℓ is even. Then, E_ℓ is spanned by all basis vectors of the form $e_\ell^A = f_{(m-\ell)/2}^A$, where A runs over all Jordan blocks of even weight $m \ge \ell$.   □

PROPOSITION A.2.2   *Let N be a nilpotent transformation of nilpotency index k. Then there exists a unique increasing filtration $W = W(N)$,*

$$\{0\} \subset W_{-k} \subset W_{-k+1} \subset \cdots \subset W_{k-1} \subset W_k = V ,$$

with the following properties:

(1) $N(W_\ell) \subset W_{\ell-2}$.

(2) For $\ell \geq 0$, $N^\ell\colon \mathrm{Gr}_\ell^W \to \mathrm{Gr}_{-\ell}^W$, where $\mathrm{Gr}_\ell^W := W_\ell/W_{\ell-1}$ is an isomorphism.

Moreover, the subspaces W_ℓ may be expressed in terms of $\ker(N^a)$ and $\mathrm{Im}(N^b)$. Hence, they are defined over $\mathbb{Q}$ (resp. over $\mathbb{R}$) if N is.

PROOF. The existence of $W(N)$ follows from Theorem A.2.1 while the uniqueness is a consequence of the uniqueness properties of the Jordan decomposition. Alternatively one may give an inductive construction of $W_\ell(N)$ as in [21, Lemma 6.4]. For an explicit construction involving kernels and images of powers of N, we refer to [22]. □

EXAMPLE A.2.3 Suppose $k = 1$. Then the weight filtration is of the form

$$\{0\} \subset W_{-1} \subset W_0 \subset W_1 = V.$$

Since $N\colon V/W_0 \to W_{-1}$ is an isomorphism it follows that

$$W_{-1}(N) = \mathrm{Im}(N) \;;\quad W_0(N) = \ker(N).$$

EXERCISE A.2.4 Prove that if $k = 2$ the weight filtration

$$\{0\} \subset W_{-2} \subset W_{-1} \subset W_0 \subset W_1 \subset W_2 = V$$

is given by

$$\{0\} \subset \mathrm{Im}(N^2) \subset \mathrm{Im}(N) \cap \ker(N) \subset \mathrm{Im}(N) + \ker(N) \subset \ker(N^2) \subset V.$$

DEFINITION A.2.5 *Let $V = \bigoplus_{\ell \in \mathbb{Z}} V_\ell$ be a finite-dimensional graded real vector space and N an endomorphism of V. We say that the pair (V, N) satisfies the **hard Lefschetz (HL) property** if and only if $N(V_\ell) \subset V_{\ell+2}$ and*

$$N^\ell\colon V_{-\ell} \to V_\ell \qquad\qquad (A.2.2)$$

is an isomorphism for all $\ell \geq 0$.

Clearly, if (V, N) satisfies HL, N is nilpotent. Moreover, for any nilpotent transformation $N \in \mathrm{End}(V)$, the pair (Gr^W, N), where Gr^W is given the opposite grading, satisfies HL.

A.3 REPRESENTATIONS OF $\mathfrak{sl}(2, \mathbb{C})$ AND LEFSCHETZ THEOREMS

We recall that the Lie algebra $\mathfrak{sl}(2, \mathbb{C})$ consists of all 2×2 complex matrices of trace zero. It has a basis consisting of

$$\mathbf{n}_+ := \begin{pmatrix} 0 & 1 \\ 0 & 0 \end{pmatrix} \;;\quad \mathbf{n}_- := \begin{pmatrix} 0 & 0 \\ 1 & 0 \end{pmatrix} \;;\quad \mathbf{y} := \begin{pmatrix} 1 & 0 \\ 0 & -1 \end{pmatrix}. \qquad (A.3.1)$$

This basis satisfies the commutation relations

$$[\mathbf{y}, \mathbf{n}_+] = 2\mathbf{n}_+ ; \quad [\mathbf{y}, \mathbf{n}_-] = -2\mathbf{n}_- ; \quad [\mathbf{n}_+, \mathbf{n}_-] = \mathbf{y} . \tag{A.3.2}$$

A *representation* ρ of $\mathfrak{sl}(2, \mathbb{C})$ on a complex vector space $V_{\mathbb{C}}$ is a Lie algebra homomorphism

$$\rho \colon \mathfrak{sl}(2, \mathbb{C}) \to \mathfrak{gl}(V_{\mathbb{C}}).$$

We denote the image of the generators of $\mathfrak{sl}(2, \mathbb{C})$ by $N_+, N_-,$ and Y. These elements satisfy commutation relations analogous to (A.3.2). Conversely, given elements $\{N_+, N_-, Y\} \subset \mathfrak{gl}(V)$ satisfying the commutation relations (A.3.2), we can define a representation $\rho \colon \mathfrak{sl}(2, \mathbb{C}) \to \mathfrak{gl}(V_{\mathbb{C}})$. We will refer to $\{N_+, N_-, Y\}$ as an $\mathfrak{sl}_2$-triple. We say that ρ is *real* if $V_{\mathbb{C}}$ is the complexification of a real vector space V and $\rho(\mathfrak{sl}(2, \mathbb{R})) \subset \mathfrak{gl}(V)$.

A representation ρ is called *irreducible* if V has no proper subspaces invariant under $\rho(\mathfrak{sl}(2, \mathbb{C}))$.

EXAMPLE A.3.1   For each $n \in \mathbb{Z}$ we may define an irreducible representation ρ_n on $V_{\mathbb{C}} = \mathbb{C}^{n+1}$. We will describe the construction for n odd and leave the even case as an exercise. If $n = 2k + 1$, we label the standard basis of V as

$$\{e_{-k}, e_{-k+2}, \ldots, e_{k-2}, e_k\}$$

and define

$$Y(e_j) := j \cdot e_j ; \quad N_-(e_j) = e_{j-2} ; \quad N_+(e_j) = \mu_j \cdot e_{j+2},$$

where the integers μ_j are the unique solution to the recursion equations

$$\mu_{j-2} - \mu_j = j ; \quad \mu_{-k-2} = 0. \tag{A.3.3}$$

It is easy to check that the first two commutation relations are satisfied. On the other hand,

$$[N_+, N_-](e_j) = N_+(e_{j-2}) - N_-(\mu_j \cdot e_{j-2}) = (\mu_{j-2} - \mu_j) \cdot e_j = j \cdot e_j = Y(e_j).$$

EXERCISE A.3.2   Solve the equations (A.3.3).

EXERCISE A.3.3   Extend the construction of the representation in Example A.3.1 to the case $n = 2k$.

We note that for the representation ρ_n we have $N_-^n = 0$ and $N_-^{n-1} \neq 0$; that is, the index of nilpotency of N_- (and of N_+) is $n - 1$. At the same time, the eigenvalues of Y range from $-n+1$ to $n-1$. We will refer to $n-1$ as the *weight* of the representation ρ_n. This notion of weight is consistent with that defined for Jordan blocks above.

The basic structure theorem about representations of $\mathfrak{sl}(2, \mathbb{C})$ follows:

THEOREM A.3.4   *Every finite-dimensional representation of* $\mathfrak{sl}(2, \mathbb{C})$ *splits as a direct sum of irreducible representations. Moreover, an irreducible representation of dimension n is isomorphic to* ρ_n.

PROOF. We refer to [35, Chapter V, Section 3] for a proof.   □

Suppose now that $\rho\colon \mathfrak{sl}(2,\mathbb{C}) \to \mathfrak{gl}(V)$ is a representation and set $N = N_-$. Let k be the nilpotency index of N. Then ρ splits as a direct sum of irreducible representations of weight at most k and we have

$$W_\ell(N) = \bigoplus_{j \leq \ell} E_j(Y). \tag{A.3.4}$$

Indeed, it is enough to check this statement for each irreducible representation and there the statement is clear.

EXERCISE A.3.5   Let $N \in \mathfrak{gl}(V)$ be nilpotent and $Y \in \mathfrak{gl}(V)$ be semisimple. Then the following are equivalent:

(1) There exists an $\mathfrak{sl}_2$-triple $\{N_+, N_-, Y\}$ with $N_- = N$.

(2) $[Y, N] = -2N$ and the weight filtration of N is given by (A.3.4).

The above exercise implies that if N is a nilpotent element, W_ℓ is its weight filtration and $\{V_\ell\}$ is a *splitting* of the filtration W in the sense that

$$W_\ell = V_\ell \oplus W_{\ell-1},$$

then if we define $Y \in \mathfrak{gl}(V)$ by $Y(v) = \ell v$ for $v \in V_\ell$, the pair $\{N, Y\}$ may be extended to an $\mathfrak{sl}_2$-triple. Moreover if N is defined over $\mathbb{Q}$ (resp. over $\mathbb{R}$) and the splitting is defined over $\mathbb{Q}$ (resp. over $\mathbb{R}$), then so is the $\mathfrak{sl}_2$-triple.

EXERCISE A.3.6   Apply Exercise A.3.5 to prove that if N is a nilpotent transformation, then there exists an $\mathfrak{sl}_2$-triple with $N = N_-$. This is a version of the Jacobson–Morosov theorem.

EXERCISE A.3.7   Let $N \in \mathfrak{gl}(V)$ be nilpotent and $Y \in \mathfrak{gl}(V)$ be semisimple. Set $V_\ell = E_\ell(Y)$. Then the following are equivalent:

(1) There exists an $\mathfrak{sl}_2$-triple $\{N_+, N_-, Y\}$ with $N_+ = N$.

(2) The pair (V, N) satisfies the hard Lefschetz property.

Thus, given a graded vector space V, the pair (V, N) satisfies HL if and only if the pair Y, N extends to an $\mathfrak{sl}_2$-triple, where $Y = \sum_\ell \ell\pi_\ell$, where $\pi_\ell\colon V \to V_\ell$ is the natural projection.

Let V be a graded real vector space and suppose (V, N) satisfies the HL property. For $\ell \leq 0$ we define

$$P_\ell := \ker\{N^{\ell+1}\colon V_\ell \to V_{\ell+2}\} \tag{A.3.5}$$

and call it the kth *primitive* space.

EXERCISE A.3.8   Suppose $N^3 = 0$ but $N^2 \neq 0$. Prove that $P_1 = V_1$ and

$$V_0 = \ker(N)/(\ker(N) \cap \operatorname{Im}(N)) \subset V_0.$$

PROPOSITION A.3.9 *Let V be a graded real vector space and suppose (V, N) satisfies the HL property. Let k be the index of nilpotency of N. Then for any $\mathfrak{sl}_2$-triple with $N_+ = N$ we have*

$$P_\ell := \ker\{N_- : V_\ell \to V_{\ell+2}\}.$$

Moreover, for every ℓ, $-k \leq \ell \leq 0$, we have

$$V_\ell = P_\ell \oplus N(V_{\ell-2}). \tag{A.3.6}$$

PROOF. Let $\{N, N_-, Y\}$ be an $\mathfrak{sl}_2$-triple with $N_+ = N$. Then V_ℓ is given by all eigenvectors of Y of eigenvalue ℓ living in the sum of irreducible components of the representation of weight ℓ and these are exactly the elements annihilated by N_+. Similarly, it suffices to verify the decomposition (A.3.6) in each irreducible component which is easy to do.   □

The decomposition (A.3.6), or more precisely, the decomposition obtained inductively from (A.3.6),

$$V_\ell = P_\ell \oplus N(P_{\ell-2}) \oplus N^2(P_{\ell-4}) + \cdots, \tag{A.3.7}$$

is called the *Lefschetz decomposition*.

EXAMPLE A.3.10 The only interesting term in the Lefschetz decomposition for $k = 2$ occurs for $\ell = 0$, where, according to Exercises A.2.4 and A.3.8 we get

$$V_0 = P_0 \oplus N(P_{-2}) = \ker(N)/(\ker(N) \cap \mathrm{Im}(N)) \oplus N(V/\ker(N^2)).$$

A.4 HODGE STRUCTURES

We will now review the basic definitions of the central object in this volume: *Hodge structures*.

DEFINITION A.4.1 *A (real) Hodge structure of weight $k \in \mathbb{Z}$ consists of*

(1) a finite-dimensional real vector space V;

(2) a decomposition of the complexification $V_{\mathbb{C}}$ as

$$V_{\mathbb{C}} = \bigoplus_{p+q=k} V^{p,q}; \quad V^{q,p} = \overline{V^{p,q}}. \tag{A.4.1}$$

We say that the Hodge structure is rational (resp. integral) if there exists a rational vector space $V_{\mathbb{Q}}$ (resp. a lattice $V_{\mathbb{Z}}$) such that $V = V_{\mathbb{Q}} \otimes_{\mathbb{Q}} \mathbb{R}$ (resp. $V = V_{\mathbb{Z}} \otimes_{\mathbb{Z}} \mathbb{R}$).

The following statement is valid for Hodge structures defined over $\mathbb{R}$, $\mathbb{Q}$, or $\mathbb{Z}$.

PROPOSITION A.4.2 *Let V and W be vector spaces with Hodge structures of weight k, ℓ, respectively. Then $\mathrm{Hom}(V, W)$ has a Hodge structure of weight $\ell - k$.*

PROOF. We set

$$\mathrm{Hom}(V,W)^{a,b} := \{X \in \mathrm{Hom}_{\mathbb{C}}(V_{\mathbb{C}}, W_{\mathbb{C}}) : X(V^{p,q}) \subset W^{p+a,q+b}\}. \qquad (A.4.2)$$

Then $\mathrm{Hom}(V,W)^{a,b} = 0$ unless $p + a + q + b = \ell$, that is, unless $a + b = \ell - k$. The rest of the verifications are left to the reader. $\qquad \square$

In particular, if we choose $W = \mathbb{R}$ with the Hodge structure $W_{\mathbb{C}} = W^{0,0}$, we see that if V has a Hodge structure of weight k, then V^* has a Hodge structure of weight $-k$. Similarly, if V and W have Hodge structures of weight k, ℓ, respectively, then $V \otimes_{\mathbb{R}} W \cong \mathrm{Hom}_{\mathbb{R}}(V^*, W)$ has a Hodge structure of weight $k + \ell$ and

$$(V \otimes W)^{a,b} = \bigoplus_{\substack{p+r=a \\ q+s=b}} V^{p,q} \otimes_{\mathbb{C}} W^{r,s}. \qquad (A.4.3)$$

Needless to say, we could take (A.4.3) as our starting point rather than (A.4.2). Note that if V has a Hodge structure of weight k then the tensor product $T^{a,b}(V)$ defined in (A.1.3) has a Hodge structure of weight $k(a - b)$.

EXAMPLE A.4.3 Let W be a real vector space with a complex structure J. Let $V_k = \bigwedge^k W^*$. Then the decomposition

$$(V_k)_{\mathbb{C}} = \oplus_{p+q=k} V^{p,q}$$

given by the bigrading (A.1.12) satisfies $V^{q,p} = \overline{V^{p,q}}$ and, therefore, defines a Hodge structure of weight k on V_k.

There are two alternative ways of describing a Hodge structure on a vector space V that will be very useful to us.

DEFINITION A.4.4 *A real (resp. rational, integral)* ***Hodge structure*** *of weight $k \in \mathbb{Z}$ consists of a real vector space V (resp. a rational vector space $V_{\mathbb{Q}}$, a lattice $V_{\mathbb{Z}}$) and a decreasing filtration*

$$\cdots F^p \subset F^{p-1} \cdots$$

of the complex vector space $V_{\mathbb{C}} = V \otimes_{\mathbb{R}} \mathbb{C}$ (resp. $V_{\mathbb{C}} = V_{\mathbb{Q}} \otimes_{\mathbb{Q}} \mathbb{C}$, $V_{\mathbb{C}} = V_{\mathbb{Z}} \otimes_{\mathbb{Z}} \mathbb{C}$) such that

$$V_{\mathbb{C}} = F^p \oplus \overline{F^{k-p+1}}. \qquad (A.4.4)$$

The equivalence of Definitions A.4.1 and A.4.4 is easy to verify. Indeed given a decomposition as in (A.4.1) we set

$$F^p = \bigoplus_{a \geq p} V^{a,k-a},$$

while given a filtration of $V_{\mathbb{C}}$ satisfying (A.4.4), the subspaces

$$V^{p,q} = F^p \cap \overline{F^q}; \quad p + q = k$$

define a decomposition of $V_{\mathbb{C}}$ satisfying (A.4.1).

In order to state the third definition of a Hodge structure, we need to recall some basic notions from representation theory. Let us denote by $\mathbb{S}(\mathbb{R})$ the real algebraic group

$$\mathbb{S}(\mathbb{R}) := \left\{ \begin{pmatrix} a & -b \\ b & a \end{pmatrix} \in \mathrm{GL}(2, \mathbb{R}) \right\}.$$

Then $\mathbb{C}^* \cong \mathbb{S}(\mathbb{R})$ via the identification

$$z = a + ib \mapsto \begin{pmatrix} a & -b \\ b & a \end{pmatrix}.$$

The circle group $S^1 = \{ z \in \mathbb{C}^* : |z| = 1 \}$ is then identified with the group of rotations $SO(2, \mathbb{R})$.

Recall that a representation of an algebraic group G defined over the field $F = \mathbb{Q}, \mathbb{R},$ or $\mathbb{C}$, on an F-vector space V_F is a group homomorphism $\varphi \colon G \to \mathrm{GL}(V_F)$.

Now, if V is a real vector space with a Hodge structure of weight k, then we may define a representation of $\mathbb{S}(\mathbb{R})$ on $V_{\mathbb{C}}$ by

$$\varphi(z)(v) := \sum_{p+q=k} z^p \, \bar{z}^q \, v_{p,q},$$

where $v = \sum v_{p,q}$ is the decomposition of v according to (A.4.1). We verify that $\varphi(z) \in \mathrm{GL}(V)$, i.e., that $\overline{\varphi(z)} = \varphi(z)$:

$$\overline{\varphi(z)}(v) = \overline{\varphi(z)(\bar{v})} = \overline{\sum_{p+q=k} z^p \, \bar{z}^q \, \overline{v_{q,p}}}$$

since $(\bar{v})_{p,q} = \overline{v_{q,p}}$. Hence $\overline{\varphi(z)}(v) = \varphi(z)(v)$. Note that $\varphi(\lambda)(v) = \lambda^k \, v$ for all $v \in V, \lambda \in \mathbb{R}^*$.

Conversely, it follows from the representation theory of $\mathbb{S}(\mathbb{R})$ that every finite-dimensional representation of $\mathbb{S}(\mathbb{R})$ on a complex vector space splits as a direct sum of one-dimensional representations, where $z \in \mathbb{S}(\mathbb{R})$ acts as multiplication by $z^p \bar{z}^q$, with $p, q \in \mathbb{Z}$. Hence, a representation $\varphi \colon \mathbb{S}(\mathbb{R}) \to \mathrm{GL}(V_{\mathbb{C}})$ defined over $\mathbb{R}$ (i.e., $\bar{\varphi} = \varphi$) decomposes $V_{\mathbb{C}}$ into subspaces $V^{p,q}$,

$$V_{\mathbb{C}} = \bigoplus V^{p,q}; \quad V^{q,p} = \overline{V^{p,q}},$$

where $\varphi(z)$ acts as multiplication by $z^p \bar{z}^q$. Note, moreover, that if $\lambda \in \mathbb{R}^* \subset \mathbb{S}(\mathbb{R})$, then $\varphi(\lambda)$ acts on $V^{p,q}$ as multiplication by λ^{p+q}. Thus, the following definition is equivalent to Definitions A.4.1 and A.4.4.

DEFINITION A.4.5 *A **real Hodge structure** of weight $k \in \mathbb{Z}$ consists of a real vector space V and a representation $\varphi \colon \mathbb{S}(\mathbb{R}) \to \mathrm{GL}(V)$ such that $\varphi(\lambda)(v) = \lambda^k \, v$ for all $v \in V$ and all $\lambda \in \mathbb{R}^* \subset \mathbb{S}(\mathbb{R})$.*

Given a Hodge structure φ of weight k on V, the linear operator $\varphi(i)\colon V_{\mathbb{C}} \to V_{\mathbb{C}}$ is called the *Weil operator* and denoted by C. Note that on $V^{p,q}$ the Weil operator acts as multiplication by i^{p-q} and, consequently, if J is a complex structure on V then for the Hodge structure of weight k defined on the exterior product $\bigwedge^k(V^*)$, the Weil operator agrees with the natural extension of J^* to $\bigwedge^k(V^*)$.

EXERCISE A.4.6   Prove that if (V, φ), (V', φ') are Hodge structures of weight k and k', respectively, then (V^*, φ^*) and $(V \otimes V', \varphi \otimes \varphi')$ are the natural Hodge structures on V^* and $V \otimes V'$ defined above. Here $\varphi^*\colon \mathbb{S}(\mathbb{R}) \to GL(V^*)$ is the representation $\varphi^*(z) := (\varphi(z))^*$ and, similarly, $(\varphi \otimes \varphi')(z) := \varphi(z) \otimes \varphi'(z)$.

DEFINITION A.4.7   *Let (V, φ) be a real Hodge structure of weight k. A **polarization** of (V, φ) is a real[2] bilinear form $Q\colon V \times V \to \mathbb{R}$ such that*

(i) $Q(u, v) = (-1)^k Q(v, u)$; i.e., Q is symmetric or skew symmetric depending on whether k is even or odd;

(ii) The Hodge decomposition is orthogonal relative to the Hermitian form H on $V_{\mathbb{C}}$ defined by

$$H(w_1, w_2) := Q(C\, w_1, \bar{w}_2), \qquad\qquad (A.4.5)$$

where $C = \varphi(i)$ is the Weil operator;

(iii) H is positive definite.

Remark.   Note that if k is even, then the Weil operator acts on $V^{p,q}$ as multiplication by ± 1 and then it is clear that the form H defined by (A.4.5) is Hermitian. A similar statement holds if k is odd since in this case C acts on $V^{p,q}$ as multiplication by $\pm i$. We also note that (ii) and (iii) above may be restated as follows:

(ii') $Q(V^{p,q}, V^{p',q'}) = 0$ if $p' \neq k - p$;

(iii') $i^{p-q} Q(w, \bar{w}) > 0$ for all $0 \neq w \in V^{p,q}$.

The statements (ii') and (iii') correspond to the *Hodge–Riemann bilinear relations* in Theorem 1.5.16.

---

[2]If (V, φ) is a rational (resp. integral) Hodge structure, then we require Q to be defined over $\mathbb{Q}$ (resp. over $\mathbb{Z}$).

Appendix B

The Kähler Identities
by Phillip Griffiths

The aim of this appendix is to give an elementary proof of the classical Kähler identities by deriving them from the analogous identities in the symplectic case, where there are no local invariants, and which may be proved by induction.[1]

B.1   SYMPLECTIC LINEAR ALGEBRA

DEFINITION B.1.1   *A **symplectic structure** on a real vector space W is an alternating bilinear form Q on W which is nondegenerate in the sense that the map $h\colon W \to W^*$ defined by $h(v) = Q(v, \bullet)$ is an isomorphism.*

We will also use the inverse $q = h^{-1}\colon W^* \to W$ determined by

$$\phi = Q(q(\phi), \bullet) \quad \forall \phi \in W^*.$$

The isomorphism q gives rise to an isomorphism $\bigwedge^* q\colon \bigwedge^* W^* \to \bigwedge^* W$ and to a bilinear form on $\bigwedge^* W^*$ also denoted by Q:

$$Q(\alpha, \beta) = \langle \alpha, {\textstyle\bigwedge}^* q(\beta) \rangle. \tag{B.1.1}$$

Note that if $\alpha \in \bigwedge^k W^*$,

$$Q(\alpha, \beta) = (-1)^k Q(\beta, \alpha).$$

We denote by ω the 2-form associated with Q:

$$\omega(v \wedge w) = Q(v, w), \tag{B.1.2}$$

and by $\Omega = \omega^d/d!$ the corresponding volume element.

EXERCISE B.1.2   Let Q be a symplectic structure on W.

(1) Prove that $\dim W$ is even.

(2) Prove that there exists a basis (*symplectic basis*) $\{v_1, \ldots, v_d, w_1, \ldots, w_d\}$ of W such that

$$Q(v_i, v_j) = Q(w_i, w_j) = 0 \quad \text{and} \quad Q(v_i, w_j) = \delta_{ij}.$$

---

[1]Many thanks to Robert Bryant and Olivier Guichard for their help in the preparation of this appendix.

(3) Let $\{v_1, \ldots, v_d, w_1, \ldots, w_d\}$ be a symplectic basis. Show that the dual basis $\{\xi_1, \ldots, \xi_d, \eta_1 \ldots, \eta_d\} \subset W^*$ is given by $\xi_i := -h(w_i)$, $\eta_i := h(v_i)$, $i = 1, \ldots, d$.

(4) Prove that for any symplectic basis of W,

$$\omega = \sum_{i=1}^{d} \xi_i \wedge \eta_i \, ; \quad \Omega = (\xi_1 \wedge \eta_1) \wedge \cdots \wedge (\xi_d \wedge \eta_d).$$

If (W, Q) is a symplectic vector space and $V = \bigwedge^* W^*$, then we may define a *symplectic star operator* $*_\sigma$ by the expression

$$*_\sigma \alpha \wedge \beta = Q(\alpha, \beta)\Omega \quad \forall \alpha, \beta \in V. \tag{B.1.3}$$

We may also consider the operator $L = L_\omega \in \text{End}(V)$ given by left-multiplication by ω and its symplectic dual Λ_σ:

$$Q(\Lambda_\sigma \alpha, \beta) = Q(\alpha, L\beta) \quad \forall \alpha, \beta \in V.$$

Let $\dim_\mathbb{R} W = 2d$ and $\pi_k \colon V \to V_k := \bigwedge^k W^*$ the natural projection; we set

$$Y := \sum_{k=0}^{2d}(k-d)\pi_k. \tag{B.1.4}$$

Then Y is a semisimple transformation whose eigenvalues are the integers between $-d$ and d.

EXAMPLE B.1.3 Suppose $d = 1$. Then given any $v_1 \in W$, there exists v_2, unique modulo $\mathbb{R}.v_1$ such that $\{v_1, v_2\}$ is a symplectic basis of (W, Q). Let $\{\xi_1, \xi_2\}$ be the dual basis. Then $\omega = \Omega = \xi_1 \wedge \xi_2$ so, in particular, the expressions $v_1 \wedge v_2$ and $\xi_1 \wedge \xi_2$ are independent of the choice of symplectic basis. In fact, ω is characterized by the fact that $Q(\omega, \omega) = 1$. For any $\alpha \in W^*$ we have $*_\sigma \alpha \wedge \alpha = Q(\alpha, \alpha)\Omega = 0$. Hence $*_\sigma \colon W^* \to W^*$ is a multiple of the identity. Checking on a symplectic basis we see that $*_\sigma$ is, in fact, the identity map on W^*. Since $Q(\omega, \omega) = 1$, it follows that

$$*_\sigma 1 = \omega \, ; \quad *_\sigma \omega = 1.$$

Hence

$$*_\sigma \circ *_\sigma = I. \tag{B.1.5}$$

Next we compute $\Lambda_\sigma \omega \in \mathbb{R}$. Since

$$Q(\Lambda_\sigma \omega, 1) = Q(\omega, L1) = Q(\omega, \omega) = 1,$$

we see that $\Lambda_\sigma \omega = 1$. Thus, we may verify directly that if $d = 1$, the linear transformations $\{L, \Lambda_\sigma, Y\}$ are an $\mathfrak{sl}_2$-triple; i.e.,

$$[Y, \Lambda_\sigma] = -2\Lambda_\sigma \, , \quad [Y, L] = 2L \, , \quad [L, \Lambda_\sigma] = Y. \tag{B.1.6}$$

Since every symplectic vector space admits a symplectic basis, it is clear that we can write any symplectic vector space as a Q-orthogonal sum of two-dimensional spaces. We next study the various operators defined above in the case of a direct sum of symplectic vector spaces. Suppose

$$W = W' \oplus W''$$

and Q', Q'' are symplectic forms on W', W'', respectively. Then $Q = Q' \oplus Q''$ defines a symplectic form on W. We denote by $q', q'', *'_\sigma, *''_\sigma, \Lambda'_\sigma, L', Y'$, etc. the corresponding operators.

EXERCISE B.1.4   Prove that the natural map

$$\textstyle\bigwedge^*(W')^* \otimes \bigwedge^*(W'')^* \to \bigwedge^* W^*, \tag{B.1.7}$$

that sends $\alpha' \otimes \alpha'' \mapsto \alpha' \wedge \alpha''$, is an isomorphism of vector spaces which may be turned into an algebra isomorphism by defining

$$(\alpha' \otimes \alpha'') \cdot (\beta' \otimes \beta'') = (-1)^{\deg(\alpha'')\deg(\beta')}(\alpha' \wedge \beta') \otimes (\alpha'' \wedge \beta'').$$

LEMMA B.1.5   *Under the identification* (B.1.7), *the following identities hold:*

(1) $\bigwedge^* q = \bigwedge^* q' \otimes \bigwedge^* q''.$

(2) $Q = Q' \otimes Q''.$

(3) $\omega = \omega' + \omega''.$

(4) $\Omega = \Omega' \wedge \Omega''.$

(5) $*_\sigma(\alpha' \otimes \alpha'') = (-1)^{\deg(\alpha')\deg(\alpha'')} *'_\sigma \alpha' \otimes *''_\sigma \alpha''.$

(6) $L = L' \otimes I_{W''} + I_{W'} \otimes L''.$

(7) $\Lambda_\sigma = \Lambda'_\sigma \otimes I_{W''} + I_{W'} \otimes \Lambda''_\sigma.$

(8) $Y = Y' \otimes I_{W''} + I_{W'} \otimes Y''.$

PROOF.   Both sides of (1) agree on $W^* = (W')^* \oplus (W'')^*$ and are morphisms of the graded algebra $\bigwedge^*(W')^* \otimes \bigwedge^*(W'')^*$. Hence they are equal. Identity (2) then follows from (1) and (B.1.1), while (3) follows from (B.1.2). Since ω' and ω'' commute we have

$$\omega^d = (\omega' + \omega'')^d = \frac{d!}{d'!d''!}(\omega')^{d'} \wedge (\omega'')^{d''},$$

since all other terms vanish. Hence (4) follows.

Next we check (5). We have

$$
\begin{aligned}
*_\sigma(\alpha' \otimes \alpha'') \wedge (\beta' \otimes \beta'')
&= Q(\alpha' \otimes \alpha'', \beta' \otimes \beta'')\Omega \\
&= Q'(\alpha', \beta')\Omega' \otimes Q''(\alpha'', \beta'')\Omega'' \\
&= *'_\sigma\alpha' \wedge \beta' \otimes *''_\sigma\alpha'' \wedge \beta'' \\
&= (-1)^{\deg(\beta')\deg(\alpha'')} *'_\sigma \alpha' \otimes *''_\sigma\alpha'' \wedge (\beta' \otimes \beta'') \\
&= (-1)^{\deg(\alpha')\deg(\alpha'')} (*'_\sigma\alpha' \otimes *'_\sigma\alpha'') \wedge (\beta' \otimes \beta''),
\end{aligned}
$$

where the last equality follows since $\alpha' \wedge *_\sigma' \beta' = 0$ unless $\deg(\alpha') = \deg(\beta')$. Hence (5) follows. The verification of (7) follows along similar lines.

The identity (6) follows directly from (3) while the last statement follows easily from the definition of Y. □

LEMMA B.1.6 *Let* $\alpha \in \bigwedge^* W^*$. *Then*

(1) $*_\sigma(*_\sigma\alpha) = \alpha$;

(2) $\Lambda_\sigma\alpha = *_\sigma L *_\sigma \alpha$.

PROOF. The first identity follows by induction from Lemma B.1.5(5) and (B.1.5), which give the identity in the dimension two case. It remains to prove Lemma B.1.5(2):

$$
\begin{aligned}
Q(\Lambda_\sigma\alpha, \beta)\Omega &= Q(\alpha, L\beta)\Omega = *_\sigma\alpha \wedge L\beta \\
&= *_\sigma\alpha \wedge \omega \wedge \beta = \omega \wedge *_\sigma\alpha \wedge \beta \\
&= L *_\sigma \alpha \wedge \beta = *_\sigma *_\sigma L *_\sigma \alpha \wedge \beta \\
&= Q(*_\sigma L *_\sigma \alpha, \beta)\Omega.
\end{aligned}
$$

Since the identity holds for all $\beta \in \bigwedge^* W^*$, the result follows. □

THEOREM B.1.7 *Let* (W, Q) *be a symplectic vector space of dimension* $2d$. *The triple of operators* $\{L, \Lambda_\sigma, Y\}$ *define an* $\mathfrak{sl}_2$-*triple.*

PROOF. This follows from induction from Lemma B.1.5(6), (7), and (8), and the fact that the relations (B.1.6) hold in the two-dimensional case. □

In particular, we have the following corollary:

COROLLARY B.1.8 *Let* (W, Q) *be a* $2d$-*dimensional symplectic vector space and let* $V = \bigwedge^* W_\mathbb{C}^*$ *graded by degree. Then* $(V[d], L)$ *satisfies the hard Lefschetz property.*

B.2 COMPATIBLE INNER PRODUCTS

If W is a real vector space with a complex structure J and $B \colon W \times W \to \mathbb{R}$ is a bilinear form on W, then we say that B is *compatible* with J if and only if

$$B(Ju, Jv) = B(u, v) \text{ for all } u, v \in W. \tag{B.2.1}$$

We shall also denote by B the bilinear extension of B to the complexification $W_\mathbb{C}$. If B is symmetric then the bilinear form

$$Q(u, v) := B(Ju, v) \tag{B.2.2}$$

is alternating and satisfies $Q(Ju, Jv) = Q(u, v)$. Moreover, Q is nondegenerate if and only if B is nondegenerate. The corresponding 2-form ω is real and lies in $\bigwedge^{1,1} W_\mathbb{C}^*$. We may recover B from Q by the formula $B(u, v) = Q(u, Jv)$.

We collect these observations in the following:

THEOREM B.2.1   *Let W be a real vector space with a complex structure J. Then the following data are equivalent:*

(1) a symmetric bilinear form B on W compatible with J

(2) an alternating bilinear form Q on W compatible with J

(3) an element $\omega \in \bigwedge^{1,1}(W_{\mathbb{C}}^*) \cap \bigwedge^2 W^*$

Suppose now that (W, J) is a vector space with a complex structure and a compatible positive-definite bilinear form B. Then we have an isomorphism $b \colon W^* \to W$ defined by $\phi = B(b(\phi), \bullet)$. We extend B to $V = \bigwedge^* W^*$ by

$$B(\alpha, \beta) := \alpha(\textstyle\bigwedge^* b(\beta)).$$

This defines a positive-definite symmetric form on V.

PROPOSITION B.2.2   *Let (W, J, B) and V be as above. By dualizing and taking the exterior product we consider the associated map $J \colon V_{\mathbb{C}} \to V_{\mathbb{C}}$. Then*

(1) $b = J \circ q$;

(2) let $\alpha, \beta \in V_{\mathbb{C}}$; then
$$B(\alpha, \beta) = Q(J\alpha, \beta). \tag{B.2.3}$$

PROOF.   For $\phi \in W^*$, $w \in W$ we have

$$\phi(w) \;=\; Q(q(\phi), w) \;=\; B(J(q(\phi)), w).$$

Hence (1) holds. Now,

$$B(\alpha, \beta) \;=\; \alpha(\textstyle\bigwedge^* b(\beta)) \;=\; \alpha(J(\textstyle\bigwedge^* q(\beta))) \;=\; (J\alpha)(\textstyle\bigwedge^* q(\beta)) \;=\; Q(J\alpha, \beta).$$

$\square$

We define a *star operator* $*$ on V by

$$*\alpha \wedge \beta \;=\; B(\alpha, \beta)\Omega \quad \forall \alpha, \beta \in V,$$

where, as before, $\Omega = \omega^d/d!$. Let Λ denote the B-adjoint of L,

$$B(\Lambda\alpha, \beta) = B(\alpha, L\beta); \quad \alpha, \beta \in V_{\mathbb{R}}.$$

We extend $*$ and Λ to $V_{\mathbb{C}}$ by linearity.

The inner product B on V may also be extended to a Hermitian inner product H on $V_{\mathbb{C}}$ by

$$H(\alpha_1 + i\alpha_2, \beta_1 + i\beta_2) := B(\alpha_1, \beta_1) + iB(\alpha_2, \beta_1) - iB(\alpha_1, \beta_2) + B(\alpha_2, \beta_2),$$

where $\alpha_i, \beta_i \in V$.

EXERCISE B.2.3   Let $\alpha, \beta \in V_{\mathbb{C}}$. Prove that

$$*\alpha \wedge \bar{\beta} = H(\alpha, \beta)\Omega.$$

Deduce that

$$*(V^{p,q}) \subset V^{n-q,n-p}. \tag{B.2.4}$$

LEMMA B.2.4   *The following relations hold:*

$$[L, J] = 0; \quad [\Lambda, J] = 0; \quad \Lambda = \Lambda_\sigma.$$

PROOF. The first identity is clear since $\omega \in \bigwedge^{1,1} W^*$ and the second follows by adjunction. Now, because of Proposition B.2.2 we have

$$
\begin{aligned}
B(\Lambda\alpha, \beta) &= B(\alpha, L\beta) = Q(J\alpha, L\beta) \\
&= (-1)^{\deg(\alpha)} Q(\alpha, JL\beta) = (-1)^{\deg(\alpha)} Q(\alpha, LJ\beta) \\
&= (-1)^{\deg(\alpha)} Q(\Lambda_\sigma\alpha, J\beta) = Q(J\Lambda_\sigma\alpha, \beta) \\
&= B(\Lambda_\sigma\alpha, \beta).
\end{aligned}
$$

$\square$

COROLLARY B.2.5   *Let (W, Q) be a symplectic vector space, $\dim Q = 2d$, J a complex structure on W such that Q is compatible with J and the associated symmetric form B is positive definite. Then $\{L, \Lambda, Y\}$ is an $\mathfrak{sl}_2$-triple and, if $V = \bigwedge^* W_{\mathbb{C}}^*$, then $(V[d], L)$ satisfies the hard Lefschetz property.*

PROOF. This follows from Lemma B.2.4 and Corollary B.1.8. $\square$

B.3   SYMPLECTIC MANIFOLDS

In (B.1.3) we defined a symplectic star operator on the exterior algebra of a vector space carrying a symplectic structure. We will now extend this construction to a symplectic manifold. Let (M, ω) be a symplectic manifold. We denote by $Q_p(v, w) = \omega_p(v \wedge w)$ the symplectic structure on $T_p(M)$ for $p \in M$. We extend the bilinear form Q to $\mathcal{A}^*(M)$, pointwise, and we denote by $\Omega = \omega^d/d!$ the associated orientation form. Let $\mathcal{A}_0^k(M)$ denote the exterior k-forms on M with compact support. Given $\alpha \in \mathcal{A}^k(M)$, $\beta \in \mathcal{A}_0^k(M)$, we define a *symplectic pairing*:

$$(\alpha, \beta)_\sigma := \int_M Q(\alpha, \beta)\,\Omega. \tag{B.3.1}$$

The symplectic star operator $*_\sigma$ on $\mathcal{A}^*(M)$ is then defined as

$$*_\sigma\alpha \wedge \beta = (\alpha, \beta)_\sigma\,\Omega \quad \text{for} \quad \alpha \in \mathcal{A}^*(M), \ \beta \in \mathcal{A}_0^*(M). \tag{B.3.2}$$

Let δ_σ, Λ_σ denote the adjoints of d and L relative to the symplectic pairing. Thus, we have

$$(\delta_\sigma\alpha, \beta)_\sigma = (\alpha, d\beta)_\sigma; \quad (\Lambda_\sigma\alpha, \beta)_\sigma = (\alpha, L\beta)_\sigma.$$

Note that for $\alpha \in \mathcal{A}^{k-1}(M), \beta \in \mathcal{A}_0^k(M)$,

$$(d\alpha, \beta)_\sigma = (-1)^k (\beta, d\alpha)_\sigma = (-1)^k (\delta_\sigma \beta, \alpha)_\sigma = -(\alpha, \delta_\sigma \beta). \tag{B.3.3}$$

EXERCISE B.3.1 The operators L, $*_\sigma$, and Λ_σ are tensorial, i.e., indicating by the subscript p the operators induced at the tangent space level we have

$$(L\alpha)_p = L_p \alpha_p ; \quad (*_\sigma \alpha)_p = *_{\sigma,p} \alpha_p ; \quad (\Lambda_\sigma \alpha)_p = \Lambda_{\sigma,p} \alpha_p$$

LEMMA B.3.2 *The following identities hold:*

$$\Lambda_\sigma = *_\sigma L *_\sigma ; \quad \delta_\sigma \alpha = -(-1)^{\deg(\alpha)} *_\sigma d *_\sigma \alpha. \tag{B.3.4}$$

PROOF. The first assertion follows from Exercise B.3.1 and the corresponding statement for symplectic vector spaces proved in Lemma B.1.6. The second assertion follows from the Stokes theorem:

$$
\begin{aligned}
(\delta_\sigma \alpha, \beta)_\sigma &= (\alpha, d\beta)_\sigma \\
&= \int_M Q(\alpha, d\beta)\, \Omega = \int_M *_\sigma \alpha \wedge d\beta \\
&= (-1)^{\deg(*_\sigma \alpha)} \left(\int_M d(*_\sigma \alpha \wedge \beta) - \int_M d(*_\sigma \alpha) \wedge \beta \right) \\
&= -(-1)^{\deg(*_\sigma \alpha)} \int_M *_\sigma (*_\sigma d *_\sigma \alpha) \wedge \beta \\
&= -(-1)^{\deg(*_\sigma \alpha)} (*_\sigma d *_\sigma \alpha, \beta)_\sigma.
\end{aligned}
$$

$\square$

EXAMPLE B.3.3 Consider $\mathbb{R}^2$ with the standard symplectic structure; i.e., $w = dx \wedge dy$. From Lemma B.3.2 and Example B.1.3 it follows that $*_\sigma$ acts as the identity on $\mathcal{A}^1(M)$, $*_\sigma 1 = \omega$, and $*_\sigma \omega = 1$. Also, $\Lambda_\sigma \omega = 1$.

Lemma B.3.2 gives

$$\delta_\sigma(f\omega) = -*_\sigma d *_\sigma f\omega = -df,$$

$$\delta_\sigma(f\, dx + g\, dy) = *_\sigma d *_\sigma (f\, dx + g\, dy) = *_\sigma d(f\, dx + g\, dy) = (g_x - f_y).$$

B.4   THE KÄHLER IDENTITIES

Let (M, J, ω) be a compact Kähler manifold and, as before, let g denote the associated Riemannian metric. As in Section 1.4.2 we consider the L^2 inner product $(\bullet, \bullet)$ on $\mathcal{A}^*(M)$ defined in (1.4.3). Let $*$ be the corresponding star operator and $\delta = -*\, d *$ the adjoint of d. We extend these operators by linearity to $\mathcal{A}^*(M, \mathbb{C})$. We have from (1.4.8) that

$$*(\mathcal{A}^{p,q}(M)) \subset \mathcal{A}^{n-q,n-p}(M).$$

As in (1.4.9) we have

$$\delta = - * d * = - * \bar{\partial} * - * \partial *,$$

with

$$\partial^* := - * \bar{\partial} * ; \quad \bar{\partial}^* := - * \partial * .$$

The identity (B.2.3) in Proposition B.2.2 allows us to relate the inner product $(\bullet, \bullet)$ and the bilinear form $(\bullet, \bullet)_\sigma$:

EXERCISE B.4.1 Prove that for $\alpha, \beta \in \mathcal{A}^k(M)$:

$$(\alpha, \beta) = (J\alpha, \beta)_\sigma. \tag{B.4.1}$$

PROPOSITION B.4.2 *Let δ and δ_σ be the adjoints of d. Then, for $\alpha \in \mathcal{A}^k(M, \mathbb{C})$,*

$$\delta\alpha = (-1)^{k-1} J \delta_\sigma J\alpha \text{ and} \tag{B.4.2}$$

$$\delta_\sigma \alpha = (-i\partial^* + i\bar{\partial}^*)\alpha. \tag{B.4.3}$$

PROOF. Let $\alpha \in \mathcal{A}^k(M, \mathbb{C})$. Then for all $\beta \in \mathcal{A}^{k-1}(M, \mathbb{C})$ we have

$$
\begin{aligned}
(\delta\alpha, \beta) &= (\alpha, d\beta) = (J\alpha, d\beta)_\sigma \\
&= (\delta_\sigma J\alpha, \beta)_\sigma \\
&= (-1)^{k-1}(J^2 \delta_\sigma J\alpha, \beta)_\sigma \\
&= (-1)^{k-1}(J \delta_\sigma J\alpha, \beta).
\end{aligned}
$$

Hence, (B.4.2) holds and inverting we get

$$\delta_\sigma \alpha = (-1)^k J \delta J \alpha.$$

Suppose now that $\alpha \in \mathcal{A}^{p,q}$, $p + q = k$. Then writing $\delta = \partial^* + \bar{\partial}^*$ we have

$$
\begin{aligned}
\delta_\sigma \alpha &= (-1)^k i^{p-q} J (\partial^*\alpha + \bar{\partial}^*\alpha) \\
&= (-1)^k i^{p-q}(i^{p-1-q}\partial^*\alpha + i^{p-q+1}\bar{\partial}^*\alpha) \\
&= -i\partial^*\alpha + i\bar{\partial}^*\alpha.
\end{aligned}
$$

$\square$

Let $L: \mathcal{A}^k(M) \to \mathcal{A}^{k+2}(M)$ be the operator $L\alpha = \omega \wedge \alpha$ and Λ its adjoint relative to the positive-definite inner product $(\bullet, \bullet)$. Since $\omega \in \mathcal{A}^{1,1}(M)$ we have

$$LJ = JL. \tag{B.4.4}$$

PROPOSITION B.4.3 *Let (M, J, ω) be a compact Kähler manifold. Then $\Lambda = \Lambda_\sigma$.*

PROOF. For each $p \in M$ we have $\Lambda_p = \Lambda_{\sigma,p}$ by Lemma B.2.4. On the other hand, it follows from Exercise B.3.1 that Λ_σ is defined pointwise. Since a similar result holds for Λ, the proposition follows. $\square$

THEOREM B.4.4 (Kähler identities)   *Let (M, J, ω) be a compact Kähler manifold. Then the following identities hold:*

(1) $[L, \partial] = 0$; $[L, \bar{\partial}] = 0$; $[\Lambda, \partial^*] = 0$; $[\Lambda, \bar{\partial}^*] = 0$.

(2) $[L, \bar{\partial}^*] = -i\partial$; $[L, \partial^*] = i\bar{\partial}$; $[\Lambda, \partial] = i\bar{\partial}^*$; $[\Lambda, \bar{\partial}] = -i\partial^*$.

PROOF. We note first of all that it suffices to prove only one of the identities in each line. The others then follow by conjugation or adjunction relative to the Hermitian form $(\bullet, \bullet)^h$. Now, since $\omega \in \mathcal{A}^{1,1}(M)$ is a closed form, we have $\partial\omega = 0$ and, consequently, $[L, \partial] = 0$. Hence, it suffices to prove the identities in (2).

We begin by showing that the following symplectic identity implies the desired result:

$$[\Lambda_\sigma, d] = \delta_\sigma. \tag{B.4.5}$$

Indeed, it follows from Proposition B.4.3 that $\Lambda_\sigma = \Lambda$; hence (B.4.3) gives

$$[\Lambda, d] = \delta_\sigma = -i\partial^* + i\bar{\partial}^*.$$

Since Λ is an operator of pure type $(-1, -1)$ we can decompose $d = \partial + \bar{\partial}$ and compare types in the above equation to get

$$[\Lambda, \partial] = i\bar{\partial}^* ; \quad [\Lambda, \bar{\partial}] = -i\partial^*.$$

Now, since the operators d and δ_σ are local, it suffices to prove (B.4.5) in a coordinate neighborhood of M. But, by Darboux's theorem, M is locally symplectomorphic to an open set in $\mathbb{R}^{2d}$ with the standard symplectic structure. Hence, we may argue as in the previous section and prove the desired identity by induction. In order to do that we observe that if the symplectic vector bundle (TM, Q) is symplectically isomorphic to a direct sum

$$(TM, Q) \cong (W', Q') \oplus (W'', Q''),$$

then

$$A^*(M) = C^\infty(M, \bigwedge{}^* TM) \cong C^\infty(M, \bigwedge{}^* W') \otimes_{C^\infty(M)} C^\infty(M, \bigwedge{}^* W''),$$

and this is an isomorphism of graded $C^\infty(M)$-algebras.

Now, in our case, $C^\infty(M, \bigwedge^* W')$ is invariant by d and we will denote by d' the restriction of d and similarly for W''. One has

$$d = d' \otimes \mathrm{id} + \mathrm{id} \otimes d''.$$

This may be verified by checking that both sides agree on functions and satisfy the Leibniz rule. Consequently,

$$\delta_\sigma = \delta'_\sigma \otimes \mathrm{id} + \mathrm{id} \otimes \delta''_\sigma.$$

Hence

$$[\Lambda_\sigma, d] = [\Lambda'_\sigma, d'] \otimes \mathrm{id} + \mathrm{id} \otimes [\Lambda''_\sigma, d'']$$

and it suffices to verify the two-dimensional case which follows directly from the computations in Example B.3.3. $\square$

Bibliography

[1] William M. Boothby. *An introduction to differentiable manifolds and Riemannian geometry*, volume 120 of *Pure and Applied Mathematics*. Academic Press, Orlando, FL, second edition, 1986.

[2] Raoul Bott and Loring W. Tu. *Differential forms in algebraic topology*, volume 82 of *Graduate Texts in Mathematics*. Springer, New York, 1982.

[3] Paul Bressler and Valery A. Lunts. Hard Lefschetz theorem and Hodge–Riemann relations for intersection cohomology of nonrational polytopes. *Indiana Univ. Math. J.*, 54(1):263–307, 2005.

[4] Yu. D. Burago and V. A. Zalgaller. *Geometric inequalities*, volume 285 of *Grundlehren der Mathematischen Wissenschaften [Fundamental Principles of Mathematical Sciences]*. Springer, Berlin, 1988. Translated from the Russian by A. B. Sosinskiĭ, Springer Series in Soviet Mathematics.

[5] Eduardo Cattani. Mixed Lefschetz theorems and Hodge–Riemann bilinear relations. *Int. Math. Res. Not.*, (10):Art. ID rnn025, 20, 2008.

[6] Shiing-shen Chern. *Complex manifolds without potential theory (with an appendix on the geometry of characteristic classes)*. Universitext. Springer, New York, second edition, 1995.

[7] Pierre Deligne. The Hodge conjecture. In *The millennium prize problems*, pages 45–53. Clay Math. Inst., Cambridge, MA, 2006.

[8] Jean-Pierre Demailly. *Complex analytic and algebraic geometry*. OpenContent Book available from http://www-fourier.ujf-grenoble.fr/~demailly/manuscripts/agbook.pdf.

[9] Tien-Cuong Dinh and Viêt-Anh Nguyên. The mixed Hodge–Riemann bilinear relations for compact Kähler manifolds. *Geom. Funct. Anal.*, 16(4):838–849, 2006.

[10] Phillip Griffiths and Joseph Harris. *Principles of algebraic geometry*. Wiley Classics Library. John Wiley & Sons, New York, 1994. Reprint of the 1978 original.

[11] M. Gromov. Convex sets and Kähler manifolds. In *Advances in differential geometry and topology*, pages 1–38. World Sci., Teaneck, NJ, 1990.

[12] W. V. D. Hodge. The topological invariants of algebraic varieties. In *Proceedings of the International Congress of Mathematicians, Cambridge, Mass., 1950, vol. 1*, pages 182–192, Providence, RI, 1952. Amer. Math. Soc.

[13] Daniel Huybrechts. *Complex geometry*. Universitext. Springer, Berlin, 2005.

[14] Kalle Karu. Hard Lefschetz theorem for nonrational polytopes. *Invent. Math.*, 157(2):419–447, 2004.

[15] Shoshichi Kobayashi. *Differential geometry of complex vector bundles*, volume 15 of *Publications of the Mathematical Society of Japan*. Princeton University Press, Princeton, NJ, 1987. Kanô Memorial Lectures, 5.

[16] Shoshichi Kobayashi and Katsumi Nomizu. *Foundations of differential geometry. Vol. II*. Wiley Classics Library. John Wiley & Sons, New York, 1996. Reprint of the 1969 original, Wiley-Interscience

[17] Kunihiko Kodaira. *Complex manifolds and deformation of complex structures*. Classics in Mathematics. Springer, Berlin, English edition, 2005. Translated from the 1981 Japanese original by Kazuo Akao.

[18] John M. Lee. *Introduction to smooth manifolds*, volume 218 of *Graduate Texts in Mathematics*. Springer, New York, 2003.

[19] James Morrow and Kunihiko Kodaira. *Complex manifolds*. AMS Chelsea Publishing, Providence, RI, 2006. Reprint of the 1971 edition with errata.

[20] A. Newlander and L. Nirenberg. Complex analytic coordinates in almost complex manifolds. *Ann. of Math. (2)*, 65:391–404, 1957.

[21] Wilfried Schmid. Variation of Hodge structure: the singularities of the period mapping. *Invent. Math.*, 22:211–319, 1973.

[22] Joseph Steenbrink and Steven Zucker. Variation of mixed Hodge structure. I. *Invent. Math.*, 80(3):489–542, 1985.

[23] Shlomo Sternberg. *Lectures on differential geometry*. Prentice-Hall, Englewood Cliffs, NJ, 1964.

[24] Bernard Teissier. Variétés toriques et polytopes. In *Bourbaki Seminar, Vol. 1980/81*, volume 901 of *Lecture Notes in Math.*, pages 71–84. Springer, Berlin, 1981.

[25] W. P. Thurston. Some simple examples of symplectic manifolds. *Proc. Amer. Math. Soc.*, 55(2):467–468, 1976.

[26] V. A. Timorin. Mixed Hodge–Riemann bilinear relations in a linear context. *Funktsional. Anal. i Prilozhen.*, 32(4):63–68, 96, 1998.

[27] V. A. Timorin. An analogue of the Hodge–Riemann relations for simple convex polyhedra. *Uspekhi Mat. Nauk*, 54(2(326)):113–162, 1999.

[28] Loring W. Tu. *An introduction to manifolds*, second edition. Universitext. Springer, New York, 2011.

[29] Claire Voisin. A counterexample to the Hodge conjecture extended to Kähler varieties. *Int. Math. Res. Not.*, (20):1057–1075, 2002.

[30] Claire Voisin. *Hodge theory and complex algebraic geometry. I*, volume 76 of *Cambridge Studies in Advanced Mathematics*. Cambridge University Press, Cambridge, English edition, 2007. Translated from the French by Leila Schneps.

[31] Claire Voisin. *Hodge theory and complex algebraic geometry. II*, volume 77 of *Cambridge Studies in Advanced Mathematics*. Cambridge University Press, Cambridge, English edition, 2007. Translated from the French by Leila Schneps.

[32] Claire Voisin. Hodge structures on cohomology algebras and geometry. *Math. Ann.*, 341(1):39–69, 2008.

[33] Frank W. Warner. *Foundations of differentiable manifolds and Lie groups*, volume 94 of *Graduate Texts in Mathematics*. Springer, New York, 1983. Corrected reprint of the 1971 edition.

[34] André Weil. *Introduction à l'étude des variétés kählériennes*. Publications de l'Institut de Mathématique de l'Université de Nancago, VI. Actualités Sci. Ind. no. 1267. Hermann, Paris, 1958.

[35] Raymond O. Wells, Jr. *Differential analysis on complex manifolds*, volume 65 of *Graduate Texts in Mathematics*. Springer, New York, third edition, 2008. With a new appendix by Oscar Garcia-Prada.

[36] Steven Zucker. The Hodge conjecture for cubic fourfolds. *Compositio Math.*, 34(2):199–209, 1977.

Chapter Two

From Sheaf Cohomology to the Algebraic de Rham
Theorem
by Fouad El Zein and Loring W. Tu

INTRODUCTION

The concepts of homology and cohomology trace their origin to the work of Poincaré in
the late nineteenth century. They attach to a topological space algebraic structures such
as groups or rings that are topological invariants of the space. There are actually many
different theories, for example, simplicial, singular, and de Rham theories. In 1931,
Georges de Rham proved a conjecture of Poincaré on a relationship between cycles and
smooth differential forms, which establishes for a smooth manifold an isomorphism
between singular cohomology with real coefficients and de Rham cohomology.

   More precisely, by integrating smooth forms over singular chains on a smooth manifold M, one obtains a linear map

$$\mathcal{A}^k(M) \to S^k(M, \mathbb{R})$$

from the vector space $\mathcal{A}^k(M)$ of smooth k-forms on M to the vector space $S^k(M, \mathbb{R})$
of real singular k-cochains on M. The theorem of de Rham asserts that this linear map
induces an isomorphism

$$H_{\mathrm{dR}}^*(M) \overset{\sim}{\to} H^*(M, \mathbb{R})$$

between the de Rham cohomology $H_{\mathrm{dR}}^*(M)$ and the singular cohomology $H^*(M, \mathbb{R})$,
under which the wedge product of classes of closed smooth differential forms corresponds to the cup product of classes of cocycles. Using complex coefficients, there is
similarly an isomorphism

$$h^*\big(\mathcal{A}^\bullet(M, \mathbb{C})\big) \overset{\sim}{\to} H^*(M, \mathbb{C}),$$

where $h^*\big(\mathcal{A}^\bullet(M, \mathbb{C})\big)$ denotes the cohomology of the complex $\mathcal{A}^\bullet(M, \mathbb{C})$ of smooth
$\mathbb{C}$-valued forms on M.

   By an algebraic variety, we will mean a reduced separated scheme of finite type
over an algebraically closed field [14, Vol. 2, Ch. VI, Sec. 1.1, p. 49]. In fact, the field
throughout the article will be the field of complex numbers. For those not familiar with

the language of schemes, there is no harm in taking an algebraic variety to be a quasi-projective variety; the proofs of the algebraic de Rham theorem are exactly the same in the two cases.

Let X be a smooth complex algebraic variety with the Zariski topology. A *regular* function on an open set $U \subset X$ is a rational function that is defined at every point of U. A differential k-form on X is *algebraic* if locally it can be written as $\sum f_I \, dg_{i_1} \wedge \cdots \wedge dg_{i_k}$ for some regular functions f_I, g_{i_j}. With the complex topology, the underlying set of the smooth variety X becomes a complex manifold X_{an}. By de Rham's theorem, the singular cohomology $H^*(X_{\mathrm{an}}, \mathbb{C})$ can be computed from the complex of smooth $\mathbb{C}$-valued differential forms on X_{an}. Grothendieck's algebraic de Rham theorem asserts that the singular cohomology $H^*(X_{\mathrm{an}}, \mathbb{C})$ can in fact be computed from the complex $\Omega^{\bullet}_{\mathrm{alg}}$ of sheaves of algebraic differential forms on X. Since algebraic de Rham cohomology can be defined over any field, Grothendieck's theorem lies at the foundation of Deligne's theory of absolute Hodge classes (see Chapter 11 in this volume).

In spite of its beauty and importance, there does not seem to be an accessible account of Grothendieck's algebraic de Rham theorem in the literature. Grothendieck's paper [7], invoking higher direct images of sheaves and a theorem of Grauert–Remmert, is quite difficult to read. An impetus for our work is to give an elementary proof of Grothendieck's theorem, elementary in the sense that we use only tools from standard textbooks as well as some results from Serre's groundbreaking FAC and GAGA papers ([12] and [13]).

This article is in two parts. In Part I, comprising Sections 1 through 6, we prove Grothendieck's algebraic de Rham theorem more or less from scratch for a smooth complex projective variety X, namely, that there is an isomorphism

$$H^*(X_{\mathrm{an}}, \mathbb{C}) \simeq \mathbb{H}^*(X, \Omega^{\bullet}_{\mathrm{alg}})$$

between the complex singular cohomology of X_{an} and the hypercohomology of the complex $\Omega^{\bullet}_{\mathrm{alg}}$ of sheaves of algebraic differential forms on X. The proof, relying mainly on Serre's GAGA principle and the technique of hypercohomology, necessitates a discussion of sheaf cohomology, coherent sheaves, and hypercohomology, and so another goal is to give an introduction to these topics. While Grothendieck's theorem is valid as a ring isomorphism, to keep the account simple, we prove only a vector space isomorphism. In fact, we do not even discuss multiplicative structures on hypercohomology. In Part II, comprising Sections 7 through 10, we develop more machinery, mainly the Čech cohomology of a sheaf and the Čech cohomology of a complex of sheaves, as tools for computing hypercohomology. We prove that the general case of Grothendieck's theorem is equivalent to the affine case, and then prove the affine case.

The reason for the two-part structure of our article is the sheer amount of background needed to prove Grothendieck's algebraic de Rham theorem in general. It seems desirable to treat the simpler case of a smooth projective variety first, so that the reader can see a major landmark before being submerged in yet more machinery. In fact, the projective case is not necessary to the proof of the general case, although the tools developed, such as sheaf cohomology and hypercohomology, are indispensable to the general proof. A reader who is already familiar with these tools can go directly to Part II.

Of the many ways to define sheaf cohomology, for example as Čech cohomology, as the cohomology of global sections of a certain resolution, or as an example of a right-derived functor in an abelian category, each has its own merit. We have settled on Godement's approach using his canonical resolution [5, Sec. 4.3, p. 167]. It has the advantage of being the most direct. Moreover, its extension to the hypercohomology of a complex of sheaves gives at once the E_2 terms of the standard spectral sequences converging to the hypercohomology.

What follows is a more detailed description of each section. In Part I, we recall in Section 1 some of the properties of sheaves. In Section 2, sheaf cohomology is defined as the cohomology of the complex of global sections of Godement's canonical resolution. In Section 3, the cohomology of a sheaf is generalized to the hypercohomology of a complex of sheaves. Section 4 defines coherent analytic and algebraic sheaves and summarizes Serre's GAGA principle for a smooth complex projective variety. Section 5 proves the holomorphic Poincaré lemma and the analytic de Rham theorem for any complex manifold, and Section 6 proves the algebraic de Rham theorem for a smooth complex projective variety.

In Part II, we develop in Sections 7 and 8 the Čech cohomology of a sheaf and of a complex of sheaves. Section 9 reduces the algebraic de Rham theorem for an algebraic variety to a theorem about affine varieties. Finally, in Section 10 we treat the affine case.

We are indebted to George Leger for his feedback and to Jeffrey D. Carlson for helpful discussions and detailed comments on the many drafts of the article. Loring Tu is also grateful to the Tufts University Faculty Research Award Committee for a New Directions in Research Award and to the National Center for Theoretical Sciences Mathematics Division (Taipei Office) in Taiwan for hosting him during part of the preparation of this manuscript.

PART I. SHEAF COHOMOLOGY, HYPERCOHOMOLOGY, AND THE PROJECTIVE CASE

2.1 SHEAVES

We assume a basic knowledge of sheaves as in [9, Ch. II, Sec. 1, pp. 60–69].

2.1.1 The Étalé Space of a Presheaf

Associated to a presheaf $\mathcal{F}$ on a topological space X is another topological space $E_{\mathcal{F}}$, called the *étalé space* of $\mathcal{F}$. Since the étalé space is needed in the construction of Godement's canonical resolution of a sheaf, we give a brief discussion here. As a set, the étalé space $E_{\mathcal{F}}$ is the disjoint union $\coprod_{p \in X} \mathcal{F}_p$ of all the stalks of $\mathcal{F}$. There is a natural projection map $\pi \colon E_{\mathcal{F}} \to X$ that maps $\mathcal{F}_p$ to p. A *section* of the étalé space $\pi \colon E_{\mathcal{F}} \to X$ over $U \subset X$ is a map $s \colon U \to E_{\mathcal{F}}$ such that $\pi \circ s = \mathrm{id}_U$, the identity map on U. For any open set $U \subset X$, element $s \in \mathcal{F}(U)$, and point $p \in U$, let $s_p \in \mathcal{F}_p$ be the germ of s at p. Then the element $s \in \mathcal{F}(U)$ defines a section $\tilde{s}$ of the étalé space over U,

$$\tilde{s} \colon U \to E_{\mathcal{F}},$$
$$p \mapsto s_p \in \mathcal{F}_p.$$

The collection

$$\{\tilde{s}(U) \mid U \text{ open in } X, \ s \in \mathcal{F}(U)\}$$

of subsets of $E_{\mathcal{F}}$ satisfies the conditions to be a basis for a topology on $E_{\mathcal{F}}$. With this topology, the étalé space $E_{\mathcal{F}}$ becomes a topological space. By construction, the topological space $E_{\mathcal{F}}$ is locally homeomorphic to X. For any element $s \in \mathcal{F}(U)$, the function $\tilde{s} \colon U \to E_{\mathcal{F}}$ is a continuous section of $E_{\mathcal{F}}$. A section t of the étalé space $E_{\mathcal{F}}$ is continuous if and only if every point $p \in X$ has a neighborhood U such that $t = \tilde{s}$ on U for some $s \in \mathcal{F}(U)$.

Let $\mathcal{F}^+$ be the presheaf that associates to each open subset $U \subset X$ the abelian group

$$\mathcal{F}^+(U) := \{\text{continuous sections } t \colon U \to E_{\mathcal{F}}\}.$$

Under pointwise addition of sections, the presheaf $\mathcal{F}^+$ is easily seen to be a sheaf, called the **sheafification** or the **associated sheaf** of the presheaf $\mathcal{F}$. There is an obvious presheaf morphism $\theta \colon \mathcal{F} \to \mathcal{F}^+$ that sends a section $s \in \mathcal{F}(U)$ to the section $\tilde{s} \in \mathcal{F}^+(U)$.

EXAMPLE 2.1.1   For each open set U in a topological space X, let $\mathcal{F}(U)$ be the group of all *constant* real-valued functions on U. At each point $p \in X$, the stalk $\mathcal{F}_p$ is $\mathbb{R}$. The étalé space $E_{\mathcal{F}}$ is thus $X \times \mathbb{R}$, but not with its usual topology. A basis for $E_{\mathcal{F}}$ consists of open sets of the form $U \times \{r\}$ for an open set $U \subset X$ and a number $r \in \mathbb{R}$. Thus, the topology on $E_{\mathcal{F}} = X \times \mathbb{R}$ is the product topology of the given topology on X and the discrete topology on $\mathbb{R}$. The sheafification $\mathcal{F}^+$ is the sheaf $\mathbb{R}$ of *locally constant* real-valued functions.

EXERCISE 2.1.2   *Prove that if $\mathcal{F}$ is a sheaf, then $\mathcal{F} \simeq \mathcal{F}^+$. (Hint: the two sheaf axioms say precisely that for every open set U, the map $\mathcal{F}(U) \to \mathcal{F}^+(U)$ is one-to-one and onto.)*

2.1.2   Exact Sequences of Sheaves

From now on, we will consider only sheaves of abelian groups. A sequence of morphisms of sheaves of abelian groups

$$\cdots \longrightarrow \mathcal{F}^1 \xrightarrow{d_1} \mathcal{F}^2 \xrightarrow{d_2} \mathcal{F}^3 \xrightarrow{d_3} \cdots$$

on a topological space X is said to be **exact** at $\mathcal{F}^k$ if $\operatorname{Im} d_{k-1} = \ker d_k$; the sequence is said to be **exact** if it is exact at every $\mathcal{F}^k$. The exactness of a sequence of morphisms of sheaves on X is equivalent to the exactness of the sequence of stalk maps at every point $p \in X$ (see [9, Exer. 1.2, p. 66]). An exact sequence of sheaves of the form

$$0 \to \mathcal{E} \to \mathcal{F} \to \mathcal{G} \to 0 \tag{2.1.1}$$

is said to be a ***short exact sequence***.

It is not too difficult to show that the exactness of the sheaf sequence (2.1.1) over a topological space X implies the exactness of the sequence of sections

$$0 \to \mathcal{E}(U) \to \mathcal{F}(U) \to \mathcal{G}(U) \tag{2.1.2}$$

for every open set $U \subset X$, but that the last map $\mathcal{F}(U) \to \mathcal{G}(U)$ need not be surjective. In fact, as we will see in Theorem 2.2.8, the cohomology $H^1(U, \mathcal{E})$ is a measure of the nonsurjectivity of the map $\mathcal{F}(U) \to \mathcal{G}(U)$ of sections.

Fix an open subset U of a topological space X. To every sheaf $\mathcal{F}$ of abelian groups on X, we can associate the abelian group $\Gamma(U, \mathcal{F}) := \mathcal{F}(U)$ of sections over U and to every sheaf map $\varphi \colon \mathcal{F} \to \mathcal{G}$, the group homomorphism $\varphi_U \colon \Gamma(U, \mathcal{F}) \to \Gamma(U, \mathcal{G})$. This makes $\Gamma(U, \ )$ a functor from sheaves of abelian groups on X to abelian groups.

A functor F from the category of sheaves of abelian groups on X to the category of abelian groups is said to be ***exact*** if it maps a short exact sequence of sheaves

$$0 \to \mathcal{E} \to \mathcal{F} \to \mathcal{G} \to 0$$

to a short exact sequence of abelian groups

$$0 \to F(\mathcal{E}) \to F(\mathcal{F}) \to F(\mathcal{G}) \to 0.$$

If instead one has only the exactness of

$$0 \to F(\mathcal{E}) \to F(\mathcal{F}) \to F(\mathcal{G}), \tag{2.1.3}$$

then F is said to be a ***left-exact functor***. The sections functor $\Gamma(U, \ )$ is left exact but not exact. (By Proposition 2.2.2 and Theorem 2.2.8, the next term in the exact sequence (2.1.3) is the first cohomology group $H^1(U, \mathcal{E})$.)

2.1.3   Resolutions

Recall that $\mathbb{R}$ is the sheaf of locally constant functions with values in $\mathbb{R}$ and $\mathcal{A}^k$ is the sheaf of smooth k-forms on a manifold M. The exterior derivative $d \colon \mathcal{A}^k(U) \to \mathcal{A}^{k+1}(U)$, as U ranges over all open sets in M, defines a morphism of sheaves $d \colon \mathcal{A}^k \to \mathcal{A}^{k+1}$.

PROPOSITION 2.1.3   *On any manifold M of dimension n, the sequence of sheaves*

$$0 \to \underline{\mathbb{R}} \to \mathcal{A}^0 \xrightarrow{d} \mathcal{A}^1 \xrightarrow{d} \cdots \xrightarrow{d} \mathcal{A}^n \to 0 \tag{2.1.4}$$

is exact.

PROOF. Exactness at $\mathcal{A}^0$ is equivalent to the exactness of the sequence of stalk maps $\mathbb{R}_p \to \mathcal{A}_p^0 \xrightarrow{d} \mathcal{A}_p^1$ for all $p \in M$. Fix a point $p \in M$. Suppose $[f] \in \mathcal{A}_p^0$ is the germ of a C^∞ function $f \colon U \to \mathbb{R}$, where U is a neighborhood of p, such that $d[f] = [0]$ in $\mathcal{A}_p^1$. Then there is a neighborhood $V \subset U$ of p on which $df \equiv 0$. Hence, f is locally constant on V and $[f] \in \mathbb{R}_p$. Conversely, if $[f] \in \mathbb{R}_p$, then $d[f] = 0$. This proves the exactness of the sequence (2.1.4) at $\mathcal{A}^0$.

Next, suppose $[\omega] \in \mathcal{A}_p^k$ is the germ of a smooth k-form ω on some neighborhood of p such that $d[\omega] = 0 \in \mathcal{A}_p^{k+1}$. This means there is a neighborhood V of p on which $d\omega \equiv 0$. By making V smaller, we may assume that V is contractible. By the Poincaré lemma [3, Cor. 4.1.1, p. 35], ω is exact on V, say $\omega = d\tau$ for some $\tau \in \mathcal{A}^{k-1}(V)$. Hence, $[\omega] = d[\tau]$ in $\mathcal{A}_p^k$. This proves the exactness of the sequence (2.1.4) at $\mathcal{A}^k$ for $k > 0$.   □

In general, an exact sequence of sheaves

$$0 \to \mathcal{A} \to \mathcal{F}^0 \to \mathcal{F}^1 \to \mathcal{F}^2 \to \cdots$$

on a topological space X is called a **resolution** of the sheaf $\mathcal{A}$. On a complex manifold M of complex dimension n, the analogue of the Poincaré lemma is the $\bar{\partial}$-Poincaré lemma [6, p. 25], from which it follows that for each fixed integer $p \geq 0$, the sheaves $\mathcal{A}^{p,q}$ of smooth (p, q)-forms on M give rise to a resolution of the sheaf Ω^p of holomorphic p-forms on M:

$$0 \to \Omega^p \to \mathcal{A}^{p,0} \xrightarrow{\bar{\partial}} \mathcal{A}^{p,1} \xrightarrow{\bar{\partial}} \cdots \xrightarrow{\bar{\partial}} \mathcal{A}^{p,n} \to 0. \qquad (2.1.5)$$

The cohomology of the **Dolbeault complex**

$$0 \to \mathcal{A}^{p,0}(M) \xrightarrow{\bar{\partial}} \mathcal{A}^{p,1}(M) \xrightarrow{\bar{\partial}} \cdots \xrightarrow{\bar{\partial}} \mathcal{A}^{p,n}(M) \to 0$$

of smooth (p, q)-forms on M is by definition the **Dolbeault cohomology** $H^{p,q}(M)$ of the complex manifold M. (For (p, q)-forms on a complex manifold, see [6] or Chapter 1.)

2.2 SHEAF COHOMOLOGY

The **de Rham cohomology** $H_{\mathrm{dR}}^*(M)$ of a smooth n-manifold M is defined to be the cohomology of the **de Rham complex**

$$0 \to \mathcal{A}^0(M) \to \mathcal{A}^1(M) \to \mathcal{A}^2(M) \to \cdots \to \mathcal{A}^n(M) \to 0$$

of C^∞-forms on M. De Rham's theorem for a smooth manifold M of dimension n gives an isomorphism between the real singular cohomology $H^k(M, \mathbb{R})$ and the de Rham cohomology of M (see [3, Th. 14.28, p. 175; Th. 15.8, p. 191]). One obtains the de Rham complex $\mathcal{A}^\bullet(M)$ by applying the global sections functor $\Gamma(M, \ )$ to the resolution

$$0 \to \mathbb{R} \to \mathcal{A}^0 \to \mathcal{A}^1 \to \mathcal{A}^2 \to \cdots \to \mathcal{A}^n \to 0$$

of $\mathbb{R}$, but omitting the initial term $\Gamma(M, \mathbb{R})$. This suggests that the cohomology of a sheaf $\mathcal{F}$ might be defined as the cohomology of the complex of global sections of a certain resolution of $\mathcal{F}$. Now every sheaf has a canonical resolution: its *Godement resolution*. Using the Godement resolution, we will obtain a well-defined cohomology theory of sheaves.

2.2.1 Godement's Canonical Resolution

Let $\mathcal{F}$ be a sheaf of abelian groups on a topological space X. In Section 2.1.1, we defined the étalé space $E_{\mathcal{F}}$ of $\mathcal{F}$. By Exercise 2.1.2, for any open set $U \subset X$, the group $\mathcal{F}(U)$ may be interpreted as

$$\mathcal{F}(U) = \mathcal{F}^+(U) = \{\text{continuous sections of } \pi \colon E_{\mathcal{F}} \to X\}.$$

Let $C^0\mathcal{F}(U)$ be the group of all (not necessarily continuous) sections of the étalé space $E_{\mathcal{F}}$ over U; in other words, $C^0\mathcal{F}(U)$ is the direct product $\prod_{p \in U} \mathcal{F}_p$. In the literature, $C^0\mathcal{F}$ is often called the sheaf of ***discontinuous sections*** of the étalé space $E_{\mathcal{F}}$ of $\mathcal{F}$. Then $\mathcal{F}^+ \simeq \mathcal{F}$ is a subsheaf of $C^0\mathcal{F}$ and there is an exact sequence

$$0 \to \mathcal{F} \to C^0\mathcal{F} \to \mathcal{Q}^1 \to 0, \tag{2.2.1}$$

where $\mathcal{Q}^1$ is the quotient sheaf $C^0\mathcal{F}/\mathcal{F}$. Repeating this construction yields exact sequences

$$0 \to \mathcal{Q}^1 \to C^0\mathcal{Q}^1 \to \mathcal{Q}^2 \to 0, \tag{2.2.2}$$
$$0 \to \mathcal{Q}^2 \to C^0\mathcal{Q}^2 \to \mathcal{Q}^3 \to 0, \tag{2.2.3}$$
$$\cdots .$$

The short exact sequences (2.2.1) and (2.2.2) can be spliced together to form a longer exact sequence

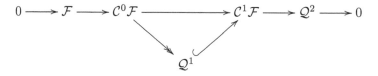

with $C^1\mathcal{F} := C^0\mathcal{Q}^1$. Splicing together all the short exact sequences (2.2.1), (2.2.2), (2.2.3), $\ldots$, and defining $C^k\mathcal{F} := C^0\mathcal{Q}^k$ results in the long exact sequence

$$0 \to \mathcal{F} \to C^0\mathcal{F} \to C^1\mathcal{F} \to C^2\mathcal{F} \to \cdots ,$$

called the ***Godement canonical resolution*** of $\mathcal{F}$. The sheaves $C^k\mathcal{F}$ are called the ***Godement sheaves*** of $\mathcal{F}$. (The letter "C" stands for "canonical.")

Next we show that the Godement resolution $\mathcal{F} \to C^\bullet\mathcal{F}$ is functorial: a sheaf map $\varphi \colon \mathcal{F} \to \mathcal{G}$ induces a morphism $\varphi_* \colon C^\bullet\mathcal{F} \to C^\bullet\mathcal{G}$ of their Godement resolutions satisfying the two functorial properties: preservation of the identity and of composition.

A sheaf morphism (sheaf map) $\varphi \colon \mathcal{E} \to \mathcal{F}$ induces a sheaf morphism

$$
\begin{array}{ccc}
\mathcal{C}^0\varphi \colon & \mathcal{C}^0\mathcal{E} \longrightarrow & \mathcal{C}^0\mathcal{F} \\
& \| & \| \\
& \prod \mathcal{E}_p & \prod \mathcal{F}_p
\end{array}
$$

and therefore a morphism of quotient sheaves

$$
\begin{array}{ccc}
\mathcal{C}^0\mathcal{E}/\mathcal{E} \longrightarrow & \mathcal{C}^0\mathcal{F}/\mathcal{F}\,, \\
\| & \| \\
\mathcal{Q}^1_{\mathcal{E}} & \mathcal{Q}^1_{\mathcal{F}}
\end{array}
$$

which in turn induces a sheaf morphism

$$
\begin{array}{ccc}
\mathcal{C}^1\varphi \colon & \mathcal{C}^0\mathcal{Q}^1_{\mathcal{E}} \longrightarrow & \mathcal{C}^0\mathcal{Q}^1_{\mathcal{F}}\,. \\
& \| & \| \\
& \mathcal{C}^1\mathcal{E} & \mathcal{C}^1\mathcal{F}
\end{array}
$$

By induction, we obtain $\mathcal{C}^k\varphi \colon \mathcal{C}^k\mathcal{E} \to \mathcal{C}^k\mathcal{F}$ for all k. It can be checked that each $\mathcal{C}^k(\ )$ is a functor from sheaves to sheaves, called the *kth Godement functor*.

Moreover, the induced morphisms $\mathcal{C}^k\varphi$ fit into a commutative diagram

$$
\begin{array}{ccccccccc}
0 \longrightarrow & \mathcal{E} & \longrightarrow & \mathcal{C}^0\mathcal{E} & \longrightarrow & \mathcal{C}^1\mathcal{E} & \longrightarrow & \mathcal{C}^2\mathcal{E} & \longrightarrow \cdots \\
& \downarrow & & \downarrow & & \downarrow & & \downarrow & \\
0 \longrightarrow & \mathcal{F} & \longrightarrow & \mathcal{C}^0\mathcal{F} & \longrightarrow & \mathcal{C}^1\mathcal{F} & \longrightarrow & \mathcal{C}^2\mathcal{F} & \longrightarrow \cdots\,,
\end{array}
$$

so that collectively $(\mathcal{C}^k\varphi)_{k=0}^{\infty}$ is a morphism of Godement resolutions.

PROPOSITION 2.2.1 *If*

$$0 \to \mathcal{E} \to \mathcal{F} \to \mathcal{G} \to 0$$

is a short exact sequence of sheaves on a topological space X and $\mathcal{C}^k(\ )$ is the kth Godement sheaf functor, then the sequence of sheaves

$$0 \to \mathcal{C}^k\mathcal{E} \to \mathcal{C}^k\mathcal{F} \to \mathcal{C}^k\mathcal{G} \to 0$$

is exact.

We say that the Godement functors $\mathcal{C}^k(\ )$ are *exact functors* from sheaves to sheaves.

PROOF. For any point $p \in X$, the stalk $\mathcal{E}_p$ is a subgroup of the stalk $\mathcal{F}_p$ with quotient group $\mathcal{G}_p = \mathcal{F}_p / \mathcal{E}_p$. Interpreting $C^0 \mathcal{E}(U)$ as the direct product $\prod_{p \in U} \mathcal{E}_p$ of stalks over U, it is easy to verify that for any open set $U \subset X$,

$$0 \to C^0 \mathcal{E}(U) \to C^0 \mathcal{F}(U) \to C^0 \mathcal{G}(U) \to 0 \tag{2.2.4}$$

is exact. In general, the direct limit of exact sequences is exact [2, Ch. 2, Exer. 19, p. 33]. Taking the direct limit of (2.2.4) over all neighborhoods of a point $p \in X$, we obtain the exact sequence of stalks

$$0 \to (C^0 \mathcal{E})_p \to (C^0 \mathcal{F})_p \to (C^0 \mathcal{G})_p \to 0$$

for all $p \in X$. Thus, the sequence of sheaves

$$0 \to C^0 \mathcal{E} \to C^0 \mathcal{F} \to C^0 \mathcal{G} \to 0$$

is exact.

Let $\mathcal{Q}_\mathcal{E}$ be the quotient sheaf $C^0 \mathcal{E} / \mathcal{E}$, and similarly for $\mathcal{Q}_\mathcal{F}$ and $\mathcal{Q}_\mathcal{G}$. Then there is a commutative diagram

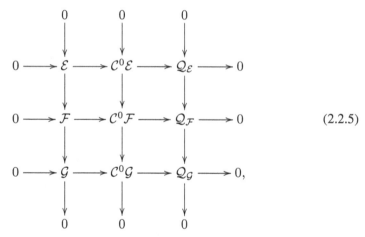

$$\tag{2.2.5}$$

in which the three rows and the first two columns are exact. It follows by the nine lemma that the last column is also exact.[1] Taking $C^0(\ )$ of the last column, we obtain an exact sequence

$$0 \longrightarrow C^0 \mathcal{Q}_\mathcal{E} \longrightarrow C^0 \mathcal{Q}_\mathcal{F} \longrightarrow C^0 \mathcal{Q}_\mathcal{G} \longrightarrow 0.$$
$$\quad\quad \| \quad\quad\quad\quad \| \quad\quad\quad\quad \|$$
$$\quad C^1 \mathcal{E} \quad\quad\quad C^1 \mathcal{F} \quad\quad\quad C^1 \mathcal{G}$$

The Godement resolution is created by alternately taking C^0 and taking quotients. We have shown that each of these two operations preserves exactness. Hence, the proposition follows by induction.   □

---

[1]To prove the nine lemma, view each column as a differential complex. Then the diagram (2.2.5) is a short exact sequence of complexes. Since the cohomology groups of the first two columns are zero, the long exact cohomology sequence of the short exact sequence implies that the cohomology of the third column is also zero [15, Th. 25.6, p. 285].

2.2.2 Cohomology with Coefficients in a Sheaf

Let $\mathcal{F}$ be a sheaf of abelian groups on a topological space X. What is so special about the Godement resolution of $\mathcal{F}$ is that it is completely canonical. For any open set U in X, applying the sections functor $\Gamma(U, \ )$ to the Godement resolution of $\mathcal{F}$ gives a complex

$$0 \to \mathcal{F}(U) \to \mathcal{C}^0\mathcal{F}(U) \to \mathcal{C}^1\mathcal{F}(U) \to \mathcal{C}^2\mathcal{F}(U) \to \cdots . \qquad (2.2.6)$$

In general, the kth **cohomology** of a complex

$$0 \to K^0 \xrightarrow{d} K^1 \xrightarrow{d} K^2 \to \cdots$$

will be denoted by

$$h^k(K^\bullet) := \frac{\ker(d \colon K^k \to K^{k+1})}{\mathrm{Im}(d \colon K^{k-1} \to K^k)}.$$

We sometimes write a complex $(K^\bullet, d)$ not as a sequence, but as a direct sum $K^\bullet = \bigoplus_{k=0}^\infty K^k$, with the understanding that $d \colon K^k \to K^{k+1}$ increases the degree by 1 and $d \circ d = 0$. The **cohomology of U with coefficients in the sheaf $\mathcal{F}$**, or the **sheaf cohomology of $\mathcal{F}$ on U**, is defined to be the cohomology of the complex $\mathcal{C}^\bullet\mathcal{F}(U) = \bigoplus_{k \geq 0} \mathcal{C}^k\mathcal{F}(U)$ of sections of the Godement resolution of $\mathcal{F}$ (with the initial term $\mathcal{F}(U)$ dropped from the complex (2.2.6)):

$$H^k(U, \mathcal{F}) := h^k(\mathcal{C}^\bullet\mathcal{F}(U)).$$

PROPOSITION 2.2.2 *Let $\mathcal{F}$ be a sheaf on a topological space X. For any open set $U \subset X$, we have $H^0(U, \mathcal{F}) = \Gamma(U, \mathcal{F})$.*

PROOF. If

$$0 \to \mathcal{F} \to \mathcal{C}^0\mathcal{F} \to \mathcal{C}^1\mathcal{F} \to \mathcal{C}^2\mathcal{F} \to \cdots$$

is the Godement resolution of $\mathcal{F}$, then by definition

$$H^0(U, \mathcal{F}) = \ker\left(d \colon \mathcal{C}^0\mathcal{F}(U) \to \mathcal{C}^1\mathcal{F}(U)\right).$$

In the notation of the preceding subsection, $d \colon \mathcal{C}^0\mathcal{F}(U) \to \mathcal{C}^1\mathcal{F}(U)$ is induced from the composition of sheaf maps

$$\mathcal{C}^0\mathcal{F} \twoheadrightarrow \mathcal{Q}^1 \hookrightarrow \mathcal{C}^1\mathcal{F}.$$

Thus, $d \colon \mathcal{C}^0\mathcal{F}(U) \to \mathcal{C}^1\mathcal{F}(U)$ is the composition of

$$\mathcal{C}^0\mathcal{F}(U) \to \mathcal{Q}^1(U) \hookrightarrow \mathcal{C}^1\mathcal{F}(U).$$

Note that the second map $\mathcal{Q}^1(U) \hookrightarrow \mathcal{C}^1\mathcal{F}(U)$ is injective, because $\Gamma(U, \ )$ is a left-exact functor. Hence,

$$H^0(U, \mathcal{F}) = \ker\left(\mathcal{C}^0\mathcal{F}(U) \to \mathcal{C}^1\mathcal{F}(U)\right)$$
$$= \ker\left(\mathcal{C}^0\mathcal{F}(U) \to \mathcal{Q}^1(U)\right).$$

But from the exactness of

$$0 \to \mathcal{F}(U) \to \mathcal{C}^0 \mathcal{F}(U) \to \mathcal{Q}^1(U),$$

we see that

$$\Gamma(U, \mathcal{F}) = \mathcal{F}(U) = \ker\left(\mathcal{C}^0 \mathcal{F}(U) \to \mathcal{Q}^1(U)\right) = H^0(U, \mathcal{F}).$$

$\square$

2.2.3 Flasque Sheaves

Flasque sheaves are a special kind of sheaf with vanishing higher cohomology. All Godement sheaves turn out to be flasque sheaves.

DEFINITION 2.2.3 *A sheaf $\mathcal{F}$ of abelian groups on a topological space X is **flasque** (French for "flabby") if for every open set $U \subset X$, the restriction map $\mathcal{F}(X) \to \mathcal{F}(U)$ is surjective.*

For any sheaf $\mathcal{F}$, the Godement sheaf $\mathcal{C}^0 \mathcal{F}$ is clearly flasque because $\mathcal{C}^0 \mathcal{F}(U)$ consists of all discontinuous sections of the étalé space $E_{\mathcal{F}}$ over U. In the notation of the preceding subsection, $\mathcal{C}^k \mathcal{F} = \mathcal{C}^0 \mathcal{Q}^k$, so all Godement sheaves $\mathcal{C}^k \mathcal{F}$ are flasque.

PROPOSITION 2.2.4   *(i) In a short exact sequence of sheaves*

$$0 \to \mathcal{E} \xrightarrow{i} \mathcal{F} \xrightarrow{j} \mathcal{G} \to 0 \tag{2.2.7}$$

over a topological space X, if $\mathcal{E}$ is flasque, then for any open set $U \subset X$, the sequence of abelian groups

$$0 \to \mathcal{E}(U) \to \mathcal{F}(U) \to \mathcal{G}(U) \to 0$$

is exact.

(ii) If $\mathcal{E}$ and $\mathcal{F}$ are flasque in (2.2.7), then $\mathcal{G}$ is flasque.

(iii) If

$$0 \to \mathcal{E} \to \mathcal{L}^0 \to \mathcal{L}^1 \to \mathcal{L}^2 \to \cdots \tag{2.2.8}$$

is an exact sequence of flasque sheaves on X, then for any open set $U \subset X$ the sequence of abelian groups of sections

$$0 \to \mathcal{E}(U) \to \mathcal{L}^0(U) \to \mathcal{L}^1(U) \to \mathcal{L}^2(U) \to \cdots \tag{2.2.9}$$

is exact.

PROOF.   (i) To simplify the notation, we will use i to denote $i_U : \mathcal{E}(U) \to \mathcal{F}(U)$ for all U; similarly, $j = j_U$. As noted in Section 2.1.2, the exactness of

$$0 \to \mathcal{E}(U) \xrightarrow{i} \mathcal{F}(U) \xrightarrow{j} \mathcal{G}(U) \tag{2.2.10}$$

is true in general, whether $\mathcal{E}$ is flasque or not. To prove the surjectivity of j for a flasque $\mathcal{E}$, let $g \in \mathcal{G}(U)$. Since $\mathcal{F} \to \mathcal{G}$ is surjective as a sheaf map, all stalk maps $\mathcal{F}_p \to \mathcal{G}_p$ are surjective. Hence, every point $p \in U$ has a neighborhood $U_\alpha \subset U$ on which there exists a section $f_\alpha \in \mathcal{F}(U_\alpha)$ such that $j(f_\alpha) = g|_{U_\alpha}$.

Let V be the largest union $\bigcup_\alpha U_\alpha$ on which there is a section $f_V \in \mathcal{F}(V)$ such that $j(f_V) = g|_V$. We claim that $V = U$. If not, then there are a set U_α not contained in V and $f_\alpha \in \mathcal{F}(U_\alpha)$ such that $j(f_\alpha) = g|_{U_\alpha}$. On $V \cap U_\alpha$, writing j for $j_{V \cap U_\alpha}$, we have

$$j(f_V - f_\alpha) = 0.$$

By the exactness of the sequence (2.2.10) at $\mathcal{F}(V \cap U_\alpha)$,

$$f_V - f_\alpha = i(e_{V,\alpha}) \text{ for some } e_{V,\alpha} \in \mathcal{E}(V \cap U_\alpha).$$

Since $\mathcal{E}$ is flasque, one can find a section $e_U \in \mathcal{E}(U)$ such that $e_U|_{V \cap U_\alpha} = e_{V,\alpha}$. On $V \cap U_\alpha$,

$$f_V = i(e_{V,\alpha}) + f_\alpha.$$

If we modify f_α to

$$\bar{f}_\alpha = i(e_U) + f_\alpha \quad \text{on } U_\alpha,$$

then $f_V = \bar{f}_\alpha$ on $V \cap U_\alpha$, and $j(\bar{f}_\alpha) = g|_{U_\alpha}$. By the gluing axiom for the sheaf $\mathcal{F}$, the elements f_V and $\bar{f}_\alpha$ piece together to give an element $f \in \mathcal{F}(V \cup U_\alpha)$ such that $j(f) = g|_{V \cup U_\alpha}$. This contradicts the maximality of V. Hence, $V = U$ and $j \colon \mathcal{F}(U) \to \mathcal{G}(U)$ is onto.

(ii) Since $\mathcal{E}$ is flasque, for any open set $U \subset X$ the rows of the commutative diagram

$$
\begin{array}{ccccccccc}
0 & \longrightarrow & \mathcal{E}(X) & \longrightarrow & \mathcal{F}(X) & \overset{j_X}{\longrightarrow} & \mathcal{G}(X) & \longrightarrow & 0 \\
& & \downarrow{\alpha} & & \downarrow{\beta} & & \downarrow{\gamma} & & \\
0 & \longrightarrow & \mathcal{E}(U) & \longrightarrow & \mathcal{F}(U) & \overset{j_U}{\longrightarrow} & \mathcal{G}(U) & \longrightarrow & 0
\end{array}
$$

are exact by (i), where α, β, and γ are the restriction maps. Since $\mathcal{F}$ is flasque, the map $\beta \colon \mathcal{F}(X) \to \mathcal{F}(U)$ is surjective. Hence,

$$j_U \circ \beta = \gamma \circ j_X \colon \mathcal{F}(X) \to \mathcal{G}(X) \to \mathcal{G}(U)$$

is surjective. Therefore, $\gamma \colon \mathcal{G}(X) \to \mathcal{G}(U)$ is surjective. This proves that $\mathcal{G}$ is flasque.

(iii) The long exact sequence (2.2.8) is equivalent to a collection of short exact sequences

$$0 \to \mathcal{E} \to \mathcal{L}^0 \to \mathcal{Q}^0 \to 0, \tag{2.2.11}$$

$$0 \to \mathcal{Q}^0 \to \mathcal{L}^1 \to \mathcal{Q}^1 \to 0, \tag{2.2.12}$$

$$\cdots .$$

In (2.2.11), the first two sheaves are flasque, so $\mathcal{Q}^0$ is flasque by (ii). Similarly, in (2.2.12), the first two sheaves are flasque, so $\mathcal{Q}^1$ is flasque. By induction, all the sheaves $\mathcal{Q}^k$ are flasque.

By (i), the functor $\Gamma(U, \ )$ transforms the short exact sequences of sheaves into short exact sequences of abelian groups

$$0 \to \mathcal{E}(U) \to \mathcal{L}^0(U) \to \mathcal{Q}^0(U) \to 0,$$
$$0 \to \mathcal{Q}^0(U) \to \mathcal{L}^1(U) \to \mathcal{Q}^1(U) \to 0,$$
$$\cdots .$$

These short exact sequences splice together into the long exact sequence (2.2.9). $\qquad\square$

COROLLARY 2.2.5 *Let $\mathcal{E}$ be a flasque sheaf on a topological space X. For every open set $U \subset X$ and every $k > 0$, the cohomology $H^k(U, \mathcal{E}) = 0$.*

PROOF. Let

$$0 \to \mathcal{E} \to \mathcal{C}^0 \mathcal{E} \to \mathcal{C}^1 \mathcal{E} \to \mathcal{C}^2 \mathcal{E} \to \cdots$$

be the Godement resolution of $\mathcal{E}$. It is an exact sequence of flasque sheaves. By Proposition 2.2.4(iii), the sequence of groups of sections

$$0 \to \mathcal{E}(U) \to \mathcal{C}^0 \mathcal{E}(U) \to \mathcal{C}^1 \mathcal{E}(U) \to \mathcal{C}^2 \mathcal{E}(U) \to \cdots$$

is exact. It follows from the definition of sheaf cohomology that

$$H^k(U, \mathcal{E}) = \begin{cases} \mathcal{E}(U) & \text{for } k = 0, \\ 0 & \text{for } k > 0. \end{cases}$$

$\qquad\square$

A sheaf $\mathcal{F}$ on a topological space X is said to be ***acyclic*** on $U \subset X$ if $H^k(U, \mathcal{F}) = 0$ for all $k > 0$. Thus, a flasque sheaf on X is acyclic on every open set of X.

EXAMPLE 2.2.6 Let X be an irreducible complex algebraic variety with the Zariski topology. Recall that the constant sheaf $\mathbb{C}$ over X is the sheaf of locally constant functions on X with values in $\mathbb{C}$. Because any two open sets in the Zariski topology of X have a nonempty intersection, the only continuous sections of the constant sheaf $\mathbb{C}$ over any open set U are the constant functions. Hence, $\mathbb{C}$ is flasque. By Corollary 2.2.5, $H^k(X, \mathbb{C}) = 0$ for all $k > 0$.

COROLLARY 2.2.7 *Let U be an open subset of a topological space X. The kth Godement sections functor $\Gamma(U, \mathcal{C}^k(\ ))$, which assigns to a sheaf $\mathcal{F}$ on X the group $\Gamma(U, \mathcal{C}^k \mathcal{F})$ of sections of $\mathcal{C}^k \mathcal{F}$ over U, is an exact functor from sheaves on X to abelian groups.*

PROOF. Let

$$0 \to \mathcal{E} \to \mathcal{F} \to \mathcal{G} \to 0$$

be an exact sequence of sheaves. By Proposition 2.2.1, for any $k \geq 0$,

$$0 \to \mathcal{C}^k \mathcal{E} \to \mathcal{C}^k \mathcal{F} \to \mathcal{C}^k \mathcal{G} \to 0$$

is an exact sequence of sheaves. Since $\mathcal{C}^k \mathcal{E}$ is flasque, by Proposition 2.2.4(i),

$$0 \to \Gamma(U, \mathcal{C}^k \mathcal{E}) \to \Gamma(U, \mathcal{C}^k \mathcal{F}) \to \Gamma(U, \mathcal{C}^k \mathcal{G}) \to 0$$

is an exact sequence of abelian groups. Hence, $\Gamma(U, \mathcal{C}^k(\ ))$ is an exact functor from sheaves to groups. $\qquad\square$

Although we do not need it, the following theorem is a fundamental property of sheaf cohomology.

THEOREM 2.2.8 *A short exact sequence*

$$0 \to \mathcal{E} \to \mathcal{F} \to \mathcal{G} \to 0$$

of sheaves of abelian groups on a topological space X induces a long exact sequence in sheaf cohomology,

$$\cdots \to H^k(X, \mathcal{E}) \to H^k(X, \mathcal{F}) \to H^k(X, \mathcal{G}) \to H^{k+1}(X, \mathcal{E}) \to \cdots .$$

PROOF. Because the Godement sections functor $\Gamma(X, \mathcal{C}^k(\ ))$ is exact, from the given short exact sequence of sheaves one obtains a short exact sequence of complexes of global sections of Godement sheaves

$$0 \to \mathcal{C}^{\bullet} \mathcal{E}(X) \to \mathcal{C}^{\bullet} \mathcal{F}(X) \to \mathcal{C}^{\bullet} \mathcal{G}(X) \to 0.$$

The long exact sequence in cohomology [15, Sec. 25] associated to this short exact sequence of complexes is the desired long exact sequence in sheaf cohomology. $\qquad\square$

2.2.4   Cohomology Sheaves and Exact Functors

As before, a sheaf will mean a sheaf of abelian groups on a topological space X. A *complex of sheaves* $\mathcal{L}^{\bullet}$ on X is a sequence of sheaves

$$0 \to \mathcal{L}^0 \xrightarrow{d} \mathcal{L}^1 \xrightarrow{d} \mathcal{L}^2 \xrightarrow{d} \cdots$$

on X such that $d \circ d = 0$. Denote the kernel and image sheaves of $\mathcal{L}^{\bullet}$ by

$$\mathcal{Z}^k := \mathcal{Z}^k(\mathcal{L}^{\bullet}) := \ker\left(d \colon \mathcal{L}^k \to \mathcal{L}^{k+1}\right),$$
$$\mathcal{B}^k := \mathcal{B}^k(\mathcal{L}^{\bullet}) := \operatorname{Im}\left(d \colon \mathcal{L}^{k-1} \to \mathcal{L}^k\right).$$

Then the *cohomology sheaf* $\mathcal{H}^k := \mathcal{H}^k(\mathcal{L}^{\bullet})$ of the complex $\mathcal{L}^{\bullet}$ is the quotient sheaf

$$\mathcal{H}^k := \mathcal{Z}^k / \mathcal{B}^k.$$

For example, by the Poincaré lemma, the complex

$$0 \to \mathcal{A}^0 \to \mathcal{A}^1 \to \mathcal{A}^2 \to \cdots$$

of sheaves of C^∞-forms on a manifold M has cohomology sheaves

$$\mathcal{H}^k = \mathcal{H}^k(\mathcal{A}^\bullet) = \begin{cases} \mathbb{R} & \text{for } k = 0, \\ 0 & \text{for } k > 0. \end{cases}$$

PROPOSITION 2.2.9   *Let* $\mathcal{L}^\bullet$ *be a complex of sheaves on a topological space* X. *The stalk of its cohomology sheaf* $\mathcal{H}^k$ *at a point* p *is the kth cohomology of the complex* $\mathcal{L}_p^\bullet$ *of stalks.*

PROOF. Since

$$\mathcal{Z}_p^k = \ker \left(d_p \colon \mathcal{L}_p^k \to \mathcal{L}_p^{k+1} \right) \text{ and } \mathcal{B}_p^k = \operatorname{Im} \left(d_p \colon \mathcal{L}_p^{k-1} \to \mathcal{L}_p^k \right)$$

(see [9, Ch. II, Exer. 1.2(a), p. 66]), one can also compute the stalk of the cohomology sheaf $\mathcal{H}^k$ by computing

$$\mathcal{H}_p^k = (\mathcal{Z}^k/\mathcal{B}^k)_p = \mathcal{Z}_p^k/\mathcal{B}_p^k = h^k(\mathcal{L}_p^\bullet),$$

the cohomology of the sequence of stalk maps of $\mathcal{L}^\bullet$ at p. $\square$

Recall that a ***morphism*** $\varphi \colon \mathcal{F}^\bullet \to \mathcal{G}^\bullet$ of complexes of sheaves is a collection of sheaf maps $\varphi^k \colon \mathcal{F}^k \to \mathcal{G}^k$ such that $\varphi^{k+1} \circ d = d \circ \varphi^k$ for all k. A morphism $\varphi \colon \mathcal{F}^\bullet \to \mathcal{G}^\bullet$ of complexes of sheaves induces morphisms $\varphi^k \colon \mathcal{H}^k(\mathcal{F}^\bullet) \to \mathcal{H}^k(\mathcal{G}^\bullet)$ of cohomology sheaves. The morphism $\varphi \colon \mathcal{F}^\bullet \to \mathcal{G}^\bullet$ of complexes of sheaves is called a ***quasi-isomorphism*** if the induced morphisms $\varphi^k \colon \mathcal{H}^k(\mathcal{F}^\bullet) \to \mathcal{H}^k(\mathcal{G}^\bullet)$ of cohomology sheaves are isomorphisms for all k.

PROPOSITION 2.2.10   *Let* $\mathcal{L}^\bullet = \bigoplus_{k \geq 0} \mathcal{L}^k$ *be a complex of sheaves on a topological space* X. *If* T *is an exact functor from sheaves on* X *to abelian groups, then it commutes with cohomology:*

$$T\left(\mathcal{H}^k(\mathcal{L}^\bullet)\right) = h^k\left(T(\mathcal{L}^\bullet)\right).$$

PROOF. We first prove that T commutes with cocycles and coboundaries. Applying the exact functor T to the exact sequence

$$0 \to \mathcal{Z}^k \to \mathcal{L}^k \xrightarrow{d} \mathcal{L}^{k+1}$$

results in the exact sequence

$$0 \to T(\mathcal{Z}^k) \to T(\mathcal{L}^k) \xrightarrow{d} T(\mathcal{L}^{k+1}),$$

which proves that

$$Z^k\left(T(\mathcal{L}^\bullet)\right) := \ker \left(T(\mathcal{L}^k) \xrightarrow{d} T(\mathcal{L}^{k+1}) \right) = T(\mathcal{Z}^k).$$

(By abuse of notation, we write the differential of $T(\mathcal{L}^\bullet)$ also as d, instead of $T(d)$.)

The differential $d\colon \mathcal{L}^{k-1} \to \mathcal{L}^k$ factors into a surjection $\mathcal{L}^{k-1} \twoheadrightarrow \mathcal{B}^k$ followed by an injection $\mathcal{B}^k \hookrightarrow \mathcal{L}^k$:

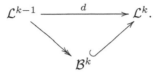

Since an exact functor preserves surjectivity and injectivity, applying T to the diagram above yields a commutative diagram

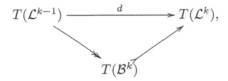

which proves that

$$B^k\left(T(\mathcal{L}^\bullet)\right) := \operatorname{Im}\left(T(\mathcal{L}^{k-1}) \xrightarrow{d} T(\mathcal{L}^k)\right) = T(\mathcal{B}^k).$$

Applying the exact functor T to the exact sequence of sheaves

$$0 \to \mathcal{B}^k \to \mathcal{Z}^k \to \mathcal{H}^k \to 0$$

gives the exact sequence of abelian groups

$$0 \to T(\mathcal{B}^k) \to T(\mathcal{Z}^k) \to T(\mathcal{H}^k) \to 0.$$

Hence,

$$T\left(\mathcal{H}^k(\mathcal{L}^\bullet)\right) = T(\mathcal{H}^k) = \frac{T(\mathcal{Z}^k)}{T(\mathcal{B}^k)} = \frac{Z^k\left(T(\mathcal{L}^\bullet)\right)}{B^k\left(T(\mathcal{L}^\bullet)\right)} = h^k\left(T(\mathcal{L}^\bullet)\right).$$

$\square$

2.2.5 Fine Sheaves

We have seen that flasque sheaves on a topological space X are acyclic on any open subset of X. Fine sheaves constitute another important class of such sheaves.

A sheaf map $f\colon \mathcal{F} \to \mathcal{G}$ over a topological space X induces at each point $x \in X$ a group homomorphism $f_x\colon \mathcal{F}_x \to \mathcal{G}_x$ of stalks. The **support** of the sheaf morphism f is defined to be

$$\operatorname{supp} f = \{x \in X \mid f_x \neq 0\}.$$

If two sheaf maps over a topological space X agree at a point, then they agree in a neighborhood of that point, so the set where two sheaf maps agree is open in X. Since the complement $X - \operatorname{supp} f$ is the subset of X where the sheaf map f agrees with the zero sheaf map, it is open and therefore $\operatorname{supp} f$ is closed.

DEFINITION 2.2.11   *Let $\mathcal{F}$ be a sheaf of abelian groups on a topological space X and $\{U_\alpha\}$ a locally finite open cover of X. A **partition of unity** of $\mathcal{F}$ subordinate to $\{U_\alpha\}$ is a collection $\{\eta_\alpha \colon \mathcal{F} \to \mathcal{F}\}$ of sheaf maps such that*

(i) $\operatorname{supp} \eta_\alpha \subset U_\alpha$;

(ii) for each point $x \in X$, the sum $\sum \eta_{\alpha,x} = \operatorname{id}_{\mathcal{F}_x}$, the identity map on the stalk $\mathcal{F}_x$.

Note that although α may range over an infinite index set, the sum in (ii) is a finite sum, because x has a neighborhood that meets only finitely many of the U_α's and $\operatorname{supp} \eta_\alpha \subset U_\alpha$.

DEFINITION 2.2.12   *A sheaf $\mathcal{F}$ on a topological space X is said to be **fine** if for every locally finite open cover $\{U_\alpha\}$ of X, the sheaf $\mathcal{F}$ admits a partition of unity subordinate to $\{U_\alpha\}$.*

PROPOSITION 2.2.13   *The sheaf $\mathcal{A}^k$ of smooth k-forms on a manifold M is a fine sheaf on M.*

PROOF.   Let $\{U_\alpha\}$ be a locally finite open cover of M. Then there is a C^∞ partition of unity $\{\rho_\alpha\}$ on M subordinate to $\{U_\alpha\}$ [15, App. C, p. 346]. (This partition of unity $\{\rho_\alpha\}$ is a collection of smooth $\mathbb{R}$-valued functions, not sheaf maps.) For any open set $U \subset M$, define $\eta_{\alpha,U} \colon \mathcal{A}^k(U) \to \mathcal{A}^k(U)$ by

$$\eta_{\alpha,U}(\omega) = \rho_\alpha \omega.$$

If $x \notin U_\alpha$, then x has a neighborhood U disjoint from $\operatorname{supp} \rho_\alpha$. Hence, ρ_α vanishes identically on U and $\eta_{\alpha,U} = 0$, so that the stalk map $\eta_{\alpha,x} \colon \mathcal{A}_x^k \to \mathcal{A}_x^k$ is the zero map. This proves that $\operatorname{supp} \eta_\alpha \subset U_\alpha$.

For any $x \in M$, the stalk map $\eta_{\alpha,x}$ is multiplication by the germ of ρ_α, so $\sum_\alpha \eta_{\alpha,x}$ is the identity map on the stalk $\mathcal{A}_x^k$. Hence, $\{\eta_\alpha\}$ is a partition of unity of the sheaf $\mathcal{A}^k$ subordinate to $\{U_\alpha\}$. $\square$

Let $\mathcal{R}$ be a sheaf of commutative rings on a topological space X. A sheaf $\mathcal{F}$ of abelian groups on X is called a **sheaf of $\mathcal{R}$-modules** (or simply an $\mathcal{R}$-**module**) if for every open set $U \subset X$, the abelian group $\mathcal{F}(U)$ has an $\mathcal{R}(U)$-module structure and moreover, for all $V \subset U$, the restriction map $\mathcal{F}(U) \to \mathcal{F}(V)$ is compatible with the module structure in the sense that the diagram

$$
\begin{array}{ccc}
\mathcal{R}(U) \times \mathcal{F}(U) & \longrightarrow & \mathcal{F}(U) \\
\downarrow & & \downarrow \\
\mathcal{R}(V) \times \mathcal{F}(V) & \longrightarrow & \mathcal{F}(V)
\end{array}
$$

commutes.

A **morphism** $\varphi \colon \mathcal{F} \to \mathcal{G}$ of sheaves of $\mathcal{R}$-modules over X is a sheaf morphism such that for each open set $U \subset X$, the group homomorphism $\varphi_U \colon \mathcal{F}(U) \to \mathcal{G}(U)$ is an $\mathcal{R}(U)$-module homomorphism.

If $\mathcal{A}^0$ is the sheaf of C^∞ functions on a manifold M, then the sheaf $\mathcal{A}^k$ of smooth k-forms on M is a sheaf of $\mathcal{A}^0$-modules. By a proof analogous to that of Proposition 2.2.13, any sheaf of $\mathcal{A}^0$-modules over a manifold is a fine sheaf. In particular, the sheaves $\mathcal{A}^{p,q}$ of smooth (p,q)-forms on a complex manifold are all fine sheaves.

2.2.6 Cohomology with Coefficients in a Fine Sheaf

A topological space X is **paracompact** if every open cover of X admits a locally finite open refinement. In working with fine sheaves, one usually has to assume that the topological space is paracompact, in order to be assured of the existence of a locally finite open cover. A common and important class of paracompact spaces is the class of topological manifolds [16, Lem. 1.9, p. 9].

A fine sheaf is generally not flasque. For example, $f(x) = \sec x$ is a C^∞ function on the open interval $U =]-\pi/2, \pi/2[$ that cannot be extended to a C^∞ function on $\mathbb{R}$. This shows that $\mathcal{A}^0(\mathbb{R}) \to \mathcal{A}^0(U)$ is not surjective. Thus, the sheaf $\mathcal{A}^0$ of C^∞ functions is a fine sheaf that is not flasque.

While flasque sheaves are useful for defining cohomology, fine sheaves are more prevalent in differential topology. Although fine sheaves need not be flasque, they share many of the properties of flasque sheaves. For example, on a manifold, Proposition 2.2.4 and Corollary 2.2.5 remain true if the sheaf $\mathcal{E}$ is fine instead of flasque.

PROPOSITION 2.2.14   *(i) In a short exact sequence of sheaves*

$$0 \to \mathcal{E} \to \mathcal{F} \to \mathcal{G} \to 0 \tag{2.2.13}$$

of abelian groups over a paracompact space X, if $\mathcal{E}$ is fine, then the sequence of abelian groups of global sections

$$0 \to \mathcal{E}(X) \xrightarrow{i} \mathcal{F}(X) \xrightarrow{j} \mathcal{G}(X) \to 0$$

is exact.
In (ii) and (iii), assume that every open subset of X is paracompact (a manifold is an example of such a space X).

(ii) If $\mathcal{E}$ is fine and $\mathcal{F}$ is flasque in (2.2.13), then $\mathcal{G}$ is flasque.

(iii) If

$$0 \to \mathcal{E} \to \mathcal{L}^0 \to \mathcal{L}^1 \to \mathcal{L}^2 \to \cdots$$

is an exact sequence of sheaves on X in which $\mathcal{E}$ is fine and all the $\mathcal{L}^k$ are flasque, then for any open set $U \subset X$, the sequence of abelian groups

$$0 \to \mathcal{E}(U) \to \mathcal{L}^0(U) \to \mathcal{L}^1(U) \to \mathcal{L}^2(U) \to \cdots$$

is exact.

PROOF. To simplify the notation, $i_U : \mathcal{E}(U) \to \mathcal{F}(U)$ will generally be denoted by i. Similarly, "f_α on $U_{\alpha\beta}$" will mean $f_\alpha|_{U_{\alpha\beta}}$. As in Proposition 2.2.4(i), it suffices to show that if $\mathcal{E}$ is a fine sheaf, then $j \colon \mathcal{F}(X) \to \mathcal{G}(X)$ is surjective. Let $g \in \mathcal{G}(X)$. Since $\mathcal{F}_p \to \mathcal{G}_p$ is surjective for all $p \in X$, there exist an open cover $\{U_\alpha\}$ of X and elements $f_\alpha \in \mathcal{F}(U_\alpha)$ such that $j(f_\alpha) = g|_{U_\alpha}$. By the paracompactness of X, we may assume that the open cover $\{U_\alpha\}$ is locally finite. On $U_{\alpha\beta} := U_\alpha \cap U_\beta$,

$$j(f_\alpha|_{U_{\alpha\beta}} - f_\beta|_{U_{\alpha\beta}}) = j(f_\alpha)|_{U_{\alpha\beta}} - j(f_\beta)|_{U_{\alpha\beta}} = g|_{U_{\alpha\beta}} - g|_{U_{\alpha\beta}} = 0.$$

By the exactness of the sequence

$$0 \to \mathcal{E}(U_{\alpha\beta}) \overset{i}{\to} \mathcal{F}(U_{\alpha\beta}) \overset{j}{\to} \mathcal{G}(U_{\alpha\beta}),$$

there is an element $e_{\alpha\beta} \in \mathcal{E}(U_{\alpha\beta})$ such that on $U_{\alpha\beta}$,

$$f_\alpha - f_\beta = i(e_{\alpha\beta}).$$

Note that on the triple intersection $U_{\alpha\beta\gamma} := U_\alpha \cap U_\beta \cap U_\gamma$, we have

$$i(e_{\alpha\beta} + e_{\beta\gamma}) = f_\alpha - f_\beta + f_\beta - f_\gamma = i(e_{\alpha\gamma}).$$

Since $\mathcal{E}$ is a fine sheaf, it admits a partition of unity $\{\eta_\alpha\}$ subordinate to $\{U_\alpha\}$. We will now view an element of $\mathcal{E}(U)$ for any open set U as a continuous section of the étalé space $E_{\mathcal{E}}$ over U. Then the section $\eta_\gamma(e_{\alpha\gamma}) \in \mathcal{E}(U_{\alpha\gamma})$ can be extended by zero to a continuous section of $E_{\mathcal{E}}$ over U_α:

$$\overline{\eta_\gamma e_{\alpha\gamma}}(p) = \begin{cases} (\eta_\gamma e_{\alpha\gamma})(p) & \text{for } p \in U_{\alpha\gamma}, \\ 0 & \text{for } p \in U_\alpha - U_{\alpha\gamma}. \end{cases}$$

(Proof of the continuity of $\overline{\eta_\gamma e_{\alpha\gamma}}$: On $U_{\alpha\gamma}$, $\eta_\gamma e_{\alpha\gamma}$ is continuous. If $p \in U_\alpha - U_{\alpha\gamma}$, then $p \notin U_\gamma$, so $p \notin \operatorname{supp} \eta_\gamma$. Since $\operatorname{supp} \eta_\gamma$ is closed, there is an open set V containing p such that $V \cap \operatorname{supp} \eta_\gamma = \varnothing$. Thus, $\overline{\eta_\gamma e_{\alpha\gamma}} = 0$ on V, which proves that $\overline{\eta_\gamma e_{\alpha\gamma}}$ is continuous at p.)

To simplify the notation, we will omit the overbar and write $\eta_\gamma e_{\alpha\gamma} \in \mathcal{E}(U_\alpha)$ also for the extension by zero of $\eta_\gamma e_{\alpha\gamma} \in \mathcal{E}(U_{\alpha\gamma})$. Let e_α be the locally finite sum

$$e_\alpha = \sum_\gamma \eta_\gamma e_{\alpha\gamma} \in \mathcal{E}(U_\alpha).$$

On the intersection $U_{\alpha\beta}$,

$$i(e_\alpha - e_\beta) = i\left(\sum_\gamma \eta_\gamma e_{\alpha\gamma} - \sum_\gamma \eta_\gamma e_{\beta\gamma}\right) = i\left(\sum_\gamma \eta_\gamma(e_{\alpha\gamma} - e_{\beta\gamma})\right)$$

$$= i\left(\sum_\gamma \eta_\gamma e_{\alpha\beta}\right) = i(e_{\alpha\beta}) = f_\alpha - f_\beta.$$

Hence, on $U_{\alpha\beta}$,

$$f_\alpha - i(e_\alpha) = f_\beta - i(e_\beta).$$

By the gluing sheaf axiom for the sheaf $\mathcal{F}$, there is an element $f \in \mathcal{F}(X)$ such that $f|_{U_\alpha} = f_\alpha - i(e_\alpha)$. Then

$$j(f)|_{U_\alpha} = j(f_\alpha) = g|_{U_\alpha} \text{ for all } \alpha.$$

By the uniqueness sheaf axiom for the sheaf $\mathcal{G}$, we have $j(f) = g \in \mathcal{G}(X)$. This proves the surjectivity of $j \colon \mathcal{F}(X) \to \mathcal{G}(X)$.

(ii), (iii) Assuming that every open subset U of X is paracompact, we can apply (i) to U. Then the proofs of (ii) and (iii) are the same as in Proposition 2.2.4(ii), (iii). $\qquad\square$

The analogue of Corollary 2.2.5 for $\mathcal{E}$ a fine sheaf then follows as before. The upshot is the following theorem.

THEOREM 2.2.15 *Let X be a topological space in which every open subset is para-compact. Then a fine sheaf on X is acyclic on every open subset U.*

REMARK 2.2.16 Sheaf cohomology can be characterized uniquely by a set of axioms [16, Def. 5.18, pp. 176–177]. Both the sheaf cohomology in terms of Godement's resolution and the Čech cohomology of a paracompact Hausdorff space satisfy these axioms [16, pp. 200–204], so at least on a paracompact Hausdorff space, sheaf cohomology is isomorphic to Čech cohomology. Since the Čech cohomology of a triangularizable space with coefficients in the constant sheaf $\mathbb{Z}$ is isomorphic to its singular cohomology with integer coefficients [3, Th. 15.8, p. 191], the sheaf cohomology $H^k(M, \mathbb{Z})$ of a manifold M is isomorphic to the singular cohomology $H^k(M, \mathbb{Z})$. In fact, the same argument shows that one may replace $\mathbb{Z}$ by $\mathbb{R}$ or by $\mathbb{C}$.

2.3 COHERENT SHEAVES AND SERRE'S GAGA PRINCIPLE

Given two sheaves $\mathcal{F}$ and $\mathcal{G}$ on X, it is easy to show that the presheaf $U \mapsto \mathcal{F}(U) \oplus \mathcal{G}(U)$ is a sheaf, called the **direct sum** of $\mathcal{F}$ and $\mathcal{G}$ and denoted by $\mathcal{F} \oplus \mathcal{G}$. We write the direct sum of p copies of $\mathcal{F}$ as $\mathcal{F}^{\oplus p}$. If U is an open subset of X, the **restriction** $\mathcal{F}|_U$ of the sheaf $\mathcal{F}$ to U is the sheaf on U defined by $(\mathcal{F}|_U)(V) = \mathcal{F}(V)$ for every open subset V of U. Let $\mathcal{R}$ be a sheaf of commutative rings on a topological space X. A sheaf $\mathcal{F}$ of $\mathcal{R}$-modules on M is **locally free of rank** p if every point $x \in M$ has a neighborhood U on which there is a sheaf isomorphism $\mathcal{F}|_U \simeq \mathcal{R}|_U^{\oplus p}$.

Given a complex manifold M, let $\mathcal{O}_M$ be its sheaf of holomorphic functions. When understood from the context, the subscript M is usually suppressed and $\mathcal{O}_M$ is simply written $\mathcal{O}$. A sheaf of $\mathcal{O}$-modules on a complex manifold is also called an **analytic sheaf**.

EXAMPLE 2.3.1 On a complex manifold M of complex dimension n, the sheaf Ω^k of holomorphic k-forms is an analytic sheaf. It is locally free of rank $\binom{n}{k}$, with local frame $\{dz_{i_1} \wedge \cdots \wedge dz_{i_k}\}$ for $1 \le i_1 < \cdots < i_k \le n$.

EXAMPLE 2.3.2 The sheaf $\mathcal{O}^*$ of nowhere-vanishing holomorphic functions with pointwise multiplication on a complex manifold M is *not* an analytic sheaf, since multiplying a nowhere-vanishing function $f \in \mathcal{O}^*(U)$ by the zero function $0 \in \mathcal{O}(U)$ will result in a function not in $\mathcal{O}^*(U)$.

Let $\mathcal{R}$ be a sheaf of commutative rings on a topological space X, let $\mathcal{F}$ be a sheaf of $\mathcal{R}$-modules on X, and let $f_1, \ldots, f_n$ be sections of $\mathcal{F}$ over an open set U in X. For any $r_1, \ldots, r_n \in \mathcal{R}(U)$, the map

$$\mathcal{R}^{\oplus n}(U) \to \mathcal{F}(U),$$

$$(r_1, \ldots, r_n) \mapsto \sum r_i f_i$$

defines a sheaf map $\varphi \colon \mathcal{R}^{\oplus n}|_U \to \mathcal{F}|_U$ over U. The kernel of φ is a subsheaf of $(\mathcal{R}|_U)^{\oplus n}$ called the **sheaf of relations** among $f_1, \ldots, f_n$, denoted by $\mathcal{S}(f_1, \ldots, f_n)$. We say that $\mathcal{F}|_U$ is **generated by** $f_1, \ldots, f_n$ if $\varphi \colon \mathcal{R}^{\oplus n} \to \mathcal{F}$ is surjective over U.

A sheaf $\mathcal{F}$ of $\mathcal{R}$-modules on X is said to be **of finite type** if every $x \in X$ has a neighborhood U on which $\mathcal{F}$ is generated by finitely many sections $f_1, \ldots, f_n \in \mathcal{F}(U)$. In particular, then, for every $y \in U$, the values $f_{1,y}, \ldots, f_{n,y} \in \mathcal{F}_y$ generate the stalk $\mathcal{F}_y$ as an $\mathcal{R}_y$-module.

DEFINITION 2.3.3   *A sheaf $\mathcal{F}$ of $\mathcal{R}$-modules on a topological space X is **coherent** if*

 (i) *$\mathcal{F}$ is of finite type; and*

 (ii) *for any open set $U \subset X$ and any collection of sections $f_1, \ldots, f_n \in \mathcal{F}(U)$, the sheaf $\mathcal{S}(f_1, \ldots, f_n)$ of relations among $f_1, \ldots, f_n$ is of finite type over U.*

THEOREM 2.3.4   (i) *The direct sum of finitely many coherent sheaves is coherent.*

 (ii) *The kernel, image, and cokernel of a morphism of coherent sheaves are coherent.*

PROOF. For a proof, see Serre [12, Subsec. 13, Ths. 1 and 2, pp. 208–209].   □

A sheaf $\mathcal{F}$ of $\mathcal{R}$-modules on a topological space X is said to be **locally finitely presented** if every $x \in X$ has a neighborhood U on which there is an exact sequence of the form

$$\mathcal{R}|_U^{\oplus q} \to \mathcal{R}|_U^{\oplus p} \to \mathcal{F}|_U \to 0;$$

in this case, we say that $\mathcal{F}$ has a **finite presentation** or that $\mathcal{F}$ is **finitely presented** on U. If $\mathcal{F}$ is a coherent sheaf of $\mathcal{R}$-modules on X, then it is locally finitely presented.

Remark.   Having a finite presentation locally is a consequence of coherence, but is not equivalent to it. Having a finite presentation means that for *one set* of generators of $\mathcal{F}$, the sheaf of relations among them is finitely generated. Coherence is a stronger condition in that it requires the sheaf of relations among *any set* of elements of $\mathcal{F}$ to be finitely generated.

A sheaf $\mathcal{R}$ of rings on X is clearly a sheaf of $\mathcal{R}$-modules of finite type. For it to be coherent, for any open set $U \subset X$ and any sections $f_1, \ldots, f_n$, the sheaf $\mathcal{S}(f_1, \ldots, f_n)$ of relations among $f_1, \ldots, f_n$ must be of finite type.

EXAMPLE 2.3.5   If $\mathcal{O}_M$ is the sheaf of holomorphic functions on a complex manifold M, then $\mathcal{O}_M$ is a coherent sheaf of $\mathcal{O}_M$-modules (Oka's theorem [4, Sec. 5]).

EXAMPLE 2.3.6 If $\mathcal{O}_X$ is the sheaf of regular functions on an algebraic variety X, then $\mathcal{O}_X$ is a coherent sheaf of $\mathcal{O}_X$-modules (Serre [12, Sec. 37, Prop. 1]).

A sheaf of $\mathcal{O}_X$-modules on an algebraic variety is called an **algebraic sheaf**.

EXAMPLE 2.3.7 On a smooth variety X of dimension n, the sheaf Ω^k of algebraic k-forms is an algebraic sheaf. It is locally free of rank $\binom{n}{k}$ [14, Ch. III, Th. 2, p. 200].

THEOREM 2.3.8 *Let $\mathcal{R}$ be a coherent sheaf of rings on a topological space X. Then a sheaf $\mathcal{F}$ of $\mathcal{R}$-modules on X is coherent if and only if it is locally finitely presented.*

PROOF. ($\Rightarrow$) True for any coherent sheaf of $\mathcal{R}$-modules, whether $\mathcal{R}$ is coherent or not.

($\Leftarrow$) Suppose there is an exact sequence

$$\mathcal{R}^{\oplus q} \to \mathcal{R}^{\oplus p} \to \mathcal{F} \to 0$$

on an open set U in X. Since $\mathcal{R}$ is coherent, by Theorem 2.3.4 so are $\mathcal{R}^{\oplus p}$, $\mathcal{R}^{\oplus q}$, and the cokernel $\mathcal{F}$ of $\mathcal{R}^{\oplus q} \to \mathcal{R}^{\oplus p}$. □

Since the structure sheaves $\mathcal{O}_X$ or $\mathcal{O}_M$ of an algebraic variety X or of a complex manifold M are coherent, an algebraic or analytic sheaf is coherent if and only if it is locally finitely presented.

EXAMPLE 2.3.9 A locally free analytic sheaf $\mathcal{F}$ over a complex manifold M is automatically coherent since every point p has a neighborhood U on which there is an exact sequence of the form

$$0 \to \mathcal{O}_U^{\oplus p} \to \mathcal{F}|_U \to 0,$$

so that $\mathcal{F}|_U$ is finitely presented.

For our purposes, we define a **Stein manifold** to be a complex manifold that is biholomorphic to a closed submanifold of $\mathbb{C}^N$ (this is not the usual definition, but is equivalent to it [11, p. 114]). In particular, a complex submanifold of $\mathbb{C}^N$ defined by finitely many holomorphic functions is a Stein manifold. One of the basic theorems about coherent analytic sheaves is Cartan's theorem B.

THEOREM 2.3.10 (Cartan's theorem B) *A coherent analytic sheaf $\mathcal{F}$ is acyclic on a Stein manifold M, i.e., $H^q(M, \mathcal{F}) = 0$ for all $q \geq 1$.*

For a proof, see [8, Th. 14, p. 243].

Let X be a smooth quasi-projective variety defined over the complex numbers and endowed with the Zariski topology. The underlying set of X with the complex topology is a complex manifold X_{an}. Similarly, if U is a Zariski open subset of X, let U_{an} be the underlying set of U with the complex topology. Since Zariski open sets are open in the complex topology, U_{an} is open in X_{an}.

Denote by $\mathcal{O}_{X_{an}}$ the sheaf of holomorphic functions on X_{an}. If $\mathcal{F}$ is a coherent algebraic sheaf on X, then X has an open cover $\{U\}$ by Zariski open sets such that on each open set U there is an exact sequence

$$\mathcal{O}_U^{\oplus q} \to \mathcal{O}_U^{\oplus p} \to \mathcal{F}|_U \to 0$$

of algebraic sheaves. Moreover, $\{U_{\text{an}}\}$ is an open cover of X_{an} and the morphism $\mathcal{O}_U^{\oplus q} \to \mathcal{O}_U^{\oplus p}$ of algebraic sheaves induces a morphism $\mathcal{O}_{U_{\text{an}}}^{\oplus q} \to \mathcal{O}_{U_{\text{an}}}^{\oplus p}$ of analytic sheaves. Hence, there is a coherent analytic sheaf $\mathcal{F}_{\text{an}}$ on the complex manifold X_{an} defined by

$$\mathcal{O}_{U_{\text{an}}}^{\oplus q} \to \mathcal{O}_{U_{\text{an}}}^{\oplus p} \to \mathcal{F}_{\text{an}}|_{U_{\text{an}}} \to 0.$$

(Rename the open cover $\{U_{\text{an}}\}$ as $\{U_\alpha\}_{\alpha \in A}$. A section of $\mathcal{F}_{\text{an}}$ over an open set $V \subset X_{\text{an}}$ is a collection of sections $s_\alpha \in (\mathcal{F}_{\text{an}}|_{U_\alpha})(U_\alpha \cap V)$ that agree on all pairwise intersections $(U_\alpha \cap V) \cap (U_\beta \cap V)$.)

In this way one obtains a functor $(\ )_{\text{an}}$ from the category of smooth complex quasi-projective varieties and coherent algebraic sheaves to the category of complex manifolds and analytic sheaves. Serre's GAGA ("Géométrie algébrique et géométrie analytique") principle [13] asserts that for smooth complex projective varieties, the functor $(\ )_{\text{an}}$ is an equivalence of categories and moreover, for all q, there are isomorphisms of cohomology groups

$$H^q(X, \mathcal{F}) \simeq H^q(X_{\text{an}}, \mathcal{F}_{\text{an}}), \tag{2.3.1}$$

where the left-hand side is the sheaf cohomology of $\mathcal{F}$ on X endowed with the Zariski topology and the right-hand side is the sheaf cohomology of $\mathcal{F}_{\text{an}}$ on X_{an} endowed with the complex topology.

When X is a smooth complex quasi-projective variety, to distinguish between sheaves of algebraic and sheaves of holomorphic forms, we write Ω^p_{alg} for the sheaf of algebraic p-forms on X and Ω^p_{an} for the sheaf of holomorphic p-forms on X_{an} (for the definition of algebraic forms, see the introduction to this chapter). If $z_1, \ldots, z_n$ are local parameters for X [14, Ch. II, Sec. 2.1, p. 98], then both Ω^p_{alg} and Ω^p_{an} are locally free with frame $\{dz_{i_1} \wedge \cdots \wedge dz_{i_p}\}$, where $I = (i_1, \ldots, i_p)$ is a strictly increasing multiindex between 1 and n inclusive. (For the algebraic case, see [14, Vol. 1, Ch. III, Sec. 5.4, Th. 4, p. 203].) Hence, locally there are sheaf isomorphisms

$$0 \to \mathcal{O}_U^{\binom{n}{p}} \to \Omega^p_{\text{alg}}|_U \to 0 \quad \text{and} \quad 0 \to \mathcal{O}_{U_{\text{an}}}^{\binom{n}{p}} \to \Omega^p_{\text{an}}|_{U_{\text{an}}} \to 0,$$

which show that Ω^p_{an} is the coherent analytic sheaf associated to the coherent algebraic sheaf Ω^p_{alg}.

Let k be a field. An **affine closed set** in k^N is the zero set of finitely many polynomials on k^N, and an **affine variety** is an algebraic variety biregular to an affine closed set. The algebraic analogue of Cartan's theorem B is the following vanishing theorem of Serre for an affine variety [12, Sec. 44, Cor. 1, p. 237].

THEOREM 2.3.11 (Serre) *A coherent algebraic sheaf $\mathcal{F}$ on an affine variety X is acyclic on X, i.e., $H^q(X, \mathcal{F}) = 0$ for all $q \geq 1$.*

2.4 THE HYPERCOHOMOLOGY OF A COMPLEX OF SHEAVES

This section requires some knowledge of double complexes and their associated spectral sequences. One possible reference is [3, Chs. 2 and 3]. The hypercohomology of a complex $\mathcal{L}^\bullet$ of sheaves of abelian groups on a topological space X generalizes the cohomology of a single sheaf. To define it, first form the double complex of global sections of the Godement resolutions of the sheaves $\mathcal{L}^q$:

$$K = \bigoplus_{p,q} K^{p,q} = \bigoplus_{p,q} \Gamma(X, \mathcal{C}^p \mathcal{L}^q).$$

This double complex comes with two differentials: a horizontal differential

$$\delta \colon K^{p,q} \to K^{p+1,q}$$

induced from the Godement resolution and a vertical differential

$$d \colon K^{p,q} \to K^{p,q+1}$$

induced from the complex $\mathcal{L}^{\bullet}$. Since the differential $d \colon \mathcal{L}^q \to \mathcal{L}^{q+1}$ induces a morphism of complexes $\mathcal{C}^{\bullet} \mathcal{L}^q \to \mathcal{C}^{\bullet} \mathcal{L}^{q+1}$, where $\mathcal{C}^{\bullet}$ is the Godement resolution, the vertical differential in the double complex K commutes with the horizontal differential. The **hypercohomology** $\mathbb{H}^*(X, \mathcal{L}^{\bullet})$ of the complex $\mathcal{L}^{\bullet}$ is the total cohomology of the double complex, i.e., the cohomology of the associated single complex

$$K^{\bullet} = \bigoplus_k K^k = \bigoplus_k \bigoplus_{p+q=k} K^{p,q}$$

with differential $D = \delta + (-1)^p d$:

$$\mathbb{H}^k(X, \mathcal{L}^{\bullet}) = H_D^k(K^{\bullet}).$$

If the complex of sheaves $\mathcal{L}^{\bullet}$ consists of a single sheaf $\mathcal{L}^0 = \mathcal{F}$ in degree 0,

$$0 \to \mathcal{F} \to 0 \to 0 \to \cdots,$$

then the double complex $\bigoplus K^{p,q} = \bigoplus \Gamma(X, \mathcal{C}^p \mathcal{L}^q)$ has nonzero entries only in the zeroth row, which is simply the complex of sections of the Godement resolution of $\mathcal{F}$:

$$K = \quad \begin{array}{c|c|c|c}
q \uparrow & & & \\
0 & 0 & 0 & \\
0 & 0 & 0 & \\
\hline
\Gamma(X, \mathcal{C}^0 \mathcal{F}) & \Gamma(X, \mathcal{C}^1 \mathcal{F}) & \Gamma(X, \mathcal{C}^2 \mathcal{F}) & \\
\hline
0 & 1 & 2 & p
\end{array}$$

In this case, the associated single complex is the complex $\Gamma(X, \mathcal{C}^{\bullet} \mathcal{F})$ of global sections of the Godement resolution of $\mathcal{F}$, and the hypercohomology of $\mathcal{L}^{\bullet}$ is the sheaf cohomology of $\mathcal{F}$:

$$\mathbb{H}^k(X, \mathcal{L}^{\bullet}) = h^k\big(\Gamma(X, \mathcal{C}^{\bullet} \mathcal{F})\big) = H^k(X, \mathcal{F}). \qquad (2.4.1)$$

It is in this sense that hypercohomology generalizes sheaf cohomology.

2.4.1 The Spectral Sequences of Hypercohomology

Associated to any double complex (K, d, δ) with commuting differentials d and δ are two spectral sequences converging to the total cohomology $H_D^*(K)$. One spectral sequence starts with $E_1 = H_d$ and $E_2 = H_\delta H_d$. By reversing the roles of d and δ, we obtain a second spectral sequence with $E_1 = H_\delta$ and $E_2 = H_d H_\delta$ (see [3, Ch. II]). By the *usual* spectral sequence of a double complex, we will mean the first spectral sequence, with the vertical differential d as the initial differential.

In the category of groups, the E_∞ term is the associated graded group of the total cohomology $H_D^*(K)$ relative to a canonically defined filtration and is not necessarily isomorphic to $H_D^*(K)$ because of the extension phenomenon in group theory.

Fix a nonnegative integer p and let $T = \Gamma(X, \mathcal{C}^p(\ ))$ be the Godement sections functor that associates to a sheaf $\mathcal{F}$ on a topological space X the group of sections $\Gamma(X, \mathcal{C}^p \mathcal{F})$ of the Godement sheaf $\mathcal{C}^p \mathcal{F}$. Since T is an exact functor by Corollary 2.2.7, by Proposition 2.2.10 it commutes with cohomology:

$$h^q\big(T(\mathcal{L}^\bullet)\big) = T\big(\mathcal{H}^q(\mathcal{L}^\bullet)\big), \tag{2.4.2}$$

where $\mathcal{H}^q := \mathcal{H}^q(\mathcal{L}^\bullet)$ is the qth cohomology sheaf of the complex $\mathcal{L}^\bullet$ (see Section 2.2.4). For the double complex $K = \bigoplus \Gamma(X, \mathcal{C}^p \mathcal{L}^q)$, the E_1 term of the first spectral sequence is the cohomology of K with respect to the vertical differential d. Thus, $E_1^{p,q} = H_d^{p,q}$ is the qth cohomology of the pth column $K^{p,\bullet} = \Gamma\big(X, \mathcal{C}^p(\mathcal{L}^\bullet)\big)$ of K:

$$\begin{aligned}
E_1^{p,q} = H_d^{p,q} = h^q(K^{p,\bullet}) &= h^q\big(\Gamma(X, \mathcal{C}^p \mathcal{L}^\bullet)\big) \\
&= h^q\big(T(\mathcal{L}^\bullet)\big) && \text{(definition of } T) \\
&= T\big(\mathcal{H}^q(\mathcal{L}^\bullet)\big) && \text{(by (2.4.2))} \\
&= \Gamma(X, \mathcal{C}^p \mathcal{H}^q) && \text{(definition of } T).
\end{aligned}$$

Hence, the E_2 term of the first spectral sequence is

$$E_2^{p,q} = H_\delta^{p,q}(E_1) = H_\delta^{p,q} H_d^{\bullet,\bullet} = h_\delta^p(H_d^{\bullet,q}) = h_\delta^p\big(\Gamma(X, \mathcal{C}^\bullet \mathcal{H}^q)\big) = \boxed{H^p(X, \mathcal{H}^q)}. \tag{2.4.3}$$

Note that the qth row of the double complex $\bigoplus K^{p,q} = \bigoplus \Gamma(X, \mathcal{C}^p \mathcal{L}^q)$ calculates the sheaf cohomology of $\mathcal{L}^q$ on X. Thus, the E_1 term of the second spectral sequence is

$$E_1^{p,q} = H_\delta^{p,q} = h_\delta^p(K^{\bullet,q}) = h_\delta^p\big(\Gamma(X, \mathcal{C}^\bullet \mathcal{L}^q)\big) = \boxed{H^p(X, \mathcal{L}^q)} \tag{2.4.4}$$

and the E_2 term is

$$E_2^{p,q} = H_d^{p,q}(E_1) = H_d^{p,q} H_\delta^{\bullet,\bullet} = h_d^q(H_\delta^{p,\bullet}) = h_d^q\big(H^p(X, \mathcal{L}^\bullet)\big).$$

THEOREM 2.4.1 *A quasi-isomorphism $\mathcal{F}^\bullet \to \mathcal{G}^\bullet$ of complexes of sheaves of abelian groups over a topological space X (see p. 84) induces a canonical isomorphism in hypercohomology:*

$$\mathbb{H}^*(X, \mathcal{F}^\bullet) \xrightarrow{\sim} \mathbb{H}^*(X, \mathcal{G}^\bullet).$$

PROOF. By the functoriality of the Godement sections functors, a morphism $\mathcal{F}^\bullet \to \mathcal{G}^\bullet$ of complexes of sheaves induces a homomorphism $\Gamma(X, \mathcal{C}^p \mathcal{F}^q) \to \Gamma(X, \mathcal{C}^p \mathcal{G}^q)$ that commutes with the two differentials d and δ and hence induces a homomorphism $\mathbb{H}^*(X, \mathcal{F}^\bullet) \to \mathbb{H}^*(X, \mathcal{G}^\bullet)$ in hypercohomology.

Since the spectral sequence construction is functorial, the morphism $\mathcal{F}^\bullet \to \mathcal{G}^\bullet$ also induces a morphism $E_r(\mathcal{F}^\bullet) \to E_r(\mathcal{G}^\bullet)$ of spectral sequences and a morphism of the filtrations

$$F_p\big(H_D(K_{\mathcal{F}^\bullet})\big) \to F_p\big(H_D(K_{\mathcal{G}^\bullet})\big)$$

on the hypercohomology of $\mathcal{F}^\bullet$ and $\mathcal{G}^\bullet$. We will shorten the notation $F_p\big(H_D(K_{\mathcal{F}^\bullet})\big)$ to $F_p(\mathcal{F}^\bullet)$.

By definition, the quasi-isomorphism $\mathcal{F}^\bullet \to \mathcal{G}^\bullet$ induces an isomorphism of cohomology sheaves $\mathcal{H}^*(\mathcal{F}^\bullet) \overset{\sim}{\to} \mathcal{H}^*(\mathcal{G}^\bullet)$, and by (2.4.3) an isomorphism of the E_2 terms of the first spectral sequences of $\mathcal{F}^\bullet$ and of $\mathcal{G}^\bullet$:

$$E_2^{p,q}(\mathcal{F}^\bullet) = H^p\big(X, \mathcal{H}^q(\mathcal{F}^\bullet)\big) \overset{\sim}{\to} H^p\big(X, \mathcal{H}^q(\mathcal{G}^\bullet)\big) = E_2^{p,q}(\mathcal{G}^\bullet).$$

An isomorphism of the E_2 terms induces an isomorphism of the E_∞ terms:

$$\bigoplus_p \frac{F_p(\mathcal{F}^\bullet)}{F_{p+1}(\mathcal{F}^\bullet)} = E_\infty(\mathcal{F}^\bullet) \overset{\sim}{\to} E_\infty(\mathcal{G}^\bullet) = \bigoplus_p \frac{F_p(\mathcal{G}^\bullet)}{F_{p+1}(\mathcal{G}^\bullet)}.$$

We claim that in fact, the canonical homomorphism $\mathbb{H}^*(X, \mathcal{F}^\bullet) \to \mathbb{H}^*(X, \mathcal{G}^\bullet)$ is an isomorphism. Fix a total degree k and let $F_p^k(\mathcal{F}^\bullet) = F_p(\mathcal{F}^\bullet) \cap \mathbb{H}^k(X, \mathcal{F}^\bullet)$. Since

$$K^{\bullet,\bullet}(\mathcal{F}^\bullet) = \bigoplus \Gamma(X, \mathcal{C}^p \mathcal{F}^q)$$

is a first-quadrant double complex, the filtration $\{F_p^k(\mathcal{F}^\bullet)\}_p$ on $\mathbb{H}^k(X, \mathcal{F}^\bullet)$ is finite in length:

$$\mathbb{H}^k(X, \mathcal{F}^\bullet) = F_0^k(\mathcal{F}^\bullet) \supset F_1^k(\mathcal{F}^\bullet) \supset \cdots \supset F_k^k(\mathcal{F}^\bullet) \supset F_{k+1}^k(\mathcal{F}^\bullet) = 0.$$

A similar finite filtration $\{F_p^k(\mathcal{G}^\bullet)\}_p$ exists on $\mathbb{H}^k(X, \mathcal{G}^\bullet)$.

Suppose $F_p^k(\mathcal{F}^\bullet) \to F_p^k(\mathcal{G}^\bullet)$ is an isomorphism. We will prove that $F_{p-1}^k(\mathcal{F}^\bullet) \to F_{p-1}^k(\mathcal{G}^\bullet)$ is an isomorphism. In the commutative diagram

$$
\begin{array}{ccccccccc}
0 & \longrightarrow & F_p^k(\mathcal{F}^\bullet) & \longrightarrow & F_{p-1}^k(\mathcal{F}^\bullet) & \longrightarrow & F_{p-1}^k(\mathcal{F}^\bullet)/F_p^k(\mathcal{F}^\bullet) & \longrightarrow & 0 \\
& & \downarrow & & \downarrow & & \downarrow & & \\
0 & \longrightarrow & F_p^k(\mathcal{G}^\bullet) & \longrightarrow & F_{p-1}^k(\mathcal{G}^\bullet) & \longrightarrow & F_{p-1}^k(\mathcal{G}^\bullet)/F_p^k(\mathcal{G}^\bullet) & \longrightarrow & 0,
\end{array}
$$

the two outside vertical maps are isomorphisms, by the induction hypothesis and because $\mathcal{F}^\bullet \to \mathcal{G}^\bullet$ induces an isomorphism of the associated graded groups. By the five lemma, the middle vertical map $F_{p-1}^k(\mathcal{F}^\bullet) \to F_{p-1}^k(\mathcal{G}^\bullet)$ is also an isomorphism. By induction on the filtration subscript p, as p moves from $k+1$ to 0, we conclude that

$$\mathbb{H}^k(X, \mathcal{F}^\bullet) = F_0^k(\mathcal{F}^\bullet) \to F_0^k(\mathcal{G}^\bullet) = \mathbb{H}^k(X, \mathcal{G}^\bullet)$$

is an isomorphism. □

THEOREM 2.4.2 *If $\mathcal{L}^\bullet$ is a complex of acyclic sheaves of abelian groups on a topological space X, then the hypercohomology of $\mathcal{L}^\bullet$ is isomorphic to the cohomology of the complex of global sections of $\mathcal{L}^\bullet$:*

$$\mathbb{H}^k(X, \mathcal{L}^\bullet) \simeq h^k(\mathcal{L}^\bullet(X)),$$

where $\mathcal{L}^\bullet(X)$ denotes the complex

$$0 \to \mathcal{L}^0(X) \to \mathcal{L}^1(X) \to \mathcal{L}^2(X) \to \cdots .$$

PROOF. Let K be the double complex $K = \bigoplus K^{p,q} = \bigoplus C^p \mathcal{L}^q(X)$. Because each $\mathcal{L}^q$ is acyclic on X, in the second spectral sequence of K, by (2.4.4) the E_1 term is

$$E_1^{p,q} = H^p(X, \mathcal{L}^q) = \begin{cases} \mathcal{L}^q(X) & \text{for } p = 0, \\ 0 & \text{for } p > 0. \end{cases}$$

$$
E_1 = H_\delta = \quad
\begin{array}{c|ccc}
q \\
& \mathcal{L}^2(X) & 0 & 0 \\
& \mathcal{L}^1(X) & 0 & 0 \\
& \mathcal{L}^0(X) & 0 & 0 \\
\hline
& 0 & 1 & 2 \qquad p
\end{array}
$$

Hence,

$$E_2^{p,q} = H_d^{p,q} H_\delta = \begin{cases} h^q(\mathcal{L}^\bullet(X)) & \text{for } p = 0, \\ 0 & \text{for } p > 0. \end{cases}$$

Therefore, the spectral sequence degenerates at the E_2 term and

$$\mathbb{H}^k(X, \mathcal{L}^\bullet) \simeq E_2^{0,k} = h^k(\mathcal{L}^\bullet(X)).$$

□

2.4.2 Acyclic Resolutions

Let $\mathcal{F}$ be a sheaf of abelian groups on a topological space X. A resolution

$$0 \to \mathcal{F} \to \mathcal{L}^0 \to \mathcal{L}^1 \to \mathcal{L}^2 \to \cdots$$

of $\mathcal{F}$ is said to be **acyclic** on X if each sheaf $\mathcal{L}^q$ is acyclic on X, i.e., $H^k(X, \mathcal{L}^q) = 0$ for all $k > 0$.

If $\mathcal{F}$ is a sheaf on X, we will denote by $\mathcal{F}^\bullet$ the complex of sheaves such that $\mathcal{F}^0 = \mathcal{F}$ and $\mathcal{F}^k = 0$ for $k > 0$.

THEOREM 2.4.3 *If $0 \to \mathcal{F} \to \mathcal{L}^\bullet$ is an acyclic resolution of the sheaf $\mathcal{F}$ on a topological space X, then the cohomology of $\mathcal{F}$ can be computed from the complex of global sections of $\mathcal{L}^\bullet$:*

$$H^k(X, \mathcal{F}) \simeq h^k(\mathcal{L}^\bullet(X)).$$

PROOF. The resolution $0 \to \mathcal{F} \to \mathcal{L}^\bullet$ may be viewed as a quasi-isomorphism of the two complexes

$$
\begin{array}{ccccccccc}
0 & \longrightarrow & \mathcal{F} & \longrightarrow & 0 & \longrightarrow & 0 & \longrightarrow & \cdots \\
& & \downarrow & & \downarrow & & \downarrow & & \\
0 & \longrightarrow & \mathcal{L}^0 & \longrightarrow & \mathcal{L}^1 & \longrightarrow & \mathcal{L}^2 & \longrightarrow & \cdots ,
\end{array}
$$

since

$$\mathcal{H}^0(\text{top row}) = \mathcal{H}^0(\mathcal{F}^\bullet) = \mathcal{F} \simeq \operatorname{Im}(\mathcal{F} \to \mathcal{L}^0) = \ker(\mathcal{L}^0 \to \mathcal{L}^1) = \mathcal{H}^0(\text{bottom row})$$

and the higher cohomology sheaves of both complexes are zero. By Theorem 2.4.1, there is an induced morphism in hypercohomology

$$\mathbb{H}^k(X, \mathcal{F}^\bullet) \simeq \mathbb{H}^k(X, \mathcal{L}^\bullet).$$

The left-hand side is simply the sheaf cohomology $H^k(X, \mathcal{F})$ by (2.4.1). By Theorem 2.4.2, the right-hand side is $h^k(\mathcal{L}^\bullet(X))$. Hence,

$$H^k(X, \mathcal{F}) \simeq h^k\big(\mathcal{L}^\bullet(X)\big).$$

$\square$

So in computing sheaf cohomology, any acyclic resolution of $\mathcal{F}$ on a topological space X can take the place of the Godement resolution.

Using acyclic resolutions, we can give simple proofs of de Rham's and Dolbeault's theorems.

EXAMPLE 2.4.4 (De Rham's theorem) By the Poincaré lemma ([3, Sec. 4, p. 33], [6, p. 38]), on a C^∞ manifold M the sequence of sheaves

$$0 \to \mathbb{R} \to \mathcal{A}^0 \to \mathcal{A}^1 \to \mathcal{A}^2 \to \cdots \qquad (2.4.5)$$

is exact. Since each $\mathcal{A}^k$ is fine and hence acyclic on M, (2.4.5) is an acyclic resolution of $\mathbb{R}$. By Theorem 2.4.3,

$$H^*(M, \mathbb{R}) \simeq h^*\big(\mathcal{A}^\bullet(M)\big) = H^*_{\mathrm{dR}}(M).$$

Because the sheaf cohomology $H^*(M, \mathbb{R})$ of a manifold is isomorphic to the real singular cohomology of M (Remark 2.2.16), de Rham's theorem follows.

EXAMPLE 2.4.5 (Dolbeault's theorem) According to the $\bar\partial$-Poincaré lemma [6, pp. 25 and 38], on a complex manifold M the sequence of sheaves

$$0 \to \Omega^p \to \mathcal{A}^{p,0} \xrightarrow{\bar\partial} \mathcal{A}^{p,1} \xrightarrow{\bar\partial} \mathcal{A}^{p,2} \to \cdots$$

is exact. As in the previous example, because each sheaf $\mathcal{A}^{p,q}$ is fine and hence acyclic, by Theorem 2.4.3,

$$H^q(M, \Omega^p) \simeq h^q\big(\mathcal{A}^{p,\bullet}(M)\big) = H^{p,q}(M).$$

This is the Dolbeault isomorphism for a complex manifold M.

2.5   THE ANALYTIC DE RHAM THEOREM

The analytic de Rham theorem is the analogue of the classical de Rham theorem for a complex manifold, according to which the singular cohomology with $\mathbb{C}$ coefficients of any complex manifold can be computed from its sheaves of holomorphic forms. Because of the holomorphic Poincaré lemma, the analytic de Rham theorem is far easier to prove than its algebraic counterpart.

2.5.1   The Holomorphic Poincaré Lemma

Let M be a complex manifold and Ω^k_{an} the sheaf of holomorphic k-forms on M. Locally, in terms of complex coordinates $z_1, \ldots, z_n$, a holomorphic form can be written as $\sum a_I \, dz_{i_1} \wedge \cdots \wedge dz_{i_n}$, where the a_I are holomorphic functions. Since for a holomorphic function a_I,

$$
da_I = \partial a_I + \bar{\partial} a_I = \sum_i \frac{\partial a_I}{\partial z_i} \, dz_i + \sum_i \frac{\partial a_I}{\partial \bar{z}_i} d\bar{z}_i = \sum_i \frac{\partial a_I}{\partial z_i} dz_i,
$$

the exterior derivative d maps holomorphic forms to holomorphic forms. Note that a_I is holomorphic if and only if $\bar{\partial} a_I = 0$.

THEOREM 2.5.1 (Holomorphic Poincaré lemma)   *On a complex manifold M of complex dimension n, the sequence*

$$
0 \to \mathbb{C} \to \Omega^0_{\mathrm{an}} \xrightarrow{d} \Omega^1_{\mathrm{an}} \xrightarrow{d} \cdots \to \Omega^n_{\mathrm{an}} \to 0
$$

of sheaves is exact.

PROOF.   We will deduce the holomorphic Poincaré lemma from the smooth Poincaré lemma and the $\bar{\partial}$-Poincaré lemma by a double complex argument. The double complex $\bigoplus \mathcal{A}^{p,q}$ of sheaves of smooth (p,q)-forms has two differentials ∂ and $\bar{\partial}$. These differentials anticommute because

$$
\begin{aligned}
0 = d \circ d &= (\partial + \bar{\partial})(\partial + \bar{\partial}) = \partial^2 + \bar{\partial}\partial + \partial\bar{\partial} + \bar{\partial}^2 \\
&= \bar{\partial}\partial + \partial\bar{\partial}.
\end{aligned}
$$

The associated single complex $\bigoplus \mathcal{A}^k_{\mathbb{C}}$, where $\mathcal{A}^k_{\mathbb{C}} = \bigoplus_{p+q=k} \mathcal{A}^{p,q}$ with differential $d = \partial + \bar{\partial}$, is simply the usual complex of sheaves of smooth $\mathbb{C}$-valued differential forms on M. By the smooth Poincaré lemma,

$$
\mathcal{H}^k_d(\mathcal{A}^\bullet_{\mathbb{C}}) = \begin{cases} \mathbb{C} & \text{for } k = 0, \\ 0 & \text{for } k > 0. \end{cases}
$$

By the $\bar{\partial}$-Poincaré lemma, the sequence

$$
0 \to \Omega^p_{\mathrm{an}} \to \mathcal{A}^{p,0} \xrightarrow{\bar{\partial}} \mathcal{A}^{p,1} \xrightarrow{\bar{\partial}} \cdots \to \mathcal{A}^{p,n} \to 0
$$

is exact for each p and so the E_1 term of the usual spectral sequence of the double complex $\bigoplus \mathcal{A}^{p,q}$ is

$$E_1 = H_{\bar{\partial}} = \begin{array}{c} q \\ \left|\begin{array}{cccc} 0 & 0 & 0 & \\ 0 & 0 & 0 & \\ \Omega_{an}^0 & \Omega_{an}^1 & \Omega_{an}^2 & \\ \hline 0 & 1 & 2 & p \end{array}\right. \end{array}$$

Hence, the E_2 term is given by

$$E_2^{p,q} = \begin{cases} \mathcal{H}_d^p(\Omega_{an}^{\bullet}) & \text{for } q = 0, \\ 0 & \text{for } q > 0. \end{cases}$$

Since the spectral sequence degenerates at the E_2 term,

$$\mathcal{H}_d^k(\Omega_{an}^{\bullet}) = E_2 = E_{\infty} \simeq \mathcal{H}_d^k(\mathcal{A}_{\mathbb{C}}^{\bullet}) = \begin{cases} \mathbb{C} & \text{for } k = 0, \\ 0 & \text{for } k > 0, \end{cases}$$

which is precisely the holomorphic Poincaré lemma.   □

2.5.2   The Analytic de Rham Theorem

THEOREM 2.5.2   *Let Ω_{an}^k be the sheaf of holomorphic k-forms on a complex manifold M. Then the singular cohomology of M with complex coefficients can be computed as the hypercohomology of the complex $\Omega_{an}^{\bullet}$:*

$$H^k(M, \mathbb{C}) \simeq \mathbb{H}^k(M, \Omega_{an}^{\bullet}).$$

PROOF.   Let $\mathbb{C}^{\bullet}$ be the complex of sheaves that is $\mathbb{C}$ in degree 0 and zero otherwise. The holomorphic Poincaré lemma may be interpreted as a quasi-isomorphism of the two complexes

$$\begin{array}{ccccccccc} 0 & \longrightarrow & \mathbb{C} & \longrightarrow & 0 & \longrightarrow & 0 & \longrightarrow & \cdots \\ & & \downarrow & & \downarrow & & \downarrow & & \\ 0 & \longrightarrow & \Omega_{an}^0 & \longrightarrow & \Omega_{an}^1 & \longrightarrow & \Omega_{an}^2 & \longrightarrow & \cdots, \end{array}$$

since

$$\begin{aligned} \mathcal{H}^0(\mathbb{C}^{\bullet}) = \mathbb{C} &\simeq \text{Im}(\mathbb{C} \to \Omega_{an}^0) \\ &= \ker(\Omega_{an}^0 \to \Omega_{an}^1) \quad \text{(by the holomorphic Poincaré lemma)} \\ &= \mathcal{H}^0(\Omega_{an}^{\bullet}) \end{aligned}$$

and the higher cohomology sheaves of both complexes are zero.

By Theorem 2.4.1, the quasi-isomorphism $\mathbb{C}^{\bullet} \simeq \Omega_{an}^{\bullet}$ induces an isomorphism

$$\mathbb{H}^*(M, \mathbb{C}^{\bullet}) \simeq \mathbb{H}^*(M, \Omega_{an}^{\bullet}) \tag{2.5.1}$$

in hypercohomology. Since $\mathbb{C}^\bullet$ is a complex of sheaves concentrated in degree 0, by (2.4.1) the left-hand side of (2.5.1) is the sheaf cohomology $H^k(M, \mathbb{C})$, which is isomorphic to the singular cohomology $H^k(M, \mathbb{C})$ by Remark 2.2.16. $\quad\square$

In contrast to the sheaves $\mathcal{A}^k$ and $\mathcal{A}^{p,q}$ in de Rham's theorem and Dolbeault's theorem, the sheaves $\Omega^\bullet_{\mathrm{an}}$ are generally neither fine nor acyclic, because in the analytic category there is no partition of unity. However, when M is a Stein manifold, the complex $\Omega^\bullet_{\mathrm{an}}$ is a complex of acyclic sheaves on M by Cartan's theorem B. It then follows from Theorem 2.4.2 that

$$\mathbb{H}^k(M, \Omega^\bullet_{\mathrm{an}}) \simeq h^k\big(\Omega^\bullet_{\mathrm{an}}(M)\big).$$

This proves the following corollary of Theorem 2.5.2.

COROLLARY 2.5.3 *The singular cohomology of a Stein manifold M with coefficients in $\mathbb{C}$ can be computed from the holomorphic de Rham complex:*

$$H^k(M, \mathbb{C}) \simeq h^k\big(\Omega^\bullet_{\mathrm{an}}(M)\big).$$

2.6 THE ALGEBRAIC DE RHAM THEOREM FOR A PROJECTIVE VARIETY

Let X be a smooth complex algebraic variety with the Zariski topology. The underlying set of X with the complex topology is a complex manifold X_{an}. Let Ω^k_{alg} be the sheaf of algebraic k-forms on X, and Ω^k_{an} the sheaf of holomorphic k-forms on X_{an}. According to the holomorphic Poincaré lemma (Theorem 2.5.1), the complex of sheaves

$$0 \to \mathbb{C} \to \Omega^0_{\mathrm{an}} \xrightarrow{d} \Omega^1_{\mathrm{an}} \xrightarrow{d} \Omega^2_{\mathrm{an}} \xrightarrow{d} \cdots \tag{2.6.1}$$

is exact. However, there is no Poincaré lemma in the algebraic category; the complex

$$0 \to \mathbb{C} \to \Omega^0_{\mathrm{alg}} \to \Omega^1_{\mathrm{alg}} \to \Omega^2_{\mathrm{alg}} \to \cdots$$

is in general not exact.

THEOREM 2.6.1 (Algebraic de Rham theorem for a projective variety) *If X is a smooth complex projective variety, then there is an isomorphism*

$$H^k(X_{\mathrm{an}}, \mathbb{C}) \simeq \mathbb{H}^k(X, \Omega^\bullet_{\mathrm{alg}})$$

between the singular cohomology of X_{an} with coefficients in $\mathbb{C}$ and the hypercohomology of X with coefficients in the complex $\Omega^\bullet_{\mathrm{alg}}$ of sheaves of algebraic differential forms on X.

PROOF. By Theorem 2.4.1, the quasi-isomorphism $\mathbb{C}^\bullet \to \Omega^\bullet_{\mathrm{an}}$ of complexes of sheaves induces an isomorphism in hypercohomology

$$\mathbb{H}^*(X_{\mathrm{an}}, \mathbb{C}^\bullet) \simeq \mathbb{H}^*(X_{\mathrm{an}}, \Omega^\bullet_{\mathrm{an}}). \tag{2.6.2}$$

In the second spectral sequence converging to $\mathbb{H}^*(X_{\mathrm{an}}, \Omega_{\mathrm{an}}^\bullet)$, by (2.4.4) the E_1 term is

$$E_{1,\mathrm{an}}^{p,q} = H^p(X_{\mathrm{an}}, \Omega_{\mathrm{an}}^q).$$

By (2.4.4) the E_1 term in the second spectral sequence converging to the hypercohomology $\mathbb{H}^*(X, \Omega_{\mathrm{alg}}^\bullet)$ is

$$E_{1,\mathrm{alg}}^{p,q} = H^p(X, \Omega_{\mathrm{alg}}^q).$$

Since X is a smooth complex projective variety, Serre's GAGA principle (2.3.1) applies and gives an isomorphism

$$H^p(X, \Omega_{\mathrm{alg}}^q) \simeq H^p(X_{\mathrm{an}}, \Omega_{\mathrm{an}}^q).$$

The isomorphism $E_{1,\mathrm{alg}} \overset{\sim}{\to} E_{1,\mathrm{an}}$ induces an isomorphism in E_∞. Hence,

$$\mathbb{H}^*(X, \Omega_{\mathrm{alg}}^\bullet) \simeq \mathbb{H}^*(X_{\mathrm{an}}, \Omega_{\mathrm{an}}^\bullet). \tag{2.6.3}$$

Combining (2.4.1), (2.6.2), and (2.6.3) gives

$$H^*(X_{\mathrm{an}}, \mathbb{C}) \simeq \mathbb{H}^*(X_{\mathrm{an}}, \mathbb{C}^\bullet) \sim \mathbb{H}^*(X_{\mathrm{an}}, \Omega_{\mathrm{an}}^\bullet) \simeq \mathbb{H}^*(X, \Omega_{\mathrm{alg}}^\bullet).$$

Finally, by the isomorphism between sheaf cohomology and singular cohomology (Remark 2.2.16), we may replace the sheaf cohomology $H^*(X_{\mathrm{an}}, \mathbb{C})$ by a singular cohomology group:

$$H^*(X_{\mathrm{an}}, \mathbb{C}) \simeq \mathbb{H}^*(X, \Omega_{\mathrm{alg}}^\bullet).$$

$\square$

PART II. ČECH COHOMOLOGY AND THE ALGEBRAIC DE RHAM THEOREM IN GENERAL

The algebraic de Rham theorem (Theorem 2.6.1) in fact does not require the hypothesis of projectivity on X. In this section we will extend it to an arbitrary smooth algebraic variety defined over $\mathbb{C}$. In order to carry out this extension, we will need to develop two more machineries: the Čech cohomology of a sheaf and the Čech cohomology of a complex of sheaves. Čech cohomology provides a practical method for computing sheaf cohomology and hypercohomology.

2.7   ČECH COHOMOLOGY OF A SHEAF

Čech cohomology may be viewed as a generalization of the Mayer–Vietoris sequence from two open sets to arbitrarily many open sets.

2.7.1 Čech Cohomology of an Open Cover

Let $\mathfrak{U} = \{U_\alpha\}_{\alpha \in A}$ be an open cover of the topological space X indexed by a linearly ordered set A, and $\mathcal{F}$ a presheaf of abelian groups on X. To simplify the notation, we will write the $(p+1)$-fold intersection $U_{\alpha_0} \cap \cdots \cap U_{\alpha_p}$ as $U_{\alpha_0 \cdots \alpha_p}$. Define the **group of Čech p-cochains** on $\mathfrak{U}$ with values in the presheaf $\mathcal{F}$ to be the direct product

$$\check{C}^p(\mathfrak{U}, \mathcal{F}) := \prod_{\alpha_0 < \cdots < \alpha_p} \mathcal{F}(U_{\alpha_0 \cdots \alpha_p}).$$

An element ω of $\check{C}^p(\mathfrak{U}, \mathcal{F})$ is then a function that assigns to each finite set of indices $\alpha_0, \ldots, \alpha_p$ an element $\omega_{\alpha_0 \ldots \alpha_p} \in \mathcal{F}(U_{\alpha_0 \ldots \alpha_p})$. We will write $\omega = (\omega_{\alpha_0 \ldots \alpha_p})$, where the subscripts range over all $\alpha_0 < \cdots < \alpha_p$. In particular, the subscripts $\alpha_0, \ldots, \alpha_p$ must all be distinct. Define the **Čech coboundary operator**

$$\delta = \delta_p \colon \check{C}^p(\mathfrak{U}, \mathcal{F}) \to \check{C}^{p+1}(\mathfrak{U}, \mathcal{F})$$

by the alternating sum formula

$$(\delta\omega)_{\alpha_0 \ldots \alpha_{p+1}} = \sum_{i=0}^{p+1} (-1)^i \omega_{\alpha_0 \cdots \widehat{\alpha_i} \cdots \alpha_{p+1}},$$

where $\widehat{\alpha_i}$ means to omit the index α_i; moreover, the restriction of $\omega_{\alpha_0 \cdots \widehat{\alpha_i} \cdots \alpha_{p+1}}$ from $U_{\alpha_0 \cdots \widehat{\alpha_i} \cdots \alpha_{p+1}}$ to $U_{\alpha_0 \ldots \alpha_{p+1}}$ is suppressed in the notation.

PROPOSITION 2.7.1 *If δ is the Čech coboundary operator, then $\delta^2 = 0$.*

PROOF. Basically, this is true because in $(\delta^2\omega)_{\alpha_0 \cdots \alpha_{p+2}}$, we omit two indices α_i, α_j twice with opposite signs. To be precise,

$$
\begin{aligned}
(\delta^2\omega)_{\alpha_0 \cdots \alpha_{p+2}} &= \sum (-1)^i (\delta\omega)_{\alpha_0 \cdots \widehat{\alpha_i} \cdots \alpha_{p+2}} \\
&= \sum_{j<i} (-1)^i (-1)^j \omega_{\alpha_0 \cdots \widehat{\alpha_j} \cdots \widehat{\alpha_i} \cdots \alpha_{p+2}} \\
&\quad + \sum_{j>i} (-1)^i (-1)^{j-1} \omega_{\alpha_0 \cdots \widehat{\alpha_i} \cdots \widehat{\alpha_j} \cdots \alpha_{p+2}} \\
&= 0.
\end{aligned}
$$

$\square$

It follows from Proposition 2.7.1 that $\check{C}^\bullet(\mathfrak{U}, \mathcal{F}) := \bigoplus_{p=0}^\infty \check{C}^p(\mathfrak{U}, \mathcal{F})$ is a cochain complex with differential δ. The cohomology of the complex $(\check{C}^*(\mathfrak{U}, \mathcal{F}), \delta)$,

$$\check{H}^p(\mathfrak{U}, \mathcal{F}) := \frac{\ker \delta_p}{\operatorname{Im} \delta_{p-1}} = \frac{\{p\text{-cocycles}\}}{\{p\text{-coboundaries}\}},$$

is called the **Čech cohomology** of the open cover $\mathfrak{U}$ with values in the presheaf $\mathcal{F}$.

2.7.2 Relation Between Čech Cohomology and Sheaf Cohomology

In this subsection we construct a natural map from the Čech cohomology of a sheaf on an open cover to its sheaf cohomology. This map is based on a property of flasque sheaves.

LEMMA 2.7.2 *Suppose $\mathcal{F}$ is a flasque sheaf of abelian groups on a topological space X, and $\mathfrak{U} = \{U_\alpha\}$ is an open cover of X. Then the augmented Čech complex*

$$0 \to \mathcal{F}(X) \to \prod_\alpha \mathcal{F}(U_\alpha) \to \prod_{\alpha<\beta} \mathcal{F}(U_{\alpha\beta}) \to \cdots$$

is exact.

In other words, for a flasque sheaf $\mathcal{F}$ on X,

$$\check{H}^k(\mathfrak{U}, \mathcal{F}) = \begin{cases} \mathcal{F}(X) & \text{for } k = 0, \\ 0 & \text{for } k > 0. \end{cases}$$

PROOF. [5, Th. 5.2.3(a), p. 207]. □

Now suppose $\mathcal{F}$ is any sheaf of abelian groups on a topological space X and $\mathfrak{U} = \{U_\alpha\}$ is an open cover of X. Let $K^{\bullet,\bullet} = \bigoplus K^{p,q}$ be the double complex

$$K^{p,q} = \check{C}^p(\mathfrak{U}, \mathcal{C}^q \mathcal{F}) = \prod_{\alpha_0 < \cdots < \alpha_p} \mathcal{C}^q \mathcal{F}(U_{\alpha_0 \cdots \alpha_p}).$$

We augment this complex with an outside bottom row ($q = -1$) and an outside left column ($p = -1$):

$$(2.7.1)$$

Note that the qth row of the double complex $K^{\bullet,\bullet}$ is the Čech cochain complex of the Godement sheaf $\mathcal{C}^q \mathcal{F}$ and the pth column is the complex of groups for computing the sheaf cohomology $\prod_{\alpha_0 < \cdots < \alpha_p} H^*(U_{\alpha_0 \cdots \alpha_p}, \mathcal{F})$.

By Lemma 2.7.2, each row of the augmented double complex (2.7.1) is exact. Hence, the E_1 term of the second spectral sequence of the double complex is

$$
E_1 = H_\delta =
\begin{array}{c|ccc}
q \\
\hline
\mathcal{C}^2 \mathcal{F}(X) & 0 & 0 \\
\mathcal{C}^1 \mathcal{F}(X) & 0 & 0 \\
\mathcal{C}^0 \mathcal{F}(X) & 0 & 0 \\
\hline
0 & 1 & 2 & p
\end{array}
$$

and the E_2 term is

$$
E_2 = H_d H_\delta =
\begin{array}{c|ccc}
q \\
\hline
H^2(X, \mathcal{F}) & 0 & 0 \\
H^1(X, \mathcal{F}) & 0 & 0 \\
H^0(X, \mathcal{F}) & 0 & 0 \\
\hline
0 & 1 & 2 & p
\end{array}
$$

So the second spectral sequence of the double complex (2.7.1) degenerates at the E_2 term and the cohomology of the associated single complex $K^\bullet$ of $\bigoplus K^{p,q}$ is

$$
H_D^k(K^\bullet) \simeq H^k(X, \mathcal{F}).
$$

In the augmented complex (2.7.1), by the construction of Godement's canonical resolution, the Čech complex $\check{C}^\bullet(\mathfrak{U}, \mathcal{F})$ injects into the complex $K^\bullet$ via a cochain map

$$
\epsilon \colon \check{C}^k(\mathfrak{U}, \mathcal{F}) \to K^{k,0} \hookrightarrow K^k,
$$

which gives rise to an induced map

$$
\epsilon^* \colon \check{H}^k(\mathfrak{U}, \mathcal{F}) \to H_D^k(K^\bullet) = H^k(X, \mathcal{F}) \tag{2.7.2}
$$

in cohomology.

DEFINITION 2.7.3 *A sheaf $\mathcal{F}$ of abelian groups on a topological space X is **acyclic on an open cover** $\mathfrak{U} = \{U_\alpha\}$ of X if the cohomology*

$$
H^k(U_{\alpha_0 \cdots \alpha_p}, \mathcal{F}) = 0
$$

for all $k > 0$ and all finite intersections $U_{\alpha_0 \cdots \alpha_p}$ of open sets in $\mathfrak{U}$.

THEOREM 2.7.4 *If a sheaf $\mathcal{F}$ of abelian groups is acyclic on an open cover $\mathfrak{U} = \{U_\alpha\}$ of a topological space X, then the induced map $\epsilon^* \colon \check{H}^k(\mathfrak{U}, \mathcal{F}) \to H^k(X, \mathcal{F})$ is an isomorphism.*

PROOF. Because $\mathcal{F}$ is acyclic on each intersection $U_{\alpha_0 \cdots \alpha_p}$, the cohomology of the pth column of (2.7.1) is $\prod H^0(U_{\alpha_0 \cdots \alpha_p}, \mathcal{F}) = \prod \mathcal{F}(U_{\alpha_0 \cdots \alpha_p})$, so that the E_1 term of the usual spectral sequence is

$$E_1 = H_d = \begin{array}{c|c|c|c}
q & & & \\
\hline
0 & 0 & 0 \\
0 & 0 & 0 \\
\prod \mathcal{F}(U_{\alpha_0}) & \prod \mathcal{F}(U_{\alpha_0\alpha_1}) & \prod \mathcal{F}(U_{\alpha_0\alpha_1\alpha_2}) \\
\hline
0 & 1 & 2 & p
\end{array},$$

and the E_2 term is

$$E_2 = H_\delta H_d = \begin{array}{c|c|c|c}
q & & & \\
\hline
0 & 0 & 0 \\
0 & 0 & 0 \\
\check{H}^0(\mathfrak{U}, \mathcal{F}) & \check{H}^1(\mathfrak{U}, \mathcal{F}) & \check{H}^2(\mathfrak{U}, \mathcal{F}) \\
\hline
0 & 1 & 2 & p
\end{array}.$$

Hence, the spectral sequence degenerates at the E_2 term and there is an isomorphism

$$\epsilon^* \colon \check{H}^k(\mathfrak{U}, \mathcal{F}) \simeq H_D^k(K^\bullet) \simeq H^k(X, \mathcal{F}).$$

$\square$

Remark. Although we used a spectral sequence argument to prove Theorem 2.7.4, in the proof there is no problem with the extension of groups in the E_∞ term, since along each antidiagonal $\bigoplus_{p+q=k} E_\infty^{p,q}$ there is only one nonzero box. For this reason, Theorem 2.7.4 holds for sheaves of abelian groups, not just for sheaves of vector spaces.

2.8   ČECH COHOMOLOGY OF A COMPLEX OF SHEAVES

Just as the cohomology of a sheaf can be computed using a Čech complex on an open cover (Theorem 2.7.4), the hypercohomology of a complex of sheaves can also be computed using the Čech method.

Let $(\mathcal{L}^\bullet, d_\mathcal{L})$ be a complex of sheaves on a topological space X, and $\mathfrak{U} = \{U_\alpha\}$ an open cover of X. To define the Čech cohomology of $\mathcal{L}^\bullet$ on $\mathfrak{U}$, let $K = \bigoplus K^{p,q}$ be the double complex

$$K^{p,q} = \check{C}^p(\mathfrak{U}, \mathcal{L}^q)$$

with its two commuting differentials δ and $d_\mathcal{L}$. We will call K the **Čech–sheaf double complex**. The **Čech cohomology** $\check{H}^*(\mathfrak{U}, \mathcal{L}^\bullet)$ of $\mathcal{L}^\bullet$ is defined to be the cohomology of the single complex

$$K^\bullet = \bigoplus K^k, \quad \text{where } K^k = \bigoplus_{p+q=k} \check{C}^p(\mathfrak{U}, \mathcal{L}^q) \text{ and } d_K = \delta + (-1)^p d_\mathcal{L},$$

associated to the Čech–sheaf double complex.

2.8.1   The Relation Between Čech Cohomology and Hypercohomology

There is an analogue of Theorem 2.7.4 that allows us to compute hypercohomology using an open cover.

THEOREM 2.8.1 *If $\mathcal{L}^\bullet$ is a complex of sheaves of abelian groups on a topological space X such that each sheaf $\mathcal{L}^q$ is acyclic on the open cover $\mathfrak{U} = \{U_\alpha\}$ of X, then there is an isomorphism $\check{H}^k(\mathfrak{U}, \mathcal{L}^\bullet) \simeq \mathbb{H}^k(X, \mathcal{L}^\bullet)$ between the Čech cohomology of $\mathcal{L}^\bullet$ on the open cover $\mathfrak{U}$ and the hypercohomology of $\mathcal{L}^\bullet$ on X.*

The Čech cohomology of the complex $\mathcal{L}^\bullet$ is the cohomology of the associated single complex of the double complex $\bigoplus_{p,q} \check{C}^p(\mathfrak{U}, \mathcal{L}^q) = \bigoplus_{p,q} \prod_\alpha \mathcal{L}^q(U_{\alpha_0 \cdots \alpha_p})$, where $\alpha = (\alpha_0 < \cdots < \alpha_p)$. The hypercohomology of the complex $\mathcal{L}^\bullet$ is the cohomology of the associated single complex of the double complex $\bigoplus_{q,r} \mathcal{C}^r \mathcal{L}^q(X)$. To compare the two, we form the triple complex with terms

$$N^{p,q,r} = \check{C}^p(\mathfrak{U}, \mathcal{C}^r \mathcal{L}^q)$$

and three commuting differentials: the Čech differential $\delta_{\check{C}}$, the differential $d_\mathcal{L}$ of the complex $\mathcal{L}^\bullet$, and the Godement differential $d_\mathcal{C}$.

Let $N^{\bullet,\bullet,\bullet}$ be any triple complex with three commuting differentials d_1, d_2, and d_3 of degrees $(1,0,0)$, $(0,1,0)$, and $(0,0,1)$, respectively. Summing $N^{p,q,r}$ over p and q, or over q and r, one can form two double complexes from $N^{\bullet,\bullet,\bullet}$:

$$N^{k,r} = \bigoplus_{p+q=k} N^{p,q,r}$$

with differentials

$$\delta = d_1 + (-1)^p d_2, \quad d = d_3,$$

and

$$N'^{p,\ell} = \bigoplus_{q+r=\ell} N^{p,q,r}$$

with differentials

$$\delta' = d_1, \quad d' = d_2 + (-1)^q d_3.$$

PROPOSITION 2.8.2 *For any triple complex $N^{\bullet,\bullet,\bullet}$, the two associated double complexes $N^{\bullet,\bullet}$ and $N'^{\bullet,\bullet}$ have the same associated single complex.*

PROOF. Clearly, the groups

$$N^n = \bigoplus_{k+r=n} N^{k,r} = \bigoplus_{p+q+r=n} N^{p,q,r}$$

and

$$N'^n = \bigoplus_{p+\ell=n} N'^{p,\ell} = \bigoplus_{p+q+r=n} N^{p,q,r}$$

are equal. The differential D for $N^\bullet = \bigoplus_n N^n$ is

$$D = \delta + (-1)^k d = d_1 + (-1)^p d_2 + (-1)^{p+q} d_3.$$

The differential D' for $N'^\bullet = \bigoplus_n N'^n$ is

$$D' = \delta' + (-1)^p d' = d_1 + (-1)^p \big(d_2 + (-1)^q d_3\big) = D.$$

$\square$

Thus, any triple complex $N^{\bullet,\bullet,\bullet}$ has an associated single complex $N^\bullet$ whose cohomology can be computed in two ways, either from the double complex $(N^{\bullet,\bullet}, D)$ or from the double complex $(N'^{\bullet,\bullet}, D')$.

We now apply this observation to the Čech–Godement–sheaf triple complex

$$N^{\bullet,\bullet,\bullet} = \bigoplus \check{C}^p(\mathfrak{U}, \mathcal{C}^r \mathcal{L}^q)$$

of the complex $\mathcal{L}^\bullet$ of sheaves. The kth column of the double complex $N^{\bullet,\bullet} = \bigoplus N^{k,r}$ is

$$\bigoplus_{p+q=k} \prod_{\alpha_0 < \cdots < \alpha_p} \mathcal{C}^{r+1} \mathcal{L}^q(U_{\alpha_0 \cdots \alpha_p})$$

$$\uparrow$$

$$\bigoplus_{p+q=k} \prod_{\alpha_0 < \cdots < \alpha_p} \mathcal{C}^r \mathcal{L}^q(U_{\alpha_0 \cdots \alpha_p})$$

$$\uparrow$$

$$\vdots$$

$$\uparrow$$

$$\bigoplus_{p+q=k} \prod_{\alpha_0 < \cdots < \alpha_p} \mathcal{C}^0 \mathcal{L}^q(U_{\alpha_0 \cdots \alpha_p}),$$

where the vertical differential d is the Godement differential d_C. Since $\mathcal{L}^\bullet$ is acyclic on the open cover $\mathfrak{U} = \{U_\alpha\}$, this column is exact except in the zeroth row, and the zeroth row of the cohomology H_d is

$$\bigoplus_{p+q=k} \prod_{\alpha_0 < \cdots < \alpha_p} \mathcal{L}^q(U_{\alpha_0 \cdots \alpha_p}) = \bigoplus_{p+q=k} \check{C}^p(\mathfrak{U}, \mathcal{L}^q) = \bigoplus_{p+q=k} K^{p,q} = K^k,$$

the associated single complex of the Čech–sheaf double complex. Thus, the E_1 term of the first spectral sequence of $N^{\bullet,\bullet}$ is

$$E_1 = H_d = \begin{array}{c|ccc}
r \\
& 0 & 0 & 0 \\
& 0 & 0 & 0 \\
& K^0 & K^1 & K^2 \\
\hline
& 0 & 1 & 2 & k
\end{array}$$

and so the E_2 term is

$$E_2 = H_\delta(H_d) = H^*_{d_K}(K^\bullet) = \check{H}^*(\mathfrak{U}, \mathcal{L}^\bullet).$$

Although we are working with abelian groups, there are no extension issues, because each antidiagonal in E_∞ contains only one nonzero group. Thus, the E_∞ term is

$$H^*_D(N^\bullet) \simeq E_2 = \check{H}^*(\mathfrak{U}, \mathcal{L}^\bullet). \tag{2.8.1}$$

On the other hand, the ℓth row of $N'^{\bullet,\bullet}$ is

$$0 \to \bigoplus_{q+r=\ell} \check{C}^0(\mathfrak{U}, \mathcal{C}^r \mathcal{L}^q) \to \cdots \to \bigoplus_{q+r=\ell} \check{C}^p(\mathfrak{U}, \mathcal{C}^r \mathcal{L}^q) \to \bigoplus_{q+r=\ell} \check{C}^{p+1}(\mathfrak{U}, \mathcal{C}^r \mathcal{L}^q) \to \cdots,$$

which is the Čech cochain complex of the flasque sheaf $\bigoplus_{q+r=\ell} \mathcal{C}^r \mathcal{L}^q$ with differential $\delta' = \delta_{\check{C}}$. Thus, each row of $N'^{\bullet,\bullet}$ is exact except in the zeroth column, and the kernel of $N'^{0,\ell} \to N'^{1,\ell}$ is $M^\ell = \bigoplus_{q+r=\ell} \mathcal{C}^r \mathcal{L}^q(X)$. Hence, the E_1 term of the second spectral sequence is

$$E_1 = H_{\delta'} = \begin{array}{c|cccc|}
\ell & & & & \\
\hline
M^2 & 0 & 0 & 0 \\
M^1 & 0 & 0 & 0 \\
M^0 & 0 & 0 & 0 \\
\hline
0 & 1 & 2 & 3 & p
\end{array}$$

The E_2 term is

$$E_2 = H_{d'}(H_{\delta'}) = H^*_{d_M}(M^\bullet) = \mathbb{H}^*(X, \mathcal{L}^\bullet).$$

Since this spectral sequence for $N^{\bullet,\bullet}$ degenerates at the E_2 term and converges to $H^*_{D'}(N'^\bullet)$, there is an isomorphism

$$E_\infty = H^*_{D'}(N'^\bullet) \simeq E_2 = \mathbb{H}^*(X, \mathcal{L}^\bullet). \tag{2.8.2}$$

By Proposition 2.8.2, the two groups in (2.8.1) and (2.8.2) are isomorphic. In this way, one obtains an isomorphism between the Čech cohomology and the hypercohomology of the complex $\mathcal{L}^\bullet$:

$$\check{H}^*(\mathfrak{U}, \mathcal{L}^\bullet) \simeq \mathbb{H}^*(X, \mathcal{L}^\bullet).$$

2.9 REDUCTION TO THE AFFINE CASE

Grothendieck proved his general algebraic de Rham theorem by reducing it to the special case of an affine variety. This section is an exposition of his ideas in [7].

THEOREM 2.9.1 (Algebraic de Rham theorem) *Let X be a smooth algebraic variety defined over the complex numbers, and X_{an} its underlying complex manifold. Then the singular cohomology of X_{an} with $\mathbb{C}$ coefficients can be computed as the hypercohomology of the complex $\Omega^\bullet_{\mathrm{alg}}$ of sheaves of algebraic differential forms on X with its Zariski topology:*

$$H^k(X_{\mathrm{an}}, \mathbb{C}) \simeq \mathbb{H}^k(X, \Omega^\bullet_{\mathrm{alg}}).$$

By the isomorphism $H^k(X_{an}, \mathbb{C}) \simeq \mathbb{H}^k(X_{an}, \Omega_{an}^\bullet)$ of the analytic de Rham theorem, Grothendieck's algebraic de Rham theorem is equivalent to an isomorphism in hypercohomology

$$\mathbb{H}^k(X, \Omega_{alg}^\bullet) \simeq \mathbb{H}^k(X_{an}, \Omega_{an}^\bullet).$$

The special case of Grothendieck's theorem for an affine variety is especially interesting, since it does not involve hypercohomology.

COROLLARY 2.9.2 (The affine case) *Let X be a smooth affine variety defined over the complex numbers and $(\Omega_{alg}^\bullet(X), d)$ the complex of algebraic differential forms on X. Then the singular cohomology with $\mathbb{C}$ coefficients of its underlying complex manifold X_{an} can be computed as the cohomology of its complex of algebraic differential forms:*

$$H^k(X_{an}, \mathbb{C}) \simeq h^k(\Omega_{alg}^\bullet(X)).$$

It is important to note that the left-hand side is the singular cohomology of the complex manifold X_{an}, not of the affine variety X. In fact, in the Zariski topology, a constant sheaf on an irreducible variety is always flasque (Example 2.2.6), and hence acyclic (Corollary 2.2.5), so that $H^k(X, \mathbb{C}) = 0$ for all $k > 0$ if X is irreducible.

2.9.1 Proof that the General Case Implies the Affine Case

Assume Theorem 2.9.1. It suffices to prove that for a smooth affine complex variety X, the hypercohomology $\mathbb{H}^k(X, \Omega_{alg}^\bullet)$ reduces to the cohomology of the complex $\Omega_{alg}^\bullet(X)$. Since $\Omega_{alg}^\bullet$ is a coherent algebraic sheaf, by Serre's vanishing theorem for an affine variety (Theorem 2.3.11), Ω_{alg}^q is acyclic on X. By Theorem 2.4.2,

$$\mathbb{H}^k(X, \Omega_{alg}^\bullet) \simeq h^k(\Omega_{alg}^\bullet(X)).$$

2.9.2 Proof that the Affine Case Implies the General Case

Assume Corollary 2.9.2. The proof is based on the facts that every algebraic variety X has an **affine open cover**, an open cover $\mathfrak{U} = \{U_\alpha\}$ in which every U_α is an affine open set, and that the intersection of two affine open sets is affine open. The existence of an affine open cover for an algebraic variety follows from the elementary fact that every quasi-projective variety has an affine open cover; since an algebraic variety by definition has an open cover by quasi-projective varieties, it necessarily has an open cover by affine varieties.

Since $\Omega_{alg}^\bullet$ is a complex of locally free and hence coherent algebraic sheaves, by Serre's vanishing theorem for an affine variety (Theorem 2.3.11), $\Omega_{alg}^\bullet$ is acyclic on an affine open cover. By Theorem 2.8.1, there is an isomorphism

$$\check{H}^*(\mathfrak{U}, \Omega_{alg}^\bullet) \simeq \mathbb{H}^*(X, \Omega_{alg}^\bullet) \tag{2.9.1}$$

between the Čech cohomology of $\Omega_{alg}^\bullet$ on the affine open cover $\mathfrak{U}$ and the hypercohomology of $\Omega_{alg}^\bullet$ on X. Similarly, by Cartan's theorem B (because a complex affine

variety with the complex topology is Stein) and Theorem 2.8.1, the corresponding statement in the analytic category is also true: if $\mathfrak{U}_{an} := \{(U_\alpha)_{an}\}$, then

$$\check{H}^*(\mathfrak{U}_{an}, \Omega_{an}^\bullet) \simeq \mathbb{H}^*(X_{an}, \Omega_{an}^\bullet). \tag{2.9.2}$$

The Čech cohomology $\check{H}^*(\mathfrak{U}, \Omega_{alg}^\bullet)$ is the cohomology of the single complex associated to the double complex $\bigoplus K_{alg}^{p,q} = \bigoplus \check{C}^p(\mathfrak{U}, \Omega_{alg}^q)$. The E_1 term of the usual
spectral sequence of this double complex is

$$\begin{aligned}
E_{1,alg}^{p,q} &= H_d^{p,q} = h_d^q(K^{p,\bullet}) = h_d^q\big(\check{C}^p(\mathfrak{U}, \Omega_{alg}^\bullet)\big) \\
&= h_d^q\bigg(\prod_{\alpha_0 < \cdots < \alpha_p} \Omega_{alg}^\bullet(U_{\alpha_0 \cdots \alpha_p}) \bigg) \\
&= \prod_{\alpha_0 < \cdots < \alpha_p} h_d^q\big(\Omega_{alg}^\bullet(U_{\alpha_0 \cdots \alpha_p})\big) \\
&= \prod_{\alpha_0 < \cdots < \alpha_p} H^q(U_{\alpha_0 \cdots \alpha_p, an}, \mathbb{C}) \qquad \text{(by Corollary 2.9.2).}
\end{aligned}$$

A completely similar computation applies to the usual spectral sequence of the
double complex $\bigoplus K_{an}^{p,q} = \bigoplus_{p,q} \check{C}^p(\mathfrak{U}_{an}, \Omega_{an}^q)$ converging to the Čech cohomology
$\check{H}^*(\mathfrak{U}_{an}, \Omega_{an}^\bullet)$: the E_1 term of this spectral sequence is

$$\begin{aligned}
E_{1,an}^{p,q} &= \prod_{\alpha_0 < \cdots < \alpha_p} h_d^q\big(\Omega_{an}^\bullet(U_{\alpha_0 \cdots \alpha_p, an})\big) \\
&= \prod_{\alpha_0 < \cdots < \alpha_p} H^q(U_{\alpha_0 \cdots \alpha_p, an}, \mathbb{C}) \qquad \text{(by Corollary 2.5.3).}
\end{aligned}$$

The isomorphism in E_1 terms,

$$E_{1,alg} \xrightarrow{\sim} E_{1,an},$$

commutes with the Čech differential $d_1 = \delta$ and induces an isomorphism in E_∞ terms,

$$\begin{array}{ccc}
E_{\infty,alg} & \xrightarrow{\quad\sim\quad} & E_{\infty,an} \\
\| & & \| \\
\check{H}^*(\mathfrak{U}, \Omega_{alg}^\bullet) & & \check{H}^*(\mathfrak{U}_{an}, \Omega_{an}^\bullet).
\end{array}$$

Combined with (2.9.1) and (2.9.2), this gives

$$\mathbb{H}^*(X, \Omega_{alg}^\bullet) \simeq \mathbb{H}^*(X_{an}, \Omega_{an}^\bullet),$$

which, as we have seen, is equivalent to the algebraic de Rham theorem (Theorem 2.9.1)
for a smooth complex algebraic variety.

2.10 THE ALGEBRAIC DE RHAM THEOREM FOR AN AFFINE VARIETY

It remains to prove the algebraic de Rham theorem in the form of Corollary 2.9.2 for a smooth affine complex variety X. This is the most difficult case and is in fact the heart of the matter. We give a proof that is different from Grothendieck's in [7].

A **normal crossing divisor** on a smooth algebraic variety is a divisor that is locally the zero set of an equation of the form $z_1 \cdots z_k = 0$, where $z_1, \ldots, z_N$ are local parameters. We first describe a standard procedure by which any smooth affine variety X may be assumed to be the complement of a normal crossing divisor D in a smooth complex projective variety Y. Let $\bar{X}$ be the projective closure of X; for example, if X is defined by polynomial equations

$$f_i(z_1, \ldots, z_N) = 0$$

in $\mathbb{C}^N$, then $\bar{X}$ is defined by the equations

$$f_i\left(\frac{Z_1}{Z_0}, \ldots, \frac{Z_N}{Z_0}\right) = 0$$

in $\mathbb{C}P^N$, where $Z_0, \ldots, Z_N$ are the homogeneous coordinates on $\mathbb{C}P^N$ and $z_i = Z_i/Z_0$. In general, $\bar{X}$ will be a singular projective variety. By Hironaka's resolution of singularities, there is a surjective regular map $\pi \colon Y \to \bar{X}$ from a smooth projective variety Y to $\bar{X}$ such that $\pi^{-1}(\bar{X} - X)$ is a normal crossing divisor D in Y and $\pi|_{Y-D} \colon Y - D \to X$ is an isomorphism. Thus, we may assume that $X = Y - D$, with an inclusion map $j \colon X \hookrightarrow Y$.

Let $\Omega^k_{Y_{an}}(*D)$ be the sheaf of meromorphic k-forms on Y_{an} that are holomorphic on X_{an} with poles of any order ≥ 0 along D_{an} (order 0 means no poles) and let $\mathcal{A}^k_{X_{an}}$ be the sheaf of C^∞ complex-valued k-forms on X_{an}. By abuse of notation, we use j also to denote the inclusion $X_{an} \hookrightarrow Y_{an}$. The **direct image sheaf** $j_* \mathcal{A}^k_{X_{an}}$ is by definition the sheaf on Y_{an} defined by

$$\left(j_* \mathcal{A}^k_{X_{an}}\right)(V) = \mathcal{A}^k_{X_{an}}(V \cap X_{an})$$

for any open set $V \subset Y_{an}$. Since a section of $\Omega^k_{Y_{an}}(*D)$ over V is holomorphic on $V \cap X_{an}$ and therefore smooth there, the sheaf $\Omega^k_{Y_{an}}(*D)$ of meromorphic forms is a subsheaf of the sheaf $j_* \mathcal{A}^k_{X_{an}}$ of smooth forms. The main lemma of our proof, due to Hodge and Atiyah [10, Lem. 17, p. 77], asserts that the inclusion

$$\Omega^\bullet_{Y_{an}}(*D) \hookrightarrow j_* \mathcal{A}^\bullet_{X_{an}} \tag{2.10.1}$$

of complexes of sheaves is a quasi-isomorphism. This lemma makes essential use of the fact that D is a normal crossing divisor. Since the proof of the lemma is quite technical, in order not to interrupt the flow of the exposition, we postpone it to the end of the chapter.

By Theorem 2.4.1, the quasi-isomorphism (2.10.1) induces an isomorphism

$$\mathbb{H}^k\left(Y_{an}, \Omega^\bullet_{Y_{an}}(*D)\right) \simeq \mathbb{H}^k\left(Y_{an}, j_* \mathcal{A}^\bullet_{X_{an}}\right) \tag{2.10.2}$$

in hypercohomology. If we can show that the right-hand side is $H^k(X_{\text{an}}, \mathbb{C})$ and the left-hand side is $h^k(\Omega_{\text{alg}}^{\bullet}(X))$, the algebraic de Rham theorem for the affine variety X (Corollary 2.9.2), $h^k(\Omega_{\text{alg}}^{\bullet}(X)) \simeq H^k(X_{\text{an}}, \mathbb{C})$, will follow.

2.10.1 The Hypercohomology of the Direct Image of a Sheaf of Smooth Forms

To deal with the right-hand side of (2.10.2), we prove a more general lemma valid on any complex manifold.

LEMMA 2.10.1 *Let M be a complex manifold and U an open submanifold, with $j \colon U \hookrightarrow M$ the inclusion map. Denote the sheaf of smooth $\mathbb{C}$-valued k-forms on U by $\mathcal{A}_U^k$. Then there is an isomorphism*

$$\mathbb{H}^k(M, j_* \mathcal{A}_U^{\bullet}) \simeq H^k(U, \mathbb{C}).$$

PROOF. Let $\mathcal{A}^0$ be the sheaf of smooth $\mathbb{C}$-valued functions on the complex manifold M. For any open set $V \subset M$, there is an $\mathcal{A}^0(V)$-module structure on $(j_* \mathcal{A}_U^k)(V) = \mathcal{A}_U^k(U \cap V)$:

$$\mathcal{A}^0(V) \times \mathcal{A}_U^k(U \cap V) \to \mathcal{A}_U^k(U \cap V),$$
$$(f, \omega) \mapsto f \cdot \omega.$$

Hence, $j_* \mathcal{A}_U^k$ is a sheaf of $\mathcal{A}^0$-modules on M. As such, $j_* \mathcal{A}_U^k$ is a fine sheaf on M (Section 2.2.5).

Since fine sheaves are acyclic, by Theorem 2.4.2,

$$
\begin{aligned}
\mathbb{H}^k(M, j_* \mathcal{A}_U^{\bullet}) &\simeq h^k\big((j_* \mathcal{A}_U^{\bullet})(M)\big) \\
&= h^k\big(\mathcal{A}_U^{\bullet}(U)\big) && \text{(definition of } j_* \mathcal{A}_U^{\bullet}) \\
&= H^k(U, \mathbb{C}) && \text{(by the smooth de Rham theorem).}
\end{aligned}
$$

$\square$

Applying the lemma to $M = Y_{\text{an}}$ and $U = X_{\text{an}}$, we obtain

$$\mathbb{H}^k(Y_{\text{an}}, j_* \mathcal{A}_{X_{\text{an}}}^{\bullet}) \simeq H^k(X_{\text{an}}, \mathbb{C}).$$

This takes care of the right-hand side of (2.10.2).

2.10.2 The Hypercohomology of Rational and Meromorphic Forms

Throughout this subsection, the smooth complex affine variety X is the complement of a normal crossing divisor D in a smooth complex projective variety Y. Let $\Omega_{Y_{\text{an}}}^q(nD)$ be the sheaf of meromorphic q-forms on Y_{an} that are holomorphic on X_{an} with poles of order $\leq n$ along D_{an}. As before, $\Omega_{Y_{\text{an}}}^q(*D)$ is the sheaf of meromorphic q-forms on Y_{an} that are holomorphic on X_{an} with at most poles (of any order) along D. Similarly, $\Omega_Y^q(*D)$ and $\Omega_Y^q(nD)$ are their algebraic counterparts, the sheaves of rational

q-forms on Y that are regular on X with poles along D of arbitrary order or order $\leq n$ respectively. Then

$$\Omega^q_{Y_{\mathrm{an}}}(*D) = \varinjlim_n \Omega^q_{Y_{\mathrm{an}}}(nD) \quad \text{and} \quad \Omega^q_Y(*D) = \varinjlim_n \Omega^q_Y(nD).$$

Let Ω^q_X and Ω^q_Y be the sheaves of regular q-forms on X and Y, respectively; they are what would be written Ω^q_{alg} if there is only one variety. Similarly, let $\Omega^q_{X_{\mathrm{an}}}$ and $\Omega^q_{Y_{\mathrm{an}}}$ be sheaves of holomorphic q-forms on X_{an} and Y_{an}, respectively. There is another description of the sheaf $\Omega^q_Y(*D)$ that will prove useful. Since a regular form on $X = Y - D$ that is not defined on D can have at most poles along D (no essential singularities), if $j \colon X \to Y$ is the inclusion map, then

$$j_*\Omega^q_X = \Omega^q_Y(*D).$$

Note that the corresponding statement in the analytic category is not true: if $j \colon X_{\mathrm{an}} \to Y_{\mathrm{an}}$ now denotes the inclusion of the corresponding analytic manifolds, then in general

$$j_*\Omega^q_{X_{\mathrm{an}}} \neq \Omega^q_{Y_{\mathrm{an}}}(*D)$$

because a holomorphic form on X_{an} that is not defined along D_{an} may have an essential singularity on D_{an}.

Our goal now is to prove that the hypercohomology $\mathbb{H}^*\left(Y_{\mathrm{an}}, \Omega^\bullet_{Y_{\mathrm{an}}}(*D)\right)$ of the complex $\Omega^\bullet_{Y_{\mathrm{an}}}(*D)$ of sheaves of meromorphic forms on Y_{an} is computable from the algebraic de Rham complex on X:

$$\mathbb{H}^k\left(Y_{\mathrm{an}}, \Omega^\bullet_{Y_{\mathrm{an}}}(*D)\right) \simeq h^k\left(\Gamma(X, \Omega^\bullet_{\mathrm{alg}})\right).$$

This will be accomplished through a series of isomorphisms.

First, we prove something akin to a GAGA principle for hypercohomology. The proof requires commuting direct limits and cohomology, for which we shall invoke the following criterion. A topological space is said to be **noetherian** if it satisfies the descending chain condition for closed sets: any descending chain $Y_1 \supset Y_2 \supset \cdots$ of closed sets must terminate after finitely many steps. As shown in a first course in algebraic geometry, affine and projective varieties are noetherian [9, Exa. 1.4.7, p. 5; Exer.1.7(b), p. 8; Exer. 2.5(a), p. 11].

PROPOSITION 2.10.2 (Commutativity of direct limit with cohomology)   *Let $(\mathcal{F}_\alpha)$ be a direct system of sheaves on a topological space Z. The natural map*

$$\varinjlim H^k(Z, \mathcal{F}_\alpha) \to H^k(Z, \varinjlim \mathcal{F}_\alpha)$$

is an isomorphism if

  (i) Z is compact; or

  (ii) Z is noetherian.

PROOF. For (i), see [10, Lem. 4, p. 61]. For (ii), see [9, Ch. III, Prop. 2.9, p. 209] or [5, Ch. II, remark after Th. 4.12.1, p. 194]. $\qquad\square$

PROPOSITION 2.10.3 *In the notation above, there is an isomorphism in hypercohomology*

$$\mathbb{H}^*\big(Y, \Omega_Y^\bullet(*D)\big) \simeq \mathbb{H}^*\big(Y_{\mathrm{an}}, \Omega_{Y_{\mathrm{an}}}^\bullet(*D)\big).$$

PROOF. Since Y is a projective variety and each $\Omega_Y^\bullet(nD)$ is locally free, we can apply Serre's GAGA principle (2.3.1) to get an isomorphism

$$H^p\big(Y, \Omega_Y^q(nD)\big) \simeq H^p\big(Y_{\mathrm{an}}, \Omega_{Y_{\mathrm{an}}}^q(nD)\big).$$

Next, take the direct limit of both sides as $n \to \infty$. Since the projective variety Y is noetherian and the complex manifold Y_{an} is compact, by Proposition 2.10.2, we obtain

$$H^p\big(Y, \varinjlim_n \Omega_Y^q(nD)\big) \simeq H^p\big(Y_{\mathrm{an}}, \varinjlim_n \Omega_{Y_{\mathrm{an}}}^q(nD)\big),$$

which is

$$H^p\big(Y, \Omega_Y^q(*D)\big) \simeq H^p\big(Y_{\mathrm{an}}, \Omega_{Y_{\mathrm{an}}}^q(*D)\big).$$

Now the two cohomology groups $H^p\big(Y, \Omega_Y^q(*D)\big)$ and $H^p\big(Y_{\mathrm{an}}, \Omega_{Y_{\mathrm{an}}}^q(*D)\big)$ are the E_1 terms of the second spectral sequences of the hypercohomologies of $\Omega_Y^\bullet(*D)$ and $\Omega_{Y_{\mathrm{an}}}^\bullet(*D)$, respectively (see (2.4.4)). An isomorphism of the E_1 terms induces an isomorphism of the E_∞ terms. Hence,

$$\mathbb{H}^*\big(Y, \Omega_Y^\bullet(*D)\big) \simeq \mathbb{H}^*\big(Y_{\mathrm{an}}, \Omega_{Y_{\mathrm{an}}}^\bullet(*D)\big).$$

$\qquad\square$

PROPOSITION 2.10.4 *In the notation above, there is an isomorphism*

$$\mathbb{H}^k\big(Y, \Omega_Y^\bullet(*D)\big) \simeq \mathbb{H}^k(X, \Omega_X^\bullet)$$

for all $k \geq 0$.

PROOF. If V is an affine open set in Y, then V is noetherian and so by Proposition 2.10.2(ii), for $p > 0$,

$$
\begin{aligned}
H^p\big(V, \Omega_Y^q(*D)\big) &= H^p\big(V, \varinjlim_n \Omega_Y^q(nD)\big) \\
&\simeq \varinjlim_n H^p\big(V, \Omega_Y^q(nD)\big) \\
&= 0,
\end{aligned}
$$

the last equality following from Serre's vanishing theorem (Theorem 2.3.11), since V is affine and $\Omega_Y^q(nD)$ is locally free and therefore coherent. Thus, the complex of sheaves $\Omega_Y^\bullet(*D)$ is acyclic on any affine open cover $\mathfrak{U} = \{U_\alpha\}$ of Y. By Theorem 2.8.1, its hypercohomology can be computed from its Čech cohomology:

$$\mathbb{H}^k\big(Y, \Omega_Y^\bullet(*D)\big) \simeq \check{H}^k\big(\mathfrak{U}, \Omega_Y^\bullet(*D)\big).$$

Recall that if $j\colon X \to Y$ is the inclusion map, then $\Omega_Y^\bullet(*D) = j_*\Omega_X^\bullet$. By definition, the Čech cohomology $\check{H}^k\big(\mathfrak{U}, \Omega_Y^\bullet(*D)\big)$ is the cohomology of the associated single complex of the double complex

$$K^{p,q} = \check{C}^p\big(\mathfrak{U}, \Omega_Y^q(*D)\big) = \check{C}^p(\mathfrak{U}, j_*\Omega_X^q)$$

$$= \prod_{\alpha_0 < \cdots < \alpha_p} \Omega^q(U_{\alpha_0\cdots\alpha_p} \cap X). \tag{2.10.3}$$

Next we compute the hypercohomology $\mathbb{H}^k(X, \Omega_X^\bullet)$. The restriction $\mathfrak{U}|_X := \{U_\alpha \cap X\}$ of $\mathfrak{U}$ to X is an affine open cover of X. Since Ω_X^q is locally free [14, Ch. III, Th. 2, p. 200], by Serre's vanishing theorem for an affine variety again,

$$H^p(U_\alpha \cap X, \Omega_X^q) = 0 \quad \text{for all } p > 0.$$

Thus, the complex of sheaves $\Omega_X^\bullet$ is acyclic on the open cover $\mathfrak{U}|_X$ of X. By Theorem 2.8.1,

$$\mathbb{H}^k(X, \Omega_X^\bullet) \simeq \check{H}^k(\mathfrak{U}|_X, \Omega_X^\bullet).$$

The Čech cohomology $\check{H}^k(\mathfrak{U}|_X, \Omega_X^\bullet)$ is the cohomology of the single complex associated to the double complex

$$K^{p,q} = \check{C}^p(\mathfrak{U}|_X, \Omega_X^q)$$

$$= \prod_{\alpha_0 < \cdots < \alpha_p} \Omega^q(U_{\alpha_0\cdots\alpha_p} \cap X). \tag{2.10.4}$$

Comparing (2.10.3) and (2.10.4), we get an isomorphism

$$\mathbb{H}^k\big(Y, \Omega_Y^\bullet(*D)\big) \simeq \mathbb{H}^k(X, \Omega_X^\bullet)$$

for every $k \geq 0$.   □

Finally, because Ω_X^q is locally free, by Serre's vanishing theorem for an affine variety still again, $H^p(X, \Omega_X^q) = 0$ for all $p > 0$. Thus, $\Omega_X^\bullet$ is a complex of acyclic sheaves on X. By Theorem 2.4.2, the hypercohomology $\mathbb{H}^k(X, \Omega_X^\bullet)$ can be computed from the complex of global sections of $\Omega_X^\bullet$:

$$\mathbb{H}^k(X, \Omega_X^\bullet) \simeq h^k\big(\Gamma(X, \Omega_X^\bullet)\big) = h^k\big(\Omega_{\mathrm{alg}}^\bullet(X)\big). \tag{2.10.5}$$

Putting together Propositions 2.10.3 and 2.10.4 with (2.10.5), we get the desired interpretation

$$\mathbb{H}^k\big(Y_{\mathrm{an}}, \Omega_{Y_{\mathrm{an}}}^\bullet(*D)\big) \simeq h^k\big(\Omega_{\mathrm{alg}}^\bullet(X)\big)$$

of the left-hand side of (2.10.2). Together with the interpretation of the right-hand side of (2.10.2) as $H^k(X_{\mathrm{an}}, \mathbb{C})$, this gives Grothendieck's algebraic de Rham theorem for an affine variety,

$$H^k(X_{\mathrm{an}}, \mathbb{C}) \simeq h^k\big(\Omega_{\mathrm{alg}}^\bullet(X)\big).$$

2.10.3   Comparison of Meromorphic and Smooth Forms

It remains to prove that (2.10.1) is a quasi-isomorphism. We will reformulate the lemma in slightly more general terms. Let M be a complex manifold of complex dimension n, let D be a normal crossing divisor in M, and let $U = M - D$ be the complement of D in M, with $j\colon U \hookrightarrow M$ the inclusion map. Denote by $\Omega_M^q(*D)$ the sheaf of meromorphic q-forms on M that are holomorphic on U with at most poles along D, and by $\mathcal{A}_U^q := \mathcal{A}_U^q(\,,\mathbb{C})$ the sheaf of smooth $\mathbb{C}$-valued q-forms on U. For each q, the sheaf $\Omega_M^q(*D)$ is a subsheaf of $j_*\mathcal{A}_U^q$.

LEMMA 2.10.5 (Fundamental lemma of Hodge and Atiyah ([10, Lem. 17, p. 77]))
The inclusion $\Omega_M^\bullet(*D) \hookrightarrow j_*\mathcal{A}_U^\bullet$ of complexes of sheaves is a quasi-isomorphism.

PROOF.   We remark first that this is a *local* statement. Indeed, the main advantage of using sheaf theory is to reduce the global statement of the algebraic de Rham theorem for an affine variety to a local result. The inclusion $\Omega_M^\bullet(*D) \hookrightarrow j_*\mathcal{A}_U^\bullet$ of complexes induces a morphism of cohomology sheaves $\mathcal{H}^*(\Omega_M^\bullet(*D)) \to \mathcal{H}^*(j_*\mathcal{A}_U^\bullet)$. It is a general fact in sheaf theory that a morphism of sheaves is an isomorphism if and only if its stalk maps are all isomorphisms [9, Prop. 1.1, p. 63], so we will first examine the stalks of the sheaves in question. There are two cases: $p \in U$ and $p \in D$. For simplicity, let $\Omega_p^q := (\Omega_M^q)_p$ be the stalk of Ω_M^q at $p \in M$ and let $\mathcal{A}_p^q := (\mathcal{A}_U^q)_p$ be the stalk of $\mathcal{A}_U^q$ at $p \in U$.

   Case 1: At a point $p \in U$, the stalk of $\Omega_M^q(*D)$ is Ω_p^q, and the stalk of $j_*\mathcal{A}_U^q$ is $\mathcal{A}_p^q$. Hence, the stalk maps of the inclusion $\Omega_M^\bullet(*D) \hookrightarrow j_*\mathcal{A}_U^\bullet$ at p are

$$
\begin{array}{ccccccc}
0 & \longrightarrow & \Omega_p^0 & \longrightarrow & \Omega_p^1 & \longrightarrow & \Omega_p^2 & \longrightarrow & \cdots \\
& & \downarrow & & \downarrow & & \downarrow & & \\
0 & \longrightarrow & \mathcal{A}_p^0 & \longrightarrow & \mathcal{A}_p^1 & \longrightarrow & \mathcal{A}_p^2 & \longrightarrow & \cdots .
\end{array}
\tag{2.10.6}
$$

Being a chain map, (2.10.6) induces a homomorphism in cohomology. By the holomorphic Poincaré lemma (Theorem 2.5.1), the cohomology of the top row of (2.10.6) is

$$
h^k(\Omega_p^\bullet) = \begin{cases} \mathbb{C} & \text{for } k = 0, \\ 0 & \text{for } k > 0. \end{cases}
$$

By the complex analogue of the smooth Poincaré lemma ([3, Sec. 4, p. 33] and [6, p. 38]), the cohomology of the bottom row of (2.10.6) is

$$
h^k(\mathcal{A}_p^\bullet) = \begin{cases} \mathbb{C} & \text{for } k = 0, \\ 0 & \text{for } k > 0. \end{cases}
$$

Since the inclusion map (2.10.6) takes $1 \in \Omega_p^0$ to $1 \in \mathcal{A}_p^0$, it is a quasi-isomorphism.
   By Proposition 2.2.9, for $p \in U$,

$$
\mathcal{H}^k\left(\Omega_M^\bullet(*D)\right)_p \simeq h^k\left((\Omega_M^\bullet(*D))_p\right) = h^k(\Omega_p^\bullet)
$$

and

$$\mathcal{H}^k(j_*\mathcal{A}_U^\bullet)_p \simeq h^k\big((j_*\mathcal{A}_U^\bullet)_p\big) = h^k(\mathcal{A}_p^\bullet).$$

Therefore, by the preceding paragraph, at $p \in U$ the inclusion $\Omega_M^\bullet(*D) \hookrightarrow j_*\mathcal{A}_U^\bullet$ induces an isomorphism of stalks

$$\mathcal{H}^k\big(\Omega_M^\bullet(*D)\big)_p \simeq \mathcal{H}^k(j_*\mathcal{A}_U^\bullet)_p \qquad (2.10.7)$$

for all $k > 0$.

Case 2: Similarly, we want to show that (2.10.7) holds for $p \notin U$, i.e., for $p \in D$. Note that to show the stalks of these sheaves at p are isomorphic, it is enough to show the spaces of sections are isomorphic over a neighborhood basis of polydisks.

Choose local coordinates $z_1, \ldots, z_n$ so that $p = (0, \ldots, 0)$ is the origin and D is the zero set of $z_1 \cdots z_k = 0$ on some coordinate neighborhood of p. Let P be the polydisk $P = \Delta^n := \Delta \times \cdots \times \Delta$ (n times), where Δ is a small disk centered at the origin in $\mathbb{C}$, say of radius ϵ for some $\epsilon > 0$. Then $P \cap U$ is the **polycylinder**

$$
\begin{aligned}
P^* := P \cap U &= \Delta^n \cap (M - D) \\
&= \{(z_1, \ldots, z_n) \in \Delta^n \mid z_i \neq 0 \text{ for } i = 1, \ldots, k\} \\
&= (\Lambda^*)^k \times \Delta^{n-k},
\end{aligned}
$$

where Δ^* is the punctured disk $\Delta - \{0\}$ in $\mathbb{C}$. Note that P^* has the homotopy type of the torus $(S^1)^k$. For $1 \leq i \leq k$, let γ_i be a circle wrapping once around the ith Δ^*. Then a basis for the homology of P^* is given by the submanifolds $\prod_{i \in J} \gamma_i$ for all the various subsets $J \subset [1, k]$.

Since on the polydisk P,

$$(j_*\mathcal{A}_U^\bullet)(P) = \mathcal{A}_U^\bullet(P \cap U) = \mathcal{A}^\bullet(P^*),$$

the cohomology of the complex $(j_*\mathcal{A}_U^\bullet)(P)$ is

$$
\begin{aligned}
h^*\big((j_*\mathcal{A}_U^\bullet)(P)\big) &= h^*\big(\mathcal{A}^\bullet(P^*)\big) \\
&= H^*(P^*, \mathbb{C}) \simeq H^*\big((S^1)^k, \mathbb{C}\big) \\
&= \bigwedge\left(\left[\frac{dz_1}{z_1}\right], \ldots, \left[\frac{dz_k}{z_k}\right]\right), \qquad (2.10.8)
\end{aligned}
$$

the free exterior algebra on the k generators $[dz_1/z_1], \ldots, [dz_k/z_k]$. Up to a constant factor of $2\pi i$, this basis is dual to the homology basis cited above, as we can see by integrating over products of loops.

For each q, the inclusion $\Omega_M^q(*D) \hookrightarrow j_*\mathcal{A}_U^q$ of sheaves induces an inclusion of groups of sections over a polydisk P:

$$\Gamma\big(P, \Omega_M^q(*D)\big) \hookrightarrow \Gamma(P, j_*\mathcal{A}_U^q).$$

As q varies, the inclusion of complexes

$$i \colon \Gamma\big(P, \Omega_M^\bullet(*D)\big) \to \Gamma(P, j_*\mathcal{A}_U^\bullet)$$

induces a homomorphism in cohomology

$$i^*: h^*\big(\Gamma(P,\Omega_M^\bullet(*D))\big) \to h^*\big(\Gamma(P, j_*\mathcal{A}_U^\bullet)\big) = \bigwedge\left(\left[\frac{dz_1}{z_1}\right], \ldots, \left[\frac{dz_k}{z_k}\right]\right).$$
(2.10.9)

Since each dz_j/z_j is a closed meromorphic form on P with poles along D, it defines a cohomology class in $h^*\big(\Gamma(P,\Omega_M^\bullet(*D))\big)$. Therefore, the map i^* is surjective. If we could show i^* were an isomorphism, then by taking the direct limit over all polydisks P containing p, we would obtain

$$\mathcal{H}^*\big(\Omega_M^\bullet(*D)\big)_p \simeq \mathcal{H}^*(j_*\mathcal{A}_U^\bullet)_p \quad \text{for } p \in D, \tag{2.10.10}$$

which would complete the proof of the fundamental lemma (Lemma 2.10.5).

We now compute the cohomology of the complex $\Gamma\big(P, \Omega_M^\bullet(*D)\big)$.

PROPOSITION 2.10.6 *Let P be a polydisk Δ^n in $\mathbb{C}^n$, and D the normal crossing divisor defined in P by $z_1 \cdots z_k = 0$. The cohomology ring $h^*\big(\Gamma(P,\Omega^\bullet(*D))\big)$ is generated by $[dz_1/z_1], \ldots, [dz_k/z_k]$.*

PROOF. The proof is by induction on the number k of irreducible components of the singular set D. When $k = 0$, the divisor D is empty and meromorphic forms on P with poles along D are holomorphic. By the holomorphic Poincaré lemma,

$$h^*\big(\Gamma(P,\Omega^\bullet)\big) = H^*(P,\mathbb{C}) = \mathbb{C}.$$

This proves the base case of the induction.

The induction step is based on the following lemma.

LEMMA 2.10.7 *Let P be a polydisk Δ^n, and D the normal crossing divisor defined by $z_1 \cdots z_k = 0$ in P. Let $\varphi \in \Gamma\big(P, \Omega^q(*D)\big)$ be a closed meromorphic q-form on P that is holomorphic on $P^* := P - D$ with at most poles along D. Then there exist closed meromorphic forms $\varphi_0 \in \Gamma\big(P,\Omega^q(*D)\big)$ and $\alpha_1 \in \Gamma\big(P, \Omega^{q-1}(*D)\big)$, which have no poles along $z_1 = 0$, such that their cohomology classes satisfy the relation*

$$[\varphi] = [\varphi_0] + \left[\frac{dz_1}{z_1}\right] \wedge [\alpha_1].$$

PROOF. Our proof is an elaboration of the proof of Hodge and Atiyah [10, Lem. 17, p. 77]. We can write φ in the form

$$\varphi = dz_1 \wedge \alpha + \beta,$$

where the meromorphic $(q-1)$-form α and the q-form β do not involve dz_1. Next, we expand α and β as Laurent series in z_1:

$$\alpha = \alpha_0 + \alpha_1 z_1^{-1} + \alpha_2 z_1^{-2} + \cdots + \alpha_r z_1^{-r},$$
$$\beta = \beta_0 + \beta_1 z_1^{-1} + \beta_2 z_1^{-2} + \cdots + \beta_r z_1^{-r},$$

where α_i and β_i for $1 \le i \le r$ do not involve z_1 or dz_1 and are meromorphic in the other variables, and α_0, β_0 are holomorphic in z_1, are meromorphic in the other variables, and do not involve dz_1. Then

$$\varphi = (dz_1 \wedge \alpha_0 + \beta_0) + \left(dz_1 \wedge \sum_{i=1}^{r} \alpha_i z_1^{-i} + \sum_{i=1}^{r} \beta_i z_1^{-i} \right).$$

Set $\varphi_0 = dz_1 \wedge \alpha_0 + \beta_0$. By comparing the coefficients of $z_1^{-i} dz_1$ and z_1^{-i}, we deduce from the condition $d\varphi = 0$,

$$d\alpha_1 = d\alpha_2 + \beta_1 = d\alpha_3 + 2\beta_2 = \cdots = r\beta_r = 0,$$
$$d\beta_1 = d\beta_2 \quad = d\beta_3 \quad = \cdots = d\beta_r = 0,$$

and $d\varphi_0 = 0$.

We can write

$$\varphi = \varphi_0 + \frac{dz_1}{z_1} \wedge \alpha_1 + \left(dz_1 \wedge \sum_{i=2}^{r} \alpha_i z_1^{-i} + \sum_{i=1}^{r} \beta_i z_1^{-i} \right). \tag{2.10.11}$$

It turns out that the term within the parentheses in (2.10.11) is $d\theta$ for

$$\theta = -\frac{\alpha_2}{z_1} - \frac{\alpha_3}{2z_1^2} - \cdots - \frac{\alpha_r}{(r-1)z_1^{r-1}}.$$

In (2.10.11), both φ_0 and α_1 are closed. Hence, the cohomology classes satisfy the relation

$$[\varphi] = [\varphi_0] + \left[\frac{dz_1}{z_1} \right] \wedge [\alpha_1].$$

$\square$

Since φ_0 and α_1 are meromorphic forms which do not have poles along $z_1 = 0$, their singularity set is contained in the normal crossing divisor defined by $z_2 \cdots z_k = 0$, which has $k-1$ irreducible components. By induction, the cohomology classes of φ_0 and α_1 are generated by $[dz_2/z_2], \ldots, [dz_k/z_k]$. Hence, $[\varphi]$ is a graded-commutative polynomial in $[dz_1/z_1], \ldots, [dz_k/z_k]$. This completes the proof of Proposition 2.10.6.

$\square$

PROPOSITION 2.10.8 *Let P be a polydisk Δ^n in $\mathbb{C}^n$, and D the normal crossing divisor defined by $z_1 \cdots z_k = 0$ in P. Then there is a ring isomorphism*

$$h^*\left(\Gamma(P, \Omega^\bullet(*D))\right) \simeq \bigwedge \left(\left[\frac{dz_1}{z_1} \right], \ldots, \left[\frac{dz_k}{z_k} \right] \right).$$

PROOF. By Proposition 2.10.6, the graded-commutative algebra $h^*\left(\Gamma(P, \Omega^\bullet(*D))\right)$ is generated by $[dz_1/z_1], \ldots, [dz_k/z_k]$. It remains to show that these generators satisfy no algebraic relations other than those implied by graded commutativity. Let $\omega_i = dz_i/z_i$ and $\omega_I := \omega_{i_1 \cdots i_r} := \omega_{i_1} \wedge \cdots \wedge \omega_{i_r}$. Any linear relation among the cohomology classes $[\omega_I]$ in $h^*\big(\Gamma(P,\Omega^\bullet(*D))\big)$ would be, on the level of forms, of the form

$$\sum c_I \omega_I = d\xi \tag{2.10.12}$$

for some meromorphic form ξ with at most poles along D. But by restriction to $P - D$, this would give automatically a relation in $\Gamma(P, j_*\mathcal{A}_U^q)$. Since $h^*\big(\Gamma(P, j_*\mathcal{A}_U^q)\big) = \bigwedge\big([\omega_1],\ldots,[\omega_k]\big)$ is freely generated by $[\omega_1],\ldots,[\omega_k]$ (see (2.10.8)), the only possible relations (2.10.12) are all implied by graded commutativity. □

Since the inclusion $\Omega_M^\bullet(*D) \hookrightarrow j_*\mathcal{A}_U^*$ induces an isomorphism

$$\mathcal{H}^*\big(\Omega_M^\bullet(*D)\big)_p \simeq \mathcal{H}^*(j_*\mathcal{A}_U^\bullet)_p$$

of stalks of cohomology sheaves for all p, the inclusion $\Omega_M^\bullet(*D) \hookrightarrow j_*\mathcal{A}_U^*$ is a quasi-isomorphism. This completes the proof of Lemma 2.10.5. □

Bibliography

[1] Alberto Arabia and Zoghman Mebkhout. Introduction à la cohomologie de de Rham. Preprint, 260 pages, 1997.

[2] Michael F. Atiyah and Ian G. Macdonald. *Introduction to Commutative Algebra*. Addison–Wesley, Reading, Massachusetts, 1969.

[3] Raoul Bott and Loring W. Tu. *Differential Forms in Algebraic Topology*, third corrected printing. Springer, New York, 1995.

[4] Henri Cartan. Variétés analytiques complexes et cohomologie. *Colloque de Bruxelles*, 41–55, 1953.

[5] Roger Godement. *Topologie algébrique et théorie des faisceaux*. Hermann, Paris, 1958.

[6] Phillip Griffiths and Joe Harris. *Principles of Algebraic Geometry*. John Wiley, New York, 1978.

[7] Alexandre Grothendieck. On the de Rham cohomology of algebraic varieties. *Publ. Math. IHES*, 29: 95–103, 1966.

[8] Robert Gunning and Hugo Rossi. *Analytic Functions of Several Complex Variables*. Prentice-Hall, Englewood Cliffs, NJ, 1965.

[9] Robin Hartshorne. *Algebraic Geometry*, volume 52 of *Graduate Texts in Mathematics*. Springer, New York, 1977.

[10] W. V. D. Hodge and Michael F. Atiyah. Integrals of the second kind on an algebraic variety. *Annals of Math.*, 62:56–91, 1955.

[11] Lars Hörmander. *An Introduction to Complex Analysis in Several Variables*, third edition. North-Holland, New York, 1990.

[12] Jean-Pierre Serre. Faisceaux algébriques cohérents. *Annals of Math.*, 61:197–278, 1955.

[13] Jean-Pierre Serre. Géométrie analytique et géométrie algébrique. *Ann. Inst. Fourier*, 6:1–42, 1956.

[14] Igor R. Shafarevich. *Basic Algebraic Geometry*, two volumes, second edition. Springer, New York, 1994.

[15] Loring W. Tu. *An Introduction to Manifolds*, second edition. Universitext. Springer, New York, 2011.

[16] Frank W. Warner. *Foundations of Differentiable Manifolds and Lie groups*, volume 94 of *Graduate Texts in Mathematics*. Springer, New York, 1983.

Chapter Three

Mixed Hodge Structures
by Fouad El Zein and Lê Dũng Tráng

INTRODUCTION

We assume that the reader is familiar with the basic theory of manifolds, basic algebraic geometry as well as cohomology theory. For instance, we shall refer to Chapter 1 and Chapter 2 in this book and we recommend to the reader books like e.g., [45], [3], or [40], and the beginning of [28] and, for complementary reading on complex algebraic and analytic geometry, [43, 39, 47].

According to Deligne, the cohomology space $H^n(X, \mathbb{C})$ of a complex algebraic variety X carries two finite filtrations by complex subvector spaces: the rationally defined weight filtration W and the Hodge filtration F defining a mixed Hodge structure (MHS) (see [7, 8]).

For a nonsingular compact complex algebraic variety, the weight filtration W is trivial, while the Hodge filtration F and its conjugate $\overline{F}$, with respect to the rational cohomology, define a Hodge decomposition. The structure in linear algebra defined on a complex vector space by such a decomposition is called a Hodge structure (HS) (see [33, 47]).

On a nonsingular variety X, the weight filtration W and the Hodge filtration F reflect properties of the normal crossing divisor (NCD) at infinity of an adequate completion of the variety obtained by resolution of singularities (see [30]), but are independent of the choice of the normal crossing divisor.

Inspired by the properties of étale cohomology of varieties over fields with positive characteristic, constructed by Grothendieck and Artin, Deligne established the existence of an MHS on the cohomology of complex algebraic varieties, depending naturally on algebraic morphisms but not on continuous maps. The theory has been fundamental in the study of topological properties of complex algebraic varieties. At the end of this introduction, we recall the topological background of homology theory and Poincaré duality, then we refer to Chapter 1 in this volume to state the Hodge decomposition on the cohomology of Kähler manifolds.

The theory involves a significant change in the method and technique from the construction of harmonic forms, (see Section 1.4.2) which is dependent on the metric, while the results here are stated on cohomology which is a topological invariant and the cohomology subspaces of type (p, q). The course is organized as follows:

Section 3.1. The abstract category of Hodge structures is defined in the first section and spectral sequences are introduced. The decomposition on the cohomology of Kähler manifolds is used to prove the degeneration at rank 1 of the spectral sequence defined by the filtration F on the de Rham complex in the projective nonsingular case. An important result here is the degeneration at rank 1 of the spectral sequence, since this result extends to all complex algebraic varieties. The section ends with the proof of the existence of a Hodge structure on the cohomology of a nonsingular compact complex algebraic variety.

Section 3.2. In the second section, we introduce an abstract definition of MHS as an object of interest in linear algebra, following closely Deligne [7]. The algebraic properties of MHS are developed on objects with three opposite filtrations in an abelian category. Then we prove Deligne's lemma on the two filtrations and explain how it can be applied to construct an MHS on the cohomology of a normal crossing divisor.

Section 3.3. In the third section, we need to develop algebraic homology techniques on filtered complexes up to filtered quasi-isomorphisms of complexes. Since an MHS involves rational cohomology constructed via different techniques than the ones used in de Rham cohomology, when we define a morphism of mixed Hodge structures, we ask for compatibility with the filtrations as well as the identification of cohomology via the various constructions. For this reason it is convenient to work constantly at the level of complexes in the language of filtered and bifiltered derived categories to ensure that the MHS constructed on cohomology are canonical and do not depend on particular resolutions used in the definition of cohomology. The main contribution by Deligne was to fix the axioms of a mixed Hodge complex (MHC) and to prove the existence of an MHS on its cohomology. This central result is applied in Section 3.4.

Section 3.4. We give here the construction of the MHS on any algebraic variety. On a noncompact nonsingular algebraic variety X, we need to introduce Deligne's logarithmic de Rham complex to take into account the properties at infinity, that is, the properties of the NCD, which is the complement of the variety in an adequate compactification of X. If X is singular, we introduce a smooth simplicial covering to construct the MHC. We also indicate an alternative construction.

As applications, let us mention deformations of nonsingular proper analytic families which define a linear invariant called variation of Hodge structure (VHS) introduced by Griffiths (see Chapters 7 and 4), and limit MHS adapted to study the degeneration of variation of Hodge structure. Variations of mixed Hodge structure, which arise in the case of any algebraic deformation, are the topic of the lectures in Chapters 7 and 8.

Finally we stress that we only introduce the necessary language to express the statements and their implications in Hodge theory, but we encourage mathematicians to look into the foundational work of Deligne in references [7, 8] to discover his dense and unique style, since intricate proofs in Hodge theory and spectral sequences are often just surveyed here.

Topological Background

The theory of homology and cohomology traces its origin to the work of Henri Poincaré in the late nineteenth century. There are actually many different theories, for example, simplicial and singular (see, e.g., [29]). In 1931, Georges de Rham proved a conjecture of Poincaré on a relationship between cycles and differential forms that establishes for a compact orientable manifold an isomorphism between singular cohomology with real coefficients and what is known now as de Rham cohomology. The result has had a great impact on the theory since it shows the diversity of applications of cohomology theory.

A basic result that we also mention is the isomorphism established by William Hodge between the space of harmonic forms on a Riemannian compact manifold and cohomology (see Section 1.4). While the space of harmonic forms depends on the choice of the metric, the cohomology space appears as an invariant of the topology. These results on de Rham cohomology, as well Čech cohomology, shifted the attention to the construction of cohomology by sheaf theory.

Our subject starts with the Hodge decomposition on the cohomology of a compact Kähler manifold. For an accessible introduction to the subject covering the fundamental statements that we need here, see the notes of Cattani in Chapter 1. For an excellent full account one may also refer to [47].

There are various applications of Hodge decomposition. On one hand, the work of Phillip Griffiths on the variation of Hodge structures [19] depends on the variation of the analytic (or algebraic) structure of a nonsingular projective variety in a family (by comparison, cohomology can only detect the underlying topological structure and it is locally constant with respect to the parameter space of the family). On the other hand, we cite the work of André Weil [46] and his conjecture, which motivated Alexander Grothendieck and his theory of motives and the work of Pierre Deligne (who solved the conjecture) on mixed Hodge structures described here.

Fundamental Class

The idea that homology should represent the classes of subvarieties has probably been at the origin of homology theory, although the effective construction of homology groups is different and more elaborate. The simplest way to construct homology groups (in fact, assumed with compact support) is to use singular simplexes, but the definition of homology groups $H_j(X, \mathbb{Z})$ of a triangulated space is the most natural. Indeed, the sum of oriented triangles of highest dimensions of an oriented triangulated topological space X of real dimension n, defines a homology class $[X] \in H_n(X, \mathbb{Z})$ (see [20, Ch. 0, Para. 4]).

We take as granted here that a compact complex algebraic variety X of dimension m can be triangulated ([31]) such that the sum of its oriented triangles of highest dimension defines a class in the homology group $H_{2m}(X, \mathbb{Z})$.

Cap Product

Cohomology groups of a topological space $H^i(X, \mathbb{Z})$ are dually defined, and there exists a topological operation on homology and cohomology, called the cap product:

$$\cap : H^q(X, \mathbb{Z}) \otimes H_p(X, \mathbb{Z}) \to H_{p-q}(X, \mathbb{Z})$$

We can now state the duality theorem of Poincaré, frequently used in geometry.

THEOREM 3.0.1 (Poincaré isomorphism) *Let X be a compact oriented topological manifold of dimension n. The cap product with the fundamental class $[X] \in H_n(X, \mathbb{Z})$,*

$$D_X : H^j(X, \mathbb{Z}) \xrightarrow{\cap [X]} H_{n-j}(X, \mathbb{Z}),$$

defines an isomorphism, for all j, $0 \le j \le n$.

Intersection Product in Topology and Geometry

On a triangulated space, cycles are defined as a sum of triangles of the same dimension with boundary zero, and homology is defined by classes of cycles modulo boundaries. It is always possible to represent two homology classes of degree $n - p$ and $n - q$ by two cycles of codimension p and q in "transversal position" on an oriented topological manifold, so that their intersection is defined as a cycle of codimension $p+q$. Moreover, for two representations by transversal cycles, the intersection cycles are homologous (see e.g., [18] or [15, 2.8]). Then, on an oriented topological manifold, a theory of intersection product on homology can be deduced:

$$H_{n-p}(X, \mathbb{Z}) \otimes H_{n-q}(X, \mathbb{Z}) \xrightarrow{\cap} H_{n-p-q}(X, \mathbb{Z}).$$

In geometry, two closed submanifolds V_1 and V_2 of a compact oriented manifold M can be isotopically deformed into a transversal position so that their intersection can be defined as a submanifold W with a sign depending on the orientation (see [25, Ch. 2]); then the homology class $[W]$ of W is up to sign $[V_1] \cap [V_2]$.

Poincaré Duality in Homology (see [20, p. 53])

On a compact oriented manifold X of dimension n, the intersection pairing

$$H_j(X, \mathbb{Z}) \otimes H_{n-j}(X, \mathbb{Z}) \xrightarrow{\cap} H_0(X, \mathbb{Z}) \xrightarrow{\text{degree}} \mathbb{Z}$$

for $0 \le j \le n$ is unimodular: the induced morphism

$$H_j(X, \mathbb{Z}) \to \text{Hom}(H_{n-j}(X, \mathbb{Z}), \mathbb{Z})$$

is surjective and its kernel is the torsion subgroup of $H_j(X, \mathbb{Z})$.

Cup Product and the Trace Morphism

The cup product is a topological product defined on the cohomology of a topological space with coefficients in $\mathbb{Z}$ [29]. In de Rham cohomology of a differentiable manifold, the cup product is defined by the exterior product of differential forms.

On an oriented topological compact manifold X of dimension n, a trace map Tr : $H^n(X, \mathbb{Z}) \to \mathbb{Z}$ is defined. In de Rham cohomology, the trace map is defined by integrating a differential form of degree n on the oriented differentiable manifold X.

Generalizations of the trace morphism will be present in various forms, including a definition on the level of complexes [42, (2.3.4)] and [27, (III.10), (VI,4)].

Poincaré Duality in Cohomology

On a compact oriented manifold X of dimension n, the composition of the trace with the cup product,

$$H^j(X, \mathbb{Z}) \otimes H^{n-j}(X, \mathbb{Z}) \xrightarrow{\cup} H^n(X, \mathbb{Z}) \xrightarrow{\text{Tr}} \mathbb{Z},$$

defines a unimodular pairing for $0 \leq j \leq n$ inducing an isomorphism

$$H^j(X, \mathbb{Q}) \xrightarrow{\sim} \text{Hom}(H^{n-j}(X, \mathbb{Q}), \mathbb{Q}).$$

Poincaré Isomorphism Transforms the Intersection Pairing into the Cup Product

The following result is proved in [20, p. 59] in the case $k' = n - k$:

Let σ be a k-cycle on an oriented manifold X of real dimension n and τ a k'-cycle on X with Poincaré duals $\eta_\sigma \in H^{n-k}(X)$ and $\eta_\tau \in H^{n-k'}(X)$; then

$$\eta_\sigma \cup \eta_\tau = \eta_{\sigma \cap \tau} \in H^{n-k-k'}(X).$$

Hodge Decomposition

Our starting point in this section is the Hodge decomposition explained in Chapter 1, Theorem 1.5.5. Let $\mathcal{E}^*(X)$ denote the de Rham complex of differential forms with complex values on a complex manifold X and $\mathcal{E}^{p,q}(X)$ the differentiable forms with complex coefficients of type (p, q), i.e., of the form

$$\omega = \sum c_{i_1,\ldots,i_p,j_1,\ldots,j_q}(z, \bar{z}) dz_{i_1} \wedge \cdots \wedge dz_{i_p} \wedge d\bar{z}_{j_1} \wedge \cdots \wedge d\bar{z}_{j_q}$$

(see formula 1.2.4, and Example A.4.3 in the appendix to Chapter 1). The cohomology subspaces of type (p, q) are defined as the space of cohomology classes represented by a form of type (p, q):

$$H^{p,q}(X) = \frac{Z_d^{p,q}(X)}{d\mathcal{E}^*(X) \cap Z_d^{p,q}(X)}, \quad \text{where} \quad Z_d^{p,q}(X) = \text{Ker}\, d \cap \mathcal{E}^{p,q}(X).$$

THEOREM 3.0.2 (Hodge decomposition)   *Let X be a compact Kähler manifold. There exists a decomposition of the complex cohomology spaces into a direct sum of complex subspaces of type (p, q):*

$$H^i(X, \mathbb{C}) = \oplus_{p+q=i} H^{p,q}(X), \quad satisfying \quad H^{p,q}(X) = \overline{H^{q,p}(X)}. \quad (3.0.1)$$

See Chapter 1, Theorem 1.5.5.

Since a complex nonsingular projective variety is Kähler (see Example 1.3.10 and Theorem 1.3.12), we deduce the following corollary:

COROLLARY 3.0.3   *There exists a Hodge decomposition on the cohomology of a complex nonsingular projective variety.*

We remark that the decomposition is canonical on the cohomology level, although it is obtained from the decomposition of the space of harmonic forms which depends on the metric. The above Hodge decomposition theorem uses successive fundamental concepts from Hermitian geometry including the definition and the properties of the *Laplacian* on the Kähler manifold with its associated *fundamental closed $(1,1)$-form*, analytic geometry on the underlying complex analytic manifold such as the type of a form and Dolbeault cohomology, and Riemannian geometry on the underlying Riemannian manifold such as harmonic forms. Besides Chapter 1, an extended exposition of the theory with full proofs, including the subtle linear algebra of Hermitian metrics needed here, can be found in [47], see also ([39, Vol. 2, Ch. IX]) and for original articles see [33]. The aim of the next section is to define a structure that extends to algebraic geometry.

3.1   HODGE STRUCTURE ON A SMOOTH COMPACT COMPLEX VARIETY

In this section we shift our attention from harmonic forms to a new structure defined on the cohomology of nonsingular compact complex algebraic varieties with underlying analytic structure not necessarily Kähler. In this setting the Hodge filtration F on cohomology plays an important role since it is obtained directly from a filtration on the de Rham complex. In this context, it is natural to introduce the spectral sequence defined by F, then the proof will consist in its degeneration at rank 1, although the ultimate argument will be a reduction to the decomposition on a Kähler manifold.

3.1.1   Hodge Structure (HS)

It is rewarding to introduce the Hodge decomposition as a formal structure in linear algebra without any reference to its construction.

DEFINITION 3.1.1 (HS1)   *A Hodge structure of weight n is defined by the data*

   (i) a finitely generated abelian group $H_{\mathbb{Z}}$;

(ii) a decomposition by complex subspaces:

$$H_{\mathbb{C}} := H_{\mathbb{Z}} \otimes_{\mathbb{Z}} \mathbb{C} = \oplus_{p+q=n} H^{p,q} \quad satisfying \quad H^{p,q} = \overline{H^{q,p}}.$$

The conjugation on $H_{\mathbb{C}}$ makes sense with respect to $H_{\mathbb{Z}}$.

A subspace $V \subset H_{\mathbb{C}} := H_{\mathbb{Z}} \otimes_{\mathbb{Z}} \mathbb{C}$ satisfying $\overline{V} = V$ has a real structure, that is, $V = (V \cap H_{\mathbb{R}}) \otimes_{\mathbb{R}} \mathbb{C}$. In particular, $H^{p,p} = (H^{p,p} \cap H_{\mathbb{R}}) \otimes_{\mathbb{R}} \mathbb{C}$. We may suppose that $H_{\mathbb{Z}}$ is a free abelian group (which is called the lattice of the HS), if we are only interested in its image in $H_{\mathbb{Q}} := H_{\mathbb{Z}} \otimes_{\mathbb{Z}} \mathbb{Q}$.

With such an abstract definition we can perform linear algebra operations on Hodge structures and define morphisms of HS.

We can replace the data (i) above by "a finite-dimensional vector space $H_{\mathbb{Q}}$ over the rational numbers $\mathbb{Q}$." Equivalently we may consider a finite-dimensional vector space over the real numbers $\mathbb{R}$. In these cases we speak of rational Hodge structure or real Hodge structure. The main point is to have a conjugation on $H_{\mathbb{C}}$ with respect to $H_{\mathbb{Q}}$ (or $H_{\mathbb{R}}$).

3.1.1.1   The Hodge Filtration

To study variations of HS, Griffiths introduced the Hodge filtration which varies holomorphically with parameters. Given a Hodge decomposition $(H_{\mathbb{Z}}, H^{p,q})$ of weight n, we define a decreasing filtration F by subspaces of $H_{\mathbb{C}}$:

$$F^p H_{\mathbb{C}} := \oplus_{r \geq p} H^{r,n-r}.$$

Then, the following decomposition is satisfied:

$$H_{\mathbb{C}} = F^p H_{\mathbb{C}} \oplus \overline{F^{n-p+1} H_{\mathbb{C}}},$$

since $\overline{F^{n-p+1} H_{\mathbb{C}}} = \oplus_{r \geq n-p+1} \overline{H^{r,n-r}} = \oplus_{i \leq p-1} H^{i,n-i}$ while $F^p H_{\mathbb{C}} = \oplus_{i \geq p} H^{i,n-i}$.

The Hodge decomposition may be recovered from the filtration by the formula:

$$H^{p,q} = F^p H_{\mathbb{C}} \cap \overline{F^q H_{\mathbb{C}}}, \quad for \ p+q = n.$$

Hence, we obtain an equivalent definition of Hodge decompositions which play an important role in the development of Hodge theory, since the Hodge filtration exists naturally on the cohomology of a smooth compact complex algebraic variety X and it is defined by a natural filtration on the algebraic de Rham complex.

DEFINITION 3.1.2 (HS2)   *A Hodge structure of weight n is defined equivalently by the data*

(i) a finitely generated abelian group $H_{\mathbb{Z}}$;

(ii) a filtration F by complex subspaces $F^p H_{\mathbb{C}}$ of $H_{\mathbb{C}}$ satisfying

$$H_{\mathbb{C}} = F^p H_{\mathbb{C}} \oplus \overline{F^{n-p+1} H_{\mathbb{C}}};$$

then

$$H^{p,q} = F^p H_{\mathbb{C}} \cap \overline{F^q H_{\mathbb{C}}}, \quad for \ p+q = n.$$

3.1.1.2 Linear Algebra Operations on Hodge Structures

Classical linear algebra operations may be carried on the above abstract definition of HS.

DEFINITION 3.1.3 *A morphism $f : H = (H_{\mathbb{Z}}, H^{p,q}) \to H' = (H'_{\mathbb{Z}}, H'^{p,q})$ of Hodge structures of the same weight n, is a homomorphism of abelian groups $f : H_{\mathbb{Z}} \to H'_{\mathbb{Z}}$ such that $f_{\mathbb{C}} : H_{\mathbb{C}} \to H'_{\mathbb{C}}$ is compatible with the decompositions, i.e., for any p, q, the $\mathbb{C}$-linear map $f_{\mathbb{C}}$ induces a $\mathbb{C}$-linear map from $H^{p,q}$ into $H'^{p,q}$.*

We have the following important result:

PROPOSITION 3.1.4 *The Hodge structures of the same weight n form an abelian category.*

In particular, the decomposition on the kernel of a morphism $\varphi : H \to H'$ is induced by the decomposition of H, while on the cokernel it is the image of the decomposition of H'.

Remark.

(1) A morphism of HS of distinct weights is torsion. Therefore, HS form an abelian category whose objects are HS and morphisms are morphisms of HS.

(2) On a general subvector space V of H endowed with an HS, the induced subspaces $H^{p,q} \cap V$ do not define an HS decomposition on V.

3.1.1.3 Tensor Product and Hom

Let H and H' be two HS of weight n and n', respectively.

(1) A Hodge structure on $H \otimes H'$ of weight $n + n'$ is defined as

    (i) $(H \otimes H')_{\mathbb{Z}} = H_{\mathbb{Z}} \otimes H'_{\mathbb{Z}}$;

    (ii) the bigrading of $(H \otimes H')_{\mathbb{C}} = H_{\mathbb{C}} \otimes H'_{\mathbb{C}}$ is the tensor product of the bigradings of $H_{\mathbb{C}}$ and $H'_{\mathbb{C}}$:

$$(H \otimes H')^{a,b} := \oplus_{p+p'=a, q+q'=b} H^{p,q} \otimes H'^{p',q'}.$$

(2) A Hodge structure on $\text{Hom}(H, H')$ of weight $n' - n$ is defined as

    (i) $\text{Hom}(H, H')_{\mathbb{Z}} := \text{Hom}_{\mathbb{Z}}(H_{\mathbb{Z}}, H'_{\mathbb{Z}})$;

    (ii) the components of the decomposition of

$$\text{Hom}(H, H')_{\mathbb{C}} := \text{Hom}_{\mathbb{Z}}(H_{\mathbb{Z}}, H'_{\mathbb{Z}}) \otimes \mathbb{C} \simeq \text{Hom}_{\mathbb{C}}(H_{\mathbb{C}}, H'_{\mathbb{C}})$$

    are defined by

$$\text{Hom}(H, H')^{a,b} := \oplus_{p'-p=a, q'-q=b} \text{Hom}_{\mathbb{C}}(H^{p,q}, H'^{p',q'}).$$

In particular, the dual H^* to H is an HS of weight $-n$.

EXAMPLE 3.1.5 (Tate–Hodge structure) $\mathbb{Z}(1)$ is an HS of weight -2 defined by

$$H_{\mathbb{Z}} = 2i\pi\mathbb{Z} \subset \mathbb{C}, \quad H_{\mathbb{C}} = H^{-1,-1}.$$

It is purely bigraded of type $(-1, -1)$, of rank 1. The m-tensor product $\mathbb{Z}(1) \otimes \cdots \otimes$ $\mathbb{Z}(1)$ is an HS of weight $-2m$ denoted by $\mathbb{Z}(m)$:

$$H_{\mathbb{Z}} = (2i\pi)^m\mathbb{Z} \subset \mathbb{C}, \quad H_{\mathbb{C}} = H^{-m,-m}.$$

Let $H = (H_{\mathbb{Z}}, \oplus_{p+q=n}H^{p,q})$ be an HS of weight n; its m-twist is an HS of weight $n - 2m$ denoted $H(m)$ and defined by

$$H(m)_{\mathbb{Z}} := H_{\mathbb{Z}} \otimes (2i\pi)^m\mathbb{Z}, \quad H(m)^{p,q} := H^{p+m,q+m}.$$

Remark.   The group of morphisms of HS is called the internal morphism group of the category of HS and is denoted by $\mathrm{Hom}_{\mathrm{HS}}(H, H')$; it is the subgroup of $\mathrm{Hom}_{\mathbb{Z}}(H_{\mathbb{Z}}, H'_{\mathbb{Z}})$ of elements of type $(0,0)$ in the HS on $\mathrm{Hom}(H, H')$.

A homomorphism of type (r, r) is a morphism of the HS $H \to H'(-r)$.

3.1.1.4   *Equivalent Definition of HS*

Let $\mathbb{S}(\mathbb{R})$ denote the subgroup

$$\mathbb{S}(\mathbb{R}) = \left\{ M(u,v) = \begin{pmatrix} u & -v \\ v & u \end{pmatrix} \in \mathrm{GL}(2,\mathbb{R}), \quad u,v \in \mathbb{R} \right\}.$$

It is isomorphic to $\mathbb{C}^*$ via the group homomorphism $M(u, v) \mapsto z = u + iv$. The interest in this isomorphism is to give a structure of a real algebraic group on $\mathbb{C}^*$; indeed the set $\mathbb{S}(\mathbb{R})$ of matrices $M(u, v)$ is a real algebraic subgroup of $\mathrm{GL}(2, \mathbb{R})$.

DEFINITION 3.1.6   *A rational Hodge structure of weight $m \in \mathbb{Z}$ is defined by a $\mathbb{Q}$-vector space H and a representation of real algebraic groups $\varphi : \mathbb{S}(\mathbb{R}) \to \mathrm{GL}(H_{\mathbb{R}})$ such that for $t \in \mathbb{R}^*$, $\varphi(t)(v) = t^m v$ for all $v \in H_{\mathbb{R}}$.*

See the proof of the equivalence with the action of the group $\mathbb{S}(\mathbb{R})$ in between the definitions A.4.4 and A.4.5 in the appendix to Chapter 1. It is based on the following lemma:

LEMMA 3.1.7   *Let $(H_{\mathbb{R}}, F)$ be a real HS of weight m, defined by the decomposition $H_{\mathbb{C}} = \oplus_{p+q=m}H^{p,q}$; then the action of $\mathbb{S}(\mathbb{R}) = \mathbb{C}^*$ on $H_{\mathbb{C}}$, defined by*

$$\left(z, v = \sum_{p+q=m} v_{p,q} \right) \mapsto \sum_{p+q=m} z^p\overline{z}^q v_{p,q},$$

corresponds to a real representation $\varphi : \mathbb{S}(\mathbb{R}) \to \mathrm{GL}(H_{\mathbb{R}})$ satisfying $\varphi(t)(v) = t^m v$ for $t \in \mathbb{R}^*$.

Remark.

(i) The complex points of

$$
\mathbb{S}(\mathbb{C}) = \left\{ M(u,v) = \begin{pmatrix} u & -v \\ v & u \end{pmatrix} \in \mathrm{GL}(2,\mathbb{C}), \quad u, v \in \mathbb{C} \right\}
$$

are the matrices with complex coefficients $u, v \in \mathbb{C}$ with determinant $u^2 + v^2 \neq 0$. Let $z = u + iv$, $z' = u - iv$, then $zz' = u^2 + v^2 \neq 0$ such that $\mathbb{S}(\mathbb{C}) \xrightarrow{\sim} \mathbb{C}^* \times \mathbb{C}^* : (u,v) \mapsto (z, z')$ is an isomorphism satisfying $z' = \overline{z}$, for $u, v \in \mathbb{R}$; in particular, $\mathbb{R}^* \hookrightarrow \mathbb{C}^* \times \mathbb{C}^* : t \mapsto (t, t)$.

(ii) We can write $\mathbb{C}^* \simeq S^1 \times \mathbb{R}^*$ as the product of the real points of the unitary subgroup $U(1)$ of $\mathbb{S}(\mathbb{C})$ defined by $u^2 + v^2 = 1$ and of the multiplicative subgroup $G_m(\mathbb{R})$ defined by $v = 0$ in $\mathbb{S}(\mathbb{R})$, and $\mathbb{S}(\mathbb{R})$ is the semiproduct of S^1 and $\mathbb{R}^*$. Then the representation gives rise to a representation of $\mathbb{R}^*$ called the scaling since it fixes the weight, and of $U(1)$ which fixes the decomposition into $H^{p,q}$ and on which $\varphi(z)$ acts as multiplication by z^{p-q}.

3.1.1.5 Polarized Hodge Structure

We add the additional structure of polarization since such structure will be satisfied, in particular, by the primitive cohomology of nonsingular complex projective varieties (see Proposition 3.1.20).

Hermitian product. To define a polarization on a Hodge structure $(H_{\mathbb{Z}}, (H_{\mathbb{C}} \simeq \oplus_{p+q=n} H^{p,q}))$, we need to introduce the Weil operator C defined on $H_{\mathbb{C}}$ as follows:

$$
C.\alpha = i^{p-q}\alpha, \quad \forall \alpha \in H^{p,q}.
$$

Notice that C is a real operator since for a real vector $v = \sum_{p+q=j} v^{p,q}$, $v^{q,p} = \overline{v^{p,q}}$, hence $\overline{Cv} = \sum \overline{i^{p-q}v^{p,q}} = \sum \overline{i}^{p-q} v^{q,p} = \sum i^{q-p} v^{q,p} = C\overline{v}$, as $\overline{i}^{p-q} = i^{q-p}$. It depends on the decomposition; in particular, for a varying Hodge structure H_t^{p-q} with parameters t, the action $C = C_t$ depends on t.

DEFINITION 3.1.8 *Let $H := (H_{\mathbb{Z}}, H_{\mathbb{C}} \simeq \oplus_{p+q=n} H^{p,q})$ be a Hodge structure of weight n. A non-degenerate bilinear form Q on $H_{\mathbb{Z}}$ defines a polarization on H if the following conditions are satisfied:*

(i) *$Q(\alpha, \beta) = (-1)^n Q(\beta, \alpha)$ (Q is alternating if n is odd, and symmetric if n is even).*

(ii) *Distinct terms $H^{p,q}$ of the Hodge decomposition are orthogonal to each other relative to the Hermitian form $\tilde{Q}$ defined on $H_{\mathbb{C}}$ by: $\tilde{Q}(\alpha, \beta) := Q(C.\alpha, \overline{\beta})$ for $\alpha, \beta \in H_{\mathbb{C}}$.*

(iii) *$\tilde{Q}$ is positive definite: $\tilde{Q}(\alpha, \alpha) := i^{p-q}Q(\alpha, \overline{\alpha})$ for all non vanishing elements $\alpha \neq 0 \in H^{p,q}$.*

In particular, we use $\overline{Q(\alpha, \beta)} = Q(\overline{\alpha}, \overline{\beta})$ since Q is real, to check for $\alpha, \beta \in H^{p,q}$:

$$\overline{\tilde{Q}(\alpha, \beta)} = \overline{Q(i^{p-q}\alpha, \overline{\beta})} = \overline{i}^{p-q}Q(\overline{\alpha}, \beta) = (-1)^j i^{p-q}Q(\overline{\alpha}, \beta)$$
$$= (-1)^{2j} i^{p-q}Q(\beta, \overline{\alpha}) = \tilde{Q}(\beta, \alpha).$$

3.1.2 Spectral Sequence of a Filtered Complex

The techniques used in the construction of the mixed Hodge structure on the cohomology of algebraic variety are based on the use of spectral sequences. Since we need to prove later an important result in this setting (the lemma on two filtrations), it is necessary to review here the theory using Deligne's notation.

3.1.2.1 Spectral Sequence Defined by a Filtered Complex (K, F)

Let $\mathbb{A}$ be an abelian category (for a first reading, we may suppose $\mathbb{A}$ is the category of vector spaces over a field). A complex K is defined by a family $(K^j)_{j \in \mathbb{Z}}$ of objects of $\mathbb{A}$ and morphisms $d_j : K^j \to K^{j+1}$ in $\mathbb{A}$ satisfying $d_{j+1} \circ d_j = 0$. A filtration F of the complex is defined by a family of subobjects $F^i \subset K^j$ satisfying $d_j(F^i K^j) \subset F^i K^{j+1}$. A complex K endowed with a filtration F is called a filtered complex. We consider decreasing filtrations $F^{i+1} \subset F^i$. In the case of an increasing filtration W_i, we obtain the terms of the spectral sequence from the decreasing case by a change of the sign of the indices of the filtration, which transforms the increasing filtration into a decreasing one F with $F^i = W_{-i}$.

DEFINITION 3.1.9 *Let K be a complex of objects of an abelian category $\mathbb{A}$, with a decreasing filtration by subcomplexes F. It induces a filtration F on the cohomology $H^*(K)$, defined by*

$$F^i H^j(K) = \mathrm{Im}(H^j(F^i K) \to H^j(K)) \quad \forall i, j \in \mathbb{Z}.$$

Let $F^i K / F^j K$ for $i < j$ denote the complex $(F^i K^r / F^j K^r, d_r)_{r \in \mathbb{Z}}$ with induced filtrations; in particular, we write $\mathrm{Gr}_F^p(K)$ for $F^p K / F^{p+1} K$. The associated graded object is $\mathrm{Gr}_F^*(K) = \oplus_{p \in \mathbb{Z}} \mathrm{Gr}_F^p(K)$. We define similarly $\mathrm{Gr}_F^i H^j(K)$ and

$$\mathrm{Gr}_F^* H^j(K) := \oplus_{i \in \mathbb{Z}} \mathrm{Gr}_F^i H^j(K) = \oplus_{j \in \mathbb{Z}} F^i H^j(K) / F^{i+1} H^j(K).$$

The spectral sequence defined by the filtered complex (K, F) gives a method to compute the graded object $\mathrm{Gr}_F^* H^*(K)$ out of the cohomology $H^*(F^i K / F^j K)$ for various indices $i > j$ of the filtration. The spectral sequence $E_r^{p,q}(K, F)$ associated to F ([5, 7]) leads for large r and under mild conditions, to the graded object $\mathrm{Gr}_F^* H^j(K)$. It consists of indexed objects of $\mathbb{A}$ endowed with differentials (see below explicit definitions):

(1) terms $E_r^{p,q}$ for $r > 0, p, q \in \mathbb{Z}$

(2) differentials $d_r : E_r^{p,q} \to E_r^{p+r,q-r+1}$ such that $d_r \circ d_r = 0$

(3) isomorphisms:

$$E_{r+1}^{p,q} \simeq H(E_r^{p-r,q+r-1} \xrightarrow{d_r} E_r^{p,q} \xrightarrow{d_r} E_r^{p+r,q-r+1})$$

of the (p, q) term of index $r + 1$ with the corresponding cohomology of the sequence with index r. To avoid ambiguity we may write $_F E_r^{p,q}$ or $E_r^{p,q}(K, F)$. The first term is defined as

$$E_1^{p,q} = H^{p+q}(\mathrm{Gr}_F^p(K)).$$

The aim of the spectral sequence is to compute the term

$$E_\infty^{p,q} := \mathrm{Gr}_F^p(H^{p+q}(K)).$$

These terms $\mathrm{Gr}_F^p(H^{p+q}(K))$ are called the limit of the spectral sequence.
   The spectral sequence is said to degenerate if

$$\forall p, q, \ \exists \ r_0(p, q) \ \text{such that} \ \forall r \geq r_0, \quad E_r^{p,q} \simeq E_\infty^{p,q} := \mathrm{Gr}_F^p H^{p+q}(K).$$

3.1.2.2   *Formulas for the Terms of the Spectral Sequence*

It will be convenient to set for $r = 0$, $E_0^{p,q} = \mathrm{Gr}_F^p(K^{p+q})$.
   To define the spectral terms $E_r^{p,q}$ with respect to F for $r > 1$, we put for $r > 0$, $p, q \in \mathbb{Z}$,

$$Z_r^{p,q} = \mathrm{Ker}(d : F^p K^{p+q} \to K^{p+q+1}/F^{p+r} K^{p+q+1}),$$
$$B_r^{p,q} = F^{p+1} K^{p+q} + d(F^{p-r+1} K^{p+q-1}).$$

Such formulas still makes sense for $r = \infty$ if we set, for a filtered object (A, F), $F^{-\infty}(A) = A$ and $F^\infty(A) = 0$,

$$Z_\infty^{p,q} = \mathrm{Ker}(d : F^p K^{p+q} \to K^{p+q+1}),$$
$$B_\infty^{p,q} = F^{p+1} K^{p+q} + d(K^{p+q+1}).$$

We set by definition,

$$E_r^{p,q} = Z_r^{p,q}/(B_r^{p,q} \cap Z_r^{p,q}), \quad E_\infty^{p,q} = Z_\infty^{p,q}/B_\infty^{p,q} \cap Z_\infty^{p,q}.$$

The notation is similar to [7] but different from [17].

Remark.   Given a statement on spectral sequences, we may deduce a dual statement if we use the following definition of B_r^{pq} and $B_\infty^{p,q}$, dual to the definition of $Z_r^{p,q}$ and $Z_\infty^{p,q}$:

$$K^{p+q}/B_r^{p,q} = \mathrm{coker}(d : F^{p-r+1} K^{p+q-1} \to K^{p+q}/F^{p+1}(K^{p+q})),$$
$$K^{p+q}/B_\infty^{p,q} = \mathrm{coker}(d : K^{p+q+1} \to K^{p+q}/F^{p+1} K^{p+q}),$$
$$E_r^{p,q} = \mathrm{Im}(Z_r^{p,q} \to K^{p+q}/B_r^{p,q}) = \mathrm{Ker}(K^{p+q}/B_r^{p,q} \to K^{p+q}/(Z_r^{p,q} + B_r^{p,q})).$$

LEMMA 3.1.10 *For each r, the differential d_r on the terms $E_r^{p,q}$ is induced by the differential $d : Z_r^{p,q} \to Z_r^{p+r,q-r+1}$ and has the following property:*

$$E_{r+1}^{p,q} \simeq H(E_r^{p-r,q+r-1} \xrightarrow{d_r} E_r^{p,q} \xrightarrow{d_r} E_r^{p+r,q-r+1}).$$

Also, $d_r \circ d_r = 0$ and the cohomology at the term $E_r^{p,q}$ is isomorphic to $E_{r+1}^{p,q}$.

The first term may be written as

$$E_1^{p,q} = H^{p+q}(\mathrm{Gr}_F^p(K))$$

so that the differentials d_1 are obtained as connecting morphisms defined by the short exact sequences of complexes

$$0 \to \mathrm{Gr}_F^{p+1} K \to F^p K / F^{p+2} K \to \mathrm{Gr}_F^p K \to 0.$$

DEFINITION 3.1.11 *The decreasing filtration on the complex K is biregular if it induces a finite filtration on K^n for each degree n.*

Then, for each (p,q), there exists an integer $r_0 - r_0(p,q)$ such that for all $r \geq r_0$, $Z_r^{p,q} = \mathrm{Ker}(d : F^p K^{p+q} \to K^{p+q+1})$, $B_r^{p,q} = F^{p+1} K^{p+q} + dK^{p+q-1}$; hence the spectral sequence degenerates:

$$Z_r^{p,q} = Z_\infty^{p,q}, \quad B_r^{p,q} = B_\infty^{p,q}, \quad E_r^{p,q} = E_\infty^{p,q}, \quad \forall r \geq r_0.$$

The spectral sequence degenerates at rank r (independent of (p,q)) if the differentials d_i of E_i^{pq} vanish for $i \geq r$ for a fixed r.

Most known applications are in the case of degenerate spectral sequences [20].

We give a formula for $E_r^{p,q}$ below in Lemma 3.2.7, but there is no obvious general result on the degeneration of the spectral sequence at a specific rank as various examples show (see [17, 4.6] or [5, Prop. 5.6]).

Note that in some cases, it is not satisfactory to get only the graded cohomology and this is one motivation to be unhappy with spectral sequences and to prefer to keep the complex, as we shall do in the later section on the derived category.

We emphasize that in Deligne–Hodge theory, the spectral sequences studied later (see sections 3.1.3, 3.2.4 and lemma 3.3.22) degenerate at rank 1 or 2; we will impose sufficient conditions to imply specifically in our case that $d_r = 0$ for $r > 1$, hence the terms $E_r^{p,q}$ are identical for all $r > 1$.

3.1.2.3 The Simple Complex Associated to a Double Complex and its Filtrations

A double complex is a bigraded object of an abelian category $\mathbb{A}$ with two differentials:

$$K^{\bullet,\bullet} := (K^{i,j})_{i,j \in \mathbb{N}}, \quad d' : K^{i,j} \to K^{i+1,j}, \quad d'' : K^{i,j} \to K^{i,j+1},$$

satisfying $d' \circ d' = 0, d'' \circ d'' = 0$ and $d'' \circ d' + d' \circ d'' = 0$. The associated complex is defined as

$$(s(K^{\bullet,\bullet}), d) : s(K^{\bullet,\bullet})^n := \oplus_{i+j=n} K^{i,j}, \quad d = d' + d''.$$

There exist two natural decreasing filtrations F' and F'' on $s(K^{\bullet,\bullet})$ defined by

$$(F')^p s(K^{\bullet,\bullet})^n := \oplus_{i+j=n, i \geq p} K^{i,j}, \quad (F'')^p s(K^{\bullet,\bullet})^n := \oplus_{i+j=n, j \geq p} K^{i,j}.$$

The corresponding terms of the spectral sequences are

- $'E_1^{p,q} := H^{p+q}(K^{p,\bullet}[-p], d'') = H^q(K^{p,\bullet}, d'');$

- $'E_2^{p,q} := H^p[H^q(K^{\bullet,\bullet}, d''), d'];$

- $''E_1^{p,q} := H^{p+q}(K^{\bullet,p}[-p], d') = H^q(K^{\bullet,p}, d');$

- $''E_2^{p,q} := H^p[H^q(K^{\bullet,\bullet}, d'), d''];$

where q and p are the partial degrees with opposition to the total degree. See Section 2.4 for examples.

3.1.2.4   Morphisms of Spectral Sequences

A morphism of filtered complexes of objects of an abelian category $\mathbb{A}$,

$$f : (K, F) \to (K', F'),$$

compatible with the filtrations $f(F^i(K) \subset F^i(K'))$, induces a morphism of the corresponding spectral sequences.

DEFINITION 3.1.12   *(i) A morphism $f : K \xrightarrow{\approx} K'$ of complexes of objects of $\mathbb{A}$ is a quasi-isomorphism denoted by $\approx$ if the induced morphisms on cohomology $H^*(f) : H^*(K) \xrightarrow{\sim} H^*(K')$ are isomorphisms for all degrees.*

(ii) A morphism $f : (K, F) \xrightarrow{\approx} (K', F)$ of complexes with biregular filtrations is a filtered quasi-isomorphism if it is compatible with the filtrations and induces a quasi-isomorphism on the graded object $\mathrm{Gr}_F^*(f) : \mathrm{Gr}_F^*(K) \xrightarrow{\approx} \mathrm{Gr}_F^*(K').$

In case (ii) we call (K', F) a filtered resolution of (K, F), while in (i) the complex K' is just a resolution of K.

PROPOSITION 3.1.13   *Let $f : (K, F) \to (K', F')$ be a filtered morphism with biregular filtrations; then the following assertions are equivalent:*

(i) f is a filtered quasi-isomorphism.

(ii) $E_1^{p,q}(f) : E_1^{p,q}(K, F) \to E_1^{p,q}(K', F')$ is an isomorphism for all p, q.

(iii) $E_r^{p,q}(f) : E_r^{p,q}(K, F) \to E_r^{p,q}(K', F')$ is an isomorphism for all $r \geq 1$ and all p, q.

By definition of the terms $E_1^{p,q}$, (ii) is equivalent to (i). We deduce (iii) from (ii) by induction: if we suppose the isomorphism in (iii) is satisfied for $r \leq r_0$, then the isomorphism for $r_0 + 1$ follows since $E_{r_0}^{p,q}(f)$ is compatible with d_{r_0}.

3.1.3 Hodge Structure on the Cohomology of Nonsingular Compact Complex Algebraic Varieties

We consider here not only the case of nonsingular projective varieties which are Kähler, but also the case of nonsingular compact algebraic varieties which may be not Kähler. The technique of proof is based on the spectral sequence defined by the trivial filtration F on the de Rham complex of sheaves of analytic differential forms Ω_X^*.

The new idea here is to observe the degeneracy of the spectral sequence of the filtered complex (Ω_X^*, F) at rank 1 and deduce the definition of the Hodge filtration on cohomology from the trivial filtration F on the complex without any reference to harmonic forms, although the proof of the decomposition is given via a reduction to the case of a projective variety, hence a compact Kähler manifold and the results on harmonic forms in this case.

Classically, we use distinct notation for X with Zariski topology and X^{an} for the associated analytic manifold; then the filtration F is defined on the algebraic de Rham hypercohomology groups and the comparison theorem [23] (see Chapter 2) is compatible with the filtrations $\mathbb{H}^i(X, F^p\Omega_X^*) \simeq \mathbb{H}^i(X^{\mathrm{an}}, F^p\Omega_{X^{\mathrm{an}}}^*)$. Hence, in this chapter we consider only the analytic structure on X and we still write X instead of X^{an}. The study of Hodge theory via the spectral sequence defined by F is already present in [16]. The degeneration in the case of characteristic 0 or $p > 0$ with a lifting hypothesis is proved in [10].

THEOREM 3.1.14 (Deligne [6]) *Let X be a smooth compact complex algebraic variety; then the filtration F by subcomplexes of the de Rham complex*

$$F^p\Omega_X^* := \Omega_X^{*\geq p} = 0 \to 0 \cdots 0 \to \Omega_X^p \to \Omega_X^{p+1} \to \cdots \to \Omega_X^n \to 0$$

induces a Hodge filtration of a Hodge structure on the cohomology of X.

DEFINITION 3.1.15 *The Hodge filtration F is defined on de Rham cohomology as follows:*

$$F^p H^i(X, \mathbb{C}) = F^p\mathbb{H}^i(X, \Omega_X^*) := \mathrm{Im}(\mathbb{H}^i(X, F^p\Omega_X^*) \to \mathbb{H}^i(X, \Omega_X^*)),$$

where the first isomorphism is defined by the holomorphic Poincaré lemma on the resolution of the constant sheaf $\mathbb{C}$ by the analytic de Rham complex Ω_X^*.

The proof is based on the degeneration at rank 1 of the spectral sequence with respect to F defined as follows:

$$_F E_1^{p,q} := \mathbb{H}^{p+q}(X, \mathrm{Gr}_F^p \Omega_X^*) \simeq H^q(X, \Omega_X^p) \Rightarrow \mathrm{Gr}_F^p \mathbb{H}^{p+q}(X, \Omega_X^*).$$

We distinguish first the projective case which follows from the underlying structure of a Kähler variety, from which the general case is deduced.

(1) For X projective, the Dolbeault resolution (Section 1.2.3) of the subcomplex $F^p\Omega_X^*$ defines a double complex $(\mathcal{E}_X^{r,*})_{r\geq p}$ of smooth forms with the differentials ∂ acting on the first degree r and $\bar\partial$ acting on the second degree $*$. The simple associated complex

denoted $s(\mathcal{E}_X^{*,*})$ is defined in degree m by summing over the terms in degrees $p+q=m$ (see Section 3.1.2.3 above), on which the filtration F is defined as follows:

$$(\mathcal{E}_X^*, d) = s(\mathcal{E}_X^{*,*}, \overline{\partial}, \partial), \quad (F^p\mathcal{E}_X^*, d) = s(\mathcal{E}_X^{r,*}, \overline{\partial}, \partial)_{r \geq p}$$

is a fine resolution of de Rham complex

$$\mathbb{H}^n(X, F^p\Omega_X^*) \simeq H^n(X, F^p\mathcal{E}^*).$$

For $p = 0$, the complex $(\mathcal{E}_X^*, d)$ contains the subcomplex of harmonic forms on which the differentials ∂ and $\overline{\partial}$ vanish. On the Kähler manifold, the terms $E_1^{p,q} = H^q(X, \Omega_X^p)$ are isomorphic to the vector subspace $\mathcal{H}^{p,q}(X)$ of $\mathbb{H}^{p+q}(X, \Omega_X^*) \simeq H^{p+q}(X, \mathcal{E}_X^*) \simeq H^{p+q}(X, \mathbb{C})$ consisting of harmonic forms of type (p, q). Since the spectral sequence $(_F E_r^{p,q}, d_r)$ degenerates to the space $\mathrm{Gr}_F^p \mathbb{H}^{p+q}(X, \Omega_X^*) \simeq \mathcal{H}^{p,q}(X)$, and since already $E_1^{p,q} = \mathrm{Gr}_F^p \mathbb{H}^{p+q}(X, \Omega_X^*)$ have the same dimension, we deduce for all $r \geq 1$ that $\dim {}_F E_1^{p,q} = \dim {}_F E_r^{p,q} = \dim \mathcal{H}^{p,q}(X)$. Hence $d_r = 0$ for $r \geq 1$, otherwise the dimension would drop, which means that the spectral sequence $(_F E_r^{p,q}, d_r)$ degenerates at rank 1, and the HS is defined by the subspaces $\mathcal{H}^{p,q}$ lifting the subspaces $\mathrm{Gr}_F^p \mathbb{H}^{p+q}(X, \Omega_X^*)$ into $\mathbb{H}^{p+q}(X, \Omega_X^*)$.

(2) If X is not projective, there exists a projective variety and a projective birational morphism $f : X' \to X$, by Chow's lemma (see [39, p. 69]). By Hironaka's desingularization ([30]) we can suppose X' smooth projective, hence X' is a Kähler manifold. We continue to write f for the associated analytic morphism defined by the algebraic map f. By the general duality theory [42], there exists for all integers p a trace map $\mathrm{Tr}(f) : Rf_*\Omega_{X'}^p \to \Omega_X^p$ inducing a map on cohomology $\mathrm{Tr}(f) : H^q(X', \Omega_{X'}^p) \to H^q(X, \Omega_X^p)$, since $\mathbb{H}^q(X, Rf_*\Omega_{X'}^p) \simeq H^q(X', \Omega_{X'}^p)$ ([6, Sec. 4], [27, VI, 4]). In our case, since f is birational, the trace map is defined on the level of de Rham complexes as follows. Let U be an open subset of X and $V \subset U$ an open dense subset of U such that f induces an isomorphism $f^{-1}(V) \xrightarrow{\sim} V$. A differential form $\omega' \in \Gamma(f^{-1}(U), \Omega_{X'}^p)$ induces a form ω on V which has the property to extend to a unique holomorphic form on U; then we define $\mathrm{Tr} f(\omega') := \omega \in \Gamma(U, \Omega_X^p)$ ([13, II, 2.1]), from which we deduce that the composition morphism with the canonical reciprocal morphism f^* is the identity

$$\mathrm{Tr}(f) \circ f^* = \mathrm{Id} : H^q(X, \Omega_X^p) \xrightarrow{f^*} H^q(X', \Omega_{X'}^p) \xrightarrow{\mathrm{Tr}(f)} H^q(X, \Omega_X^p).$$

In particular, f^* is injective. Since the map f^* is compatible with the filtrations F on de Rham complexes on X and X', we deduce a map of spectral sequences:

$$E_1^{p,q} = H^q(X, \Omega_X^p) \xrightarrow{f^*} E_1'^{p,q} = H^q(X', \Omega_{X'}^p), \quad f^* : E_r^{p,q}(X) \hookrightarrow E_r^{p,q}(X'),$$

which commutes with the differentials d_r and is injective on all terms. The differential d_1 vanishes on X', hence it must vanish on X, then the terms for $r = 2$ coincide with the terms for $r = 1$. The proof continues by induction on r as we can repeat the argument for all r.

The degeneration of the Hodge spectral sequence on X at rank 1 follows, and it is equivalent to the isomorphism

$$\mathbb{H}^n(X, F^p\Omega_X^*) \xrightarrow{\sim} F^p\mathbb{H}^n(X, \Omega_X^*).$$

Equivalently, the dimension of the hypercohomology of the de Rham complex $\mathbb{H}^j(X, \Omega_X^*)$ is equal to the dimension of Hodge cohomology $\oplus_{p+q=j}H^q(X, \Omega_X^p)$.

However, we still need to lift Dolbeault cohomology $H^q(X, \Omega_X^p)$ into subspaces of $\mathbb{H}^{p+q}(X, \Omega_X^*)$. Using conjugation, we deduce from the Hodge filtration, the definition of the following subspaces:

$$H^{p,q}(X) := F^pH^n(X, \mathbb{C}) \cap \overline{F^qH^n(X, \mathbb{C})}, \quad \text{for } p + q = n,$$

satisfying $H^{p,q}(X) = \overline{H^{q,p}(X)}$. Now we check the decomposition

$$H^n(X, \mathbb{C}) = \oplus_{p+q=n}H^{p,q}(X)$$

and deduce that $H^{p,q}(X) \simeq H^q(X, \Omega_X^p)$. Since f^* is injective we have

$$F^pH^n(X) \cap \overline{F^{n-p+1}H^n(X)} \subset F^pH^n(X') \cap \overline{F^{n-p+1}H^n(X')} = 0.$$

This shows that $F^pH^n(X) + \overline{F^{n-p+1}H^n(X)}$ is a direct sum. We want to prove that this sum equals $H^n(X)$.

Let $h^{p,q} = \dim H^q(X, \Omega_X^p)$; since the spectral sequence degenerates at rank 1, we have

$$\dim F^pH^n(X) = \sum_{i \geq p} h^{i,n-i}, \quad \dim \overline{F^{n-p+1}H^n(X)} = \sum_{i \geq n-p+1} h^{i,n-i};$$

then

$$\sum_{i \geq p} h^{i,n-i} + \sum_{i \geq n-p+1} h^{i,n-i} \leq \dim H^n(X) = \sum_i h^{i,n-i},$$

from which we deduce the inequality $\sum_{i \geq p} h^{i,n-i} \leq \sum_{i \leq n-p} h^{i,n-i}$.

By Serre duality ([28, Cor. 7.13]) on X of dimension N, we have

$$H^j(X, \Omega_X^i)^* \simeq H^{N-j}(X, \Omega_X^{N-i});$$

hence $h^{i,j} = h^{N-i,N-j}$, which transforms the inequality into

$$\sum_{N-i \leq N-p} h^{N-i,N-n+i} \leq \sum_{N-i \geq N-n+p} h^{N-i,N-n+i},$$

from which, by setting $j = N - i, q = N - n + p$, we deduce the opposite inequality on $H^m(X)$ for $m = 2N - n$:

$$\sum_{j \geq q} h^{j,m-j} \geq \sum_{j \leq m-q} h^{j,m-j}$$

for all q and m. In particular, setting $j = i$ and $m = n$,

$$\sum_{i \geq p} h^{i,n-i} \geq \sum_{i \leq n-p} h^{i,n-i}, \quad \text{hence} \quad \sum_{i \geq p} h^{i,n-i} = \sum_{i \leq n-p} h^{i,n-i}.$$

This implies $\dim F^p + \dim \overline{F^{n-p+1}} = \dim H^n(X)$. Hence

$$H^n(X) = F^p H^n(X) \oplus \overline{F^{n-p+1}} H^n(X),$$

which, in particular, induces a decomposition

$$F^{n-1} H^n(X) = F^p H^n(X) \oplus H^{n-1,n-p+1}(X).$$

COROLLARY 3.1.16 ([7, Cor. 5.4]) *On a nonsingular complex algebraic variety,*

(i) *a cohomology class a is of type (p,q) ($a \in H^{p,q}(X)$) if and only if it can be represented by a closed form α of type (p,q);*

(ii) *a cohomology class a may be represented by a form α satisfying $\overline{\partial}(\alpha) = 0$ and $\partial \alpha = 0$;*

(iii) *if a form α satisfies $\overline{\partial}(\alpha) = 0$ and $\partial \alpha = 0$, then the following four conditions are equivalent:*

   (1) *There exists β such that $\alpha = d\beta$.*

   (2) *There exists β such that $\alpha = \partial \beta$.*

   (3) *There exists β such that $\alpha = \overline{\partial} \beta$.*

   (4) *There exists β such that $\alpha = \partial \overline{\partial} \beta$.*

3.1.3.1 Compatibility of Poincaré Duality with HS

On compact oriented differentiable manifolds, the wedge product of differential forms defines the cup product on de Rham cohomology [20] and the integration of the form of maximal degree defines the trace, so that we can deduce the compatibility of Poincaré duality with Hodge structure.

The Trace Map. On a compact oriented manifold X of dimension n, we already mentioned that the cup product on de Rham cohomology is defined by the wedge product on the level of differential forms:

$$H^i_{\mathrm{deRham}}(X) \otimes H^j_{\mathrm{deRham}}(X) \xrightarrow{\cup} H^{i+j}_{\mathrm{deRham}}(X).$$

By the Stokes theorem, the integral over X of differential forms ω of highest degree n depends only on the class of ω modulo boundaries; hence it defines a map called the trace:

$$\mathrm{Tr} : H^n(X, \mathbb{C}) \to \mathbb{C}, \quad [\omega] \mapsto \int_X \omega.$$

It is convenient to extend the trace to a map on the de Rham complex inducing zero in degree different than n:

$$\text{Tr} : \Omega_X^*[n] \to \mathbb{C}, \quad [\omega] \mapsto \int_X \omega,$$

which is a special case of the definition of the trace on the dualizing complex needed for a general duality theory. On de Rham cohomology, we restate the following theorem:

THEOREM 3.1.17 (Poincaré duality) *Let X be a compact oriented manifold of dimension n. The composition of the trace map with the cup product*

$$H^j(X, \mathbb{C}) \otimes H^{n-j}(X, \mathbb{C}) \xrightarrow{\cup} H^n(X, \mathbb{C}) \xrightarrow{\text{Tr}} \mathbb{C}$$

defines an isomorphism

$$H^j(X, \mathbb{C}) \xrightarrow{\sim} \text{Hom}(H^{n-j}(X, \mathbb{C}), \mathbb{C}).$$

Then we define the trace as a map of HS,

$$H^{2n}(X, \mathbb{C}) \xrightarrow{\sim} \mathbb{C}(-n), \quad \omega \to \frac{1}{(2i\pi)^n} \int_X \omega,$$

such that Poincaré duality is compatible with HS:

$$H^{n-i}(X, \mathbb{C}) \simeq \text{Hom}(H^{n+i}(X, \mathbb{C}), \mathbb{C}(-n)),$$

where the duality between $H^{p,q}$ and $H^{n-p,n-q}$ is Serre duality [28, Cor. 7.13]. The HS on homology is defined by duality:

$$(H_i(X, \mathbb{C}), F) \simeq \text{Hom}((H^i(X, \mathbb{C}), F), \mathbb{C}),$$

where $\mathbb{C}$ is an HS of weight 0, hence $H_i(X, \mathbb{Z})$ is of weight $-i$. Then, Poincaré duality becomes an isomorphism of HS: $H^{n+i}(X, \mathbb{C}) \simeq H_{n-i}(X, \mathbb{C})(-n)$.

3.1.3.2 Gysin Morphism

Let $f : X \to Y$ be an algebraic morphism of nonsingular compact algebraic varieties with $\dim X = n$ and $\dim Y = m$. Since $f^* : H^i(Y, \mathbb{Q}) \to H^i(X, \mathbb{Q})$ is compatible with HS, its Poincaré dual

$$\text{Gysin}(f) : H^j(X, \mathbb{Q}) \to H^{j+2(m-n)}(Y, \mathbb{Q})(m-n)$$

is compatible with HS after a shift by $-2(m-n)$ on the right.

3.1.4 Lefschetz Decomposition and Polarized Hodge Structure

We define one more specific structure on cohomology of compact Kähler manifolds, namely, the Lefschetz decomposition and Riemann bilinear relations on primitive cohomology subspaces, which leads to the abstract definition of polarized Hodge structures. The original proof Lefschetz [35] was incomplete and based on topology instead of Hodge theory.

3.1.4.1   Lefschetz Decomposition and Primitive Cohomology

The class of the fundamental form $[\omega] \in H^2(X, \mathbb{R})$ of Hodge type $(1, 1)$ defined by the Hermitian metric of the underlying Kähler structure on X acts on cohomology by repeated cup product with $[\omega]$ and defines morphisms

$$L : H^q(X, \mathbb{R}) \to H^{q+2}(X, \mathbb{R}), \quad L : H^q(X, \mathbb{C}) \to H^{q+2}(X, \mathbb{C}) : [\varphi] \mapsto [\omega] \wedge [\varphi].$$

Referring to de Rham cohomology, the action of L is represented on the level of forms as $\varphi \mapsto \omega \wedge \varphi$ (since ω is closed, the image of a closed form (resp. a boundary) is closed (resp. a boundary)), hence L depends on the class $[\omega]$.

DEFINITION 3.1.18   *Let $n = \dim X$. The primitive cohomology subspaces are defined for $i \geq 0$ as*

$$H_{\mathrm{prim}}^{n-i}(X, \mathbb{R}) := \mathrm{Ker}(L^{i+1} : H^{n-i}(X, \mathbb{R}) \to H^{n+i+2}(X, \mathbb{R}))$$

and similarly for complex coefficients $H_{\mathrm{prim}}^{n-i}(X, \mathbb{C}) \simeq H_{\mathrm{prim}}^{n-i}(X, \mathbb{R}) \otimes_{\mathbb{R}} \mathbb{C}$.

The operator L is of Hodge type $(1, 1)$ since it sends the subspace $H^{p,q}$ to $H^{p+1,q+1}$. Hence, the action $L^{i+1} : H^{n-i}(X, \mathbb{C}) \to H^{n+i+2}(X, \mathbb{C})$ is a morphism of Hodge type $(i + 1, i + 1)$, and the kernel is endowed with an induced Hodge decomposition which is a strong condition on the primitive subspaces

$$H_{\mathrm{prim}}^{p,q} := H_{\mathrm{prim}}^{p+q} \cap H^{p,q}(X, \mathbb{C}), \quad H_{\mathrm{prim}}^i(X, \mathbb{C}) = \oplus_{p+q=i} H_{\mathrm{prim}}^{p,q}.$$

The following isomorphism, referred to as the hard Lefschetz theorem, puts a strong condition on the cohomology of projective, and more generally compact Kähler, manifolds and gives rise to a decomposition of the cohomology in terms of primitive subspaces:

THEOREM 3.1.19   *Let X be a compact Kähler manifold.*

  (i) *Hard Lefschetz theorem. The iterated linear operator L induces isomorphisms for each i, $1 \leq i \leq n$:*

$$L^i : H^{n-i}(X, \mathbb{R}) \xrightarrow{\sim} H^{n+i}(X, \mathbb{R}), \quad L^i : H^{n-i}(X, \mathbb{C}) \xrightarrow{\sim} H^{n+i}(X, \mathbb{C}).$$

  (ii) *Lefschetz decomposition. The cohomology decomposes into a direct sum of images of primitive subspaces by L^r, $r \geq 0$:*

$$H^q(X, \mathbb{R}) = \oplus_{r \geq 0} L^r H_{\mathrm{prim}}^{q-2r}(\mathbb{R}), \quad H^q(X, \mathbb{C}) = \oplus_{r \geq 0} L^r H_{\mathrm{prim}}^{q-2r}(\mathbb{C}).$$

  Moreover, the Lefschetz decomposition is compatible with Hodge decomposition.

  (iii) *If X is projective, then the action of L is defined with rational coefficients and the decomposition applies to rational cohomology.*

We refer to the proof in Section 1.5.3 of Chapter 1 and for more details to [43, Section 6.2.2] and [47].

3.1.4.2   Hermitian Product on Cohomology

We deduce from the isomorphism in the hard Lefschetz theorem and Poincaré duality, a scalar product on cohomology of smooth complex projective varieties compatible with HS and satisfying relations known as Hodge–Riemann relations leading to a polarization of the primitive cohomology, which is an additional structure characteristic of such varieties. Representing cohomology classes by differential forms, we define a bilinear form,

$$Q(\alpha, \beta) = (-1)^{\frac{j(j-1)}{2}} \int_X \alpha \wedge \beta \wedge \omega^{n-j} \quad \forall [\alpha], [\beta] \in H^j(X, \mathbb{C}),$$

where ω is in the Kähler class, the product of α with ω^{n-j} represents the action of L^{n-j} and the integral of the product with β represents Poincaré duality.

Properties of the Product

The above product $Q(\alpha, \beta)$ depends on j and only on the class of α and β. The following properties are satisfied:

  (i)   The product Q is real (it takes real values on real forms) since ω is real; in other words, the matrix of Q is real, skew symmetric if j is odd and symmetric if j is even.

 (ii)   It is non-degenerate (by Lefschetz isomorphism and Poincaré duality).

(iii)   By consideration of type, the Hodge and Lefschetz decompositions satisfy, with respect to Q, the relations

$$Q(H^{p,q}, H^{p',q'}) = 0 \quad \text{unless } p = q', q = p'.$$

On projective varieties, the Kähler class is in the integral lattice defined by cohomology with coefficients in $\mathbb{Z}$, hence the product is defined on rational cohomology and preserves the integral lattice. In this case we have more precise positivity relations in terms of the primitive component $H^{p,q}_{\mathrm{prim}}(X, \mathbb{C})$ of the cohomology $H^{n+q}(X, \mathbb{C})$.

PROPOSITION 3.1.20 (Hodge–Riemann bilinear relations)   *Let X be smooth projective, then the product $\tilde{Q}$ defined by Q and the Weil operator C on the Hodge structure $H^j(X, \mathbb{C})$ (see def. 3.1.8):*

$$\tilde{Q}(\alpha, \beta) = Q(C.\alpha, \overline{\beta}), \quad \forall [\alpha], [\beta] \in H^j(X, \mathbb{C})$$

is Hermitian nondegenerate. Precisely, the Hodge decomposition is orthogonal with respect to $\tilde{Q}$, and the product $i^{p-q} Q(\alpha, \overline{\alpha})$ is positive definite on the primitive component $H^{p,q}_{\mathrm{prim}}$:

$$i^{p-q} Q(\alpha, \overline{\alpha}) > 0 \quad \forall \alpha \in H^{p,q}_{\mathrm{prim}}, \ \alpha \neq 0.$$

We refer to the proof in Section 1.5.3 of Chapter 1.

Remark.   When the class $[\omega] \in H^j(X, \mathbb{Z})$ is integral, which is the case for projective varieties, the product Q is integral, i.e., with integral value on integral classes.

3.1.4.3 Projective Case

On a smooth projective variety $X \subset \mathbb{P}^n_{\mathbb{C}}$, we can choose the first Chern class $c_1(\mathcal{L})$ of the line bundle $\mathcal{L}$ defined by the restriction of the hyperplane line bundle $\mathcal{O}(1)$ on $\mathbb{P}^n_{\mathbb{C}}$ to represent the Kähler class $[\omega]$ defined by the Hermitian metric on the underlying Kähler variety X: $c_1(\mathcal{L}) = [\omega]$ (see Section 1.3.2). Hence, we have an integral representative of the class $[\omega]$ in the image of $H^2(X, \mathbb{Z}) \to H^2(X, \mathbb{C})$, which has the following consequence: the operator $L : H^q(X, \mathbb{Q}) \to H^{q+2}(X, \mathbb{Q})$ acts on rational cohomology. This fact characterizes projective varieties among compact Kähler manifolds since a theorem of Kodaira (see [47, Ch. VI]) states that a Kähler variety with an integral class $[\omega]$ is projective, i.e., it can be embedded as a closed analytic subvariety in a projective space; hence by the Chow lemma it is necessarily a projective subvariety.

Topological interpretation   In the projective case, the class $[\omega]$ corresponds to the homology class of a hyperplane section $[H] \in H_{2n-2}(X, \mathbb{Z})$, so that the operator L corresponds to the intersection with $[H]$ in X and the result is an isomorphism:

$$H_{n+k}(X) \xrightarrow{(\cap [H])^k} H_{n-k}(X).$$

The primitive cohomology $H^{n-k}_{\mathrm{prim}}(X)$ corresponds to the image of

$$H_{n-k}(X - H, \mathbb{Q}) \to H_{n-k}(X, \mathbb{Q}).$$

COROLLARY 3.1.21   *The cohomology of a projective complex smooth variety carries a polarized Hodge structure defined by the composition of Poincaré duality with the Lefschetz operator L.*

3.1.5 Examples

We list now some known examples of HS, mainly on tori.

3.1.5.1 Cohomology of Projective Spaces

The HS on cohomology is polarized by the first Chern class of the canonical line bundle $H = c_1(\mathcal{O}_{\mathbb{P}^n}(1))$ dual to the homology class of a hyperplane (Section 1.3.2).

PROPOSITION 3.1.22 $H^i(\mathbb{P}^n, \mathbb{Z}) = 0$ *for i odd and $H^i(\mathbb{P}^n, \mathbb{Z}) = \mathbb{Z}$ for i even, with generator $[H]^i$ equal to the cup product to the power i of the cohomology class of a hyperplane $[H]$; hence $H^{2r}(\mathbb{P}^n, \mathbb{Z}) = \mathbb{Z}(-r)$ as HS.*

3.1.5.2 Hodge Decomposition on Complex Tori

Let T_Λ be a complex torus of dimension r defined by a lattice $\Lambda \subset \mathbb{C}^r$ as in (see Example 1.1.27 for $r = 2$). The cohomology of a complex torus T_Λ is easy to compute by the Künneth formula, since it is diffeomorphic to a product of circles: $T_\Lambda \simeq (S^1)^{2r}$. Hence $H^1(T_\Lambda, \mathbb{Z}) \simeq \mathbb{Z}^{2r}$ and $H^j(T_\Lambda, \mathbb{Z}) \simeq \wedge^j H^1(T_\Lambda, \mathbb{Z})$.

The cohomology with complex coefficients may be computed by de Rham cohomology. In this case, since the complex tangent space is trivial, we have natural cohomology elements given by the translation-invariant forms, hence with constant coefficients. The finite complex vector space T_0^* of constant holomorphic 1-forms is isomorphic to $\mathbb{C}^r$ and generated by $dz_j, j \in [1, r]$ and the Hodge decomposition reduces to prove $H^j(X, \mathbb{C}) \simeq \oplus_{p+q=j} \wedge^p T_0^* \otimes \wedge^q \overline{T_0^*}$, $p \geq 0, q \geq 0$, which is a consequence of the above computation of the cohomology in this case.

3.1.5.3 Moduli Space of Complex Tori

We may parametrize all lattices as follows:

- The group $\mathrm{GL}(2r, \mathbb{R})$ acts transitively on the set of all lattices of $\mathbb{C}^r$.

  We choose a basis $\tau = (\tau_1, \ldots, \tau_{2i-1}, \tau_{2i}, \ldots, \tau_{2r})$, $i \in [1, r]$ of a lattice L, then it defines a basis of $\mathbb{R}^{2r}$ over $\mathbb{R}$. An element φ of $\mathrm{GL}(2r, \mathbb{R})$ is given by the linear transformation which sends τ into the basis $\varphi(\tau) = \tau'$ of $\mathbb{R}^{2r}$ over $\mathbb{R}$. The element φ of $\mathrm{GL}(2r, \mathbb{R})$ carries the lattice L onto the lattice L' defined by the basis τ'.

- The isotopy group of elements with neutral action is $\mathrm{GL}(2r, \mathbb{Z})$ since τ and τ' define the same lattice if and only if $\varphi \in \mathrm{GL}(2r, \mathbb{Z})$.

  Hence the space of lattices is the quotient group $\mathrm{GL}(2r, \mathbb{R}) / \mathrm{GL}(2r, \mathbb{Z})$.

- Two tori defined by the lattice L and L' are analytically isomorphic if and only if there is an element of $\mathrm{GL}(r, \mathbb{C})$ which transforms the lattice L into the lattice L' (see Example 1.1.27).

It follows that the parameter space of complex tori is the quotient

$$\mathrm{GL}(2r, \mathbb{Z}) \backslash \mathrm{GL}(2r, \mathbb{R}) / \mathrm{GL}(r, \mathbb{C}),$$

where $\mathrm{GL}(r, \mathbb{C})$ is embedded naturally in $\mathrm{GL}(2r, \mathbb{R})$ as a complex linear map is $\mathbb{R}$-linear.

For $r = 1$, the quotient $\mathrm{GL}(2, \mathbb{R}) / \mathrm{GL}(1, \mathbb{C})$ is isomorphic to $\mathbb{C} - \mathbb{R}$ since, up to complex isomorphisms, a lattice is generated by $1, \tau \in \mathbb{C}$ independent over $\mathbb{R}$, and hence is completely determined by $\tau \in \mathbb{C} - \mathbb{R}$. The moduli space is the orbit space of the action of $\mathrm{GL}(2, \mathbb{Z})$ on the space $\mathrm{GL}(2, \mathbb{R}) / \mathrm{GL}(1, \mathbb{C}) = \mathbb{C} - \mathbb{R}$. Since $\mathrm{GL}(2, \mathbb{Z})$ is the disjoint union of $\mathrm{SL}(2, \mathbb{Z})$ and the integral 2×2 matrices of determinant equal to -1, that orbit space is the one of the action of $\mathrm{SL}(2, \mathbb{Z})$ on the upper half plane:

$$\left(\left(\begin{array}{cc} a & b \\ c & d \end{array} \right), z \right) \mapsto \frac{az + b}{cz + d}.$$

The Hodge structures of the various complex tori define a variation of Hodge structures on the moduli space of all complex tori.

3.1.5.4   Polarized Hodge Structures of Dimension 2 and Weight 1

Let H be a real vector space of dimension 2 endowed with a skew-symmetric quadratic form and (e_1, e_2) a basis in which the matrix of Q is

$$Q = \begin{pmatrix} 0 & 1 \\ -1 & 0 \end{pmatrix}, \quad Q(u, v) = {}^t v Q u, \quad u, v \in \mathbb{Q}^2;$$

then Hodge decomposition $H_{\mathbb{C}} = H^{1,0} \oplus H^{0,1}$ is defined by the one-dimensional subspace $H^{1,0}$ with generator v of coordinates $(v_1, v_2) \in \mathbb{C}^2$. While $Q(v, v) = 0$ since Q is skew symmetric, the Hodge–Riemann positivity condition is written as $iQ(v, \bar{v}) = -i(v_1 \bar{v}_2 - \bar{v}_1 v_2) > 0$, hence $v_2 \neq 0$, so we divide by v_2 to get a unique representative of $H^{1,0}$ by a vector of the form $v = (\tau, 1)$ with $\operatorname{Im}(\tau) > 0$. Hence the Poincaré half plane $\{z \in \mathbb{C} : \operatorname{Im} z > 0\}$ is a classifying space for polarized Hodge structures of dimension 2. This will applies to the cohomology $H := H^1(T, \mathbb{R})$ of a complex torus of dimension 1.

In particular, we remark that the torus T is a projective variety. We write $T := \mathbb{C}/L$ as the quotient of $\mathbb{C}$ with a nondegenerate lattice L and $T*$ for the complement of the reference point represented by the class of the lattice. To construct an embedding, we use the Weierstrass elliptic function $\mathcal{P}(z)$ ([28, Ch. 4, Sec. 4]) defined as a sum of a series over the elements of L in $\mathbb{C}$; it defines with its derivative a holomorphic map $(\mathcal{P}(z), \mathcal{P}'(z)) : T^* \to \mathbb{C}^2 \subset \mathbb{P}^2_{\mathbb{C}}$, meromorphic on T, which extends to an embedding of the torus onto a smooth elliptic curve in the projective space.

3.1.5.5   Polarized Hodge Structures of Weight 1 and Abelian Varieties

Given a Hodge structure $(H_{\mathbb{Z}}, H^{1,0} \oplus H^{0,1})$, the projection on $H^{0,1}$ induces an isomorphism of $H_{\mathbb{R}}$ onto $H^{0,1}$ as real vector spaces:

$$H_{\mathbb{R}} \to H_{\mathbb{C}} = H^{1,0} \oplus H^{0,1} \to H^{0,1}$$

since $\overline{H^{0,1}} = H^{1,0}$, hence $H_{\mathbb{R}} \cap H^{1,0} = 0$. Then we deduce that $H_{\mathbb{Z}}$ projects to a lattice in the complex space $H^{0,1}$, and the quotient $T := H^{0,1}/H_{\mathbb{Z}}$ is a complex torus.

This HS appears in geometry in the case of a complex manifold X. The exact sequence of sheaves defined by $f \mapsto e^{2i\pi f}$,

$$0 \to \mathbb{Z} \to \mathcal{O}_X \to \mathcal{O}_X^* \to 1,$$

where 1 on the right is the neutral element of the multiplicative group structure on $\mathcal{O}_X^*$, and its associated long exact sequence

$$\to H^1(X, \mathbb{Z}) \to H^1(X, \mathcal{O}_X) \to H^1(X, \mathcal{O}_X^*) \to H^2(X, \mathbb{Z}) \to,$$

have the following geometric interpretation. Since the space $H^1(X, \mathcal{O}_X^*)$ is the group of isomorphism classes of line bundles on X, the last morphism defines the Chern class of the line bundle. We deduce the isomorphism

$$T := \frac{H^1(X, \mathcal{O}_X)}{\operatorname{Im} H^1(X, \mathbb{Z})} \simeq \operatorname{Pic}^0(X) := \operatorname{Ker}(H^1(X, \mathcal{O}_X^*) \xrightarrow{c_1} H^2(X, \mathbb{Z}))$$

of the torus T with the Picard variety $\mathrm{Pic}^0(X)$ parametrizing the holomorphic line bundles $\mathcal{L}$ on X with first Chern class equal to zero: $c_1(\mathcal{L}) = 0$. The Picard variety of a smooth projective variety is an abelian variety. (Hint: define a Kähler form with integral class on $\mathrm{Pic}^0(X)$ and apply Kodaira's embedding theorem.)

3.1.5.6 Polarized Hodge Structures of Weight 2

The polarized Hodge structures of weight 2 are given by the data

$$(H_{\mathbb{Z}}, \; H^{2,0} \oplus H^{1,1} \oplus H^{0,2}; \; H^{0,2} = \overline{H^{2,0}}, \; H^{1,1} = \overline{H^{1,1}}; \; Q)$$

where the intersection form Q is symmetric and the product $\tilde{Q}(\alpha, \beta) = Q(\alpha, \overline{\beta})$ is Hermitian, positive definite on $H^{1,1}$ and negative definite on $H^{2,0}$ and $H^{0,2}$; moreover the Hodge decomposition is orthogonal with respect to the Hermitian form $\tilde{Q}$.

Hence the HS is determined by the subspace $H^{2,0} \subset H_{\mathbb{C}}$ which is totally isotropic for Q on which F is negative definite on $H^{2,0}$.

Then $H^{0,2}$ is defined by conjugation and $H^{1,1} = (H^{2,0} \oplus H^{0,2})^{\perp}$ by orthogonality. The signature of Q is $(h^{1,1}, 2h^{2,0})$.

3.1.6 Cohomology Class of a Subvariety and Hodge Conjecture

An oriented compact topological variety V of dimension n has a fundamental class in its homology group $H_{2n}(V, \mathbb{Z})$, which plays an important role in Poincaré duality. By duality, the class corresponds to a class in the cohomology group $H^0(V, \mathbb{Z})$.

In algebraic and analytic geometry, the existence of the fundamental class extends to all subvarieties since their singular subset has codimension 2. We construct here the class of a closed complex algebraic subvariety (resp. complex analytic subspace) of codimension p in de Rham cohomology $H^{2p}_{\mathrm{DR}}(V)$ of a smooth complex projective variety V (resp. compact Kähler manifold) and we show it is rational of Hodge type (p, p); then we state the Hodge conjecture.

LEMMA 3.1.23 *Let X be a complex manifold and Z a compact complex analytic subspace of dimension m in X. The integral of the restriction of a form ω on X to the smooth open subset Z_{smooth} of Z, is convergent and defines a linear map on forms of degree $2m$. It induces a linear map on cohomology*

$$\mathrm{cl}(Z) : H^{2m}(X, \mathbb{C}) \to \mathbb{C}, \quad [\omega] \mapsto \int_{Z_{\mathrm{smooth}}} \omega_{|Z}.$$

Moreover, if X is compact Kähler, $\mathrm{cl}(Z)$ vanishes on all components of the Hodge decomposition which are distinct from $H^{m,m}$.

If Z is compact and smooth, the integral is well defined on the class $[\omega]$ since for $\omega = d\eta$ the integral vanishes by the Stokes theorem, as the integral of η on the boundary $\partial Z = \emptyset$ of Z vanishes.

If Z is not smooth, the easiest proof is to use a desingularization $\pi : Z' \to Z$ inducing an isomorphism over Z_{smooth} (see [30]) which implies the convergence of the integral of the restriction $\omega_{|Z}$ and the equality of the integrals of $\pi^*(\omega_{|Z})$ and $\omega_{|Z}$. In particular, the integral is independent of the choice of Z'. The restriction $\omega_{|Z}$ of degree $2m$ vanishes unless it is of type (m,m) since Z_{smooth} is an analytic manifold of dimension m.

If the compact complex analytic space Z is of codimension p in the smooth compact complex manifold X, its class $\text{cl}(Z) \in H^{2n-2p}(X,\mathbb{C})^*$ corresponds by Poincaré duality on X, to a fundamental cohomology class $[\eta_Z] \in H^{2p}(X,\mathbb{C})$. Then we have by definition the following lemma:

LEMMA AND DEFINITION 3.1.24 *For a compact complex manifold X, the fundamental cohomology class $[\eta_Z] \in H^{p,p}(X,\mathbb{C})$ of a closed complex submanifold Z of codimension p satisfies the following relation:*

$$\int_X \varphi \wedge \eta_Z = \int_Z \varphi_{|Z} \quad \forall \varphi \in \mathcal{E}^{n-p,n-p}(X).$$

LEMMA 3.1.25 *For a compact Kähler manifold X,*

$$H^{p,p}(X,\mathbb{C}) \neq 0 \quad \text{for } 0 \leq p \leq \dim X.$$

In fact, the integral of the volume form $\int_X \omega^n > 0$. It follows that the cohomology class $[\omega]^n \neq 0 \in H^{2n}(X,\mathbb{C})$; hence the cohomology class $[\omega]^p \neq 0 \in H^{p,p}(X,\mathbb{C})$ since its cup product with $[\omega]^{n-p}$ is not zero.

LEMMA 3.1.26 *For a compact Kähler manifold X, the cohomology class of a compact complex analytic closed submanifold Z of codimension p is a nonzero element $[\eta_Z] \neq 0 \in H^{p,p}(X,\mathbb{C})$, for $0 \leq p \leq \dim X$.*

PROOF. For a compact Kähler manifold X, with a Kähler form ω, the integral on Z of the restriction $\omega_{|Z}$ is positive since it is a Kähler form on Z; hence $[\eta_Z] \neq 0$ since by Poincaré duality,

$$\int_X (\wedge^{n-p}\omega) \wedge \eta_Z = \int_Z \wedge^{n-p}(\omega_{|Z}) > 0.$$

$\square$

This class is compatible with the fundamental class naturally defined in homology. Indeed, the homology class $i_*[Z]$ of a compact complex analytic subspace $i : Z \to X$ of codimension p in the smooth compact complex manifold X of dimension n corresponds, by the inverse of the Poincaré duality isomorphism, to a fundamental cohomology class

$$[\eta_Z]^{\text{top}} \in H^{2p}(X,\mathbb{Z}).$$

LEMMA 3.1.27 *The canonical morphism $H_{2n-2p}(X,\mathbb{Z}) \to H^{2n-2p}(X,\mathbb{C})^*$ carries the topological class $[Z]$ of an analytic subspace Z of codimension p in X into the fundamental class $\text{cl}(Z)$. Similarly, the morphism $H^{2p}(X,\mathbb{Z}) \to H^{2p}(X,\mathbb{C})$ carries the topological class $[\eta_Z]^{\text{top}}$ to $[\eta_Z]$.*

3.1.6.1   Hodge Conjecture

A natural question is to find what conditions can be made on cohomology classes representing classes of algebraic subvarieties, including the classes in cohomology with coefficients in $\mathbb{Z}$. The Hodge type (p, p) of the fundamental class of analytic compact submanifolds of codimension p is an analytic condition. The search for properties characterizing classes of cycles has been motivated by the Hodge conjecture:

DEFINITION 3.1.28 (Hodge classes)   *For each integer $p \in \mathbb{N}$, let $H^{p,p}(X)$ denote the subspace of $H^{2p}(X, \mathbb{C})$ of type (p, p); the group of rational (p, p) cycles*

$$H^{p,p}(X, \mathbb{Q}) := H^{2p}(X, \mathbb{Q}) \cap H^{p,p}(X) = H^{2p}(X, \mathbb{Q}) \cap F^p H^{2p}(X, \mathbb{C})$$

is called the group of rational Hodge classes of type (p, p).

DEFINITION 3.1.29   *An r-cycle of an algebraic variety X is a formal finite linear combination $\sum_{i \in [1,h]} m_i Z_i$ of closed irreducible subvarieties Z_i of dimension r with integer coefficients m_i. The group of r-cycles is denoted by $\mathcal{Z}_r(X)$.*

On a compact complex algebraic manifold, the class of closed irreducible subvarieties of codimension p extends into a linear morphism:

$$\mathrm{cl}_{\mathbb{Q}} : \mathcal{Z}_p(X) \otimes \mathbb{Q} \to H^{p,p}(X, \mathbb{Q}) : \sum_{i \in [1,h]} m_i Z_i \mapsto \sum_{i \in [1,h]} m_i \eta_{Z_i} \quad \forall m_i \in \mathbb{Q}.$$

The elements of the image of $\mathrm{cl}_{\mathbb{Q}}$ are called rational algebraic Hodge classes of type (p, p).

HODGE CONJECTURE ([32])   *On a nonsingular complex projective variety, any rational Hodge class of type (p, p) is algebraic, i.e., in the image of $\mathrm{cl}_{\mathbb{Q}}$.*

Originally, the Hodge conjecture was stated with $\mathbb{Z}$-coefficients.

Let $\varphi : H^{2p}(X, \mathbb{Z}) \to H^{2p}(X, \mathbb{C})$ denote the canonical map and define the group of integral Hodge classes of type (p, p) as

$$H^{p,p}(X, \mathbb{Z}) := \{x \in H^{2p}(X, \mathbb{Z}) : \varphi(x) \in H^{p,p}(X, \mathbb{C})\};$$

then, is any integral Hodge class of type (p, p) algebraic?

However, there are torsion elements which cannot be represented by algebraic cycles (see [1]). Also, there exist compact Kähler complex manifolds not containing enough analytic subspaces to represent all Hodge cycles (see [44]). See [11] for a summary of results related to the Hodge conjecture.

Absolute Hodge cycle   Deligne added another property for algebraic cycles by introducing the notion of the absolute Hodge cycle (see [9] and Chapter 11). An algebraic cycle Z is necessarily defined over a field extension K of finite type over $\mathbb{Q}$. Then its cohomology class in the de Rham cohomology of X over the field K defines, for each embedding $\sigma : K \to \mathbb{C}$, a cohomology class $[Z]_\sigma$ of type (p, p) in the cohomology of the complex manifold X_σ^{an}. We refer to Chapter 11 for a comprehensive study of this theory.

Grothendieck construction of the fundamental class   For an algebraic subvariety Z of codimension p in a variety X of dimension n, the fundamental class of Z can be defined as an element of the group $\mathrm{Ext}^p(\mathcal{O}_Z, \Omega_X^p)$ (see [22, 27]). Let U be an affine open subset and suppose that $Z \cap U$ is defined as a complete intersection by p regular functions f_i, if we use the Koszul resolution of $\mathcal{O}_{Z \cap U}$ defined by the elements f_i to compute the extension groups; then the fundamental class is defined by a symbol:

$$\left[\begin{array}{c} df_1 \wedge \cdots \wedge df_p \\ f_1 \cdots f_p \end{array} \right] \in \mathrm{Ext}^p(\mathcal{O}_{Z \cap U}, \Omega_U^p).$$

This symbol is independent of the choice of generators of $\mathcal{O}_{Z \cap U}$, and it is the restriction of a unique class defined over Z which defines the cohomology class of Z in the de Rham cohomology group $\mathbb{H}_Z^{2p}(X, \Omega_X^*)$ with support in Z ([12], [13, IV, Prop. 5]). The extension groups and cohomology groups with support in closed algebraic subvarieties form the basic idea of constructing the dualizing complex ω_X of X as part of Grothendieck duality theory (see [27]).

3.2   MIXED HODGE STRUCTURES (MHS)

Motivated by conjectural properties of cohomology of varieties in positive characteristic as stated by Weil, Deligne imagined the correct structure to put on cohomology of any complex algebraic variety, possibly open or singular (later Deligne solved Weil's conjecture).

Since the knowledge of the linear algebra structure underlying MHS is supposed to help the reader before being confronted with their construction, we first introduce the category of mixed Hodge structures (MHS) consisting of vector spaces endowed with weight W and Hodge F filtrations by subvector spaces satisfying adequate axioms (see [7]). The striking result to remember is that morphisms of MHS are *strict* (see Section 3.2.1.3) for both filtrations W and F.

Only, linear algebra is needed for the proofs in the abstract setting. The corresponding theory in an abelian category is developed for objects with opposite filtrations.

- We start with a formal study of filtrations needed in the definitions of MHS. Since we are essentially concerned by filtrations of vector spaces, it is not more difficult to describe this notion in the terminology of abelian categories.

- Next we define MHS and prove that they form an abelian category with morphisms which are strict with respect to W and F (see Section 3.2.18). The proof is based on a canonical decomposition of an MHS (see Section 3.2.2.3).

- We end this section with a result on spectral sequences which is essential for the construction of MHS (Deligne's lemma on two filtrations; see Section 3.2.3.1).

3.2.1 Filtrations

Given a morphism in an additive category, the isomorphism between the image and coimage is one of the conditions to define an abelian category. In the additive category of filtered vector spaces endowed with finite filtrations by vector subspaces, given a morphism $f : (H, F) \to (H', F')$ compatible with the filtrations $f(F^j) \subset F'^j$, the filtration obtained from F on the coimage does not coincide in general with the filtration induced by F' on the image. The morphism is called strict if they coincide (see Section 3.2.1.3).

This kind of problem will occur for various repeated restrictions of filtrations. Here we need to define with precision the properties of induced filtrations, since this is at the heart of the subject of MHS.

On a subquotient B/C of a filtered vector space A, and in general of a filtered object of an abelian category, there are two ways to restrict the filtration: first on the subobject B, then on the quotient B/C, or on the quotient A/C, then on the subobject B/C. On an object A with two filtrations W and F, we can restrict F on each Gr_W^n, or W on each Gr_F^m and we may get different objects if we start with W then F or we reverse the order. We need to know precise relations between the induced filtrations in these various ways, and to know the precise behavior of a linear morphism with respect to such induced filtrations.

As a main application, we will indicate (see Section 3.2.3.1) three different ways to induce a filtration on the terms of a spectral sequence. A central result is to give conditions on the filtered complex such that the three induced filtrations coincide.

First we recall some preliminaries on filtrations in an abelian category $\mathbb{A}$.

DEFINITION 3.2.1 *A decreasing (resp. increasing) filtration F of an object A of $\mathbb{A}$ is a family of subobjects of A, satisfying*

$$\forall n, m, \qquad n \le m \Longrightarrow F^m(A) \subset F^n(A) \qquad (resp. \ n \le m \Longrightarrow F_n(A) \subset F_m(A)).$$

The pair (A, F) will be called a filtered object.

If F is a decreasing filtration (resp. W an increasing filtration), a shifted filtration $F[n]$ (resp. $W[n]$) by an integer n is defined as

$$(F[n])^p(A) = F^{p+n}(A), \quad (W[n])_p(A) = W_{p-n}(A).$$

Decreasing filtrations F are considered for general study. Statements for increasing filtrations W are deduced by the change of indices $W_n(A) = F^{-n}(A)$. A filtration is finite if there exist integers n and m such that $F^n(A) = A$ and $F^m(A) = 0$.

3.2.1.1 Induced Filtration

A filtered object (A, F) induces a filtration on a subobject $i : B \to A$ of A defined by $F^n(B) = B \cap F^n(A)$. Dually, the quotient filtration on A/B, with canonical projection $p : A \to A/B$, is defined by

$$F^n(A/B) = p(F^n(A)) = (B + F^n(A))/B \simeq F^n(A)/(B \cap F^n(A)).$$

DEFINITION 3.2.2   *The graded object associated to* (A, F) *is defined as*

$$\mathrm{Gr}_F(A) = \oplus_{n \in \mathbb{Z}} \mathrm{Gr}_F^n(A) \quad where \quad \mathrm{Gr}_F^n(A) = F^n(A)/F^{n+1}(A).$$

3.2.1.2   Filtered Morphisms and Induced Filtration on Cohomology

A morphism of filtered objects $(A, F) \xrightarrow{f} (B, F)$ is a morphism $A \xrightarrow{f} B$ compatible with the filtrations $f(F^n(A)) \subset F^n(B)$ for all $n \in \mathbb{Z}$. The cohomology of a sequence of filtered morphisms

$$(A, F) \xrightarrow{f} (B, F) \xrightarrow{g} (C, F)$$

satisfying $g \circ f = 0$ is defined as $H = \mathrm{Ker}\, g/\mathrm{Im}\, f$; it is *filtered* and endowed with the quotient filtration of the induced filtration on $\mathrm{Ker}\, g$. It is equal to the induced filtration on H by the quotient filtration on $B/\mathrm{Im}\, f$ where $H \subset (B/\mathrm{Im}\, f)$.

   Filtered objects (resp. objects with a finite filtration) form an additive category with the existence of *kernel, cokernel, image,* and *coimage* of a morphism with natural induced filtrations. However, the image and coimage will not be necessarily filtered isomorphic.

   This is the main obstruction to obtaining an abelian category. To get around this obstruction, we are lead to define the notion of strictness for compatible morphisms.

3.2.1.3   Strictness

For filtered modules over a ring, a morphism of filtered objects $f : (A, F) \to (B, F)$ is called strict if the relation

$$f(F^n(A)) = f(A) \cap F^n(B)$$

is satisfied; that is, any element $b \in F^n(B) \cap \mathrm{Im}\, A$ is already in $\mathrm{Im}\, F^n(A)$. Next, we consider an additive category where we suppose the existence of a subobject of B, still denoted $f(A) \cap F^n(B)$, containing all common subobjects of $f(A)$ and $F^n(B)$.

DEFINITION 3.2.3   *A filtered morphism in an additive category*

$$f : (A, F) \to (B, F)$$

is called strict, or strictly compatible with the filtrations, if it induces a filtered isomorphism

$$(\mathrm{Coim}(f), F) \to (\mathrm{Im}(f), F),$$

from the coimage to the image of f *with their induced filtrations. Equivalently,*

$$f(F^n(A)) = f(A) \cap F^n(B)$$

for all n.

   This concept is basic to the theory, so we mention the following criteria:

PROPOSITION 3.2.4   *(i) A filtered morphism $f : (A, F) \to (B, F)$ of objects with finite filtrations is strict if and only if the following exact sequence of graded objects is exact:*

$$0 \to \mathrm{Gr}_F(\mathrm{Ker}\, f) \to \mathrm{Gr}_F(A) \to \mathrm{Gr}_F(B) \to \mathrm{Gr}_F(\mathrm{Coker}\, f) \to 0.$$

(ii) Let $S : (A, F) \xrightarrow{f} (B, F) \xrightarrow{g} (C, F)$ be a sequence S of strict morphisms such that $g \circ f = 0$; then the cohomology H with induced filtration satisfies

$$H(\mathrm{Gr}_F(S)) \simeq \mathrm{Gr}_F(H(S)).$$

3.2.1.4 Degeneration of a Spectral Sequence and Strictness

The following proposition gives a remarkable criterion of degeneration at rank 1 of the spectral sequence with respect to the filtration F.

PROPOSITION 3.2.5 ([7, Prop. 1.3.2])   *Let K be a complex with a biregular filtration F. The following conditions are equivalent:*

   (i) The spectral sequence defined by F degenerates at rank 1 ($E_1 = E_\infty$).

   (ii) The morphism $H^i(F^p(K)) \to F^p H^i(K)$ is an isomorphism for all p.

   (iii) The differentials $d : K^i \to K^{i+1}$ are strictly compatible with the filtrations.

3.2.1.5 Two Filtrations

Let A be an object of $\mathbb{A}$ with two filtrations F and G. By definition, $\mathrm{Gr}_F^n(A)$ is a quotient of a subobject of A, and as such, it is endowed with an induced filtration by G. Its associated graded object defines a bigraded object $\mathrm{Gr}_G^n \mathrm{Gr}_F^m(A)_{n,m \in \mathbb{Z}}$. We refer to [7] for the following lemma:

LEMMA 3.2.6 (Zassenhaus' lemma)   *The objects $\mathrm{Gr}_G^n \mathrm{Gr}_F^m(A)$ and $\mathrm{Gr}_F^m \mathrm{Gr}_G^n(A)$ are isomorphic.*

Remark.   Let H be a third filtration on A. It induces a filtration on $\mathrm{Gr}_F(A)$, hence on $\mathrm{Gr}_G \mathrm{Gr}_F(A)$. It induces also a filtration on $\mathrm{Gr}_F \mathrm{Gr}_G(A)$. These filtrations do not correspond in general under the above isomorphism. In the formula $\mathrm{Gr}_H \mathrm{Gr}_G \mathrm{Gr}_F(A)$, G and H have symmetric roles, but not F and G.

3.2.1.6 Hom and Tensor Functors

If A and B are two filtered modules on some ring, we define a filtration on Hom:

$$F^k \mathrm{Hom}(A, B) = \{f : A \to B : \forall n, f(F^n(A)) \subset F^{n+k}(B)\}.$$

Hence

$$\mathrm{Hom}((A, F), (B, F)) = F^0(\mathrm{Hom}(A, B)).$$

If A and B are two filtered modules on some ring, we define

$$F^k(A \otimes B) = \sum_m \operatorname{Im}(F^m(A) \otimes F^{k-m}(B) \to A \otimes B).$$

3.2.1.7 Multifunctor

In general, if $H : \mathbb{A}_1 \times \ldots \times \mathbb{A}_n \to \mathbb{B}$ is a right exact multiadditive functor, we define

$$F^k(H(A_1, \ldots, A_n)) = \sum_{\sum k_i = k} \operatorname{Im}(H(F^{k_1}A_1, \ldots, F^{k_n}A_n) \to H(A_1, \ldots, A_n))$$

and dually if H is left exact,

$$F^k(H(A_1, \ldots, A_n)) = \bigcap_{\sum k_i = k} \operatorname{Ker}(H(A_1, \ldots, A_n) \to H(A_1/F^{k_1}A_1, \ldots, A_n/F^{k_n}A_n)).$$

If H is exact, both definitions are equivalent.

3.2.1.8 Spectral Sequence for Increasing Filtrations

To illustrate the notation of induced filtrations, we give the formulas of the spectral sequence with respect to an increasing filtration W on a complex K, where the preceding formulas (see Section 3.1.2.1) are applied to the decreasing filtration F deduced from W by the change of indices $F^i = W_{-i}$. We set for all j, $n \le m$ and $n \le i \le m$,

$$W_i H^j(W_m K/W_n K) = \operatorname{Im}(H^j(W_i K/W_n K) \to H^j(W_m K/W_n K));$$

then the terms for all $r \ge 1, p$ and q are written as follows:

LEMMA 3.2.7 *The terms of the spectral sequence for (K, W) are equal to*

$$E_r^{p,q}(K, W) = \operatorname{Gr}_{-p}^W H^{p+q}(W_{-p+r-1}K/W_{-p-r}K).$$

PROOF. Let (K_r^p, W) denote the quotient complex $K_r^p := W_{-p+r-1}K/W_{-p-r}K$ with the induced filtration by subcomplexes W; we put

$$Z_\infty^{p,q}(K_r^p, W) := \operatorname{Ker}(d : (W_{-p}K^{p+q}/W_{-p-r}K^{p+q})$$
$$\to (W_{-p+r-1}K^{p+q+1}/W_{-p-r}K^{p+q+1})),$$
$$B_\infty^{p,q}(K_r^p, W) := (W_{-p-1}K^{p+q} + dW_{-p+r-1}K^{p+q-1})/W_{-p-r}K^{p+q}.$$

They coincide, *up to the quotient by $W_{-p-r}K^{p+q}$*, with $Z_r^{p,q}(K, W)$ (resp. $B_r^{p,q}(K, W)$) with

$$Z_r^{p,q} := \operatorname{Ker}(d : W_{-p}K^{p+q} \to K^{p+q+1}/W_{-p-r}K^{p+q+1}),$$
$$B_r^{p,q} := W_{-p-1}K^{p+q} + dW_{-p+r-1}K^{p+q-1};$$

then, we define

$$E_\infty^{p,q}(K_r^p, W) = \frac{Z_\infty^{p,q}(K_r^p, W)}{B_\infty^{p,q}(K_r^p, W) \cap Z_\infty^{p,q}(K_r^p, W)}$$
$$= \mathrm{Gr}_{-p}^W H^{p+q}(W_{-p+r-1}K/W_{-p-r}K)$$

and find

$$E_r^{p,q}(K, W) = Z_r^{p,q}/(B_r^{p,q} \cap Z_r^{pq})$$
$$= Z_\infty^{p,q}(K_r^p, W)/(B_\infty^{p,q}(K_r^p, W) \cap Z_\infty^{p,q}(K_r^p, W))$$
$$= E_\infty^{p,q}(K_r^p, W) = \mathrm{Gr}_{-p}^W H^{pq}(W_{-p+r-1}K/W_{-p-r}K).$$

To define the differential d_r, we consider the exact sequence

$$0 \to W_{-p-r}K/W_{-p-2r}K \to W_{-p+r-1}K/W_{-p-2r}K$$
$$\to W_{-p+r-1}K/W_{-p-r}K \to 0,$$

and the connecting morphism

$$H^{p+q}(W_{-p+r-1}K/W_{-p-r}K) \overset{\partial}{\to} H^{p+q+1}(W_{-p-r}K/W_{-p-2r}K).$$

The injection $W_{-p-r}K \to W_{-p-1}K$ induces a morphism

$$\varphi: H^{p+q+1}(W_{-p-r}K/W_{-p-2r}K) \to W_{-p-r}H^{p+q+1}(W_{-p-1}K/W_{-p-2r}K).$$

Let π denote the projection on the right-hand term below, equal to $E_r^{p+r,q-r+1}$:

$$W_{-p-r}H^{p+q+1}(W_{-p-1}K/W_{-p-2r}K) \overset{\pi}{\to} \mathrm{Gr}_{-p-r}^W H^{p+q+1}(W_{-p-1}K/W_{-p-2r}K).$$

The composition of morphisms $\pi \circ \varphi \circ \partial$ restricted to $W_{-p}H^{p+q}(W_{-p+r-1}K/W_{-p-r}K)$ induces the differential

$$d_r : E_r^{p,q} \to E_r^{p+r,q-r+1},$$

while the injection $W_{-p+r-1} \to W_{-p+r}K$ induces the isomorphism

$$H(E_r^{p,q}, d_r) \overset{\sim}{\to} E_{r+1}^{pq} = \mathrm{Gr}_{-p}^W H^{p+q}(W_{-p+r}K/W_{-p-r-1}K).$$

$\square$

3.2.1.9 n-Opposite Filtrations

The linear algebra of HS applies to an abelian category $\mathbb{A}$ if we use the following definition where no conjugation map appears.

DEFINITION 3.2.8 (n-opposite filtrations) *Two finite filtrations F and G on an object A of an abelian category $\mathbb{A}$ are n-opposite if*

$$\mathrm{Gr}_F^p \mathrm{Gr}_G^q(A) = 0 \quad \text{for } p + q \neq n.$$

Hence, the Hodge filtration F on a Hodge structure of weight n is n-opposite to its conjugate $\overline{F}$.

EXAMPLE 3.2.9 Let $A^{p,q}$ be a bigraded object of $\mathbb{A}$ such that $A^{p,q} = 0$ for $p + q \neq n$ and $A^{p,q} = 0$ for all but a finite number of pairs (p,q); then we define two n-opposite filtrations F and G on $A := \bigoplus_{p,q} A^{p,q}$ by

$$F^p(A) = \bigoplus_{p' \geq p} A^{p',q'}, \quad G^q(A) = \bigoplus_{q' \geq q} A^{p',q'}.$$

We have $\mathrm{Gr}_F^p \mathrm{Gr}_G^q(A) = A^{p,q}$.

PROPOSITION 3.2.10   *(i) Two finite filtrations F and G on an object A are n-opposite if and only if*

$$\forall p, q, \quad p + q = n + 1 \Rightarrow F^p(A) \oplus G^q(A) \simeq A.$$

(ii) If F and G are n-opposite, and if we put $A^{p,q} = F^p(A) \cap G^q(A)$, for $p + q = n$, $A^{p,q} = 0$ for $p + q \neq n$, then A is a direct sum of $A^{p,q}$.

The above constructions define an equivalence of categories between objects of $\mathbb{A}$ with two n-opposite filtrations and bigraded objects of $\mathbb{A}$ of the type described in the example; moreover, F and G can be deduced from the bigraded object $A^{p,q}$ of A by the above procedure.

3.2.1.10   Opposite Filtrations on an HS

The previous definitions of HS may be stated in terms of induced filtrations. For any A-module H_A where A is a subring of $\mathbb{R}$, the complex conjugation extends to a conjugation on the space $H_{\mathbb{C}} = H_A \otimes_A \mathbb{C}$. A filtration F on $H_{\mathbb{C}}$ has a conjugate filtration $\overline{F}$ such that $(\overline{F})^j H_{\mathbb{C}} = \overline{F^j H_{\mathbb{C}}}$.

DEFINITION 3.2.11 (HS3)   *An A-Hodge structure H of weight n consists of*

(i) an A-module of finite type H_A;

(ii) a finite filtration F on $H_{\mathbb{C}}$ (the Hodge filtration) such that F and its conjugate $\overline{F}$ satisfy the relation

$$\mathrm{Gr}_F^p \mathrm{Gr}_{\overline{F}}^q(H_{\mathbb{C}}) = 0, \quad \text{for } p + q \neq n;$$

equivalently F is opposite to its conjugate $\overline{F}$.

The HS is called real when $A = \mathbb{R}$, rational when $A = \mathbb{Q}$ and integral when $A = \mathbb{Z}$; then the module H_A or its image in $H_{\mathbb{Q}}$ is called the lattice.

3.2.1.11   Complex Hodge Structure

For some arguments in analysis, we do not need a real substructure.

DEFINITION 3.2.12   *A complex HS of weight n on a complex vector space H is given by a pair of n-opposite filtrations F and $\overline{F}$; hence a decomposition into a direct sum of subspaces*

$$H = \oplus_{p+q=n} H^{p,q}, \quad \text{where } H^{p,q} = F^p \cap \overline{F}^q.$$

The two n-opposite filtrations F and $\overline{F}$ on a complex HS of weight n can be recovered from the decomposition by the formula

$$F^p = \oplus_{i \geq p} H^{i,n-i}, \quad \overline{F}^q = \oplus_{i \leq n-q} H^{i,n-i}.$$

Here we do not assume the existence of conjugation although we keep the notation $\overline{F}$. An A-Hodge structure of weight n defines a complex HS on $H = H_{\mathbb{C}}$.

3.2.1.12   Polarized Complex Hodge Structure

To define polarization, we recall that the conjugate space $\overline{H}$ of a complex vector space H, is the same group H with a different complex structure. The identity map on the group H defines a real linear map $\sigma : H \to \overline{H}$ and the product by scalars satisfying the relation

$$\vee \lambda \in \mathbb{C}, \, v \in H : \lambda \times_{\overline{H}} o(v) := o(\overline{\lambda} \times_H v),$$

where $\lambda \times_{\overline{H}} \sigma(v)$ (resp. $(\overline{\lambda} \times_H v)$) is the product with a scalar in $\overline{H}$ (resp. H). Then, the complex structure on $\overline{H}$ is unique. On the other hand, a complex linear morphism $f : V \to V'$ defines a complex linear conjugate morphism $\overline{f} : \overline{V} \to \overline{V}'$ satisfying $\overline{f}(\sigma(v)) = \sigma(f(v))$.

DEFINITION 3.2.13   *A polarization of a complex HS of weight n is a bilinear morphism $S : H \otimes \overline{H} \to \mathbb{C}$ such that*

$$S(x, \sigma(y)) = (-1)^n \overline{S(y, \sigma(x))} \text{ for } x, y \in L \text{ and } S(F^p, \sigma(\overline{F}^q)) = 0 \text{ for } p+q > n.$$

Moreover, $S(C(H)u, \sigma(v))$ is a positive-definite Hermitian form on H where $C(H)$ denotes the Weil action on H ($C(H)u := i^{p-q}u$ for $u \in H^{p,q}$).

EXAMPLE 3.2.14   A complex HS of weight 0 on a complex vector space H is given by a decomposition into a direct sum of subspaces $H = \oplus_{p \in \mathbb{Z}} H^p$ with $F^p = \oplus_{i \geq p} H^p$ and $\overline{F}^q = \oplus_{i \leq -q} H^i$; hence $F^p \cap \overline{F}^{-p} = H^p$.

A polarization is a Hermitian form on H for which the decomposition is orthogonal and whose restriction to H^p is definite for p even and negative definite for odd p.

3.2.2   Mixed Hodge Structures (MHS)

Let $A = \mathbb{Z}, \mathbb{Q}$, or $\mathbb{R}$, and define $A \otimes \mathbb{Q}$ as $\mathbb{Q}$ if $A = \mathbb{Z}$ or $\mathbb{Q}$ and $\mathbb{R}$ if $A = \mathbb{R}$ (according to [8, III.0.3], we may suppose A to be a noetherian subring of $\mathbb{R}$ such that $A \otimes \mathbb{R}$ is a field). For an A-module H_A of finite type, we write $H_{A \otimes \mathbb{Q}}$ for the $(A \otimes \mathbb{Q})$-vector space $(H_A) \otimes_A (A \otimes \mathbb{Q})$. It is a rational space if $A = \mathbb{Z}$ or $\mathbb{Q}$ and a real space if $A = \mathbb{R}$.

DEFINITION 3.2.15 (MHS)   *An A-mixed Hodge structure H consists of*

(1) an A-module of finite type H_A;

(2) a finite increasing filtration W of the $A \otimes \mathbb{Q}$-vector space $H_{A \otimes \mathbb{Q}}$ called the weight filtration;

(3) a finite decreasing filtration F of the $\mathbb{C}$-vector space $H_{\mathbb{C}} = H_A \otimes_A \mathbb{C}$, called the Hodge filtration, and $W^{\mathbb{C}} := W \otimes \mathbb{C}$, such that the systems

$$\mathrm{Gr}_n^W H := (\mathrm{Gr}_n^W (H_{A \otimes \mathbb{Q}}), \mathrm{Gr}_n^W (H_{A \otimes \mathbb{Q}}) \otimes \mathbb{C} \simeq \mathrm{Gr}_n^{W^{\mathbb{C}}} H_{\mathbb{C}}, (\mathrm{Gr}_n^{W^{\mathbb{C}}} H_{\mathbb{C}}, F))$$

are $A \otimes \mathbb{Q}$-HS of weight n.

Recall the definition of the induced filtration

$$F^p \mathrm{Gr}_n^{W^{\mathbb{C}}} H_{\mathbb{C}} := ((F^p \cap W_n^{\mathbb{C}}) + W_{n-1}^{\mathbb{C}})/W_{n-1}^{\mathbb{C}} \subset W_n^{\mathbb{C}}/W_{n-1}^{\mathbb{C}}.$$

The MHS is called real if $A = \mathbb{R}$, rational if $A = \mathbb{Q}$, and integral if $A = \mathbb{Z}$.

3.2.2.1 Three Opposite Filtrations

Most of the proofs on the algebraic structure of MHS may be carried for three filtrations in an abelian category defined as follows:

DEFINITION 3.2.16 (Opposite filtrations)   *Three finite filtrations W (increasing), F, and G on an object A of $\mathbb{A}$ are opposite if*

$$\mathrm{Gr}_F^p \mathrm{Gr}_G^q \mathrm{Gr}_n^W (A) = 0, \quad \text{for } p + q \neq n.$$

This condition is symmetric in F and G. It means that F and G induce on the quotient $W_n(A)/W_{n-1}(A)$ two n-opposite filtrations; then $\mathrm{Gr}_n^W (A)$ is bigraded:

$$W_n(A)/W_{n-1}(A) = \oplus_{p+q=n} A^{p,q} \quad \text{where } A^{p,q} = \mathrm{Gr}_F^p \mathrm{Gr}_G^q \mathrm{Gr}_{p+q}^W (A).$$

EXAMPLE 3.2.17   (i) A bigraded object $A = \oplus A^{p,q}$ of finite bigrading has the following three opposite filtrations:

$$W_n = \oplus_{p+q \leq n} A^{p,q}, \quad F^p = \oplus_{p' \geq p} A^{p',q'}, \quad G^q = \oplus_{q' \geq q} A^{p',q'}.$$

(ii) In the definition of an A-MHS, the filtration $W_{\mathbb{C}}$ on $H_{\mathbb{C}}$ obtained from W by scalar extension, the filtration F, and its complex conjugate, form a system of three opposite filtrations $(W_{\mathbb{C}}, F, \overline{F})$.

3.2.2.2 Morphism of Mixed Hodge Structures

A morphism $f : H \to H'$ of MHS is a morphism $f : H_A \to H'_A$ whose extension to $H_{\mathbb{Q}}$ (resp. $H_{\mathbb{C}}$) is compatible with the filtration W, i.e., $f(W_j H_A) \subset W_j H'_A$ (resp. F, i.e., $f(F^j H_A) \subset F^j H'_A$), which implies that it is also compatible with $\overline{F}$.

These definitions allow us to speak of the category of MHS. The main result of this section is the following theorem:

THEOREM 3.2.18 (Deligne)   *The category of mixed Hodge structures is abelian.*

The proof relies on the following decomposition.

3.2.2.3 Canonical Decomposition of the Weight Filtration

While there is an equivalence for an HS between the Hodge filtration and the Hodge decomposition, there is no such result for the weight filtration of an MHS. In the category of MHS, the short exact sequence $0 \to \mathrm{Gr}_{n-1}^{W} \to W_n/W_{n-2} \to \mathrm{Gr}_n^{W} \to 0$ is a nonsplit extension of the two Hodge structures Gr_n^{W} and Gr_{n-1}^{W}.

The construction of the decomposition relies on the existence, for each pair of integers (p, q), of the following canonical subspaces of $H_{\mathbb{C}}$:

$$I^{p,q} = (F^p \cap W_{p+q}) \cap (\overline{F^q} \cap W_{p+q} + \overline{F^{q-1}} \cap W_{p+q-2} + \overline{F^{q-2}} \cap W_{p+q-3} + \cdots) \quad (3.2.1)$$

By construction they are related for $p + q = n$ to the components $H^{p,q}$ of the Hodge decomposition $\mathrm{Gr}_n^{W} H \simeq \oplus_{n=p+q} H^{p,q}$:

PROPOSITION 3.2.19   *The projection, for each (p, q),*

$$\varphi : W_{p+q} \to \mathrm{Gr}_{p+q}^{W} H \simeq \oplus_{p'+q'=p+q} H^{p',q'},$$

induces an isomorphism $I^{p,q} \xrightarrow{\sim} H^{p,q}$. Moreover,

$$W_n = \oplus_{p+q \leq n} I^{p,q}, \qquad F^p = \oplus_{p' \geq p} I^{p',q'}.$$

Remark.   The proof by induction is based on the formula for $i > 0$,

$$F^{p_i} H \oplus \overline{F^{q_i}} H \simeq \mathrm{Gr}_{n-i}^{W} H \quad \forall p_i, q_i : p_i + q_i = n - i + 1,$$

used to construct for $i > 0$ a decreasing family (p_i, q_i), starting with $p_0 = p, q_0 = q$, $p + q = n$. We choose here a sequence of the type $(p_0, q_0) = (p, q)$, and for $i > 0$, $p_i = p, q_i = q - i + 1$, which explains the asymmetry in the formula defining $I^{p,q}$.

(i) In general $I^{p,q} \neq \overline{I^{q,p}}$; we have only $I^{p,q} \equiv \overline{I^{q,p}}$ modulo W_{p+q-2}.

(ii) A morphism of MHS is necessarily compatible with this decomposition, which is the main ingredient later to prove the strictness (see Section 3.2.1.3) with respect to W and F.

3.2.2.4 Proof of the Proposition

The restriction of φ to $I^{p,q}$ is an isomorphism:

(i) *Injectivity of φ.* Let $n = p + q$. We deduce from the formula modulo W_{n-1} for $I^{p,q}$ that $\varphi(I^{p,q}) \subset H^{p,q} = (F^p \cap \overline{F^q})(\mathrm{Gr}_n^{W} H)$. Let $v \in I^{p,q}$ such that $\varphi(v) = 0$; then $v \in F^p \cap W_{n-1}$ and, since the class $\mathrm{cl}(v) \in (F^p \cap \overline{F^q})(\mathrm{Gr}_{n-1}^{W} H) = 0$ as $p + q > n - 1$, $\mathrm{cl}(v)$ must vanish; so we deduce that $v \in F^p \cap W_{n-2}$. This is a step in an inductive argument based on the formula $F^p \oplus \overline{F^{q-r+1}} \simeq \mathrm{Gr}_{n-r}^{W} H$.

We want to prove $v \in F^p \cap W_{n-r}$ for all $r > 0$. We just proved this for $r = 2$. We write

$$v \in \overline{F^q} \cap W_n + \sum_{r-1 \geq i \geq 1} \overline{F^{q-i}} \cap W_{n-i-1} + \sum_{i > r-1 \geq 1} \overline{F^{q-i}} \cap W_{n-i-1}.$$

Since $W_{n-i-1} \subset W_{n-r-1}$ for $i > r - 1$, the right term is in W_{n-r-1}, and since $\overline{F}$ is decreasing, we deduce $v \in F^p \cap \overline{F^{q-r+1}} \cap W_{n-r}$ modulo W_{n-r-1}. As $(F^p \cap \overline{F^{q-r+1}}) \operatorname{Gr}_{n-r}^W H = 0$ for $r > 0$, the class $\operatorname{cl}(v) = 0 \in \operatorname{Gr}_{n-r}^W H$; then $v \in F^p \cap W_{n-r-1}$. Finally, as $W_{n-r} = 0$ for large r, we deduce $v = 0$.

(ii) *Surjectivity of φ.* Let $\alpha \in H^{p,q}$; there exists $v_0 \in F^p \cap W_n$ and $u_0 \in F^q \cap W_n$ such that $\varphi(v_0) = \alpha = \varphi(\overline{u}_0)$; hence $v_0 = \overline{u}_0 + w_0$ with $w_0 \in W_{n-1}$. Applying the formula $F^p \oplus \overline{F^q} \simeq \operatorname{Gr}_{n-1}^W H$, the class of w_0 decomposes as $\operatorname{cl}(w_0) = \operatorname{cl}(v') + \operatorname{cl}(\overline{u}')$ with $v' \in F^p \cap W_{n-1}$ and $u' \in F^q \cap W_{n-1}$; hence there exists $w_1 \in W_{n-2}$ such that $v_0 = \overline{u}_0 + v' + \overline{u}' + w_1$. Let $v_1 := v_0 - v'$ and $u_1 := u_0 + u'$; then

$$v_1 = \overline{u}_1 + w_1, \quad \text{where} \quad u_1 \in F^q \cap W_n, \; v_1 \in F^p \cap W_n, \; w_1 \in W_{n-2}.$$

By an inductive argument on k, we apply the formula $F^p \oplus \overline{F^{q-k+1}} \simeq \operatorname{Gr}_{n-k}^W H$ to find vectors v_k, u_k, w_k satisfying

$$v_k \in F^p \cap W_n, \; w_k \in W_{n-1-k}, \; \varphi(v_k) = \alpha, \; v_k = \overline{u}_k + w_k,$$

$$u_k \in F^q \cap W_n + F^{q-1} \cap W_{n-2} + F^{q-2} \cap W_{n-3} + \cdots + F^{q+1-k} \cap W_{n-k};$$

then, we decompose the class of w_k in $\operatorname{Gr}_{n-k-1}^W H$ in the inductive step as above. For large k, $W_{n-1-k} = 0$; hence we find $v_k = \overline{u}_k \in I^{p,q}$ and $\varphi(v_k) = \alpha$. Moreover $W_n = W_{n-1} \oplus (\oplus_{p+q=n} I^{p,q})$; hence, by induction on n, W_n is a direct sum of $I^{p,q}$ for $p + q \leq n$.

Next we suppose, by induction, the formula for F^p satisfied for all $v \in W_{n-1} \cap F^p$. The image of an element $v \in F^p \cap W_n$ in $\operatorname{Gr}_n^W H$ decomposes into Hodge components of type $(i, n - i)$ with $i \geq p$ since $v \in F^p \cap W_n$. Hence, the decomposition of v may be written as $v = v_1 + v_2$ with $v_1 \in \oplus_{i<p} I^{i,n-i}$ and $v_2 \in \oplus_{i \geq p} I^{i,n-i}$ with $v_1 \in W_{n-1}$ since its image vanishes in $\operatorname{Gr}_n^W H$. The formula for F^p follows by induction.

3.2.2.5   *Proof of the Theorem: Abelianness of the Category of MHS and Strictness*

A remarkable feature of MHS is that every morphism of MHS is necessarily strict for both filtrations, W and F, in consequence the category of MHS is abelian.

LEMMA 3.2.20   *The kernel (resp. cokernel) of a morphism f of mixed Hodge structure $H \to H'$ is a mixed Hodge structure K with underlying module K_A equal to the kernel (resp. cokernel) of $f_A : H_A \to H'_A$; moreover, $K_{A \otimes \mathbb{Q}}$ and $K_{A \otimes \mathbb{C}}$ are endowed with induced filtrations (resp. quotient filtrations) by W on $H_{A \otimes \mathbb{Q}}$ (resp. $H'_{A \otimes \mathbb{Q}}$) and F on $H_\mathbb{C}$ (resp. $H'_\mathbb{C}$).*

PROOF. A morphism compatible with the filtrations is necessarily compatible with the canonical decomposition of the MHS into $\oplus I^{p,q}$. It is enough to check the statement on $K_{\mathbb{C}}$ (this is why in what follows we drop the index $\mathbb{C}$ in the notation). We consider on $K = \mathrm{Ker}(f)$ the induced filtrations from H. The morphism $\mathrm{Gr}^W K \to \mathrm{Gr}^W H$ is injective, since it is injective on the corresponding terms $I^{p,q}$; moreover, the filtration F (resp. $\overline{F}$) of K induces on $\mathrm{Gr}^W K$ the inverse image of the filtration F (resp. $\overline{F}$) on $\mathrm{Gr}^W H$:

$$\mathrm{Gr}^W K = \oplus_{p,q} (\mathrm{Gr}^W K) \cap H^{p,q}(\mathrm{Gr}^W H),$$

where $H^{p,q}(\mathrm{Gr}^W K) = (\mathrm{Gr}^W K) \cap H^{p,q}(\mathrm{Gr}^W H)$. Hence the filtrations W, F on K define an MHS on K which is a kernel of f in the category of MHS. The statement on the cokernel follows by duality. □

We still need to prove that for a morphism f of MHS, the canonical morphism $\mathrm{Coim}(f) \to \mathrm{Im}(f)$ is an isomorphism of MHS. Since by the above lemma, $\mathrm{Coim}(f)$ and $\mathrm{Im}(f)$ are endowed with natural MHS, the result follows from the following statement:

A morphism of MHS which induces an isomorphism on the lattices, is an isomorphism of MHS.

COROLLARY 3.2.21   *(i) Each morphism $f : H \to H'$ is strictly compatible with the filtrations W on $H_{A \otimes \mathbb{Q}}$ and $H'_{A \otimes \mathbb{Q}}$ as well as the filtrations F on $H_{\mathbb{C}}$ and $H'_{\mathbb{C}}$. The induced morphism $\mathrm{Gr}_n^W(f) : \mathrm{Gr}_n^W H_{A \otimes \mathbb{Q}} \to \mathrm{Gr}_n^W H'_{A \otimes \mathbb{Q}}$ is compatible with the $A \otimes \mathbb{Q}$ HS, and the induced morphism $\mathrm{Gr}_F^p(f) : \mathrm{Gr}_F^p(H_{\mathbb{C}}) \to \mathrm{Gr}_F^p(H'_{\mathbb{C}})$ is strictly compatible with the induced filtration by $W^{\mathbb{C}}$.*

(ii) The functor Gr_n^W from the category of MHS to the category $A \otimes \mathbb{Q}$ HS of weight n is exact and the functor Gr_F^p is also exact.

Remark. The above result shows that any exact sequence of MHS gives rise to various exact sequences which have, in the case of MHS on cohomology of algebraic varieties that we are going to construct, interesting geometrical interpretation. In fact, we deduce from each long exact sequence of MHS

$$H'^i \to H^i \to H''^i \to H'^{i+1},$$

various exact sequences

$$\mathrm{Gr}_n^W H'^i \to \mathrm{Gr}_n^W H^i \to \mathrm{Gr}_n^W H''^i \to \mathrm{Gr}_n^W H'^{i+1}$$

for $\mathbb{Q}$ or $\mathbb{C}$ coefficients, and similarly exact sequences

$$\mathrm{Gr}_F^n H'^i \to \mathrm{Gr}_F^n H^i \to \mathrm{Gr}_F^n H''^i \to \mathrm{Gr}_F^n H'^{i+1},$$

$$\mathrm{Gr}_F^m \mathrm{Gr}_n^W H'^i \to \mathrm{Gr}_F^m \mathrm{Gr}_n^W H^i \to \mathrm{Gr}_F^m \mathrm{Gr}_n^W H''^i \to \mathrm{Gr}_F^m \mathrm{Gr}_n^W H'^{i+1}.$$

3.2.2.6 Hodge Numbers

Let H be an MHS; set

$$H^{pq} = \operatorname{Gr}_F^p \operatorname{Gr}_{\overline{F}}^q \operatorname{Gr}_{p+q}^W H_{\mathbb{C}} = (\operatorname{Gr}_{p+q}^W H_{\mathbb{C}})^{p,q}.$$

The Hodge numbers of H are the Hodge numbers of the Hodge structure on $\operatorname{Gr}_{p+q}^W H$, that is, the set of integers $h^{pq} = \dim_{\mathbb{C}} H^{pq}$.

In fact, the proof by Deligne is in terms of opposite filtrations in an abelian category:

THEOREM 3.2.22 (Deligne [7]) *Let $\mathbb{A}$ be an abelian category and $\mathbb{A}'$ the category of objects of $\mathbb{A}$ endowed with three opposite filtrations W (increasing), F and G. The morphisms of $\mathbb{A}'$ are morphisms in $\mathbb{A}$ compatible with the three filtrations. Then*

  (i) $\mathbb{A}'$ is an abelian category;

  (ii) the kernel (resp. cokernel) of a morphism $f : A \to B$ in $\mathbb{A}'$ is the kernel (resp. cokernel) of f in $\mathbb{A}$, endowed with the three induced filtrations from A (resp. quotient of the filtrations on B);

  (iii) any morphism $f : A \to B$ in $\mathbb{A}'$ is strictly compatible with the filtrations $W, F,$ and G; the morphism $\operatorname{Gr}^W(f)$ is compatible with the bigradings of $\operatorname{Gr}^W(A)$ and $\operatorname{Gr}^W(B)$; the morphisms $\operatorname{Gr}_F(f)$ and $\operatorname{Gr}_G(f)$ are strictly compatible with the induced filtration by W;

  (iv) the forget-filtration functors, as well as the following functors from $\mathbb{A}'$ to $\mathbb{A}$, are exact: $\operatorname{Gr}^W, \operatorname{Gr}_F, \operatorname{Gr}_G, \operatorname{Gr}^W \operatorname{Gr}_F \simeq \operatorname{Gr}_F \operatorname{Gr}^W, \operatorname{Gr}_F \operatorname{Gr}_G \operatorname{Gr}^W, \text{ and } \operatorname{Gr}_G \operatorname{Gr}^W \simeq \operatorname{Gr}^W \operatorname{Gr}_G.$

EXAMPLE 3.2.23 (1) A Hodge structure H of weight n is an MHS with a trivial weight filtration

$$W_i(H_{\mathbb{Q}}) = 0 \quad \text{for} \quad i < n \quad \text{and} \quad W_i(H_{\mathbb{Q}}) = H_{\mathbb{Q}} \quad \text{for} \quad i \geq n;$$

then, it is called pure of weight n.

(2) Let (H^i, F_i) be a finite family of A-HS of weight $i \in \mathbb{Z}$; then $H = \oplus_i H^i$ is endowed with the following MHS:

$$H_A = \oplus_i H_A^i, \quad W_n = \oplus_{i \leq n} H_A^i \otimes \mathbb{Q}, \quad F^p = \oplus_i F_i^p.$$

(3) Let $H_{\mathbb{Z}} = i\mathbb{Z}^n \subset \mathbb{C}^n$; then we consider the isomorphism $H_{\mathbb{Z}} \otimes \mathbb{C} \simeq \mathbb{C}^n$ defined with respect to the canonical basis e_j of $\mathbb{Z}^n$ by

$$H_{\mathbb{C}} \xrightarrow{\sim} \mathbb{C}^n : ie_j \otimes (a_j + ib_j) \mapsto i(a_j + ib_j)e_j = (-b_j + ia_j)e_j;$$

hence, the conjugation $\sigma(ie_j \otimes (a_j + ib_j)) = ie_j \otimes (a_j - ib_j)$ on $H_{\mathbb{C}}$ corresponds to the following conjugation on $\mathbb{C}^n$: $\sigma(-b_j + ia_j)e_j = (b_j + ia_j)e_j$.

(4) Let $H = (H_{\mathbb{Z}}, F, W)$ be an MHS; its m-twist is an MHS denoted by $H(m)$ and defined by $H(m)_{\mathbb{Z}} := H_{\mathbb{Z}} \otimes (2i\pi)^m \mathbb{Z}$, $W_r H(m) := (W_{r+2m} H_{\mathbb{Q}}) \otimes (2i\pi)^m \mathbb{Q}$, and $F^r H(m) := F^{r+m} H_{\mathbb{C}}$.

3.2.2.7 Tensor Product and Hom

Let H and H' be two MHS.

(1) The MHS tensor product $H \otimes H'$ is defined by applying the general rules of filtrations:

   (i) $(H \otimes H')_{\mathbb{Z}} = H_{\mathbb{Z}} \otimes H'_{\mathbb{Z}}$.

   (ii) $W_r(H \otimes H')_{\mathbb{Q}} := \sum_{p+p'=r} W_p H_{\mathbb{Q}} \otimes W_{p'} H'_{\mathbb{Q}}$.

   (iii) $F^r(H \otimes H')_{\mathbb{C}} := \sum_{p+p'=r} F^p H_{\mathbb{C}} \otimes F^{p'} H'_{\mathbb{C}}$.

(2) The MHS $\mathrm{Hom}(H, H')$, called the internal Hom (to distinguish it from the group of morphisms of the two MHS), is defined as follows:

   (i) $\mathrm{Hom}(H, H')_{\mathbb{Z}} := \mathrm{Hom}_{\mathbb{Z}}(H_{\mathbb{Z}}, H'_{\mathbb{Z}})$.

   (ii) $W_r \mathrm{Hom}(H, H')_{\mathbb{Q}} := \{f : \mathrm{Hom}_{\mathbb{Q}}(H_{\mathbb{Q}}, H'_{\mathbb{Q}}) : \forall n, f(W_n H) \subset W_{n+2r} H'\}$.

   (iii) $F^r \mathrm{Hom}(H, H')_{\mathbb{C}} := \{f : \mathrm{Hom}_{\mathbb{C}}(H_{\mathbb{C}}, H'_{\mathbb{C}}) : \forall n, f(F^n H) \subset F^{n+r} H'\}$.

In particular, the dual H^* of a mixed Hodge structure H is an MHS.

3.2.2.8 Complex Mixed Hodge Structures

Although the cohomology of algebraic varieties carries an MHS defined over $\mathbb{Z}$, we may need to work in analysis without the $\mathbb{Z}$-lattice.

DEFINITION 3.2.24 *A complex mixed Hodge structure shifted by n on a complex vector space H is given by an increasing filtration W and two decreasing filtrations F and G such that $(\mathrm{Gr}^W_k H, F, G)$, with the induced filtrations, is a complex HS of weight $n + k$.*

This complex MHS shifted by n is sometimes called of weight n. For $n = 0$, we recover the definition of a complex MHS.

3.2.2.9 Variation of Complex Mixed Hodge Structures

The structure which appears in deformation theory on the cohomology of the fibers of a morphism of algebraic varieties leads to the introduction of the concept of variation of MHS.

DEFINITION 3.2.25 *(i) A variation of complex Hodge structures (VHS) on a complex manifold X of weight n is given by the data $(\mathcal{H}, F, \overline{F})$, where $\mathcal{H}$ is a complex local system, F (resp. $\overline{F}$) is a decreasing filtration by subbundles of the vector bundle $\mathcal{O}_X \otimes_{\mathbb{C}} \mathcal{H}$ varying holomorphically (resp. $\mathcal{O}_{\overline{X}} \otimes_{\mathbb{C}} \mathcal{H}$ on the conjugate variety $\overline{X}$ with antiholomorphic structural sheaf, i.e., varying antiholomorphically on X) such that for each point $x \in X$, the data $(\mathcal{H}(x), F(x), \overline{F}(x))$ form a Hodge structure of weight n. Moreover, the connection ∇ defined by the local* *system satisfies Griffiths transversality: for tangent vectors v holomorphic and $\bar{u}$ antiholomorphic,*

$$(\nabla_v F^p) \subset F^{p-1}, \quad (\nabla_{\bar{u}} \overline{F}^p) \subset \overline{F}^{p-1}.$$

(ii) A variation of complex mixed Hodge structures (VMHS) of weight n on X is given by the data

$$(\mathcal{H}, W, F, \overline{F}),$$

where $\mathcal{H}$ is a complex local system, W is an increasing filtration by sublocal systems, F (resp. $\overline{F}$) is a decreasing filtration varying holomorphically (resp. antiholomorphically) satisfying Griffiths transversality:

$$(\nabla_v F^p) \subset F^{p-1}, \quad (\nabla_{\bar{u}} \overline{F}^p) \subset \overline{F}^{p-1}$$

such that $(\mathrm{Gr}_k^W \mathcal{H}, F, \overline{F})$, with the induced filtrations, is a complex VHS of weight $n + k$.

For $n = 0$ we just say "a complex variation of MHS." Let $\overline{\mathcal{H}}$ be the conjugate local system of $\mathcal{L}$. A linear morphism $S : \mathcal{H} \otimes_{\mathbb{C}} \overline{\mathcal{H}} \to \mathbb{C}_X$ defines a polarization of a VHS if it defines a polarization at each point $x \in X$. A complex MHS shifted by n is graded polarizable if $(\mathrm{Gr}_k^W \mathcal{H}, F, \overline{F})$ is a polarized variation Hodge structure. For a study of the degeneration of variations of MHS, see Chapters 7 and 8.

3.2.3 Induced Filtrations on Spectral Sequences

To construct an MHS, we start from a bifiltered complex (K, F, W) satisfying conditions which will be introduced later under the terminology of mixed Hodge complex.

The fact that the two filtrations induce an MHS on the cohomology of the complex is based on a delicate study of the induced filtration by F on the spectral sequence defined by W. This study by Deligne, known as the two filtrations lemma, is presented here.

To explain the difficulty, imagine for a moment that we want to give a proof by induction on the length of W. Suppose that the weights of a mixed Hodge complex (K, W, F) vary from $W_0 = 0$ to $W_l = K$ and suppose we did construct the mixed Hodge structure on the cohomology of W_{l-1}; then we consider the long exact sequence of cohomology

$$H^{i-1}(\mathrm{Gr}_l^W K) \to H^i(W_{l-1}K) \to H^i(W_l K) \to H^i(\mathrm{Gr}_l^W K) \to H^{i+1}(W_{l-1}K).$$

The result would imply that the morphisms of the sequence are strict; hence the difficulty is a question of relative positions of the subspaces W_p and F^q on $H^i(W_l K)$ with respect to $\mathrm{Im}\, H^i(W_{l-1}K)$ and the projection on $H^i(\mathrm{Gr}_l^W K)$.

3.2.3.1 Deligne's Two Filtrations Lemma

This section relates results on various induced filtrations on terms of a spectral sequence, contained in [7, 8]. Let (K, F, W) be a bifiltered complex of objects of an abelian category. What we have in mind is to find axioms to define an MHS with induced filtrations W and F on the cohomology of K. The filtration W by subcomplexes defines a spectral sequence $E_r(K, W)$. The second filtration F induces filtrations on the terms of $E_r(K, W)$ in three different ways. A detailed study will show that these filtrations coincide under adequate hypotheses.

Later, the induced filtration will have an interpretation as a Hodge structure on the terms of the spectral sequence under a suitable hypothesis on the bifiltered complex. Since the proof is technical but difficult, we emphasize here the main ideas as a guide to Deligne's proof.

3.2.3.2 Direct and Recurrent Filtrations

Let (K, F, W) be a bifiltered complex of objects of an abelian category, bounded below. The filtration F, assumed to be biregular, induces on the terms $E_r^{p,q}$ of the spectral sequence $E(K, W)$ various filtrations as follows:

DEFINITION 3.2.26 (Direct filtrations: F_d, F_{d^*}) *Let $(E_r(K, W), d_r)$ denote the graded complex consisting of the terms $E_r^{p,q}$. The first direct filtration on $E_r(K, W)$ is the filtration F_d defined for r finite or $r = \infty$, by the image*

$$F_d^p(E_r(K, W)) = \mathrm{Im}(E_r(F^p K, W) \to E_r(K, W)).$$

Dually, the second direct filtration F_{d^*} on $E_r(K, W)$ is defined by the kernel

$$F_{d^*}^p(E_r(K, W)) = \mathrm{Ker}(E_r(K, W) \to E_r(K/F^p K, W)).$$

The filtrations F_d, F_{d^*} are naturally induced by F, hence compatible with the differentials d_r.

They coincide on $E_r^{p,q}$ for $r = 0, 1$, since $B_r^{p,q} \subset Z_r^{p,q}$.

LEMMA 3.2.27 $F_d = F_{d^*}$ on $E_0^{p,q} = \mathrm{Gr}_F^p(K^{p+q})$ and on $E_1^{p,q} = H^{p+q}(\mathrm{Gr}_p^W K)$.

DEFINITION 3.2.28 (Recurrent filtration: F_{rec}) *The recurrent filtration F_{rec} on $E_r^{p,q}$ is defined by induction on r as follows:*

(i) *On $E_0^{p,q}$, $F_{\mathrm{rec}} = F_d = F_{d^*}$.*

(ii) *The recurrent filtration F_{rec} on $E_r^{p,q}$ induces a filtration on $\mathrm{Ker}\, d_r$, which induces in its turn the recurrent filtration F_{rec} on $E_{r+1}^{p,q}$ as a quotient of $\mathrm{Ker}\, d_r$.*

3.2.3.3 Comparison of $F_d, F_{\mathrm{rec}}, F_{d^*}$

The preceding definitions of direct filtrations apply to $E_\infty^{p,q}$ as well. They are compatible with the isomorphism $E_r^{p,q} \simeq E_\infty^{p,q}$ for large r, from which we deduce also a recurrent filtration F_{rec} on $E_\infty^{p,q}$. The filtrations F and W induce each a filtration on $H^{p+q}(K)$. We want to prove that the isomorphism $E_\infty^{p,q} \simeq \text{Gr}_W^{-p} H^{p+q}(K)$ is compatible with F_{rec} on $E_\infty^{p,q}$ and F on the right-hand term.

In general we have only the following inclusions:

PROPOSITION 3.2.29   *(i) On $E_r^{p,q}$, we have the inclusions*

$$F_d(E_r^{p,q}) \subset F_{\text{rec}}(E_r^{p,q}) \subset F_{d^*}(E_r^{p,q}).$$

(ii) On $E_\infty^{p,q}$, the filtration induced by the filtration F on $H^*(K)$ satisfies

$$F_d(E_\infty^{p,q}) \subset F(E_\infty^{p,q}) \subset F_{d^*}(E_\infty^{p,q}).$$

(iii) The differential d_r is compatible with F_d and F_{d^*}.

We want to introduce conditions on the bifiltered complex in order that these three filtrations coincide. For this we need to know the compatibility of d_r with F_{rec}. Deligne proves an intermediary statement.

THEOREM 3.2.30 (Two filtrations (Deligne [7, 1.3.16], [8, 7.2]))   *Let K be a complex with two filtrations W and F. We suppose W biregular and for a fixed integer $r_0 \geq 0$:*
$(*r_0)$ For each nonnegative integer $r < r_0$, the differentials d_r of the graded complex $E_r(K, W)$ are strictly compatible with F_{rec}.
    Then, we have the following results:

(i) For $r \leq r_0$ the sequence

$$0 \to E_r(F^p K, W) \to E_r(K, W) \to E_r(K/F^p K, W) \to 0$$

is exact, and for $r = r_0 + 1$, the sequence

$$E_r(F^p K, W) \to E_r(K, W) \to E_r(K/F^p K, W)$$

is exact. In particular, for $r \leq r_0 + 1$, the two direct and the recurrent filtration on $E_r(K, W)$ coincide: $F_d = F_{\text{rec}} = F_{d^*}$.

(ii) For $a < b$ and $r < r_0$, the differentials d_r of the graded complex $E_r(F^a K/F^b K, W)$ are strictly compatible with F_{rec}.

(iii) If the above condition $(*r_0)$ is satisfied for all r_0, then the filtrations $F_d, F_{\text{rec}}, F_{d^*}$ agree and the isomorphism $E_\infty^{p,q} \simeq \text{Gr}_W^{-p} H^{p+q}(K)$ is compatible with the induced common filtration F on the right.

Moreover, we have an isomorphism of spectral sequences:

$$\text{Gr}_F^p E_r(K, W) \simeq E_r(\text{Gr}_F^p K, W)$$

and the spectral sequence $E(K, F)$ (with respect to F) degenerates at rank 1 ($E_1 = E_\infty$).

PROOF. This surprising statement looks natural only if we have in mind the degeneration of $E(K, F)$ at rank 1 and the strictness in the category of mixed Hodge structures.

For fixed p, we consider the following property:

(P_r) $E_i(F^p K, W)$ injects into $E_i(K, W)$ for $i \leq r$ and its image is F_{rec}^p for $i \leq r+1$.

We already noted that (P_0) is satisfied. The proof by induction on r will apply as long as r remains $\leq r_0$. Suppose $r < r_0$ and (P_s) is true for all $s \leq r$; we prove (P_{r+1}). The sequence

$$E_r(F^p K, W) \xrightarrow{d_r} E_r(F^p K, W) \xrightarrow{d_r} E_r(F^p K, W)$$

injects into

$$E_r(K, W) \xrightarrow{d_r} E_r(K, W) \xrightarrow{d_r} E_r(K, W)$$

with image $F_d = F_{\mathrm{rec}}$; then the image of F_{rec}^p in E_{r+1},

$$F_{\mathrm{rec}}^p E_{r+1} = \mathrm{Im}[\mathrm{Ker}(F_{\mathrm{rec}}^p E_r(K, W) \xrightarrow{d_r} E_r(K, W)) \to E_{r+1}(K, W)],$$

coincides with the image of F_d^p, which is by definition

$$\mathrm{Im}[E_{r+1}(F^p K, W) \to E_{r+1}(K, W)].$$

Since d_r is strictly compatible with F_{rec}, we have

$$d_r E_r(K, W) \cap E_r(F^p K, W) = d_r E_r(F^p K, W),$$

which means that $E_{r+1}(F^p K, W)$ injects into $E_{r+1}(K, W)$; hence we deduce the injectivity for $r + 1$. Since $\mathrm{Ker}\, d_r$ on F_{rec}^p is equal to $\mathrm{Ker}\, d_r$ on $E_{r+1}(F^p K, W)$, we deduce $F_{\mathrm{rec}}^p = F_d^p$ on $E_{r+2}(K, W)$, which proves (P_{r+1}).

Then, (i) follows from a dual statement applied to F_{d*} and (ii) follows, because we have an exact sequence

$$0 \to E_r(F^b K, W) \to E_r(F^a K, W) \to E_r(F^a K/F^b K, W) \to 0.$$

(iii) We deduce the next exact sequence and its dual from (i) and (ii):

$$0 \to E_r(F^{p+1} K, W) \to E_r(F^p K, W) \to E_r(\mathrm{Gr}_F^p K, W) \to 0,$$
$$0 \leftarrow E_r(K/F^p K, W) \leftarrow E_r(F^{p+1} K, W) \leftarrow E_r(\mathrm{Gr}_F^p K, W) \leftarrow 0.$$

In view of the injections in (i) and the coincidence of $F_d = F_{\mathrm{rec}} = F_{d*}$ we have a unique filtration F; the quotient of the first two terms in the first exact sequence is isomorphic to $\mathrm{Gr}_F^p E_r(K, W)$, hence we deduce an isomorphism

$$\mathrm{Gr}_F^p E_r(K, W) \simeq E_r(\mathrm{Gr}_F^p K, W)$$

compatible with d_r and autodual. If the hypothesis is now true for all r, we deduce an exact sequence

$$0 \to E_\infty(F^p K, W) \to E_\infty(K, W) \to E_\infty(K/F^p K, W) \to 0,$$

which is identical to

$$0 \to \mathrm{Gr}_W H^*(F^p K) \to \mathrm{Gr}_W H^*(K) \to \mathrm{Gr}_W H^*(K/F^p K) \to 0,$$

from which we deduce, for all i,

$$0 \to H^i(F^p K) \to H^i(K) \to H^i(K/F^p K) \to 0.$$

Hence the spectral sequence with respect to F degenerates at rank 1 and the filtrations W induced on $H^i(F^p K)$ from $(F^p K, W)$ and from $(H^i(K), W)$ coincide. □

A category of complexes called mixed Hodge complexes satisfying the condition $(*r_0)$ for all r_0 will be introduced later.

The condition $(*r_0)$ apply inductively for a category of complexes called mixed Hodge complexes which will be introduced later. It is applied in the next case as an example.

3.2.4   MHS of a Normal Crossing Divisor (NCD)

An algebraic subvariety Y of a complex smooth algebraic variety is called a normal crossing divisor (NCD) if at each point $y \in Y$, there exists a neighborhood for the transcendental topology U_y and coordinates $z = (z_1, \ldots, z_n) : U_y \to D^n$ to a product of the complex disk such that the image of $Y \cap U_y$ is defined by the equation $f(z) = z_1 \cdots z_p = 0$ for some $p \leq n$. We consider a closed algebraic subvariety Y with NCD in a compact complex smooth algebraic variety X such that its irreducible components $(Y_i)_{i \in I}$ are smooth, and we put an order on the set of indices I of the components of Y.

3.2.4.1   Mayer–Vietoris Resolution

The singular subset of Y is defined locally as the subset of points $\{z : df(z) = 0\}$. Let S_q denote the set of strictly increasing sequences $\sigma = (\sigma_0, \ldots, \sigma_q)$ on the ordered set of indices I, $Y_\sigma = Y_{\sigma_0} \cap \cdots \cap Y_{\sigma_q}$, $Y_{\underline{q}} = \coprod_{\sigma \in S_q} Y_\sigma$ is the disjoint union, and for all $j \in [0, q], q \geq 1$, let $\lambda_{j,\underline{q}} : Y_{\underline{q}} \to Y_{\underline{q-1}}$ denote the map inducing for each σ the embedding $\lambda_{j,\sigma} : Y_\sigma \to Y_{\sigma(\widehat{j})}$, where $\sigma(\widehat{j}) = (\sigma_0, \ldots, \widehat{\sigma_j}, \ldots, \sigma_q)$ is obtained by deleting σ_j. Let $\Pi_q : Y_{\underline{q}} \to Y$ denote the canonical projection and $\lambda_{j,\underline{q}}^* : \Pi_* \mathbb{Z}_{Y_{\underline{q-1}}} \to \Pi_* \mathbb{Z}_{Y_{\underline{q}}}$ the restriction map defined by $\lambda_{j,\underline{q}}$ for $j \in [0, q]$. The various images of Π_q (or simply Π) define a natural stratification on X of dimension n:

$$X \supset Y = \Pi(Y_{\underline{0}}) \supset \cdots \supset \Pi(Y_{\underline{q}}) \supset \Pi(Y_{\underline{q+1}}) \supset \cdots \supset \Pi(Y_{\underline{n-1}}) \supset \emptyset$$

with smooth strata formed by the connected components of $\Pi_q(Y_{\underline{q}}) - \Pi_{q+1}(Y_{\underline{q+1}})$ of dimension $n - q - 1$.

LEMMA AND DEFINITION 3.2.31 (Mayer–Vietoris resolution of $\mathbb{Z}_Y$)   *The canonical morphism $\mathbb{Z}_Y \to \Pi_* \mathbb{Z}_{Y_{\underline{0}}}$ defines a quasi-isomorphism with the following complex of sheaves $\Pi_* \mathbb{Z}_{Y_\bullet}$:*

$$0 \to \Pi_* \mathbb{Z}_{Y_{\underline{0}}} \to \Pi_* \mathbb{Z}_{Y_{\underline{1}}} \to \cdots \to \Pi_* \mathbb{Z}_{Y_{\underline{q-1}}} \overset{\delta_{q-1}}{\to} \Pi_* \mathbb{Z}_{Y_{\underline{q}}} \to \cdots,$$

where $\delta_{q-1} = \sum_{j \in [0,q]} (-1)^j \lambda_{j,q}^*.$

This resolution is associated to a hypercovering of Y by topological spaces in the following sense. Consider the diagram of spaces over Y:

$$Y_\bullet = \left(Y_{\underline{0}} \;\overset{\leftarrow}{\underset{\leftarrow}{\overset{\leftarrow}{}}}\; Y_{\underline{1}} \;\overset{\leftarrow}{\leftarrow}\; \cdots \quad Y_{\underline{q-1}} \;\overset{\overset{\lambda_{j,q}}{\leftarrow}}{\underset{\longleftarrow}{\vdots}}\; Y_{\underline{q}} \cdots \right) \overset{\Pi}{\to} Y.$$

This diagram is the strict simplicial scheme associated in [7] to the normal crossing divisor Y, called hereafter Mayer–Vietoris. The Mayer–Vietoris complex is canonically associated as direct image by Π of the sheaf $\mathbb{Z}_{Y_\bullet}$ equal to $\mathbb{Z}_{Y_{\underline{i}}}$ on $Y_{\underline{i}}$. The generalization of such a resolution is the basis of the later construction of mixed Hodge structure using simplicial covering of an algebraic variety.

3.2.4.2 *The Cohomological Mixed Hodge Complex of an NCD*

The weight filtration W on $\Pi_* \mathbb{Q}_{Y_\bullet}$ (it will define the weight of an MHS on the hypercohomology) is defined by

$$W_{-q}(\Pi_* \mathbb{Q}_{Y_\bullet}) = \sigma_{\bullet \geq q} \Pi_* \mathbb{Q}_{Y_\bullet} = \Pi_* \sigma_{\bullet \geq q} \mathbb{Q}_{Y_\bullet}, \quad \mathrm{Gr}_{-q}^W (\Pi_* \mathbb{Q}_{Y_\bullet}) \simeq \Pi_* \mathbb{Q}_{Y_{\underline{q}}}[-q].$$

To define the filtration F, we introduce the complexes $\Omega_{Y_{\underline{i}}}^*$ of differential forms on $Y_{\underline{i}}$. The simple complex $s(\Omega_{Y_\bullet}^*)$ (see Section 3.1.2.3) is associated to the double complex $\Pi_* \Omega_{Y_\bullet}^*$ with the exterior differential d of forms and the differential δ_* defined by $\delta_{q-1} = \sum_{j \in [0,q]} (-1)^j \lambda_{j,q}^*$ on $\Pi_* \Omega_{Y_{\underline{q-1}}}^*$. Then, the weight W and Hodge F filtrations are defined as

$$W_{-q} = s(\sigma_{\bullet \geq q} \Omega_{Y_\bullet}^*) = s(0 \to \cdots 0 \to \Pi_* \Omega_{Y_{\underline{q}}}^* \to \Pi_* \Omega_{Y_{\underline{q+1}}}^* \to \cdots),$$
$$F^p = s(\sigma_{* \geq p} \Omega_{Y_\bullet}^*) = s(0 \to \cdots 0 \to \Pi_* \Omega_{Y_\bullet}^p \to \Pi_* \Omega_{Y_\bullet}^{p+1} \to \cdots).$$

We have an isomorphism of complexes of sheaves of abelian groups on Y compatible with the filtration

$$(\mathrm{Gr}_{-q}^W s(\Omega_{Y_\bullet}^*), F) \simeq (\Pi_* \Omega_{Y_{\underline{q}}}^* [-q], F)$$

inducing quasi-isomorphisms

$$(\Pi_* \mathbb{Q}_{Y_\bullet}, W) \otimes \mathbb{C} = (\mathbb{C}_{Y_\bullet}, W) \overset{\approx}{\underset{\alpha}{\longrightarrow}} (s(\Omega_{Y_\bullet}^*), W),$$

$$\mathrm{Gr}_{-q}^W (\Pi_* \mathbb{Q}_{Y_\bullet}) \otimes \mathbb{C} = \mathrm{Gr}_{-q}^W (\Pi_* \mathbb{C}_{Y_\bullet}) \simeq \Pi_* \mathbb{C}_{Y_{\underline{q}}}[-q] \overset{\approx}{\underset{\alpha_q}{\longrightarrow}} \Pi_* \Omega_{Y_{\underline{q}}}^* [-q] \simeq \mathrm{Gr}_{-q}^W s(\Omega_{Y_\bullet}^*).$$

The above situation is a model of the future constructions which lead to the **MHS** on cohomology of any algebraic variety. It is summarized by the construction of a system of filtered complexes with compatible quasi-isomorphisms

$$\mathbb{K} = [\mathbb{K}_\mathbb{Z}; (\mathbb{K}_\mathbb{Q}, W_\mathbb{Q}), \mathbb{K}_\mathbb{Z} \otimes \mathbb{Q} \simeq \mathbb{K}_\mathbb{Q}; (\mathbb{K}_\mathbb{C}, W, F), (\mathbb{K}_\mathbb{Q}, W_\mathbb{Q}) \otimes \mathbb{C} \simeq (\mathbb{K}_\mathbb{C}, W)]$$

defined in our case by

$$\mathbb{Z}_Y, \ (\Pi_*\mathbb{Q}_{Y_\bullet}, W); \ \mathbb{Q}_Y \xrightarrow{\approx} \Pi_*\mathbb{Q}_{Y_\bullet}; \ (s(\Omega_{Y_\bullet}^*), W, F); \ (\Pi_*\mathbb{Q}_{Y_\bullet}, W) \otimes \mathbb{C} \xrightarrow{\approx} (s(\Omega_{Y_\bullet}^*), W).$$

We extract the characteristic property of the system that we need by the remark that $\mathrm{Gr}^W(\mathbb{K})$ with the induced filtration by F,

$$\mathrm{Gr}_{-q}^W(\Pi_*\mathbb{Q}_{Y_\bullet}), \quad \mathrm{Gr}_{-q}^W(\Pi_*\mathbb{Q}_{Y_\bullet}) \otimes \mathbb{C} \simeq \mathrm{Gr}_{-q}^W s(\Omega_{Y_\bullet}^*), \quad (\mathrm{Gr}_{-q}^W s(\Omega_{Y_\bullet}^*), F)$$

is a shifted cohomological Hodge complex. Indeed, this system is defined by the non-singular compact complex case of the various intersections $Y_{\underline{i}}$.

In terms of Dolbeault resolutions $(s(\mathcal{E}_{Y_\bullet}^{*,*}), W, F)$, the above conditions induce on the complex of global sections $\Gamma(Y, s(\mathcal{E}_{Y_\bullet}^{*,*}), W, F) := (\mathbb{R}\Gamma(Y, \mathbb{C}), W, F)$ a structure called a mixed Hodge complex in the sense that $\mathbb{R}\Gamma(Y, \mathrm{Gr}^W \mathbb{K})$ have the structure of Hodge complexes with shifted weights

$$(\mathrm{Gr}_{-i}^W(\mathbb{R}\Gamma(Y, \mathbb{C})), F) := (\Gamma(Y, W_{-i}s(\mathcal{E}_{Y_\bullet}^{*,*}))/\Gamma(Y, W_{-i-1}s(\mathcal{E}_{Y_\bullet}^{*,*})), F)$$
$$\simeq (\Gamma(Y, \mathrm{Gr}_{-i}^W s(\mathcal{E}_{Y_\bullet}^{*,*})), F) \simeq (\mathbb{R}\Gamma(Y_{\underline{i}}, \Omega_{Y_{\underline{i}}}^*[-i]), F)$$

The mth graded complex with respect to W has a structure called a Hodge complex of weight m in the sense that

$$(H^n(\mathrm{Gr}_{-i}^W \mathbb{R}\Gamma(Y, \mathbb{C})), F) \simeq (H^{n-i}(Y_{\underline{i}}, \mathbb{C}), F)$$

is an HS of weight $n - i$.

The terms of the spectral sequence $E_1(K, W)$ of (K, W) are written as

$$_W E_1^{p,q} = \mathbb{H}^{p+q}(Y, \mathrm{Gr}_{-p}^W(s\Omega_{Y_\bullet}^*)) \simeq \mathbb{H}^{p+q}(Y, \Pi_*\Omega_{Y_p}^*[-p]) \simeq H^q(Y_{\underline{p}}, \mathbb{C}).$$

They carry the HS of weight q on the cohomology of the space Y_p. The differential is a combinatorial restriction map inducing a morphism of Hodge structures

$$d_1 = \sum_{j \leq p+1} (-1)^j \lambda_{j,p+1}^* : H^q(Y_{\underline{p}}, \mathbb{C}) \to H^q(Y_{\underline{p+1}}, \mathbb{C}).$$

As morphisms of HS, the differentials are strict with respect to the filtration F, equal to F_{rec} (see Theorem 3.2.30). Hence, we can apply the theorem to deduce the condition $(*, 3)$ and obtain that the differential d_2 on the induced HS of weight q on the term $_W E_2^{p,q}$ is compatible with the HS, but since $d_2 : {}_W E_2^{p,q} \to {}_W E_2^{p+2,q-1}$ is a morphism of HS of different weights, it is strict and must vanish. Then, the argument applies inductively for $r \geq 2$ to show that the spectral sequence degenerates at rank 2: ($E_2 = E_\infty$). Finally, we deduce the following proposition:

PROPOSITION 3.2.32 *The system $\mathbb{K}$ associated to a normal crossing divisor Y, with smooth proper irreducible components, defines a mixed Hodge structure on the cohomology $H^i(Y, \mathbb{Q})$, with weights varying between 0 and i.*

COROLLARY 3.2.33 *The Hodge structure on* $\mathrm{Gr}_q^W H^{p+q}(Y, \mathbb{C})$ *is the cohomology of the complex of Hodge structures defined by* $(H^q(Y_\bullet, \mathbb{C}), d_1)$ *equal to* $H^q(Y_{\underline{p}}, \mathbb{C})$ *in degree* $p \geq 0$:

$$(\mathrm{Gr}_q^W H^{p+q}(Y, \mathbb{C}), F) \simeq (H^p(H^q(Y_\bullet, \mathbb{C}), d_1), F).$$

In particular, the weight of $H^i(Y, \mathbb{C})$ *varies in the interval* $[0, i]$: $\mathrm{Gr}_q^W H^i(Y, \mathbb{C}) = 0$ *for* $q \notin [0, i]$.

We will see that the last condition on the weight is true for all complete varieties.

3.3 MIXED HODGE COMPLEX

The construction of mixed Hodge structures on the cohomology of algebraic varieties is similar to the case of a normal crossing divisor. For each algebraic variety we need to construct a system of filtered complexes

$$\mathbb{K} = [\mathbb{K}_\mathbb{Z}; (\mathbb{K}_\mathbb{Q}, W_\mathbb{Q}), \mathbb{K}_\mathbb{Z} \otimes \mathbb{Q} \simeq \mathbb{K}_\mathbb{Q}; (\mathbb{K}_\mathbb{C}, W, F), (\mathbb{K}_\mathbb{Q}, W_\mathbb{Q}) \otimes \mathbb{C} \simeq (\mathbb{K}_\mathbb{C}, W)]$$

with a filtration W on the rational level and a filtration F on the complex level satisfying the following condition:

the cohomology groups $\mathbb{H}^j(\mathrm{Gr}_i^W(\mathbb{K}))$ *with the induced filtration* F *are HS of weight* $j + i$.

The two filtrations lemma (Section 3.2.3.1) on spectral sequences is used to prove that the jth cohomology $(H^j(K), W[j], F)$ of the bifiltered complex (K, W, F) carries a mixed Hodge structure (the weight filtration of the MHS is deduced from W by adding j to the index). Such a system is called a mixed Hodge complex (MHC).

The topological techniques used to construct W on the rational level are different from the geometrical techniques represented by the de Rham complex used to construct the filtration F on the complex level. Comparison morphisms between the rational and complex levels must be added in order to obtain a satisfactory functorial theory of mixed Hodge structures with respect to algebraic morphisms.

However, the comparison between the rational and the complex filtrations W may not be defined by a direct morphism of complexes as in the previous NCD case but by a diagram of morphisms of one of the types

$$(K_1, W_1) \xleftarrow[g_1]{\approx} (K_1', W_1') \xrightarrow{f_1} (K_2, W_2), \qquad (K_1, W_1) \xrightarrow{f_2} (K_2', W_2') \xleftarrow[g_2]{\approx} (K_2, W_2),$$

where g_1 and g_2 are filtered quasi-isomorphisms as, for example, in the case of the logarithmic complex in the next section.

This type of diagram of morphisms appears in the derived category of complexes of an abelian category constructed by Verdier [41, 42, 34, 7]. Defining the system $\mathbb{K}$ in a similar category called the filtered derived category ensures the correct identification of the cohomology with its filtrations defining an MHS independently of the choice of acyclic resolutions.

In such a category, a diagram of morphisms of filtered complexes induces a morphism of the corresponding spectral sequences, but the reciprocal statement is not true: the existence of a diagram of quasi-isomorphisms is stronger than the existence of an isomorphism of spectral sequences.

Derived filtered categories described below have been used extensively in the more recent theory of perverse sheaves (see [2]).

Finally, to put a mixed Hodge structure on the relative cohomology, we discuss the technique of the mixed cone which associates a new MHC to a morphism of an MHC. The morphism must be defined at the level of the category of complexes and not up to homotopy; the mixed Hodge structure obtained depends on the homotopy between various resolutions (see [14]).

3.3.1 Derived Category

The hypercohomology of a functor of abelian categories $F : \mathbb{A} \to \mathbb{B}$ on an abelian category $\mathbb{A}$ with enough injectives is defined at a complex K of objects of $\mathbb{A}$ by considering an injective, or in general acyclic (see below) resolution $I(K)$ of K [24]. It is the cohomology object $H^i(F(I(K)))$ of $\mathbb{B}$.

If $I'(K)$ is a distinct resolution, there exists a unique isomorphism

$$\phi_i(K) : H^i(F(I(K))) \to H^i(F(I'(K))).$$

Hence, we can choose an injective resolution and there is no ambiguity in the definition with respect to the choice.

We remark however that the complexes $F(I(K))$ and $F(I'(K))$ are not necessarily isomorphic. By taking the hypercohomology, the information on the complex is lost. The idea of Grothendieck is to construct a category where the various resolutions are isomorphic (not only their cohomologies are isomorphic).

Verdier gives the construction of such a category in [41] in two steps. In the first step he constructs the homotopy category where the morphisms are classes of morphisms of complexes defined up to homotopy (see [34, 27]), and in the second step, a process of inverting all quasi-isomorphisms called localization is carried out by a calculus of fractions similar to the process of inverting a multiplicative system in a ring, although in this case the system of quasi-isomorphisms is not commutative.

We give here the minimum needed to understand the mechanism in Deligne's definition of a mixed Hodge structure. A full account may be found in [41, 34, 27]. Recently, the formalism of the derived category has been fundamental in the study of perverse sheaves [2]. To check various statements given here without proof, we recommend [34].

3.3.1.1   The homotopy category $K(\mathbb{A})$

Let $\mathbb{A}$ be an abelian category and let $C(\mathbb{A})$ (resp. $C^+(\mathbb{A})$, $C^-(\mathbb{A})$, $C^b(\mathbb{A})$) denote the abelian category of complexes of objects in $\mathbb{A}$ (resp. complexes $X^\bullet$ satisfying $X^j = 0$ for $j << 0$, for $j >> 0$, both conditions, i.e., for j outside a finite interval).

A *homotopy* between two morphisms of complexes $f, g : X^\bullet \to Y^\bullet$ is a family of morphisms $h^j : X^j \to Y^{j-1}$ in $\mathbb{A}$ satisfying $f^j - g^j = d_Y^{j-1} \circ h^j + h^{j+1} \circ d_X^j$. Homotopy defines an equivalence relation on the additive group $\operatorname{Hom}_{C(\mathbb{A})}(X^\bullet, Y^\bullet)$.

DEFINITION 3.3.1 *The category $K(\mathbb{A})$ has the same object as the category of complexes $C(\mathbb{A})$, while the additive group of morphisms $\operatorname{Hom}_{K(\mathbb{A})}(X^\bullet, Y^\bullet)$ is the group of morphisms of the complexes of $\mathbb{A}$ modulo the homotopy equivalence relation.*

We define $K^+(\mathbb{A})$, $K^-(\mathbb{A})$, and $K^b(\mathbb{A})$ similarly.

3.3.1.2   Injective Resolutions

An abelian category $\mathbb{A}$ is said to have enough injectives if each object $A \in \mathbb{A}$ is embedded in an injective object of $\mathbb{A}$.

Any complex X of $\mathbb{A}$ bounded below is quasi-isomorphic to a complex of injective objects $I^\bullet(X)$ called its injective resolution ([34, Thm. 6.1]).

PROPOSITION 3.3.2 *Given a morphism $f : A_1 \to A_2$ in $C^+(\mathbb{A})$ and two injective resolutions $A_i \xrightarrow{\approx} I^\bullet(A_i)$ of A_i, there exists an extension of f as a morphism of resolutions $I^\bullet(f) : I^\bullet(A_1) \to I^\bullet(A_2)$; moreover, two extensions of f are homotopic.*

See [34, Secs. I.6 and I.7] or [27, Lem. 4.7]. Hence, an injective resolution of an object in $\mathbb{A}$ becomes unique up to a unique isomorphism in the category $K^+(\mathbb{A})$.

The category $K^+(\mathbb{A})$ is additive but not necessarily abelian, even if $\mathbb{A}$ is abelian. Although we keep the same objects of $C^+(\mathbb{A})$, the transformation on Hom makes an important difference since a homotopy equivalence between two complexes (i.e., $f : X \to Y$ and $g : Y \to X$ such that $g \circ f$ and $f \circ g$ are homotopic to the identity) becomes an isomorphism.

Remark. In the category $\mathbb{A}$ of abelian sheaves on a topological space V, the ith group of cohomology of a sheaf $\mathcal{F}$ is defined, up to an isomorphism, as the cohomology of the space of global sections of an injective resolution $H^i(I^\bullet(\mathcal{F})(V))$.

The complex of global sections $I^\bullet(\mathcal{F})(V)$ is defined, up to a homotopy, in the category of groups $C^+(\mathbb{Z})$. Hence, in the homotopy category of groups $K^+(\mathbb{Z})$, the complex $I^\bullet(\mathcal{F})(V)$ is defined, up to an isomorphism. It is called the higher direct image of $\mathcal{F}$ by the global section functor Γ and denoted $R\Gamma(V, \mathcal{F})$.

3.3.1.3   The Derived Category $D(\mathbb{A})$

For any two resolutions of a complex K, defined by quasi-isomorphisms $\phi_1 : K \xrightarrow{\approx} K_1$ and $\phi_2 : K \xrightarrow{\approx} K_2$, there exists a common injective resolution $\psi_1 : K_1 \xrightarrow{\approx} I^\bullet$ and $\psi_2 : K_2 \xrightarrow{\approx} I^\bullet$ inducing on K homotopic resolutions $\psi_i \circ \phi_i$, $i = 1, 2$. In this case, in classical homological algebra, the ith hypercohomology of a left exact functor F is defined by $H^i(F(I^\bullet))$.

The idea of Verdier is to construct in general, by inverting quasi-isomorphisms of complexes, a new category $D(\mathbb{A})$ where all quasi-isomorphisms are isomorphic, without any reference to injective resolutions. The category $D(\mathbb{A})$ has the same objects as

$K(\mathbb{A})$ but with a different additive group of morphisms. We describe now the additive group of two objects $\mathrm{Hom}_{D(\mathbb{A})}(X, Y)$.

Let I_Y denote the category whose objects are quasi-isomorphisms $s' : Y \xrightarrow{\approx} Y'$ in $K(\mathbb{A})$. Let $s'' : Y \xrightarrow{\approx} Y''$ be another object. A morphism $h : s' \to s''$ in I_Y is defined by a morphism $h : Y' \to Y''$ satisfying $h \circ s' = s''$. The key property is that we can take limits in $K(\mathbb{A})$; hence we define

$$\mathrm{Hom}_{D(\mathbb{A})}(X, Y) := \varinjlim_{I_Y} \mathrm{Hom}_{K(\mathbb{A})}(X, Y').$$

A morphism $f : X \to Y$ in $D(\mathbb{A})$ is represented in the inductive limit by a diagram of morphisms: $X \xrightarrow{f'} Y' \xleftarrow[s']{\approx} Y$ where s' is a quasi-isomorphism in $K(\mathbb{A})$. Two diagrams $X \xrightarrow{f'} Y' \xleftarrow[s']{\approx} Y$ and $X \xrightarrow{f''} Y'' \xleftarrow[s'']{\approx} Y$ represent the same morphism f if and only if there exists a diagram

$$
\begin{array}{ccccc}
 & & X & & \\
 & \swarrow f' & \downarrow & f'' \searrow & \\
Y' & \xrightarrow[u]{\approx} & Y''' & \xleftarrow[v]{\approx} & Y''
\end{array}
$$

such that $u \circ s' = v \circ s'' \in \mathrm{Hom}(Y, Y''')$ and $u \circ f' = v \circ f'' \in \mathrm{Hom}(X, Y''')$. In this case, the morphism f may be represented by a symbol $s'^{-1} \circ f'$ and this representation is not unique since in the above limit, $s'^{-1} \circ f' = s''^{-1} \circ f''$. The construction of $D^+(\mathbb{A})$ is similar.

When there are enough injectives, the Hom of two objects A_1, A_2 in $D^+(\mathbb{A})$ is defined by their injective resolutions:

COROLLARY 3.3.3 *With the notation of the above proposition,*

$$\mathrm{Hom}_{D^+(\mathbb{A})}(A_1, A_2) \simeq \mathrm{Hom}_{D^+(\mathbb{A})}(I^\bullet(A_1), I^\bullet(A_2)) \simeq \mathrm{Hom}_{K^+(\mathbb{A})}(I^\bullet(A_1), I^\bullet(A_2)).$$

See [41, Ch. II, Sec. 2, Thm. 2.2, p. 304]. This is [27, Prop. 4.7, based on Lems. 4.4, 4.5]; see also [34, Sec. I.6]. In particular, all resolutions of a complex are isomorphic in the derived category.

Remark. We can equivalently consider the category J_X whose objects are quasi-isomorphisms $s' : X' \xrightarrow{\approx} X$ in $K(\mathbb{A})$ and define

$$\mathrm{Hom}_{D(\mathbb{A})}(X, Y) := \varinjlim_{J_X} \mathrm{Hom}_{K(\mathbb{A})}(X', Y);$$

hence a morphism $f : X \to Y$ in $D(\mathbb{A})$ is represented by a diagram of morphisms in the inductive limit: $X \xleftarrow[s']{\approx} X' \xrightarrow{f'} Y$.

3.3.1.4   The Mapping Cone Construction

We define the translate of a complex (K, d_K), denoted by TK or $K[1]$, by shifting the degrees:

$$(TK)^i = K^{i+1}, \quad d_{TK} = -d_K.$$

Let $u : K \to L$ be a morphism of complexes in $C^+\mathbb{A}$; the mapping cone $C(u)$ is the complex $TK \oplus L$ with the differential

$$d : C(u)^k := K^{k+1} \oplus L^k \to C(u)^{k+1} := K^{k+2} \oplus L^{k+1}$$

defined by

$$(a, b) \mapsto (-d_K(a), u(a) + d_L(b)).$$

The exact sequence associated to $C(u)$ is

$$0 \to L \xrightarrow{I} C(u) \to TK \to 0.$$

Remark.   Let h denote a homotopy between two morphisms $u, u' : K \to L$; we define an isomorphism $I_h : C(u) \xrightarrow{\sim} C(u')$ by the matrix $\begin{pmatrix} \mathrm{Id} & 0 \\ h & \mathrm{Id} \end{pmatrix}$ acting on $TK \oplus L$, which commute with the injections of L in $C(u)$ and $C(u')$, and with the projections on TK.

   Let h and h' be two homotopies of u to u'. A second homotopy of h to h', that is, a family of morphisms $k^{j+2} : K^{j+2} \to L^j$ for $j \in \mathbb{Z}$, satisfying $h - h' = d_L \circ k - k \circ d_K$, defines a homotopy of I_h to $I_{h'}$.

3.3.1.5   Distinguished Triangles

We write the exact sequence of the mapping cone u as

$$K \xrightarrow{u} L \xrightarrow{I} C(u) \xrightarrow{+1} .$$

It is called a distinguished triangle since the last map may be continued to the same exact sequence shifted by T. A distinguished triangle in $K(\mathbb{A})$ is a sequence of complexes isomorphic to the image in $K(\mathbb{A})$ of a distinguished triangle associated to a cone in $C(\mathbb{A})$.

   Triangles are defined in $K(\mathbb{A})$ by short exact sequences of complexes which split in each degree (see [41, 2-4, p. 272], [34, I.4, Def. 4.7], and [2, 1.1.2]). We remark that

(1) the cone over the identity morphism of a complex X is homotopic to zero;

(2) using the construction of the mapping cylinder over a morphism of complexes $u : X \to Y$ (see [34, I.4]), one can transform u, up to a homotopy equivalence, into an injective morphism of complexes.

A distinguished triangle *in the derived category* $D(\mathbb{A})$ is a sequence of complexes isomorphic to the image in $D(\mathbb{A})$ of a distinguished triangle in $K(\mathbb{A})$. Long exact sequences of cohomologies are associated to triangles.

Remark. Each short exact sequence of complexes $0 \xrightarrow{u} X \xrightarrow{v} Y \to Z \to 0$ is isomorphic to the distinguished triangle in $D^+\mathbb{A}$ defined by the cone $C(u)$ over u. The morphism $C(u) \to Z$ is defined by v and we use the connection morphism in the associated long exact sequence to check it is a quasi-isomorphism (see [41, Ch. II 1.5, p. 295], [34, 6.8, 6.9], and [2, (1.1.3)]).

3.3.1.6 Derived Functor

Let $F : \mathbb{A} \to \mathbb{B}$ be a functor of abelian categories. We denote also by $F : C^+\mathbb{A} \to C^+\mathbb{B}$ the corresponding functor on complexes, and by $Q_\mathbb{A} : C^+\mathbb{A} \to D^+\mathbb{A}$, $Q_\mathbb{B} : C^+\mathbb{B} \to D^+\mathbb{B}$ the canonical localizing functors. If *the category $\mathbb{A}$ has enough injective objects*, a derived functor,

$$RF : D^+\mathbb{A} \to D^+\mathbb{B}$$

satisfying $RF \circ Q_\mathbb{A} = Q_\mathbb{B} \circ F$ is defined as follows:

(a) Given a complex K in $D^+(\mathbb{A})$, we start by choosing an injective resolution of K, that is, a quasi-isomorphism $i : K \xrightarrow{\approx} I(K)$, where the components of I are injectives in each degree (see [34, 7.9] or [27, Lem. 4.6, p. 42]).

(b) We define $RF(K) = F(I(K))$.

A morphism $f : K \to K'$ gives rise to a morphism $RF(K) \to RF(K')$ functorially, since f can be extended to a morphism $I(f) : F(I(K)) \to F(I(K'))$, defined on the injective resolutions uniquely up to homotopy.

In particular, for a different choice of an injective resolution $J(K)$ of K, we have an isomorphism $F(I(K)) \simeq F(J(K))$ in $D^+(\mathbb{B})$.

Remark. The basic idea is that a functor F need not carry a quasi-isomorphism $\phi : K \xrightarrow{\approx} K'$ into a quasi-isomorphism $F(\phi) : F(K) \xrightarrow{\approx} F(K')$, but $F(\phi)$ will be a quasi-isomorphism if the complexes are injective, since then ϕ has an inverse up to homotopy.

DEFINITION 3.3.4   (i) *The cohomology $H^j(RF(K))$ is called the jth hypercohomology $R^jF(K)$ of F at K.*

(ii) *An object $A \in \mathbb{A}$ is F-acyclic if $R^jF(A) = 0$ for $j > 0$.*

Remark.

(i) We often add the condition that the functor $F : \mathbb{A} \to \mathbb{B}$ is left exact. In this case $R^0F \simeq F$ and we recover the theory of satellite functors in [5].

(ii) If $K \xrightarrow{\approx} K'$ is a quasi-isomorphism of complexes, hence an isomorphism in the derived category, the morphism $RF(K) \to RF(K')$ is a quasi-isomorphism, hence an isomorphism in the derived category, while the morphism $FK \to FK'$ which is not a quasi-isomorphism in general, is not well defined on the derived category.

(iii) It is important to know that we can use acyclic objects to compute RF: for any resolution $A(K)$ of a complex K by acyclic objects, $K \xrightarrow{\approx} A(K)$, $F(A(K))$ is isomorphic to the complex $RF(K)$.

For example, the hypercohomology of the global section functor Γ in the case of sheaves on a topological space, is equal to the cohomology defined via flasque resolutions or any "acyclic" resolution.

(iv) The dual construction defines the left-derived functor LF of a functor F if there exist enough projectives in the category $\mathbb{A}$.

(v) Verdier defines a derived functor even if there are not enough injectives (see [41, Ch. II, Sec. 2, p. 301]) and gives a construction of the derived functor in [41, Ch. II Sec. 2, p. 304] under suitable conditions.

3.3.1.7 Extensions

Fix a complex $B^\bullet$ of objects in $\mathbb{A}$. We consider the covariant functor $\mathrm{Hom}^\bullet(B^\bullet, *)$ from the category of complexes of objects in $\mathbb{A}$ to the category of complexes of abelian groups defined for a complex $A^\bullet$ by

$$(\mathrm{Hom}^\bullet(B^\bullet, A^\bullet))^n = \prod_{p \in \mathbb{Z}} \mathrm{Hom}_{\mathbb{A}}(B^p, A^{n+p}),$$

with the differential $d^n f$ of f in degree n defined by

$$[d^n f]^p = d_{A^\bullet}^{n+p} \circ f^p + (-1)^{n+1} f^{p+1} \circ d_{B^\bullet}^p.$$

The associated derived functor is $\mathrm{RHom}^\bullet(B^\bullet, *)$.

Suppose there are enough projectives and injectives in $\mathbb{A}$ and the complexes are bounded. If $P^\bullet \to B^\bullet$ is a projective resolution of the complex $B^\bullet$, and $A^\bullet \to I^\bullet$ is an injective resolution of $A^\bullet$, then, in $D^b(\mathbb{A})$, $\mathrm{RHom}^\bullet(B^\bullet, A^\bullet) := \mathrm{Hom}^\bullet(B^\bullet, I^\bullet)$ is isomorphic to $\mathrm{Hom}^\bullet(P^\bullet, A^\bullet)$.

The cycles (resp. boundaries) of $\mathrm{Hom}^\bullet(B^\bullet, I^\bullet)$ in degree n are the morphisms of complexes $\mathrm{Hom}(B^\bullet, I^\bullet[n])$ (resp. consist of morphisms homotopic to zero). Hence, the cohomology group $H^0(\mathrm{RHom}^\bullet(B^\bullet, A^\bullet)) \simeq H^0(\mathrm{RHom}^\bullet(B^\bullet, I^\bullet))$ is naturally isomorphic as a group to the group of morphisms from $B^\bullet$ to $A^\bullet$ in the derived category

$$H^0(\mathrm{RHom}^\bullet(B^\bullet, A^\bullet)) \simeq \mathrm{Hom}_{D^b(\mathbb{A})}(B^\bullet, A^\bullet).$$

Since for all k, $\mathrm{RHom}^\bullet(B^\bullet, A^\bullet[k]) = \mathrm{RHom}^\bullet(B^\bullet, A^\bullet)[k]$, the kth extension group is defined by

$$\mathrm{Ext}^k(B^\bullet, A^\bullet) := H^k(\mathrm{RHom}^\bullet(B^\bullet, A^\bullet)) := H^0(\mathrm{RHom}^\bullet(B^\bullet, A^\bullet[k])).$$

LEMMA 3.3.5 (Extension groups)   *When the abelian category $\mathbb{A}$ has enough injectives, the group $\operatorname{Hom}_{D(\mathbb{A})}(X^{\bullet}, Y^{\bullet})$ of morphisms of two complexes in the derived category $D(\mathbb{A})$ has an interpretation as an extension group*

$$\operatorname{Hom}_{D(\mathbb{A})}(X^{\bullet}, Y^{\bullet}[n]) = \operatorname{Ext}^{n}(X^{\bullet}, Y^{\bullet}),$$

where the groups on the right-hand side are derived from the Hom *functor. In general the group*

$$\operatorname{Hom}_{D(\mathbb{A})}(A, B[n])$$

of two objects in the category $\mathbb{A}$ may be interpreted as the Yoneda n-extension group [34, (XI.4)].

3.3.1.8   Filtered Homotopy Categories $K^{+}F(\mathbb{A}), K^{+}F_{2}(\mathbb{A})$

For an abelian category $\mathbb{A}$, let $F\mathbb{A}$ (resp. $F_{2}\mathbb{A}$) denote the category of filtered objects (resp. bifiltered) of $\mathbb{A}$ with finite filtration(s), and $C^{+}F\mathbb{A}$ (resp. $C^{+}F_{2}\mathbb{A}$) the category of complexes of $F\mathbb{A}$ (resp. $F_{2}\mathbb{A}$) bounded on the left (zero in degrees near $-\infty$) with morphisms of complexes respecting the filtration(s).

Two morphisms $u, u' : (K, F, W) \to (K', F, W)$, where W (resp. F) denotes uniformly increasing (resp. decreasing) filtrations on K or K', are homotopic if there exists a homotopy from u to u' compatible with the filtrations; then it induces a homotopy on each term $k^{i+1} : F^{j}K^{i+1} \to F^{j}K'^{i}$ (resp. for W) and, in particular, $\operatorname{Gr}_{F}(u - u')$ (resp. $\operatorname{Gr}_{F}\operatorname{Gr}_{W}(u - u')$) is homotopic to 0.

The homotopy category whose objects are bounded below complexes of filtered (resp. bifiltered) objects of $\mathbb{A}$, and whose morphisms are equivalence classes modulo homotopy compatible with the filtration(s), is denoted by $K^{+}F\mathbb{A}$ (resp. $K^{+}F_{2}\mathbb{A}$).

Filtered resolutions. In the presence of two filtrations by subcomplexes F and W on a complex K of objects of an abelian category $\mathbb{A}$, the filtration F induces by restriction a new filtration F on the terms $W^{i}K$, which also induces a quotient filtration F on $\operatorname{Gr}_{W}^{i} K$. In this way, we define the graded complexes $\operatorname{Gr}_{F} K$, $\operatorname{Gr}_{W} K$, and $\operatorname{Gr}_{F} \operatorname{Gr}_{W} K$.

DEFINITION 3.3.6   *A morphism $f : (K, F, W) \xrightarrow{\approx} (K', F, W)$ of complexes with biregular filtrations F and W is a bifiltered quasi-isomorphism if $\operatorname{Gr}_{F}^{*} \operatorname{Gr}_{W}^{*}(f)$ is a quasi-isomorphism.*

3.3.1.9   Derived Filtered Categories $D^{+}F(\mathbb{A}), D^{+}F_{2}(\mathbb{A})$

They are deduced from $K^{+}F\mathbb{A}$ (resp. $K^{+}F_{2}\mathbb{A}$) by inverting filtered quasi-isomorphisms (resp. bifiltered quasi-isomorphisms). The objects of $D^{+}F\mathbb{A}$ (resp. $D^{+}F_{2}\mathbb{A}$) are complexes of filtered objects of $\mathbb{A}$ as of $K^{+}F\mathbb{A}$ (resp. $K^{+}F_{2}\mathbb{A}$). Hence, the morphisms are represented by diagrams with filtered (resp. bifiltered) quasi-isomorphisms.

3.3.1.10 Triangles

The complex $T(K, F, W)$ and the cone $C(u)$ of a morphism

$$u : (K, F, W) \to (K', F, W)$$

are endowed naturally with filtrations F and W. The exact sequence associated to $C(u)$ is compatible with the filtrations. A filtered homotopy h of morphisms u and u' defines a filtered isomorphism of cones $I_h : C(u) \xrightarrow{\sim} C(u')$.

Distinguished (or exact) triangles are defined similarly in $K^+F\mathbb{A}$ and $K^+F_2\mathbb{A}$ as well as in $D^+F\mathbb{A}$ and $D^+F_2\mathbb{A}$. Long filtered (resp. bifiltered) exact sequences of cohomologies are associated to triangles.

Remark. The morphisms of exact sequences are not necessarily strict for the induced filtrations on cohomology; however, they will be strict in the case of the class of mixed Hodge complexes giving rise to MHS that we want to define.

3.3.2 Derived Functor on a Filtered Complex

Let $T : \mathbb{A} \to \mathbb{B}$ be a left exact functor of abelian categories with enough injectives in $\mathbb{A}$. To construct a derived functor $RT : D^+F\mathbb{A} \to D^+F\mathbb{B}$ (resp. $RT : D^+F_2\mathbb{A} \to D^+F_2\mathbb{B}$), we need to introduce the concept of T-acyclic filtered resolutions. Given a filtered complex with biregular filtration(s) we define first the image of the filtrations via acyclic filtered resolutions. Then, we remark that a filtered quasi-isomorphism $\phi : (K, F) \xrightarrow{\approx} (K', F')$ of complexes filtered by subcomplexes of T-acyclic sheaves has as its image by T, a filtered quasi-isomorphism $T(\phi) : (TK, TF) \xrightarrow{\approx} (TK', TF')$; therefore the construction factors through the derived filtered category by RT.

3.3.2.1 Image of a Filtration by a Left Exact Functor

Let (A, F) be a filtered object in $\mathbb{A}$, with a finite filtration. Since T is left exact, a filtration TF of TA is defined by the subobjects $TF^p(A)$.

If $\mathrm{Gr}_F(A)$ is T-acyclic, the objects $F^p(A)$ are T-acyclic as successive extensions of T-acyclic objects. Hence, the image by T of the sequence of acyclic objects is exact:

$$0 \to F^{p+1}(A) \to F^p(A) \to \mathrm{Gr}_F^p(A) \to 0.$$

LEMMA 3.3.7 *If* $\mathrm{Gr}_F(A)$ *is a* T-*acyclic object, we have* $\mathrm{Gr}_{TF}(TA) \simeq T\,\mathrm{Gr}_F(A)$.

3.3.2.2 Filtered T-Acyclic Objects

Let A be an object with two finite filtrations F and W such that $\mathrm{Gr}_F \mathrm{Gr}_W A$ is T-acyclic; then the objects $\mathrm{Gr}_F A$ and $\mathrm{Gr}_W A$ are T-acyclic, as well as $F^q(A) \cap W^p(A)$. As a consequence of acyclicity, the sequences

$$0 \to T(F^q \cap W^{p+1}) \to T(F^q \cap W^p) \to T((F^q \cap W^p)/(F^q \cap W^{p+1})) \to 0$$

are exact, and $T(F^q(\mathrm{Gr}_W^p A))$ is the image in $T(\mathrm{Gr}_W^p(A))$ of $T(F^q \cap W^p)$. Moreover, the isomorphism $\mathrm{Gr}_{TW} TA \simeq T(\mathrm{Gr}_W A)$ transforms the filtration $\mathrm{Gr}_{TW}(TF)$ on $\mathrm{Gr}_{TW}(TA)$ into the filtration $T(\mathrm{Gr}_W(F))$ on $T(\mathrm{Gr}_W A)$.

3.3.2.3   The Derived Filtered Functor $RT : D^+F(\mathbb{A}) \to D^+F(\mathbb{B})$

Let F be a biregular filtration of K. A filtered T-acyclic resolution of K is given by a filtered quasi-isomorphism $i : (K, F) \to (K', F')$ to a complex with a biregular filtration such that for all integers p and n, $\mathrm{Gr}_F^p(K^n)$ is acyclic for T.

LEMMA AND DEFINITION 3.3.8 (Filtered derived functor of a left exact functor $T : \mathbb{A} \to \mathbb{B}$)   *Suppose we are given functorially for each filtered complex (K, F) a filtered T-acyclic resolution $i : (K, F) \to (K', F')$. We define $T' : C^+F(\mathbb{A}) \to D^+F(\mathbb{B})$ by the formula $T'(K, F) = (TK', TF')$. A filtered quasi-isomorphism $f : (K_1, F_1) \to (K_2, F_2)$ induces an isomorphism in $D^+F(\mathbb{B})$,*

$$T'(f) : T'(K_1, F_1) \simeq T'(K_2, F_2).$$

Hence T' factors through a derived functor $RT : D^+F(\mathbb{A}) \to D^+F(\mathbb{B})$ such that $RT(K, F) = (TK', TF')$, and we have $\mathrm{Gr}_F RT(K) \simeq RT(\mathrm{Gr}_F K)$, where

$$FRT(K) := T(F'(K')).$$

In particular, for a different choice (K'', F'') of (K', F'), we have an isomorphism $(TK'', TF'') \simeq (TK', TF')$ in $D^+F(\mathbb{B})$ and

$$RT(\mathrm{Gr}_F K) \simeq \mathrm{Gr}_{TF'} T(K') \simeq \mathrm{Gr}_{TF''} T(K'').$$

Remark.   Due to the above properties, a bifiltered quasi-isomorphism of bifiltered complexes induces a bifiltered isomorphism on their hypercohomology.

EXAMPLE 3.3.9 (Godement resolution)   In the particular case of interest, where $\mathbb{A}$ is the category of sheaves of A-modules on a topological space X, and where T is the global section functor Γ, an example of a filtered T-acyclic resolution of K is the simple complex $\mathcal{G}^*(K)$, associated to the double complex defined by Godement resolution $\mathcal{G}^*$ in each degree of K (see Chapter 2 of this volume, [34, Ch. II, Sec. 3.6, p.95], or [17, Ch. II, Sec. 4.3, p.167]) filtered by $\mathcal{G}^*(F^pK)$.

This example will apply to the next result for bifiltered complexes (K, W, F) with resolutions $(\mathcal{G}^*K, \mathcal{G}^*W, \mathcal{G}^*F)$ satisfying

$$\mathrm{Gr}_{\mathcal{G}^*F} \mathrm{Gr}_{\mathcal{G}^*W}(\mathcal{G}^*K) \simeq \mathcal{G}^*(\mathrm{Gr}_F \mathrm{Gr}_W K).$$

3.3.2.4   The Filtered Derived Functor $RT : D^+F_2(\mathbb{A}) \to D^+F_2(\mathbb{B})$

Let F and W be two biregular filtrations of K. A bifiltered T-acyclic resolution of K is a bifiltered quasi-isomorphism $f : (K, W, F) \to (K', W', F')$ such that W' and F' are biregular filtrations on K' and for all p, q, n: $\mathrm{Gr}_F^p \mathrm{Gr}_W^q(K'^n)$ is acyclic for T.

LEMMA AND DEFINITION 3.3.10   *Let (K, F, W) be a bifiltered complex, $T : \mathbb{A} \to \mathbb{B}$ a left exact functor and $i : (K, F, W) \to (K', F', W')$ a functorial bifiltered T-acyclic resolution. We define $T' : C^+F(\mathbb{A}) \to D^+F(\mathbb{B})$ by the formula $T'(K, F, W) = (TK', TF', TW')$. A bifiltered quasi-isomorphism*

$$f : (K_1, F_1, W_1) \to (K_2, F_2, W_2)$$

induces an isomorphism $T'(f) : T'(K_1, F_1, W_1) \simeq T'(K_2, F_2, W_2)$ *in* $D^+F_2(\mathbb{B})$. *Hence* T' *factors through a derived functor* $RT : D^+F_2(\mathbb{A}) \to D^+F_2(\mathbb{B})$ *such that*

$$RT(K, F, W) = (TK', TF', TW'),$$

and we have

$$\mathrm{Gr}_F \, \mathrm{Gr}_W \, RT(K) \simeq RT(\mathrm{Gr}_F \, \mathrm{Gr}_W \, K).$$

In particular, for a different choice (K'', F'', W'') of (K', F', W') we have an iso-morphism $(TK'', TF'', TW'') \simeq (TK', TF', TW')$ in $D^+F_2(\mathbb{B})$ and

$$RT(\mathrm{Gr}_F \, \mathrm{Gr}_W \, K) \simeq \mathrm{Gr}_{TF'} \, \mathrm{Gr}_{TW'} \, T(K') \simeq \mathrm{Gr}_{TF''} \, \mathrm{Gr}_{TW''} \, T(K'').$$

3.3.2.5 *Hypercohomology Spectral Sequence*

An object of $D^+F(\mathbb{A})$ defines a spectral sequence functorial with respect to mor-phisms. Let $T : \mathbb{A} \to \mathbb{B}$ be a left exact functor of abelian categories, (K, F) an object of $D^+F\mathbb{A}$ and $RT(K, F) : D^+F\mathbb{A} \to D^+F\mathbb{B}$ its derived functor. Since $\mathrm{Gr}_F \, RT(K) \simeq RT(\mathrm{Gr}_F \, K)$, the spectral sequence defined by the filtered complex $RT(K, F)$ is written as

$$_F E_1^{p,q} = R^{p+q} T(\mathrm{Gr}_F^p \, K) \Rightarrow \mathrm{Gr}_F^p \, R^{p \mid q} T(K).$$

Indeed, $H^{p+q}(\mathrm{Gr}_F^p \, RT(K)) \simeq H^{p+q}(RT(\mathrm{Gr}_F^p(K)))$. This is the hypercohomology spectral sequence of the filtered complex K with respect to the functor T. The spectral sequence depends functorially on K and a filtered quasi-isomorphism induces an iso-morphism of spectral sequences. The differentials d_1 of this spectral sequence are the image by T of the connecting morphisms defined by the short exact sequences

$$0 \to \mathrm{Gr}_F^{p+1} \, K \to F^p K / F^{p+2} K \to \mathrm{Gr}_F^p \, K \to 0.$$

For an increasing filtration W on K, we have

$$_W E_1^{p,q} = R^{p+q} T(\mathrm{Gr}_{-p}^W) \Rightarrow \mathrm{Gr}_{-p}^W \, R^{p+q} T(K).$$

EXAMPLE 3.3.11 (The τ and σ filtrations)   (1) Let K be a complex; the canonical filtration by truncation τ is the increasing filtration by subcomplexes

$$\tau_p K^\bullet := (\cdots \to K^{p-1} \to \mathrm{Ker} \, d_p \to 0 \to \cdots \to 0)$$

such that

$$\mathrm{Gr}_p^\tau \, K \xrightarrow{\approx} H^p(K)[-p], \quad H^i(\tau_{\leq p}(K)) = H^i(K) \quad \text{if} \quad i \leq p,$$

while $H^i(\tau_{\leq p}(K)) = 0$ if $i > p$.

(2) The subcomplexes of K,

$$\sigma_p K^\bullet := K^{\bullet \geq p} := (0 \to \cdots \to 0 \to K^p \to K^{p+1} \to \cdots),$$

define a decreasing biregular filtration, called the trivial filtration σ of K such that $\mathrm{Gr}_\sigma^p K = K^p[-p]$ (it coincides with the filtration F on the de Rham complex).

A quasi-isomorphism $f : K \to K'$ is necessarily a filtered quasi-isomorphism for both filtrations τ and σ. The hypercohomology spectral sequences of a left exact functor attached to both filtrations of K are the two natural hypercohomology spectral sequences of K.

3.3.2.6 Construction of the Hypercohomology Spectral Sequence and the Filtration L

Let $K^\bullet := ((K^i)_{i\in\mathbb{Z}}, d)$ be a complex of objects in an abelian category and F a left exact functor with values in the category of abelian groups (for example, $K^\bullet$ is a complex of sheaves on a topological space X and Γ is the functor of global sections). To construct the hypercohomology spectral sequence, we consider F-acyclic resolutions $(K^{i,*}, d'')$ of K^i forming a double complex $K^{i,j}$ with differentials $d' : K^{i,j} \to K^{i+1,j}$ and $d'' : K^{i,j} \to K^{i,j+1}$ such that the kernels $Z^{i,j}$ of d' (resp. the image $B^{i,j}$ of d', the cohomology $H^{i,j}$) form an acyclic resolution with varying index j, of the kernel Z_i of d on K^i (resp. the image B^i of d, the cohomology $H^i(K)$). The decreasing filtration L by subcomplexes of the simple complex $sFK := s(FK^{*,*})$ associated to the double complex, is defined by

$$L^r(sF(K))^n := \oplus_{i+j=n,\, j\geq r} F(K^{i,j}).$$

The associated spectral sequence starts with the terms

$$E_0^{p,q} := \mathrm{Gr}_L^p(sF(K)^{p+q}) := F(K^{q,p}), \quad d' : F(K^{q,p}) \to F(K^{p+1,p}),$$

$$E_1^{p,q} := H^{p+q}(\mathrm{Gr}_L^p(sF(K))) = H^{p+q}(F(K^{*,p})[-p], d') = H^q(F(K^{*,p}), d'),$$

where $F(K^{*,p})$ has degree $*+p$ in the complex $F(K^{*,p})[-p]$, and the terms $E_1^{p,q}$ form a complex for varying p with differential induced by $d'' : E_1^{p,q} \to E_1^{p+1,q}$.

It follows that $E_2^{p,q} = H^p(H^q(F(K^{*,*}), d'), d'')$.

Since the cohomology groups $(H^q(K^{p,*}, d'), d'')$ for various p and induced differential d'' form an acyclic resolution of $H^q(K^\bullet)$, the cohomology for d'' is $E_2^{p,q} = R^p F(H^q(K))$; hence we have a spectral sequence

$$E_2^{p,q} := R^p F(H^q K) \Longrightarrow R^{p+q} F(K).$$

3.3.2.7 Comparison Lemma for L, τ, and the Filtration $\mathrm{Dec}(L)$

We have a comparison lemma for the above filtration L and τ on $s(FK^{*,*})$:

LEMMA 3.3.12 *On the cohomology group $R^n F(K) := H^n(s(FK^{*,*}))$, the induced filtrations by L and τ coincide up to a change in indices, $\tau_{-p} = L^{n+p}$.*

Since τ is increasing and $\mathrm{Gr}_{-p}^\tau K = H^{-p}(K)[p]$, its associated spectral sequence is

$$E_1^{p,q}(\tau) = R^{p+q} F(\mathrm{Gr}_{-p}^\tau K) = R^{p+q} F(H^{-p} K[p]) = R^{2p+q} F(H^{-p} K).$$

The filtration τ is related to the filtration L by a process of decalage described in [7, (1.3.3)].

3.3.2.8 The Filtration $\mathrm{Dec}\, F$

Let F be a decreasing filtration on a complex K; the filtration $\mathrm{Dec}\, F$ is defined as

$$(\mathrm{Dec}\, F)^p K^n := \mathrm{Ker}(F^{p+n} K^n \to K^{n+1}/(F^{p+n+1} K^{n+1})).$$

Then $E_0^{p,q}(\mathrm{Dec}\, F) := \mathrm{Gr}_{\mathrm{Dec}\, F}^p K^{p+q}$ and we have a morphism in degree $p+q$,

$$(E_0^{p,q}(\mathrm{Dec}\, F), d_0) \to (E_1^{2p+q,-p}(F), d_1)$$

inducing an isomorphism in rank $r \geq 1$: $(E_r^{p,q}(\mathrm{Dec}\, F), d_r) \simeq (E_{r+1}^{2p+q,-p}(F), d_{r+1})$ (see [7, Prop. 1.3.4]). Hence, by definition,

$$(\mathrm{Dec}\, L)^p(s(FK))^n = (\oplus_{i+j=n, j>p+n} F(K^{i,j}))$$

$$\oplus \mathrm{Ker}[F(K^{-p,p+n}) \xrightarrow{d'} F(K^{-p+1,p+n})].$$

By construction, the sum of the double complex $\tau_p K^{*,j}$ for varying j defines an acyclic resolution of $\tau_p K$; then $RF(K, \tau)$ is the complex $s(FK^{*,*})$ filtered by

$$(F\tau)_p := s(\tau_p K^{*,j})_{j \geq 0}.$$

Hence, we have a morphism

$$(s(FK^{*,*}), F\tau_{-p}) \to (s(FK^{*,*}), (\mathrm{Dec}\, L)^p)$$

inducing isomorphisms for $r > 0$,

$$E_r^{p,q}(\tau) \simeq E_r^{p,q}(\mathrm{Dec}\, L) \simeq E_{r+1}^{2p+q,-p}(L)$$

and

$$E_1^{p,q}(\tau) \simeq E_2^{2p+q,-p}(L) = R^{2p+q} F(H^{-p} K).$$

3.3.2.9 Leray's Spectral Sequence

Let $f : X \to V$ be a continuous map of topological spaces and $\mathcal{F}$ be a sheaf of abelian groups on X. To construct $Rf_* \mathcal{F}$, we use a flasque resolution of $\mathcal{F}$. We consider the filtrations τ and L on $Rf_* \mathcal{F}$ as above. If we apply the functor Γ of global sections on V, we deduce from the above statement the following result on the cohomology of X, since $\mathbb{H}^p(V, Rf_* \mathcal{F}) \simeq \mathbb{H}^p(X, \mathcal{F})$.

LEMMA AND DEFINITION 3.3.13   (i) *There exists a filtration L on the cohomology $H^n(X, \mathcal{F})$ and a convergent spectral sequence defined by f, starting at rank 2:*

$$E_2^{p,q} = \mathbb{H}^p(V, R^q f_* \mathcal{F}) \Rightarrow E_\infty^{p,q} = \mathrm{Gr}_L^p H^{p+q}(X, \mathcal{F}).$$

   (ii) *There exists a filtration τ on the cohomology $H^n(X, \mathcal{F})$ and a convergent spectral sequence defined by f, starting at rank 1:*

$$E_1^{p,q} = \mathbb{H}^{2p+q}(V, R^{-p} f_* \mathcal{F}) \Rightarrow E_\infty^{p,q} = \mathrm{Gr}_{-p}^\tau H^{p+q}(X, \mathcal{F})$$

with isomorphisms for $r > 0$, $E_r^{p,q}(\tau) \simeq E_{r+1}^{2p+q,-p}(L)$.

   On the cohomology group $H^n(X, \mathcal{F})$, the term of the induced filtration τ_p coincides with the term of the induced filtration L^{n-p}.

3.3.3 Mixed Hodge Complex (MHC)

Now we give sufficient conditions on the filtrations of a bifiltered complex (K, W, F) in order to obtain a mixed Hodge structure with the induced filtrations by W and F on the cohomology of K. The structure defined on K by these conditions is called a mixed Hodge complex (MHC).

Let Γ denote the global sections functor on an algebraic variety V; we construct on V a bifiltered complex of sheaves $(K_{\mathbb{C}}, F, W)$, where the filtration W is rationally defined, called a cohomological MHC, such that its image by the bifiltered derived functor $R\Gamma(V, K, W, F)$ is an MHC.

Then, by the two filtrations lemma (see Section 3.2.3.1), the filtrations induced by W and F on the hypercohomology $\mathbb{H}^i(V, K_{\mathbb{C}})$ defines an MHS.

This result is so powerful that the rest of the theory will consist of the construction of a cohomological MHC for all algebraic varieties. Hence, the theoretical path to construct an MHS on a variety follows the pattern

$$\text{cohomological MHC} \Rightarrow \text{MHC} \Rightarrow \text{MHS}.$$

The logarithmic complex in the next section gives the basic example of cohomological MHC. It is true that a direct study of the logarithmic complex by Griffiths and Schmid [21] is very attractive, but the above pattern in the initial work of Deligne is easy to apply, flexible and helps to go beyond this case toward a general theory.

The de Rham complex of a smooth compact complex variety is a special case of a mixed Hodge complex, called a Hodge complex (HC) with the characteristic property that it induces a Hodge structure on its hypercohomology. We start by rewriting the Hodge theory of the first section with this terminology since it is suited to generalization to MHC.

3.3.3.1 Definitions

Let A denote $\mathbb{Z}$, $\mathbb{Q}$, or $\mathbb{R}$ as in Section 3.2.2, and let $D^+(\mathbb{Z})$ (resp. $D^+(\mathbb{C})$) denote the derived category of $\mathbb{Z}$-modules (resp. $\mathbb{C}$-vector spaces). The corresponding derived category of sheaves on V are denoted by $D^+(V, \mathbb{Z})$ and $D^+(V, \mathbb{C})$.

DEFINITION 3.3.14 (Hodge complex (HC)) *A Hodge A-complex K of weight n consists of*

(1) a complex K_A of A-modules such that $H^k(K_A)$ is an A-module of finite type for all k;

(2) a filtered complex $(K_{\mathbb{C}}, F)$ of $\mathbb{C}$-vector spaces;

(3) an isomorphism $\alpha : K_A \otimes \mathbb{C} \simeq K_{\mathbb{C}}$ in $D^+(\mathbb{C})$.

The following axioms must be satisfied:

(1) (HC 1) The differential $d^i : (K_A^i, F) \to (K_A^{i+1}, F)$ is strict for all i (we say that the differential d of $K_{\mathbb{C}}$ is strictly compatible with the filtration F).

(2) (HC 2) For all k, the induced filtration F on $H^k(K_{\mathbb{C}}) \simeq H^k(K_A) \otimes \mathbb{C}$ defines an A-Hodge structure of weight $n + k$ on $H^k(K_A)$.

The condition (HC1) is equivalent to the degeneration of the spectral sequence defined by $(K_{\mathbb{C}}, F)$ at rank 1, i.e., $E_1 = E_{\infty}$ (see Section 3.2.5). By (HC2) the filtration F is $(n + k)$-opposed to its complex conjugate (conjugation makes sense since $A \subset \mathbb{R}$).

DEFINITION 3.3.15 *Let X be a topological space. An A-cohomological Hodge complex K of weight n on X consists of*

(1) a complex of sheaves K_A of A-modules on X;

(2) a filtered complex of sheaves $(K_{\mathbb{C}}, F)$ of $\mathbb{C}$-vector spaces on X;

(3) an isomorphism $\alpha : K_A \otimes \mathbb{C} \xrightarrow{\approx} K_{\mathbb{C}}$ in $D^+(X, \mathbb{C})$ of $\mathbb{C}$-sheaves on X.

Moreover, the following axiom must be satisfied:

(4) (CHC) The triple $(R\Gamma(X, K_A), R\Gamma(X, K_{\mathbb{C}}, F), R\Gamma(\alpha))$ is a Hodge complex of weight n.

If (K, F) is an HC (resp. cohomological HC) of weight n, then $(K[m], F[p])$ is a Hodge complex (resp. cohomological HC) of weight $n + m - 2p$.

EXAMPLE 3.3.16 The Hodge decomposition theorem may be stated as follows:
Let X be a compact complex algebraic manifold and consider

(i) the complex $K_{\mathbb{Z}}$ reduced to a constant sheaf $\mathbb{Z}$ on X in degree 0;

(ii) the analytic de Rham complex $K_{\mathbb{C}} := \Omega_X^*$ with its trivial filtration by subcomplexes $F^p := \Omega_X^{* \geq p}$,

$$F^p \Omega_X^* := 0 \to \cdots \to 0 \to \Omega_X^p \to \Omega_X^{p+1} \to \cdots \to \Omega_X^n \to 0;$$

(iii) the quasi-isomorphism $\alpha : K_{\mathbb{Z}} \otimes \mathbb{C} \xrightarrow{\approx} \Omega_X^*$ (Poincaré lemma).

Then $(K_{\mathbb{Z}}, (K_{\mathbb{C}}, F), \alpha)$ is a cohomological HC on X of weight 0; its hypercohomology on X is isomorphic to the cohomology of X and carries an HS with Hodge filtration induced by F.

Now, we define the structure including two filtrations W and F needed on a complex, in order to define an MHS on its cohomology.

DEFINITION 3.3.17 (MHC) *An A-mixed Hodge complex (MHC) K consists of*

(i) a complex K_A of A-modules such that $H^k(K_A)$ is an A-module of finite type for all k;

(ii) a filtered complex $(K_{A \otimes \mathbb{Q}}, W)$ of $(A \otimes \mathbb{Q})$-vector spaces with an increasing filtration W;

(iii) *an isomorphism $K_A \otimes \mathbb{Q} \xrightarrow{\sim} K_{A \otimes \mathbb{Q}}$ in $D^+(A \otimes \mathbb{Q})$;*

(iv) *a bifiltered complex $(K_{\mathbb{C}}, W, F)$ of $\mathbb{C}$-vector spaces with an increasing (resp. decreasing) filtration W (resp. F) and an isomorphism in $D^+ F(\mathbb{C})$,*

$$\alpha : (K_{A \otimes \mathbb{Q}}, W) \otimes \mathbb{C} \xrightarrow{\sim} (K_{\mathbb{C}}, W).$$

Moreover, the following axiom (MHC) is satisfied: for all n, the system consisting of

- *the complex $\mathrm{Gr}_n^W(K_{A \otimes \mathbb{Q}})$ of $(A \otimes \mathbb{Q})$-vector spaces;*

- *the complex $\mathrm{Gr}_n^W(K_{\mathbb{C}}, F)$ of $\mathbb{C}$-vector spaces with induced filtration F;*

- *the isomorphism $\mathrm{Gr}_n^W(\alpha) : \mathrm{Gr}_n^W(K_{A \otimes \mathbb{Q}}) \otimes \mathbb{C} \xrightarrow{\sim} \mathrm{Gr}_n^W(K_{\mathbb{C}})$*

form an $A \otimes \mathbb{Q}$-Hodge complex of weight n.

The above structure has a corresponding structure on a complex of sheaves on X called a cohomological MHC:

DEFINITION 3.3.18 (Cohomological MHC)   *An A-cohomological mixed Hodge complex K on a topological space X consists of*

(i) *a complex K_A of sheaves of A-modules on X such that $H^k(X, K_A)$ are A-modules of finite type;*

(ii) *a filtered complex $(K_{A \otimes \mathbb{Q}}, W)$ of sheaves of $(A \otimes \mathbb{Q})$-vector spaces on X with an increasing filtration W and an isomorphism $K_A \otimes \mathbb{Q} \simeq K_{A \otimes \mathbb{Q}}$ in $D^+(X, A \otimes \mathbb{Q})$;*

(iii) *a bifiltered complex of sheaves $(K_{\mathbb{C}}, W, F)$ of $\mathbb{C}$-vector spaces on X with an increasing (resp. decreasing) filtration W (resp. F) and an isomorphism in $D^+ F(X, \mathbb{C})$,*

$$\alpha : (K_{A \otimes \mathbb{Q}}, W) \otimes \mathbb{C} \xrightarrow{\sim} (K_{\mathbb{C}}, W).$$

Moreover, the following axiom is satisfied: for all n, the system consisting of

- *the complex $\mathrm{Gr}_n^W(K_{A \otimes \mathbb{Q}})$ of sheaves of $(A \otimes \mathbb{Q})$-vector spaces on X;*

- *the complex $\mathrm{Gr}_n^W(K_{\mathbb{C}}, F)$ of sheaves of $\mathbb{C}$-vector spaces on X with induced F;*

- *the isomorphism $\mathrm{Gr}_n^W(\alpha) : \mathrm{Gr}_n^W(K_{A \otimes \mathbb{Q}}) \otimes \mathbb{C} \xrightarrow{\sim} \mathrm{Gr}_n^W(K_{\mathbb{C}})$*

is an $A \otimes \mathbb{Q}$-cohomological HC on X of weight n.

If (K, W, F) is an MHC (resp. cohomological MHC), then for all m and $n \in \mathbb{Z}$, $(K[m], W[m - 2n], F[n])$ is an MHC (resp. cohomological MHC).

In fact, any HC or MHC described here is obtained from de Rham complexes with modifications (at infinity) as the logarithmic complex in the next section. A new construction of HC was introduced later with the theory of differential modules and perverse sheaves [2] and [38] following the theory of intersection complexes but it is not covered in this lecture.

Now we first explain how to deduce an MHC from a cohomological MHC, and then an MHS from an MHC.

PROPOSITION 3.3.19 *Let* $K = (K_A, (K_{A \otimes \mathbb{Q}}, W), (K_\mathbb{C}, W, F), \alpha)$ *be an A-cohomological MHC with a compatibility isomorphism* α; *then*

$$R\Gamma K = (R\Gamma K_A, R\Gamma(K_{A \otimes \mathbb{Q}}, W), R\Gamma(K_\mathbb{C}, W, F), R\Gamma(\alpha))$$

with the compatibility isomorphism $R\Gamma(\alpha)$ *is an A-MHC.*

3.3.3.2   MHS on the Cohomology of an MHC

The main result of Deligne in [7] and [8] can be briefly stated:

THEOREM 3.3.20 (Deligne)   *The cohomology of a mixed Hodge complex carries a mixed Hodge structure.*

The proof of this result requires a detailed study of spectral sequences based on the two filtrations lemma (Section 3.2.3.1). We give first the properties of the various spectral sequences which may be of interest as independent results.

Precisely, the weight spectral sequence of an MHC is in the category of HS. So, the graded group Gr^W of the MHS on cohomology is approached step by step by HS on the terms of the weight spectral sequence $_W E_r^{p,q}$ of $(K_\mathbb{C}, W)$.

However, the big surprise is that the spectral sequence degenerates quickly, at rank 2 for W (and at rank 1 for F): the terms $_W E_1^{p,q}$ and $_W E_2^{p,q}$ are all that is needed in the computation.

We show first that the first terms $E_1^{p,q}$ of the spectral sequence with respect to W carry an HS of weight q defined by the induced filtration by F. Moreover, the differentials d_1 are morphisms of HS; hence the terms $E_2^{p,q}$ carry an HS of weight q.

Then the proof based on Section 3.2.3.1 consists in showing that d_r is compatible with the induced HS, but for $r > 1$ it is a morphism between two HS of different weight; hence it must vanish.

PROPOSITION 3.3.21 (MHS on the cohomology of an MHC)   *Let K be an A-MHC.*

(i) *The filtration $W[n]$ of $H^n(K_A) \otimes \mathbb{Q} \simeq H^n(K_{A \otimes \mathbb{Q}})$,*

$$(W[n])_q(H^n(K_{A \otimes \mathbb{Q}})) := \mathrm{Im}(H^n(W_{q-n} K_{A \otimes \mathbb{Q}}) \to H^n(K_{A \otimes \mathbb{Q}})),$$

and the filtration F on $H^n(K_\mathbb{C}) \simeq H^n(K_A) \otimes_A \mathbb{C}$,

$$F^p(H^n(K_\mathbb{C})) := \mathrm{Im}(H^n(F^p K_\mathbb{C}) \to H^n(K_\mathbb{C})),$$

define on $H^n(K)$ an A-mixed Hodge structure,

$$(H^n(K_A), (H^n(K_{A \otimes \mathbb{Q}}), W), (H^n(K_\mathbb{C}), W, F)).$$

(ii) *On the terms $_W E_r^{p,q}(K_\mathbb{C}, W)$, the recurrent filtration and the two direct filtrations coincide $F_d = F_{\mathrm{rec}} = F_{d^*}$ and define the filtration F of a Hodge structure of weight q, moreover d_r is compatible with F.*

(iii) *The morphisms $d_1 : {}_W E_1^{p,q} \to {}_W E_1^{p+1,q}$ are strictly compatible with F.*

(iv) *The spectral sequence of* $(K_{A\otimes\mathbb{Q}}, W)$ *degenerates at rank* 2 $({}_W E_2 = {}_W E_\infty)$.

(v) *The spectral sequence of* $(K_\mathbb{C}, F)$ *degenerates at rank* 1 $({}_F E_1 = {}_F E_\infty)$.

(vi) *The spectral sequence of the complex* $\mathrm{Gr}_F^p(K_\mathbb{C})$, *with the induced filtration* W, *degenerates at rank* 2.

Note that the indices of the weight filtration are not given by the indices of the induced filtration W on cohomology, but are shifted by n. One should remember that the weight of the HS on the terms ${}_W E_r^{p,q}$ is always q, hence the weight of $\mathrm{Gr}_{-p}^W H^{p+q}(K)$ is q, i.e., the induced term W_{-p} is considered with index q: $W_{-p} = (W[p+q])_q$.

3.3.3.3 Proof of the Existence of an MHS on the Cohomology of an MHC

We need to check that the hypothesis $(*r_0)$ in Theorem 3.2.30 applies to MHC, which is done by induction on r_0. If we assume that the filtrations $F_d = F_{\mathrm{rec}} = F_{d^*}$ coincide for $r < r_0$ and moreover define the same filtration F of a Hodge structure of weight q on $E_r^{p,q}(K,W)$, and $d_r : E_r^{p,q} \to E_r^{p+r,q-r+1}$ is compatible with such a Hodge structure, then, in particular, d_r is strictly compatible with F; hence the induction applies.

LEMMA 3.3.22 *For* $r \geq 1$, *the differentials* d_r *of the spectral sequence* ${}_W E_r$ *are strictly compatible with the recurrent filtration* $F = F_{\mathrm{rec}}$. *For* $r \geq 2$, *they vanish.*

The initial statement applies for $r = 1$ by definition of an MHC since the complex $\mathrm{Gr}_{-p}^W K$ is an HC of weight $-p$. Hence, the two direct filtrations and the recurrent filtration F_{rec} coincide with the Hodge filtration F on ${}_W E_1^{p,q} = H^{p+q}(\mathrm{Gr}_{-p}^W K)$. The differential d_1 is compatible with the direct filtrations, hence with F_{rec}, and commutes with complex conjugation since it is defined on $A \otimes \mathbb{Q}$; hence it is compatible with $\overline{F}_{\mathrm{rec}}$. Then it is strictly compatible with the Hodge filtration $F = F_{\mathrm{rec}}$.

The filtration F_{rec} defined in this way is q-opposed to its complex conjugate and defines an HS of weight q on ${}_W E_2^{pq}$.

We suppose by induction that the two direct filtrations and the recurrent filtration coincide on ${}_W E_s (s \leq r+1) : F_d = F_{\mathrm{rec}} = F_{d^*}$ and ${}_W E_r = {}_W E_2$. On ${}_W E_2^{p,q} = {}_W E_r^{p,q}$, the filtration $F_{\mathrm{rec}} := F$ is compatible with d_r and q-opposed to its complex conjugate. Hence the morphism $d_r : {}_W E_r^{p,q} \to {}_W E_r^{p+r,q-r+1}$ is a morphism of an HS of weight q to an HS of weight $q - r + 1$ and must vanish for $r > 1$. In particular, we deduce that the weight spectral sequence degenerates at rank 2.

The filtration on ${}_W E_\infty^{p,q}$ induced by the filtration F on $H^{p+q}(K)$ coincides with the filtration F_{rec} on ${}_W E_2^{p,q}$.

3.3.4 Relative Cohomology and the Mixed Cone

The notion of a morphism of an MHC involves compatibility between the rational level and complex level. They are stated in the derived category to give some freedom in the choice of resolutions while keeping track of this compatibility. This is particularly interesting in the proof of functoriality of an MHS.

However, to put an MHS on the relative cohomology it is natural to use the cone construction over a morphism u. If u is given as a class $[u]$ up to homotopy, the cone construction depends on the choice of the representative u as we have seen.

To define a mixed Hodge structure on the relative cohomology, we must define the notion of the mixed cone with respect to a representative of the morphism on the level of complexes.

The isomorphism between two structures obtained for two representatives depends on the choice of a homotopy; hence it is not naturally defined.

Nevertheless, this notion is interesting in applications since in general the MHC used to define an MHS on a variety X is in fact defined in $C^+F(X, \mathbb{Q})$ and $C^+F^2(X, \mathbb{C})$.

3.3.4.1   The Shift on the Weight

Let (K, W) be a complex of objects of an abelian category $\mathbb{A}$ with an increasing filtration W. We denote by $(T_M K, W)$ or $(K[1], W[1])$ the complex shifted by a translation on degrees of K and of W such that $(W[1]_n K[1]) := (W_{n-1}K)[1]$ or $W_n(T_M K) = W_{n-1}TK$, i.e., $(W[1]_n K[1])^i := (W_{n-1}K)^{i+1}$. Then

$$\mathrm{Gr}_n^W (K[1], W[1]) = (\mathrm{Gr}_{n-1}^W K)[1]$$

and if (K, W, F) is an MHC, then $H^i(\mathrm{Gr}_n^W (K[1], W[1]), F) = H^{i+1}(\mathrm{Gr}_{n-1}^W K, F)$ is an HS of weight $n + i$; in other words, $(K[1], W[1], F)$ is an MHC.

DEFINITION 3.3.23 (Mixed cone)   *Let* $u : (K, W, F) \to (K', W', F')$ *be a morphism of complexes in* $C^+F(\mathbb{A})$ *(resp.* $C^+F_2(\mathbb{A})$*) with increasing filtrations* W, W' *(resp. decreasing filtrations* F, F'*). The structure of the mixed cone* $C_M(u)$ *is defined on the cone complex* $C(u) := K[1] \oplus K'$ *with the filtrations* $W[1] \oplus W'$ *(resp.* $F \oplus F'$*).*

The definition is not in the derived category but on the level of filtered complexes. In particular, the mixed cone of an MHC is an MHC since

$$(\mathrm{Gr}_n^W C_M(u), F) = ((\mathrm{Gr}_{n-1}^W K)[1], F) \oplus (\mathrm{Gr}_n^{W'} K', F')$$

is an HC of weight n.

3.3.4.2   Morphisms of MHC

A morphism $u : K \to K'$ of an MHC (resp. cohomological MHC) consists of morphisms

$u_A : K_A \to K'_A$ in $D^+A($ resp. $D^+(X, A))$,

$u_{A \otimes \mathbb{Q}} : (K_{A \otimes \mathbb{Q}}, W) \to (K'_{A \otimes \mathbb{Q}}, W)$ in $D^+F(A \otimes \mathbb{Q})$ (resp. $D^+F(X, A \otimes \mathbb{Q}))$,

$u_{\mathbb{C}} : (K_{\mathbb{C}}, W, F) \to (K'_{\mathbb{C}}, W, F)$ in $D^+F_2\mathbb{C}$ (resp. $D^+F_2(X, \mathbb{C}))$,

and commutative diagrams,

$$
\begin{array}{ccc}
K_{A \otimes \mathbb{Q}} & \xrightarrow{u_{A \otimes \mathbb{Q}}} & K'_{A \otimes \mathbb{Q}} \\
\wr\downarrow \alpha & & \wr\downarrow \alpha' \\
K_A \otimes \mathbb{Q} & \xrightarrow{u_A \otimes \mathbb{Q}} & K'_A \otimes \mathbb{Q},
\end{array}
\qquad
\begin{array}{ccc}
(K_{A \otimes \mathbb{Q}}, W) \otimes \mathbb{C} & \xrightarrow{u_{A \otimes \mathbb{Q}} \otimes \mathbb{C}} & (K'_{A \otimes \mathbb{Q}}, W) \otimes \mathbb{C} \\
\wr\downarrow \beta & & \wr\downarrow \beta' \\
(K_{\mathbb{C}}, W) & \xrightarrow{u_{\mathbb{C}}} & (K'_{\mathbb{C}}, W)
\end{array}
$$

in $D^+(A \otimes \mathbb{Q})$ on the left (resp. $D^+(X, A \otimes \mathbb{Q})$ on the right) compatible with W in $D^+F(\mathbb{C})$ (resp. $D^+F(X, \mathbb{C})$).

Remark. Let $u : K \to K'$ be a morphism of an MHC. There exists a quasi-isomorphism $v = (v_A, v_{A \otimes \mathbb{Q}}, v_{\mathbb{C}}) : \tilde{K} \xrightarrow{\approx} K$ and a morphism $\tilde{u} = (\tilde{u}_A, \tilde{u}_{A \otimes \mathbb{Q}}, \tilde{u}_{\mathbb{C}}) : \tilde{K} \to K'$ of an MHC such that v and $\tilde{u}$ are defined successively in C^+A, $C^+F(A \otimes \mathbb{Q})$ and $C^+F_2\mathbb{C}$; i.e., we can find, by definition, diagrams

$$K_A \xleftarrow{\approx} \tilde{K}_A \to K'_A, \quad K_{A \otimes \mathbb{Q}} \xleftarrow{\approx} \tilde{K}_{A \otimes \mathbb{Q}} \to K'_{A \otimes \mathbb{Q}}, \quad K_{\mathbb{C}} \xleftarrow{\approx} \tilde{K}_{\mathbb{C}} \to K'_{\mathbb{C}},$$

or in short $K \xleftarrow[v]{\approx} \tilde{K} \xrightarrow{\tilde{u}} K'$ (or equivalently $K \xrightarrow{\tilde{u}} \tilde{K}' \xleftarrow[v]{\approx} K'$) representing u.

3.3.4.3 Dependence on Homotopy

Consider a morphism $u : K \to K'$ of an MHC, represented by a morphism of complexes $\tilde{u} : \tilde{K} \to K'$.

To define the mixed cone $C_M(\tilde{u})$ out of

(i) the cones $C(\tilde{u}_A) \in C^+(A)$, $C_M(\tilde{u}_{A \otimes \mathbb{Q}}) \in C^+F(A \otimes \mathbb{Q})$, $C_M(\tilde{u}_{\mathbb{C}}) \in C^+F_2(\mathbb{C})$, we still need to define compatibility isomorphisms

$$\gamma_1 : C_M(\tilde{u}_{A \otimes \mathbb{Q}}) \simeq C(\tilde{u}_A) \otimes \mathbb{Q}, \quad \gamma_2 : (C_M(\tilde{u}_{\mathbb{C}}), W) \simeq (C_M(\tilde{u}_{A \otimes \mathbb{Q}}), W) \otimes \mathbb{C},$$

successively in $C^+(A \otimes \mathbb{Q})$ and $C^+F(\mathbb{C})$. With the notation of Section 3.3.4.2, the choice of isomorphisms $C_M(\tilde{\alpha}, \tilde{\alpha}')$ and $C_M(\tilde{\beta}, \tilde{\beta}')$ representing the compatibility isomorphisms in K and K' does not define compatibility isomorphisms for the cone since the diagrams of compatibility are commutative only up to homotopy, that is, there exist homotopies h_1 and h_2 such that

$$\tilde{\alpha}' \circ (\tilde{u}_{A \otimes \mathbb{Q}}) - (\tilde{u}_A \otimes \mathbb{Q}) \circ \tilde{\alpha} = h_1 \circ d + d \circ h_1$$

and

$$\tilde{\beta}' \circ \tilde{u}_{\mathbb{C}} - (\tilde{u}_{A \otimes \mathbb{Q}} \otimes \mathbb{C}) \circ \tilde{\beta} = h_2 \circ d + d \circ h_2.$$

(ii) Then we can define the compatibility isomorphism as

$$C_M(\tilde{\alpha}, \tilde{\alpha}', h_1) := \begin{pmatrix} \tilde{\alpha} & 0 \\ h_1 & \tilde{\alpha}' \end{pmatrix} : C_M(\tilde{u}_{A \otimes \mathbb{Q}}) \xrightarrow{\sim} C(\tilde{u}_A) \otimes \mathbb{Q}$$

and a similar formula for $C_M(\tilde{\beta}, \tilde{\beta}', h_2)$.

LEMMA AND DEFINITION 3.3.24 *Let $u : K \to K'$ be a morphism of MHC. The mixed cone $C_M(\tilde{u}, h_1, h_2)$ constructed above depends on the choices of the homotopies (h_1, h_2), the choice of a representative $\tilde{u}$ of u, and satisfies the relation*

$$\mathrm{Gr}_n^W(C_M(\tilde{u}), F) \simeq (\mathrm{Gr}_{n-1}^W(T\tilde{K}), F) \oplus (\mathrm{Gr}_n^W K', F)$$

is an HC of weight n; hence $C_M(\tilde{u}, h_1, h_2)$ is an MHC.

Remark. The MHC used in the case of a projective NCD case and its complement are naturally defined in $C^+F(X, \mathbb{Q})$ and $C^+F_2(X, \mathbb{C})$.

3.4 MHS ON THE COHOMOLOGY OF A COMPLEX ALGEBRAIC VARIETY

THEOREM 3.4.1 (Deligne) *The cohomology of complex algebraic varieties is endowed with a mixed Hodge structure functorial with respect to algebraic morphisms.*

The aim of this section is to prove the above theorem. The uniqueness follows easily once we have fixed the case of compact normal crossing divisors which, in particular, includes the nonsingular compact case. All that we need is to construct explicitly on each algebraic variety X a cohomological MHC, to which we apply the previous abstract algebraic study to define the MHS on the cohomology groups of X.

First, on a smooth complex variety X containing a normal crossing divisor (NCD) Y with smooth irreducible components, we shall construct the complex of sheaves of differential forms with logarithmic singularities along Y denoted $\Omega_X^*(\operatorname{Log} Y)$, or $\Omega_X^*\langle Y \rangle$, whose hypercohomology on X is isomorphic to the cohomology of $X - Y$ with coefficients in $\mathbb{C}$. We shall endow this complex with two filtrations W and F.

When X is also compact algebraic, the bifiltered complex $(\Omega_X^*(\operatorname{Log} Y), W, F)$ underlies the structure of a cohomological MHC which defines a mixed Hodge structure on the cohomology of $X - Y$. In other terms, to construct the MHS of a smooth variety V, we have to consider a compactification of V by a compact algebraic variety X, which always exists by a result of Nagata [36]. Moreover, by Hironaka's desingularization theorem [30], we can suppose X smooth and the complement $Y = X - V$ to be an NCD with smooth irreducible components. Then, the MHS on the cohomology of $V = X - Y$ will be deduced from the logarithmic complex $(\Omega_X^*(\operatorname{Log} Y), W, F)$. It is not difficult to show that such an MHS does not depend on the compactification X and will be referred to as the MHS on (the cohomology groups of) V. In some sense it depends on the asymptotic properties at infinity of V. The weights of the MHS on the cohomology $H^i(V)$ of a smooth variety V, i.e., the weights of the HS on Gr_j^W are $j \geq i$ and to be precise $W_{i-1} = 0, W_{2i} = H^i(V)$.

The hypercohomology on X of the Poincaré–Verdier dual of the logarithmic complex is the cohomology with compact support $H_c^*(X - Y, \mathbb{C})$ of $X - Y$ (see [14]) is more natural to construct as it can be deduced from the mixed cone construction (shifted by $[-1]$) over the restriction map $\Omega_X^* \to \Omega_{Y_\bullet}^*$ from the de Rham complex on X to the MHC on the normal crossing divisor Y described in Section 3.2. It is associated to the natural morphism $\mathbb{Z}_X \to \mathbb{Z}_{Y_\bullet}$.

Let Z be a sub-NCD of Y, the union of some components of Y; then the complement $Y - Z$ is an open NCD and its cohomology may be described by a double complex combination of the open case and the NCD case [14]. It is a model for the simplicial general case in this section. In this case, the weights of the MHS on the cohomology $H^j(Y - Z)$ vary from 0 to $2j$ which is valid for all algebraic varieties.

For any algebraic variety, the construction is based on a diagram of algebraic varieties

$$
X_\bullet = \left(X_0 \; \overset{\leftarrow}{\underset{\leftarrow}{\overset{\leftarrow}{}}} \; X_1 \; \overset{\leftarrow}{\underset{\leftarrow}{\overset{\leftarrow}{}}} \; \cdots \; X_{q-1} \quad \overset{X_*(\delta_i)}{\underset{\longleftarrow}{\vdots}} \quad X_q \cdots \right),
$$

similar to the model in the case of an NCD. Here the $X_*(\delta_i)$ are called the face maps (see Definition 3.4.11), one for each $i \in [0, q]$, and satisfy commutativity relations under composition.

In this case, we consider diagrams of complexes of sheaves and the resolutions of such sheaves are defined with compatibility relations with respect to the maps $X_*(\delta_i)$.

We are interested in such diagrams when they form a simplicial hypercovering of an algebraic variety X by nonsingular varieties; in other words, when the diagram defines a resolution of the constant sheaf $\mathbb{Z}_X$ on X by the direct images of constant sheaves on the various nonsingular X_i. Using a general simplicial technique combined with Hironaka's desingularization at the various steps, Deligne shows the existence of such simplicial resolutions [8].

In the case of a noncompact variety X, we can embed $X_\bullet$ into a corresponding diagram $V_\bullet$ of compact smooth complex varieties such that $V_\bullet - X_\bullet$ is a normal crossing divisor in each index. Then we can use the logarithmic complexes on the terms of $V_\bullet$ to construct an associated cohomological MHC on the variety $V_\bullet$. Thus we can develop a process to deduce an MHC defining an MHS on the cohomology of X which generalizes the construction of the MHC in the case of an NCD and more generally an open NCD (the difference of two NCD). The construction of the weight filtration is based on a diagonal process and it is similar to a repeated mixed cone construction without the ambiguity of the choice of homotopy, since resolutions of simplicial complexes of sheaves are functorial in the simplicial derived category.

In particular, we should view the simplicial category as a set of diagrams and the construction is carried out with respect to such diagrams. In fact, there exists another construction based on a set of diagrams defined by cubical schemes [26, 37].

At the end of this section, we give an alternative construction for embedded varieties with diagrams of four edges only (see [14]), which shows that the ambiguity in the mixed mapping cone construction may be overcome.

In all cases, the mixed Hodge structure is constructed first for smooth varieties and normal crossing divisors, then it is deduced for general varieties. The uniqueness follows from the compatibility of the MHS with Poincaré duality and classical exact sequences on cohomology as we will see.

3.4.1 MHS on the Cohomology of Smooth Algebraic Varieties

As we have already said, to construct the mixed Hodge structure on the cohomology of a smooth complex algebraic variety V, we use a result of Nagata [36] to embed V as an open Zariski subset of a complete variety Z (here we need the algebraic structure on V). Then the singularities of Z are included in $D := Z - V$. Since Hironaka's desingularization process in characteristic 0 (see [30]) is carried out by blowing up smooth centers above D, there exists a variety $X \to Z$ above Z such that the inverse image of D is a normal crossing divisor Y with smooth components in X such that $X - Y \simeq Z - D$.

Hence, we may start with the hypothesis that $V = X^* := X - Y$ is the complement of a normal crossing divisor Y in a smooth compact algebraic variety X. The construction of the mixed Hodge structure is reduced to this situation, if it is proved that it does not depend on the choice of X.

We introduce now the logarithmic complex underlying the structure of a cohomological mixed Hodge complex on X which computes the cohomology of V.

3.4.1.1   The Logarithmic Complex

Let X be a complex manifold and Y be an NCD in X. We denote by $j : X^* \to X$ the embedding of $X^* := X - Y$ into X. We say that a meromorphic form ω has a pole of order at most 1 along Y if at each point $y \in Y$, $f\omega$ is holomorphic for some local equation f of Y at y. Let $\Omega_X^*(*Y)$ denote the subcomplex of $j_*\Omega_{X^*}^*$ defined by meromorphic forms along Y, holomorphic on X^* (excluding forms with essential singularities in $j_*\Omega_{X^*}^*$). The idea of a logarithmic complex probably originated in the work of Atiyah and Hodge and the algebraic de Rham theorem of Grothendieck as explained at the end of Chapter 2.

DEFINITION 3.4.2 (Logarithmic complex)   *The logarithmic de Rham complex of X along a normal crossing divisor Y is the subcomplex $\Omega_X^*(\mathrm{Log}\, Y)$ of the complex $\Omega_X^*(*Y)$ defined by the sections ω such that ω and $d\omega$ both have a pole of order at most 1 along Y.*

By definition, at each point $y \in Y$, there exist local coordinates $(z_i)_{i\in[1,n]}$ on X and a subset $I_y \subset [1, n]$ depending on $y \in Y$ such that Y is defined at y by the equation $\Pi_{i\in I_y} z_i = 0$. Then ω and $d\omega$ have logarithmic poles along Y if and only if ω can be written locally as

$$\omega = \sum_{i_1,\dots,i_r \in I_y} \varphi_{i_1,\dots,i_r} \frac{dz_{i_1}}{z_{i_1}} \wedge \cdots \wedge \frac{dz_{i_r}}{z_{i_r}}, \quad \text{where} \quad \varphi_{i_1,\dots,i_r} \text{ is holomorphic.}$$

Indeed, $d(1/z_i) = -dz_i/(z_i)^2$ has a pole of order 2, and $d\omega$ will have a pole along $z_i = 0$ of order 2, unless ω is divisible by dz_i/z_i, i.e., $\omega = \omega' \wedge (dz_i/z_i)$.

This formula may be used as a definition; then we prove the independence of the choice of coordinates, that is, ω may be written in this form with respect to any set of local coordinates at y.

The $\mathcal{O}_X$-module $\Omega_X^1(\mathrm{Log}\, Y)$ is locally free at $y \in Y$ with basis $(dz_i/z_i)_{i\in I_y}$ and $(dz_j)_{j\in[1,n]-I_y}$ and $\Omega_X^p(\mathrm{Log}\, Y) \simeq \wedge^p \Omega_X^1(\mathrm{Log}\, Y)$.

Let $f : X_1 \to X_2$ be a morphism of complex manifolds, with normal crossing divisors Y_i in X_i for $i = 1, 2$ such that $f^{-1}(Y_2) = Y_1$. Then, the reciprocal morphism $f^* : f^*(j_{2*}\Omega_{X_2^*}^*) \to j_{1*}\Omega_{X_1^*}^*$ induces a morphism on logarithmic complexes:

$$f^* : f^*\Omega_{X_2}^*(\mathrm{Log}\, Y_2) \to \Omega_{X_1}^*(\mathrm{Log}\, Y_1).$$

3.4.1.2   The Weight Filtration W

Let $Y = \cup_{i \in I} Y_i$ be the union of smooth irreducible divisors. We put an order on I. Let S^q denote the set of strictly increasing sequences $\sigma = (\sigma_1, \ldots, \sigma_q)$ in the set of indices I, such that $Y_\sigma \neq \emptyset$, where $Y_\sigma = Y_{\sigma_1 \cdots \sigma_q} = Y_{\sigma_1} \cap \cdots \cap Y_{\sigma_q}$. Set $Y^q = \coprod_{\sigma \in S^q} Y_\sigma$, the disjoint union of Y_σ, $Y^0 = X$, and let $\Pi : Y^q \to Y$ be the canonical projection. An increasing filtration W of the logarithmic complex, called the weight filtration, is defined as follows:

$$W_m(\Omega_X^p(\operatorname{Log} Y)) = \sum_{\sigma \in S^m} \Omega_X^{p-m} \wedge dz_{\sigma_1}/z_{\sigma_1} \wedge \cdots \wedge dz_{\sigma_m}/z_{\sigma_m}.$$

The sub-$\mathcal{O}_X$-module $W_m(\Omega_X^p(\operatorname{Log} Y)) \subset \Omega_X^p(\operatorname{Log} Y)$ is the smallest submodule stable by exterior multiplication with local sections of Ω_X^* and containing the products $dz_{i_1}/z_{i_1} \wedge \cdots \wedge dz_{i_k}/z_{i_k}$ for $k \leq m$ for local equations z_j of the components of Y.

3.4.1.3   The Residue Isomorphism

We define now Poincaré residue isomorphisms:

$$\operatorname{Res} : \operatorname{Gr}_m^W(\Omega_X^p(\operatorname{Log} Y)) \to \Pi_* \Omega_{Y^m}^{p-m}, \quad \operatorname{Res} : \operatorname{Gr}_m^W(\Omega_X^*(\operatorname{Log} Y)) \xrightarrow{\sim} \Pi_* \Omega_{Y^m}^*[-m].$$

Locally, at a point y on the intersection of an ordered set of m components $Y_{i_1}, \ldots, Y_{i_m}$ of Y, we choose a set of local equations z_i for $i \in I_y$ of the components of Y at y and an order of the indices $i \in I_y$, then the Poincaré residue, defined on W_m by

$$\operatorname{Res}_{Y_{i_1}, \ldots, Y_{i_m}}(\alpha \wedge dz_{i_1}/z_{i_1} \wedge \cdots \wedge dz_{i_m}/z_{i_m}) = \alpha_{|Y_{i_1, \ldots, i_m}},$$

vanishes on W_{m-1}, hence it induces on Gr_m^W a morphism independent of the choice of the equations and compatible with the differentials.

To prove the isomorphism, we construct its inverse. For each sequence of indices $\sigma = (i_1, \ldots, i_m)$, we consider the morphism $\rho_\sigma : \Omega_X^p \to \operatorname{Gr}_m^W(\Omega_X^{p+m}(\operatorname{Log} Y))$, defined locally as

$$\rho_\sigma(\alpha) = \alpha \wedge dz_{\sigma_1}/z_{\sigma_1} \wedge \cdots \wedge dz_{i_m}/z_{i_m}.$$

It does not depend on the choice of z_i, since for another choice of coordinates z_i', z_i/z_i' are holomorphic and the difference $(dz_i/z_i) - (dz_i'/z_i') = d(z_i/z_i')/(z_i/z_i')$ is holomorphic; then

$$\rho_\sigma(\alpha) - \alpha \wedge dz_{i_1}'/z_{i_1}' \wedge \cdots \wedge dz_{i_m}'/z_{i_m}' \in W_{m-1}\Omega_X^{p+m}(\operatorname{Log} Y),$$

and successively $\rho_\sigma(\alpha) - \rho_\sigma'(\alpha) \in W_{m-1}\Omega_X^{p+m}(\operatorname{Log} Y)$.

We have $\rho_\sigma(z_{i_j} \cdot \beta) = 0$ and $\rho_\sigma(dz_{i_j} \wedge \beta') = 0$ for sections β of Ω_X^p and β' of Ω_X^{p-1}, hence ρ_σ factors by $\overline{\rho}_\sigma$ on $\Pi_* \Omega_{Y_\sigma}^p$. The local definitions glue together to define a global construction of a morphism of complexes on X:

$$\overline{\rho}_\sigma : \Pi_* \Omega_{Y_\sigma}^p \to \operatorname{Gr}_m^W(\Omega_X^{p+m}(\operatorname{Log} Y)), \quad \overline{\rho} : \Pi_* \Omega_{Y^m}^*[-m] \to \operatorname{Gr}_m^W \Omega_X^*(\operatorname{Log} Y).$$

LEMMA 3.4.3 *We have the following isomorphisms of sheaves:*

(i) $H^i(\mathrm{Gr}_m^W \Omega_X^*(\mathrm{Log}\,Y)) \simeq \Pi_* \mathbb{C}_{Y^m}$ *for $i = m$ and 0 for $i \neq m$.*

(ii) $H^i(W_r \Omega_X^*(\mathrm{Log}\,Y)) \simeq \Pi_* \mathbb{C}_{Y^i}$ *for $i \leq r$ and $H^i(W_r \Omega_X^*(\mathrm{Log}\,Y)) = 0$ for $i > r$; in particular, $H^i(\Omega_X^*(\mathrm{Log}\,Y)) \simeq \Pi_* \mathbb{C}_{Y^i}$.*

PROOF. The statement in (i) follows from the residue isomorphism.
The statement in (ii) follows easily by induction on r, from (i) and the long exact sequence associated to the short exact sequence $0 \to W_r \to W_{r+1} \to \mathrm{Gr}_{r+1}^W \to 0$, written as

$$\cdots \to H^i(W_r) \to H^i(W_{r+1}) \to H^i(\mathrm{Gr}_{r+1}^W) \to H^{i+1}(W_r) \to \cdots .$$

$\square$

PROPOSITION 3.4.4 (Rational weight filtration W) *The morphisms of filtered complexes*

$$(\Omega_X^*(\mathrm{Log}\,Y), W) \overset{\alpha}{\leftarrow} (\Omega_X^*(\mathrm{Log}\,Y), \tau) \overset{\beta}{\to} (j_* \Omega_{X^*}^*, \tau),$$

where τ is the truncation filtration, are filtered quasi-isomorphisms.

PROOF. We deduce the quasi-isomorphism α from the lemma. The morphism j is Stein, since for each polydisk $U(y)$ in X centered at a point $y \in Y$, the inverse image $X^* \cap U(y)$ is Stein as the complement of a hypersurface; hence j is acyclic for coherent sheaves, that is, $Rj_* \Omega_{X^*}^* \simeq j_* \Omega_{X^*}^*$. By the Poincaré lemma, $\mathbb{C}_{X^*} \simeq \Omega_{X^*}^*$, so that $Rj_* \mathbb{C}_{X^*} \simeq j_* \Omega_{X^*}^*$; hence it is enough to prove $\mathrm{Gr}_i^\tau Rj_* \mathbb{C}_{X^*} \simeq \Pi_* \mathbb{C}_{Y^i}$, which is a local statement.

For each polydisk $U(y) = U$, the open subset $U^* = U - U \cap Y \simeq (D^*)^m \times D^{n-m}$ is homotopic to an i-dimensional torus $(S^1)^m$. Hence the hypercohomology $\mathbb{H}^i(U, Rj_* \mathbb{C}_{X^*}) = H^i(U^*, \mathbb{C})$ can be computed by the Künneth formula and is equal to $\wedge^i H^1(U^*, \mathbb{C}) \simeq \wedge^i \Gamma(U, \Omega_U^1(\mathrm{Log}\,Y))$, where the wedge products of $dz_i/z_i, i \in [1, m]$ form a basis dual (up to powers of $2i\pi$) to the homology basis defined by products of homology classes of S^1 (the duality is obtained by integrating a product of k-forms on the product of k circles with value $(2i\pi)^k$ in the case of coincidence of the indices, or 0 otherwise. $\square$

COROLLARY 3.4.5 *The weight filtration is rationally defined.*

The main point here is that the τ filtration is defined with rational coefficients as $(Rj_* \mathbb{Q}_{X^*}, \tau) \otimes \mathbb{C}$, which gives the rational definition for W.

3.4.1.4 Hodge Filtration F

It is defined by the formula $F^p = \Omega_X^{* \geq p}(\mathrm{Log}\,Y)$, which includes all forms of type (p', q') with $p' \geq p$. We have

$$\mathrm{Res} : F^p(\mathrm{Gr}_m^W \Omega_X^*(\mathrm{Log}\,Y)) \simeq \Pi_* F^{p-m} \Omega_{Y^m}^*[-m];$$

hence a filtered isomorphism

$$\mathrm{Res}: (\mathrm{Gr}_m^W \, \Omega_X^*(\mathrm{Log}\, Y), F) \simeq (\Pi_* \Omega_{Y^m}^*[-m], F[-m]).$$

COROLLARY 3.4.6 *The system $\mathbb{K}$ such that*

(1) $(\mathbb{K}^\mathbb{Q}, W) = (Rj_* \mathbb{Q}_{X^*}, \tau) \in \mathrm{Ob}\, D^+ F(X, \mathbb{Q})$;

(2) $(\mathbb{K}^\mathbb{C}, W, F) = (\Omega_X^*(\mathrm{Log}\, Y), W, F) \in \mathrm{Ob}\, D^+ F_2(X, \mathbb{C})$;

(3) the isomorphism $(\mathbb{K}^\mathbb{Q}, W) \otimes \mathbb{C} \simeq (\mathbb{K}^\mathbb{C}, W)$ in $D^+ F(X, \mathbb{C})$;

is a cohomological MHC on X.

Remark. There exist functorial resolutions of the above system $\mathbb{K}$, such as Godement resolutions, for example, denoted below by $\mathcal{G}^\bullet$, which define the morphisms of filtered complexes in $C^+ F(X, \mathbb{C})$:

$$\mathcal{G}^\bullet(\Omega_X^*(\mathrm{Log}\, Y), W) \xleftarrow{\alpha} \mathcal{G}^\bullet(\Omega_X^*(\mathrm{Log}\, Y), \tau) \xrightarrow{\beta} j_* \mathcal{G}^\bullet(\Omega_{X^*}^*, \tau) \xleftarrow{I} j_* \mathcal{G}^\bullet(\mathbb{C}_{X^*}, \tau).$$

Then, the mixed cone construction applies to the system. For example, the mixed cone over the mapping $(\Omega_X^*, F) \to (\Omega_X^*(\mathrm{Log}\, Y), W, F)$ from the HC on X of weight 0 is quasi-isomorphic to $R\Gamma_Y \mathbb{C}[1]$. It puts an MHS on the cohomology with support in Y. Also, the quotient complex $(\Omega_X^*(\mathrm{Log}\, Y)/\Omega_X^*, W, F)$ defines the same MHS.

THEOREM 3.4.7 (MHS of a smooth variety (Deligne)) *The system $K = R\Gamma(X, \mathbb{K})$ is a mixed Hodge complex. It endows the cohomology of $X^* = X - Y$ with a canonical mixed Hodge structure.*

PROOF. The result follows directly from the general theory of cohomological MHC. Nevertheless, it is interesting to understand what is needed for a direct proof and to compute the weight spectral sequence at rank 1:

$$_W E_1^{p,q}(R\Gamma(X, \Omega_X^*(\mathrm{Log}\, Y))) = \mathbb{H}^{p+q}(X, \mathrm{Gr}_{-p}^W \Omega_X^*(\mathrm{Log}\, Y)) \simeq \mathbb{H}^{p+q}(X, \Pi_* \Omega_{Y-p}^*[p])$$

$$\simeq H^{2p+q}(Y^{-p}, \mathbb{C}) \Rightarrow \mathrm{Gr}_q^W H^{p+q}(X^*, \mathbb{C}),$$

where the double arrow means that the spectral sequence degenerates to the cohomology graded with respect to the filtration W induced by the weight on the complex level, shifted by $p + q$. We recall the proof: the differential d_1,

$$d_1 = \sum_{j=1}^{-p} (-1)^{j+1} G(\lambda_{j,-p}) = G : H^{2p+q}(Y^{-p}, \mathbb{C}) \longrightarrow H^{2p+q+2}(Y^{-p-1}, \mathbb{C}),$$

where $\lambda_{j,-p}$ is defined as in Section 3.2.4.1, is equal to an alternate Gysin morphism, Poincaré dual to the alternate restriction morphism

$$\rho = \sum_{j=1}^{-p} (-1)^{j+1} \lambda_{j,-p}^* : H^{2n-q}(Y^{-p-1}, \mathbb{C}) \to H^{2n-q}(Y^{-p}, \mathbb{C}).$$

Therefore, the first term,

$$(_W E_1^{p,q}, d_1)_{p\in\mathbb{Z}} = (H^{2p+q}(Y^{-p}, \mathbb{C}), d_1)_{p\in\mathbb{Z}}$$

is viewed as a complex in the category of HS of weight q. It follows that the terms

$$_W E_2^{p,q} = H^p(_W E_1^{*,q}, d_1)$$

are endowed with an HS of weight q. We need to prove that the differential d_2 is compatible with the induced Hodge filtration. For this we introduced the direct filtrations compatible with d_2 and proved that they coincide with the induced Hodge filtration. The differential d_2 is necessarily zero since it is a morphism of HS of different weights: the HS of weight q on $E_2^{p,q}$ and the HS of weight $q-1$ on $E_2^{p+2,q-1}$. The proof is the same for any MHC and consists of a recurrent argument to show in this way that the differentials d_i for $i \geq 2$ are zero (see Theorem 3.2.30).   □

3.4.1.5 Independence of the Compactification and Functoriality

Let U be a smooth complex algebraic variety, X (resp. X') a compactification of U by a normal crossing divisor Y (resp. Y') at infinity, $j : U \to X$ (resp. $j' : U \to X'$) the open embedding; then $j \times j' : U \to X \times X'$ is a locally closed embedding, with closure V. By desingularizing V outside the image of U, we are reduced to the case where we have a smooth variety $X'' \xrightarrow{f} X$ such that $Y'' := f^{-1}(Y)$ is an NCD and $U \simeq X'' - Y''$; then we have an induced morphism f^* on the corresponding logarithmic complexes, compatible with the structure of MHC. It follows that the induced morphism f^* on hypercohomology is compatible with the MHS. As it is an isomorphism on the underlying hypercohomology groups compatible with the filtration F and W, it is necessarily an isomorphism of MHS.

Functoriality. Let $f : U \to V$ be a morphism of smooth varieties, and let X (resp. Z) be smooth compactifications of U (resp. V) by NCD at infinity; then taking the closure of the graph of f in $X \times Z$ and desingularizing, we are reduced to the case where there exists a compactification X with an extension $\overline{f} : X \to Z$ inducing f on U. The induced morphism $\overline{f}^*$ on the corresponding logarithmic complexes is compatible with the filtrations W and F and with the structure of MHC; hence it is compatible with the MHS on hypercohomology.

PROPOSITION 3.4.8 *Let U be a smooth complex algebraic variety.*

(i) *The Hodge numbers $h^{p,q} := \dim H^{p,q}(\mathrm{Gr}_{p+q}^W H^i(U, \mathbb{C}))$ vanish for $p, q \notin [0, i]$. In particular, the weight of the cohomology $H^i(U, \mathbb{C})$ varies from i to $2i$.*

(ii) *let X be a smooth compactification of U; then*

$$W_i H^i(U, \mathbb{Q}) = \mathrm{Im}(H^i(X, \mathbb{Q}) \to H^i(U, \mathbb{Q})).$$

PROOF.   (i) The space $\mathrm{Gr}_r^W H^i(U, \mathbb{Q})$ is isomorphic to the term $E_2^{i-r,r}$ of the spectral sequence with a Hodge structure of weight r; hence it is a subquotient of $E_1^{i-r,r} = H^{2i-r}(Y^{r-i}, \mathbb{Q})(i-r)$ of the twisted MHS on Y^{r-i}, which gives the following relation with the Hodge numbers $h^{p,q}$ and $h^{p',q'}$ on Y^{r-i}: $h^{p,q} = h^{p',q'}$ for $(p,q) = (p'+r-i, q'+r-i)$. Since $h^{p',q'}(H^{2i-r}(Y^{r-i}, \mathbb{Q})) = 0$ unless $r-i \geq 0$ ($Y^{r-i} \neq \emptyset$) and $2i - r \geq 0$ (the degree of cohomology), we deduce $i \leq r \leq 2i$. Moreover, $p', q' \in [0, 2i-r]$; hence $p, q \in [r-i, i]$.

(ii) Suppose first that U is the complement of an NCD and let $j : U \to X$ be the inclusion. By definition, $W_i H^i(U, \mathbb{Q}) = \mathrm{Im}(\mathbb{H}^i(X, \tau_{\leq 0} Rj_* \mathbb{Q}_U) \to H^i(U, \mathbb{Q}))$; hence it is equal to the image of $H^i(X, \mathbb{Q})$ since: $\mathbb{Q} \simeq j_* \mathbb{Q}_U \simeq \tau_{\leq 0} Rj_* \mathbb{Q}_U$.

If $X - U$ is not an NCD, there exists a desingularization $\pi : X' \to X$ with an embedding of U onto $U' \subset X'$ as the complement of an NCD; then we use the trace map $\mathrm{Tr}\,\pi : H^i(X', \mathbb{Q}) \to H^i(X, \mathbb{Q})$ satisfying $(\mathrm{Tr}\,\pi) \circ \pi^* = \mathrm{Id}$ and compatible with Hodge structures. In fact the trace map is defined as a morphism of sheaves $R\pi_* \mathbb{Q}_{X'} \to \mathbb{Q}_X$ (see [42, (2.3.4)] and [27, (III.10), (VI,4)]) and hence commutes with the restriction to U. In particular, the images of both cohomology groups $H^i(X', \mathbb{Q})$ and $H^i(X, \mathbb{Q})$ coincide in $H^i(U, \mathbb{Q})$. $\square$

EXAMPLE 3.4.9 (Riemann Surface)   Let $\overline{C}$ be a connected compact Riemann surface of genus g, $Y = \{x_1, \ldots, x_m\}$ a subset of m points, and $C = \overline{C} - Y$ the open surface with $m > 0$ points in $\overline{C}$ deleted. The long exact sequence

$$0 \to H^1(\overline{C}, \mathbb{Z}) \to H^1(C, \mathbb{Z}) \to H^2_Y(\overline{C}, \mathbb{Z}) = \oplus_{i=1}^{i=m} \mathbb{Z} \to H^2(\overline{C}, \mathbb{Z}) \to H^2(C, \mathbb{Z}) = 0$$

reduces to the following short exact sequence of mixed Hodge structures:

$$0 \to H^1(\overline{C}, \mathbb{Z}) \to H^1(C, \mathbb{Z}) \to \mathbb{Z}^{m-1} \simeq \mathrm{Ker}(\oplus_{i=1}^{i=m} \mathbb{Z} \to \mathbb{Z}) \to 0,$$

where $H^1(C, \mathbb{Z}) = W_2 H^1(C, \mathbb{Z})$ is of rank $2g + m - 1$, $W_1 H^1(C, \mathbb{Z}) = H^1(\overline{C}, \mathbb{Z})$ is of rank $2g$ and $\mathrm{Gr}_2^W H^1(C, \mathbb{Z}) \simeq \mathbb{Z}^{m-1}$.

The Hodge filtration is given by

- $F^0 H^1(C, \mathbb{C}) = H^1(C, \mathbb{C})$ is of rank $2g + m - 1$;

- $F^1 H^1(C, \mathbb{C})$ has dimension $g + m - 1$, where

$$F^1 H^1(C, \mathbb{C}) \simeq \mathbb{H}^1(\overline{C}, (0 \to \Omega^1_{\overline{C}}(\mathrm{Log}\{x_1, \ldots, x_m\})))$$
$$\simeq H^0(\overline{C}, \Omega^1_{\overline{C}}(\mathrm{Log}\{x_1, \ldots, x_m\})).$$

  Indeed,
$$\mathrm{Gr}_F^0 H^1(C, \mathbb{C}) \simeq H^1(\overline{C}, \mathrm{Gr}_F^0 \Omega^*_{\overline{C}}(\mathrm{Log}\{x_1, \ldots, x_m\})) \simeq H^1(\overline{C}, \mathcal{O}_{\overline{C}})$$
  is of rank g,

- $F^2 H^1(C, \mathbb{C}) = 0$.

The exact sequence $0 \to \Omega^1_{\overline{C}} \to \Omega^1_{\overline{C}}(\mathrm{Log}\{x_1, \ldots, x_m\}) \to \mathcal{O}_{\{x_1, \ldots, x_m\}} \to 0$ is defined by the residue morphism and has the associated long exact sequence

$$0 \to H^0(\overline{C}, \Omega^1_{\overline{C}}) \simeq \mathbb{C}^g \to H^0(\overline{C}, \Omega^1_{\overline{C}}(\mathrm{Log}\{x_1, \ldots, x_m\})) \to H^0(\overline{C}, \mathcal{O}_{\{x_1, \ldots, x_m\}}) \simeq$$
$$\mathbb{C}^m \to H^1(\overline{C}, \Omega^1_{\overline{C}}) \simeq \mathbb{C} \to H^1(\overline{C}, \Omega^1_{\overline{C}}(\mathrm{Log}\{x_1, \ldots, x_m\})) \simeq 0.$$

EXAMPLE 3.4.10 (Hypersurfaces) Let $i : Y \hookrightarrow P$ be a smooth hypersurface in a projective variety P. To describe the cohomology of the affine open set $U = P - Y$, we may use

(i) the complex of rational algebraic forms on P regular on U denoted $\Omega^*(U) = \Omega_P^*(*Y)$ (this follows from Grothendieck's result on algebraic de Rham cohomology; see Chapter 2);

(ii) the complex of forms on the analytic projective space meromorphic along Y, holomorphic on U denoted by $\Omega_{P_{an}}^*(*Y)$, where the Hodge filtration is described by the order of the pole (see [7, Prop. 3.1.11]) (the trivial filtration F described above on $\Omega^*(U)$ does not induce the correct Hodge filtration if U is not compact);

(iii) the complex of forms with logarithmic singularities denoted by the logarithmic complex $\Omega_{P_{an}}^*(\text{Log } Y)$ with its natural filtration F.

(1) For example, in the case of a curve Y in a plane $P = \mathbb{P}^2$, the global holomorphic forms are all rational by Serre's result on the cohomology of coherent sheaves. The residue along Y fits into an exact sequence of sheaves:

$$0 \to \Omega_P^2 \to \Omega_P^2(\text{Log } Y) \to i_*\Omega_Y^1 \to 0.$$

Since $H^0(P, \Omega_P^2) = H^1(P, \Omega_P^2) = 0$ ($h^{2,0} = h^{2,1} = 0$), we deduce from the associated long exact sequence, the isomorphism

$$\text{Res} : H^0(P, \Omega_P^2(\text{Log } Y)) \xrightarrow{\sim} H^0(Y, \Omega_Y^1).$$

Hence, the 1-forms on Y are residues of rational 2-forms on P with a simple pole along the curve.

In homogeneous coordinates, let $F = 0$ be a homogeneous equation of Y of degree d. Then, the 1-forms on Y are residues along Y of rational forms

$$\frac{A(z_0 dz_1 \wedge dz_2 - z_1 dz_0 \wedge dz_2 + z_2 dz_0 \wedge dz_1)}{F},$$

where A is homogeneous of degree $d - 3$ (see [4, Exa. 3.2.8]).

(2) The exact sequence defined by the relative cohomology (or cohomology with support in Y)

$$H^{k-1}(U) \xrightarrow{\partial} H_Y^k(P) \to H^k(P) \xrightarrow{j^*} H^k(U)$$

is transformed by Thom's isomorphism into

$$H^{k-1}(U) \xrightarrow{r} H^{k-2}(Y) \xrightarrow{i_*} H^k(P) \xrightarrow{j^*} H^k(U),$$

where r is the topological Leray residue map, dual to the map $\tau : H_{k-2}(Y) \to H_{k-1}(U)$, associating to a cycle c its inverse image in the boundary of a tubular neighborhood, which is a fibration by circles over c, and i_* is a Gysin map that is Poincaré dual to the map i^* in cohomology.

For $P = \mathbb{P}^{n+1}$ and n odd, the map r below is an isomorphism:

$$H^{n-1}(Y) \simeq H^{n+1}(P) \to H^{n+1}(U) \xrightarrow{r} H^n(Y) \xrightarrow{i_*} H^{n+2}(P) = 0 \xrightarrow{j^*} H^{n+2}(U).$$

For n even the map r is injective and surjective onto the primitive cohomology $H^n_{\mathrm{prim}}(X)$, defined as the kernel of i_*:

$$H^{n+1}(P) = 0 \to H^{n+1}(U) \xrightarrow{r} H^n(Y) \xrightarrow{i_*} H^{n+2}(P) = \mathbb{Q} \xrightarrow{j^*} H^{n+2}(U).$$

3.4.2 MHS on Cohomology of Simplicial Varieties

To construct a natural mixed Hodge structure on the cohomology of an algebraic variety S, not necessarily smooth or compact, Deligne considers a *simplicial smooth variety* $\pi : U_\bullet \to S$ which is a *cohomological resolution* of the original variety S in the sense that the direct image $\pi_* \mathbb{Z}_{U_\bullet}$ is a resolution of $\mathbb{Z}_S$ (descent theorem; see Theorem 3.4.16).

If S is not compact, it is possible to embed $U_\bullet$, into a simplicial smooth compact complex variety $X_\bullet$ such that $X_\bullet - U_\bullet$ consists of a normal crossing divisor in each term; then the various logarithmic complexes are connected by functorial relations and form a *simplicial cohomological MHC* giving rise to a cohomological MHC, defining a mixed Hodge structure on the cohomology of $U_\bullet$, which is transported on the cohomology of S by the augmentation map. Although such a construction is technically elaborate, the above abstract development of MHC leads easily to the result without further difficulty.

3.4.2.1 Simplicial Category

The simplicial category Δ is defined as follows:

(i) The objects of Δ are the subsets $\Delta_n := \{0, 1, \ldots, n\}$ of integers for $n \in \mathbb{N}$.

(ii) The set of morphisms of Δ are the sets $H_{p,q}$ of increasing mappings from Δ_p to Δ_q for integers $p, q \geq 0$, with the natural composition of mappings : $H_{pq} \times H_{qr} \to H_{\mathrm{pr}}$.

Notice that $f : \Delta_p \to \Delta_q$ is increasing (in the nonstrict sense) if for all $i < j$, $f(i) \leq f(j)$.

DEFINITION 3.4.11 *We define, for $0 \leq i \leq n + 1$, the ith face map, as the unique strictly increasing mapping $\delta_i : \Delta_n \to \Delta_{n+1}$ such that $i \notin \delta_i(\Delta_n)$.*

The semisimplicial category $\Delta_>$ is obtained when we only consider the strictly increasing morphisms in Δ. In what follows we could restrict the constructions to semisimplicial spaces which underlie the simplicial spaces and work only with such spaces, as we use only the face maps.

DEFINITION 3.4.12   *A simplicial (resp. cosimplicial) object* $X_\bullet := (X_n)_{n \in \mathbb{N}}$ *of a category* $\mathcal{C}$ *is a contravariant (resp. covariant) functor* F *from* Δ *to* $\mathcal{C}$.

A morphism $a : X_\bullet \to Y_\bullet$ *of simplicial (resp. cosimplicial) objects is defined by its components* $a_n : X_n \to Y_n$ *compatible with the various maps* $F(f)$ *for each simplicial map* $f \in H_{pq}$ *for all* $p, q \in \mathbb{N}$ *i.e.; the maps* a_p *and* a_q *commute with* $X(f)$ *and* $Y(f)$.

The functor $T : \Delta \to \mathcal{C}$ is defined by $T(\Delta_n) := X_n$ and for each $f : \Delta_p \to \Delta_q$, by $T(f) : X_q \to X_p$ (resp. $T(f) : X_p \to X_q$), $T(f)$ will be denoted by $X_\bullet(f)$.

3.4.2.2   Sheaves on a Simplicial Space

If $\mathcal{C}$ is the category of topological spaces, a simplicial functor defines a simplicial topological space. A sheaf F^* on a simplicial topological space $X_\bullet$ is defined by

(1) a family of sheaves F^n on X_n;

(2) for each $f : \Delta_n \to \Delta_m$ with $X_\bullet(f) : X_m \to X_n$, an $X_\bullet(f)$-morphism $F(f)_*$ from F^n to F^m, that is, maps $X_\bullet(f)^* F^n \to F^m$ on X_m satisfying for all $g : \Delta_r \to \Delta_n$, $F(f \circ g)_* = F(f)_* \circ F(g)_*$.

A morphism $u : F^* \to G^*$ is a family of morphisms $u^n : F^n \to G^n$ such that for all $f : \Delta_n \to \Delta_m$ we have $u^m F(f)_* = G(f)_* u^n$, where

$$X_*(f)^* F^n \xrightarrow{F(f)_*} F^m \xrightarrow{u^m} G^m, \quad X_*(f)^* F^n \xrightarrow{X_*(f)^*(u_n)} X_*(f)^* G^n \xrightarrow{G(f)_*} G^m.$$

The image of the ith face map by a functor is also denoted abusively by the same symbol:

$$\delta_i : X_{n+1} \to X_n.$$

Given a ring A, we will consider the derived category of cosimplicial sheaves of A-modules.

3.4.2.3   Derived Filtered Category on a Simplicial Space

The definition of a complex of sheaves K on a simplicial topological space X_* follows from the definition of sheaves. Such a complex has two degrees $K := K^{p,q}$, where p is the degree of the complex and q is the simplicial degree; hence for each p, $K^{p,*}$ is a simplicial sheaf and for each q, $K^{*,q}$ is a complex on X_q.

A quasi-isomorphism $\gamma : K \to K'$ (resp. filtered, bifiltered) of simplicial complexes (resp. with filtrations) on $X_\bullet$, is a morphism of simplicial complexes inducing a quasi-isomorphism $\gamma^{*,q} : K^{*,q} \to K'^{*,q}$ (resp. filtered, bifiltered) for each space X_q.

The definition of the derived category (resp. filtered, bifiltered) of the abelian category of abelian sheaves of groups (resp. vector spaces) on a simplicial space is obtained by inverting the quasi-isomorphisms (resp. filtered, bifiltered).

3.4.2.4 Constant and Augmented Simplicial Space

A topological space S defines a simplicial constant space $S_\bullet$ such that $S_n = S$ for all n and $S_\bullet(f) = \mathrm{Id}$ for all $f \in H^{p,q}$.

An augmented simplicial space $\pi : X_\bullet \to S$ is defined by a family of maps $\pi_n : X_n \to S_n = S$ defining a morphism of simplicial spaces.

The structural sheaves $\mathcal{O}_{X_n}$ of a simplicial complex analytic space form a simplicial sheaf of rings. Let $\pi : X_\bullet \to S$ be an augmentation to a complex analytic space S. The various de Rham complexes of sheaves $\Omega^*_{X_n/S}$ for $n \in \mathbb{N}$ form a complex of sheaves on $X_\bullet$ denoted $\Omega^*_{X_\bullet/S}$.

A simplicial sheaf F^* on the constant simplicial space $S_\bullet$ defined by S corresponds to a cosimplicial sheaf on S; hence if F^* is abelian, it defines a complex via the face maps, with

$$d = \sum_i (-1)^i \delta_i : F^n \to F^{n+1}.$$

A complex of abelian sheaves K on $S_\bullet$, denoted by $K^{n,m}$ with m the cosimplicial degree, defines a simple complex sK (see Section 3.1.2.3):

$$(sK)^n := \oplus_{p+q=n} K^{p,q} : \ d(x^{p,q}) = d_K(x^{p,q}) + \sum_i (-1)^i \delta_i x^{p,q}.$$

The following filtration L with respect to the second degree will be useful:

$$L^r(sK) = s(K^{p,q})_{q \geq r}.$$

3.4.2.5 Direct Image in the Derived Category of Filtered or Bifiltered Abelian Sheaves

For an augmented simplicial space $a : X_\bullet \to S$, we define a functor denoted Ra_* on complexes K (resp. filtered (K,F), bifiltered (K,F,W)) of abelian sheaves on $X_\bullet$. We may view S as a constant simplicial scheme $S_\bullet$ and a as a morphism $a_\bullet : X_\bullet \to S_\bullet$. In the first step we construct a complex I (resp. (I,F), (I,F,W)) of acyclic (for example flabby) sheaves, quasi-isomorphic (resp. filtered, bifiltered) to K (resp. (K,F), (K,F,W)); we can always take Godement resolutions ([34, Ch. II, Sec. 3.6, p. 95] or [17, Ch. II, Sec. 4.3, p. 167]), for example, then in each degree p, $(a_q)_* I^{p,q}$ on $S_q = S$ defines for varying q a cosimplicial sheaf on S denoted $(a_\bullet)_* I^{p,\bullet}$.

For each p, we deduce from the various face maps δ_i a differential on $(a_\bullet)_* I^{p,\bullet}$ which defines the structure of a differential graded object on $(a_\bullet)_* I^{p,\bullet}$.

Then, we can view $(a_\bullet)_* I^{\bullet,\bullet}$ as a double complex whose associated simple complex is denoted $s(a_\bullet)_* I := Ra_* K$:

$$(Ra_* K)^n := \oplus_{p+q=n} (a_q)_* I^{p,q},$$

$$dx^{p,q} = d_I(x^{p,q}) + (-1)^p \sum_{i=0}^{q+1} (-1)^i \delta_i x^{p,q} \in (Ra_* K)^{n+1},$$

where q is the simplicial index $\delta_i(x^{p,q}) \in I^{p,q+1}$ and p is the degree of the complex I. The filtration L on $s(a_\bullet)_* I := Ra_* K$ defines a spectral sequence

$$E_1^{p,q} = R^q(a_p)_*(K_{|X_p}) := H^q(R(a_p)_*(K_{|X_p})) \Rightarrow H^{p+q}(Ra_* K) := R^{p+q}a_* K.$$

In the filtered case, the definition of $Ra_*(K, F)$ and $Ra_*(K, F, W)$ is similar.

In the case S is a point, we may introduce $R\Gamma^i K := R\Gamma(X_i, K)$, the derived global functor on each space X_i:

LEMMA AND DEFINITION 3.4.13   *(i) The cosimplicial derived global sections complex $R\Gamma^\bullet K$ is defined by the family of derived global sections $R\Gamma(X_n, K_{|X_n})$ on each space X_n.*

(ii) The hypercohomology of K is defined by taking the simple complex defined by the graded differential complex associated to the cosimplicial complex $R\Gamma^\bullet K$,

$$R\Gamma(X_\bullet, K) := sR\Gamma^\bullet K, \quad \mathbb{H}^i(X_\bullet, K) := H^i(R\Gamma(X_\bullet, K)).$$

Topological realization   Recall that a morphism of simplices $f : \Delta_n \to \Delta_m$ has a geometric realization $|f| : |\Delta_n| \to |\Delta_m|$ as the affine map defined when we identify a simplex Δ_n with the vertices of its affine realization in $\mathbb{R}^{\Delta_n}$. We construct the topological realization of a topological semisimplicial space $X_\bullet$ as the quotient of the topological space $Y = \coprod_{n \geq 0} X_n \times |\Delta_n|$ by the equivalence relation $\mathcal{R}$ generated by the identifications

$$\forall f : \Delta_n \to \Delta_m, x \in X_m, a \in |\Delta_n|, \quad (x, |f|(a)) \equiv (X_\bullet(f)(x), a).$$

The topological realization $|X_\bullet|$ is the quotient space of Y, modulo the relation $\mathcal{R}$, with its quotient topology. The construction above of the cohomology amounts to the computation of the cohomology of the topological space $|X_\bullet|$ with coefficient in an abelian group A: $H^i(X_\bullet, A) \simeq H^i(|X_\bullet|, A)$.

3.4.2.6   Cohomological Descent

Let $a : X_\bullet \to S$ be an augmented simplicial scheme; any abelian sheaf F on S, lifts to a sheaf $a^* F$ on $X_\bullet$ and we have a natural morphism

$$\varphi(a) : F \to Ra_* a^* F \quad \text{in} \quad D^+(S).$$

DEFINITION 3.4.14 (Cohomological descent)   *The morphism $a : X_\bullet \to S$ is of cohomological descent if the natural morphism $\varphi(a)$ is an isomorphism in $D^+(S)$ for all abelian sheaves F on S.*

The definition amounts to the following conditions ($\varphi(a)$ is a quasi-isomorphism):

$$F \xrightarrow{\sim} \mathrm{Ker}(a_{0*}a_0^* F \xrightarrow{\delta_1 - \delta_0} a_1^* F), \quad R^i a_* a^* F = 0 \quad \text{for} \quad i > 0.$$

In this case, for all complexes K in $D^+(S)$,

$$R\Gamma(S, K) \simeq R\Gamma(X_\bullet, a^* K)$$

and we have a spectral sequence

$$E_1^{p,q} = \mathbb{H}^q(X_p, a_p^* K) \Rightarrow \mathbb{H}^{p+q}(S, K), \quad d_1 = \sum_i (-1)^i \delta_i : E_1^{p,q} \to E_1^{p+1,q}.$$

3.4.2.7 MHS on Cohomology of Algebraic Varieties

A simplicial complex variety $X_\bullet$ is smooth (resp. compact) if every X_n is smooth (resp. compact).

DEFINITION 3.4.15 (Simplicial NCD)   *A simplicial normal crossing divisor is a family $Y_n \subset X_n$ of NCD such that the family of open subsets $U_n := X_n - Y_n$ form a simplicial subvariety $U_\bullet$ of $X_\bullet$; hence the family of filtered logarithmic complexes $((\Omega_{X_n}^*(\mathrm{Log}\, Y_n))_{n \geq 0}, W)$ form a filtered complex on $X_\bullet$.*

The following construction is the motivation of this section:

THEOREM 3.4.16 (Simplicial covering (Deligne [8, 6.2.8]))   *For each separated complex variety S,*

(i) *there exist a simplicial compact smooth complex variety $X_\bullet$ containing a simplicial normal crossing divisor $Y_\bullet$ in $X_\bullet$ and an augmentation $a : U_\bullet = (X_\bullet - Y_\bullet) \to S$ satisfying the cohomological descent property: for all abelian sheaves F on S, we have an isomorphism $F \xrightarrow{\sim} Ra_* a^* F$;*

(ii) *moreover, for each morphism $f : S \to S'$, there exists a morphism $f_\bullet : X_\bullet \to X'_\bullet$ of simplicial compact smooth complex varieties with simplicial normal crossing divisors $Y_\bullet$ and $Y'_\bullet$ and augmented complements $a : U_\bullet \to S$ and $a' : U'_\bullet \to S'$ satisfying the cohomological descent property, with $f_\bullet(U_\bullet) \subset U'_\bullet$ and $a' \circ f = a$.*

The proof is based on Hironaka's desingularization theorem and on a general construction of hypercoverings described briefly by Deligne in [8] after preliminaries on the general theory of hypercoverings. The desingularization is carried out at each step of the construction by induction.

Remark.   We can and shall assume that the normal crossing divisors have smooth irreducible components.

3.4.2.8 Cosimplicial (Cohomological) Mixed Hodge Complex

An A-cosimplicial mixed Hodge complex K consists of

(i) a complex K_A of cosimplicial A-modules;

(ii) a filtered complex $(K_{A\otimes\mathbb{Q}}, W)$ of cosimplicial $A \otimes \mathbb{Q}$ vector spaces with an increasing filtration W and an isomorphism $K_{A\otimes\mathbb{Q}} \simeq K_A \otimes \mathbb{Q}$ in the derived category of cosimplicial A-vector spaces;

(iii) a bifiltered complex $(K_\mathbb{C}, W, F)$ of cosimplicial A-vector spaces with an increasing (resp. decreasing) filtration W (resp. F) and an isomorphism

$$\alpha : (K_{A\otimes\mathbb{Q}}, W) \otimes \mathbb{C} \xrightarrow{\sim} (K_\mathbb{C}, W)$$

in the derived category of cosimplicial A-vector spaces.

Cosimplicial cohomological MHC
   Similarly, an A-cosimplicial cohomological mixed Hodge complex K on a topological simplicial space $X_\bullet$ consists of

(i) a complex K_A of sheaves of A-modules on $X_\bullet$ such that $\mathbb{H}^k(X_\bullet, K_A)$ are A-modules of finite type;

(ii) a filtered complex $(K_{A\otimes\mathbb{Q}}, W)$ of filtered sheaves of $A \otimes \mathbb{Q}$ modules on $X_\bullet$ with an increasing filtration W and an isomorphism $K_{A\otimes\mathbb{Q}} \simeq K_A \otimes \mathbb{Q}$ in the derived category on $X_\bullet$;

(iii) a bifiltered complex $(K_\mathbb{C}, W, F)$ of sheaves of complex vector spaces on $X_\bullet$ with an increasing (resp. decreasing) filtration W (resp. F) and an isomorphism in the derived category of A-modules $\alpha : (K_{A\otimes\mathbb{Q}}, W) \otimes \mathbb{C} \xrightarrow{\sim} (K_\mathbb{C}, W)$ on $X_\bullet$.

Moreover, the following axiom is satisfied:
   (CMHC) The restriction of K to each X_n is an A-cohomological MHC.

Cosimplicial MHC associated to a cosimplicial cohomological MHC
   If we apply the global section functor on each X_n to an A-cosimplicial cohomological MHC K, we get an A-cosimplicial MHC defined by the data:

(1) a cosimplicial complex $R\Gamma^\bullet K_A$ (see Section 3.4.2.5) in the derived category of cosimplicial A-modules;

(2) a filtered cosimplicial complex $R\Gamma^\bullet(K_{A\otimes\mathbb{Q}}, W)$ in the derived category of filtered cosimplicial vector spaces, and an isomorphism $(R\Gamma^\bullet K_A \otimes \mathbb{Q}) \simeq R\Gamma^\bullet K_{A\otimes\mathbb{Q}}$;

(3) a bifiltered cosimplicial complex $R\Gamma^\bullet(K_\mathbb{C}, W, F)$ in the derived category of bifiltered cosimplicial vector spaces;

(4) an isomorphism $R\Gamma^\bullet(K_{A\otimes\mathbb{Q}}, W) \otimes \mathbb{C} \simeq R\Gamma^\bullet(K_\mathbb{C}, W)$ in the derived category of filtered cosimplicial vector spaces;

   Moreover, the following axiom must be satisfied:

(5) the restriction of the system defined by K to each X_n, $R\Gamma^n K := R\Gamma(X_n, K_{|X_n})$ is an MHC.

EXAMPLE 3.4.17 (Cosimplicial logarithmic complex)   Let $X_\bullet$ be a simplicial compact smooth complex algebraic variety with $Y_\bullet$ a simplicial NCD in $X_\bullet$ such that $j_\bullet : U_\bullet = (X_\bullet - Y_\bullet) \to X_\bullet$ is an open simplicial embedding, then

$$(Rj_{\bullet*}\mathbb{Z}, (Rj_{\bullet*}\mathbb{Q}, \tau), (\Omega^*_{X_\bullet}(\text{Log } Y_\bullet), W, F))$$

is a cohomological MHC on $X_\bullet$.

3.4.2.9   Diagonal Filtration

A cosimplicial mixed Hodge complex K defines a differential graded complex which is viewed as a double complex whose associated simple complex is denoted sK. We put on sK a weight filtration by a diagonal process.

DEFINITION 3.4.18 (Differential graded A-MHC)   *(i) A (differential graded) DG^+-complex (or a complex of differential graded objects) is a bounded-below complex with two degrees: the first is defined by the degree of the complex and the second by the degree of the grading, viewed as a double complex.*

   (ii) A differential graded (filtered or bifiltered) A-MHC is defined by a system of DG^+-complex

$$K_A, (K_{A\otimes\mathbb{Q}}, W), K_A \otimes \mathbb{Q} \simeq K_{A\otimes\mathbb{Q}}, (K_{\mathbb{C}}, W, F), (K_{A\otimes\mathbb{Q}}, W) \otimes \mathbb{C} \simeq (K_{\mathbb{C}}, W)$$

such that for each degree n of the grading, the components at the A, $A\otimes\mathbb{Q}$ levels and complex level $(K^{*,n}_{\mathbb{C}}, W, F)$ form an A-MHC.

A cosimplicial MHC (K, W, F) defines a DG^+-A-MHC

$$K_A, (K_{A\otimes\mathbb{Q}}, W), K_A \otimes \mathbb{Q} \simeq K_{A\otimes\mathbb{Q}}, (K_{\mathbb{C}}, W, F), (K_{A\otimes\mathbb{Q}}, W) \otimes \mathbb{C} \simeq (K_{\mathbb{C}}, W),$$

where the degree of the grading is the cosimplicial degree and the differential of the grading is deduced from the face maps δ_i as in Section 3.4.2.5.

   The hypercohomology of an A-cohomological mixed Hodge complex K on $X_\bullet$ is defined by the simple complex $sR\Gamma^\bullet K$ associated to the DG^+-A-MHC $R\Gamma^\bullet K$.

DEFINITION 3.4.19 (Diagonal filtration)   *The diagonal $\delta(W, L)$ of the two filtrations W and L on sK is defined by*

$$\delta(W, L)_n(sK)^i := \oplus_{p+q=i} W_{n+q} K^{p,q}, \quad F^r(sK)^i := \oplus_{p+q=i} F^r K^{p,q},$$

where $L^r(sK) = s(K^{p,q})_{q \geq r}$.

3.4.2.10   Properties of $\delta(W, L)$

We have

$$\text{Gr}^{\delta(W,L)}_n(sK) \simeq \oplus_p \text{Gr}^W_{n+p} K^{*,p}[-p].$$

In the case of a DG^+-complex defined as the hypercohomology of a complex (K, W) on a simplicial space $X_\bullet$, we have

$$\mathrm{Gr}_n^{\delta(W,L)} R\Gamma K \simeq \oplus_p R\Gamma(X_p, \mathrm{Gr}_{n+p}^W K^{*,p})[-p],$$

and for a bifiltered complex with a decreasing F,

$$\mathrm{Gr}_n^{\delta(W,L)} R\Gamma(K, F) \simeq \oplus_p R\Gamma(X_p, (\mathrm{Gr}_{n+p}^W K, F))[-p].$$

THEOREM 3.4.20 (Hodge theory of simplicial varieties (Deligne [8, Thm. 8.1.15])) *Let K be the graded differential A-mixed Hodge complex defined by a cosimplicial A-mixed Hodge complex.*

(i) *Then $(sK, \delta(W, L), F)$ is an A-mixed Hodge complex. The first terms of the weight spectral sequence*

$$_{\delta(W,L)}E_1^{pq}(sK \otimes \mathbb{Q}) = \oplus_n H^{q-n}(\mathrm{Gr}_n^W K^{*,p+n})$$

form the simple complex $(_{\delta(W,L)}E_1^{pq}, d_1)$ of $A \otimes \mathbb{Q}$-Hodge structures of weight q associated to the double complex, where $m = n + p$ and E_1^{pq} is represented by the sum of the terms on the diagonal :

$$H^{q-(n+1)}(\mathrm{Gr}_{n+1}^W K^{*,m+1}) \xrightarrow{\partial} H^{q-n}(\mathrm{Gr}_n^W K^{*,m+1}) \xrightarrow{\partial} H^{q-(n-1)}(\mathrm{Gr}_{n-1}^W K^{*,m+1})$$
$$d'' \uparrow \qquad\qquad\qquad d'' \uparrow \qquad\qquad\qquad d'' \uparrow$$
$$H^{q-(n+1)}(\mathrm{Gr}_{n+1}^W K^{*,m}) \xrightarrow{\partial} H^{q-n}(\mathrm{Gr}_n^W K^{*,m}) \xrightarrow{\partial} H^{q-(n-1)}(\mathrm{Gr}_{n-1}^W K^{*,m})$$

where ∂ is a connecting morphism and d'' is simplicial.

(ii) *The terms $_L E_r$ for $r > 0$ of the spectral sequence defined by $(sK_{A \otimes \mathbb{Q}}, L)$ are endowed with a natural A-mixed Hodge structure, with differentials d_r compatible with such structures.*

(iii) *The filtration L on $H^*(sK)$ is a filtration in the category of mixed Hodge structures and*

$$(\mathrm{Gr}_L^p H^{p+q}(sK), \delta(W, L)[p+q], F) = (_L E_\infty^{pq}, W, F).$$

The proof follows directly from the general study of MHC. The statement in (iii) follows from the lemma:

LEMMA 3.4.21 *If $H = (H_A, W, F)$ is an A-mixed Hodge structure, a filtration L of H_A is a filtration of mixed Hodge structure if and only if, for all n,*

$$(\mathrm{Gr}_L^n H_A, \mathrm{Gr}_L^n(W), \mathrm{Gr}_L^n(F))$$

is an A-mixed Hodge structure.

However, the proof of the theorem is a particular case of a general theory on filtered mixed Hodge complexes useful in the study of filtered MHS (see Section 8.2.2).

3.4.2.11 Mixed Hodge Complex of a Simplicial Variety

In the case of a smooth simplicial variety $U_\bullet$, the complement of a normal crossing divisor $Y_\bullet$ in a smooth compact simplicial variety $X_\bullet$, the cohomology groups $H^n(U_\bullet, \mathbb{Z})$ are endowed with the mixed Hodge structure defined by the mixed Hodge complex

$$(R\Gamma(U_\bullet, \mathbb{Z}), \ R\Gamma(X_\bullet, Rj_*\mathbb{Q}_{U_\bullet}, \delta(\tau, L)), R\Gamma(X_\bullet, \Omega^*_{X_\bullet}(\mathrm{Log}\, Y_\bullet)), \delta(W, L), F)$$

with natural compatibility isomorphisms, satisfying

$$\mathrm{Gr}^{\delta(W,L)}_n R\Gamma(X_\bullet, \Omega^*_{X_\bullet}(\mathrm{Log}\, Y_\bullet)) \simeq \oplus_m \mathrm{Gr}^W_{n+m} R\Gamma(X_m, \Omega^*_{X_m}(\mathrm{Log}\, Y_m))[-m]$$
$$\simeq \oplus_m R\Gamma(Y^{n+m}_m, \mathbb{C})[-n - 2m],$$

where the first isomorphism is a property of the diagonal filtration and the second is a property of the weight of the logarithmic complex for the open set U_m, where Y^{n+m}_m denotes the disjoint union of intersections of $n + m$ components of the normal crossing divisor Y_m of simplicial degree m. Moreover,

$$_{\delta(W,L)}E^{p,q}_1 = \oplus_n H^{q-2n}(Y^n_{n+p}, \mathbb{Q}) \Rightarrow H^{p+q}(U_\bullet, \mathbb{Q}).$$

The filtration F induces on $_{\delta(W,L)}E^{p,q}_1$ a Hodge structure of weight q and the differentials d_1 are compatible with the Hodge structures. The term E_1 is the simple complex associated to the double complex of Hodge structure of weight q, where G is an alternating Gysin map:

$$
\begin{array}{ccccc}
H^{q-(2n+2)}(Y^{n+1}_{p+n+1}, \mathbb{Q}) & \xrightarrow{G} & H^{q-2n}(Y^n_{p+n+1}, \mathbb{Q}) & \xrightarrow{G} & H^{q-2(n-2)}(Y^{n-1}_{p+n+1}, \mathbb{Q}) \\
\sum_i(-1)^i\delta_i \uparrow & & \sum_i(-1)^i\delta_i \uparrow & & \sum_i(-1)^i\delta_i \uparrow \\
H^{q-(2n+2)}(Y^{n+1}_{p+n}, \mathbb{Q}) & \xrightarrow{G} & H^{q-2n}(Y^n_{p+n}, \mathbb{Q}) & \xrightarrow{G} & H^{q-2(n-2)}(Y^{n-1}_{p+n}, \mathbb{Q}),
\end{array}
$$

where the Hodge structures in the columns are twisted by $-n - 1$, $-n$, and $-n + 1$, respectively; the lines are defined by the logarithmic complex, while the vertical differentials are simplicial. We deduce from the general theory the following proposition:

PROPOSITION 3.4.22 (Hodge theory of simplicial varieties)   (i) *Let $U_\bullet := X_\bullet - Y_\bullet$ be the complement of a simplicial normal crossing divisor $Y_\bullet$ in a smooth compact simplicial algebraic variety $X_\bullet$ (Definition 3.4.15). The mixed Hodge structure on $H^n(U_\bullet, \mathbb{Z})$ is defined by the graded differential mixed Hodge complex associated to the simplicial MHC defined by the logarithmic complex (Example 3.4.17) on each term of $X_\bullet$. It is functorial in the couple $(U_\bullet, X_\bullet)$.*

(ii) *The rational weight spectral sequence degenerates at rank 2 and the Hodge structure on $E^{p,q}_2$ induced by $E^{p,q}_1$ is isomorphic to the HS on $\mathrm{Gr}^q_W H^{p+q}(U_\bullet, \mathbb{Q})$.*

(iii) *The Hodge numbers $h^{p,q}$ of $H^n(U_\bullet, \mathbb{Q})$ vanish for $p \notin [0, n]$ or $q \notin [0, n]$.*

(iv) *For $Y_\bullet = \emptyset$, the Hodge numbers $h^{p,q}$ of $H^n(X_\bullet, \mathbb{Q})$ vanish for $p \notin [0, n]$ or $q \notin [0, n]$ or $p + q > n$.*

DEFINITION 3.4.23 (MHS of an algebraic variety) *The mixed Hodge structure on the cohomology $H^n(X, \mathbb{Z})$ of a complex algebraic variety X is defined by a simplicial resolution $X_{\bullet}$ (Theorem 3.4.16) embedded into a simplicial compact complex algebraic variety $V_{\bullet}$ with a simplicial NCD, $V_{\bullet} - X_{\bullet}$, via the isomorphism defined by the augmentation map $X_{\bullet} \to X$.*

The mixed Hodge structure just defined does not depend on the resolution $X_{\bullet}$ or the embedding into $V_{\bullet}$ and is functorial in X.

3.4.2.12 Problems

(1) Let $i : Y \to X$ be a closed subvariety of X and $j : U := (X - Y) \to X$ the embedding of the complement. Then the two long exact sequences of cohomology,

$$\cdots \to H^i(X, X - Y, \mathbb{Z}) \to H^i(X, \mathbb{Z}) \to H^i(Y, \mathbb{Z}) \to H_Y^{i+1}(X, \mathbb{Z}) \to \cdots ,$$
$$\cdots \to H_Y^i(X, \mathbb{Z}) \to H^i(X, \mathbb{Z}) \to H^i(X - Y, \mathbb{Z}) \to H_Y^{i+1}(X, \mathbb{Z}) \to \cdots$$

underlie exact sequences of mixed Hodge structure.

The idea is to use a simplicial hypercovering of the morphism i in order to define two mixed Hodge complexes: $K(Y)$ on Y and $K(X)$ on X with a well-defined morphism on the level of complexes of sheaves $i^* : K(X) \to K(Y)$ (resp. $j^* : K(X) \to K(X - Y)$), then the long exact sequence is associated to the mixed cone $C_M(i^*)$(resp. $C_M(j^*)$).

In particular, one deduces associated long exact sequences by taking the graded spaces with respect to the filtrations F and W.

(2) *Künneth formula* [37]. Let X and Y be two algebraic varieties; then the isomorphisms of cohomology vector spaces

$$H^i(X \times Y, \mathbb{C}) \simeq \oplus_{r+s=i} H^r(X, \mathbb{C}) \otimes H^s(Y, \mathbb{C})$$

underlie isomorphisms of $\mathbb{Q}$-mixed Hodge structure. The answer is in three steps:

   (i) Consider the tensor product of two mixed Hodge complexes defining the mixed Hodge structure of X and Y from which we deduce the right-hand term, the direct sum of the tensor product of mixed Hodge structures.

   (ii) Construct a quasi-isomorphism of the tensor product with a mixed Hodge complex defining the mixed Hodge structure of $X \times Y$.

   (iii) Deduce that the cup product on the cohomology of an algebraic variety is compatible with mixed Hodge structures.

3.4.3 MHS on the Cohomology of a Complete Embedded Algebraic Variety

For embedded varieties into smooth varieties, the mixed Hodge structure on cohomology can be obtained by a simple method using exact sequences, once the mixed Hodge structure for the normal crossing divisor has been constructed, which should easily be convincing of the natural aspect of this theory. The technical ingredients consist of Poincaré duality and its dual the trace (or Gysin) morphism.

Let $p : X' \to X$ be a proper morphism of complex smooth varieties of the same dimension, Y a closed subvariety of X and $Y' = p^{-1}(Y)$. We suppose that Y' is an NCD in X' and that the restriction of p induces an isomorphism $p_{X'-Y'} : X' - Y' \xrightarrow{\sim} X - Y$:

$$
\begin{array}{ccccc}
Y' & \xrightarrow{i'} & X' & \xleftarrow{j'} & X' - Y' \\
\downarrow p_Y & & \downarrow p & & \downarrow p_{X'-Y'} \\
Y & \xrightarrow{i} & X & \xleftarrow{j} & X - Y.
\end{array}
$$

The trace morphism $\mathrm{Tr}\, p$ is defined as the Poincaré dual to the inverse image p^* on cohomology; hence $\mathrm{Tr}\, p$ is compatible with HS. It can be defined at the level of sheaf resolutions of $\mathbb{Z}_{X'}$ and $\mathbb{Z}_X$ as constructed by Verdier [42] in the derived category $\mathrm{Tr}\, p: Rp_*\mathbb{Z}_{X'} \to \mathbb{Z}_X$; hence we deduce morphisms $\mathrm{Tr}\, p : H_c^i(X' - Y', \mathbb{Z}) \to H_c^i(X - Y, \mathbb{Z})$ and by restriction morphisms depending on the embeddings of Y and Y' into X and X':

$$(\mathrm{Tr}\, p)_{|Y} : Rp_*\mathbb{Z}_{Y'} \to \mathbb{Z}_Y, \ (\mathrm{Tr}\, p)_{|Y} : H^i(Y', \mathbb{Z}) \to H^i(Y, \mathbb{Z}).$$

Remark. Let U be a neighborhood of Y in X, retract by deformation onto Y such that $U' = p^{-1}(U)$ is a retract by deformation onto Y'. Then the morphism $(\mathrm{Tr}\, p)_{|Y}$ is deduced from $\mathrm{Tr}(p_{|U})$ in the diagram

$$
\begin{array}{ccc}
H^i(Y', \mathbb{Z}) & \xleftarrow{\sim} & H^i(U', \mathbb{Z}) \\
\downarrow (\mathrm{Tr}\, p)_{|Y} & & \downarrow \mathrm{Tr}(p_{|U}) \\
H^i(Y, \mathbb{Z}) & \xleftarrow{\sim} & H^i(U, \mathbb{Z}).
\end{array}
$$

Consider now the following diagram:

$$
\begin{array}{ccccc}
R\Gamma_c(X' - Y', \mathbb{Z}) & \xrightarrow{j'_*} & R\Gamma(X', \mathbb{Z}) & \xrightarrow{i'^*} & R\Gamma(Y', \mathbb{Z}) \\
\mathrm{Tr}\, p \downarrow & & \downarrow \mathrm{Tr}\, p & & \downarrow (\mathrm{Tr}\, p)_{|Y} \\
R\Gamma_c(X - Y, \mathbb{Z}) & \xrightarrow{j_*} & R\Gamma(X, \mathbb{Z}) & \xrightarrow{i^*} & R\Gamma(Y, \mathbb{Z}).
\end{array}
$$

PROPOSITION 3.4.24 ([14])   *(i) The morphism* $p_Y^* : H^i(Y, \mathbb{Z}) \to H^i(Y', \mathbb{Z})$ *is injective with retraction* $(\mathrm{Tr}\, p)_{|Y}$.

(ii) We have a quasi-isomorphism of $i_*\mathbb{Z}_Y$ with the cone $C(i'^* - \mathrm{Tr}\, p)$ of the morphism $i'^* - \mathrm{Tr}\, p$. The long exact sequence associated to the cone splits into short exact sequences:

$$0 \to H^i(X', \mathbb{Z}) \xrightarrow{i'^* - \mathrm{Tr}\, p} H^i(Y', \mathbb{Z}) \oplus H^i(X, \mathbb{Z}) \xrightarrow{(\mathrm{Tr}\, p)_{|Y} + i^*} H^i(Y, \mathbb{Z}) \to 0.$$

Moreover $i'^* - \mathrm{Tr}\, p$ is a morphism of mixed Hodge structures. In particular, the weight of $H^i(Y, \mathbb{C})$ varies in the interval $[0, i]$ since this is true for Y' and X.

LEMMA AND DEFINITION 3.4.25 *The mixed Hodge structure of Y is defined as the cokernel of $i'^* - \operatorname{Tr} p$ via its isomorphism with $H^i(Y, \mathbb{Z})$, induced by $(\operatorname{Tr} p)_{|Y} + i^*$. It coincides with Deligne's mixed Hodge structure.*

This result shows the uniqueness of the theory of mixed Hodge structures, once the MHS of the normal crossing divisor Y' has been constructed. The above technique consists of the realization of the MHS on the cohomology of Y as relative cohomology with MHS on X, X', and Y', all smooth compact or NCD. Notice that the MHS on $H^i(Y, \mathbb{Z})$ is realized as a quotient and not as an extension.

PROPOSITION 3.4.26 *Let X, X' be compact algebraic varieties with X' nonsingular and let $p : X' \to X$ be a surjective proper morphism. Then for all integers i, we have*

$$W_{i-1} H^i(X, \mathbb{Q}) = \operatorname{Ker}(H^i(X, \mathbb{Q}) \xrightarrow{p^*} H^i(X', \mathbb{Q})).$$

In particular, this result applies to a desingularization of X.

We have trivially $W_{i-1} H^i(X, \mathbb{Q}) \subset \operatorname{Ker} p^*$ since $H^i(X', \mathbb{Q})$ is of pure weight i. Let $i : Y \to X$ be the subvariety of singular points in X and let $Y' := p^{-1}(Y)$, denote by $p_{Y'} : Y' \to Y$ the morphism induced by p and by $i' : Y' \to X'$ the injection into X'; then we have a long exact sequence:

$$H^{i-1}(Y', \mathbb{Q}) \to H^i(X, \mathbb{Q}) \xrightarrow{(p^*, -i^*)} H^i(X', \mathbb{Q}) \oplus H^i(Y, \mathbb{Q}) \xrightarrow{i'^* + p_{Y'}^*} H^i(Y', \mathbb{Q}) \to \cdots .$$

It is enough to prove $\operatorname{Ker} p^* \subset \operatorname{Ker} i^*$, since then

$$\operatorname{Ker} p^* = \operatorname{Ker} p^* \cap \operatorname{Ker} i^* - \operatorname{Im}(H^{i-1}(Y', \mathbb{Q}) \to H^i(X, \mathbb{Q})),$$

where the weight of $H^{i-1}(Y', \mathbb{Q})$ is $\leq i - 1$. By induction on the dimension of X, we may suppose that the proposition is true for Y. Let $\alpha : Y'' \to Y'$ be a desingularization of Y', $q := p_{Y'} \circ \alpha$ and $i'' := i' \circ \alpha$; then we have a commutative diagram,

$$
\begin{array}{ccc}
Y'' & \xrightarrow{i''} & X' \\
q \downarrow & & \downarrow p \\
Y & \xrightarrow{i} & X,
\end{array}
$$

where Y'' is compact and nonsingular.

Let $a \in \operatorname{Gr}_i^W H^i(X, \mathbb{Q})$ such that $\operatorname{Gr}_i^W p^*(a) = 0$; then $\operatorname{Gr}_i^W (p \circ i'')^*(a) = 0$. Hence $\operatorname{Gr}_i^W (i \circ q)^*(a) = 0$ since $\operatorname{Gr}_i^W (i \circ q)^* = \operatorname{Gr}_i^W (p \circ i'')^*$. By induction, $\operatorname{Gr}_i^W (q)^*$ is injective; then we deduce $\operatorname{Gr}_i^W (i^*)(a) = 0$.

Mixed Hodge structure on the cohomology of an embedded algebraic variety The construction still applies for nonproper varieties if we construct the MHS of an open NCD.

Hypothesis. Let $i_Z : Z \to X$ be a closed embedding and $i_X : X \to P$ a closed embedding in a projective space (or any proper smooth complex algebraic variety). By

Hironaka desingularization we construct a diagram,

$$
\begin{array}{ccccc}
Z'' & \to & X'' & \to & P'' \\
\downarrow & & \downarrow & & \downarrow \\
Z' & \to & X' & \to & P' \\
\downarrow & & \downarrow & & \downarrow \\
Z & \to & X & \to & P,
\end{array}
$$

first by blowing up centers over Z so to obtain a smooth space $p : P' \to P$ such that $Z' := p^{-1}(Z)$ is a normal crossing divisor; set $X' := p^{-1}(X)$; then

$$
p_| : X' - Z' \xrightarrow{\sim} X - Z, \quad p_| : P' - Z' \xrightarrow{\sim} P - Z
$$

are isomorphisms since the modifications are all over Z. Next, by blowing up centers over X' we obtain a smooth space $q : P'' \to P'$ such that $X'' := q^{-1}(X')$ and $Z'' := q^{-1}(Z')$ are NCD, and $q_| : P'' - X'' \xrightarrow{\sim} P' - X'$. Then, we deduce the diagram

$$
\begin{array}{ccccc}
X'' - Z'' & \xrightarrow{i''_X} & P'' - Z'' & \xleftarrow{j''} & P'' - X'' \\
\downarrow & & q_| \downarrow & & q_| \downarrow \wr \\
X' - Z' & \xrightarrow{i'_X} & P' - Z' & \xleftarrow{j'} & P' - X'.
\end{array}
$$

Since the desingularization is a sequence of blow-ups above X', we still have an isomorphism induced by q on the right-hand side of the preceding diagram. For $\dim P = d$ and all integers i, the morphism $q^*_| : H_c^{2d-i}(P'' - Z'', \mathbb{Q}) \to H_c^{2d-i}(P' - Z', \mathbb{Q})$ is well defined on cohomology with compact support since q is proper. Its Poincaré dual is called the trace morphism $\operatorname{Tr} q_| : H^i(P'' - Z'', \mathbb{Q}) \to H^i(P' - Z', \mathbb{Q})$ and satisfies the relation $\operatorname{Tr} q_| \circ q^*_| = \operatorname{Id}$. Moreover, the trace morphism is defined as a morphism of sheaves $q_{|*}\mathbb{Z}_{P''-Z''} \to \mathbb{Z}_{P'-Z'}$ (see [42]); hence an induced trace morphism $(\operatorname{Tr} q)_| : H^i(X'' - Z'', \mathbb{Q}) \to H^i(X' - Z', \mathbb{Q})$ is well defined.

PROPOSITION 3.4.27 *With the notation of the above diagram, we have short exact sequences*

$$
0 \to H^i(P'' - Z'', \mathbb{Q}) \xrightarrow{(i''_X)^* - \operatorname{Tr} q_|} H^i(X'' - Z'', \mathbb{Q}) \oplus H^i(P' - Z', \mathbb{Q})
$$
$$
\xrightarrow{(i'_X)^* - (\operatorname{Tr} q)_|} H^i(X' - Z', \mathbb{Q}) \to 0.
$$

Since we have a vertical isomorphism q on the right-hand side of the above diagram, we deduce a long exact sequence of cohomology spaces containing the sequences of the proposition; the injectivity of $(i''_X)^* - \operatorname{Tr} q_|$ and the surjectivity of $(i'_X)^* - (\operatorname{Tr} q)_|$ are deduced from $\operatorname{Tr} q_| \circ q^*_| = \operatorname{Id}$ and $(\operatorname{Tr} q)_| \circ q^*_X = \operatorname{Id}$; hence the long exact sequence splits into short exact sequences.

COROLLARY 3.4.28 *The cohomology $H^i(X - Z, \mathbb{Z})$ is isomorphic to $H^i(X' - Z', \mathbb{Z})$ since $X - Z \simeq X' - Z'$, and then carries the MHS isomorphic to the cokernel of $(i''_X)^* - \operatorname{Tr} q_|$, which is a morphism of MHS.*

In particular, the MHS on $X - Z$ is uniquely defined by the MHS on an open NCD and the compatibility of the morphism $(i'_X)^* - (\operatorname{Tr} q)_|$ with MHS.

The left-hand term carries an MHS as the special case of the complement of the normal crossing divisor Z'' into the smooth proper variety P'', while the middle term is the complement of the normal crossing divisor Z'' into the normal crossing divisor X''. Both cases can be treated by the above special cases without the general theory of simplicial varieties.

This shows that the MHS is uniquely defined by the construction on an open NCD and the compatibility with the morphism $(i''_X)^* - \operatorname{Tr} q_|$.

Bibliography

[1] M. F. Atiyah, F. Hirzebruch. *Analytic cycles on complex manifolds*. Topology 1, 1962, 25–45.

[2] A. A. Beilinson, J. Bernstein, P. Deligne. *Analyse et topologie sur les espaces singuliers*. Astérisque 100, Soc. Math. Fr., France, 1982.

[3] R. Bott, L. W. Tu. Differential forms in algebraic topology. third corrected printing, Springer, New York, 1995.

[4] J. Carlson, S. Muller-Stach, C. Peters. Period mappings and period domains. Cambridge Studies in Advanced Mathematics 85, 2003.

[5] H. Cartan, S. Eilenberg. Homological algebra. Princeton Math. series 19, Princeton University Press, 1956.

[6] P. Deligne. *Théorème de Lefschetz et critères de dégénérescence de suites spectrales*. Publ. Math. IHES 35, 1968, 107–126.

[7] P. Deligne. *Théorie de Hodge II*. Publ. Math. IHES 40, 1972, 5–57.

[8] P. Deligne. *Théorie de Hodge III*. Publ. Math. IHES 44, 1975, 6–77.

[9] P. Deligne. *Hodge cycles on abelian varieties* (notes by J. S. Milne), in Lecture Notes in Math. 900, 1982, 9–100, Springer.

[10] P. Deligne, L. Illusie. *Relèvements modulo p^2 et décomposition du complexe de de Rham*. Inv. Math. 89, 1987, 247–270.

[11] P. Deligne. *The Hodge conjecture*. The Millennium Prize Problems, Clay Math. Institute 2006, 45–53.

[12] F. El Zein. *La classe fondamentale d'un cycle*. Compositio Mathematica 29, 1974, 9–33.

[13] F. El Zein. *Complexe dualisant et applications à la classe fondamentale d'un cycle*. Mémoire de la S.M.F. 58, 1978, 1–66.

[14] F. El Zein. *Mixed Hodge structures*. Trans. AMS 275, 1983, 71–106.

[15] F. El Zein, A. Némethi. *On the weight filtration of the homology of algebraic varieties: the generalized Leray cycles*. Ann. Scuola Norm. Sup. Pisa Cl. Sci. 5, vol. I, 2002, 869–903.

[16] A. Fröhlicher. *Relations between the cohomology groups of Dolbeault and topological invariants.* Proc. Nat. Acad. Sci. USA 41, 1955, 641–644.

[17] R. Godement. Topologie algébrique et théorie des faisceaux. Paris, Hermann, 1958.

[18] M. Goresky, R. MacPherson. *Intersection homology theory.* Topology, 19, 1980.

[19] P. Griffiths. *A transcendental method in algebraic geometry.* Congrès Int. Math. 1, 1970, 113–119.

[20] P. Griffiths, J. Harris. Principles of algebraic geometry. John Wiley, 1978.

[21] P. Griffiths, W. Schmid. *Recent developments in Hodge theory.* in Discrete subgroups of Lie groups, Oxford University Press, 1973.

[22] A. Grothendieck. *The cohomology theory of abstract algebraic varieties.* Proc. Int. Cong. Math., Edinburgh, 1958, 103–118.

[23] A. Grothendieck. *On the de Rham cohomology of algebraic varieties.* Publ. Math. IHES 29, 1966, 95–103.

[24] A. Grothendieck. *Sur quelques points d'algèbre homologique.* Tôhoku Math. J. 9, 1957, 119–221.

[25] V. Guillemin, A. Pollack. Differential topology. New Jersey, 1974.

[26] F. Guillén, V. Navarro Aznar, P. Pascual-Gainza, F. Puerta. *Hyperrésolutions cubiques et descente cohomologique.* Springer Lecture Notes in Math. 1335, 1988.

[27] R. Hartshorne. Residues and duality. Springer Lecture Notes in Math. 20, 1966.

[28] R. Hartshorne. Algebraic geometry. Springer GTM 52, 1977.

[29] A. Hatcher. Algebraic topology. Cambridge University Press, 2002.

[30] H. Hironaka. *Resolution of singularities of an algebraic variety over a field of characteristic zero.* Ann. Math. 79, 1964, I: 109–203; II: 205–326.

[31] H. Hironaka. *Triangulations of algebraic sets.* Proc. Symp. Pure Math. 29, AMS, Providence, 1975, 165–185.

[32] W. V. D. Hodge. *The topological invariants of algebraic varieties.* Proc. I.C.M., 1950, 181–192.

[33] W. V. D. Hodge. The theory and applications of harmonic integrals. Cambridge University Press, New York, 1941 (second ed., 1952).

[34] B. Iversen. Cohomology of sheaves. Springer, Universitext, 1986.

[35] S. Lefschetz. L'Analysis Situs et la Géométrie Algébrique. Gauthiers-Villars, Paris, 1924.

[36] M. Nagata. *Imbedding of an abstract variety in a complete variety.* J. of Math. Kyoto 2, I, 1962, 1–10.

[37] C. Peters, J. Steenbrink. Mixed Hodge structures. Ergebnisse der Mathematik. 3. Folge, A Series of Modern Surveys in Mathematics, 2010.

[38] M. Saito. *Modules de Hodge polarisables.* Pub. RIMS 553, Kyoto University, 1986.

[39] I. R. Shafarevich. Basic algebraic geometry 2. Springer, 1994.

[40] L. W. Tu. An introduction to manifolds. second edition, Springer, Universitext, 2011.

[41] J. L. Verdier. *Catégories dérivées*, in *Cohomologie etale*. Springer Lecture Notes in Math. 569, 262–312, 1977; and *Des catégories dérivées des catégories abéliennes*, Astérisque 239, 1996.

[42] J. L. Verdier. *Dualité dans la cohomologie des espaces localement compacts.* Séminaire Bourbaki 300, 1965.

[43] C. Voisin. Hodge theory and complex algebraic geometry. Cambridge University Press, 2002.

[44] C. Voisin. A counterexample to the Hodge conjecture extended to Kähler varieties. Int. Math. Res. 20, 1057–1075, 2002.

[45] F. W. Warner. Foundations of differentiable manifolds and Lie groups. Graduate Texts in Math. 94, Springer, 1983.

[46] A. Weil. Introduction aux variétés kählériennes . Hermann, Paris, 1958.

[47] R. O. Wells Jr. Differential analysis on complex manifolds. Graduate Texts in Math. 65, Springer, 1979.

Chapter Four

Period Domains and Period Mappings

by James Carlson

INTRODUCTION AND REVIEW

The aim of these lectures is to develop a working understanding of the notions of period domain and period mapping, as well as familiarity with basic examples thereof. The fundamental references are [10] and [11]. We will not give specific references to these, but essentially everything below is contained in, or derivative of these articles. Three general references are [12], [15], and [5].

In previous lectures you have studied the notion of a polarized Hodge structure H of weight n over the integers, for which the motivating example is the primitive cohomology in dimension n of a projective algebraic manifold of the same dimension. We discuss primitivity below.

To recapitulate, H is a triple $(H_{\mathbb{Z}}, \oplus H^{p,q}, Q)$, where

(a) $H_{\mathbb{Z}}$ is a free $\mathbb{Z}$-module;

(b) $\oplus H^{p,q}$ is a direct sum decomposition of the complex vector space $H_{\mathbb{C}} = H_{\mathbb{Z}} \otimes_{\mathbb{Z}} \mathbb{C}$ satisfying $\overline{H^{p,q}} = H^{q,p}$, where $p + q = k$;

(c) $Q(x, y)$ is a nondegenerate bilinear form which is symmetric for k even and antisymmetric for k odd.

For the n-dimensional cohomology of a smooth projective variety M of dimension n, Q is, up to sign, given by the intersection form, i.e., by cup product followed by evaluation on the fundamental class. The sign is, up to sign, given by the cup product

$$Q(\alpha, \beta) = (-1)^{n(n+1)/2} \langle \alpha, \beta \rangle,$$

where

$$\langle \alpha, \beta \rangle = \int_M \alpha \wedge \beta.$$

The bilinear form is compatible with the direct sum decomposition (the *Hodge decomposition*) in the following way:

$$Q(x, y) = 0 \text{ if } x \in H^{p,q} \text{ and } y \in H^{r,s}, (r, s) \neq (q, p); \tag{4.0.1}$$

$$i^{p-q} Q(x, \bar{x}) > 0 \text{ for } x \in H^{p,q} - \{0\}. \tag{4.0.2}$$

The compatibility relations are the *Riemann bilinear relations*. The *Weil operator* is the linear transformation $C : H_{\mathbb{C}} \longrightarrow H_{\mathbb{C}}$ such that $C(x) = i^{p-q}x$. It is a real operator, that is, it restricts to a real linear transformation of $H_{\mathbb{R}}$. The expression

$$h(x, y) = Q(Cx, \bar{y}) \tag{4.0.3}$$

defines a positive Hermitian form. Said another way, the Hermitian form $Q(x, \bar{y})$ is definite on the Hodge spaces $H^{p,q}$, but the sign of the form alternates as p increases.

A Hodge structure can also be defined by a filtration. Let

$$F^p = \bigoplus_{a \geq p} H^{a,b}.$$

Then one has the decreasing filtration $\cdots \supset F^0 \supset F^1 \supset F^2 \supset \cdots$. It satisfies

$$H = F^p \oplus \overline{F^{k-p+1}}$$

for a Hodge structure of weight k. This is the *Hodge filtration*. A Hodge filtration defines a Hodge decomposition with

$$H^{p,q} = F^p \cap \bar{F}^q.$$

Let us consider these definitions in the context of a compact Riemann surface M. Let $\phi = f(z)dz$ be a 1-form of type $(1, 0)$. Since $dz \wedge dz = 0$, its exterior derivative is

$$d\phi = \frac{\partial f}{\partial \bar{z}}d\bar{z} \wedge dz. \tag{4.0.4}$$

Thus, if ϕ is a closed form then $\partial f/\partial \bar{z} = 0$, and so f is holomorphic. Conversely, if f is holomorphic, then $d\phi = 0$. The space $H^{1,0}(M)$ is the space spanned by closed 1-forms of type $(1, 0)$. By what we have just seen, this is the same as the space of holomorphic 1-forms, aka *abelian differentials*. The Hodge structure is given by this subspace and the subspace $H^{0,1}(M)$ spanned by the complex conjugates of the abelian differentials:

$$H^1(M, \mathbb{C}) = H^{1,0}(M) \oplus H^{0,1}(M) = \{f(z)dz\} \oplus \{g(\bar{z})d\bar{z}\},$$

where $g(\bar{z})$ is antiholomorphic.

Remark. The subspaces $H^{p,q}$ of the complex cohomology are defined as the space of cohomology classes represented by closed forms of type (p, q)—forms with p dz_i's and q $d\bar{z}_j$'s. The fact that these subspaces give a direct sum decomposition is hard, and requires

(a) the Hodge theorem, which states that each cohomology class has a unique harmonic representative;

(b) the fact that the Laplace operator commutes with the operation of projection onto the space of forms of type (p, q).

A harmonic form is a solution of Laplace's equation, $\Delta \alpha = 0$. The Laplacian commutes with the (p, q) projectors when the underlying manifold is a compact Kähler manifold. For an example of a complex manifold which is not Kähler, take the Hopf surface, defined as the quotient of $\mathbb{C}^2 - \{0\}$ by the group of dilations $z \mapsto 2^n z$. It is homeomorphic to $S^1 \times S^3$, so that H^1 is one-dimensional. The spaces $H^{1,0}$ and $H^{0,1}$ are defined as above, but in this case $H^{1,0} = H^{0,1}$. Thus the Hopf surface cannot carry a Kähler structure, nor can its cohomology carry a Hodge structure.

The cohomology class of a nonzero abelian differential is nonzero. To see this, consider the integral

$$\sqrt{-1} \int_M \phi \wedge \bar{\phi}. \tag{4.0.5}$$

Let $\{T_\alpha\}$ be a triangulation of M that is so fine that each closed triangle T_α is contained in a coordinate neighborhood with coordinate z_α. Thus on T_α, it follows that $\phi = f_\alpha(z_\alpha) dz_\alpha$. Therefore the above integral is a sum of terms

$$\sqrt{-1} \int_{T_\alpha} |f_\alpha(z_\alpha)|^2 dz_\alpha \wedge d\bar{z}_\alpha. \tag{4.0.6}$$

Now $\sqrt{-1} dz_\alpha \wedge d\bar{z}_\alpha = 2 dx_\alpha \wedge dy_\alpha$, where $z_\alpha = x_\alpha + \sqrt{-1} y_\alpha$. The form $dx_\alpha \wedge dy_\alpha$ is the volume form in the natural orientation determined by the complex structure: rotation by 90° counterclockwise in the x_α-y_α plane. Thus the integral (4.0.5) is a sum of positive terms (4.0.6). We conclude that if $\phi \neq 0$, then

$$\sqrt{-1} \int_M \phi \wedge \bar{\phi} > 0. \tag{4.0.7}$$

This is the *second Riemann bilinear relation*; it implies that the cohomology class of ϕ is nonzero.

Consider a second abelian differential $\psi = \{g_\alpha(z_\alpha) dz_\alpha\}$. Then

$$\sqrt{-1} \int_M \phi \wedge \psi = 0, \tag{4.0.8}$$

since the integral is a sum of integrals with integrands $f_\alpha(z_\alpha) g_\alpha(z_\alpha) dz_\alpha \wedge dz_\alpha = 0$. This is the *first Riemann bilinear relation*. We conclude that *the first cohomology group of a compact orientable Riemann surface is a polarized Hodge structure of weight 1*.

Let us now return to the issue of primitivity. Let M be a smooth projective variety of dimension n and let

$$L : H^k(M) \longrightarrow H^{k+2}(M) \tag{4.0.9}$$

be the operator such that $L(x) = \omega \wedge x$, where ω is the positive $(1,1)$-form which represents the Kähler class. A cohomology class $x \in H^k(M)$ for $k < n$ is *primitive* if $L^{n-k+1}(x) = 0$. Define the primitive cohomology, which we write as $H^k(M)_o$ or $H^k(M)_{\text{prim}}$, to be the kernel of L^{n-k+1}. Then

$$H^n(M)_o = \{x \in H^n(M) \mid L(x) = 0\}.$$

For example, for an algebraic surface, where $n = 2$, the primitive cohomology is the orthogonal complement of the hyperplane class. In general the cohomology is the direct sum (over the rational numbers) of the sub-Hodge structures $L^i H^j(M)_o$.

Algebraic surfaces

Let us study the Hermitian form

$$h(\alpha, \beta) = \int_M \alpha \wedge \bar{\beta}$$

on an algebraic surface $M \subset \mathbb{P}^n$. Let ω be the Kähler form on $\mathbb{P}^n$, that is, the form given by $\sqrt{-1}\partial\bar{\partial} \log \|Z\|^2$, where Z is the vector of homogeneous coordinates. This form is positive, meaning that if we write it as

$$\omega = \sqrt{-1} \sum_{ij} h_{ij} dz_i \wedge d\bar{z}_j,$$

where the z_i are holomorphic local coordinates, then the matrix (h_{ij}) is a positive Hermitian matrix. For such a differential form, there are holomorphic coordinates satisfying

$$\omega(p) = \sqrt{-1} \sum_i dz_i \wedge d\bar{z}_i.$$

The last two equations hold for suitable coordinates on *any* complex submanifold M of $\mathbb{P}^n$. On an algebraic surface, the Kähler form in "good" coordinates at p is given by

$$\omega(p) = \sqrt{-1}(dz_1 \wedge d\bar{z}_1 + dz_2 \wedge d\bar{z}_2). \qquad (4.0.10)$$

Then

$$\omega(p)^2 = 2\sqrt{-1}^2(dz_1 \wedge d\bar{z}_1 \wedge dz_2 \wedge d\bar{z}_2)$$

is a positive multiple of the volume form at p. Since this inequality holds at every point of M, we conclude that

$$\int_M \omega^2 > 0.$$

Since the integral is nonzero, the cohomology class of ω is nonzero.

Now consider a holomorphic 2-form ϕ, which is given in local coordinates by $f(z_1, z_1)dz_1 \wedge dz_2$. We have

$$\phi \wedge \bar{\phi} = |f|^2 dz_1 \wedge dz_2 \wedge d\bar{z}_1 \wedge d\bar{z}_2 = -|f|^2 dz_1 \wedge d\bar{z}_1 \wedge dz_2 \wedge d\bar{z}_2,$$

so that

$$\phi \wedge \bar{\phi} = (\sqrt{-1})^2 |f|^2 dz_1 \wedge d\bar{z}_1 \wedge dz_2 \wedge d\bar{z}_2 > 0.$$

The integrand $\phi \wedge \bar{\phi}$ is a *positive* multiple of the volume form at each point, and so

$$\int_M \phi \wedge \bar{\phi} > 0.$$

This is the second Riemann bilinear relation for the $(2, 0)$ space of an algebraic surface.

Consider next the harmonic representative ϕ of a primitive class of type $(1,1)$. There are local coordinates which simultaneously diagonalize the Kähler form and the form ϕ at a given point p. Then $\omega(p)$ is as above, in (4.0.10), while $\phi(p) = a_1 dz_1 \wedge d\bar{z}_1 + a_2 dz_2 \wedge d\bar{z}_2$. Now the wedge product of harmonic forms is not in general harmonic, any more than the product of harmonic functions is harmonic. However, *the product of a parallel form and a harmonic form is harmonic*, just as the product of a constant function and a harmonic function is constant. A parallel form is by definition one whose covariant derivative is zero—it is a "constant" for the covariant derivative and all operators like the Laplacian, that are derived from the covariant derivative. The Kähler form, which is a form derived from the metric, is parallel. Hence the product of this form with a harmonic form is harmonic. Consider now the form $\omega \wedge \phi$. Since the cohomology class of ϕ is zero, so is the cohomology class of $\omega \wedge \phi$. But this last form is also harmonic. Since it is the unique harmonic representative of the zero cohomology class, it is zero as a differential form. Therefore, at each point we have

$$\omega \wedge \phi(p) = \sqrt{-1}(dz_1 \wedge d\bar{z}_1 + dz_2 \wedge d\bar{z}_2) \wedge (a_1 dz_1 \wedge dz_1 + a_2 dz_2 \wedge dz_2) = 0,$$

from which we conclude that $a_1 + a_2 = 0$ at p. Write

$$\phi(p) = \lambda(dz_1 \wedge d\bar{z}_1 - dz_2 \wedge d\bar{z}_2).$$

Then

$$\phi \wedge \bar{\phi}(p) = -2(\sqrt{-1})^2 |\lambda|^2 dz_1 \wedge d\bar{z}_1 \wedge dz_2 \wedge d\bar{z}_2$$

is a *negative* multiple of the volume form at p, and so

$$\int_M \phi \wedge \bar{\phi} < 0.$$

We conclude that the Hermitian form $\langle x, \bar{y} \rangle$ given by the cup product is negative on a primitive class of type $(1,1)$. *This is the first sign of a general pattern: in primitive cohomology, the signs of $\langle x, \bar{y} \rangle$ on the spaces $H^{p,q}$ alternate as p increases.* Note also that in this case the Hermitian form takes opposite signs on the two parts of the $(1,1)$ cohomology: the primitive part on the one hand and the Kähler form on the other.

Let us return to the polarizing form Q defined at the beginning of this section. Take

$$Q(\alpha, \beta) = (-1)^{k(k+1)/2} \int_M \alpha \wedge \beta \wedge \omega^{n-k}$$

for the k-th cohomology of an n-dimensional compact Kähler manifold, where $k \leq n$. Then Q defines a polarization on the primitive part of the k-th cohomology: the form $h(x, y)$ defined in (4.0.3) is positive definite (and the signs of $\langle x, \bar{y} \rangle$ alternate).

4.1 PERIOD DOMAINS AND MONODROMY

Fix a lattice $H_\mathbb{Z}$, a weight k, a bilinear form Q, and a vector of Hodge numbers $h = (h^{p,q}) = (\dim H^{p,q})$. For example, we might take $H_\mathbb{Z} = \mathbb{Z}^{2g}$, the weight to be 1, Q to be the standard symplectic form, and $h = (g, g)$, where $h^{1,0} = g$ and $h^{0,1} = g$. This is the case of Riemann surfaces. Fix a lattice $\mathbb{Z}^n$, where $n = \mathrm{rank}\, H_{\mathbb{Z}}$, and fix a bilinear form Q_0 on $\mathbb{Z}^n$ isometric to Q. A *marked Hodge structure* (H, m) is a Hodge structure H on $H_{\mathbb{Z}}$ together with an isometry $m : \mathbb{Z}^n \longrightarrow H_{\mathbb{Z}}$. A marked Hodge structure determines a distinguished basis $\{m(e_i)\}$ of $H_{\mathbb{Z}}$, and such a distinguished basis in turn determines a marking.

Let D be the set of all marked Hodge structures on $H_{\mathbb{Z}}$ polarized by Q with Hodge numbers h. This is the *period domain* with the given data. We will show below that this set is a complex manifold. We also study important special cases of period domains, beginning with elliptic curves and Riemann surfaces of higher genus, then progressing to period domains of higher weight. For domains of higher weight, we encounter a new phenomenon, *Griffiths transversality*, which plays an important role.

Let Γ be the group of isometries of $H_{\mathbb{Z}}$ relative to Q. This is an arithmetic group—a group of integral matrices defined by algebraic equations—which acts on D. Because the action is properly discontinuous, the quotient D/Γ—the *period space*—is defined as an analytic space. While D parametrizes marked Hodge structures, D/Γ parametrizes isomorphism classes of Hodge structures. As a quotient under a properly discontinuous action, the period space has the structure of an *orbifold* or *V-manifold*. Every point of such a quotient has a neighborhood which is the quotient of an open set in $\mathbb{C}^n$ by the action of a finite group. When the group is trivial, the quotient is of course a manifold. In general, open sets of orbifolds are parametrized by open sets in $\mathbb{C}^n$. However, the parametrization may be n-to-1 with $n > 1$. In the case of dimension 1, the only possible local orbifold structure is the quotient of a disk by a cyclic group. But the quotient of the unit disk Δ by the group μ_n of n-th roots of unity is homeomorphic to the unit disk. Indeed, the map $f(z) = z^n$ identifies Δ/μ_n with Δ, and the map $f : \Delta \longrightarrow \Delta$ descends to a bijection $\Delta/\mu_n \longrightarrow \Delta$. Thus the analytic space underlying a one-dimensional orbifold is smooth.

Note that the group Γ may be viewed as the subgroup of matrices with integer entries in an orthogonal or symplectic group.

Consider now a family of algebraic varieties $\{X_s \mid s \in S\}$. Let $\Delta \subset S$ be the *discriminant locus*—the set of points in S where the fiber X_s is singular. We assume this to be a proper (Zariski) closed subset. The map f that associates to a point $s \in S - \Delta$ the class of the Hodge structure of $H^k(X_s)$ in D/Γ is called the *period map*. It is a map with quite special properties; in particular, it is holomorphic, it is the quotient of a holomorphic map of $\tilde{f} : \tilde{S} \longrightarrow D$, where $\tilde{S}$ is the universal cover, and its behavior as one approaches the discriminant locus is controlled by the *monodromy representation*. The monodromy representation is a homomorphism

$$\rho : \pi_1(S, o) \longrightarrow \mathrm{Aut}\, H^k(X_0)$$

which is equivariant in the sense that

$$\tilde{f}(\gamma x) = \rho(\gamma)\tilde{f}(x),$$

where $\tilde{f}$ is the "lift" of f.

To define the monodromy representation, consider first a family of algebraic varieties X/Δ: a map $f : X \longrightarrow \Delta$, where Δ is the disk of unit radius, and where every point of Δ except the origin is a regular value. Thus the fibers $X_s = f^{-1}(s)$ are smooth for $s \neq 0$, while X_0 is in general singular. Let $\xi = \partial/\partial\theta$ be the angular vector field. It defines a flow $\phi(\theta)$, where $\phi(\theta)$ is the diffeomorphism "rotation counterclockwise by θ radians." Note that $\phi(\theta_1)\phi(\theta_2) = \phi(\theta_1+\theta_2)$: the flow is a one-parameter family of diffeomorphisms such that $\phi(2\pi)$ is the identity. Over the punctured disk $\Delta^* = \Delta - \{0\}$, one may construct a vector field η which lifts ξ in the sense that $f_*\eta = \xi$. Let $\psi(\theta)$ be the associated flow. It satisfies $f \circ \psi(\theta) = \phi(\theta) \circ f$ and $\psi(\theta_1)\psi(\theta_2) = \psi(\theta_1 + \theta_2)$. Let $T = \psi(2\pi)$. This transformation, the monodromy transformation, is not usually the identity map, even when considered on the level of homology, which is usually how it is viewed.

To define the monodromy representation for families with an arbitrary base in place of Δ, one proceeds as follows. Let X/S be a family of varieties with discriminant locus Δ. Let γ be a loop in $S-\Delta$. It is given parametrically by a map $\gamma(t) : [0,1] \longrightarrow S-\Delta$. Consider the "cylinder" $f^{-1}(\gamma([0,1]))$ over the "circle" $\gamma([0,1])$. Let η be a vector field defined on the cylinder such that $f_*\eta = \partial/\partial t$. Let $\psi(t)$ be the corresponding flow, and let $\rho(\gamma) = \psi(1)_*$ considered in homology: using the flow, push the cycles around the cylinder from the fiber $X_{\gamma(0)}$ back to the same fiber. This map is not necessarily the identity, though it is if γ is homotopic to the identity. There results a homomorphism

$$\pi_1(S - \Delta, o) \longrightarrow \mathrm{Aut}(H^k(X_0, \mathbb{Z})).$$

This is the *monodromy representation*.

For a simple example of a monodromy representation, take a Moebius band M (see Figure 4.1). It is a bundle over the circle with fiber which can be identified with the interval $[-1, 1]$. Let $f : M \longrightarrow S^1$ be the projection. Consider also the boundary of the Moebius band, ∂M. The fiber of $f : \partial M \longrightarrow S^1$ is the two point space $f^{-1}(\theta) \cong \{-1, +1\}$. What is the monodromy representation for $\partial M \longrightarrow S^1$? It is the nontrivial map $\pi_1(S^1, 0) \longrightarrow \mathrm{Aut}\, H_0(\{-1, +1\})$, which can be identified with the natural map $\mathbb{Z} \longrightarrow \mathbb{Z}/2$. It is generated by the permutation which interchanges $+1$ and -1. As is the case here, monodromy representations are often (nearly) surjective and have large kernels. See [2] and [8].

EXERCISE 4.1.1 Consider the family $M \longrightarrow S^1$ where M is the Moebius strip. What is the monodromy on $H_1(f^{-1}(0), \partial f^{-1}(0)) \cong H_1([-1, +1], \{-1, +1\})$?

A more significant example, which we will study in more detail in the next section, concerns the family of elliptic curves $y^2 = (x^2 - t)(x - 1)$. For small, nonzero t, the fibers of this family are smooth. The fiber over the origin is a cubic curve with one node, and in this nodal case, one refers to the monodromy map T as the *Picard–Lefschetz transformation*. To compute it, consider as homology basis the positively oriented circle δ as in Figure 4.2. It encircles the two branch points at $\pm\sqrt{t}$. Let γ be the path that runs from the branch cut between 1 and ∞ to the branch cut between $-\sqrt{t}$ and $+\sqrt{t}$. When it meets the cut, it travels upward then makes a large rightward arc before traveling back to the cut from 1 to ∞. Now rotate the branch cut from $-\sqrt{t}$

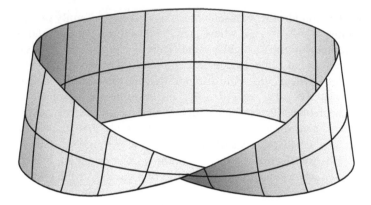

Figure 4.1: Moebius band.

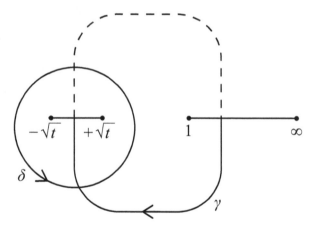

Figure 4.2: Picard–Lefschetz, $\theta = 0$

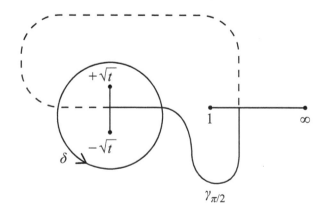

Figure 4.3: Picard–Lefschetz, $\theta = \pi/2$

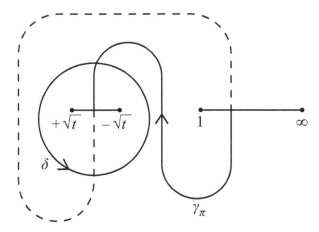

Figure 4.4: Picard–Lefschetz, $\theta = \pi$

to $+\sqrt{t}$ counterclockwise by π radians. As the branch cut rotates, so does the circular path δ, dragging γ with it as it goes. By drawing pictures at $\theta = 0, \pi/2$, and π, as in Figures 4.2, 4.3, and 4.4, one can compute the effect of this motion. Let $\gamma_{\pi/2}$ and γ_π be the loop γ after rotation of the branch cut by the indicated angle. Write

$$\gamma_\pi = a\delta + b\gamma.$$

Determine the coefficients a and b by taking intersection products with δ and γ. One finds that the Picard–Lefschetz transformation acts by $T(\delta) = \delta$ and $T(\gamma) = \gamma + \delta$. Thus the matrix of T is

$$T = \begin{pmatrix} 1 & 1 \\ 0 & 1 \end{pmatrix}. \tag{4.1.1}$$

Figure 4.5 gives an alternative view of the monodromy transformation, as a so-called *Dehn twist*.

EXERCISE 4.1.2   What is the monodromy for the family $y^2 = x(x-1)(x-t)$ near $t = 0, 1, \infty$?

4.2 ELLIPTIC CURVES

Let us now study the period domain and period space for Hodge structures of elliptic curves. This study will provide a guide to understanding period domains for arbitrary Hodge structures of weight 1. Once we do this, we will consider the nonclassical situation, that of Hodge structures of higher weight.

Let $\mathcal{E}$ be an *elliptic curve*, that is, a Riemann surface of genus 1. Such a surface can be defined by the affine equation $y^2 = p(x)$, where $p(x)$ is a cubic polynomial with distinct roots. Then $\mathcal{E}$ is a double cover of the Riemann sphere $\mathbb{CP}^1$ with branch points at the roots of p and also at infinity. A homology basis $\{\delta, \gamma\}$ for $\mathcal{E}$ is pictured in Figure 4.6.

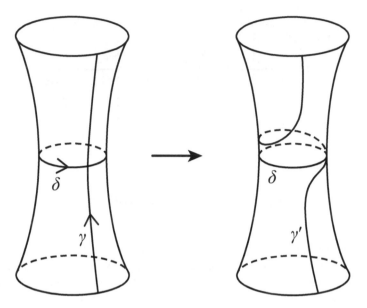

Figure 4.5: Dehn twist

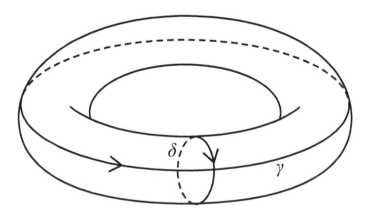

Figure 4.6: Homology basis for an elliptic curve

The intersection matrix for the indicated homology basis is the "standard symplectic form,"

$$Q_0 = \begin{pmatrix} 0 & 1 \\ -1 & 0 \end{pmatrix}.$$

Note that it is unimodular, i.e., has determinant 1. This is a reflection of Poincaré duality. Let $\{\delta^*, \gamma^*\}$ be the basis which is dual to the standard symplectic basis, i.e., $\delta^*(\delta) = \gamma^*(\gamma) = 1$, $\delta^*(\gamma) = \gamma^*(\delta) = 0$. As noted above, such a basis determines a marking of $H^1(M)$.

For a basis of $H^{1,0}(S)$, take the differential form

$$\omega = \frac{dx}{y} = \frac{dx}{\sqrt{p(x)}}.$$

It is obvious that ω is holomorphic on the part of S above the complement of the set of zeros of $p(x)$, plus the point at infinity. A calculation in local coordinates shows that ω is holomorphic at those points as well. (Exercise).

We can now give an approximate answer to the question "what is the period domain for Hodge structures of elliptic curves?" A marking $m : \mathbb{Z}^2 \longrightarrow H^1(M, \mathbb{Z})$ determines a subspace

$$m^{-1}(H^{1,0}(M)) \subset \mathbb{C}^2.$$

That is, a marked Hodge structure determines a one-dimensional subspace of $\mathbb{C}^2$. The set of one-dimensional subspaces of a two-dimensional vector space is a Grassmann variety. In this case it is just complex projective space of dimension 1. Thus the set of marked Hodge structures D of weight 1 with $h^{1,0} = 1$ can be identified with a subset of $\mathbb{CP}^1$.

We ask, is $D = \mathbb{CP}^1$? The discussion below shows that it is not. Nonetheless, $\mathbb{CP}^1$ does play a special role. Every period domain has its so-called *compact dual* $\check{D}$, and in the present case, $\check{D} = \mathbb{CP}^1$. The compact dual is a compact complex manifold with a number of special properties, and the period domain D is an open subset of it.

To answer the question of which part of $\mathbb{CP}^1$ corresponds to polarized Hodge structures, consider the vector of integrals

$$(A, B) = \left(\int_\delta \omega, \int_\gamma \omega \right).$$

These are the *periods* of ω. We refer to them as the A-period and the B-period. The periods express ω in terms of the basis $\{\delta^*, \gamma^*\}$:

$$\omega = A\delta^* + B\gamma^*.$$

The second Riemann bilinear relation is the statement

$$\sqrt{-1} \int_S \omega \wedge \overline{\omega} > 0.$$

Substituting the expression for ω in terms of the dual basis and using

$$(\delta^* \cup \gamma^*)[S] = 1,$$

we find that
$$\sqrt{-1}(A\bar{B} - B\bar{A}) > 0. \qquad (4.2.1)$$

It follows that $A \neq 0$ and $B \neq 0$. Therefore the *period ratio* $Z = B/A$ is defined. It depends only on the choice of marking and is therefore an invariant of the marked Hodge structure (H, m).

The ratio Z can be viewed as the B-period of the unique cohomology class in $H^{1,0}$ whose A-period is 1:
$$(A, B) = (1, Z).$$

From (4.2.1), it follows that Z, the *normalized period*, has positive imaginary part. Thus to the marked Hodge structure (H^1, m) is associated a point in the upper half plane,
$$\mathcal{H} = \{z \in \mathbb{C} \mid \Im z > 0\}.$$

Consequently there is a map
$$\{\text{marked Hodge structures}\} \longrightarrow \mathcal{H}. \qquad (4.2.2)$$

This map has an inverse given by
$$Z \in \mathcal{H} \mapsto \mathbb{C}(\delta^* + Z\gamma^*).$$

Thus (4.2.2) is an isomorphism:
$$D = \{\text{marked Hodge structures}\} \cong \mathcal{H}.$$

To see how D sits inside the Grassmannian $\mathbb{CP}^1$, let $[A, B]$ be homogeneous coordinates for projective space. Identify $\{[A, B] \mid A \neq 0\}$ with the complex line $\mathbb{C}$ via $[A, B] \mapsto B/A$. Then $\mathbb{CP}^1$ is identified with the one-point compactification of the complex line, where the point at infinity corresponds to the point of $\mathbb{CP}^1$ with homogeneous coordinates $[0, 1]$. The inclusion of D in $\check{D}$ can then be identified with the composition of maps
$$\mathcal{H} \longrightarrow \mathbb{C} \longrightarrow \mathbb{C} \cup \{\infty\} \cong \mathbb{CP}^1.$$

The upper half plane can be thought of as the part of the northern hemisphere strictly above the equator, which in turn can be thought of as the one-point compactification of the real line. This is no surprise, since the upper half plane is biholomorphic to the unit disk.

Having identified the period domain D with the upper half plane, let us identify the period space D/Γ. The key question is, what is the group of transformations that preserves the lattice $H^1_{\mathbb{Z}}$ and the bilinear form Q? The answer is clear: it is the group of 2×2 integral symplectic matrices. This is a group which acts transitively on markings. Let M be such a matrix, and consider the equation
$${}^t M Q_0 M = Q_0.$$

Set
$$M = \begin{pmatrix} a & b \\ c & d \end{pmatrix}.$$

The above matrix equation is equivalent to the single scalar equation $ad - bc = 1$, that is, to the condition $\det M = 1$. Thus the group Γ, which *is* an integral symplectic group, is also the group of integer matrices of determinant 1, that is, the group $\mathrm{SL}(2, \mathbb{Z})$. This group acts on complex projective space by fractional linear transformations. Indeed, we have

$$(1, Z)M = (a + cZ, b + dZ) \equiv (1, (b + dZ)/(a + cZ)),$$

so the action on normalized period matrices is by

$$Z \mapsto \frac{b + dZ}{a + cZ}.$$

The action on $\mathcal{H}$ is *properly discontinuous*: that is, for a compact set $K \subset \mathcal{H}$, there are only finitely many group elements g such that $gK \cap K \neq \emptyset$. Consequently the quotient $\mathcal{H}/\Gamma$ is a Hausdorff topological space. It is even more: an analytic manifold with a natural orbifold structure. To conclude, we have found that

$$\{\text{isomorphism classes of Hodge structures}\} \cong D/\Gamma \cong \mathcal{H}/\Gamma.$$

To see what sort of object the period space is, note that a fundamental domain for Γ is given by the set

$$\mathcal{H} = \{z \in \mathcal{H} \mid |\Re(z)| \leq 1/2, |z| \geq 1\},$$

which is pictured in Figure 4.7. The domain $\mathcal{H}$ is a triangle. One vertex is at infinity, and the other two are at ω and $-\bar{\omega}$, where

$$\omega = \frac{-1 + \sqrt{-3}}{2}$$

is a primitive cube root of unity. The group Γ is generated by the element

$$S = \begin{pmatrix} 0 & 1 \\ -1 & 0 \end{pmatrix},$$

corresponding to the fractional linear transformation

$$S(Z) = -1/Z,$$

and the element

$$T = \begin{pmatrix} 1 & 1 \\ 0 & 1 \end{pmatrix},$$

which we recognize as a Picard–Lefschetz transformation, corresponding to the fractional linear transformation

$$T(Z) = Z + 1.$$

The period space D/Γ is the same as the fundamental domain $\mathcal{H}$ modulo the identifications determined by S and T. The identification defined by T glues the right and left sides of $\mathcal{H}$ to make a cylinder infinite in one direction, identifying, for example, the points ω and $-1/\omega = -\bar{\omega}$. The map S glues a half circle on the boundary of the cylinder to the opposite half circle: the arc from ω to i is identified with the arc from $-1/\omega$ to i. Topologically, the result is a disk. As an orbifold it can be identified with the complex line $\mathbb{C}$ with two special points corresponding to ω and i, which are fixed points for the action of Γ. Better yet, there is a meromorphic function $j(z)$, the quotient of modular forms of weight 12, which is invariant under the action of Γ and which gives a bijective holomorphic map $\mathcal{H}/\Gamma \longrightarrow \mathbb{C}$. The forms are defined as follows.

$$g_2 = 60 \sum_{(m,n)\neq(0,0)} \frac{1}{(m+n\tau)^4},$$

$$g_3 = 140 \sum_{(m,n)\neq(0,0)} \frac{1}{(m+n\tau)^6},$$

and

$$\Delta = g_2^3 - 27g_3^2.$$

The function $\Delta(\tau)$ is the *discriminant*. It vanishes if and only if the elliptic curve

$$y^2 = 4x^3 - g_2 x - g_3$$

is singular. The j-invariant is the quotient

$$j(\tau) = \frac{1728g_2^3}{\Delta}.$$

Thus $j(\tau) = 0$ for the elliptic curve with $g_2 = 0$. By a change of coordinates, we may assume that the curve has equation $y^2 = x^3 - 1$. This is an elliptic curve with branch points at infinity and the cube roots of unity. It has an automorphism of order 6, given by $(x, y) \mapsto (\omega x, -y)$.

EXERCISE 4.2.1   Show that the group $G = \mathrm{SL}(2, \mathbb{R})$ acts transitively on the upper half plane by fractional linear transformations. Show that the isotropy group $K = \{g \in G \mid g \cdot i = i\}$ is isomorphic to the unitary group $U(1)$. Then show $\mathcal{H} \cong G/K$. Thus the period space is $G_{\mathbb{Z}}\backslash G/K$, where $G_{\mathbb{Z}}$ is the set of integer-valued points of G.

4.3   PERIOD MAPPINGS: AN EXAMPLE

Let us consider now the family of elliptic curves $\mathcal{E}_t$ given by

$$y^2 = x(x - 1)(x - t).$$

The parameter space for this family is the extended complex line $\mathbb{C} \cup \{\infty\} \cong \mathbb{CP}^1$. The fibers $\mathcal{E}_t$ are smooth for $t \neq 0, 1, \infty$. The set of points $\Delta = \{0, 1, \infty\}$ is the *discriminant locus* of the family. Thus the family $\{\mathcal{E}_t\}$ is smooth when restricted to $\mathbb{CP}^1 - \Delta$.

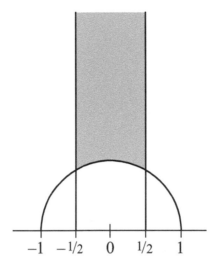

Figure 4.7: Fundamental domain

Fix a point t_0 in the complement of the discriminant locus, and let U be a coordinate disk centered at t_0 which lies in the complement of the discriminant locus. Then a homology basis $\{\delta, \gamma\}$ for $\mathcal{E}_t$ exists which lies in the inverse image of the complement of U. As we see in Figure 4.8, this is a fancy way of saying the following:

> *Fix δ and γ as in the figure to get a homology basis of $\mathcal{E}_{t_0}$. Then the same δ and γ give a homology basis for $\mathcal{E}_t$ for t sufficiently close to t_0.*

Therefore the periods for $\mathcal{E}_t$ take the form

$$A_t = \int_\delta \frac{dx}{\sqrt{x(x-1)(x-t)}}, \qquad B_t = \int_\gamma \frac{dx}{\sqrt{x(x-1)(x-t)}},$$

where the domains of integration δ and γ are fixed. Since the integrands depend holomorphically on the parameter t, so do the periods A_t and B_t, as well as the ratio $Z_t = B_t/A_t$. Thus the period map is a holomorphic function on U with values in $\mathcal{H}$.

So far our approach to the period map has been local. One way of defining the period map globally is to consider its full analytic continuation, which will be a function from the universal cover of the parameter space minus the discriminant locus to the period domain, in this case the upper half plane. The branches of the analytic continuation correspond to different markings of the fibers $\mathcal{E}_t$, that is, to different homology bases. Consider therefore the composed map

$$\{\text{universal cover of } \mathbb{CP}^1 - \Delta\} \longrightarrow \mathcal{H} \longrightarrow \Gamma \backslash \mathcal{H}.$$

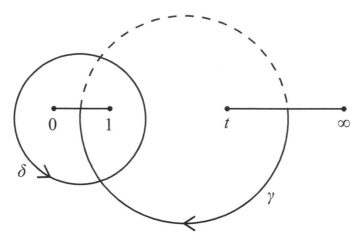

Figure 4.8: Branch cuts for $y^2 = x(x-1)(x-t)$

This map is holomorphic, and it is also invariant under the action of covering transformations on the left. Thus we obtain a holomorphic map

$$\mathbb{CP}^1 - \Delta \longrightarrow \Gamma \backslash \mathcal{H}.$$

This is the period map.

The family of elliptic curves just discussed has three singular fibers. Are there nontrivial families with fewer fibers? Consider a family with just two singular fibers, which we may take to be at zero and infinity. The parameter space for the smooth fibers is $\mathbb{C}^*$, the complex line with the origin removed. The universal cover of $\mathbb{C}^*$ is the complex line, with covering map $\mathbb{C} \longrightarrow \mathbb{C}^*$ given by $\exp 2\pi i z$. Thus the lift of the period map to the universal cover is a holomorphic map $\mathbb{C} \longrightarrow \mathcal{H}$. Now the upper half plane is biholomorphic to the unit disk (exercise: verify this). Consequently the lift is in essence a bounded entire function. Such functions are constant. This means that if s and t are nonzero complex numbers, then $\mathcal{E}_s$ and $\mathcal{E}_t$ are isomorphic.

As a related application, suppose that one has a family of elliptic curves parametrized by the unit disk minus the origin, which we write as Δ^*. Suppose further that the monodromy transformation is trivial. This means that analytic continuation defines a period map $f : \Delta^* \longrightarrow \mathcal{H}$. Again using the fact that $\mathcal{H}$ is biholomorphic to the disk, we apply the Riemann removable singularity theorem to conclude that the period map extends to a holomorphic map $f : \Delta \longrightarrow \mathcal{H}$. The point $f(0)$ corresponds to the Hodge structure of a smooth elliptic curve. From this we conclude that the family of elliptic curves on the punctured disk is the restriction of a family of elliptic curves on the disk.

Asymptotics of the period map

Let us examine the behavior of the period map for the family $y^2 = x(x-1)(x-t)$ as t approaches infinity along the ray $[2, \infty)$ on the real axis. Let δ and γ be as in

Figure 4.8. Then the A-period is given by

$$A(t) = \int_\delta \frac{dx}{\sqrt{x(x-1)(x-t)}}.$$

When δ is a circle of radius $R > 2$ and t is large compared to R, then $x(x-1) \sim x^2$ and $\sqrt{x-t} \sim \sqrt{-t}$, so that

$$A(t) \sim \int_\delta \frac{dx}{x\sqrt{-t}} \sim \frac{2\pi}{\sqrt{t}}.$$

One can give a more careful argument which yields the same approximation. By deforming the path of integration over γ, we find that the B-period is given by

$$B(t) = \int_\gamma \frac{dx}{\sqrt{x(x-1)(x-t)}} = -2\int_1^t \frac{dx}{\sqrt{x(x-1)(x-t)}}.$$

Write the integral as the sum of two terms,

$$B(t) = -2\int_1^2 \frac{dx}{\sqrt{x(x-1)(x-t)}} - 2\int_2^t \frac{dx}{\sqrt{x(x-1)(x-t)}},$$

where $|t| \gg 2$. The first term is bounded in t, while the second is asymptotic to

$$-2\int_2^t \frac{dx}{x\sqrt{(x-t)}}.$$

This integral can be computed exactly in terms of the arctangent from which one finds

$$B(t) \sim \frac{4}{\sqrt{t}}\left[1 + O\left(\frac{1}{t}\right)\right] \arctan\left(\sqrt{\frac{C}{t} - 1}\right)$$

The expression in the arctangent is asymptotic to

$$\frac{\sqrt{-1}}{2}\log\frac{C}{t},$$

so that

$$B(t) \sim \frac{2\sqrt{-1}}{\sqrt{t}}\log t.$$

Thus the period ratio Z has the asymptotic form

$$Z(t) \sim \frac{\sqrt{-1}}{\pi}\log t.$$

Behavior of this kind generalizes to arbitrary degenerations, though the result is much harder to prove. The dominant term in a period is of the form $t^a (\log t)^b$, where a is a rational number and so $\lambda(a) = \exp(2\pi i a)$ is a root of unity. The parameters a and b

are related to the monodromy T. The root of unity $\lambda(a)$ is an eigenvalue of T. If m is the least common multiple of the parameters a, then T^m is unipotent, and so $T^m - 1$ is nilpotent. The index of nilpotence is $b + 1$, that is, b is the smallest positive integer such that

$$(T^m - 1)^{b+1} = 0.$$

For algebraic curves, $(T^m - 1)^2 = 0$, for surfaces $(T^m - 1)^3 = 0$, etc. The index of nilpotence can be less than maximal in special cases. For example, in a degeneration of algebraic varieties where the central fiber consists of two smooth varieties meeting transversely in a smooth variety, the index of nilpotence is 2.

EXERCISE 4.3.1   (a) Consider the family of zero-dimensional varieties X_t defined by $z^p = t$. Describe the monodromy on $H_0(X_t)$.

   (b) (Harder) Consider the family $z^p + w^q = t$. Describe the monodromy on $H_0(X_t)$. (See also [14].)

EXERCISE 4.3.2   Consider the family of elliptic curves $\{\mathcal{E}_t\}$ defined by $y^2 = x^3 - 1$. What is the monodromy representation? What is the period map? Describe the singular fibers.

4.4   HODGE STRUCTURES OF WEIGHT 1

Let us now study the period domain for polarized Hodge structures of weight 1. One source of such Hodge structures is the first cohomology of Riemann surfaces. Another is the first cohomology of abelian varieties. An abelian variety is a compact complex torus that is also a projective algebraic variety. Elliptic curves are abelian varieties of dimension 1.

It is natural to ask if all weight-1 Hodge structures come from Riemann surfaces. This is the case for $g = h^{1,0} = 1$. The starting point for the proof of this statement is the fact that the moduli space (the space of isomorphism classes) of elliptic curves and the period space have the same dimension. For $g > 1$, the dimension of the moduli space is $3g - 3$. By the Torelli theorem, the period map for Riemann surfaces is injective. Thus the corresponding space of Hodge structures has dimension $3g - 3$. As for the period space, we show below that it has dimension $g(g + 1)/2$. Therefore the dimension of the space of Hodge structures of genus g is larger than the dimension of the space of Hodge structures coming from Riemann surfaces, except in the cases $g = 1, g = 2$.

Despite the fact that not all polarized Hodge structures are Hodge structures of algebraic curves, they all come from the first cohomology of algebraic varieties. This is because a polarized Hodge structure of weight 1 determines an abelian variety in a natural way. The underlying torus is the quotient

$$J(H) = \frac{H_{\mathbb{C}}}{H^{1,0} + H_{\mathbb{Z}}}.$$

The polarizing form Q may be viewed as an element of $\Lambda^2 H_{\mathbb{Z}}^1$, hence as an element ω_Q of $H^2(J, \mathbb{Z})$. The first Riemann bilinear relation implies that ω_Q has type $(1, 1)$. The second Riemann bilinear relation is equivalent to the statement that ω_Q is represented by a positive $(1,1)$-form. Using the exponential sheaf sequence, one shows that ω_Q is the first Chern class of a holomorphic line bundle L. Because the first Chern class is positive, a theorem of Kodaira applies to show that sections of some power of L give a projective embedding of J.

EXERCISE 4.4.1 Let H be a Hodge structure of weight 1, and let $J(H)$ be the associated torus. Show that the first cohomology of $J(H)$ is isomorphic to H as a Hodge structure.

Let us now describe the period space D for polarized Hodge structures of weight 1 and genus $g = \dim H^{1,0}$. Although the period space depends on the choice of a skew form Q, we work here under the assumption that the matrix of Q in a suitable basis takes the form

$$Q = \begin{pmatrix} 0 & I \\ -I & 0 \end{pmatrix},$$

where I is the $g \times g$ identity matrix. Let $\{\delta^j, \gamma^j\}$ be a basis for $H_{\mathbb{Z}}$, where j runs from 1 to g. Write $e_i = \delta_i$ and $e_{g+i} = \gamma_i$. Assume that the matrix of inner products (e_i, e_j) is Q. Let $\{\phi_i\}$ be a basis for $H^{1,0}$, where i runs from 1 to g. Then

$$\phi_i = \sum_j A_{ij}\delta^j + \sum_j B_{ij}\gamma^j.$$

The $g \times 2g$ matrix

$$P = (A, B)$$

is the *period matrix* of the Hodge structure with respect to the given bases. The basis for $H_{\mathbb{Z}}$ defines a marking m of the Hodge structure. Then $S = m^{-1}H^{1,0} \subset \mathbb{C}^{2g}$ is the row space of the period matrix. Therefore the period domain D is a subset of the Grassmannian of g-planes in complex $2g$-space. The Riemann bilinear relations impose restrictions on these subspaces. The first Riemann bilinear relation gives a set of quadratic equations that express the fact that $Q(v, w) = 0$ for any two vectors in S. The second Riemann bilinear relation gives a set of inequalities: $\sqrt{-1}Q(v, \bar{v}) > 0$ for any nonzero vector in S.

To understand both the equations and the inequalities, we first show that the period matrix A is nonsingular. To that end, let $\phi = \sum_m v_m \phi_m$ be an arbitrary nonzero element of $H^{1,0}$. Then

$$\sqrt{-1}Q(\phi, \bar{\phi}) > 0. \tag{4.4.1}$$

Therefore,

$$\sqrt{-1} \sum_{m,n,j} Q(v_m A_{mj}\delta^j + v_m B_{mj}\gamma^j, \bar{v}_n \bar{A}_{nk}\delta^k + \bar{v}_n \bar{B}_{nk}\gamma^k) > 0,$$

and so

$$\sqrt{-1} \sum_{m,n,j} (v_m \bar{v}_n A_{mj}\bar{B}_{nj} - v_m \bar{v}_n B_{mj}\bar{A}_{nj}) > 0 \tag{4.4.2}$$

for arbitrary choices of the quantities v_m. Let

$$H_{mn} = \sqrt{-1} \sum_j (A_{mj} \bar{B}_{nj} - B_{mj} \bar{A}_{nj}).$$

Then

$$H = \sqrt{-1}(AB^* - BA^*) > 0, \qquad (4.4.3)$$

where B^* is the Hermitian conjugate of B. Using (4.4.1), we conclude that H is positive definite. If A were singular, there would exist a vector v such that $vA = 0$, in which case we would also have $A^* \bar{v} = 0$, so that $v H \bar{v} = 0$, a contradiction. Thus A is nonsingular.

At this point we know that the period matrix can be brought to the normalized form $P = (I, Z)$, where for now, Z is an arbitrary $g \times g$ matrix. Let us apply the first Riemann bilinear relation to elements $\phi = v_m \phi_m$ and $\psi = w_n \phi_n$ of $H^{1,0}$. One finds that

$$\sum_{m,n,j} v_m w_n A_{mj} B_{nj} - v_m w_n B_{mj} A_{nj} = 0. \qquad (4.4.4)$$

Setting A to the identity matrix, this simplifies to

$$\sum_{m,n} v_m w_n B_{nm} - v_m w_n B_{mn} = 0, \qquad (4.4.5)$$

which can be written

$$v(B - {}^t B)w = 0$$

for arbitrary v and w. Therefore B is a symmetric matrix, as claimed. The fact that H is positive definite is equivalent to the statement that Z has positive-definite imaginary part. To conclude,

$$\mathcal{H}_g = \{ Z \mid Z \text{ is a complex symmetric matrix with positive imaginary part} \}.$$

Its dimension is $g(g+1)/2$.

Remark. As in the case of $\mathcal{H} = \mathcal{H}_1$, there is a group-theoretic description. Let $G = \mathrm{Sp}(g, \mathbb{R})$ be the group of real $2g \times 2g$ matrices which preserve the form Q. This is the real symplectic group. Fix a polarized Hodge structure H. Let g be an element of the isotropy group K of this "reference" Hodge structure. The element g preserves the reference structure and is determined by its restriction to $H^{1,0}$. The restriction to $H^{1,0}$ preserves the positive Hermitian form $\sqrt{-1}Q(v, \bar{w})$. Therefore K is the unitary group of $H^{1,0}$. Since G acts transitively on $\mathcal{H}_g$ (exercise!), we find that

$$\mathcal{H}_g \cong G/K = \mathrm{Sp}(g, \mathbb{R})/U(g).$$

The group which permutes the markings is $G_{\mathbb{Z}} = \mathrm{Sp}(g, \mathbb{Z})$. Thus the period space is

$$\Gamma \backslash \mathcal{H}_g = G_{\mathbb{Z}} \backslash G/K = \mathrm{Sp}(g, \mathbb{Z}) \backslash \mathrm{Sp}(g, \mathbb{R})/U(g).$$

EXERCISE 4.4.2 Compute the dimensions of the Lie groups $\mathrm{Sp}(g, \mathbb{R})$ and $U(g)$. Then compute the dimension of the Siegel upper half-space in terms of these dimensions.

Remark. Let G be a noncompact simple Lie group and let K be a maximal compact subgroup. The quotient $D = G/K$ is a *homogeneous space*—a space on which there is a transitive group action. Quotients of a noncompact simple Lie group by a maximal compact subgroup are *symmetric spaces*: they carry a G-invariant Riemannian metric, and at each point x of D there is an isometry i_x which fixes x and which acts as -1 on the tangent space. That symmetry is given by an element of K which can be identified with $-I$, where I is the identity matrix when G and K are identified with matrix groups in a suitable way. The Siegel upper half-space is an example of a *Hermitian* symmetric space. This is a symmetric space which is also a complex manifold. For such spaces the isotropy group contains a natural subgroup isomorphic to the circle group $U(1)$. Its action gives the complex structure tensor, for then isometries representing multiplication by a unit complex number are defined. In particular, an isometry representing rotation counterclockwise through an angle of $90°$ is defined at each point.

Remark. Let G/K be a Hermitian symmetric space, and let $\Gamma = G_{\mathbb{Z}}$. A theorem of Baily–Borel shows that $M = G_{\mathbb{Z}} \backslash G/K$ has a projective embedding. Thus, like the quotient of the upper half plane by the action of $\mathrm{SL}(2, \mathbb{Z})$, these spaces are quasi-projective algebraic varieties. By this we mean that they are of the form "a projective variety minus a projective subvariety."

4.5   HODGE STRUCTURES OF WEIGHT 2

New phenomena arise when one considers Hodge structures of weight greater than 1. All of these phenomena present themselves in the weight-2 case, a case that we can study without notational complications. Polarized Hodge structures of weight 2 arise in nature as the primitive second cohomology of an algebraic surface M. In this case primitivity has a simple meaning: orthogonal to the hyperplane class. Such a Hodge structure has the form

$$H_o^2 = H^{2,0} \oplus H_o^{1,1} \oplus H^{0,2}.$$

We remarked earlier that classes of type $(2, 0)$ are primitive for reasons of type. The same is of course true for classes of type $(0, 2)$. For polarized Hodge structures of weight 2, the Hodge filtration is $F^2 = H^{2,0}$, $F^1 = H^{2,0} \oplus H_o^{1,1}$, $F^0 = H_o^2$. By the first Riemann bilinear relation, the orthogonal complement of F^2 is F^1. Thus the data F^2 and Q determine the polarized Hodge structure. Let us set $p = \dim H^{2,0}$ and $q = \dim H_o^{1,1}$. Then a marked Hodge structure determines a subspace $S = m^{-1}F^2$ of dimension p in $\mathbb{C}^{2p+q}$. The period domain D whose description we seek is therefore a subset of the Grassmannian of p-planes in $(2p + q)$-space. As in the case of weight-1 structures, the first and second bilinear relations impose certain equalities and inequalities. As a result, D will be an open subset of a closed submanifold $\check{D}$ of the Grassmannian. The submanifold $\check{D}$ is the set of isotropic spaces of dimension p—the set of p-dimensional subspaces S on which Q is identically zero. The open set D is defined by the requirement that $Q(v, \bar{v}) < 0$ for all nonzero vectors in S.

As before, the period domain is a complex homogeneous space with a simple group-theoretic description. Let G be the special orthogonal group of the vector space $H_\mathbb{R}$ endowed with the symmetric bilinear form Q. This group acts transitively on marked Hodge structures. Now fix a marked Hodge structure and consider an element g of G which leaves this Hodge structure invariant. Then g restricts to the subspaces of the Hodge decomposition; since it is a real linear transformation, its restriction to $H^{2,0}$ determines its restriction to $H^{0,2}$. The restriction to $H^{2,0}$ also preserves the Hermitian form $Q(v, \bar{w})$, which is negative definite on this subspace. Similarly, the restriction of g to $H_o^{1,1}$ preserves the form $Q(v, w)$, which is positive definite on this subspace. (Recall that Q is the intersection form up to sign, and the sign $(-1)^{n(n+1)/2}$ in this case is -1.) Let V be the isotropy subgroup of G: the subgroup whose elements leave the particular marked Hodge structure fixed. We have just defined a map $V \longrightarrow U(p) \times SO(q)$ via $g \mapsto (g|H^{2,0}, g|H_o^{1,1})$. This map is an isomorphism (exercise). Thus we find that

$$D \cong G/V.$$

The easiest way to see the complex structure on D is to view it as an open set of $\check{D}$, which in turn is an algebraic submanifold of the Grassmannian. Since G acts by holomorphic transformations, the complex structure defined group-theoretically agrees with the one defined naively.

We now encounter the first major difference with classical Hodge theory. In general, the isotropy group V, while compact, is not maximal compact. Indeed, for weight-2 structures, a maximal compact subgroup K has the form $SO(2p) \times SO(q)$. An element of K is an element of G which preserves the subspace $H_o^{1,1}$ and the subspace $H^{2,0} \oplus H^{0,2}$. The restriction of Q to the real points of $H^{2,0} \oplus H^{0,2}$ is a negative-definite form of rank $2p$. Thus K is isomorphic to $S(O(2p) \times O(q))$. Only in the case $p = 1$ do we have $V = K$. In that case D is a Hermitian symmetric space, and it can be realized as a bounded domain in complex Euclidean space. In all other cases D is not Hermitian symmetric, and it is not an open subset of complex Euclidean space. Indeed, D abounds in compact complex subvarieties of positive dimension, namely, K/V and its translates by G (and its "sufficiently small" translates by $G_\mathbb{C}$). It is worth mentioning that

(a) the Hermitian symmetric case occurs for K3 surfaces, hence their prominent role in algebraic geometry; and

(b) in general the map $G/V \longrightarrow G/K$ has as target a symmetric space, with source a complex manifold.

However, the map is not holomorphic, even when G/K happens to be Hermitian symmetric.

Despite the nonclassical nature of D in higher weight, there are two facts that make period domains and period mappings useful. The first is that the period map is a holomorphic. The second is that it satisfies a differential equation which forces it, unlike an arbitrary holomorphic map into a period domain, to behave as if it were mapping into a bounded domain. This differential equation, which we shall explain presently, is *Griffiths transversality*.

Let us begin with the definition of the period map in complete generality. Let $f : X \longrightarrow S$ be a smooth family of algebraic varieties. Let $U \subset S$ be a contractible open set with distinguished point b, the base point. Then the restriction of f to $f^{-1}(U)$ is a family differentiably isomorphic to $X_b \times U \longrightarrow U$, where the map is a projection on the second factor. Let $\{\delta^i\}$ be a basis for $H^k(X_b, \mathbb{Z})_o$. Because of the product structure, this basis defines a basis for $H^k(X_s, \mathbb{Z})_o$ for all s in U. Consequently we have a family of markings $m_s : \mathbb{Z}^n \longrightarrow H^k(X_s, \mathbb{Z})$. Let $F_s^p = m_s^{-1}(F^p H^k(X_s)_o)$. Thus is defined a family of filtrations F_s^p of $\mathbb{C}^n$. This is the local period map $p : U \longrightarrow D$. If one accepts for now that the period mapping is holomorphic, one defines the period mapping from the universal cover of S to D as the full analytic continuation of such a local period map. Thus one has first

$$p : \tilde{S} \longrightarrow D$$

and subsequently the quotient map

$$p : S \longrightarrow \Gamma \backslash D.$$

There is, however, another argument. Lift the family X/S to a family $X/\tilde{S}$. Then the bundle of cohomology groups of fibers is trivial, i.e., is isomorphic to $H^k(X_b, \mathbb{Z}) \times \tilde{S}$. Therefore there is a marking for the pull-back of the cohomology of the family to the universal cover of S. Using this marking, which makes no reference to the as yet unproved holomorphicity of the period map, we construct a period map $p : \tilde{S} \longrightarrow D$.

Below, we will show that the period map is holomorphic in the special case of surfaces in $\mathbb{P}^3$. The method of proof, which relies on the Griffiths–Poincaré residue, works for the case of hypersurfaces in $\mathbb{P}^n$. Using the residue calculus, we will also establish Griffiths transversality for hypersurfaces.

EXERCISE 4.5.1 (a) Determine the groups G, V, and K for a Hodge structure of weight 3. (b) Do the same for weight 4.

4.6 POINCARÉ RESIDUES

Let X_i be homogeneous coordinates on $\mathbb{C}^{n+1}$, where i ranges from 0 to n. Consider the n-form

$$\Phi = d(X_1/X_0) \wedge \cdots \wedge d(X_n/X_0) = \frac{1}{X_0^{n+1}} \sum_{i=0}^{n} X_i dX_0 \wedge \cdots \wedge \widehat{dX_i} \wedge \cdots \wedge dX_n,$$

where we view the X_i/X_0 as affine coordinates on the open set $U_0 = \{[X] \mid X_0 \neq 0\}$. Note that this form is homogeneous of degree 0. An object is homogeneous of degree d if it is multiplied by λ^d when each X_i is multiplied by λ. Define

$$\Omega = \sum_{i=0}^{n} X_i dX_0 \wedge \cdots \wedge \widehat{dX_i} \wedge \cdots \wedge dX_n.$$

This expression, in which no variable is distinguished, is homogeneous of degree $n+1$. We can write $\Phi = \Omega/X_0^{n+1}$. It is a meromorphic form which is homogenous of degree 0 and which has a pole of order $n+1$ along the hyperplane $X_0 = 0$. More generally, we can construct meromorphic forms with poles of arbitrary poles on a hypersurface via the expression

$$\Omega_A = \frac{A\Omega}{Q^r},$$

so long as the homogeneous polynomial A is such that Ω_A is homogenous of degree 0. Consider, for example, the case of a hypersurface of degree d in $\mathbb{P}^3$. Let $Q = 0$ be the defining equation. The degree of Ω is 4, so meromorphic forms with a pole of order 1 are given by

$$\Omega_A = \frac{A\Omega}{Q},$$

where $\deg A = d - 4$. It is easy to count the dimension of the space of meromorphic forms of this kind using the formula for the dimension of the space of homogeneous polynomials of degree d in n variables, namely, the binomial coefficient $\binom{d+n-1}{n-1}$. As a mnemonic device, remember that

(a) for $d = 1$, the binomial coefficient is n, the number of homogeneous coordinates;

(b) in general the binomial coefficient is a polynomial (with integer coefficients) of degree $n-1$ in the variable d.

(Exercise: prove all these statements.)

We come now to the Poincaré residue. The forms Ω_A for fixed Q define cohomology classes of degree n on $\mathbb{P}^n - X$, where X is the locus $Q = 0$. Grothendieck's algebraic de Rham theorem applied to this case states that the cohomology of $\mathbb{P}^n - X$ is generated by the classes of the Ω_A. We assume here that X is smooth.

There is a purely topological construction

$$\text{res} : H^n(\mathbb{P}^n - X) \longrightarrow H^{n-1}(X)$$

which is defined as follows. Given a cycle γ of dimension $n-1$ on X, let $T_\epsilon(\gamma)$ denote the tube of radius ϵ around it relative to some Riemannian metric. If ϵ is sufficiently small, then the boundary of this tube lies in $\mathbb{P}^n - X$. Define

$$\text{res}(\alpha)(\gamma) = \frac{1}{2\pi i}\alpha(\partial T_\epsilon(\gamma)).$$

We call this the *topological residue* of α. For small ϵ, the region inside the tube $T_{\epsilon'}$ but outside $T_{\epsilon''}$ for $\epsilon'' < \epsilon' < \epsilon$, is bounded by smooth, nonintersecting tubes. The fact that α is closed gives (by Stokes' theorem) that

$$\alpha(\partial T_{\epsilon'}(\gamma)) = \alpha(\partial T_{\epsilon''}(\gamma)).$$

Thus the topological residue is independent of the tube chosen, so long as it is small enough. We can also write

$$\text{res}(\alpha)(\gamma) = \lim_{\epsilon \to 0} \frac{1}{2\pi i}\alpha(\partial T_\epsilon(\gamma)),$$

even though the sequence is in fact constant.

The residue map fits into the exact sequence

$$\cdots \longrightarrow H^n(\mathbb{P}^n) \longrightarrow H^n(\mathbb{P}^n - X) \xrightarrow{\text{res}} H^{n-1}(X) \xrightarrow{G} H^{n+1}(\mathbb{P}^n) \longrightarrow \cdots$$

where G is the *Gysin* map. The Gysin map is the transpose via Poincaré duality of the map $H_k(X) \longrightarrow H_k(\mathbb{P}^n)$, and the "tube sequence" is itself the transpose via Poincaré duality of the exact sequence for the homology of the pair $(\mathbb{P}^n, \mathbb{P}^n - X)$. The kernel of the Gysin map is the primitive cohomology. Thus there is a surjection

$$H^n(\mathbb{P}^n - X) \xrightarrow{\text{res}} H^{n-1}(X)_o \longrightarrow 0.$$

When α is represented by a meromorphic form Ω_A, we would like to represent its residue by a differential form on X. To this end we will define an *analytic residue* for forms with pole of order 1. We will then compare the analytic and topological residues, concluding that they are the same on the level of cohomology.

Let us consider first the local version of the analytic residue, where projective space is replaced by a coordinate neighborhood in $\mathbb{C}^n$, Ω_A is replaced by the expression $\alpha = a dz_1 \wedge \cdots \wedge dz_n/f$, and where the hypersurface is defined by the holomorphic equation $f(z_1, \ldots, z_n) = 0$. Note that

$$df = \sum_i \frac{\partial f}{\partial z_i} dz_i.$$

If $f(z) = 0$ defines a smooth hypersurface, then at least one of its partial derivatives is nonvanishing at each point of $f = 0$. Assume that the partial derivative $\partial f/\partial z_n$ is not zero at a point of $f = 0$. By shrinking the neighborhood if necessary, we may assume that this derivative is nonzero throughout the neighborhood. Multiply df by $dz_1 \wedge \cdots \wedge dz_{n-1}$. Then

$$dz_1 \wedge \cdots \wedge dz_{n-1} \wedge df = \frac{\partial f}{\partial z_n} dz_1 \wedge \cdots \wedge dz_n.$$

Thus

$$\frac{a dz_1 \wedge \cdots \wedge dz_n}{f} = \frac{a dz_1 \wedge \cdots \wedge dz_{n-1}}{\partial f/\partial z_n} \wedge \frac{df}{f}.$$

Define the analytic residue to be the coefficient of df/f, restricted to $f = 0$. Thus

$$\text{Res}\left(\frac{a dz_1 \wedge \cdots \wedge dz_n}{f}\right) = \frac{a dz_1 \wedge \cdots \wedge dz_{n-1}}{\partial f/\partial z_n}\bigg|_{f=0}.$$

Since $\partial f/\partial z_n$ is nonzero on $f = 0$, we see that the analytic residue of a meromorphic form with a pole of order 1 is a holomorphic form.

If one of the partial derivatives of f is nonvanishing at p, we may use the implicit function theorem to choose new holomorphic coordinates near p so that the submanifold $f = 0$ is given by $z_n = 0$. Then the analytic residue is given by

$$\text{Res}\left(\frac{a dz_1 \wedge \cdots \wedge dz_n}{z_n}\right) = a dz_1 \wedge \cdots \wedge dz_{n-1}\bigg|_{z_n=0}.$$

Let $\{U_\beta\}$ be a coordinate cover of a neighborhood of $T_\epsilon(X)$ which restricts to a coordinate cover of X. We may assume that X is locally given by $z_n^\beta = 0$. Let $\{\rho_\beta\}$ be a partition of unity subordinate to the given cover. Then

$$\mathrm{res}(\alpha)(\gamma) = \lim_{\epsilon \to 0} \frac{1}{2\pi i} \sum_\beta \int_{\partial T_\epsilon(\gamma)} \rho_\beta a_\beta dz_1^\beta \wedge \cdots \wedge dz_{n-1}^\beta \frac{dz_n^\beta}{z_n^\beta}.$$

Use Fubini's theorem to integrate first with respect to z_n^β and then with respect to the other variables. To evaluate the integral in z_n^β, define for any function $g(z_1^\beta, \ldots, z_n^\beta)$, the quantity

$$[g]_\epsilon(z_1^\beta, \ldots z_{n-1}^\beta) = \frac{1}{2\pi} \int_0^{2\pi} g(z_1^\beta, \ldots, z_{n-1}^\beta, \epsilon e^{2\pi i \theta_n^\beta}) d\theta_n^\beta.$$

The average value $[g]_\epsilon$ satisfies

$$\lim_{\epsilon \to 0} [g(z_1, \ldots, z_{n-1})]_\epsilon = g(z_1, \ldots, z_{n-1}, 0),$$

and so

$$\mathrm{res}(\alpha)(\gamma) = \sum_\beta \int_\gamma (\rho_\beta a_\beta)(z_1^\beta, \ldots, z_{n-1}^\beta, 0) dz_1^\beta \wedge \cdots \wedge dz_{n-1}^\beta.$$

The integral on the right is the integral of the analytic residue over γ. To summarize,

$$\mathrm{res}(\alpha)(\gamma) = \mathrm{Res}(\alpha)(\gamma).$$

Thus the two residue maps are the same on the level of cohomology classes.

Recall again what Grothendieck's theorem says: the cohomology of $H^n(\mathbb{P}^n - X)$ is spanned by the classes $A\Omega/Q^r$. Thus there is a filtration of the cohomology by order of pole: define

$$P^r H^n(\mathbb{P}^n - X) = \{\beta \mid \beta = [A\Omega/Q^k], \text{ where } k \le r\}.$$

The filtration $P^1 \subset P^2 \subset \cdots \subset P^j \subset \cdots = H^n(\mathbb{P}^n - X)$ is increasing and eventually exhausts the cohomology of the complement. In fact $P^{n+1} = H^n(\mathbb{P}^n - X)$. The filtration by order of pole maps to a filtration of the primitive cohomology. By the discussion above, we have

$$\mathrm{res}\, P^1 H^n(\mathbb{P}^n - X) = F^{n-1} H^{n-1}(X)_o.$$

More elaborate arguments show that the residue map carries the increasing filtration by order of pole to the decreasing filtration by Hodge level:

$$\mathrm{res}\, P^r H^n(\mathbb{P}^n - X) = F^{n-r} H^{n-1}(X)_o.$$

Thus one has the following fundamental result, due to Griffiths [10]:

THEOREM 4.6.1 *The residue map identifies the filtration by order of pole with the Hodge filtration.*

4.7   PROPERTIES OF THE PERIOD MAPPING

We will now establish some properties of the period mapping for hypersurfaces. All properties stated hold in general (see Eduardo Cattani's lectures in Chapter 7).

THEOREM 4.7.1   *The period mapping is holomorphic.*

To understand why the period mapping is holomorphic, we need to understand how to compute its derivative. To this end, let $\{F^p(t)\}$ be a decreasing filtration with parameter t. Let $v_i(t)$ be a basis of $F^p(t)$. To first order, $v_i(t) = v_i(0) + t\dot{v}_i(0)$. Define the homomorphism

$$\Phi_p : F^p(0) \longrightarrow H_{\mathbb{C}}/F^p(0)$$

by

$$\Phi_p(v_i(0)) = \dot{v}_i(0) \bmod F^p(0).$$

This homomorphism is the velocity vector of the curve $F^p(t)$, and it is zero if and only if $F^p(t)$ is constant to first order as a map into the Grassmannian of k-planes in $H_{\mathbb{C}}$, where $k = \dim F^p$.

The period domain D can be viewed as embedded as an open subset of a closed submanifold of a product of Grassmannians. It inherits its complex structure from this product. We show that $F^n H^n(X)$ varies holomorphically in that complex structure, where X is a hypersurface in $\mathbb{P}^{n+1}$, but make some remarks about the general case. To that end, let $Q(t) = Q + tR$ be a pencil of equations defining a pencil of hypersurfaces. Let A be a polynomial such that $A\Omega/Q(t)$ defines a meromorphic form on $\mathbb{P}^{n+1}$. Then res $A\Omega/Q(t)$ is a curve of vectors in $F^n(t)$. Let γ_i be a basis for $H_n(X_0, \mathbb{Z})$, where $Q(t) = 0$ defines X_t. For a given ϵ, there is a δ such that the tubes $\partial T_\epsilon(\gamma_i)$ form a basis for $H^{n+1}(\mathbb{P}^{n+1} - X_t)$ so long as $|t| < \delta$. Thus coordinates for res $A\Omega/Q(t)$ are given by the integrals (periods)

$$I_k(t) = \int_{\partial T_\epsilon(\gamma_i)} \frac{A\Omega}{Q(t)}.$$

Because the tubes do not vary as t varies, the domain of integration is fixed, as in the example of elliptic curves. Therefore derivatives of $I_k(t)$ can be computed by differentiation under the integral sign. Since $Q(t)$ depends holomorphically on parameters, the derivative of $I_k(t)$ with respect to $\bar{t}$ is zero, and so $F^n(t)$ is holomorphic. The same argument applies to $F^p(t)$ once one establishes compatibility of the filtrations by order of pole and by Hodge level.

The next assertion is Griffiths transversality.

THEOREM 4.7.2   *Let $\{F^p(t)\}$ be a family of Hodge filtrations coming from a family of projective algebraic manifolds. Then Φ_p takes values in F^{p-1}/F^p.*

There are two parts to the proof in the case of hypersurfaces. The first is an observation from calculus. Let

$$I_k(t) = \int_{\partial T_\epsilon(\gamma_i)} \frac{A\Omega}{Q(t)^r},$$

where here we allow poles of arbitrary order. Then

$$\frac{\partial I_k(t)}{\partial t}\bigg|_{t=0} = -(r+1)\int_{\partial T_\epsilon(\gamma_i)} \frac{\dot{Q}(0)A\Omega}{Q(0)^{r+1}}.$$

Thus differentiating a meromorphic differential moves it by just one step in the pole filtration. There is, however, a subtlety. It could happen that a meromorphic differential with a pole of order r is cohomologous to one with a pole of smaller order. In this case, the cohomology class does not have the expected level in the filtration by order of pole, i.e., it does not have the same level as the given form which represents it.

The second part of the proof is the compatibility between the filtration by order of pole and the Hodge filtration. We have established it only for hypersurfaces and only for $F^n H^n(X)$. However, given that compatibility, we see that the derivative map Φ_p takes values in F^{p-1}/F^p.

Because of the distinction between the order of pole of a meromorphic differential and the minimum order of pole of a meromorphic form in the cohomology class of a differential, it can be a somewhat delicate matter to determine whether $\Phi_p \neq 0$, that is, whether the derivative of the period mapping is nonzero. We address this issue in the next section.

Remark. Griffiths' transversality has many remarkable consequences. One is that for weight greater than 1, modulo rare exemptions such as the case $h^{2,0} = 1$, a generic Hodge structure is not the cohomology of an algebraic variety (or even a motive). This is because the image of any period map is a subvariety of the period space of smaller dimension. Therefore the set of "geometric Hodge structures" is at best a countable union of subvarieties of lower dimension in D/Γ.

4.8   THE JACOBIAN IDEAL AND THE LOCAL TORELLI THEOREM

Let us now investigate the question of whether the cohomology class of a meromorphic form can be represented by one of lower pole order. An answer to this question will lead to a proof of the following result [10], which is the local Torelli theorem of Griffiths.

THEOREM 4.8.1 *The period map for hypersurfaces of degree d in $\mathbb{P}^n$ is locally injective for $d > 2$ and $n > 1$, except for the case of cubic surfaces.*

To efficiently study differential forms of all degree on $\mathbb{P}^{n+1} - X$, we introduce a calculus which will prove convenient. Let $dV = dX_0 \wedge \cdots \wedge dX_{n+1}$ be the "holomorphic volume form" on $\mathbb{C}^{n+2}$. Let

$$E = \sum_k X_k\, i\left(\frac{\partial}{\partial X_k}\right)$$

be the Euler operator, where $i(\xi)$ is interior multiplication by a vector field ξ. It is an operator which is homogeneous of degree 0. Then

$$\Omega = i(E)dV.$$

Apply the operator $i(E)$ to the identity

$$dQ \wedge dV = 0$$

to obtain

$$(\deg Q)QdV = dQ \wedge \Omega.$$

If we use "$\equiv$" to mean "up to addition of an expression which is a multiple of Q," then this relation reads

$$dQ \wedge \Omega \equiv 0.$$

Apply $K_\ell = i(\partial/\partial X_\ell)$ to the preceding to obtain

$$Q_\ell \Omega \equiv dQ \wedge \Omega_\ell,$$

where

$$\Omega_\ell = K_\ell \Omega$$

and

$$Q_\ell = \frac{\partial Q}{\partial X_\ell}.$$

Now consider a general meromorphic n-form on $\mathbb{P}^{n+1} - X$. It can be written as a sum of terms which are homogeneous of degree 0:

$$\sum_\ell \frac{A_\ell K_\ell \Omega}{Q^r}.$$

The exterior derivative of such an expression is simple if we ignore all terms of lower pole order:

$$d \sum_\ell \frac{A_\ell K_\ell \Omega}{Q^r} \equiv -(r+1) \sum_\ell \frac{A_\ell dQ \wedge K_\ell \Omega}{Q^{r+1}}.$$

One also has the identity

$$K_\ell dQ + dQ K_\ell = Q_\ell.$$

Therefore

$$d \sum_\ell \frac{A_\ell K_\ell \Omega}{Q^r} \equiv (r+1) \sum_\ell \frac{A_\ell K_\ell dQ \wedge \Omega}{Q^{r+1}} - (r+1) \sum_\ell \frac{A_\ell Q_\ell \Omega}{Q^{r+1}}.$$

But $dQ \wedge \Omega \equiv 0$, so that

$$d \sum_\ell \frac{A_\ell K_\ell \Omega}{Q^r} \equiv -(r+1) \sum_\ell \frac{A_\ell Q_\ell \Omega}{Q^{r+1}}.$$

Define the *Jacobian ideal* to be the ideal generated by the partial derivatives of Q. Then we have the following result, which we have proved in only one direction:

PROPOSITION 4.8.2 *Let Ω_A be a meromorphic pole with a form which has a pole of order r. It is cohomologous to a form with a pole of order one less if (and also only if) A is in the Jacobian ideal of Q.*

As a first consequence of this result, we compute the dimension of $H^{n,0}(X)$. This space is spanned by cohomology classes of residues of meromorphic forms $A\Omega/Q$. The degree of the numerator is $d - (n + 2)$. The Jacobian ideal is generated in degree $d - 1 > d - (n + 2)$. Thus the cohomology class of the residue of Ω_A is nonzero if A is nonzero, just as in the case of abelian differentials. We conclude that

$$\dim H^{n,0}(X) = \binom{d-1}{n+1}.$$

This expression is a polynomial in d of degree $n + 1$ with integer coefficients.

We can now prove the local Torelli theorem of Griffiths. Consider first a hypersurface X of degree d in $\mathbb{P}^{n+1}$. The space $F^n H^n(X)$ is nonzero under the hypotheses of the theorem and is spanned by residues of meromorphic forms $A\Omega/Q$, where $\deg A = d - (n + 2)$. The derivative of such a form at $t = 0$ for the pencil $Q + tR$ has the form

$$-\frac{RA\Omega}{Q^2}.$$

The numerator has degree $2d - (n + 1)$. To proceed, we call upon an important fact from commutative algebra that holds whenever X is smooth:

PROPOSITION 4.8.3   *Let $Q(X_0, \ldots, X_{n+1})$ be a homogeneous polynomial such that the hypersurface $Q = 0$ is smooth. Let $S = \mathbb{C}[X_0, \ldots, X_{n+1}]/J$ be the quotient ring. It is a finite-dimensional graded $\mathbb{C}$-algebra. Its component of top degree has degree $t = (d - 1)(n + 2)$ and dimension 1. Let $\phi : S^t \longrightarrow \mathbb{C}$ be any nonzero map. Consider the composition*

$$S^a \times S^{t-a} \longrightarrow S^t \overset{\phi}{\longrightarrow} \mathbb{C},$$

where the first map sends classes represented by polynomials A and B to the class of AB. The composition is a perfect pairing.

To prove the local Torelli theorem, fix a hypersurface $Q = 0$, and assume that the derivative of the period map is zero in the direction R. That is, assume that the pencil $Q + tR$ is such that the derivative of every class Ω_A is zero. Then RA lies in the Jacobian ideal for all A of degree $d - (n + 2)$. Consequently RB is in the Jacobian ideal for all B of degree $t - d$. Under the stated hypotheses, the numbers $d - (n + 2)$ and $t - d$ are nonnegative. Because the pairing is perfect, it follows that R is in the Jacobian ideal. But an element of degree d in the Jacobian ideal has the form

$$R = \sum_{ij} A_{ij} X_i \frac{\partial Q}{\partial X_j}.$$

A vector field on $\mathbb{P}^n$ has the general form

$$\xi = \sum_{ij} A_{ij} X_i \frac{\partial}{\partial X_j},$$

corresponding to the one-parameter group

$$I + tA \in \mathrm{GL}(n + 2, \mathbb{C}).$$

Thus $R = \xi Q$ is tangent to the action of $\mathrm{PGL}(n+1)$, so that to first order, the pencil $Q + tR = 0$ is constant. This completes the proof.

4.9   THE HORIZONTAL DISTRIBUTION—DISTANCE-DECREASING PROPERTIES

Let us examine Griffiths transversality in more detail. To this end it is useful to consider the Hodge structure on the Lie algebra of the group G which acts transitively on the period domain. Given any set of Hodge structures, one can build new Hodge structures using the functorial operations of linear algebra: direct sum, tensor product, dual, and hom. (Exercise: reimagine the definitions.) In particular, if H is a Hodge structure, then $\mathrm{End}(H_{\mathbb{R}})$ carries a natural Hodge structure: if $\phi : H_{\mathbb{C}} \longrightarrow H_{\mathbb{C}}$ satisfies $\phi(H^{r,s}) \subset H^{r+p,s+q}$, then ϕ has type (p, q). Note that this Hodge structure is of weight 0.

Let g be the Lie algebra of G. Then $g_{\mathbb{C}}$ carries a Hodge decomposition inherited from the one on $\mathrm{End}(H_{\mathbb{C}})$. The Lie algebra of V is $g^{0,0}$. The holomorphic tangent space at H can be identified with the subalgebra

$$g = \oplus_{p<0} \, g^{p,-p},$$

and one has

$$g_{\mathbb{C}} = g^- \oplus g^{0,0} \oplus g^+,$$

where

$$g^+ = \oplus_{p>0} \, g^{p,-p} \text{ and } g^- = \oplus_{p<0} \, g^{p,-p},$$

and g^- is the complex conjugate of g^+. The space g^- is the holomorphic tangent space of D at H. This space is a V-module, and it defines a homogeneous bundle isomorphic to the holomorphic tangent bundle. The subspace $g^{-1,1}$ is also a V-module. The homogeneous subbundle of the tangent bundle which it defines is the bundle of holomorphic tangent vectors which satisfy Griffiths transversality. We denote this bundle by $T_{\mathrm{hor}}(D)$.

EXERCISE 4.9.1   Show that g^- is a Lie subalgebra. Describe it and the associated Lie group as explicitly as possible.

Another decomposition of the Lie algebra is given by

$$g_{\mathrm{even}} = \oplus_{p \text{ is even}} \, g^{-p,p}$$

and

$$g_{\mathrm{odd}} = \oplus_{p \text{ is odd}} \, g^{-p,p},$$

so that

$$g_{\mathbb{C}} = g_{\mathrm{even}} \oplus g_{\mathrm{odd}}.$$

This decomposition is in fact the decomposition

$$g_{\mathbb{C}} = \mathfrak{k}_{\mathbb{C}} \oplus \mathfrak{p}_{\mathbb{C}},$$

where $\mathfrak{k}_{\mathbb{C}}$ is the Lie algebra of the complexification of the maximal compact subgroup K of G containing V, and where $\mathfrak{p}_{\mathbb{C}}$ is the orthogonal complement of $\mathfrak{k}_{\mathbb{C}}$ with respect to the Killing form.

It is only when $K = V$, that is, when $\mathfrak{g}_{\text{even}} = \mathfrak{g}^{0,0}$, that D is a Hermitian symmetric domain. Such is the case in weight 2 if and only if $h^{2,0} = 1$ because of the "accidental" isomorphism $SO(2, \mathbb{R}) \cong U(1)$.

The bundle $T_{\text{hor}}(D)$ has special properties relative to the holomorphic sectional curvatures of D. While we do not have the time to develop the theory needed to explain this in full, here is a sketch of the needed argument. Let M be a surface in $\mathbb{R}^3$, x a point of M, and N a plane through x that contains the normal vector. The vector N cuts out a curve on M passing through x. That curve has an osculating circle—a circle which best approximates it. Let $K(N)$ be the reciprocal of the radius of that circle, taken with the correct sign: positive if the curve bends away from the normal vector, negative if it bends toward it. Call this the curvature of the curve $K(N)$. The extreme values of $K(N)$ occur for orthogonal planes. Their product is called the *Gaussian curvature* of M at x. Gauss showed that the Gaussian curvature is defined intrinsically, that is, by the metric on M. Now let M be any Riemannian manifold and P a plane in the tangent space of M at x. Let S_p be the surface consisting of geodesics emanating from x tangent to P. Let $K(P)$ be the Gauss curvature in the induced metric. There is associated to every plane in the tangent space a number, the *sectional curvature*. Finally, suppose that M is a complex manifold. Then there is an endomorphism J of the tangent bundle whose square is minus the identity. This is the complex structure tensor. It gives a coherent notion of multiplication by $\sqrt{-1}$. A plane in the tangent space is a complex tangent plane if it is invariant under the action of J. The sectional curvatures of these planes are called *holomorphic sectional curvatures*. The invariant metric on the Riemann sphere has constant holomorphic sectional curvature $+1$. For the invariant metric on the torus, the curvature is zero. On the unit disk or the upper half plane, in the invariant metric, the curvature is -1. The unit ball (complex hyperbolic space) has an invariant metric with holomorphic sectional curvature -1. However, the Riemannian sectional curvatures are variable.

For period domains, the holomorphic sectional curvatures associated to vectors in $\mathfrak{g}^{-1,1}$ are negative, while those for vectors tangent to fibers of the map $G/V \longrightarrow K/V$ are positive.

We come now to a fundamental principle of complex geometry: *if $f : M \longrightarrow N$ is a holomorphic map between complex manifolds with negative sectional curvatures, then $f^* ds_N^2 \leq ds_M^2$. That is, horizontal maps decrease distances.* The statement is NOT true for maps not tangent to the horizontal distribution.

Let us take the distance-decreasing property of period maps as a given and draw some consequences from it. The first of these, an analogue of Liouville's theorem in one complex variable, illustrates the fact that period domains act with respect to horizontal holomorphic maps as holomorphic maps do with respect to bounded domains.

PROPOSITION 4.9.2 *Let $f : \mathbb{C} \longrightarrow D$ be a holomorphic horizontal map. Then f is constant.*

For the proof, consider the Poincaré metric on the disk of radius R,

$$ds_R^2 = \frac{R^2 dz d\bar{z}}{(R^2 - |z|^2)^2}.$$

This is the metric of curvature -1. If ds_D^2 is the G-invariant metric on D, we have $f^* ds_D^2 \le ds_R^2$. Notice that

$$ds_R^2(0) = \frac{dz d\bar{z}}{R^2},$$

and that

$$f^* ds_D^2 = C dz d\bar{z}$$

for some $C > 0$. Then

$$C \le \frac{1}{R^2}$$

for all $R > 0$. For R large enough, this is a contradiction.

COROLLARY 4.9.3   *Let $f : \mathbb{C}^* \longrightarrow \Gamma \backslash D$ be a period map. Then f is constant.*

PROOF.   For the proof, note that f has a lift $f : \mathbb{C} \longrightarrow D$. By the previous proposition, the lift must be constant.   □

COROLLARY 4.9.4   *Let $X/\mathbb{P}^1$ be a family of algebraic hypersurfaces of positive dimension and degree at least 3 (or 4 in the case of two-dimensional algebraic surfaces). Then $X/\mathbb{P}^1$ has at least three singular fibers.*

THEOREM 4.9.5 (Monodromy theorem)   *Let γ be monodromy transformation for a period map $f : \Delta^* \longrightarrow \Gamma \backslash D$. Then γ is quasi-unipotent. That is, there are integers m and N such that $(\gamma^m - 1)^N = 0$.*

The theorem says that the eigenvalues of γ are roots of unity and that a suitable power of γ is nilpotent. For the proof, consider the lift $\tilde{f} : \mathcal{H} \longrightarrow D$, and use the Poincaré metric

$$ds_H^2 = \frac{dx^2 + dy^2}{y^2}.$$

The distance between $\sqrt{-1}n$ and $\sqrt{-1}n + 1$ is $1/n$, so that

$$d_D((\tilde{f}(\sqrt{-1}n), \tilde{f}(\sqrt{-1}n + 1)) \le 1/n.$$

Write $\tilde{f}(\sqrt{-1}n) = g_n V$, for some $g_n \in G$. Then the above relation reads

$$d_D(g_n V, \gamma g_n V) \le 1/n,$$

or

$$d_D(V, g_n^{-1} \gamma g_n V) \le 1/n.$$

Consequently the sequence $\{g_n^{-1} \gamma g_n\}$ converges to the compact subgroup V. If the conjugacy class of γ has a limit point in a compact group, then its eigenvalues are of absolute value 1. Because $\gamma \in G_{\mathbb{Z}}$, the eigenvalues are algebraic integers. Their conjugates are also eigenvalues. A theorem of Kronecker says that an algebraic integer, all of whose conjugates are of absolute value one, is a root of unity.   □

4.10   THE HORIZONTAL DISTRIBUTION—INTEGRAL MANIFOLDS

A subbundle of the holomorphic tangent bundle defines a *distribution*, that is, a field of subspaces at every point. An *integral manifold* of a distribution is a manifold everywhere tangent to the distribution. A distribution is said to be *involutive*, or *integrable*, if whenever X and Y are vector fields tangent to the distribution, so is the Lie bracket $[X, Y]$. A theorem of Frobenius states that every point of an involutive distribution has a neighborhood U and coordinates $z_1, \ldots, z_n$ such that the distribution, assumed to have dimension k at each point, is spanned by $\partial / \partial z_i$, $i = 1, \ldots, k$. The integral manifolds are then locally given by equations $z_i = c_i$ where $i = k + 1, \ldots, n$. Thus the integral manifolds foliate the given manifold.

A distribution can be defined as the set of tangent vectors which are annihilated by a set of 1-forms $\{\theta_i\}$. Consider the relation which defines the exterior derivative:

$$d\theta(X, Y) = X\theta(Y) - Y\theta(X) - \theta([X, Y]).$$

From it we see that a distribution is involutive if and only if the $d\theta_i$ annihilate vectors tangent to it. This statement is equivalent to the statements (a) the $d\theta_i$ are in the algebraic ideal generated by the forms θ_i, (b) if $\theta_1, \ldots, \theta_k$ are 1-forms whose common null space is the given distribution, then $d\theta \wedge \theta_1 \wedge \cdots \wedge \theta_k = 0$.

Consider now the *contact* distribution, defined by the null space of the 1-form

$$\theta = dz - \sum_{i=1}^{n} y_i dx_i. \tag{4.10.1}$$

This is the so-called *contact form*. Note that

$$d\theta = \sum_{i=1}^{n} dx_i \wedge dy_i$$

and that

$$\theta \wedge (d\theta)^n = \pm dx_1 \wedge \cdots \wedge dx_n \wedge dy_1 \wedge \cdots \wedge dy_n \wedge dz \neq 0.$$

Thus $\theta \wedge d\theta \neq 0$, and so the distribution defined by the contact form is not involutive. It follows that integral manifolds of the contact distribution are not of codimension 1. Indeed, they are of dimension n, a fact that is already clear for $n = 1$. See [1] or [13].

It is easy to exhibit an n-dimensional integral manifold of the contact distribution. For any function $f(x_1, \ldots, x_n)$, consider the manifold M parametrized by

$$(x_1, \ldots, x_n) \mapsto \left(x_1, \ldots, x_n, \frac{\partial f}{\partial x_i}, \ldots, \frac{\partial f}{\partial x_n}, f\right).$$

It satisfies the contact equation.

A space of tangent vectors annihilated by the θ_i and such that pairs of tangent vectors are annihilated by the $d\theta_i$ is called an *integral element*. The tangent space to an

integral manifold is necessarily an integral element. In the case of the contact distribution, this necessary condition is also sufficient, as the above construction implies.

The reason for considering the contact distribution here is that it provides a simple
model for the Griffiths distribution, which is noninvolutive whenever it is nontrivial.
Some natural questions are,

(a) what is the maximal dimension of an integral manifold?

(b) can one characterize them?

(c) what can one say about integral manifolds that are maximal with respect to inclusion?

(d) what can one say about "generic" integral manifolds?

We study some of these questions in the case of weight 2. To this end, choose a
basis (Hodge frame) compatible with the Hodge decomposition $H^{2,0} \oplus H^{1,1} \oplus H^{0,2}$
and such that the intersection matrix for the bilinear form Q is

$$Q = \begin{pmatrix} 0 & 0 & I \\ 0 & -I & 0 \\ I & 0 & 0 \end{pmatrix}.$$

Matrices of the form

$$g = \begin{pmatrix} I & 0 & 0 \\ X & I & 0 \\ Z & {}^tX & I \end{pmatrix},$$

where $Z + {}^tZ = {}^tXX$, satisfy ${}^tgQg = Q$. The matrices g act on Hodge frames, and
the set of such g constitutes the unipotent group $G^- = \exp \mathfrak{g}^-$.

The *Maurer–Cartan* form for G^- is the form $\omega = g^{-1}dg$ where g is in block
lower-triangular form as above. It has the form

$$\omega = \begin{pmatrix} 0 & 0 & 0 \\ dX & 0 & 0 \\ W & {}^tdX & 0 \end{pmatrix},$$

where $W = dZ - {}^tXdX$ is skew symmetric. A holomorphic tangent vector ξ is
horizontal if and only if it is annihilated by W. The equation $W_{ij} = 0$ reads

$$dZ_{ij} = \sum_k {}^tX_{ik}dX_{kj} = \sum_k X_{ki}dX_{kj}.$$

Thus the Griffiths distribution in weight 2 is given by a system of coupled contact
equations.

In the case $h^{2,0} = 2$, the Maurer–Cartan form depends on a single form $W_{12} =
-W_{21}$. If we set $h^{2,0} = 2$, $h_o^{1,1} = q$, $X_{k1} = x_k$, $X_{k2} = y_k$, and $Z_{21} = z$, then the
equation $W_{21} = 0$ reads

$$dz = \sum_i y_i dx_i.$$

In this case, the Griffiths distribution *is* the contact distribution! The maximum dimension of an integral manifold is q, and integral manifolds are given locally by $x \mapsto (x, \nabla z(x), z(x))$, as noted above.

For $h^{2,0} = p > 2$, the behavior of the Griffiths distribution is more complicated. To understand it better, consider the Maurer–Cartan form ω for G^- It is a $\mathfrak{g}^-$-valued 1-form which satisfies the integrability condition $d\omega - \omega \wedge \omega = 0$. In the weight-2 case,

$$
d\omega = d \begin{pmatrix} 0 & 0 & 0 \\ dX & 0 & 0 \\ W & {}^t dX & 0 \end{pmatrix} = \begin{pmatrix} 0 & 0 & 0 \\ 0 & 0 & 0 \\ dW & 0 & 0 \end{pmatrix},
$$

where $dW = 0$ on any integral element E. Thus $d\omega = 0$ on E. From $d\omega = \omega \wedge \omega$, it follows that

$$
\omega \wedge \omega(X, Y) = [X, Y] = 0
$$

for any vectors X and Y in E. Consequently an integral element $E \subset \mathfrak{g}^{-1,1}$ satisfies $[E, E] = 0$. That is, *integral elements are abelian subalgebras of* $\mathfrak{g}^-$ *contained in* $\mathfrak{g}^{-1,1}$. This result holds in arbitrary weight, and so the integral elements are always defined by quadratic equations.

Is every integral element tangent to an integral manifold? In general, the answer is no. However, for the Griffiths distribution, the answer is yes: *given an abelian subalgebra* $\mathfrak{a} \subset \mathfrak{g}^{-1,1}$, *the manifold* $V = \exp \mathfrak{a}$ *is an integral manifold with tangent space* $\mathfrak{a}$.

The next question is, on how many free parameters does an integral manifold depend? Two extreme answers to this question are the cases of *rigidity* and *flexibility*. In the first case, an integral manifold through x with given tangent space is completely determined. Examples are given in [6]. These are precisely the examples in weight 2, $h_o^{1,1}$ even of maximal dimension $h^{2,0} h^{1,1}/2$, with $h^{2,0} > 2$. In the second case, the integral manifolds with given tangent space depend on infinitely many parameters. The contact system is an example: the parameter is an arbitrary smooth function f.

The integral manifolds of maximal dimension of [6] are defined as follows. Fix a Hodge structure H with $\dim H^{2,0} = p$ and $\dim H^{1,1} = 2q$. Fix a complex structure J on $H_{\mathbb{R}}^{1,1}$. Let $H^{1,1} = H_+^{1,1} \oplus H_-^{1,1}$ be the eigenspace decomposition for J, where $\pm$ refers to the eigenvalue $\pm\sqrt{-1}$. Note that $H_+^{1,1}$ is totally isotropic: $Q(v, w) = 0$ for all $v, w \in H_+^{1,1}$. Let

$$
\mathfrak{a} = \operatorname{Hom}(H^{2,0}, H_+^{1,1}).
$$

Note that $\mathfrak{a}$ is a subspace of $\operatorname{Hom}(H^{2,0}, H^{1,1}) \cong \mathfrak{g}^{-1,1}$ of dimension pq. It is also an abelian subalgebra. To see this, write

$$
g(X, Z) = \begin{pmatrix} I & 0 & 0 \\ X & I & 0 \\ Z & {}^t X & 0 \end{pmatrix}.
$$

Then $[g(X, Z), g(X', Z')] = g(0, X^t X' - X'^t X)$. The matrix $X^t X'$ is a matrix of dot products of column vectors of X. The dot products vanish because $H_+^{1,1}$ is isotropic.   □

The integral element $\mathfrak{a} = \mathrm{Hom}(H^{2,0}, H^{1,1}_+)$ is tangent to the variation of Hodge structure $\exp \mathfrak{a}$. Let $V' \subset G = \mathrm{SO}(2p, 2q)$ be the group which preserves the reference Hodge structure and which commutes with J. Then $V' \cong U(p) \times U(q)$. Let $G' \cong U(p, q)$ be the subgroup of G which preserves J. The orbit of $\mathfrak{a}$ is an open subset of the orbit M of the reference Hodge structure under the action of G'. Thus M *is a Hermitian symmetric space embedded in* D as a closed, horizontal, complex manifold isomorphic to G'/V'. It is an example of a *Mumford–Tate domain*: the orbit of a reference Hodge structure under the action of a Mumford–Tate group; see [9].

As shown in [6], one can choose J artfully so as to ensure that there is an arithmetic group Γ operating on D with $G'/\Gamma \cap G'$ of finite volume and even compact. The space $H/\Gamma \cap H$ is (quasi)-projective.

NOTE 4.10.1   If $\dim H^{1,1}$ is odd, there is a variation of Hodge structure U of maximal dimension $pq+1$ which fibers over the unit disk and whose fibers are the W 's described above. However, U, which is a kind of tube domain, does not admit a discrete group action with finite covolume. *Whether there is a quasi-projective example of dimension $pq + 1$ is unknown.*

An integral element $\mathfrak{a}$ (aka infinitesimal variation of Hodge structure, aka abelian subspace) is *maximal* if whenever $\mathfrak{a}'$ is another integral element containing $\mathfrak{a}$, then $\mathfrak{a}' = \mathfrak{a}$. Integral elements of maximal dimension are maximal, but *the converse is not true.* For a geometric example, consider the integral elements that come from variations of Hodge structure of sufficiently high degree. The proof (see [3]) is based on Donagi's symmetrizer construction.

We also give a "linear algebra" example (see [7]):

THEOREM 4.10.2   *Let $\mathfrak{a}$ be an integral element for a weight-2 Hodge structure H. Suppose that $\mathfrak{a}$ is generic in the sense that there is a vector $v \in H^{2,0}$ such that $\mathfrak{a}(v) = H^{1,1}$. Then $\mathfrak{a}$ is isomorphic to $H^{1,1}$ and is maximal.*

PROOF.   Since the map $\mathfrak{a} \longrightarrow H^{1,1}$ defined by $X \mapsto X(e_1)$ is surjective, $\dim \mathfrak{a} \geq \dim H^{1,1} = q$. We may choose a basis $e_1, \ldots, e_p$ for $H^{2,0}$ such that

(1) $\mathfrak{a}(e_1) = H^{1,1}$;

(2) there is a basis $M^1, \ldots, M^q$ for $\mathfrak{a}$ and a basis $e^1, \ldots, e^q$ of $H^{1,1}$ such that $M^i(e_1) = e^j$.

Elements of $\mathfrak{a}$ are given by $q \times p$ rectangular matrices A, B, etc. The abelian subspace condition then reads $(A, B) = 0$, where $(A, B)_{ij} = A_i \cdot B_j - B_i \cdot A_j$, and where A_i denotes the ith column of A.

Let N be an element of $\mathfrak{a}$. We may subtract a linear combination of the M^i so that $N(e_1) = 0$, i.e., the first column N_1 vanishes.

On the one hand, $(M^k, N) = 0$. On the other hand,

$$(M^k, N)_{1j} = M^k_1 \cdot N_j - N_1 \cdot M^k_j = e_k \cdot N_j = N_{kj}.$$

Thus $N_{kj} = 0$ for all k, j. Thus N is in the span of the M^i.   □

THEOREM 4.10.3 *Let $\mathfrak{a}$ be a generic integral element for a weight-2 Hodge structure. Set $q = \dim H^{2,0}$. Then all integral manifolds tangent to $\mathfrak{a}$ are given by functions $f_2, \ldots, f_q$ satisfying $[H_{f_i}, H_{f_j}] = 0$ where H_f is the Hessian of H.*

The proof is elementary and uses the matrix-valued contact system described earlier:

$$dZ_{ij} = \sum_k {}^t X_{ik} dX_{kj} = \sum_k X_{ki} dX_{kj}. \tag{4.10.2}$$

Consider

$$dZ_{i1} = \sum_k X_{ki} dX_{k1}.$$

Then, as usual, Z_{i1} is a function $f_i(X_{11}, X_{21}, \ldots, X_{q1})$. The X_{ki} are determined by the contact equation

$$X_{ki} = \frac{\partial f_i}{\partial X_{ki}}.$$

Once the X_{ki} are known (see equation (4.10.2)), the Z_{ij} are determined as well. What remains is the question of whether there are systems of functions with commuting Hessians, so that the f_i solve the differential equations. Here is a partial answer:

EXAMPLE 4.10.4 Choose the f_i so that their Hessians are diagonal. Thus enforce

$$\frac{\partial^2 f_i}{\partial x_j \partial x_k} = 0, \qquad i \neq j.$$

In two variables this is the equation $\partial^2 g / \partial x \partial y = 0$, a form of the wave equation. A class of solutions is given by $g(x, y) = a(x) + b(y)$. More generally,

$$f_i(x_1, \ldots, x_s) = \sum_j h_{ij}(x_j)$$

gives a very restricted but nonetheless infinite-dimensional family of solutions.

Bibliography

[1] V.I. Arnold, *Mathematical Methods of Classical Mechanics*, Graduate Texts in Mathematics, Vol. 60, Springer-Verlag, New York, 2010.

[2] A. Beauville, Le groupe de monodromie d'hypersurfaces et d'intersections complètes, *Complex Analysis and Algebraic Geometry* Lecture Notes in Mathematics **1194** (1986), 1–18.

[3] J. Carlson and R. Donagi, Hypersurface variations are maximal, I, *Inventiones Math.* **89** (1987), 371–374.

[4] J. Carlson, A. Kasparian, and D. Toledo, Variations of Hodge structure of maximal dimension, *Duke Math. J.* **58** (1989), 669–694.

[5] J. Carlson, C. Peters, and S. Müller-Stach. *Period Mappings and Period Domains*, Cambridge Studies in Advanced Mathematics **85**, Cambridge University Press.

[6] J. Carlson and C. Simpson, Shimura varieties of weight two Hodge structures, *Hodge Theory*, Springer Lecture Notes in Mathematics **1246** (1987), 1–15.

[7] J. Carlson and D. Toledo, Generic integral manifolds for weight two period domains. *Trans. Amer. Math. Soc.* **356**(6) (2004), 2241–2249.

[8] J. Carlson and D. Toledo, Discriminant complements and kernels of monodromy representations, *Duke Math. J.* **97** (1999), 621–648. (alg-geom/9708002)

[9] M. Green, P. Griffiths, and M. Kerr, *Mumford–Tate Groups and Domains: Their Geometry and Arithmetic* (2012), Princeton University Press.

[10] P. Griffiths, Periods of rational integrals I, II, *Ann. of Math.* **90** (1969), 460–495; 498–541.

[11] P. Griffiths, Periods of rational integrals III, *Pub. Math. IHES* **38** (1970), 125–180.

[12] P. Griffiths and J. Harris, *Principles of Algebraic Geometry* (1985), Wiley, New York.

[13] R. Mayer, Coupled contact systems and rigidity of maximal variations of Hodge structure, *Trans. AMS* **352**(5) (2000), 2121–2144.

[14] M. Sebastiani and R. Thom, Un résultat sur la monodromie, *Invent. Math.* **13** (1971), 90–96.

[15] C. Voisin, *Hodge Theory and Complex Algebraic Geometry I, II*, Cambridge Studies in Advanced Mathematics, Cambridge University Press.

Chapter Five

The Hodge Theory of Maps

by Mark Andrea de Cataldo and Luca Migliorini

Lectures 1–3 by Luca Migliorini

These three lectures summarize classical results of Hodge theory concerning algebraic maps. Lectures 4 and 5, to be delivered by M. A. de Cataldo, will discuss more recent results. I will not try to trace the history of the subject nor attribute the results discussed. Coherently with this policy, the bibliography only contains textbooks and a survey, and no original papers. Furthermore, quite often the results will not be presented in their maximal generality; in particular, I'll always stick to projective maps, even though some results discussed hold more generally.

INTRODUCTION

Hodge theory gives nontrivial restrictions on the topology of a nonsingular projective variety, or, more generally, of a compact Kähler manifold: the odd Betti numbers are even, the hard Lefschetz theorem, the formality theorem, stating that the real homotopy type of such a variety is, if simply connected, determined by the cohomology ring. Similarly, Hodge theory gives nontrivial topological constraints on algebraic maps. This is, broadly speaking, what these lectures are about.

A new aspect which emerges when dealing with maps is that one is forced to deal with singularities: even assuming that the domain and target of an algebraic map are nonsingular, asking that the map is smooth is much too restrictive: there are singular fibers, and this brings into the picture the technical tools to deal with them: stratification theory and topological invariants of singular spaces, such as intersection cohomology. This latter, which turns out to be a good replacement for cohomology when dealing with singular varieties, is better understood as the hypercohomology of a complex of sheaves, and this naturally leads to considering objects in the "constructible" derived category.

The question which we plan to address can also be formulated as follows: *How is the existence of an algebraic map $f : X \to Y$ of complex algebraic varieties reflected in the topological invariants of X?* From this point of view, one is looking for a relative version of Hodge theory, classical theory corresponding to the case in which Y is a point. Hodge theory encodes the algebraic structure of X in linear algebra data on $H^*(X)$. If X is nonsingular and projective this amounts to the (p,q) decomposition

$$H^r = \bigoplus_{p+q=r} H^{p,q} \text{ with the symmetry } H^{p,q} = \overline{H^{q,p}},$$

which it is possible to enrich with polarization data after the choice of an ample line bundle.

As we will see in Lecture 2, for a general X, i.e., maybe singular and noncompact, the (p,q) decomposition is replaced by a more complicated linear algebra object, the mixed Hodge structure, consisting of two filtrations $W_\bullet$ on $H^*(X,\mathbb{Q})$, and $F^\bullet$ on $H^*(X,\mathbb{C})$ with compatibility conditions.

Similarly, given a projective map $f : X \longrightarrow Y$, with X is nonsingular, we look for a linear algebra datum encoding the datum of the map f, with the obvious requirements:

- This datum should be compatible with the Hodge structure of X.

- It should impose strong constraints of a linear algebra type.

- It should have a vivid geometric interpretation.

The theorems discussed by de Cataldo in the last lecture of the course, with their Hodge-theoretic counterparts, give some answers to these questions.

Remark.   Unless otherwise stated, all the cohomology groups are with rational coefficients.

5.1   LECTURE 1: THE SMOOTH CASE: E_2-DEGENERATION

We suppose that $f : X \to Y$ is a projective smooth map of nonsingular (connected) quasi-projective varieties, that is, f factors as $p_2 \circ i$ via a closed immersion $i : X \longrightarrow Y \times \mathbb{P}^N$ into some product $Y \times \mathbb{P}^N$, and the fibers of f are nonsingular projective manifolds. The nonsingular hypersurfaces of a fixed degree in some projective space give an interesting example. More generally we have the following:

EXAMPLE 5.1.1 (The universal hyperplane section)   Let $X \subseteq \mathbb{P}^n(\mathbb{C})$ be a nonsingular projective variety, denote by $\mathbb{P}^n(\mathbb{C})^\vee$ the dual projective space, whose points are hyperplanes in $\mathbb{P}^n(\mathbb{C})$, and define

$$\mathcal{X} := \{(x,H) \in X \times \mathbb{P}^n(\mathbb{C})^\vee \text{ such that } x \in H \cap X\},$$

with the second projection $p_2 : \mathcal{X} \longrightarrow \mathbb{P}^n(\mathbb{C})^\vee$. The fiber over the point of $\mathbb{P}^n(\mathbb{C})^\vee$ corresponding to the hyperplane H is the hyperplane section $H \cap X \subseteq X$. Since the projection $\mathcal{X} \longrightarrow X$ makes $\mathcal{X}$ into a projective bundle over X, it follows that $\mathcal{X}$ is nonsingular.

Let

$$X^\vee = \{H \in \mathbb{P}^n(\mathbb{C})^\vee \text{ such that } H \cap X \text{ is singular}\}.$$

It is an algebraic subvariety of $\mathbb{P}^n(\mathbb{C})^\vee$, called the dual variety of X.

Set

$$U_{\mathrm{reg}} := \mathbb{P}^n(\mathbb{C})^\vee \setminus X^\vee, \qquad \mathcal{X}_{\mathrm{reg}} := p_2^{-1}(U_{\mathrm{reg}}).$$

Then the restriction

$$p_{2|} : \mathcal{X}_{\mathrm{reg}} \longrightarrow U_{\mathrm{reg}}$$

is a projective smooth map. For example, let $\mathbb{P}^n(\mathbb{C}) \xrightarrow{v_d} \mathbb{P}^N(\mathbb{C})$ be the dth Veronese embedding of $\mathbb{P}^n(\mathbb{C})$: setting $X := v_d(\mathbb{P}^n(\mathbb{C}))$, the construction gives the family of degree d hypersurfaces in $\mathbb{P}^n(\mathbb{C})$.

By a classical result (the Ehresmann fibration lemma), a map as above is a C^∞-fiber bundle, namely, for some manifold F, each point $y \in Y$ has a neighborhood N in the analytic topology, such that there is a fiber-preserving diffeomorphism

$$f^{-1}(N) \xrightarrow{\simeq} N \times F,$$

$$\left\downarrow f \quad \swarrow p_1\right.$$

$$N$$

where p_1 denotes the projection on the first factor.

In particular, for every i, the ith higher direct image sheaf $R^i f_* \mathbb{Q}$, whose stalk at a point $y \in Y$ is

$$(R^i f_* \mathbb{Q})_y \simeq H^i(f^{-1}(y)),$$

is a *local system*, i.e., a locally constant sheaf of finite-dimensional $\mathbb{Q}$-vector spaces. If Y is simply connected, the $R^i f_* \mathbb{Q}$'s are in fact constant sheaves. In general, choosing a base point $y_0 \in Y$, we have associated *monodromy* representations

$$\rho^i : \pi_1(Y, y_0) \to \mathrm{Aut}(H^i(f^{-1}(y_0))). \tag{5.1.1}$$

For general reasons there is the *Leray spectral sequence*

$$E_2^{pq} = H^p(Y, R^q f_* \mathbb{Q}) \to H^{p+q}(X).$$

For a fibration of differentiable manifolds, the Leray spectral sequence can be non-trivial even if Y is simply connected; for example, in the Hopf fibration $f : S^3 \to S^2$, the differential $d_2 : E_2^{01} \to E_2^{20}$ is nonzero.

THEOREM 5.1.2 *The Leray spectral sequence associated to a smooth projective map degenerates at E_2.*

In fact a stronger statement can be proved:

THEOREM 5.1.3 *There exists an isomorphism in the bounded derived category of sheaves with constructible cohomology (see Lecture 3)*

$$Rf_* \mathbb{Q} \simeq \bigoplus_q R^q f_* \mathbb{Q}[-q].$$

In particular, if Y is simply connected, Theorem 5.1.2 gives an isomorphism of vector spaces

$$H^r(X) \simeq \bigoplus_{a+b=r} H^a(Y) \otimes H^b(f^{-1}(y_0));$$

the fibration behaves, from the point of view of additive cohomology, as if it were a product. Simple examples ($\mathbb{P}^1$-fibrations, for instance) show that in general the isomorphism cannot be made compatible with the ring structure on cohomology.

SKETCH OF THE PROOF OF THEOREM 5.1.2. Suppose that $\mathcal{L}$ is a relatively ample line bundle on X, and denote by $L \in H^2(X)$ its first Chern class. Let $n :=$ $\dim X - \dim Y$ be the relative dimension of the map. The hard Lefschetz theorem applied to the fibers of f gives isomorphisms $L^k : H^{n-k}(f^{-1}(y)) \longrightarrow H^{n+k}(f^{-1}(y))$, hence isomorphisms of local systems $L^k : R^{n-k}f_*\mathbb{Q} \longrightarrow R^{n+k}f_*\mathbb{Q}$. On the cohomology of each fiber we have a Lefschetz decomposition

$$H^i(f^{-1}(y)) = \oplus L^a P^{i-2a}(f^{-1}(y)),$$

with $P^i(f^{-1}(y)) := \operatorname{Ker} L^{n-i+1} : H^i(f^{-1}(y)) \longrightarrow H^{2n-i+2}(f^{-1}(y))$. We have a corresponding decomposition of local systems

$$R^i f_*\mathbb{Q} = \oplus L^a \mathcal{P}^{i-2a},$$

where $\mathcal{P}^r$ denotes the local system with stalk $\mathcal{P}^r_y = P^r(f^{-1}(y))$. The differentials of the Leray spectral sequence are compatible with this decomposition. Let us show for example that $d_2 = 0$. It is enough to show this on the direct summand $H^p(Y, \mathcal{P}^q) \subseteq E_2^{pq}$. We have the following:

$$
\begin{array}{ccc}
H^p(Y, \mathcal{P}^q) & \xrightarrow{\quad d_2 \quad} & H^{p+2}(Y, R^{q-1}f_*\mathbb{Q}) \\
\downarrow{\scriptstyle L^{n-q+1}} & & \downarrow{\scriptstyle L^{n-q+1}} \\
H^p(Y, R^{2n-q+2}f_*\mathbb{Q}) & \xrightarrow{\quad d_2 \quad} & H^{p+2}(Y, R^{2n-q+1}f_*\mathbb{Q}).
\end{array}
$$

The left vertical arrow is the zero map by the definition of the primitive local system $\mathcal{P}^q$, while the right vertical arrow is an isomorphism by the hard Lefschetz theorem applied to the fibers of f; hence $d_2 = 0$. $\qquad\square$

The stronger statement about the splitting in the derived category is obtained in a similar way, considering the spectral sequence associated with the functors, one for each q,

$$\operatorname{Hom}_{\mathcal{D}(Y)}(R^q f_*\mathbb{Q}[-q], -).$$

There is a spectral sequence converging to $\operatorname{Hom}_{\mathcal{D}(Y)}(R^q f_*\mathbb{Q}[-q], Rf_*\mathbb{Q})$. Exactly the same argument used for the Leray spectral sequence gives the degeneration to E_2 of this one; in particular, the maps

$$\operatorname{Hom}_{\mathcal{D}(Y)}(R^q f_*\mathbb{Q}[-q], Rf_*\mathbb{Q}) \longrightarrow \operatorname{Hom}_{\mathcal{D}(Y)}(R^q f_*\mathbb{Q}, R^q f_*\mathbb{Q}).$$

are surjective. For every q, we can lift the identity map $R^q f_* \mathbb{Q} \longrightarrow R^q f_* \mathbb{Q}$ to a map $R^q f_* \mathbb{Q}[-q] \longrightarrow Rf_* \mathbb{Q}$, so as to obtain a map

$$\bigoplus_q R^q f_* \mathbb{Q}[-q] \longrightarrow Rf_* \mathbb{Q},$$

which induces the identity on all cohomology sheaves, so it is, in particular, a quasi-isomorphism.

Remark. For singular maps, the Leray spectral sequence is seldom degenerate. If $f : X \to Y$ is a resolution of the singularities of a projective variety Y whose cohomology has a mixed Hodge structure (see Lecture 2) which is not pure, then f^* cannot be injective, and in view of the edge sequence the Leray spectral sequence cannot degenerate at E_2.

5.2   LECTURE 2: MIXED HODGE STRUCTURES

5.2.1   Mixed Hodge Structures on the Cohomology of Algebraic Varieties

As the following two elementary examples show, one cannot expect that Hodge theory extends to singular or noncompact varieties.

EXAMPLE 5.2.1   Consider the projective plane curve $\mathcal{C}$ of equation

$$Y^2 Z - X^2 (X - Z) = 0.$$

It is immediately seen that $\dim H^1(\mathcal{C}) = 1$, hence there cannot be a (p, q)-decomposition $H^1(\mathcal{C}, \mathbb{C}) = H^{1,0} \oplus \overline{H^{1,0}}$ on this vector space.

EXAMPLE 5.2.2   Consider $\mathbb{C}^* \subseteq \mathbb{P}^1(\mathbb{C})$. Clearly $\dim H^1(\mathbb{C}^*) = 1$, and again there cannot be a (p, q)-decomposition $H^1(\mathbb{C}^*, \mathbb{C}) = H^{1,0} \oplus \overline{H^{1,0}}$ on this vector space.

Basically, there are two possibilities:

1. Allow linear algebra structures which are more complicated than the "simple" (p, q) decomposition.

2. Consider different topological invariants.

Both possibilities turn out to have remarkable consequences. In this lecture we will consider the first option. The second possibility, leading to the definition of *intersection cohomology*, will be considered in de Cataldo's Lecture 4 (in Section 6.1).

DEFINITION 5.2.3 (Mixed Hodge structure)   *A (rational) mixed Hodge structure (MHS) consists of the following datum:*

1. *A vector space $V_{\mathbb{Q}}$ over $\mathbb{Q}$ with a finite increasing filtration* (the weight filtration)

$$\{0\} = W_a \subseteq W_{a+1} \subseteq \cdots \subseteq W_b = V_{\mathbb{Q}}$$

2. *A finite decreasing filtration* (the Hodge filtration) *on $V_{\mathbb{C}} := V_{\mathbb{Q}} \otimes \mathbb{C}$,*

$$V_{\mathbb{C}} = F^q \supseteq F^{q+1} \supseteq \cdots \supseteq F^m \supseteq F^{m+1} = \{0\}$$

which, for every k, induces on $\mathrm{Gr}_k^W V_{\mathbb{C}} = (W_k/W_{k-1}) \otimes \mathbb{C}$, a pure Hodge structure of weight k, namely,

$$\mathrm{Gr}_k^W V_{\mathbb{C}} = \bigoplus_{p+q=k} \left(\mathrm{Gr}_k^W V_{\mathbb{C}}\right)^{pq},$$

where

$$\left(\mathrm{Gr}_k^W V_{\mathbb{C}}\right)^{pq} := F^p \, \mathrm{Gr}_k^W V_{\mathbb{C}} \cap \overline{F^q \, \mathrm{Gr}_k^W V_{\mathbb{C}}}.$$

Let us recall the following definition.

DEFINITION 5.2.4 (Filtered and strict filtered maps)   *Let $(V, G^\bullet)$, $(V', G'^\bullet)$ be two vector spaces endowed with increasing filtrations, and $f : V \longrightarrow V'$ a filtered map, namely, a linear map such that $f(G^a V) \subseteq G'^a V'$. The map f is said to be* strict *if, for every a,*

$$f(G^a V) = \mathrm{Im}\, f \cap G'_a V'.$$

An analogous definition holds for decreasing filtrations. Morphisms of mixed Hodge structures are just what one expects them to be:

DEFINITION 5.2.5 (Morphism of mixed Hodge structures)   *A map of mixed Hodge structures $f : (V_{\mathbb{Q}}, W_\bullet, F^\bullet) \longrightarrow (V'_{\mathbb{Q}}, W'_\bullet, F'^\bullet)$ is a linear map $f : V_{\mathbb{Q}} \longrightarrow V'_{\mathbb{Q}}$ filtered with respect to $W_\bullet$, $W'_\bullet$, such that $f_{\mathbb{C}} : V_{\mathbb{C}} \longrightarrow V'_{\mathbb{C}}$ is filtered with respect to $F^\bullet$, $F'^\bullet$.*

One can similarly define morphisms of mixed Hodge structures of type (k, k); they become just morphisms of mixed Hodge structures after an appropriate Tate twist.

The remarkable formal properties of mixed Hodge structures will be treated in other courses; we list just some of them here:

THEOREM 5.2.6   *Mixed Hodge structures with morphisms of mixed Hodge structures form an abelian category. A morphism f of mixed Hodge structures is strict with respect to $W_\bullet$, and $f_{\mathbb{C}}$ is strict with respect to $F^\bullet$.*

We have the following remarkable theorem:

THEOREM 5.2.7   *The cohomology groups $H^i(Y)$ of a complex algebraic variety Y have a functorial mixed Hodge structure. Furthermore we have the following restrictions on the weights:*

1. $W_a H^i(Y) = \{0\}$ *for $a < 0$ and $W_a H^i(Y) = H^i(Y)$ for $a \geq 2i$.*

2. *If Y is nonsingular, then $W_a H^i(Y) = \{0\}$ for $a < i$, and*

$$W_i H^i(Y) = \mathrm{Im}\ H^i(\overline{Y}) \longrightarrow H^i(Y),$$

where $\overline{Y}$ is any compactification of Y.

3. *If Y is complete, then $W_a H^i(Y) = H^i(Y)$ for $a \geq i$, and*

$$W_{i-1} H^i(Y) = \mathrm{Ker}\ H^i(Y) \longrightarrow H^i(\widetilde{Y}),$$

where $\widetilde{Y} \longrightarrow Y$ is a resolution of singularities of Y.

Functoriality here means that the pull-back map $f^* : H^i(Y) \longrightarrow H^i(X)$ associated with an algebraic map $f : X \longrightarrow Y$ is a morphism of mixed Hodge structures.

Remark. In the case in which Y is nonsingular and projective, the weight filtration is trivial, namely, $W_{i-1} H^i(Y) = \{0\}$ and $W_i H^i(Y) = H^i(Y)$, and $F^\bullet$ is the standard filtration associated with the Hodge decomposition. In the example of the nodal curve, $W_{-1} H^1(Y) = \{0\}$ and $W_0 H^1(Y) = H^1(Y)$, and every class has "type" $(0,0)$. In the example of the punctured line, $W_1 H^1(Y) = \{0\}$ and $W_2 H^1(Y) = H^1(Y)$, and every class has "type" $(1,1)$.

5.2.2 The Global Invariant Cycle Theorem

The following theorem is a consequence of the restrictions on the weights of the cohomology groups of an algebraic variety.

THEOREM 5.2.8 (Weight principle) *Let $Z \subseteq U \subseteq X$ be inclusions, where X is a nonsingular projective variety, $U \subseteq X$ is a Zariski dense open subset and $Z \subseteq U$ is a closed subvariety of X. Then the images in the cohomology of Z of the restriction maps from X and from U coincide, namely, setting*

$$H^l(X) \xrightarrow{a} H^l(U) \xrightarrow{b} H^l(Z),$$

we have $\mathrm{Im}\ b = \mathrm{Im}\ b \circ a.$

SKETCH OF PROOF. The maps a and b are strictly compatible with the weight filtration. Hence, it follows from Theorem 5.2.7 that $\mathrm{Im}\ b = \mathrm{Im}\ W_l\ H^l(U)$. The statement follows since, again by Theorem 5.2.7, $\mathrm{Im}\ a = W_l\ H^l(U)$. $\qquad\square$

Remark. Despite its innocent appearance, this is an extremely strong statement, imposing nontrivial constraints on the topology of algebraic maps. For contrast look at the real picture,

$$Z = S^1 \subseteq U = \mathbb{C}^* \subseteq X = \mathbb{P}^1(\mathbb{C}).$$

The restriction map $H^1(\mathbb{P}^1(\mathbb{C})) \longrightarrow H^1(S^1)$ is zero, while $H^1(\mathbb{C}^*) \longrightarrow H^1(S^1)$ is an isomorphism.

The following is the global invariant cycle theorem, which follows quite directly from the weight principle above:

THEOREM 5.2.9 (The global invariant cycle theorem)   *Suppose $f : X \to Y$ is a smooth projective map, with Y connected, and let $\overline{X}$ be a nonsingular compactification of X. Then, for $y_0 \in Y$,*

$$H^l(f^{-1}(y_0))^{\pi_1(Y, y_0)} = \mathrm{Im}\,\{H^l(\overline{X}) \longrightarrow H^l(f^{-1}(y_0))\}.$$

Remark.   The previous theorem is most often used when we have a projective map, not necessarily smooth, $\overline{f} : \overline{X} \longrightarrow \overline{Y}$, with $\overline{X}$ nonsingular. There is a dense Zariski open subset $Y \subseteq \overline{Y}$ such that $X := \overline{f}^{-1}(Y) \longrightarrow Y$ is a smooth map. Then Theorem 5.2.9 states that the monodromy invariants in the cohomology of a generic fiber are precisely the classes obtained by restriction from the total space of the family.

Remark also that, while it is clear that a cohomology class in X restricts to a monodromy invariant class in the cohomology of the fiber at y_0, the converse is by no means obvious and is in fact specific to algebraic maps: let us consider again the Hopf fibration and identify a fiber with S^1; the generator of $H^1(S^1)$ is clearly monodromy invariant, as the monodromy of the Hopf fibration is trivial, but it is not the restriction of a class in S^3, as $H^1(S^3) = 0$.

SKETCH OF PROOF OF THE GLOBAL INVARIANT CYCLE THEOREM (5.2.9).   It follows from Theorem 5.1.2 proved in Lecture 1, that the Leray spectral sequence for f degenerates at E_2, in particular, for all l, the map

$$H^l(X) \longrightarrow E_2^{0l} = H^0(Y, R^l f_* \mathbb{Q})$$

is surjective. We have the natural identification

$$H^0(Y, R^l f_* \mathbb{Q}) \simeq H^l(f^{-1}(y_0))^{\pi_1(Y, y_0)} \subseteq H^l(f^{-1}(y_0)),$$

and the composition

$$H^l(X) \longrightarrow H^0(Y, R^l f_* \mathbb{Q}) \xrightarrow{\simeq} H^l(f^{-1}(y_0))^{\pi_1(Y, y_0)} \longrightarrow H^l(f^{-1}(y_0))$$

is a map of (mixed) Hodge structures. By Theorem 5.2.8 we have

$$\mathrm{Im}\, H^l(\overline{X}) \to H^l(f^{-1}(y_0)) = \mathrm{Im}\, H^l(X) \to H^l(f^{-1}(y_0)) = H^l(f^{-1}(y_0))^{\pi_1(Y, y_0)}.$$

$$\square$$

5.2.3   Semisimplicity of Monodromy

A consequence of the global invariant cycle theorem (Theorem 5.2.9) is that the subspace

$$H^i(f^{-1}(y_0))^{\pi_1(Y, y_0)} \subseteq H^i(f^{-1}(y_0))$$

of monodromy invariants is a (pure) sub-Hodge structure of $H^i(f^{-1}(y_0))$. It is interesting to compare this fact with the local situation:

EXAMPLE 5.2.10 Consider the family of degenerating elliptic curves C_t of equations

$$\{Y^2 Z - X(X - tZ)(X - Z) = 0\} \subseteq \mathbb{P}^2(\mathbb{C}) \times \Delta,$$

where t is a coordinate on the disk Δ. The monodromy operator is a length 2 Jordan block, and the subspace of monodromy invariants is one-dimensional, spanned by the vanishing cycle of the degeneration. Hence the subspace cannot be a sub-Hodge structure of the weight 1 Hodge structure $H^1(C_{t_0})$ for $t \neq 0$. As we will see in Lecture 3, the local invariant cycle theorem with the associated Clemens–Schmid exact sequence deals with this kind of setup, defining a *mixed Hodge structure* on $H^1(C_{t_0})$ whose weight filtration is related to the monodromy.

The fact that the space of monodromy invariants is a sub-Hodge structure of the cohomology of a fiber can be refined as follows (recall that a representation is said to be *irreducible* if it has no nontrivial invariant subspace):

THEOREM 5.2.11 (Semisimplicity theorem) *Suppose $f : X \to Y$ is a smooth projective map of quasi-projective manifolds. Then the monodromy representations ρ^i defined in (5.1.1) are semisimple, namely, they split as a direct sum of irreducible representations.*

Again one can compare with the local setup of a family over a disk which is smooth outside 0. In this case, by the *monodromy theorem*, the monodromy operator T is *quasi-unipotent*, namely, $(T^a - I)^b = 0$ (compare with Example 5.2.10). Again, the semisimplicity of global monodromy is a specific property of algebraic geometry. For examples there are Lefschetz pencils on symplectic varieties of dimension 4 with non-semi-simple monodromy. We will not discuss the proof of Theorem 5.2.11, but let us summarize what we have so far.

THEOREM 5.2.12 *Suppose $f : X \to Y$ is a smooth projective map of quasi-projective manifolds of relative dimension n, $\mathcal{L}$ a relatively ample line bundle, and L its first Chern class.*

- **Decomposition:** *there is an isomorphism in $\mathcal{D}_Y$,*

$$Rf_*\mathbb{Q} \simeq \bigoplus_q R^q f_*\mathbb{Q}[-q].$$

- **Hard Lefschetz along the fibers:** *cupping with L defines isomorphisms of local systems on Y,*
$$L^k : R^{n-q} f_*\mathbb{Q} \longrightarrow R^{n+q} f_*\mathbb{Q}.$$

- **Semisimplicity:** *the local systems $R^q f_*\mathbb{Q}$ are, for every q, semisimple local systems.*

As will be explained in the next lectures, the three statements above will generalize to any projective map, once the local systems are replaced by *intersection cohomology complexes of local systems*. More precisely, the category of local systems is replaced by a different category, that of *perverse sheaves*, which has strikingly similar formal properties.

5.3   LECTURE 3: TWO CLASSICAL THEOREMS ON SURFACES AND THE LOCAL INVARIANT CYCLE THEOREM

5.3.1   Homological Interpretation of the Contraction Criterion and Zariski's Lemma

Let $f : X \longrightarrow Y$ be a birational proper map with X a nonsingular surface, and Y normal; let $y \in Y$ be a singular point. Set $C := f^{-1}(y)$ and let C_i, for $i = 1, \ldots, k$, be its irreducible components. Then the intersection numbers $(C_i \cdot C_j)$ give a symmetric matrix, hence a symmetric bilinear form, called the *intersection form*, on the vector space generated by the C_i's.

EXAMPLE 5.3.1   Let $C \subseteq \mathbb{P}^n$ be a nonsingular projective curve, and let $\mathcal{C} \subseteq \mathbb{C}^{n+1}$ be the affine cone over C, with vertex o. Blowing up the vertex we get a nonsingular surface $\widetilde{\mathcal{C}}$, the total space of the line bundle $\mathcal{O}(-1)_{|C}$, with the blow-down map $\beta :$ $\widetilde{\mathcal{C}} \longrightarrow \mathcal{C}$. We have $\beta^{-1}(o) = C$, where C is embedded in $\widetilde{\mathcal{C}}$ as the zero section. In this case the intersection form amounts to the self-intersection number $C^2 = -\deg C$.

We have the classical theorem:

THEOREM 5.3.2   *The* intersection form $(C_i \cdot C_j)$ *associated to an exceptional curve* $C = \bigcup C_i$ *is negative definite.*

There is a similar statement for surfaces mapping to a curve: the setup is as follows: $f : X \longrightarrow Y$ is a projective map, with X a nonsingular surface, and Y a nonsingular curve. Let $y \in Y$, and $C := f^{-1}(y)$ and let C_i, for $i = 1, \ldots, k$, be its irreducible components. As before, the intersection numbers $(C_i \cdot C_j)$ define a symmetric bilinear form, the *intersection form*, on the vector space $\mathcal{V}$ generated by the C_i's. If $f^{-1}(y) = \sum n_i C_i$, clearly its intersection with every element in $\mathcal{V}$ vanishes, being algebraically equivalent to every other fiber. The following is known as *Zariski's lemma*:

THEOREM 5.3.3   *The intersection form is negative semidefinite. Its radical is spanned by the class* $\sum n_i C_i$ *of the fiber of* y.

Clearly the statement above is empty whenever the fiber over y is irreducible.

We are going to give an interpretation of these two classical results in terms of splittings of the derived direct image sheaf $Rf_*\mathbb{Q}$. To do this we need to introduce the constructible derived category (see [1, 2, 4, 5]).

Recall that a *stratification* Σ of an algebraic variety Y is a decomposition $Y = \coprod_{l \geq 0} S_l$, where the $S_l \subseteq Y$ are locally closed and nonsingular subvarieties of pure complex dimension l. A sheaf on Y is said to be *constructible* if there exists a stratification of the variety such that the restriction of the sheaf to each stratum is a locally constant sheaf of finite rank. The category $\mathcal{D}_Y$ that we are interested in has

- *objects*: bounded complexes of sheaves

$$K^\bullet : \cdots \longrightarrow K^i \xrightarrow{d^i} K^{i+1} \longrightarrow \cdots$$

of $\mathbb{Q}$-vector spaces, such that their cohomology sheaves

$$\mathcal{H}^i(K^\bullet) := \operatorname{Ker} d^i / \operatorname{Im} d^{i-1}$$

are constructible;

- *morphisms*: a map $\phi : K^\bullet \longrightarrow L^\bullet$ between two objects is defined by a diagram

$$K^\bullet \xrightarrow{\phi'} \tilde{L}^\bullet \xleftarrow{u} L^\bullet,$$

where u is a quasi-isomorphism, namely, a map of complexes of sheaves which induces isomorphisms

$$\mathcal{H}^i(u) : \mathcal{H}^i(L^\bullet) \longrightarrow \mathcal{H}^i(\tilde{L}^\bullet)$$

for every i. There is a natural equivalence relation on such diagrams which we will not discuss. A map in $\mathcal{D}_Y$ is an equivalence class of such diagrams.

The category $\mathcal{D}_Y$ is remarkably stable with respect to many operations, most notably Verdier's duality, to be explained in a future lecture by M. A. de Cataldo. Furthermore, algebraic maps $f : X \longrightarrow Y$ define several functors between $\mathcal{D}_Y$ and $\mathcal{D}_Y$, such as the derived direct image Rf_*, the proper direct image $Rf_!$, the pull-back functors f^*, and $f^!$, related by a rich formalism. For details see the texts in the bibliography. We will use this formalism freely. Recall also that the *(standard) truncation functors* $\tau_{\le}, \tau_{\ge}$ defined on complexes send quasi-isomorphisms to quasi-isomorphisms and obviously preserve constructibility of cohomology sheaves; hence they define functors in $\mathcal{D}_Y$.

Let $f : X \to Y$ be the resolution of singularities of the normal surface Y. For simplicity, let us suppose that Y has a unique singular point y, and as before, let us set $C := f^{-1}(y)$ and $C = \bigcup C_i$.

Let us study the complex $Rf_*\mathbb{Q}_X$.

There is the commutative diagram with Cartesian squares

$$
\begin{array}{ccccc}
C & \xrightarrow{I} & X & \xleftarrow{J} & U \\
\downarrow & & \downarrow{\scriptstyle f} & & \Vert{\scriptstyle =} \\
y & \xrightarrow{i} & Y & \xleftarrow{j} & U
\end{array}
$$

where $U := Y \setminus y = X \setminus Z$.

The distinguished attaching triangle associated to the restriction from Y to U gives

$$i_! i^! Rf_*\mathbb{Q}_X \longrightarrow Rf_*\mathbb{Q}_X \longrightarrow Rj_* j^* Rf_*\mathbb{Q}_X \simeq Rj_*\mathbb{Q}_{Y \setminus \{y\}} \xrightarrow{[1]}.$$

Let us consider the long exact sequence of the cohomology sheaves at y. Note that $R^i f_*\mathbb{Q} = 0$ for $i > 2$, $R^1 f_*\mathbb{Q}$, $R^2 f_*\mathbb{Q}$ are concentrated at y, while $R^0 f_*\mathbb{Q} = \mathbb{Q}$ by force of our hypotheses.

There is a fundamental system of Stein neighborhoods of y such that if N is any neighborhood in the family and $N' := N \setminus \{y\}$ we have

$$\mathcal{H}^a(i_! i^! Rf_*\mathbb{Q}_X)_y = H^a(f^{-1}(N), f^{-1}(N')) \simeq H_{4-a}(C),$$

which vanishes for $a \neq 2, 3, 4$,

$$\mathcal{H}^a(Rf_*\mathbb{Q}_X)_y = H^a(f^{-1}(N)) \simeq H^a(C),$$

which vanishes for $a \neq 0, 1, 2$, and

$$\mathcal{H}^a(Rj_* j^* Rf_*\mathbb{Q}_X)_y = \mathcal{H}^a(Rj_*\mathbb{Q}_{N'})_y \simeq H^a(N').$$

There is the adjunction map $\mathbb{Q}_Y \to Rf_* f^*\mathbb{Q}_Y = Rf_*\mathbb{Q}_X$. This map does not split. We study the obstruction to this failure. As already noted, since $\dim_{\mathbb{C}} f^{-1}(y) = 1$, we have $\tau_{\leq 2} Rf_*\mathbb{Q}_X \simeq Rf_*\mathbb{Q}_X$. The truncation functors yield a map

$$\widetilde{u} \; : \; Rf_*\mathbb{Q}_X \longrightarrow \tau_{\leq 2} Rj_*\mathbb{Q}_{N'}.$$

Consider the truncation distinguished triangle

$$\tau_{\leq 1} Rj_*\mathbb{Q}_{N'} \longrightarrow \tau_{\leq 2} Rj_*\mathbb{Q}_{N'} \longrightarrow \mathcal{H}^2(Rj_*\mathbb{Q}_{N'})[-2] \xrightarrow{\;[1]\;}.$$

Note that the last complex is in fact reduced to the skyscraper complex $H^2(N')_y[-2]$. Apply the cohomological functor $\mathrm{Hom}_{\mathcal{D}_Y}(Rf_*\mathbb{Q}_X, -)$ to the triangle and take the associated long exact sequence

$$\begin{aligned}
0 \;\; &\to \mathrm{Hom}(Rf_*\mathbb{Q}_X, \tau_{\leq 1} Rj_*\mathbb{Q}_{N'}) \to \mathrm{Hom}(Rf_*\mathbb{Q}_X, \tau_{\leq 2} Rj_*\mathbb{Q}_{N'}) \\
&\to \mathrm{Hom}(Rf_*\mathbb{Q}_X, H^2(N')[-2]).
\end{aligned}$$

The map $\widetilde{u}$ maps to a map in

$$\begin{aligned}
\mathrm{Hom}(Rf_*\mathbb{Q}_X, H^2(N')[-2]) \;\; &= \mathrm{Hom}(R^2 f_*\mathbb{Q}_X, H^2(N')) \\
&= \mathrm{Hom}(H^2(f^{-1}(N)), H^2(f^{-1}(N'))).
\end{aligned}$$

This map is just the restriction map, which fits in the long exact sequence

$$\cdots \longrightarrow H_2(C) \xrightarrow{\;I\;} H^2(C) \simeq H^2(f^{-1}(N)) \longrightarrow H^2(f^{-1}(N')) \longrightarrow \cdots,$$

where $I \in \mathrm{Hom}(H_2(C), H^2(C)) \simeq H^2(C) \otimes H^2(C)$ is just the intersection matrix discussed above. This being nondegenerate, we have that I is an isomorphism, and the restriction map $H^2(f^{-1}(N)) \longrightarrow H^2(f^{-1}(N'))$ vanishes. This means that there exists a (unique) lift $\widetilde{v} : Rf_*\mathbb{Q}_X \to \tau_{\leq 1} Rj_*\mathbb{Q}_U$.

Taking the cone $C(\widetilde{v})$ of this map, one obtains a distinguished triangle

$$\to C(\widetilde{v}) \to f_*\mathbb{Q}_X \to \tau_{\leq 1} j_*\mathbb{Q}_U \xrightarrow{\;[1]\;}.$$

An argument similar to the previous one, shows that $\widetilde{v}$ admits a canonical splitting so that there is a canonical isomorphism in $\mathcal{D}_Y$:

$$Rf_*\mathbb{Q}_X \simeq \tau_{\leq 1}Rj_*\mathbb{Q}_U \oplus H^2(C)_y[-2].$$

Thus, $\mathbb{Q}_Y \to Rf_*\mathbb{Q}_X$ does not split: however, in looking for the nonexistent map $Rf_*\mathbb{Q}_X \to \mathbb{Q}_Y$, we find the remarkable and more interesting splitting map $\widetilde{v}$.

Why is this interesting? Because, up to shift, the complex $\tau_{\leq 1}Rj_*\mathbb{Q}_{N'}$ is, by definition, the *intersection cohomology complex* (see the next lecture) $\mathcal{IC}_Y$ of Y. The complex $H^2(C)_y[-2]$ is, up to shift, the intersection cohomology complex of y with multiplicity $b_2(Z)$. We may rewrite the splitting as

$$Rf_*\mathbb{Q}_X[2] \simeq \mathcal{IC}_Y \oplus \mathcal{IC}_y^{\oplus b_2(Z)},$$

thus obtaining a first nontrivial example of the decomposition theorem.

We can do something similar for a projective map $f : X \longrightarrow Y$ with X a nonsingular surface, and Y a nonsingular curve. For simplicity, we assume that the generic fiber is connected. Set $j : Y' \longrightarrow Y$, the open embedding of the regular value locus of f. The restriction $f' : X' := f^{-1}(Y') \longrightarrow Y'$ is a smooth family of curves, and by Theorem 5.1.2 we have a splitting in $\mathcal{D}_{Y'}$,

$$Rf'_*\mathbb{Q}_{X'} \simeq R^0 f'_*\mathbb{Q}_{X'} \oplus R^1 f'_*\mathbb{Q}_{X'}[-1] \oplus R^2 f'_*\mathbb{Q}_{X'}[-2], \qquad (5.3.1)$$

which we may rewrite as

$$Rf'_*\mathbb{Q}_{X'} \simeq \mathbb{Q}_{Y'} \oplus R^1 f'_*\mathbb{Q}_{X'}[-1] \oplus \mathbb{Q}_{Y'}[-2], \qquad (5.3.2)$$

as we have $R^0 f'_*\mathbb{Q}_{X'} = R^2 f'_*\mathbb{Q}_{X'} = \mathbb{Q}_{Y'}$, since the fibers over Y' are nonsingular and connected.

If we try to investigate whether we can extend the splitting (5.3.2) to Y, we see that this time, at the crucial step, we may use Zariski's lemma, to conclude that there is an isomorphism

$$Rf_*\mathbb{Q}_X \simeq \mathbb{Q}_Y \oplus j_*R^1 f'_*\mathbb{Q}_{X'}[-1] \oplus \mathbb{Q}[-2] \oplus (\oplus V_{y_i}), \qquad (5.3.3)$$

where V_{y_i} is a skyscraper sheaf concentrated at the points y_i where the fiber is reducible; the dimension of V_{y_i} equals the number of irreducible components of the fiber over y_i minus one.

Again, it turns out that the nonderived direct image sheaf $j_*R^1 f'_*\mathbb{Q}_{Y'}$, defined by

$$j_*R^1 f'_*\mathbb{Q}_{Y'} = \tau_{\geq 0}Rj_*R^1 f'_*\mathbb{Q}_{Y'},$$

is the *intersection cohomology complex* of the local system $R^1 f'_*\mathbb{Q}_{Y'}$. The two examples discussed above turn out to be special cases of the general theorems which will be discussed in the last lecture of this course.

5.3.2 The Local Invariant Cycle Theorem, the Limit Mixed Hodge Structure, and the Clemens–Schmid Exact Sequence ([3, 6])

We consider again the family of curves $f : X \longrightarrow Y$ and, specifically, the sheaf-theoretic decomposition (5.3.3); let $y \in Y \setminus Y'$, and take the stalk at y of the first cohomology sheaf $\mathcal{H}^1$. Pick a disk N around y, and let $y_0 \in N \setminus \{y\}$. We have a monodromy operator $T : H^1(f^{-1}(y_0)) \longrightarrow H^1(f^{-1}(y_0))$. Recall that if $\mathcal{V}$ is a local system on the punctured disk, with monodromy T, then $(j_*\mathcal{V})_0 = H^0(\Delta^*, \mathcal{V}) \simeq \mathrm{Ker}(T - I)$. We find

$$H^1(f^{-1}(y)) = (R^1 f_* \mathbb{Q})_y = (j_* R^1 f'_* \mathbb{Q}_{Y'})_y = \mathrm{Ker}(T - I) \subseteq H^1(f^{-1}(y_0)).$$

This is, in this particular case, the content of the *local invariant cycle theorem*, whose general statement follows:

THEOREM 5.3.4 (Local invariant cycle theorem) *Let $f : X \to \Delta$ be a projective flat map, smooth over the punctured disk Δ^*, and assume X is nonsingular. Denote by $X_0 := f^{-1}(0)$ the central fiber, let $t_0 \in \Delta^*$ be a fixed base point, and $X_{t_0} := f^{-1}(t_0)$. Let $T : H^k(X_{t_0}) \longrightarrow H^k(X_{t_0})$ be the monodromy around 0. Then, for every k, the sequence*

$$H^k(X_0) \longrightarrow H^k(X_{t_0}) \xrightarrow{T-I} H^k(X_{t_0})$$

is exact.

Note that the degeneration at E_2 theorem (Theorem 5.1.2) implies that every monodromy invariant cohomology class on a generic fiber X_{t_0} is the restriction of a class on $X \setminus X_0$. The local invariant cycle theorem however says much more: if Δ is small enough, there is a homotopy equivalence $X \simeq X_0$ (the retraction on the central fiber). Theorem 5.3.4 states that every monodromy-invariant cohomology class on X_{t_0} comes in fact from a cohomology class on the total space X.

The statement of Theorem 5.3.4 can be refined by introducing a rather nonintuitive mixed Hodge structure on $H^k(X_{t_0})$, the *limit mixed Hodge structure*, which we now describe. Let $f : X^* \to \Delta^*$ be a projective smooth map over the punctured disk Δ^*.

The monodromy operators $T : H^k(X_{t_0}) \longrightarrow H^k(X_{t_0})$ are quasi-unipotent, namely, $(T^a - I)^b = 0$. Taking base change by $\zeta^a : \Delta^* \longrightarrow \Delta^*$ has the effect of replacing the monodromy T with its power T^a; hence we can suppose T unipotent. We define

$$N := \log T = \sum \frac{1}{k}(T - I)^k,$$

noting that the sum is finite. N is nilpotent: suppose $N^b = 0$ (it actually follows from the monodromy theorem that we can take $b = k + 1$ if we are considering the monodromy of $H^k(X_{t_0})$). The following linear algebra result is basically a (very useful) reformulation of the Jordan form theorem for a nilpotent endomorphism of a vector space:

THEOREM 5.3.5 (Monodromy weight filtration)   *There is an increasing filtration of* $\mathbb{Q}$-*subspaces of* $H^k(X_{t_0})$:

$$\{0\} \subseteq W_0 \subseteq W_1 \subseteq \cdots \subseteq W_{2k-1} \subseteq W_{2k} = H^k(X_{t_0})$$

such that

- $N(W_l) \subseteq W_{l-2}$ *for every* l;
- $N^l : \mathrm{Gr}_W^{k+l} \longrightarrow \mathrm{Gr}_W^{k-l}$ *is an isomorphism for every* l.

We have the following surprising result:

THEOREM 5.3.6 (The limit mixed Hodge structure)   *For every* k, *there exists a decreasing filtration* $F_{\mathrm{lim}}^\bullet$ *on* $H^k(X_{t_0}, \mathbb{C})$ *such that* $(H^k(X_{t_0}, \mathbb{C}), W_\bullet, F_{\mathrm{lim}}^\bullet)$, *where* $W_\bullet$ *is the filtration associated with the endomorphism* $\log T$ *as described in Theorem 5.3.5, is a mixed Hodge structure. Furthermore, the map* $\log T : H^k(X_{t_0}, \mathbb{Q}) \longrightarrow H^k(X_{t_0}, \mathbb{Q})$ *becomes a map of type* $(-1, -1)$ *with respect to this limit mixed Hodge structure.*

Even more remarkably, this limit mixed Hodge structure, which, we emphasize, is constructed just from the family over the punctured disk, is related to the mixed Hodge structure of the central fiber, thus giving a delicate interplay between the monodromy properties of the smooth family and the geometry of the central fiber. This is the content of the Clemens–Schmid exact sequence theorem.

Suppose $f : \overline{X} \longrightarrow \Delta$ is a projective map, smooth outside 0, such that the central fiber is a reduced divisor with global normal crossing, namely, in the irreducible components decomposition

$$f^{-1}(0) = \bigcup X_\alpha,$$

the X_α are nonsingular and meet transversely. The *semistable reduction theorem* states that any degeneration over the disk may be brought in semistable form after a finite base change ramified at 0 and a birational modification. The mixed Hodge structure of the cohomology of a normal crossing is easily expressed by a spectral sequence in the category of mixed Hodge structures involving the cohomology groups of the X_α's and their intersections. We have the (co)specialization map, defined as the composition $H^k(X_0, \mathbb{Q}) \xrightarrow{\simeq} H^k(X, \mathbb{Q}) \longrightarrow H^k(X_{t_0}, \mathbb{Q})$.

THEOREM 5.3.7 (Clemens–Schmid exact sequence)   *The specialization map is a map of mixed Hodge structures if we consider on* $H^k(X_{t_0}, \mathbb{Q})$ *the limit mixed Hodge structure. There is an exact sequence of mixed Hodge structures (with appropriate Tate twists)*

$$\cdots \longrightarrow H_{2\dim X - k}(X_0) \longrightarrow H^k(X_0) \longrightarrow H^k(X_{t_0}) \xrightarrow{N} H^k(X_{t_0})$$

$$\longrightarrow H_{2\dim X - k - 2}(X_0) \longrightarrow H^{k+2}(X_0) \longrightarrow H^{k+2}(X_{t_0}) \xrightarrow{N} \cdots.$$

Bibliography

[1] Mark A. de Cataldo and Luca Migliorini. "The decomposition theorem and the topology of algebraic maps," *Bull. Amer. Math. Soc.*, vol. 46, n. 4 (2009), 535–633.

[2] Alexandru Dimca. *Sheaves in topology*, Universitext. Springer, Berlin, 2004.

[3] Phillip Griffiths, ed. *Topics in transcendental algebraic geometry*, Ann. of Math. Stud. **106**, Princeton University Press, Princeton, 1984.

[4] Birger Iversen. *Cohomology of sheaves*, Universitext. Springer, Berlin, Heidelberg, 1986.

[5] Masaki Kashiwara and Pierre Schapira. *Sheaves on manifolds*, volume 292 of *Grundlehren der mathematischen Wissenschaften*. Springer, Berlin, Heidelberg, 1990.

[6] Chris A. M. Peters and Joseph H. M. Steenbrink. *Mixed Hodge structures*, volume 52 of *Ergebnisse der Mathematik und ihrer Grenzgebiete. 3. Folge. A Series of Modern Surveys in Mathematics [Results in Mathematics and Related Areas. 3rd Series]*. Springer, Berlin, 2008.

Chapter Six

The Hodge Theory of Maps
by Mark Andrea de Cataldo and Luca Migliorini
Lectures 4–5 by Mark Andrea de Cataldo

These are the lecture notes from my two lectures 4 and 5. To get an idea of what you will find in them, parse the table of contents at the beginning of the book. The lectures had a very informal flavor to them and, by choice, the notes reflect this fact. There are plenty of exercises and some references so you can start looking things up on your own. My book [5] contains some of the notions discussed here, as well as some amplifications.

6.1 LECTURE 4

6.1.1 Sheaf Cohomology and All That (A Minimalist Approach)

1. We say that a sheaf of abelian groups I on a topological space X is *injective* if

   the abelian-group-valued functor on sheaves $\mathrm{Hom}(-, I)$ is exact.

   See [4, 10, 13, 11].

   Of course, the notion of injectivity makes sense in any abelian category, so we may speak of injective abelian groups, modules over a ring, etc.

2. **Exercise**

   (a) Verify that for every sheaf F, the functor $\mathrm{Hom}(-, F)$ is exact on one side (which one?), but, in general, not on the other.

   (b) The injectivity of I is equivalent to the following: for every injection $F \to G$ and every map $F \to I$ there is a map $G \to I$ making the "obvious" diagram (part of the exercise is to identify this diagram) commutative.

   (c) A short exact sequence $0 \to I \to A \to B \to 0$, with I injective, splits.

   (d) If $0 \to A \to B \to C \to 0$ is exact and A is injective, then B is injective if and only if C is.

   (e) A vector space over a field k is an injective k-module.

(f) **By** reversing the arrows, you can define the notion of *projectivity* (for sheaves, modules over a ring, etc.). Show that free implies projective.

3. It is a fact that every abelian group can be embedded into an injective abelian group. Obviously, this is true in the category of vector spaces!

4. **Exercise**

   Deduce from the embedding statement above that every sheaf F can be embedded into an injective sheaf. (Hint: consider the direct product sheaf $\Pi_{x \in X} F_x$ on X and work stalk by stalk using 3.)

5. By iteration of the embedding result established in Exercise 4, it is easy to show that given every sheaf F, there is an *injective resolution of F*, i.e., a long exact sequence

$$0 \longrightarrow F \xrightarrow{e} I^0 \xrightarrow{d^0} I^1 \xrightarrow{d^1} I^2 \xrightarrow{d^2} \cdots$$

   such that each I is injective.

6. The resolution is not unique, but it is so in the homotopy category. Let us not worry about this; see [4] (part of the work to be done by the young (at heart) reader, is to dig out the relevant statement from the references given here!). Under suitable assumptions, usually automatically verified when working with algebraic varieties, the injective resolution can be chosen to be bounded, i.e., $I^k = 0$, for $k \gg 0$; see [13].

7. Let $f : X \to Y$ be a continuous map of topological spaces and F be a sheaf on X.

   The *direct image sheaf $f_* F$* on Y is the sheaf

$$Y \overset{\text{open}}{\supseteq} U \longmapsto F(f^{-1}(U)).$$

   You should check that the above definition yields a sheaf, not just a presheaf.

8. A *complex of sheaves K* is a diagram of sheaves and maps of sheaves:

$$\cdots \longrightarrow K^i \xrightarrow{d^i} K^{i+1} \xrightarrow{d^{i+1}} \cdots$$

   with $d^2 = 0$.

   We have the *cohomology sheaves*

$$\mathcal{H}^i(K) := \operatorname{Ker} d^i / \operatorname{Im} d^{i-1};$$

   recall that everything is first defined as a presheaf and you must take the associated sheaf; the only exception is the kernel: the kernel presheaf of a map of sheaves is automatically a sheaf (check this).

   A *map of complexes $f : K \to L$* is a compatible system of maps $f^i : K^i \to L^i$. Compatible means that the "obvious" diagrams are commutative.

There are the induced maps of sheaves $\mathcal{H}^i(f) : \mathcal{H}^i(K) \to \mathcal{H}^i(L)$ for all $i \in \mathbb{Z}$.

A *quasi-isomorphism (qis)* $f : K \to L$ is a map inducing isomorphisms on all cohomology sheaves.

The *translated complex* $K[l]$ has $(K[l])^i := K^{l+i}$ with the same differentials (up to the sign $(-1)^l$).

Note that $K[1]$ means moving the entries one step to the *left* and taking $-d$.

An exact sequence of complexes is the "obvious" thing (make this explicit).

Later, I will mention *distinguished triangles*:

$$K \longrightarrow L \longrightarrow M \xrightarrow{+} K[1].$$

You can mentally replace this with a short exact sequence

$$0 \longrightarrow K \longrightarrow L \longrightarrow M \longrightarrow 0$$

and this turns out to be ok.

9. The *direct image complex* Rf_*F associated with (F, f) is "the" complex of sheaves on Y
$$Rf_*F := f_*I,$$
where $F \to I$ is an injective resolution as above.

This is well defined up to unique isomorphism in the homotopy category. This is easy to verify (check it). For the basic definitions and a proof of this fact see [4] (note that there are no sheaves in this reference, the point is the use of the properties of injective objects).

10. If C is a *bounded below* complex of sheaves on X, i.e., with $\mathcal{H}^i(K) = 0$ for all $i \ll 0$ (and we assume this from now on), then C admits a *bounded below injective resolution*, i.e., a qis $C \to I$, where each entry I^j is injective, and I is bounded below.

Again, this is well defined up to unique isomorphism in the homotopy category.

Rf_* is a "derived functor." However, this notion and the proof of this fact require plunging into the derived category, which we do not do in these notes. See [7].

11. We can thus define the *derived direct image complex* of a bounded below complex of sheaves C on X by first choosing a bounded below injective resolution $C \to I$ and then by setting
$$Rf_*C := f_*I;$$
this is a bounded below complex of sheaves on Y.

12. Define the *(hyper)cohomology groups of $(X$ with coefficients in) C* as follows:

- Take the unique map $c : X \to p$ (a point).

- Take the complex of global sections $Rc_*C = c_*I = I(X)$.

- Set (the right-hand side, being the cohomology of a complex of abelian groups, is an abelian group)

$$H^i(X, C) := H^i(I(X)).$$

13. **Exercise** (As mentioned earlier, from now on complexes are assumed to be bounded below.)

Use the homotopy statements to formulate and prove that these groups are well defined (typically, this means unique up to unique isomorphism; make this precise).

14. The *direct image sheaves on Y with respect to f* of the bounded below complex C of sheaves on X are

$$R^i f_*C := \mathcal{H}^i(Rf_*C) := \mathcal{H}^i(f_*I), \qquad i \in \mathbb{Z}.$$

These are well defined (see Exercise 13).

By boundedness, they are zero for $i \ll 0$ (depending on C).

If C is a sheaf, then they are zero for $i < 0$.

15. **Exercise**

Observe that if $C = F$ is a sheaf, then $R^0 f_*F = f_*F$ (as defined earlier).

Prove that the sheaf $R^i f_*C$ is the sheaf associated with the *presheaf*

$$U \longmapsto H^i\left(f^{-1}(U), C\right). \tag{6.1.1}$$

(See [11].) This fact is very important in order to build an intuition for higher direct images. You should test it against the examples that come to your mind (including all those appearing in these notes).

Note that even if C is a sheaf, then, in general, (6.1.1) above defines a presheaf. Give many examples of this fact.

Recall that while a presheaf and the associated sheaf can be very different, they have canonically isomorphic stalks! It follows that (6.1.1) can be used to try and determine the stalks of the higher direct image sheaves. Compute these stalks in many examples.

Remark that for every $y \in Y$ there is a natural map (it is called the base change map)

$$(R^i f_*C)_y \longrightarrow H^i(X_y, C_{|X_y}) \tag{6.1.2}$$

between the stalk of the direct image and the cohomology of the fiber $X_y := f^{-1}(y)$.

Give examples where this map is not an isomorphism/injective/surjective.

16. Given a sheaf G on Y, the *pull-back* f^*G is the sheaf associated with the *presheaf* (the limit below is the direct limit over the directed set of open sets $W \subseteq Y$ containing $f(U)$):

$$U \longmapsto \lim_{W \supseteq f(U)} G(W).$$

This presheaf is not a sheaf even when $f : X \to Y$ is the obvious map from a set with two elements to a set with one element (both with the discrete topology) and G is constant.

The pull-back defined above should not be confused with the pull-back of a quasi-coherent sheaf with respect to a map of algebraic varieties. (This is discussed very well in [11]).

In [8], you will find a very beautiful discussion of the étale space associated with a presheaf, and hence with a sheaf. This is all done in the general context of sheaves of sets; it is very worthwhile to study sheaves of sets, i.e., sheaves without the additional algebraic structures (sometimes the additional structure may hinder some of the basic principles).

The first, important surprise is that every map of sets yields a sheaf on the target: the sheaf of the local section of the map.

For example, a local homeomorphism, which can fail to be surjective (by way of contrast, the étale space of a sheaf of abelian groups on a space always surjects onto the space due to the obvious fact that we always have the zero section!) yields a sheaf on the target whose étale space is canonically isomorphic with the domain.

Ask yourself: can I view a 2:1 covering space as a sheaf? Yes, see above. Can I view the same covering as a sheaf of abelian groups? No, unless the covering is trivial (a sheaf of abelian groups always has the zero section!).

Whereas the definition of direct image f_*F is easy, the étale space of f_*F may bear very little resemblance to the one of F. On the other hand, while the definition of f^*G is a bit more complicated, the étale space $|f^*G|$ of f^*G is canonically isomorphic with the fiber product over Y of X with the étale space $|G|$ of G:

$$|f^*G| = |G| \times_Y X.$$

17. It is a fact that if I on X is injective, then f_*I on Y is injective.

A nice proof of this fact uses the fact that the pull-back functor f^* on sheaves is the left adjoint to f_*, i.e., (cf. [7])

$$\mathrm{Hom}(f^*F, G) = \mathrm{Hom}(F, f_*G).$$

18. **Exercise**

Use the adjunction property in (17) to prove that I injective implies f_*I injective.

Observe that the converse does not hold.

Observe that if J is injective on Y, then, in general, the pull-back f^*J is not injective on X.

Find classes of maps $f : X \to Y$ for which J injective on Y implies f^*J injective on X.

19. **Exercise**

Use that f_* preserves injectives to deduce that

$$H^i(X, C) = H^i(Y, Rf_*C).$$

20. It is a fact that on a good space X the cohomology defined above with coefficients in the constant sheaf $\mathbb{Z}_X$ is the same as the one defined by using singular and Cech cohomologies (see [14, 15]):

$$H^i(X, \mathbb{Z}_X) = H^i(X, \mathbb{Z}) = \check{H}^i(X, \mathbb{Z}).$$

21. **Exercise** (For a different perspective on what follows, see Lecture 1.)

(a) Let $j : \mathbb{R}^n - \{0\} \longrightarrow \mathbb{R}^n$ be the open immersion. Determine the sheaves $R^q j_* \mathbb{Z}$.

(b) (This is the very first occurrence of the decomposition theorem (DT) in these notes!) Let $X = Y = \mathbb{C}$, $X^* = Y^* = \mathbb{C}^*$, let $f : \mathbb{C} \to \mathbb{C}$ be the holomorphic map $z \mapsto z^2$, and let $g : \mathbb{C}^* \to \mathbb{C}^*$ be the restriction of f to $\mathbb{C}^* := \mathbb{C} - \{0\}$.

Show that $R^i f_* \mathbb{Z}_X = 0$ for all $i > 0$. Ditto for g.

Show that there is a split short exact sequence of sheaves of vector spaces (if you use $\mathbb{Z}$-coefficients, there is no splitting)

$$0 \longrightarrow \mathbb{Q}_Y \longrightarrow f_* \mathbb{Q}_X \longrightarrow Q \longrightarrow 0$$

and determine the stalks of Q.

Ditto for g and observe that what you obtain for g is the restriction to the open set Y^* of what you obtain for f on Y. (This is a general fact that you may find in the literature as "the base change theorem holds for an open immersion.")

The short exact sequence above, when restricted to Y^*, is one of locally constant sheaves (recall that a locally constant sheaf of abelian groups—you can guess the definition in the case of sheaves of sets with stalk a fixed set—with stalk a group L is a sheaf that is locally isomorphic to the constant sheaf with stalk L) and the restriction Q^* of Q to Y^* is the locally constant sheaf with stalk $\mathbb{Q}$ at a point $y \in Y^*$ endowed with the automorphism multiplication by -1 (explain what this must mean).

The locally constant sheaf Q^* on Y^* (and thus on the unit circle) is a good example of a nonconstant sheaf with stalk $\mathbb{Q}$, or $\mathbb{Z}$. Another good example is the sheaf of orientations of a nonorientable manifold: the stalk is a set given by two points; this is not a sheaf of groups! If the manifold is orientable, then the choice of an orientation turns the sheaf of sets into a locally constant sheaf of abelian groups with stalk $\mathbb{Z}/2\mathbb{Z}$.

(c) Show that on a good connected space X, a locally constant sheaf L (we often call such an object a *local system*) yields a representation of the fundamental group $\pi_1(X, x)$ in the group $A(L_x)$ of automorphisms of the stalk L_x at a pre-fixed point $x \in X$, and vice versa. (Hint: consider the quotient $(\tilde{X} \times L_x)/\pi_1(X, x)$ under a suitable action; here $\tilde{X}$ is a universal covering of X.)

(d) Use the principle of analytic continuation and the monodromy theorem (cf. [12]) to prove that every local system on a simply connected space is constant (trivial representation).

(e) Give an example of a local system that is not semisimple.

(The relevant definitions are simple := irreducible := no nontrivial subobject; semisimple := direct sum of simples.)

(Hint: consider, for example, the standard 2×2 unipotent matrix.)

The matrix in the hint given above is the one of the Picard–Lefschetz transformation associated with the degeneration of a one-parameter family of elliptic plane cubic curves to a rational cubic curve with a node; in other words it is the monodromy of the associated nontrivial! fiber bundle over a punctured disk with fiber $S^1 \times S^1$.

(f) Given a fiber bundle, e.g., a smooth proper map (see the Ehresmann fibration lemma, e.g., in [16]) $f : X \to Y$, with fiber X_y, prove that the direct image sheaf is locally constant with typical stalk

$$(R^i f_* \mathbb{Z}_X)_y = H^i(X_y, \mathbb{Z}).$$

(g) Show that the Hopf bundle $h : S^3 \to S^2$, with fiber S^1, is not (isomorphic to) a trivial bundle. Though the bundle is not trivial, the local systems $R^i h_* \mathbb{Z}_{S^3}$ are trivial on the simply connected S^2.

Do the same for $k : S^1 \times S^3 \to S^2$. Verify that you can turn the above into a proper holomorphic submersion of compact complex manifolds $k : S \to \mathbb{CP}^1$ (see the Hopf surface in [1]).

Show that the Deligne theorem (see Lecture 1) on the degeneration for smooth projective maps cannot hold for the Hopf map above. Deduce that this is an example of a map in complex geometry for which the decomposition theorem (DT) does not hold.

(h) Show that if a map f is proper and with finite fibers (e.g., a finite topological covering, a branched covering, the normalization of a complex space, for example of a curve, the embedding of a closed subvariety, etc.), then $R^i f_* F = 0$ for every $i > 0$ and every sheaf F.

Give explicit examples of finite maps and compute $f_* \mathbb{Z}$ in those examples.

22. **Some examples of maps $f : X \to Y$ to play with (some have already appeared above)**

   (a) $f : (0,1) \to [0,1]$:
   $f_* \mathbb{Z}_X = R f_* \mathbb{Z}_X = \mathbb{Z}_Y$;
   the base change map (6.1.2) is zero at $0 \in X$.

   (b) $f : \Delta^* \to \Delta$ (immersion of punctured unit disk into the unit disk in $\mathbb{C}$):
   $f_* \mathbb{Z}_X = \mathbb{Z}_Y$;
   $R^1 f_* \mathbb{Z}_X = \mathbb{Z}_o = H^1(X, \mathbb{Z}_X)$, $o \in \Delta$ the puncture;
   there is a nonsplit exact sequence

   $$0 \longrightarrow \mathbb{Z}_Y \longrightarrow R f_* \mathbb{Z}_X \longrightarrow \mathbb{Z}_o[-1] \longrightarrow 0.$$

   (c) $f : \Delta \to \Delta$, $z \mapsto z^2$:
   $R^0 f_* = f_*$; $R^i f_* = 0$; $f_* = R f_*$;
   the natural short exact sequence

   $$0 \longrightarrow R_Y \longrightarrow f_* R_X \longrightarrow Q(R) \longrightarrow 0$$

   does not split for $R = \mathbb{Z}$, but it splits if 2 is invertible in R.

   (d) $f : \Delta^* \to \Delta^*$, $z \mapsto z^2$:
   $R^0 f_* = f_*$; $R^i f_* = 0$; $f_* = R f_*$;
   the natural short exact sequence

   $$0 \longrightarrow R_Y \longrightarrow f_* R_X \longrightarrow Q(R) \longrightarrow 0$$

   does not split for $R = \mathbb{Z}$, but it splits if 2 is invertible in R;
   the stalk $Q(R)_p$ at $p := 1/4 \in \Delta^*$ (the target) is a rank-1 free R-module generated by the equivalence class $[(1, -1)]$ in $R^2/R = (f_* R_X)_p/(R_Y)_p$, modulo the equivalence relation $(a, b) \sim (a', b')$ if and only if $(a - a' = b - b')$; here (a, b) is viewed as a constant R-valued function in the preimage of a small connected neighborhood of p, this preimage being the disconnected union of two small connected neighborhoods of $\pm 1/2 \in \Delta^*$ (the domain);
   if we circuit once (e.g., counterclockwise) the origin of the target Δ^* starting at $1/4$ and returning to it, then the pair $(1, -1)$ is turned into the pair $(-1, 1)$; this is the monodromy representation on the stalk $Q(R)_p$;
   we see that in order to split $R_Y \to f_* R_X$, or equivalently, $f_* R_X \to Q(R)$, we need to be able to divide by 2.

   In this example and in the previous one, the conclusion of the decomposition theorem (DT) (see Section 6.2.1) holds, provided we use coefficients in a field of characteristic 0.

The DT already fails for integer and for $\mathbb{Z}/2\mathbb{Z}$ coefficients in these simple examples.

(e) $f : \mathbb{R} \to Y := \mathbb{R}/\sim$, where Y is obtained by identifying $\pm 1 \in \mathbb{R}$ to one point o (this can be visualized as the real curve $y^2 = x^2 - x^3$ inside $\mathbb{R}^2$, with o the origin):
$f_*\mathbb{Z}_X = Rf_*\mathbb{Z}_X; (f_*\mathbb{Z}_X)_o \simeq \mathbb{Z}^2;$
there is the natural nonsplit short exact sequence

$$0 \longrightarrow \mathbb{Z}_Y \longrightarrow f_*\mathbb{Z}_X \longrightarrow \mathbb{Z}_o \longrightarrow 0.$$

Let $j : U := Y - \{o\} \to Y$ be the open immersion $j_*\mathbb{Z} = Rj_*\mathbb{Z}; (j_*\mathbb{Z})_o \simeq \mathbb{Z}^4;$
there is the natural nonsplit short exact sequence

$$0 \longrightarrow \mathbb{Z}_Y \longrightarrow j_*\mathbb{Z}_U \longrightarrow \mathbb{Z}_o^3 \longrightarrow 0;$$

note that there is a natural nonsplit short exact sequence

$$0 \longrightarrow f_*\mathbb{Z}_X \longrightarrow j_*\mathbb{Z}_U \longrightarrow \mathbb{Z}_o^2 \longrightarrow 0. \qquad (6.1.3)$$

(f) $f : \mathbb{C} \to Y := \mathbb{C}/\sim$, where Y is obtained by identifying $\pm 1 \in \mathbb{C}$ to a point o and let $j : U = Y - \setminus\{o\} \to Y$ (this can be visualized as the complex curve $y^2 = x^2 - x^3$ inside $\mathbb{C}^2$, with o the origin):
this is analogous to the previous example, but it has an entirely different flavor:
$$Rf_*\mathbb{Z}_X = f_*\mathbb{Z} = j_*\mathbb{Z}_U. \qquad (6.1.4)$$

This is another example where the DT holds (in fact here it holds with $\mathbb{Z}$-coefficients).

(g) $f : S^3 \longrightarrow S^3$, the famous Hopf S^1-bundle; it is a map of real algebraic varieties for which the conclusion of Deligne's theorem Section 6.2.1(3) does not hold: we have the trivial local systems

$$R^0 f_*\mathbb{Z}_X = R^1 f_*\mathbb{Z}_X = \mathbb{Z}_Y, \qquad R^i f_*\mathbb{Z}_X = 0 \ \forall i \geq 2$$

and a nonsplit (even if we replace $\mathbb{Z}$ with $\mathbb{Q}$) short exact sequence

$$0 \longrightarrow \mathbb{Z}_Y \longrightarrow Rf_*\mathbb{Z}_X \longrightarrow \mathbb{Z}_Y[-1] \longrightarrow 0$$

(n.b. if it did split, then the first Betti number $0 = b_1(S^3) = 1!$).

(h) Consider the action of the group $\mathbb{Z}$ on $X' := \mathbb{C}^2 - \{(0,0)\}$ given by $(z, w) \mapsto (2z, 2w)$.
There are no fixed points and the (punctured complex lines through the origin, $w = mz$, are preserved).

One shows that $X := X'/\mathbb{Z}$ is a compact complex surface (a Hopf surface, see [1]) endowed with a proper holomorphic submersion (i.e., with differential everywhere of maximal rank) $f : X \to Y = \mathbb{CP}^1$.

After dividing by the $\mathbb{Z}$-action, each line $w = mz$ turns into a compact Riemann surface of genus 1, which in turn is the fiber $f^{-1}(m)$. Of course, $m = \infty$ corresponds to the line $z = 0$.

If we take the unit 3-sphere in $\mathbb{C}^2$, then, $f_{|S^3} : S^3 \to \mathbb{CP}^1 = S^2$ is the Hopf bundle of the previous example.

There is a natural filtration of $Rf_*\mathbb{Z}_X$:

$$0 = K^{-1} \subseteq K^0 \subseteq K^1 \subseteq K^2 = Rf_*\mathbb{Z}_X$$

into subcomplexes with

$$K^0/K^{-1} = \mathbb{Z}_Y, \quad K^1/K^0 = \mathbb{Z}_Y^2, \quad K^2/K_1 = \mathbb{Z}_Y.$$

As in the previous example, we cannot have a splitting

$$Rf_*\mathbb{Z}_Y \simeq \mathbb{Z}_Y \oplus \mathbb{Z}_Y^2[-1] \oplus \mathbb{Z}_Y$$

(not even replacing $\mathbb{Z}$ with $\mathbb{Q}$) in view of the fact that this would imply that $1 = b_1(X) = 2$.

This is an example of a proper holomorphic submersion, where the fibers and the target are projective varieties, but for which the conclusion of Deligne's theorem Section 6.2.1(3) does not hold.

(i) Let $C \subseteq \mathbb{CP}^2$ be a nonsingular complex algebraic curve (it is also a compact Riemann surface), let $\mathcal{U}$ be the universal holomorphic line bundle on $\mathbb{CP}^2$ (the fiber at a point is naturally the complex line parametrized by the point), let X be the complex surface total space of the line bundle $\mathcal{U}_{|C}$, let $Y \subseteq \mathbb{C}^3$ be the affine cone over C; it is a singular surface with an isolated point at the vertex (origin) $o \in Y$.

The blow-up of Y at the vertex coincides with X (check this).

Let $f : X \to Y$ be the natural map (it contracts the zero section of X).

Let $j : U := Y - \{o\} \to Y$ be the open immersion.

We have the first (for us) example of the DT for a nonfinite map (for details see [6]):

$$Rf_*\mathbb{Q} \simeq \tau_{\leq 1}Rj_*\mathbb{Q}_U \oplus \mathbb{Q}_o[-2] \qquad (6.1.5)$$

(given a complex K, its standard truncated subcomplex $\tau_{\leq i}K$ is the complex L with $L^j = K^j$ for every $j < i$, $L^i := \mathrm{Ker}\, d_K^i$, $K^j = 0$ for every $j > i$; its most important property is that it has the same cohomology sheaves $\mathcal{H}^j(L)$ as K for every $j \leq i$ and $\mathcal{H}^j(L) = 0$ for every $j > i$).

The most important aspect of the splitting (6.1.5) is that the right-hand side does not contain the symbol f denoting the map! This is in striking similarity with (6.1.4), another example of DT.

The relevant direct image sheaves for f are

$$R^0 f_* \mathbb{Q}_X = \mathbb{Q}_Y, \quad R^1 f_* \mathbb{Q}_X = \mathbb{Q}_o^{2g}, \quad R^2 f_* \mathbb{Q}_X = \mathbb{Q}_o$$

(the map is proper, the proper base change theorem holds, see [12], or [13], so that the base change map (6.1.2) is an iso).
The relevant direct image sheaves for j are

$$Rj_* \mathbb{Q}_U = \mathbb{Q}_Y, \quad R^1 j_* \mathbb{Q}_U = \mathbb{Q}_o^{2g}, \quad R^2 j_* \mathbb{Q}_U = \mathbb{Q}_o^{2g}, \quad R^3 j_* \mathbb{Q}_U = \mathbb{Q}_o;$$

this requires a fair amount of work (as a by-product, you will appreciate the importance of the base change theorem for proper maps, which you cannot use here!):
$j_* \mathbb{Q}_U = \mathbb{Q}_Y$ is because U is connected;
the computation on the higher $R^i j_* \mathbb{Q}_U$ boils down to determining the groups $H^i(U, \mathbb{Q}_U)$ (see (6.1.1));
on the other hand, $U \to C$ is the $\mathbb{C}^*$-bundle of the line bundle $\mathcal{U}_{|C}$ and this calculation is carried out in [3] (in fact, it is carried out for the associated oriented S^1-bundle) (be warned that [3] uses the Leray spectral sequence: this is a perfect chance to learn about it without being overwhelmed by the indices and by being shown very clearly how everything works; an alternative without spectral sequences is, for example, any textbook in algebraic topology covering the Wang sequence (i.e., the long exact sequence of an oriented S^1-bundle; by the way, it can be recovered using the Leray spectral sequence!).
Note that if we replace $\mathbb{Q}$ with $\mathbb{Z}$ we lose the splitting (6.1.5) due to torsion phenomena.
Note that there is a nonsplit short exact sequence

$$0 \longrightarrow \mathbb{Q}_Y \longrightarrow \tau_{\leq 1} Rj_* \mathbb{Q}_U \longrightarrow \mathbb{Q}_o^{2g} \longrightarrow 0.$$

A direct proof that this splitting cannot occur is a bit technical (omitted).
For us it is important to note that $\tau_{\leq 1} Rj_* \mathbb{Q}_U$ is the intersection complex of I_Y of Y (see Section 6.1.2) and intersection complexes I_Y never split nontrivially into a direct sum of complexes.

(j) $f : X = \mathbb{C} \times C \to Y$, where C is a compact Riemann surface as in the previous example, where Y is obtained from X by identifying $\Gamma := \{0\} \times C$ to a point $o \in Y$ and leaving the rest of X unchanged. Let $U := Y - \{o\} = X - \Gamma$.
Note that Γ defines the trivial class in $H^2(X, \mathbb{Z})$, because you can send it to infinity!, i.e., view it as the boundary of $\mathbb{R}^{\geq 0} \times C$.
The actual generator for $H^2(X, \mathbb{Z}) = H^2(C, \mathbb{Z})$ is given by the class of a complex line $\mathbb{C} \times c, c \in C$.
You should contrast what is above with the previous example given by the total space of a line bundle with negative degree. Of course, here X is the total space of the trivial line bundle on C.

The map f is not algebraic, not even holomorphic, in fact Y is not a complex space.

The DT cannot hold for f: the relevant cohomology sheaves for Rf_* are

$$f_*\mathbb{Q}_X = \mathbb{Q}_Y, \quad , R^1 f_*\mathbb{Q}_X = \mathbb{Q}_o^{2g}, \quad R^2 f_*\mathbb{Q} = \mathbb{Q}_o;$$

the relevant cohomology sheaves for $\tau_{\leq 1} Rj_*\mathbb{Q}_U$ are

$$j_*\mathbb{Q}_U = \mathbb{Q}_Y, \qquad R^1 j_*\mathbb{Q}_U = \mathbb{Q}_o^{2g+1};$$

it follows that (6.1.5), and hence the DT, do not hold in this case.

For more details and a discussion relating the first Chern classes of the trivial and of the negative line bundle to the DT see [6], which also explains (see also [5]) how to use Borel–Moore homology cycles to describe cohomology, as we have suggested above.

6.1.2   The Intersection Cohomology Complex

We shall limit ourselves to define and "calculate" the intersection complex I_X of a variety of dimension d with one isolated singularity:

$$Y = Y_{\text{reg}} \coprod Y_{\text{sing}}, \qquad U := Y_{\text{reg}}, \qquad Y_{\text{sing}} = \{p\},$$

$$U \xrightarrow{\ j\ } Y \xleftarrow{\ i\ } p.$$

This is done for ease of exposition only. Of course, the intersection cohomology complex I_Y, and its variants $I_Y(L)$ with twisted coefficients, can be defined for any variety Y, regardless of the singularities.

1. Recall that given a complex K the ath *truncated complex* $\tau_{\leq a} K$ is the subcomplex C with the following entries:

    $$C^b = K^b \ \forall b < a, \qquad C^a = \text{Ker } d^a, \qquad C^b = 0 \ \forall b > a.$$

    The single most important property is that

    $$\mathcal{H}^b(\tau_{\leq a} K) = \mathcal{H}^b(K) \qquad \forall b \leq a, \qquad \text{zero otherwise.}$$

2. Let Y be as above. Define the *intersection cohomology complex* (with coefficients in $\mathbb{Z}$, for example) as follows:

    $$I_Y := \tau_{\leq d-1} Rj_*\mathbb{Z}_U.$$

3. **Toy model**

    What follows is related to Section 6.1.1, Exercise 22(i).

    Let $Y \subseteq \mathbb{C}^3$ be the affine cone over an elliptic curve $E \subseteq \mathbb{CP}^2$.

$R^0 j_* \mathbb{Z}_U = \mathbb{Z}_Y$ (recall that we always have $R^0 f_* = f_*$).

As to the others we observe that U is the $\mathbb{C}^*$-bundle of the hyperplane line bundle H on E, i.e., the one induced by the hyperplane bundle on $\mathbb{CP}^2$. By choosing a metric, we get the unit sphere (here S^1) bundle U' over E. Note that U' and U have the same homotopy type. The bundle $U' \to E$ is automatically an oriented S^1-bundle. The associated Euler class $e \in H^2(E, \mathbb{Z})$ is the first Chern class $c_1(H)$.

4. **Exercise**

(You will find all you need in [3].) Use the spectral sequence for this oriented bundle (here it is just the Wang sequence) to compute the groups

$$H^i(U', \mathbb{Z}) = H^i(U, \mathbb{Z}).$$

Answer: (Caution: the answer below is for $\mathbb{Q}$-coefficients only!; work this situation out in the case of $\mathbb{Z}$-coefficients and keep track of the torsion.)

$$H^0(U) = H^0(E), \quad H^1(U) = H^1(E), \quad H^2(U) = H^1(E), \quad H^3(U) = H^2(E).$$

Deduce that, with $\mathbb{Q}$-coefficients (work out the $\mathbb{Z}$ case as well), we have that I_Y has only two nonzero cohomology sheaves

$$\mathcal{H}^0(I_Y) = \mathbb{Q}_Y, \qquad \mathcal{H}^1(I_Y) = H^1(E)_p \text{ (skyscraper at } p).$$

5. **Exercise**

Compute I_Y for $Y = \mathbb{C}^d$, with p the origin.

Answer: $I_Y = \mathbb{Q}_Y$ (here $\mathbb{Z}$-coefficients are ok).

6. The above result is general:

if Y is nonsingular, then $I_Y = \mathbb{Q}_Y$ ($\mathbb{Z}$ ok);

if Y is the quotient of a nonsingular variety by a finite group action, then $I_Y = \mathbb{Q}_Y$ ($\mathbb{Z}$ coefficients, KO!).

7. Let L be a local system on U. Define

$$I_Y(L) := \tau_{\leq d-1} Rj_* L.$$

Note that (this is a general fact)

$$\mathcal{H}^0(I_Y(L)) = j_* L.$$

8. Useful notation: $j_! L$ is the sheaf on Y which agrees with L on U and has stalk zero at p.

9. **Exercise**

(a) Let C be a singular curve. Compute I_C.

Answer: let $f : \hat{C} \to C$ be the normalization. Then $I_C = f_* \mathbb{Z}_{\hat{C}}$.

(b) Let things be as in Section 6.1.1, Exercise 21(b). Let $L = (f_* \mathbb{Z}_X)_{|Y^*}$ and $M := Q_{|Y^*}$. Compute

$$I_Y(L), \qquad I_Y(M).$$

(c) Let U be as in the toy model Exercise 3. Determine $\pi_1(U)$. Classify local systems of low ranks on U. Find some of their $I_Y(L)$'s.

(d) Let $f : C \to D$ be a branched cover of nonsingular curves. Let $f^o : C^o \to D^o$ be the corresponding topological covering space, obtained by removing the branching points and their preimages.

Prove that $L := f^o_* \mathbb{Q}_{C^o}$ is semisimple ($\mathbb{Z}$-coefficients is KO!, even for the identity: $\mathbb{Z}$ is not a simple $\mathbb{Z}$-module!).

Determine $I_D(L)$ and describe its stalks. (Try the case when C is replaced by a surface, threefold, etc.)

6.1.3   Verdier Duality

For ease of exposition, we work with rational coefficients.

1. Let M^m be an oriented manifold. We have Poincaré duality:

$$H^i(M, \mathbb{Q}) \simeq H_c^{m-i}(M, \mathbb{Q})^*. \tag{6.1.6}$$

2. **Exercise**

Find compact and noncompact examples of the failure of Poincaré duality for singular complex varieties.

(The easiest way to do this is to find nonmatching Betti numbers.)

3. Verdier duality (which we do not define here; see [7]) is the culmination of a construction that achieves the following generalization of Poincaré duality to the case of complexes of sheaves on locally compact spaces.

Given a complex of sheaves K on Y, its *Verdier dual* K^* is a canonically defined complex on Y such that for every open $U \subseteq Y$,

$$H^i(U, K^*) = H_c^{-i}(U, K)^*. \tag{6.1.7}$$

Note that $H_c^i(Y, K)$ is defined the same way as $H^i(Y, K)$, except that we take global sections with compact supports.

The formation of K^* is contravariantly functorial in K:

$$K \longrightarrow L, \qquad K^* \longleftarrow L^*,$$

and satisfies

$$K^{**} = K, \qquad (K[l])^* = K^*[-l].$$

4. **Exercise**

Recall the definition of the translation functor $[m]$ on complexes (see Section 6.1.1) and those of H^i and H^i_c and show directly that

$$H^i(Y, K[l]) = H^{i+l}(Y, K), \qquad H^i_c(Y, K[l]) = H^{i+l}_c(Y, K).$$

5. It is a fact that, for the oriented manifold M^m, the chosen orientation determines an isomorphism

$$\mathbb{Q}^*_Y = \mathbb{Q}_Y[m]$$

so that we get Poincaré duality. Verify this!; that is, verify that $(6.1.7) \Longrightarrow (6.1.6)$

(do not take it for granted, you will see what duality means over a point!).

If M is not oriented, then you get something else. See [3] (look for "densities"), see [13] (look for "sheaf of orientations"), see [12] (look for "Borel–Moore chains"), and the resulting complex of sheaves (see also [2]).

6. One of the most important properties of I_Y is its self-duality, which we express as follows (the translation by d is for notational convenience): first set

$$\mathrm{IC}_Y := I_Y[d]$$

(we have translated the complex I_Y, which had nonzero cohomology sheaves only in degrees $[0, d-1]$, to the *left* by d units, so that the corresponding interval is now $[-d, -1]$); then we have that

$$\mathrm{IC}^*_Y = \mathrm{IC}_Y.$$

7. **Exercise**

Use the toy model to verify that the equality holds (in that case) at the level of cohomology sheaves by verifying that (here V is a "typical" neighborhood of p)

$$\mathcal{H}^i(\mathrm{IC}_Y)_p = H^i(V, \mathrm{IC}_V) = H^{-i}_c(V, \mathrm{IC}_V)^*.$$

(To do this, you will need to compute $H^i_c(U)$ as you did $H^i(U)$; be careful though about using homotopy types and H_c!) You will find the following distinguished triangle useful—recall we can view them as short exact sequences, and as such, yielding a long exact sequence of cohomology groups, with or without supports:

$$\mathcal{H}^0(I_Y) \longrightarrow I_Y \longrightarrow \mathcal{H}^1(I_Y)[-1] \xrightarrow{+};$$

you will also find useful the long exact sequence

$$\cdots \longrightarrow H^a_c(U) \longrightarrow H^a_c(Y) \longrightarrow H^a_c(p) \longrightarrow H^{a+1}_c(U) \longrightarrow \cdots.$$

8. Define the *intersection cohomology groups of Y* as

$$\mathrm{IH}^i(Y) = H^i(Y, I_Y), \qquad \mathrm{IH}^i_c(Y) = H^i_c(Y, I_Y).$$

The original definition is more geometric and involves chains and boundaries, like in the early days of homology; see [2].

9. Since $\mathrm{IC}^*_Y = \mathrm{IC}_Y$, we get

$$H^i(Y, \mathrm{IC}_Y) = H^{-i}_c(Y, \mathrm{IC}_Y)^*.$$

Using $\mathrm{IC}_Y = I_Y[d]$, Verdier duality implies that

$$H^i(Y, I_Y) = H^{2n-i}_c(Y, I_Y)^*,$$

and we immediately deduce *Poincaré duality for intersection cohomology groups* on an arbitrarily singular complex algebraic variety (or complex space):

$$\mathrm{IH}^i(Y, I_Y) = \mathrm{IH}^{2d-i}_c(Y, I_Y)^*.$$

10. Variant for twisted coefficients.

If $Y^o \subseteq Y_{\mathrm{reg}} \subseteq Y$, L is a local system on a nonempty open set Y^o and L^* is the dual local system, then we have $I_Y(L)$, its translated $\mathrm{IC}_Y(L)$, and we have a canonical isomorphism

$$\mathrm{IC}_Y(L)^* = \mathrm{IC}_Y(L^*).$$

There is the corresponding duality statement for the groups $\mathrm{IH}^i(Y, I_Y(L))$, etc.:

$$\mathrm{IH}^i(Y, I_Y(L^*)) = \mathrm{IH}^{2d-i}_c(Y, I_Y(L))^*.$$

11. **Exercise**

Define the dual local system L^* of a local system L as the sheaf of germs of sheaf maps $L \to \mathbb{Q}_Y$.

(a) Show that it is a local system and that there is a pairing (map of sheaves)

$$L \otimes_{\mathbb{Q}_Y} L^* \longrightarrow \mathbb{Q}_Y$$

inducing identifications

$$(L_y)^* = (L^*)_y.$$

(Recall that the tensor product is defined by taking the sheaf associated with the presheaf tensor product (because of local constancy of all the players, in this case the presheaf is a sheaf): $U \mapsto L(U) \otimes_{\mathbb{Q}_U(U)} L^*(U)$.)

(b) If L is given by the representation $r : \pi_1(Y, y) \to A(L_y)$ (see Section 6.1.1, Exercise 21(c)), find an expression for a representation associated with L^*. (Hint: inverse–transpose.)

12. Verdier duality and Rf_* for a proper map.

It is a fact that if f is proper, then

$$(Rf_*C)^* = Rf_*(C^*).$$

We apply this to $\mathrm{IC}_Y(L)^* = \mathrm{IC}_Y(L^*)$ and get

$$(Rf_* \mathrm{IC}_Y(L))^* = Rf_* \mathrm{IC}_Y(L^*).$$

In particular, $Rf_* \mathrm{IC}_Y$ is self-dual.

6.2 LECTURE 5

6.2.1 The Decomposition Theorem (DT)

1. Let $f : X \to Y$ be a *proper* map of algebraic varieties and L be a *semisimple* (= direct sum of simples; simple = no nontrivial subobject) local system with $\mathbb{Q}$-coefficients (most of what follows fails with coefficients not in a field of characteristic 0)

on a Zariski dense open set $X^o \subseteq X_{\mathrm{reg}} \subseteq X$.

Examples include

- X is nonsingular, $L = \mathbb{Q}_X$, then $I_X(L) = I_X = \mathbb{Q}_X$;
- X is singular, $L = \mathbb{Q}_{X_{\mathrm{reg}}}$, then $I_X(L) = I_X$.

2. **Decomposition theorem**

The following statement is the deepest known fact concerning the homology of algebraic varieties.

There is a splitting in the derived category of sheaves on Y:

$$Rf_* I_X(L) \simeq \bigoplus_{b \in B} I_{\overline{Z_b}}(L_b)[l_b], \tag{6.2.1}$$

where

- *B is a finite set of indices;*
- *$Z_b \subseteq Y$ is a collection of locally closed nonsingular subvarieties;*
- *L_b is a semisimple local system on Z_b; and*
- *$l_b \in \mathbb{Z}$.*

What does it mean to have a splitting in the derived category?

Well, I did not define what a derived category is (and I will not). Still, we can deduce immediately from (6.2.1) that the intersection cohomology groups of the domain split into a direct sum of intersection cohomology groups on the target.

3. The case where we take $I_X = I_X(L)$ is already important.

Even if X and Y are smooth, we must deal with I_Z's on Y, i.e., we cannot have a direct sum of shifted sheaves for example.

Deligne's theorem (1968), including the semisimplicity statement (1972) for proper smooth maps of smooth varieties (see Lectures 1 and 2) is a special case and it reads as follows:

$$Rf_* \mathbb{Q}_X \simeq \bigoplus_{i \geq 0} R^i f_* \mathbb{Q}_X[-i], \qquad I_Y(R^i f_* \mathbb{Q}_X) = R^i f_* \mathbb{Q}_X.$$

4. **Exercise**

By using the self-duality of IC_Y, the rule $(K[l])^* = K^*[-l]$, the DT above, and the fact that $IC_T = I_T[\dim T]$, show that (6.2.1) can be rewritten in the following more symmetric form, where r is a uniquely determined nonnegative integer:

$$Rf_* IC_X \simeq \bigoplus_{i=-r}^{r} P^i[-i],$$

where each P^i is a direct sum of some of the $IC_{\overline{Z_b}}$ appearing above, *without translations* $[-]!$, and

$$(P^i)^* = P^{-i} \qquad \forall i \in \mathbb{Z}.$$

Try this first in the case of smooth proper maps, where $Rf_* \mathbb{Q}_X = \oplus R^i f_* \mathbb{Q}_X[-i]$. This may help to get used to the change of indexing scheme as you go from I_Y to $IC_Y = I_Y[d]$.

5. **Exercise**

(a) Go back to all the examples we met earlier and determine, in the cases where the DT is applicable, the summands appearing on the left of (6.2.1).

(b) (See [6].) Let $f : X \to C$ be a proper algebraic map with connected fibers, X a nonsingular algebraic surface, C a nonsingular algebraic curve.

Let C^o be the set of regular values, $\Sigma := C \setminus C^o$ (it is a fact that it is finite). Let $f^o : X^o \to C^o$ and $j : C^o \to C$ be the obvious maps.

Deligne's theorem applies to f^o and is a statement on C^o; show that it takes the following form:

$$Rf_*^o \mathbb{Q}_{X^o} \simeq \mathbb{Q}_{C^o} \oplus R^i f_*^o \mathbb{Q}_{X^o}[-1] \oplus \mathbb{Q}_{C^o}[-2].$$

Show that the DT on C must take the form (let $R^1 := R^1 f_*^o \mathbb{Q}_{X^o}$)

$$Rf_* \mathbb{Q}_X \simeq \mathbb{Q}_C \oplus j_* R^1[-1] \oplus \mathbb{Q}_C[-2] \oplus V_\Sigma[-2],$$

where V_Σ is the skyscraper sheaf on the finite set Σ with stalk at each $\sigma \in \Sigma$ a vector space V_σ of rank equal to the number of irreducible components of $f^{-1}(\sigma)$ *minus 1*.

Find a more canonical description of V_σ as a quotient of $H^2(f^{-1}(\sigma))$.

Note that this splitting contains quite a lot of information. Extract it:

- The only feature of $f^{-1}(\sigma)$ that contributes to $H^*(X)$ is its number of irreducible components; if this is 1, there is no contribution, no matter how singular (including multiplicities) the fiber is.
- Let $c \in C$, let Δ be a small disk around c, let $\eta \in \Delta^*$ be a regular value.

  We have the bundle $f^* : X_{\Delta^*} \to \Delta^*$ with typical fiber $X_\eta := f^{-1}(\eta)$. We have the (local) monodromy for this bundle; i.e., R^i is a local system; i.e., $\pi_1(\Delta^*) = \mathbb{Z}$ acts on $H^i(X_\eta)$.

  Denote by $R^{1\pi_1} \subseteq R^1_\eta$ the invariants of this (local) action.

  Show the following general fact: for local systems L on a good connected space Z and for a point $z \in Z$ we have that the invariants of the local system $L_z^{\pi_1(Z,z)} = H^0(Z, L)$.

  Let $X_c := f^{-1}(c)$ be the central fiber; there are the natural restriction maps

$$H^1(X_\eta) \supseteq H^1(X_\eta)^{\pi_1} \xleftarrow{r} H^1(f^{-1}(\Delta)) \xrightarrow{\simeq} H^1(X_c).$$

Use the DT above to deduce that r is *surjective*—this is the celebrated *local invariant cycle theorem*: all local invariant classes come from X_Δ; it comes *for free* from the DT.

Finally observe, that in this case, we indeed have $Rf_*\mathbb{Q}_X \simeq \oplus R^i f_*\mathbb{Q}[-i]$ (but you should view this as a coincidence due to the low dimension).

(c) Write down the DT for a projective bundle over a smooth variety.

(d) Ditto for the blowing up of a nonsingular subvariety of a nonsingular variety.

(e) Let Y be a threefold with an isolated singularity at $p \in Y$. Let $f : X \to Y$ be a resolution of the singularities of Y: X is nonsingular, f is proper and it is an isomorphism over $Y - \{p\}$.

  (i) Assume $\dim f^{-1}(p) = 2$; show, using the symmetries expressed by Exercise 4, that the DT takes the form

$$Rf_*\mathbb{Q}_X = IC_Y \oplus V_p[-2] \oplus W_p[-4],$$

  where $V_p \simeq W_p^*$ are skyscraper sheaves with dual stalks.
  Hint: use $H_4(X_p) \neq 0$ (why is this true?) to infer, using that $\mathcal{H}^4(I_X) = 0$, that one must have a summand contributing to $R^4 f_*\mathbb{Q}$.
  Deduce that the irreducible components of top dimension 2 of X_p yield linearly independent cohomology classes in $H^2(X)$.

  (ii) Assume $\dim f^{-1}(p) \leq 1$. Show that we must have

$$Rf_*\mathbb{Q}_X = I_Y.$$

Note that this is remarkable and highlights a general principle: the proper algebraic maps are restricted by the fact that the topology of Y, impersonated by I_Y, restricts the topology of X.

As we have seen in our examples to play with at the end of Section 6.1.1, there are no such general restriction in other geometries, e.g., proper C^∞ maps, proper real algebraic maps, proper holomorphic maps.

6.2.2 The Relative Hard Lefschetz and the Hard Lefschetz for Intersection Cohomology Groups

1. Let $f : X \to Y$ be a projective smooth map of nonsingular varieties and $\ell \in H^2(X, \mathbb{Q})$ be the first Chern class of a line bundle on X which is ample (Hermitian positive) on every X_y.

   We have the iterated cup product map (how do you make this precise?)

   $$\ell^i : R^j f_* \mathbb{Q}_X \longrightarrow R^{j+2i} f_* \mathbb{Q}_X.$$

   For every fiber $X_y := f^{-1}(y)$, we have the *hard Lefschetz theorem* ([9]) for the iterated cup product action of $\ell_y \in H^*(X_y, \mathbb{Q})$; let $d = \dim X_y$.

   The hard Lefschetz theorem on the fibers of the smooth proper map f implies at once that we have the isomorphisms of sheaves

   $$\ell^i : R^{d-i} f_* \mathbb{Q}_X \xrightarrow{\simeq} R^{d+i} f_* \mathbb{Q}_X \qquad (6.2.2)$$

   and we view this fact as the *relative hard Lefschetz theorem for smooth proper maps*.

2. In an earlier exercise, you were asked to find examples of the failure of Poincaré duality. It was suggested you find examples of (necessarily singular) complex projective varieties of complex dimension d for which one does not have the symmetry predicted by Poincaré duality: $b_{d-i} = b_{d+i}$, for every $i \in \mathbb{Z}$. Since the conclusion of the hard Lefschetz theorem yields the same symmetry for the Betti numbers, we see that for these same examples, the conclusion of the hard Lefschetz theorem does not hold.

   If the hard Lefschetz theorem does not hold for singular projective varieties, the sheaf-theoretic counterpart (6.2.2) cannot hold (why?) for an arbitrary proper map, even if the domain and target are nonsingular and the map is surjective (this is due to the singularities of the fibers).

   In short, the relative hard Lefschetz does not hold if formulated in terms of an isomorphism between direct image *sheaves*.

3. Recall the symmetric form of the DT (see Section 6.2.1, Exercise 4):

   $$Rf_* \mathrm{IC}_X \simeq \bigoplus_{i=-r}^{r} P^i[-i].$$

It is a formality to show that given a map $f : X \to Y$ and a cohomology class $\ell \in H^2(X, \mathbb{Q})$ we get iterated cup product maps

$$\ell^i : P^j \to P^{j+2i}.$$

The *relative hard Lefschetz theorem* (RHL) is the statement that if f is proper and if ℓ is the first Chern class of an ample line bundle on X, or at least ample on every fiber of f, then we have that the iterated cup product maps

$$\ell^i : P^{-i} \xrightarrow{\simeq} P^i \tag{6.2.3}$$

are isomorphisms for every $i \geq 0$.

In other words, the conclusion of the RHL (6.2.2) for smooth proper maps, expressed as an isomorphism of direct image sheaves, remains valid for arbitrary proper maps provided

- we push forward IC_X, i.e., we form $Rf_* \mathrm{IC}_X$, vs. $Rf_* \mathbb{Q}_X$ for which nothing so clean holds in general; and

- we consider the complexes P^i, instead of the direct image sheaves.

In the interest of perspective, let me add that the P^i are the so-called perverse direct image complexes of IC_X with respect to f and are special perverse sheaves on Y. The circle of ideas is now closed:

RHL is a statement about the perverse direct image complexes of $Rf_* \mathrm{IC}_X$!

Note that Verdier duality shows that $P^{-i} = (P^i)^*$. Verdier duality holds in general, outside of the realm of algebraic geometry and holds, for example for the Hopf surface map $h : S \to \mathbb{CP}^1$. In the context of complex geometry, the RHL, $\ell^i : P^{-i} \simeq P^i$, is a considerably deeper statement than Poincaré duality.

4. **Exercise**

(a) Make the statement of the RHL explicit in the example of a map from a surface to a curve (see Section 6.2.1, Exercise 5(b)).

(b) Ditto for Section 6.2.1, Exercise 5(e)(i). (Hint: in this case you get $\ell : V_p \simeq W_p$.)

Interpret geometrically, i.e., in terms of intersection theory, the isomorphism $i : V_p \simeq W_p^*$ (PD) and $l : V_p \simeq W_p$ (RHL).

(*Answer*: see [6]) let D_k be the fundamental classes of the exceptional divisors (which are the surfaces in X contracted to p); interpret W_p as (equivalence classes of) topological 2-cycles w; then i sends D_k to the linear map sending w to $D_k \cdot w \in H^6(X, \mathbb{Q}) \simeq \mathbb{Q}$; the map should be viewed as the operation of intersecting with a hyperplane section H on X and it sends D_k to the 2-cycle $D_k \cap H$. Now you can word out the conclusions of PD and RHL and appreciate them.)

5. The hard Lefschetz theorem on the intersection cohomology groups $\mathrm{IH}(Y, \mathbb{Q})$ of a projective variety X of dimension d. Let us apply RHL to the proper map $X \to$ point:

   let ℓ be the first Chern class of an ample line bundle on X of dimension d, then

   $$\ell^i : \mathrm{IH}^{d-i}(X, \mathbb{Q}) \overset{\simeq}{\longrightarrow} \mathrm{IH}^{d+i}(X, \mathbb{Q}).$$

6. Hodge–Lefschetz package for intersection cohomology.

   Let X be a projective variety. Then the statements (see [9] for these statements) of the two (hard and hyperplane section) Lefschetz theorems, of the primitive Lefschetz decomposition, of the Hodge decomposition and of the Hodge–Riemann bilinear relations hold for the rational intersection cohomology group of $\mathrm{IH}(X, \mathbb{Q})$.

7. **Exercise** (Compare what follows with the first part of Lecture 3.)

   Let $f : X \to Y$ be a resolution of the singularities of a projective surface with isolated singularities (for simplicity only; after you solve this, you may want to tackle the case when the singularities are not isolated).

   Show that the DT takes the form

   $$Rf_*\mathbb{Q}_X[2] = \mathrm{IC}_Y \oplus V_\Sigma,$$

   where Σ is the set of singularities of Y and V_Σ is the skyscraper with fiber $V_\sigma = H^2(X_\sigma)$ (here $X_\sigma := f^{-1}(\sigma)$).

   Deduce that the fundamental classes E_i of the curves given by the irreducible components in the fibers are linearly independent.

   Use Poincaré duality to deduce that the intersection form (cup product) matrix $||E_i \cdot E_j||$ on these classes is nondegenerate.

   (Grauert proved a general theorem, valid in the analytic context and for an analytic germ (Y, o) that even shows that this form is negative definite.)

   Show that the contribution $\mathrm{IH}^*(Y)$ to $H^*(X)$ can be viewed as the space orthogonal, with respect to the cup product, to the span of the E_i's.

   Deduce that $\mathrm{IH}^*(Y)$ sits inside $H^*(X, \mathbb{Q})$ compatibly with the Hodge decomposition of $H^*(X, \mathbb{C})$, i.e., $\mathrm{IH}^j(Y, \mathbb{Q})$ inherits a pure Hodge structure of weight j.

Bibliography

[1] W. Barth, C. Peters, A. Van de Ven. *Compact complex surfaces, Ergebnisse der Mathematik und ihrer Grenzgebiete* (3) 4. Springer, Berlin, 1984.

[2] A. Borel et al. *Intersection cohomology*, volume 50 of *Progress in Mathematics*. Birkhäuser, Boston-Basel-Stuttgart, 1984.

[3] R. Bott, L. Tu. *Differential forms in algebraic topology*, volume 82 of *Graduate Texts in Mathematics*. Springer, New York, 1982.

[4] H. Cartan, S. Eilenberg. *Homological algebra*. Princeton University Press, Princeton, 1999.

[5] M. A. de Cataldo. *The Hodge theory of projective manifolds*. Imperial College Press, London, 2007.

[6] M. A. de Cataldo, L. Migliorini. Intersection forms, topology of algebraic maps and motivic decompositions for resolutions of threefolds. In *Algebraic Cycles and Motives*, I, volume no. 343 of the London Math. Soc., pages 102–137, Cambridge University Press, Cambridge, UK, 2007.

[7] S. Gelfand, Y. I. Manin. *Methods of homological algebra*, second edition. Springer Monographs in Mathematics. Springer, Berlin, 2003.

[8] R. Godement. *Topologie algébrique et théorie des faisceaux*. Publications de l'Institut de Mathematique de l'Université de Strasbourg, XIII. Actualités Scientifiques et Industrielles, No. 1252. Hermann, Paris, 1973.

[9] P. Griffiths, J. Harris. *Principles of algebraic geometry*. Pure and Applied Mathematics. Wiley-Interscience, New York, 1978.

[10] A. Grothendieck. Sur quelques points d' algèbre homologique. The Tohoku Mathematical Journal, Second Series 9:119–221, 1957.

[11] R. Hartshorne. *Algebraic geometry*, volume 52 of *Graduate Texts in Mathematics*. Springer, New York, 1977.

[12] B. Iversen. *Cohomology of sheaves*. Universitext, Springer, Berlin, 1986.

[13] M. Kashiwara, P. Schapira. *Sheaves on manifolds*. With a chapter in French by Christian Houzel. Corrected reprint of the 1990 original. Volume 292 of the *Grundlehren der Mathematischen Wissenschaften*. Springer, Berlin, 1994.

[14] D. Mumford. *Algebraic geometry I: complex projective varieties*. Reprint of the 1976 edition. Classics in Mathematics. Springer, Berlin, 1995.

[15] E. Spanier. *Algebraic topology*. Corrected reprint of the 1966 original. Springer, New York, 1994.

[16] C. Voisin, *Hodge theory and complex algebraic geometry, I, II*, volumes 76 and 77 of *Cambridge Studies in Advanced Mathematics*. Cambridge University Press, Cambridge, English edition, 2007. Translated from the French by Leila Schneps.

Chapter Seven

Introduction to Variations of Hodge Structure
by Eduardo Cattani

INTRODUCTION

The modern theory of variations of Hodge structure (although some authors have referred to this period as the prehistory) begins with the work of Griffiths [21, 22, 23] and continues with that of Deligne [16, 17, 18] and Schmid [35]. The basic objects of study are *period domains* which parametrize the possible polarized Hodge structures in the cohomology of a given smooth projective variety. An analytic family of such varieties gives rise to a holomorphic map with values in a period domain, satisfying an additional system of differential equations. Moreover, period domains are homogeneous quasi-projective varieties and, following Griffiths and Schmid, one can apply Lie-theoretic techniques to study these maps.

These notes are not intended as a comprehensive survey of the theory of variations of Hodge structure (VHS). We refer the reader to the surveys [14, 23, 26, 2, 28], the collections [24, 1], and the monographs [3, 34, 38, 39] for fuller accounts of various aspects of the theory. In these notes we will emphasize the theory of *abstract* variations of Hodge structure and, in particular, their asymptotic behavior. The geometric aspects are the subject of Chapter 4 and the theory of variations of mixed Hodge structures is treated in Chapter 8.

In Section 7.1, we study the basic correspondence between local systems, representations of the fundamental group, and bundles with a flat connection. The second section is devoted to the study of analytic families of smooth projective varieties, the Kodaira–Spencer map, Griffiths' period map and a discussion of its main properties: holomorphicity and horizontality. These properties motivate the notion of an abstract VHS. In Section 7.3, we define the classifying spaces for polarized Hodge structures and study some of their basic properties. The last two sections deal with the asymptotics of a period mapping with particular attention to Schmid's orbit theorems. We emphasize throughout this discussion the relationship between nilpotent and SL_2-orbits and mixed Hodge structures.

In these notes I have often drawn from previous works in collaboration with Aroldo Kaplan, Wilfried Schmid, Pierre Deligne, and Javier Fernandez. I am very grateful to all of them.

7.1 LOCAL SYSTEMS AND FLAT CONNECTIONS

In this section we will collect some basic results about local systems and bundles with flat connections that play a central role in the notion of a variation of Hodge structure. We refer to [15] and [38, Section 9.2] for details.

7.1.1 Local Systems

We recall that the constant sheaf with stalk $\mathbb{C}^n$ over a topological space X is the sheaf of locally constant functions on X with values in $\mathbb{C}^n$.

DEFINITION 7.1.1 *A sheaf $\mathcal{L}$ over B is called a **local system** of $\mathbb{C}$-vector spaces if it is locally isomorphic to the constant sheaf with stalk $\mathbb{C}^n$ for a fixed n.*

If $\mathcal{L} \to B$ is a local system, and we fix a base point $b_0 \in B$, then for any curve $\gamma \colon [0, 1] \to B$, $\gamma(0) = b_0$, $\gamma(1) = b_1$, the pull-back $\gamma^*(\mathcal{L})$ to $[0, 1]$ is locally constant, hence constant. Thus we get a $\mathbb{C}$-vector space isomorphism

$$\tau^\gamma \colon \mathcal{L}_{b_1} \to \mathcal{L}_{b_0},$$

which depends only on the homotopy class of the path γ. Taking closed loops based at b_0, we get a map

$$\rho \colon \pi_1(B, b_0) \to \mathrm{GL}(\mathcal{L}_{b_0}) \cong \mathrm{GL}(n, \mathbb{C}).$$

It is easy to check that ρ is a group homomorphism and, consequently, it defines a representation of the fundamental group $\pi_1(B, b_0)$ on $\mathcal{L}_{b_0} \cong \mathbb{C}^n$. If B is connected, as we shall assume throughout, this construction is independent, up to conjugation, of the base point b_0.

Conversely, suppose $\rho \colon \pi_1(B, b_0) \to \mathrm{GL}(n, \mathbb{C})$ is a finite-dimensional representation and let $\mathfrak{p} \colon \tilde{B} \to B$ be the universal covering space of B. The fundamental group $\pi_1(B, b_0)$ acts on $\tilde{B}$ by covering (deck) transformations[1] and we may define a holomorphic vector bundle $\mathbb{V} \to B$ by

$$\mathbb{V} := \tilde{B} \times \mathbb{C}^n / \sim,$$

where the equivalence relation $\sim$ is defined as

$$(\tilde{b}, v) \sim (\sigma(\tilde{b}), \rho(\sigma^{-1})(v)); \quad \sigma \in \pi_1(B, b_0), \tag{7.1.1}$$

and the map $\mathbb{V} \to B$ is the natural projection from $\tilde{B}$ to B.[2] Suppose $U \subset B$ is an evenly covered open set in B, that is, $\mathfrak{p}^{-1}(U)$ is a disjoint union of open sets $W_j \subset \tilde{B}$ biholomorphic to U. Let us define $\mathfrak{p}_j = \mathfrak{p}|_{W_j}$. Then, given any $v \in \mathbb{C}^n$ we have, for any choice of j, a local section

$$\hat{v}(z) = [\mathfrak{p}_j^{-1}(z), v]; \quad z \in U$$

---

[1] Since the group action on $\pi_1(B, b_0)$ is defined by concatenation of loops, the action on the universal covering space is a right action.

[2] In other words, $\mathbb{V}$ is the vector bundle associated to the principal bundle $\pi_1(B, b_0) \to \tilde{B} \to B$ by the representation ρ.

on U. We call $\hat{v}$ a *constant section* of the bundle $\mathbb{V}$ and note that this notion is well defined since the transition functions of the bundle $\mathbb{V}$ take value on the discrete group $\rho(\pi_1(B, b_0))$. We denote by $\mathcal{L}$ the sheaf of constant local sections of $\mathbb{V}$. Clearly, $\mathcal{L}$ is a locally constant sheaf, i.e., a local system. As we shall see below, every local system arises in this way from a certain class of holomorphic vector bundles.

EXAMPLE 7.1.2 Let $B = \Delta^* := \{z \in \mathbb{C} : 0 < |z| < 1\}$. For $t_0 \in \Delta^*$ we have $\pi_1(\Delta^*, t_0) \cong \mathbb{Z}$, where we choose as generator a simple loop c oriented clockwise. Let

$$\rho \colon \pi_1(\Delta^*, t_0) \cong \mathbb{Z} \to \mathbb{C}^2 \, ; \quad \rho(n) = \begin{pmatrix} 1 & n \\ 0 & 1 \end{pmatrix} \in \mathrm{GL}(2, \mathbb{C}). \tag{7.1.2}$$

Recalling that the upper half plane $H = \{z = x + iy \in \mathbb{C} : y > 0\}$ is the universal covering space of Δ^* with projection $z \mapsto \exp(2\pi i z)$, we have a commutative diagram:

$$
\begin{array}{ccc}
H \times \mathbb{C}^2 & \longrightarrow & \mathbb{V} \cong H \times \mathbb{C}^2 / \sim \\
\mathrm{pr}_1 \downarrow & & \downarrow \\
H & \xrightarrow{\exp(2\pi i \bullet)} & \Delta^*.
\end{array}
$$

Let N be the nilpotent transformation

$$N = \begin{pmatrix} 0 & 1 \\ 0 & 0 \end{pmatrix}.$$

Then, for any $v \in \mathbb{C}^2$, the map $\tilde{v} \colon \Delta^* \to \mathbb{V}$ defined by

$$\tilde{v}(t) := \left[\frac{\log t}{2\pi i} \, , \, \exp\left(\frac{\log t}{2\pi i} N \right) \cdot v \right] \in H \times \mathbb{C}^2 / \sim \tag{7.1.3}$$

is a section of the vector bundle $\mathbb{V}$. Indeed, suppose we follow a determination of log around the generator loop c, then z_0 changes to $z_0 - 1$ (i.e., $\rho(c)(z_0) = z_0 - 1$), while the second component is modified by the linear map $\exp(\ N) = \rho(c^{-1})$ as required by (7.1.1). Note that, on a contractible neighborhood U of $t \in \Delta^*$, we can write

$$\tilde{v}(t) = \exp((\log t / 2\pi i)N) \cdot \hat{v}(t),$$

where $\hat{v}(t)$ is the constant section defined on U.

This example may be generalized to any nilpotent transformation $N \in \mathfrak{gl}(V)$ of a complex vector space V if we define

$$\rho \colon \pi_1(\Delta^*, t_0) \to \mathrm{GL}(V)$$

by $\rho(c) = \exp(N)$, where c is, again, a simple loop oriented clockwise, and to commuting nilpotent transformations $\{N_1, \ldots, N_r\} \in \mathfrak{gl}(V)$ by considering $B = (\Delta^*)^r$ and

$$\rho \colon \pi_1((\Delta^*)^r, t_0) \cong \mathbb{Z}^r \to \mathrm{GL}(V),$$

the representation that maps the jth standard generator of $\mathbb{Z}^r$ to $\gamma_j = \exp N_j$.

7.1.2   Flat Bundles

As we saw in the previous section, a local system over B gives rise to a representation of the fundamental group of B which, in turn, may be used to construct a vector bundle endowed with a subspace of "distinguished" constant sections isomorphic to the original local system. Here, we want to explore what is involved in the existence of this subspace of constant sections from the point of view of the bundle itself. We refer to [15, 29] for details.

Recall that a holomorphic connection on a holomorphic vector bundle $E \to B$ is a $\mathbb{C}$-linear map

$$\nabla \colon \mathcal{O}(U, E) \to \Omega^1(U) \otimes_{\mathcal{O}(U)} \mathcal{O}(U, E) := \Omega^1(U, E),$$

where $U \subset B$ is an open set and such that

$$\nabla(f \cdot \sigma) \; = \; df \otimes \sigma + f \cdot \nabla\sigma \,; \quad f \in \mathcal{O}(U), \; \sigma \in \mathcal{O}(U, E).$$

In terms of a local holomorphic frame $\sigma_1, \ldots, \sigma_d$ of $\mathcal{O}(U, E)$, we can write

$$\nabla\sigma_j \; = \; \sum_{i=1}^d \theta_{ij} \otimes \sigma_i.$$

The holomorphic forms $\theta_{ij} \in \Omega^1(U)$ are called *connection forms*.

DEFINITION 7.1.3   *Let $E \to B$ be a bundle with a connection ∇. A section σ in $\mathcal{O}(U, E)$ is said to be **flat** if $\nabla\sigma = 0$. The connection ∇ is called **flat** if there is a trivializing cover of B for which the corresponding frame consists of flat sections.*

A connection on a holomorphic line bundle $E \to B$ allows us to differentiate holomorphic (resp. smooth) sections of E in the direction of a holomorphic vector field X on $U \subset B$. Indeed, for U small enough and a frame $\sigma_1, \ldots, \sigma_d$ of $\mathcal{O}(U, E)$ we set

$$\nabla_X \left(\sum_{j=1}^d f_j \sigma_j \right) \; := \; \sum_{i=1}^d \left(X(f_i) + \sum_{j=1}^d f_j \, \theta_{ij}(X) \right) \sigma_i \,.$$

Clearly, if the coefficients f_j are holomorphic, so is the resulting section.

EXERCISE 7.1.4   Prove that the connection forms must satisfy the following *compatibility condition*: if $\sigma_1', \ldots, \sigma_d'$ is another frame on U and

$$\sigma_j' = \sum_{i=1}^d g_{ij} \sigma_i \,; \quad g_{ij} \in \mathcal{O}(U),$$

then

$$\sum_{i=1}^d g_{ji} \, \theta_{ik}' = dg_{jk} + \sum_{i=1}^d \theta_{ji} \, g_{ik}. \tag{7.1.4}$$

Deduce that if we define the matrices: $\theta = (\theta_{ij})$, $\theta' = (\theta_{ij}')$, $g = (g_{ij})$, $dg = (dg_{ij})$, then

$$\theta' \; = \; g^{-1} \cdot dg + g^{-1} \cdot \theta \cdot g.$$

EXERCISE 7.1.5 Let $L \to M$ be a line bundle and suppose that U_α is a trivializing cover of M with transition functions $g_{\alpha\beta} \in \mathcal{O}(U_\alpha \cap U_\beta)$. Prove that a connection on M is given by a collection of holomorphic 1-forms $\theta_\alpha \in \Omega^1(U_\alpha)$ such that

$$\theta_\beta|_{U_\alpha \cap U_\beta} - \theta_\alpha|_{U_\alpha \cap U_\beta} = d(\log g_{\alpha\beta}).$$

The curvature matrix of a holomorphic connection ∇ is defined as the matrix of holomorphic 2-forms

$$\Theta_{ij} = d\theta_{ij} - \sum_{k=1}^{d} \theta_{ik} \wedge \theta_{kj},$$

or, in matrix notation,

$$\Theta = d\theta - \theta \wedge \theta.$$

EXERCISE 7.1.6 With the notation of Exercise 7.1.4, prove that

$$\Theta' = g^{-1} \cdot \Theta \cdot g. \tag{7.1.5}$$

The curvature forms measure the "failure" of the connection to be flat:

THEOREM 7.1.7 *A connection is flat if and only if the curvature forms are identically zero.*

PROOF. We note, first of all, that (7.1.5) implies that the vanishing of the curvature forms is independent of the choice of frame. On the other hand, if ∇ is flat, we can find a trivializing cover where the connection forms and, therefore, the curvature forms vanish.

Suppose $(U, z_1, \ldots, z_n)$ is a coordinate neighborhood on B with a local frame $\sigma_1, \ldots, \sigma_d \in \mathcal{O}(U, E)$. Then, we may define coordinates $\{z_1, \ldots, z_n, \xi_1, \ldots, \xi_d\}$ on E. The forms

$$d\xi_i + \sum_{j=1}^{d} \xi_j \, \theta_{ij}$$

define a distribution of dimension n on E corresponding to *flat liftings*. The existence of an n-dimensional integral manifold is equivalent to the existence of a flat local frame. The distribution is involutive if and only if the curvature forms vanish. Thus, the result follows from the Frobenius theorem. We refer to [29, Proposition 2.5] or to [38, Section 9.2.1] for a full proof. □

Suppose now that a vector bundle $\mathbb{V} \to B$ arises from a local system $\mathcal{L}$ as before. Then, the bundle V has a trivializing cover $\{U_\alpha\}$, relative to which the transition functions are constant (since they take values in a discrete subgroup of $\mathrm{GL}(n, \mathbb{C})$), and it follows from (7.1.4) that the local forms $\theta_{ij} = 0$ define a connection on $\mathbb{V}$; that is, relative to the frame $\sigma_1^\alpha, \ldots, \sigma_d^\alpha$ arising from that trivializing cover, we may define

$$\nabla \left(\sum_{i=1}^{d} f_i \, \sigma_i^\alpha \right) = \sum_{i=1}^{d} df_i \otimes \sigma_i^\alpha.$$

Since the curvature forms for ∇ vanish, it follows that ∇ is flat.

Conversely, suppose $E \to B$ is a bundle with a flat connection ∇. Then the transition functions corresponding to a covering by open sets endowed with flat frames must be constant. Consequently we can define a local system of constant sections, i.e., the flat sections.

Summarizing the results of these two subsections we can say that there is an equivalence between the following three categories:

1. Local systems over a connected, complex manifold B

2. Finite-dimensional representations of the fundamental group $\pi_1(B, b_0)$

3. Holomorphic bundles $\mathbb{V} \to B$ with a flat connection ∇

7.2   ANALYTIC FAMILIES

We will be interested in considering families of compact Kähler manifolds or smooth projective varieties varying analytically. Specifically, we consider a map

$$\varphi \colon \mathcal{X} \to B,$$

where $\mathcal{X}$ and B are complex manifolds, and φ is a proper, holomorphic submersion; i.e., φ is surjective and, for every $x \in \mathcal{X}$, the differential

$$\varphi_{*,x} \colon T_x(\mathcal{X}) \to T_{\varphi(x)}(B)$$

is also surjective.

It follows from Theorem 1.1.14 that for each $b \in B$, the fiber $X_b := \varphi^{-1}(b)$ is a complex submanifold of $\mathcal{X}$ of codimension equal to the dimension of B. Moreover, since φ is proper, X_b is compact. We think of $\{X_b; \ b \in B\}$ as an analytic family of compact complex manifolds. The following theorem asserts that $\varphi \colon \mathcal{X} \to B$ is a C^∞ fiber bundle; i.e., it is locally a product:

THEOREM 7.2.1   *For every $b_0 \in B$ there exists a polydisk $\mathcal{U}$ centered at b_0 and a C^∞ map $F \colon \varphi^{-1}(\mathcal{U}) \subset \mathcal{X} \to \mathcal{U} \times X_{b_0}$ such that the diagram*

$$\varphi^{-1}(\mathcal{U}) \xrightarrow{\quad F \quad} \mathcal{U} \times X_{b_0}, \tag{7.2.1}$$

$$\varphi \searrow \qquad \swarrow \mathrm{pr}_1$$

$$\mathcal{U}$$

where pr_1 is the projection on the first factor, commutes. Moreover, for every $x \in X_{b_0}$ the map

$$\sigma_x \colon \mathcal{U} \to \mathcal{X} ; \quad \sigma_x(b) := F^{-1}(b, x), \tag{7.2.2}$$

is holomorphic.

Remark. For the first statement to hold, it suffices to assume that $\mathcal{X}, B$, and φ are smooth. In that context it is a classical result, the Ehresmann fibration lemma. In fact, the family φ trivializes over any contractible neighborhood of $b \in B$. We refer to [38, Theorem 9.3] for a complete proof of both statements.

Given a curve $\mu: [0, 1] \to B$ such that $\mu(0) = b_0$, $\mu(1) = b_1$, we may piece together the local trivializations given by Theorem 7.2.1 to get a diffeomorphism

$$f_\mu: X = X_{b_0} \to X_{b_1}.$$

This gives rise to an isomorphism $f_\mu^*: H^k(X_{b_1}, \mathbb{A}) \to H^k(X_{b_0}, \mathbb{A})$, where $\mathbb{A} = \mathbb{Z}, \mathbb{Q}, \mathbb{R}, \mathbb{C}$, which turns out to depend only on the homotopy class of μ. Thus, we get a representation of $\pi_1(B, b_0)$ on $H^k(X_{b_0}, \mathbb{A})$. We will denote by $\mathbb{H}^k \to B$ the holomorphic vector bundle associated with the representation of $\pi_1(B, b_0)$ on $H^k(X_{b_0}, \mathbb{C})$. The fiber of $\mathbb{H}^k$ over $b \in B$ is isomorphic to $H^k(X_b, \mathbb{C})$ and the flat connection is known as the *Gauss–Manin connection*. Given a trivialization $F: \varphi^{-1}(\mathcal{U}) \to \mathcal{U} \times X_{b_0}$ with inverse G, we get well-defined isomorphisms

$$g_b^*: H^k(X_{b_0}, \mathbb{C}) \to H^k(X_b, \mathbb{C})$$

and, for any $\alpha \in H^k(X_{b_0}, \mathbb{C})$, the map $\hat{\alpha}: \mathcal{U} \to \mathbb{H}^k$ defined by $b \mapsto g_b^*(\alpha)$ is a flat section of $\mathbb{H}^k$. We often refer to $\hat{\alpha}$ as the *parallel translate* of α.

Because of Theorem 7.2.1, the local system of flat sections agrees with is the kth *direct image sheaf* $R^k\varphi_*\mathbb{A}$. Indeed, we recall that $R^k\varphi_*\mathbb{A}$ is the sheaf associated with the presheaf that assigns to an open set U the cohomology $H^k(\varphi^{-1}(U), \mathbb{A})$. In our case, we may assume, without loss of generality, that the map

$$\mathrm{pr}_2 \circ F : \varphi^{-1}(U) \to X_{b_0}$$

deduced from (7.2.1) is a deformation retract. Hence, for U contractible,

$$H^k(\varphi^{-1}(U), \mathbb{A}) \cong H^k(X_{b_0}, \mathbb{A}).$$

For later use, we state a version of the Cartan–Lie formula in this context: Suppose Ω is a differential form of degree k on $\varphi^{-1}(\mathcal{U})$ such that its restriction to X_b is closed for all $b \in \mathcal{U}$. Then the map $b \in \mathcal{U} \mapsto [\Omega|_{X_b}] \in H^k(X_b, \mathbb{C})$ is a smooth section of the bundle $\mathbb{H}^k$ which we shall denote by ω.

PROPOSITION 7.2.2 ([38, Proposition 9.14]) *Let V be a vector field on $\mathcal{U}$ and $\hat{V}$ a vector field on $\mathcal{X}$ such that $\varphi_*(\hat{V}) = V$. Then*

$$\nabla_V \omega(t) = \left[\mathrm{int}_{\hat{V}}(d\Omega|_{X_t}) \right].$$

7.2.1 The Kodaira–Spencer Map

Let $\varphi : \mathcal{X} \to B$ be a family of compact Kähler manifolds. In what follows, we will assume that B is a polydisk and that we have chosen local coordinates $(t_1, \ldots, t_r)$ in B centered at 0. We set $X = X_0$. Let G denote the inverse of the diffeomorphism

$F\colon \mathcal{X} \to B \times X$. Even though for every $t \in B$, the fiber X_t is a complex submanifold of $\mathcal{X}$, the restriction

$$g_t := G|_{\{t\}\times X}\colon X \to X_t\,, \quad t \in B$$

is, generally, only a C^∞-diffeomorphism and carries the complex structure J_{X_t} to a $(1,1)$ tensor $J_t := g_t^*(J_{X_t})$ on X satisfying $J_t^2 = -\,\mathrm{id}$. Moreover, it follows from Theorem 1.2.5 that J_t is integrable. Thus, we may, alternatively, think of $\{X_t\}$ as a family of complex structures on a fixed C^∞ manifold X. The Kodaira–Spencer map to be defined below should be seen as measuring the "derivative" of $\{X_t\}$ at $t = 0$ [30].

Let TX, $T\mathcal{X}$, and TB denote the tangent bundles of X, $\mathcal{X}$, and B, respectively. Recalling that for each $x \in X \subset \mathcal{X}$,

$$T_x(X) = \ker\{\varphi_{*,x}\colon T_x\mathcal{X} \to T_0 B\},$$

we have an exact sequence of vector bundles over X:

$$0 \to TX \hookrightarrow T\mathcal{X}|_X \xrightarrow{\varphi_*^*} X \times T_0(B) \to 0.$$

On the other hand, the fact that φ is a submersion means that we also have an exact sequence of bundles over $\mathcal{X}$:

$$0 \to T_{\mathcal{X}/B} \to T\mathcal{X} \xrightarrow{\varphi_*^*} \varphi^*(TB) \to 0,$$

where $\varphi^*(TB)$ is the pull-back bundle defined in (1.1.20) and the *relative* bundle $T_{\mathcal{X}/B}$ is defined as the kernel of φ_*.

Since φ is holomorphic, these maps are compatible with the complex structures and, therefore, we get analogous exact sequences of holomorphic tangent bundles:

$$0 \to T^h X \hookrightarrow T^h\mathcal{X}|_X \xrightarrow{\varphi_*^*} X \times T_0^h(B) \to 0, \qquad (7.2.3)$$

$$0 \to T^h_{\mathcal{X}/B} \to T^h\mathcal{X} \xrightarrow{\varphi_*^*} \varphi^*(T^h B) \to 0. \qquad (7.2.4)$$

The sequence (7.2.3) gives rise to an exact sequence of sheaves of holomorphic sections and, consequently, to a long exact sequence in cohomology yielding, in particular, a map,

$$H^0(X, \mathcal{O}(X \times T_0^h(B))) \to H^1(X, \mathcal{O}(T^h X)),$$

where $\mathcal{O}(T^h(X))$ is the sheaf of holomorphic vector fields on X.

Since X is compact, any global holomorphic function is constant and, consequently,

$$H^0(X, \mathcal{O}(X \times T_0^h(B))) \cong T_0^h(B).$$

On the other hand, it follows from the Dolbeault theorem (see [27, Corollary 2.6.25]) that

$$H^1(X, \mathcal{O}(T^h X)) \cong H^{0,1}_{\bar\partial}(X, T^h X).$$

DEFINITION 7.2.3 *The map*

$$\rho\colon T_0^h(B) \to H^1(X, \mathcal{O}(T^h X)) \cong H_{\bar{\partial}}^{0,1}(X, T^h X) \tag{7.2.5}$$

*is called the **Kodaira–Spencer map** at $t = 0$.*

We may obtain a description of ρ using the map σ_x defined in (7.2.2). Indeed, for each $v \in T_0^h(B)$, let us denote by V the constant holomorphic vector field on B whose value at 0 is v. We may regard V as a holomorphic vector field on $B \times X$ and we define a C^∞ vector field Y_v on $\mathcal{X}$ by $Y_v = G_*(V)$. Note that for $x \in X, t \in B$,

$$Y_v(\sigma_x(t)) = (\sigma_x)_{*,t}(V) \tag{7.2.6}$$

and, therefore, Y_v is a vector field of type $(1, 0)$. Moreover,

$$\phi_*(Y_v(\sigma_x(t))) = V(t),$$

and, therefore, Y_v is the unique smooth vector field of type $(1, 0)$ on $\mathcal{X}$ projecting to V. In local coordinates $(U, \{z_1^U, \dots, z_n^U\})$ the vector field $Y_v|_X$ may be written as

$$Y_v(x) = \sum_{j=1}^{n} \nu_j^U \frac{\partial}{\partial z_j^U}. \tag{7.2.7}$$

Since the coordinate changes in $T^h X$ are holomorphic one can show the following result:

EXERCISE 7.2.4 The expression

$$\alpha_v = \sum_{j-1}^{n} \bar{\partial}(\nu_j^U) \otimes \frac{\partial}{\partial z_j^U}$$

defines a global $(0, 1)$-form on X with values on the holomorphic tangent bundle $T^h X$.

Following the steps involved in the proof of the Dolbeault isomorphism it is easy to check that $[\alpha_v]$ is the cohomology class in $H_{\bar{\partial}}^{0,1}(X, T^h X)$ corresponding to $\rho(v)$ in (7.2.5).

We will give a different description of the form α_v which motivates the definition of the Kodaira–Spencer map. As noted above, the family $\varphi\colon \mathcal{X} \to B$ gives rise to a family $\{J_t : t \in B\}$ of almost-complex structures on X. As we saw in Proposition A.1.2 of Chapter 1, such an almost-complex structure is equivalent to a splitting for each $x \in X$:

$$T_{x,\mathbb{C}}(X) = (T_x)_t^+ \oplus (T_x)_t^- ; \quad (T_x)_t^- = \overline{(T_x)_t^+},$$

and where $(T_x)_0^+ = T_x^h(X)$. If t is small enough we may assume that the projection of $(T_x)_t^-$ on $(T_x)_0^-$, according to the decomposition corresponding to $t = 0$, is surjective. Hence, in a coordinate neighborhood $(U, \{z_1^U, \dots, z_n^U\})$, there is a basis of the subspace $(T_x)_t^-$, $x \in U$ of the form

$$\frac{\partial}{\partial \bar{z}_k^U} - \sum_{j=1}^{n} w_{jk}^U(z, t) \frac{\partial}{\partial z_j^U} ; \quad k = 1, \dots, n.$$

Thus, the local expression

$$\sum_{j=1}^{n} w_{jk}^{U}(z,t)\, \frac{\partial}{\partial z_{j}^{U}}$$

describes, in local coordinates, how the almost-complex structure varies with $t \in B$. Now, given $v \in T_{0}^{h}(B)$ we can "differentiate" the above expression in the direction of v to get [3]

$$\sum_{j=1}^{n} v\left(w_{jk}^{U}(z,t)\right) \frac{\partial}{\partial z_{j}^{U}}.$$

We then have the following result whose proof may be found in [38, Section 9.1.2]:

PROPOSITION 7.2.5   *For each $v \in T_{0}^{h}(B)$ the expression*

$$\sum_{j=1}^{n} \bar{\partial}\left(v\left(w_{jk}(z,t)\right)\right) \otimes \frac{\partial}{\partial z_{j}^{U}}$$

defines a $\bar{\partial}$-closed, global $(0,1)$-form on X with values on the holomorphic tangent bundle $T^{h}X$, whose cohomology class in $H_{\bar{\partial}}^{0,1}(X, T^{h}X)$ agrees with the Kodaira–Spencer class $\rho(v)$.

7.3   VARIATIONS OF HODGE STRUCTURE

7.3.1   Geometric Variations of Hodge Structure

We consider a family $\varphi \colon \mathcal{X} \to B$ and assume that $\mathcal{X} \subset \mathbb{P}^{N}$ so that each fiber X_{t}, $t \in B$, is now a smooth projective variety.[4] The Chern class of the hyperplane bundle restricted to $\mathcal{X}$ induces integral Kähler classes $\omega_{t} \in H^{1,1}(X_{t}) \cap H^{2}(X_{t}, \mathbb{Z})$ which fit together to define a section of the local system $R^{2}\varphi_{*}\mathbb{Z}$ over B. On each fiber X_{t} we have a Hodge decomposition:

$$H^{k}(X_{t}, \mathbb{C}) = \bigoplus_{p+q=k} H^{p,q}(X_{t}),$$

where $H^{p,q}(X_{t})$ is the space of de Rham cohomology classes having a representative of bidegree (p, q), and

$$H^{p,q}(X_{t}) \cong H_{\bar{\partial}}^{p,q}(X_{t}) \cong H^{q}(X_{t}, \Omega_{X_{t}}^{p}),$$

where the last term is the cohomology of X_{t} with values on the sheaf of holomorphic p-forms $\Omega_{X_{t}}^{p}$.

THEOREM 7.3.1   *The Hodge numbers $h^{p,q}(X_{t}) = \dim_{\mathbb{C}} H^{p,q}(X_{t})$ are constant.*

---

[3] Note that in what follows we could as well assume $v \in T_{0}(B)$; i.e., v need not be of type $(1, 0)$.

[4] Much of what follows holds with weaker assumptions. Indeed, it is enough to assume that a fiber X_{t} is Kähler to deduce that it will be Kähler for parameters close to t. We will not deal with this more general situation here and refer to [38, Section 9.3] for details.

PROOF. Recall that $H_{\bar{\partial}}^{p,q}(X_t) \cong \mathcal{H}^{p,q}(X_t)$, the $\bar{\partial}$-harmonic forms of bidegree (p,q). The Laplacian $\Delta_{\bar{\partial}}^{X_t}$ varies smoothly with the parameter t and consequently the dimension of its kernel is upper semicontinuous on t. This follows from the ellipticity of the Laplacian [40, Theorem 4.13]. Hence,

$$\dim_{\mathbb{C}} \mathcal{H}^{p,q}(X_t) \leq \dim_{\mathbb{C}} \mathcal{H}^{p,q}(X_{t_0}),$$

for t in a neighborhood of t_0. But, on the other hand,

$$\sum_{p+q=k} \dim_{\mathbb{C}} \mathcal{H}^{p,q}(X_t) = b^k(X_t) = b^k(X_{t_0}) = \sum_{p+q=k} \dim_{\mathbb{C}} \mathcal{H}^{p,q}(X_{t_0})$$

since X_t is diffeomorphic to X_{t_0}. Hence, $\dim_{\mathbb{C}} \mathcal{H}^{p,q}(X_t)$ must be constant.   □

Recall from Definition A.4.4 in Chapter 1 that the Hodge decomposition on the cohomology $H^k(X_t, \mathbb{C})$ may be described by the filtration

$$F^p(X_t) := \bigoplus_{a \geq p} H^{a,k-a}(X_t), \tag{7.3.1}$$

which satisfies the condition $H^k(X_t, \mathbb{C}) = F^p(X_t) \oplus \overline{F^{k-p+1}(X_t)}$. We set $f^p = \sum_{a \geq p} h^{a,k-a}$. Assume now that B is contractible and that $\mathcal{X}$ is C^∞-trivial over B. Then we have diffeomorphisms $g_t : X = X_{t_0} \to X_t$ which induce isomorphisms

$$g_t^* : H^k(X_t, \mathbb{C}) \to H^k(X, \mathbb{C}).$$

This allows us to define a map[5]

$$\mathcal{P}^p : B \to \mathcal{G}(f^p, H^k(X, \mathbb{C})) ; \quad \mathcal{P}^p(t) = g_t^*(F^p(X_t)), \tag{7.3.2}$$

where $\mathcal{G}(f^p, H^k(X, \mathbb{C}))$ denotes the Grassmannian of f^p-dimensional subspaces of $H^k(X, \mathbb{C})$.

A theorem of Kodaira (see [38, Proposition 9.22]) implies that since the dimension is constant, the spaces of harmonic forms $\mathcal{H}^{p,q}(X_t)$ vary smoothly with t. Hence the map $\mathcal{P}^p$ is smooth. In fact we have the following theorem:

THEOREM 7.3.2 *The map $\mathcal{P}^p$ is holomorphic.*

PROOF. In order to prove Theorem 7.3.2 we need to understand the differential of the map $\mathcal{P}^p$. For simplicity we will assume that $B = \Delta = \{z \in \mathbb{C} : |z| < 1\}$, though the results apply with minimal changes in the general case. Suppose then that $\varphi : \mathcal{X} \to \Delta$ is an analytic family, $X = X_0 = \varphi^{-1}(0)$, and we have a trivialization

$$F : \mathcal{X} \to \Delta \times X.$$

---

[5] Voisin [38] refers to the map $\mathcal{P}^p$ as the *period* map. To avoid confusion we will reserve this name for the map that assigns to $t \in B$ the flag of subspaces $g_t^*(F^p(X_t))$, $p = 0, \ldots, k$.

Set $G = F^{-1}$ and denote by $g_t \colon X \to X_t = \varphi^{-1}(t)$ the restriction $G|_{\{t\} \times X}$. Then

$$\mathcal{P}^p(t) = F^p(t) := g_t^*(F^p(X_t)) \subset H^k(X, \mathbb{C}),$$

and its differential at $t = 0$ is a linear map:

$$\mathcal{P}^p_{*,0} \colon T_0(\Delta) \to \mathrm{Hom}(F^p(0), H^k(X, \mathbb{C})/F^p(0)). \tag{7.3.3}$$

The assertion that $\mathcal{P}^p$ is holomorphic at $t = 0$ is equivalent to the statement that

$$\mathcal{P}^p_{*,0} \left(\frac{\partial}{\partial \bar{t}} \Big|_{t=0} \right) = 0.$$

We now describe the map (7.3.3) explicitly. Let $\alpha \in F^p(0)$. Since the subbundle $\mathbb{F}^p \subset \mathbb{H}^k$ whose fiber over t is $F^p(X_t)$ is C^∞, we can construct a smooth section σ of $\mathbb{F}^p$ over $U \subset \Delta$, $0 \in U$, so that $\sigma(0) = \alpha$. Note that for $t \in U$,

$$\sigma(t) \in F^p(X_t) \subset H^k(X_t, \mathbb{C}),$$

and, consequently, we may view $g_t^*(\sigma(t))$ as a curve in $H^k(X, \mathbb{C})$ such that $g_t^*(\sigma(t)) \in F^p(t)$. Then,

$$\left(\mathcal{P}^p_{*,0} \left(\frac{\partial}{\partial \bar{t}} \Big|_{t=0} \right) \right)(\alpha) = \left[\frac{\partial g_t^*(\sigma(t))}{\partial \bar{t}} \Big|_{t=0} \right] \quad \mathrm{mod}\ F^p(0),$$

where $\partial/\partial \bar{t}|_{t=0}$ acts on the coefficients of the forms $g_t^*(\sigma(t))$. Alternatively, we may regard this action as the pull-back of the covariant derivative

$$\nabla_{\partial/\partial \bar{t}}(\sigma) \tag{7.3.4}$$

evaluated at $t = 0$.

Kodaira's theorem [38, Proposition 9.22] also means that we can realize the cohomology classes $\sigma(t)$ as the restriction of a global form

$$\Theta \in \bigoplus_{a \geq p} \mathcal{A}^{a, k-a}(\mathcal{X}) \tag{7.3.5}$$

such that $d(\Theta|_{X_t}) = 0$ (in fact we may assume that $\Theta|_{X_t}$ is harmonic) and

$$\sigma(t) = [\Theta|_{X_t}] \in H^k(X_t, \mathbb{C}).$$

Hence, by Proposition 7.2.2 we have

$$\frac{\partial g_t^*(\sigma(t))}{\partial \bar{t}} = \mathrm{int}_{\partial/\partial \bar{t}}(dG^*(\Theta)). \tag{7.3.6}$$

Now, in view of (7.3.5), we have that at $t = 0$, ψ restricts to a closed form whose cohomology class lies in $F^p(X)$ and, therefore, (7.3.6) vanishes modulo $F^p(X)$. $\square$

The interpretation of the differential of $\mathcal{P}^p$ in terms of the Gauss–Manin connection as in (7.3.4) and Proposition 7.2.2 allow us to obtain a deeper statement:

THEOREM 7.3.3 (Griffiths horizontality)   *Let* $\varphi \colon \mathcal{X} \to B$ *be an analytic family and let* $(\mathbb{H}^k, \nabla)$ *denote the holomorphic vector bundle with the (flat) Gauss–Manin connection. Let* $\sigma \in \Gamma(B, \mathbb{F}^p)$ *be a smooth section of the holomorphic subbundle* $\mathbb{F}^p \subset \mathbb{H}^k$. *Then, for any* $(1, 0)$ *vector field* V *on* B,

$$\nabla_V(\sigma) \in \Gamma(B, \mathbb{F}^{p-1}). \tag{7.3.7}$$

PROOF. Again, for simplicity, we consider the case $B = \Delta$. Then, arguing as in the proof of Theorem 7.3.2, we have

$$\nabla_{\frac{\partial}{\partial t}}(\sigma) = \frac{\partial}{\partial t}\left(g_t^*(\sigma(t))\right).$$

But then

$$\frac{\partial}{\partial t}\left(g_t^*(\sigma(t))\right) = \operatorname{int}_{\partial/\partial t}(dG^*(\Theta))|_{t=0} - d\phi|_{t=0}.$$

Since, clearly, the right-hand side lies in $F^{p-1}(0)$, the result follows.   □

Remark.   Given the Dolbeault isomorphism $H_{\bar{\partial}}^{p,q}(X) \cong H^q(X, \Omega^p)$, we can represent the differential of $\mathcal{P}^p$,

$$\mathcal{P}_{*,0}^p \colon T_0^h(B) \to \operatorname{Hom}(H^q(X_0, \Omega^p), H^{q+1}(X_0, \Omega^{p-1})),$$

as the composition of the Kodaira–Spencer map

$$\rho \colon T_0^h(B) \to H^1(X, T^h(X_0))$$

with the map

$$H^1(X, T^h(X_0)) \to \operatorname{Hom}(H^q(X_0, \Omega^p), H^{q+1}(X_0, \Omega^{p-1}))$$

given by interior product and the product in Čech cohomology. We refer to [38, Theorem 10.4] for details.

Given a family $\varphi \colon \mathcal{X} \to B$ with $\mathcal{X} \subset \mathbb{P}^N$, the Chern class of the hyperplane bundle restricted to $\mathcal{X}$ induces integral Kähler classes $\omega_t \in H^{1,1}(X_t) \cap H^2(X_t, \mathbb{Z})$ which fit together to define a section of the local system $R^2\varphi_*\mathbb{Z}$ over B. This means that cup product by powers of the Kähler classes is a flat morphism and, consequently, the restriction of the Gauss–Manin connection to the primitive cohomology remains flat. Similarly, the polarization forms are flat and they polarize the Hodge decompositions on each fiber $H_0^k(X_t, \mathbb{C})$.

7.3.2   Abstract Variations of Hodge Structure

The geometric situation described in Section 7.3.1 may be abstracted in the following definition:

DEFINITION 7.3.4   *Let B be a connected complex manifold. A **variation of Hodge structure** of weight k (VHS) over B consists of a local system $\mathcal{V}_{\mathbb{Z}}$ of free $\mathbb{Z}$-modules[6] and a filtration of the associated holomorphic vector bundle $\mathbb{V}$,*

$$\cdots \subset \mathbb{F}^p \subset \mathbb{F}^{p-1} \subset \cdots$$

by holomorphic subbundles $\mathbb{F}^p$ satisfying

1. $\mathbb{V} = \mathbb{F}^p \oplus \overline{\mathbb{F}^{k-p+1}}$ *as C^∞ bundles, where the conjugation is taken relative to the local system of real vector spaces $\mathcal{V}_{\mathbb{R}} := \mathcal{V}_{\mathbb{Z}} \otimes \mathbb{R}$.*

2. $\nabla(\mathcal{F}^p) \subset \Omega_B^1 \otimes \mathcal{F}^{p-1}$, *where ∇ denotes the flat connection on $\mathbb{V}$, and $\mathcal{F}^p$ denotes the sheaf of holomorphic sections of $\mathbb{F}^p$.*

We will refer to the holomorphic subbundles $\mathbb{F}^p$ as the *Hodge bundles* of the variation. It follows from Section A.4 in Chapter 1 that for each $t \in B$, we have a Hodge decomposition,

$$\mathbb{V}_t = \bigoplus_{p+q=k} \mathbb{V}_t^{p,q} \; ; \quad \mathbb{V}_t^{q,p} = \overline{\mathbb{V}_t^{p,q}}, \tag{7.3.8}$$

where $\mathbb{V}^{p,q}$ is the C^∞ subbundle of $\mathbb{V}$ defined by

$$\mathbb{V}^{p,q} = \mathbb{F}^p \cap \overline{\mathbb{F}^q}.$$

We will say that a VHS $(\mathbb{V}, \nabla, \{\mathbb{F}^p\})$ is *polarized* if there exists a flat nondegenerate bilinear form $\mathcal{Q}$ of parity $(-1)^k$ on $\mathbb{V}$, defined over $\mathbb{Z}$, such that for each $t \in B$ the Hodge structure on $\mathbb{V}_t$ is polarized, in the sense of Definition 1.5.14, by $\mathcal{Q}_t$.

We note that we can define a flat Hermitian form on $\mathbb{F}$ by $\mathcal{Q}^h(\cdot, \cdot) := i^{-k} \mathcal{Q}(\cdot, \overline{\cdot})$ making the decomposition (7.3.8) orthogonal and such that $(-1)^p \mathcal{Q}^h$ is positive definite on $\mathbb{V}^{p,k-p}$. The (generally not flat) positive-definite Hermitian form on $\mathbb{V}$,

$$\mathcal{H} := \sum_{p+q=k} (-1)^p \mathcal{Q}^h|_{\mathbb{V}^{p,q}},$$

is usually called the *Hodge metric* on $\mathbb{V}$.

We may then restate Theorems 7.3.2 and 7.3.3 together with the Hodge–Riemann bilinear relations as asserting that given a family $\varphi : \mathcal{X} \to B$ of smooth projective varieties, the holomorphic bundle whose fibers are the primitive cohomology $H_0^k(X_t, \mathbb{C})$, $t \in B$, endowed with the flat Gauss–Manin connection, carry a polarized Hodge structure of weight k.[7]

---

[6] In other contexts one only assumes the existence of a local system $\mathcal{V}_{\mathbb{Q}}$ (resp. $\mathcal{V}_{\mathbb{R}}$) of vector spaces over $\mathbb{Q}$ (resp. over $\mathbb{R}$) and refers to the resulting structure as a *rational* (resp. *real*) variation of Hodge structure.

[7] The restriction to the primitive cohomology is not actually necessary since we may modify the polarization form according to the Lefschetz decomposition to obtain a polarized Hodge structure on $H^k(X_t, \mathbb{C})$ and this construction gives rise to a variation as well.

7.4   CLASSIFYING SPACES

In analogy with the case of a family of projective varieties, we may regard a variation of Hodge structure as a family of Hodge structures on a fixed vector space V_{t_0}. This is done via parallel translation relative to the flat connection ∇ and the result is well defined modulo the action of the homotopy group $\pi_1(B, t_0)$.

In what follows we will fix the following data:

1. A lattice $V_{\mathbb{Z}}$: we will denote by $V_{\mathbb{A}} = V_{\mathbb{Z}} \otimes_{\mathbb{Z}} \mathbb{A}$, for $\mathbb{A} = \mathbb{Q}, \mathbb{R}$, or $\mathbb{C}$

2. An integer k

3. A collection of *Hodge numbers* $h^{p,q}$, $p + q = k$ such that $h^{p,q} = h^{q,p}$ and $\sum h^{p,q} = \dim_{\mathbb{C}} V_{\mathbb{C}}$; we set

$$f^p = \sum_{p \geq a} h^{a, k-a}$$

4. An integral, nondegenerate bilinear form Q, of parity $(-1)^k$

DEFINITION 7.4.1   *The space* $D = D(V_{\mathbb{Z}}, Q, k, \{h^{p,q}\})$ *consisting of all Hodge structures on* $V_{\mathbb{R}}$, *of weight* k *and Hodge numbers* $h^{p,q}$, *polarized by* Q *is called the* **classifying space of Hodge structures of weight** k **and Hodge numbers** $\{h^{p,q}\}$.

We will also consider the space $\check{D}$ consisting of all decreasing filtrations of $V_{\mathbb{C}}$:

$$\cdots \subset F^p \subset F^{p-1} \subset \cdots$$

such that $\dim_{\mathbb{C}} F^p = f^p$ and

$$Q(F^p, F^{k-p+1}) = 0. \tag{7.4.1}$$

We will refer to $\check{D}$ as the *dual* of D. It is easy to check that D is an open subset of $\check{D}$.

EXAMPLE 7.4.2   A Hodge structure of weight 1 is a complex structure on $V_{\mathbb{R}}$; that is, a decomposition $V_{\mathbb{C}} = \Omega \oplus \bar{\Omega}$, $\Omega \subset V_{\mathbb{C}}$. The polarization form Q is a nondegenerate alternating form and the polarization conditions reduce to

$$Q(\Omega, \Omega) = 0 ; \quad iQ(u, \bar{u}) > 0 \text{ if } 0 \neq u \in \Omega.$$

Hence, the classifying space for Hodge structures of weight 1 is the Siegel upper half-space defined in Example 1.1.23. The dual $\check{D}$ agrees with the space M defined in that same example. Geometrically, the weight-1 case corresponds to the study of the Hodge structure in the cohomology $H^1(X, \mathbb{C})$ for a smooth algebraic curve X. This example is discussed from that point of view in Chapter 4.

The space $\check{D}$ may be regarded as the set of points in the product of Grassmannians

$$\prod_{p=1}^{k} \mathcal{G}(f^p, V_{\mathbb{C}})$$

satisfying the flag compatibility conditions and the polynomial condition (7.4.1). Hence, $\check{D}$ is a projective variety. In fact we have the following theorem:

THEOREM 7.4.3 *Both D and $\check{D}$ are smooth complex manifolds. Indeed, $\check{D}$ is a homogeneous space $\check{D} \cong G_{\mathbb{C}}/B$, where $G_{\mathbb{C}} := \mathrm{Aut}(V_{\mathbb{C}}, Q)$ is the group of elements in $\mathrm{GL}(V, \mathbb{C})$ that preserve the nondegenerate bilinear form Q, and $B \subset G_{\mathbb{C}}$ is the subgroup preserving a given flag $\{F_0^p\} \in \check{D}$. The open subset D of $\check{D}$ is an orbit of the real group $G = \mathrm{Aut}(V_{\mathbb{R}}, Q)$ and $D \cong G/K$, where $K = G \cap B$ is a compact subgroup.*

PROOF. The smoothness of $\check{D}$ follows from the homogeneity statement and since D is open in $\check{D}$, it is smooth as well. Rather than give a complete proof, which may be found in [21, Theorem 4.3], we sketch the homogeneity argument in the cases of weight 1 and 2.

For $k = 1$, $\dim_{\mathbb{C}} V_{\mathbb{C}} = 2n$, and $G_{\mathbb{C}} \cong \mathrm{Sp}(n, \mathbb{C})$. The nondegeneracy of Q implies that given any n-dimensional subspace $\Omega \in V_{\mathbb{C}}$ such that $Q(\Omega, \Omega) = 0$ (i.e., a maximal isotropic subspace of $V_{\mathbb{C}}$), there exists a basis $\{w_1, \dots, w_{2n}\}$ of $V_{\mathbb{C}}$ such that $\{w_1, \dots, w_n\}$ is a basis of Ω and, in this basis, the form Q is

$$Q = \begin{pmatrix} 0 & -iI_n \\ iI_n & 0 \end{pmatrix}. \tag{7.4.2}$$

This shows that $G_{\mathbb{C}}$ acts transitively on $\check{D}$. On the other hand, if $\Omega_0 \in D$, then we can choose our basis so that $w_{n+i} = \bar{w}_i$ and, consequently, the group of real transformations $G \cong \mathrm{Sp}(n, \mathbb{R})$ acts transitively on D. The isotropy subgroup at some point $\Omega_0 \in D$ consists of real transformations in $\mathrm{GL}(V_{\mathbb{R}}) \cong \mathrm{GL}(2n, \mathbb{R})$ that preserve a complex structure and a Hermitian form in the resulting n-dimensional complex vector space. Hence $K \cong \mathrm{U}(n)$ and

$$D \cong \mathrm{Sp}(n, \mathbb{R})/\mathrm{U}(n).$$

In the weight-2 case, $\dim_{\mathbb{C}} V_{\mathbb{C}} = 2h^{2,0} + h^{1,1}$ and Q is a nondegenerate symmetric form defined over $\mathbb{R}$ (in fact, over $\mathbb{Z}$). The complex Lie group G_C is then isomorphic to $\mathrm{O}(2h^{2,0} + h^{1,1}, \mathbb{C})$. Given a polarized Hodge structure

$$V_{\mathbb{C}} = V_0^{2,0} \oplus V_0^{1,1} \oplus V_0^{0,2}; \quad V_0^{0,2} = \overline{V_0^{2,0}},$$

the real vector space $V_{\mathbb{R}}$ decomposes as

$$V_{\mathbb{R}} = \left((V_0^{2,0} \oplus V_0^{0,2}) \cap V_{\mathbb{R}} \right) \oplus \left(V_0^{1,1} \cap V_{\mathbb{R}} \right) \tag{7.4.3}$$

and the form Q is negative definite on the first summand and positive definite on the second. Hence $G \cong \mathrm{O}(2h^{2,0}, h^{1,1})$. On the other hand, the elements in G that fix the reference Hodge structure must preserve each summand of (7.4.3). In the first summand, they must, in addition, preserve the complex structure $V_0^{2,0} \oplus V_0^{0,2}$ and a (negative-)definite Hermitian form while, on the second summand, they must preserve a positive-definite real symmetric form. Hence,

$$K \cong \mathrm{U}(h^{2,0}) \times \mathrm{O}(h^{1,1}).$$

Clearly the connected component of G acts transitively as well. These arguments generalize to arbitrary weight. $\square$

EXERCISE 7.4.4 Describe the groups G and K for arbitrary even and odd weights.

The tangent bundles of the homogeneous spaces D and $\check{D}$ may be described in terms of Lie algebras. Let $\mathfrak{g}$ (resp. $\mathfrak{g}_0$) denote the Lie algebra of $G_{\mathbb{C}}$ (resp. G). Then

$$\mathfrak{g} = \{X \in \mathfrak{gl}(V_{\mathbb{C}}) : Q(Xu, v) + Q(u, Xv) = 0 \; \forall \; u, v \in V_C\},$$

and $\mathfrak{g}_0 = \mathfrak{g} \cap \mathfrak{gl}(V_{\mathbb{R}})$. The choice of a Hodge filtration $F_0 := \{F_0^p\}$ defines a filtration on $\mathfrak{g}$:

$$F^a \mathfrak{g} := \{X \in \mathfrak{g} : X(F_0^p) \subset F_0^{p+a}\}.$$

We may assume that $F_0 \in D$; in particular, $F^a\mathfrak{g}$ defines a Hodge structure of weight 0 on $\mathfrak{g}$:

$$\mathfrak{g}^{a,-a} := \{X \in \mathfrak{g} : X(V_0^{p,q}) \subset V_0^{p+a,q-a}\}.$$

Note that $[F^p\mathfrak{g}, F^q\mathfrak{g}] \subset F^{p+q}\mathfrak{g}$ and $[\mathfrak{g}^{a,-a}, \mathfrak{g}^{b,-b}] \subset \mathfrak{g}^{a+b,-a-b}$. The Lie algebra $\mathfrak{b}$ of B is the subalgebra $F^0\mathfrak{g}$ and the Lie algebra of K is given by

$$\mathfrak{v} = \mathfrak{g}_0 \cap \mathfrak{b} = \mathfrak{g}^{0,0} \cap \mathfrak{g}_0.$$

Since $\check{D} = G_{\mathbb{C}}/B$ and B is the stabilizer of F_0, the holomorphic tangent space of $\check{D}$ at F_0 is $\mathfrak{g}/\mathfrak{h}$, while the tangent space at any other point is obtained via the action of G. More precisely, the holomorphic tangent bundle of $\check{D}$ is a homogeneous vector bundle associated to the principal bundle

$$B \to G_{\mathbb{C}} \to \check{D}$$

by the adjoint action of B on $\mathfrak{g}/\mathfrak{b}$. In other words,

$$T^h(\check{D}) \cong \check{D} \times_B \mathfrak{g}/\mathfrak{b}.$$

Since $[F^0\mathfrak{g}, F^p\mathfrak{g}] \subset F^p\mathfrak{g}$, it follows that the adjoint action of B leaves invariant the subspaces $F^p\mathfrak{g}$. In particular, we can consider the homogeneous *horizontal* subbundle $T^{-1,1}(\check{D})$ of $T^h(\check{D})$ associated with the subspace $F^{-1}\mathfrak{g} = \mathfrak{b} \oplus \mathfrak{g}^{-1,1}$. Since $D \subset \check{D}$ is open, these bundles restrict to holomorphic bundles over D.

It will be useful to unravel the definition of the horizontal bundle. We may view an element of the fiber of $T^{-1,1}(\check{D})$ as the equivalence class of a pair $(F, [X])$, where $F \in \check{D}$ and $[X] \in \mathfrak{g}/\mathfrak{b}$. Then, if $F = g \cdot F_0$, we have that $\mathrm{Ad}(g^{-1})(X) \in F^{-1}\mathfrak{g}$, and, if we regard $\mathfrak{g}$ as a Lie algebra of endomorphisms of $V_{\mathbb{C}}$, this implies

$$(g^{-1} \cdot X \cdot g)F_0^p \subset F_0^{p-1}$$

or, equivalently,

$$X(F^p) \subset F^{p-1}, \tag{7.4.4}$$

where in (7.4.4), X must be thought of as the $G_{\mathbb{C}}$-invariant vector field on $\check{D}$ defined by $[X]$.

We may now define the period map of an abstract variation of Hodge structure. Consider a polarized variation of Hodge structure $(\mathbb{V}, \nabla, \mathcal{Q}, \{\mathbb{F}^p\})$ over a connected, complex manifold B and let $b_0 \in B$. Given a curve $\mu \colon [0,1] \to B$ with $\mu(0) = b_0$ and $\mu(1) = b_1$, we have a $\mathbb{C}$-linear isomorphism

$$\mu^* \colon \mathbb{V}_{b_1} \to \mathbb{V}_{b_0}$$

defined by parallel translation relative to the flat connection ∇. These isomorphisms depend on the homotopy class of μ and, as before, we denote by

$$\rho \colon \pi_1(B, b_0) \to \mathrm{GL}(\mathbb{V}_{b_0})$$

the resulting representation. We call ρ the *monodromy representation* and the image

$$\Gamma := \rho(\pi_1(B, b_0)) \subset \mathrm{GL}(\mathbb{V}_{b_0}, \mathbb{Z}) \qquad (7.4.5)$$

the *monodromy subgroup*. We note that since $\mathcal{V}_{\mathbb{Z}}$ and $\mathcal{Q}$ are flat, the monodromy representation is defined over $\mathbb{Z}$ and preserves the bilinear form $\mathcal{Q}_{b_0}$. In particular, since V is compact, the action of Γ on D is properly discontinuous and the quotient D/Γ is an analytic variety.

Hence, we may view the polarized Hodge structures on the fibers of $\mathbb{V}$ as a family of polarized Hodge structures on $\mathbb{V}_{b_0}$ well defined up to the action of the monodromy subgroup. That is, we obtain a map

$$\Phi \colon B \to D/\Gamma, \qquad (7.4.6)$$

where D is the appropriate classifying space for polarized Hodge structures. We call Φ the *period map* of the polarized VHS.

THEOREM 7.4.5 *The period map has local liftings to D which are holomorphic. Moreover, the differential takes values on the horizontal subbundle $T^{-1,1}(D)$.*

PROOF. This is just the statement that the subbundles $\mathbb{F}^p$ are holomorphic together with condition (2) in Definition 7.3.4. $\qquad\qquad\qquad\qquad\qquad\qquad\qquad\square$

We will refer to any locally liftable map $\Phi \colon B \to D/\Gamma$ with holomorphic and horizontal local liftings as a period map.

EXAMPLE 7.4.6 In the weight-1 case, $T^{-1,1}(D) = T^h(D)$. Hence, a period map is simply a locally liftable, holomorphic map $\Phi \colon B \to D/\Gamma$, where D is the Siegel upper half-space and $\Gamma \subset \mathrm{Sp}(n, \mathbb{Z})$ is a discrete subgroup. If $\tilde{B} \to B$ is the universal covering of B, we get a global lifting

$$
\begin{array}{ccc}
\tilde{B} & \overset{\tilde{\Phi}}{\longrightarrow} & D \\
\downarrow & & \downarrow \\
B & \overset{\Phi}{\longrightarrow} & D/\Gamma
\end{array}
$$

and the map $\tilde{\Phi}$ is a holomorphic map with values in Siegel's upper half-space.

EXERCISE 7.4.7 Describe period maps in the weight-2 case.

7.5   MIXED HODGE STRUCTURES AND THE ORBIT THEOREMS

In the remainder of this chapter we will be interested in studying the asymptotic behavior of a polarized variation of Hodge structure of weight k. Geometrically, this situation arises when we have a family of smooth projective varieties $\mathcal{X} \to B$, where B is a quasi-projective variety defined as the complement of a divisor with normal crossings Y in a smooth projective variety $\bar{B}$. Then, locally on the divisor Y, we may consider the polarized variation of Hodge structure defined by the primitive cohomology over an open set $U \subset \bar{B}$ such that $U = \Delta^n$ and

$$U \cap Y = \{z \in \Delta^n : z_1 \cdots z_r = 0\}.$$

This means that

$$U \cap B = (\Delta^*)^r \times \Delta^{n-r}.$$

Thus, we will consider period maps

$$\Phi \colon (\Delta^*)^r \times \Delta^{n-r} \to D/\Gamma \tag{7.5.1}$$

and their liftings to the universal cover,

$$\tilde{\Phi} \colon H^r \times \Delta^{n-r} \to D,$$

where $H = \{z \in \mathbb{C} : \mathrm{Im}(z) > 0\}$ is the universal covering space of Δ^* as in Example 7.1.2. The map $\tilde{\Phi}$ is then holomorphic and horizontal.

We will denote by $c_1, \ldots, c_r$ the generators of $\pi_1((\Delta^*)^r)$; i.e., c_j is a clockwise loop around the origin in the jth factor Δ^*. Let $\gamma_j = \rho(c_j)$. Clearly the monodromy transformations γ_j, $j = 1, \ldots, r$, commute. We have the following theorem:

THEOREM 7.5.1 (Monodromy theorem)   *The monodromy transformations γ_j, for $j = 1, \ldots, r$, are quasi-unipotent; that is, there exist integers ν_j such that $(\gamma_j^{\nu_j} - \mathrm{id})$ is nilpotent. Moreover, the index of nilpotency of $(\gamma_j^{\nu_j} - \mathrm{id})$ is at most $k + 1$.*

PROOF.   In the geometric case this result is due to Landman [31]. The proof for (integral) variations of Hodge structure is due to Borel (cf. [35, (4.5)]). The statement on the index of nilpotency is proved in [35, (6.1)].   □

7.5.1   Nilpotent Orbits

For simplicity we will often assume that $r = n$. This will, generally, entail no loss of generality as our statements will usually hold uniformly on compact subsets of Δ^{n-r} but this will be made precise when necessary. We will also assume that the monodromy transformations γ_j are actually *unipotent*, that is, $\nu_j = 1$. This may be accomplished by lifting the period map to a finite covering of $(\Delta^*)^r$. We point out that most of the results that follow hold for real variations of Hodge structure, provided that we assume that the monodromy transformations are unipotent. In what follows we will write

$$\gamma_j = e^{N_j}; \quad j = 1, \ldots, r, \tag{7.5.2}$$

where N_j are nilpotent elements in $\mathfrak{g} \cap \mathfrak{gl}(V_{\mathbb{Q}})$ such that $N^{k+1} = 0$. We then have

$$\tilde{\Phi}(z_1, \ldots, z_j + 1, \ldots, z_r) = \exp(N_j) \cdot \tilde{\Phi}(z_1, \ldots, z_j, \ldots, z_r),$$

and the map $\Psi \colon H^r \to \check{D}$ defined by

$$\Psi(z_1, \ldots, z_r) := \exp\left(-\sum_{j=1}^r z_j N_j\right) \cdot \tilde{\Phi}(z_1, \ldots, z_r) \tag{7.5.3}$$

is the lifting of a holomorphic map $\psi \colon (\Delta^*)^r \to D$ so that

$$\psi(t_1, \ldots, t_r) = \Psi\left(\frac{\log t_1}{2\pi i}, \ldots, \frac{\log t_r}{2\pi i}\right). \tag{7.5.4}$$

EXAMPLE 7.5.2 Let $F_0 \in \check{D}$ and let $N_1, \ldots, N_r$ be commuting elements in $\mathfrak{g} \cap \mathfrak{gl}(V_{\mathbb{Q}})$ such that

$$N_j(F_0^p) \subset F_0^{p-1}. \tag{7.5.5}$$

Then the map

$$\theta \colon H^r \to \check{D}; \quad \theta(z_1, \ldots, z_r) = \exp\left(\sum_{j=1}^r z_j N_j\right) \cdot F_0$$

is holomorphic and, because of (7.5.5) and (7.4.4), its differential takes values on the horizontal subbundle. Hence, if we assume that there exists $\alpha > 0$ such that

$$\theta(z_1, \ldots, z_r) \in D; \quad \text{for} \quad \text{Im}(z_j) > \alpha,$$

the map θ is the lifting of a period map defined on a product of punctured disks Δ_ε^* of radius $\varepsilon > 0$. Such a map will be called a *nilpotent orbit*. Note that for a nilpotent orbit the map (7.5.3) is constant, equal to F_0.

THEOREM 7.5.3 (Nilpotent orbit theorem)   *Let $\Phi \colon (\Delta^*)^r \times \Delta^{n-r} \to D$ be a period map and let $N_1, \ldots, N_r$ be the monodromy logarithms. Let*

$$\psi \colon (\Delta^*)^r \times \Delta^{n-r} \to \check{D}$$

be defined in a manner analogous to (7.5.4). Then

1. *the map ψ extends holomorphically to $\Delta^r \times \Delta^{n-r}$;*

2. *for each $w \in \Delta^{n-r}$, the map $\theta \colon H^r \to \check{D}$ given by*

$$\theta(z, w) = \exp\left(\sum z_j N_j\right) . \psi(0, w)$$

   is a nilpotent orbit; moreover, if $C \subset \Delta^{n-r}$ is compact, there exists $\alpha > 0$ such that $\theta(z, w) \in D$ for $\text{Im}(z_j) > \alpha$, $1 \leq j \leq n$, $w \in C$;

3. *for any G-invariant distance d on D, there exist positive constants β, K such that, for* $\mathrm{Im}(z_j) > \alpha$,

$$d(\Phi(z,w), \theta(z,w)) \leq K \sum_j (\mathrm{Im}(z_j))^\beta \, e^{-2\pi \mathrm{Im}(z_j)} \, .$$

Moreover, the constants α, β, K depend only on the choice of d and the weight and Hodge numbers used to define D and may be chosen uniformly for w in a compact subset $C \subset \Delta^{n-r}$.

PROOF. The proof of Theorem 7.5.3, which is due to Wilfried Schmid [35], hinges upon the existence of G-invariant Hermitian metrics on D, whose holomorphic sectional curvatures along horizontal directions are negative and bounded away from zero [25]. We refer the reader to [26] for an expository account and to [36] for an enlightening proof in the case when D is Hermitian symmetric;[8] the latter is also explicitly worked out in [4] for VHS of weight 1. We should remark that the distance estimate in (3) is stronger than that in Schmid's original version [35, (4.12)] and is due to Deligne (see [11, (1.15)] for a proof).   □

The nilpotent orbit theorem may be interpreted in the context of Deligne's *canonical extension* [15]. Let $V \;\rangle\; (\Delta^*)^r \times \Delta^{n-r}$ be the flat bundle underlying a polarized VHS and pick a base point (t_0, w_0). Given $v \in V := V_{(t_0,w_0)}$, let $v^\flat$ denote the multivalued flat section of V defined by v. Then

$$\tilde{v}(t,w) := \exp\left(\sum_{j=1}^r \frac{\log t_j}{2\pi i} N_j \right) \cdot v^\flat(t,w) \tag{7.5.6}$$

is a global section of V. The canonical extension $\overline{V} \to \Delta^n$ is characterized by its being trivialized by sections of the form (7.5.6). The nilpotent orbit theorem is then equivalent to the regularity of the Gauss–Manin connection and implies that the Hodge bundles $\mathbb{F}^p$ extend to holomorphic subbundles $\overline{\mathbb{F}^p} \subset \overline{V}$. Writing the Hodge bundles in terms of a basis of sections of the form (7.5.6) yields the holomorphic map Ψ. Its constant part—corresponding to the nilpotent orbit—defines a polarized VHS as well. The connection ∇ extends to a connection on Δ^n with logarithmic poles along the divisor $\{t_1 \cdots t_r = 0\}$ and nilpotent residues.

Given a period map $\Phi \colon (\Delta^*)^r \to D/\Gamma$, we will call the value

$$F_{\mathrm{lim}} := \psi(0) \in \check{D}$$

the *limit Hodge filtration*. Note that F_{lim} depends on the choice of coordinates in $(\Delta^*)^r$. Indeed, suppose $(\hat{t}_1, \ldots, \hat{t}_r)$ is another local coordinate system compatible with the divisor structure. Then, after relabeling if necessary, we must have $(\hat{t}_1, \ldots, \hat{t}_r) =$

---

[8]In this case the nilpotent orbit theorem follows from the classical Schwarz lemma.

$(t_1 f_1(t), \ldots, t_r f_r(t))$, where f_j are holomorphic around $0 \in \Delta^r$ and $f_j(0) \neq 0$. We then have from (7.5.4),

$$\hat{\psi}(\hat{t}) = \exp\left(-\frac{1}{2\pi i}\sum_{j=1}^{r}\log(\hat{t}_j)N_j\right) \cdot \Phi(\hat{t})$$

$$= \exp\left(-\frac{1}{2\pi i}\sum_{j=1}^{r}\log(f_j)N_j\right)\exp\left(-\frac{1}{2\pi i}\sum_{j=1}^{r}\log(t_j)N_j\right) \cdot \Phi(t) \quad (7.5.7)$$

$$= \exp\left(-\frac{1}{2\pi i}\sum_{j=1}^{r}\log(f_j)N_j\right) \cdot \Psi(t),$$

and, letting $t \to 0$,

$$\hat{F}_{\lim} = \exp\left(-\frac{1}{2\pi i}\sum_{j}\log(f_j(0))N_j\right) \cdot F_{\lim}. \quad (7.5.8)$$

7.5.2 Mixed Hodge Structures

We will review some basic notions about mixed Hodge structures which will be needed in the asymptotic description of variations of Hodge structure. Chapter 3 contains an expanded account of the theory of mixed Hodge structures and the book by Peters and Steenbrink [34] serves as a comprehensive reference. We will mostly follow the notation introduced in [11].

DEFINITION 7.5.4 *Let $V_{\mathbb{Q}}$ be a vector space over $\mathbb{Q}$, $V_{\mathbb{R}} = V_{\mathbb{Q}} \otimes \mathbb{R}$, and $V_{\mathbb{C}} = V_{\mathbb{Q}} \otimes \mathbb{C}$. A mixed Hodge Structure (MHS) on $V_{\mathbb{C}}$ consists of a pair of filtrations of V, (W, F), where W is increasing and defined over $\mathbb{Q}$, and F is decreasing, such that F induces a Hodge structure of weight k on $\mathrm{Gr}_k^W := W_k/W_{k-1}$ for each k.*

The filtration W is called the *weight filtration*, while F is called the *Hodge filtration*. We point out that for many of the subsequent results, it is enough to assume that W is defined over $\mathbb{R}$. This notion is compatible with passage to the dual and with tensor products. In particular, given an MHS on $V_{\mathbb{C}}$ we may define an MHS on $\mathfrak{gl}(V_{\mathbb{C}})$ by

$$W_a \mathfrak{gl} := \{X \in \mathfrak{gl}(V_{\mathbb{C}}) : X(W_\ell) \subset W_{\ell+a}\},$$

$$F^b \mathfrak{gl} := \{X \in \mathfrak{gl}(V_{\mathbb{C}}) : X(F^p) \subset F^{p+b}\}.$$

An element $T \in W_{2a}\mathfrak{gl} \cap F^a\mathfrak{gl} \cap \mathfrak{gl}(V_{\mathbb{Q}})$ is called an (a, a)-*morphism* of (W, F).

DEFINITION 7.5.5 *A splitting of an MHS (W, F) is a bigrading*

$$V_{\mathbb{C}} = \bigoplus_{p,q} J^{p,q}$$

such that

$$W_\ell = \bigoplus_{p+q \leq \ell} J^{p,q}; \quad F^p = \bigoplus_{a \geq p} J^{a,b}. \quad (7.5.9)$$

An (a, a)-morphism T of an MHS (W, F) is said to be *compatible* with the splitting $\{J^{p,q}\}$ if $T(J^{p,q}) \subset J^{p+a,q+a}$.

Every MHS admits splittings compatible with all its morphisms. In particular, we have the following result due to Deligne[19]:

THEOREM 7.5.6   *Given an MHS (W, F) the subspaces*

$$I^{p,q} := F^p \cap W_{p+q} \cap \left(\overline{F^q} \cap W_{p+q} + \overline{U_{p+q-2}^{q-1}} \right), \qquad (7.5.10)$$

where

$$U_b^a = \sum_{j \geq 0} F^{a-j} \cap W_{b-j},$$

define a splitting of (W, F) compatible with all morphisms. Moreover, $\{I^{p,q}\}$ is uniquely characterized by the property

$$I^{p,q} \equiv \overline{I^{q,p}} \left(\mathrm{mod} \bigoplus_{a<p;\ b<q} I^{a,b} \right). \qquad (7.5.11)$$

This correspondence establishes an equivalence of categories between MHS and bi-gradings $\{I^{p,q}\}$ satisfying (7.5.11).

PROOF. We refer to [11, Theorem 2.13] for a proof.   □

DEFINITION 7.5.7   *A mixed Hodge structure (W, F) is said to **split over** $\mathbb{R}$ if it admits a splitting $\{J^{p,q}\}$ such that*

$$J^{q,p} = \overline{J^{p,q}}.$$

In this case,

$$V_{\mathbb{C}} = \bigoplus_k \left(\bigoplus_{p+q=k} J^{p,q} \right)$$

is a decomposition of $V_{\mathbb{C}}$ as a direct sum of Hodge structures.

EXAMPLE 7.5.8   The paradigmatic example of a mixed Hodge structure split over $\mathbb{R}$ is the Hodge decomposition on the cohomology of a compact Kähler manifold X discussed in Chapter 1. Let

$$V_{\mathbb{Q}} = H^*(X, \mathbb{Q}) = \bigoplus_{k=0}^{2n} H^k(X, \mathbb{Q})$$

and set

$$J^{p,q} = H^{n-p,n-q}(X).$$

Thus,

$$W_\ell = \bigoplus_{d \geq 2n-\ell} H^d(X, \mathbb{C}), \quad F^p = \bigoplus_s \bigoplus_{r \leq n-p} H^{r,s}(X). \qquad (7.5.12)$$

With this choice of indexing, the operators L_ω, where ω is a Kähler class, are $(-1,-1)$-morphisms of the MHS.

The situation described by Example 7.5.8 carries additional structure: the Lefschetz theorems and the Hodge–Riemann bilinear relations. We extend these ideas to the case of abstract mixed Hodge structures. Recall from Proposition A.2.2, Chapter 1 that given a nilpotent transformation $N \in \mathfrak{gl}(V_{\mathbb{Q}})$ there exists a unique increasing filtration defined over $\mathbb{Q}$, $W = W_{\ell}(N)$, such that

1. $N(W_{\ell}) \subset W_{\ell-2}$;

2. for $\ell \geq 0$, $N^{\ell} \colon \mathrm{Gr}_{\ell}^{W} \to \mathrm{Gr}_{-\ell}^{W}$ is an isomorphism.

DEFINITION 7.5.9 *A **polarized MHS** (PMHS) [9, (2.4)] of weight $k \in \mathbb{Z}$ on $V_{\mathbb{C}}$ consists of an MHS (W, F) on V, a $(-1, -1)$-morphism $N \in \mathfrak{g} \cap \mathfrak{gl}(V_{\mathbb{Q}})$, and a nondegenerate, rational bilinear form Q such that*

1. *$N^{k+1} = 0$;*

2. *$W = W(N)[-k]$, where $W[-k]_{\ell} := W_{\ell-k}$;*

3. *$Q(F^a, F^{k-a+1}) = 0$; and*

4. *the Hodge structure of weight $k + l$ induced by F on*

$$\ker(N^{l+1} : \mathrm{Gr}_{k+l}^{W} \to \mathrm{Gr}_{k-l-2}^{W})$$

is polarized by $Q(\cdot, N^{l}\cdot)$.

EXAMPLE 7.5.10 We continue with Example 7.5.8. We may restate the hard Lefschetz theorem (see Corollary 1.5.9 and the Hodge–Riemann bilinear relations (Theorem 1.5.16)) by saying that the mixed Hodge structure in the cohomology $H^*(X, \mathbb{C})$ of an n-dimensional compact Kähler manifold X is an MHS of weight n polarized by the rational bilinear form Q on $H^*(X, \mathbb{C})$ defined by

$$Q([\alpha], [\beta]) \; = \; (-1)^{r(r+1)/2} \int_X \alpha \wedge \beta \, ; \quad [\alpha] \in H^r(X, \mathbb{C}), [\beta] \in H^s(X, \mathbb{C})$$

and the nilpotent operator L_{ω} for any Kähler class ω. Note that Theorem 1.5.16(2) is the assertion that $L_{\omega} \in \mathfrak{g} \cap \mathfrak{gl}(V_{\mathbb{Q}})$.

There is a very close relationship between polarized mixed Hodge structures and nilpotent orbits as indicated by the following:

THEOREM 7.5.11 *Let $\theta(z) = \exp(\sum_{j=1}^{r} z_j N_j) \cdot F$ be a nilpotent orbit in the sense of Example 7.5.2; then*

1. *every element in the cone*

$$\mathcal{C} := \left\{ N = \sum_{j=1}^{r} \lambda_j \, N_j \, ; \lambda_j \in \mathbb{R}_{>0} \right\} \subset \mathfrak{g}$$

defines the same weight filtration $W(\mathcal{C})$;

2. *the pair* $(W(\mathcal{C})[-k], F)$ *defines an MHS polarized by every* $N \in \mathcal{C}$, *the* **limit mixed Hodge structure;**

3. *conversely, suppose* $\{N_1, \ldots, N_r\} \in \mathfrak{g} \cap \mathfrak{gl}(V_{\mathbb{Q}})$ *are commuting nilpotent elements with the property that the weight filtration* $W(\sum \lambda_j N_j)$ *is independent of the choice of* $\lambda_1, \ldots, \lambda_r \in \mathbb{R}_{>0}$. *Then, if* $F \in \check{D}$ *is such that* $(W(\mathcal{C})[-k], F)$ *is polarized by every*[9] *element* $N \in \mathcal{C}$, *the map* $\theta(z) = \exp(\sum_{j=1}^r z_j N_j) \cdot F$ *is a nilpotent orbit.*

PROOF. Part 1 is proved in [9, (3.3)], while part 2 was proved by Schmid [35, Theorem 6.16] as a consequence of his SL_2-orbit theorem to be discussed below. In the case of geometric variations it was also shown by Steenbrink [37] and Clemens and Schmid [13]. The converse is [11, Proposition 4.66]. $\qquad\square$

Remark. If $\{N_1, \ldots, N_r\}$ and F satisfy the conditions in Theorem 7.5.11, part 3 we will simply say that $(N_1, \ldots, N_r; F)$ is a nilpotent orbit. This notation emphasizes the fact that nilpotent orbits are (polarized) linear algebra objects.

EXAMPLE 7.5.12 We continue with the situation discussed in Examples 7.5.8 and 7.5.10. Let $\omega_1, \ldots, \omega_r \in \mathcal{K}$ be Kähler classes in the compact Kähler manifold X. Then, clearly, the nilpotent transformations $L_{\omega_1}, \ldots, L_{\omega_r}$ commute and since every positive linear combination is also a Kähler class, it follows from the hard Lefschetz theorem that the weight filtration is independent of the coefficients. Moreover, the assumptions of Theorem 7.5.11, part 3 hold and therefore the map

$$\theta(z_1, \ldots, z_r) := \exp(z_1 L_{\omega_1} + \cdots + z_r L_{\omega_r}) \cdot F,$$

where F is as in (7.5.12) is a nilpotent orbit and hence defines a variation of Hodge structure on $H^*(X, \mathbb{C})$. Note that these Hodge structures are defined in the *total* cohomology of X. The relationship between this VHS and mirror symmetry is discussed in [7, 8]. This PVHS plays a central role in the mixed Lefschetz and Hodge–Riemann bilinear relations (see [5]) discussed in Chapter 1.

7.5.3 SL_2-Orbits

Theorem 7.5.11 establishes a relationship between polarized mixed Hodge structures and nilpotent orbits. In the case of PMHS split over $\mathbb{R}$ this correspondence yields an equivalence with a particular class of nilpotent orbits equivariant under a natural action of $SL(2, \mathbb{R})$. For simplicity, we will restrict ourselves to the one-variable case and refer the reader to [11, 10] for the general case.

Let (W, F_0) be an MHS on $V_{\mathbb{C}}$, split over $\mathbb{R}$ and polarized by $N \in F_0^{-1}\mathfrak{g} \cap \mathfrak{gl}(V_{\mathbb{Q}})$. Since $W = W(N)[-k]$, the subspaces

$$V_\ell = \bigoplus_{p+q=k+\ell} I^{p,q}(W, F_0), \qquad -k \le \ell \le k$$

---

[9]In fact, it suffices to assume that this holds for *some* $N \in \mathcal{C}$.

constitute a grading of $W(N)$ defined over $\mathbb{R}$. Let $Y = Y(W, F_0)$ denote the real semisimple endomorphism of $V_\mathbb{C}$ which acts on V_ℓ as multiplication by the integer ℓ. Since $NV_\ell \subset V_{\ell-2}$,

$$[Y, N] = -2N. \tag{7.5.13}$$

Because N polarizes the MHS (W, F_0), we have $Y \in \mathfrak{g}_0$ and there exists $N^+ \in \mathfrak{g}_0$ such that $[Y, N^+] = 2N^+, [N^+, N] = Y$ (cf. [11, (2.7)]).

Therefore, there is a Lie algebra homomorphism $\rho \colon \mathfrak{sl}(2, \mathbb{C}) \to \mathfrak{g}$ defined over $\mathbb{R}$ such that, for the standard generators $\{\mathbf{y}, \mathbf{n}_+, \mathbf{n}_-\}$ defined in (A.3.1) of Chapter 1,

$$\rho(\mathbf{y}) = Y, \quad \rho(\mathbf{n}_-) = N, \quad \rho(\mathbf{n}_+) = N^+. \tag{7.5.14}$$

The Lie algebra $\mathfrak{sl}(2, \mathbb{C})$ carries a Hodge structure of weight 0:

$$(\mathfrak{sl}(2, \mathbb{C}))^{-1,1} = \overline{(\mathfrak{sl}(2, \mathbb{C}))^{1,-1}} = \mathbb{C}(i\,\mathbf{y} + \mathbf{n}_- + \mathbf{n}_+),$$

$$(\mathfrak{sl}(2, \mathbb{C}))^{0,0} = \mathbb{C}(\mathbf{n}_+ - \mathbf{n}_-).$$

A homomorphism $\rho \colon \mathfrak{sl}(2, \mathbb{C}) \to \mathfrak{g}$ is said to be *Hodge* at $F \in D$, if it is a morphism of Hodge structures: that defined above on $\mathfrak{sl}(2, \mathbb{C})$ and the one determined by $F\mathfrak{g}$ in $\mathfrak{g}$. The lifting $\tilde{\rho} \colon \mathrm{SL}(2, \mathbb{C}) \to G_\mathbb{C}$ of such a morphism induces a horizontal, equivariant embedding $\hat{\rho} \colon \mathbb{P}^1 \longrightarrow \check{D}$ by $\hat{\rho}(g \cdot i) = \rho(g) \cdot F$, $g \in \mathrm{SL}(2, \mathbb{C})$. Moreover,

1. $\tilde{\rho}(\mathrm{SL}(2, \mathbb{R})) \subset G_\mathbb{R}$ and therefore $\hat{\rho}(H) \subset D$, where H is the upper half plane;

2. $\hat{\rho}(z) = (\exp zN)(\exp(-iN)) \cdot F$;

3. $\hat{\rho}(z) = (\exp xN)(\exp(-1/2)\log y\, Y) \cdot F$ for $z = x + iy \in H$.

Every polarized MHS split over $\mathbb{R}$ gives rise to a Hodge representation:

THEOREM 7.5.13   *Let (W, F_0) be an MHS split over $\mathbb{R}$ polarized by $N \in \mathfrak{g}$. Then*

1. *the filtration $F_{\sqrt{-1}} := \exp i\, N \cdot F_0$ lies in D;*

2. *the homomorphism $\rho \colon \mathfrak{sl}(2, \mathbb{C}) \to \mathfrak{g}$ defined by (7.5.14) is Hodge at $F_{\sqrt{-1}}$.*

Conversely, if a homomorphism $\rho \colon \mathfrak{sl}(2, \mathbb{C}) \to \mathfrak{g}$ is Hodge at $F \in D$, then

$$(W(\rho(\mathbf{n}_-))[-k], \exp(-i\rho(\mathbf{n}_-)) \cdot F)$$

is an MHS, split over $\mathbb{R}$ and polarized by $\rho(\mathbf{n}_-)$.

The following is a simplified version of Schmid's SL_2-orbit theorem. We refer to [35] for a full statement and proof.

THEOREM 7.5.14 (SL_2-orbit theorem)   *Let $z \mapsto \exp zN \cdot F$ be a nilpotent orbit. There exists*

1. *a filtration $F_{\sqrt{-1}} \in D$;*

2. *a homomorphism $\rho \colon \mathfrak{sl}(2, \mathbb{C}) \to \mathfrak{g}$ Hodge at $F_{\sqrt{-1}}$;*

3. *a real analytic, $G_{\mathbb{R}}$-valued function $g(y)$, defined for $y \gg 0$;*

such that

1. $N = \rho(\mathbf{n}_-)$;

2. *for $y \gg 0$, $\exp(iyN) \cdot F = g(y) \exp(iyN) \cdot F_0$, where $F_0 = \exp(-iN) \cdot F_{\sqrt{-1}}$;*

3. *both $g(y)$ and $g(y)^{-1}$ have convergent power series expansions around $y = \infty$, of the form $1 + \sum_{n=1}^{\infty} A_n \, y^{-n}$, with $A_n \in W_{n-1}\mathfrak{g} \cap \ker(\mathrm{ad}N)^{n+1}$.*

We may regard the SL_2-orbit theorem as associating to any given nilpotent orbit a distinguished nilpotent orbit, whose corresponding limit mixed Hodge structure splits over $\mathbb{R}$, together with a very fine description of the relationship between the two orbits. In particular, it yields the fact that nilpotent orbits are equivalent to PMHS and, given this, it may be interpreted as associating to any PMHS another one which splits over $\mathbb{R}$. One may reverse this process and take as a starting point the existence of the limit MHS associated with a nilpotent orbit. It is then possible to characterize functorially the PMHS corresponding to the SL_2-orbit. We refer to [11] for a full discussion. It is also possible to define other functorial real splittings of an MHS. One such is due to Deligne [19] (see also [11, Proposition 2.20]) and is central to the several-variable arguments in [11]. For applications of the one and several variables SL_2-orbit theorems see [35, 11, 12, 6]. For generalizations to variations of MHS see Chapter 8.

7.6   ASYMPTOTIC BEHAVIOR OF A PERIOD MAPPING

In this section we will study the asymptotic behavior of a period map. Much of this material is taken from [10, 7, 8]. Our setting is the same as in the previous section; i.e., we consider a period map

$$\Phi \colon (\Delta^*)^r \times \Delta^{n-r} \to D/\Gamma$$

and its lifting to the universal cover $\tilde{\Phi} \colon H^r \times \Delta^{n-r} \to D$.

The map $\tilde{\Phi}$ is, thus, holomorphic and horizontal. We assume that the monodromy transformations $\gamma_1, \ldots, \gamma_r$ are unipotent and let $N_1, \ldots, N_r \in \mathfrak{g} \cap \mathfrak{gl}(V_{\mathbb{Q}})$ denote the monodromy logarithms. Let $F_{\mathrm{lim}}(w)$, $w \in \Delta^{n-r}$ be the limit Hodge filtration. Then, for each $w \in \Delta^{n-r}$ we have a nilpotent orbit $(N_1, \ldots, N_r; F_{\mathrm{lim}}(w))$. Moreover, the nilpotent orbit theorem implies that we may write

$$\tilde{\Phi}(z, w) \; = \; \exp\left(\sum_{j=1}^{r} z_j N_j \right) \cdot \psi(t, w), \tag{7.6.1}$$

where $t_j = \exp(2\pi i z_j)$, and $\psi(t, w)$ is a holomorphic map on Δ^n with values on $\check{D}$ and $\psi(0, w) = F_{\mathrm{lim}}(w)$.

Since $\check{D}$ is a homogeneous space of the Lie group $G_{\mathbb{C}}$, we can obtain holomorphic liftings of ψ to $G_{\mathbb{C}}$. We describe a lifting adapted to the limit mixed Hodge structure.

Let $W = W(\mathcal{C})[-k]$ denote the shifted weight filtration of any linear combination of $N_1, \ldots, N_r$ with positive real coefficients, and let $F_0 = F_{\lim}(0)$. We let $\{I^{p,q}\}$ denote the canonical bigrading of the mixed Hodge structure (W, F_0) (see Theorem 7.5.6). The subspaces

$$I^{a,b}\mathfrak{g} := \{X \in \mathfrak{g} : X(I^{p,q}) \subset I^{p+a,q+b}\} \qquad (7.6.2)$$

define the canonical bigrading of the mixed Hodge structure defined by $(W\mathfrak{g}, F_0\mathfrak{g})$ on $\mathfrak{g}$. We note that

$$\left[I^{a,b}\mathfrak{g}, I^{a',b'}\mathfrak{g}\right] \subset I^{a+a',b+b'}\mathfrak{g}.$$

Set

$$\mathfrak{p}_a := \bigoplus_q I^{a,q}\mathfrak{g} \quad \text{and} \quad \mathfrak{g}_- := \bigoplus_{a \leq -1} \mathfrak{p}_a. \qquad (7.6.3)$$

Since, by (7.5.9),

$$F_0^0(\mathfrak{g}) = \bigoplus_{p \geq 0} I^{p,q}\mathfrak{g},$$

it follows that $\mathfrak{g}_-$ is a nilpotent subalgebra of $\mathfrak{g}$ complementary to $\mathfrak{b} = F_0^0(\mathfrak{g})$, the Lie algebra of the isotropy subgroup B of $G_{\mathbb{C}}$ at F_0. Hence, in a neighborhood of the origin in Δ^n, we may write

$$\psi(t, w) = \exp(\Gamma(t, w)) \cdot F_0,$$

where

$$\Gamma \colon U \subset \Delta^n \to \mathfrak{g}_-$$

is holomorphic in an open set U around the origin, and $\Gamma(0) = 0$. Consequently, we may rewrite (7.6.1) as

$$\tilde{\Phi}(t, w) = \exp\left(\sum_{j=1}^r z_j N_j\right) \cdot \exp(\Gamma(t, w)) \cdot F_0. \qquad (7.6.4)$$

Since $N_j \in I^{-1,-1}\mathfrak{g} \subset \mathfrak{g}_-$, the product

$$E(z, w) := \exp\left(\sum_{j=1}^r z_j N_j\right) \cdot \exp(\Gamma(t, w))$$

lies in the group $\exp(\mathfrak{g}_-)$ and, hence we may write $E(z, w) := \exp(X(z, w))$, with $X(z, w) \in \mathfrak{g}_-$. It follows from (7.4.4) that the horizontality of $\tilde{\Phi}$ implies that

$$E^{-1} dE \in \mathfrak{p}_{-1} \otimes T^*(H^r \times \Delta^{n-r}).$$

Hence, writing

$$X(z, w) = \sum_{j \leq -1} X_j(z, w); \quad X_j \in \mathfrak{p}_j,$$

we have

$$
\begin{aligned}
E^{-1}\,dE &= \exp(-X(z,w))\,d(\exp(X(z,w))) \\
&= \left(I - X + \frac{X^2}{2} - \cdots\right)(dX_{-1} + dX_{-2} + \cdots) \\
&\equiv dX_{-1}\left(\bmod \left(\bigoplus_{a\leq -2}\mathfrak{p}_a\right)\otimes T^*(H^r \times \Delta^{n-r})\right),
\end{aligned}
$$

and therefore we must have

$$E^{-1}\,dE = dX_{-1}. \tag{7.6.5}$$

Note that, in particular, it follows from (7.6.5) that $dE^{-1}\wedge dE = 0$, and equating terms according to the decomposition of $\mathfrak{g}_-$ it follows that

$$dX_{-1} \wedge dX_{-1} = 0. \tag{7.6.6}$$

THEOREM 7.6.1   *Let* $(N_1,\dots,N_r,F)$ *be a nilpotent orbit and let*

$$\Gamma\colon \Delta^r \times \Delta^{n-r} \to \mathfrak{g}_-$$

be a holomorphic map with $\Gamma(0,0) = 0$.

1. If the map

$$\tilde{\Phi}\colon H^r \times \Delta^{n-r} \to \check{D}$$

is horizontal then it lies in D *for* $\mathrm{Im}(z_j) > \alpha$, *where the constant* α *may be chosen uniformly on compact subsets of* Δ^{n-r}. *In other words,* $\tilde{\Phi}$ *is the lifting of a period map defined in a neighborhood of* $0 \in \Delta^n$.

2. Let $R\colon \Delta^r \times \Delta^{n-r} \in \mathfrak{p}_{-1}$ *be a holomorphic map with* $R(0,0) = 0$ *and set*

$$X_{-1}(z,w) = \sum_{j=1}^{r} z_j\,N_j + R(t,w)\,;\quad t_j = \exp(2\pi i z_j).$$

Then, if X_{-1} *satisfies the differential equation* (7.6.6), *there exists a unique period map* Φ *defined in a neighborhood of* $0 \in \Delta^n$ *and such that* $R = \Gamma_{-1}$.

PROOF.   The first statement is [10, Theorem 2.8] and is a consequence of the several-variables asymptotic results in [11]. The second statement is [7, Theorem 2.7]. Its proof consists in showing that the differential equation (7.6.6) is the integrability condition required for finding a (unique) solution of (7.6.5). The result then follows from part 1. A proof in another context may be found in [33].   □

Theorem 7.6.1 means that, asymptotically, a period map consists of linear-algebraic and analytic data. The linear-algebraic data is given by the nilpotent orbit or, equivalently, the polarized mixed Hodge structure. The analytic data is given by the holomorphic, $\mathfrak{p}_{-1}$-valued map Γ_{-1}. We conclude this chapter with two examples illustrating the asymptotic behavior of period maps and the meaning of some of the objects defined above.

EXAMPLE 7.6.2 Consider a PVHS over Δ^* of Hodge structures of weight 1 on the $2n$-dimensional vector space V polarized by the form Q. Let $\Phi\colon \Delta^* \to D/\operatorname{Sp}(V_{\mathbb{Z}}, Q)$ be the corresponding period map. The monodromy logarithm N satisfies $N^2 = 0$ and, by Example A.2.3 in Chapter 1, its weight filtration is

$$W_{-1}(N) = \operatorname{Im}(N)\,; \quad W_0(N) = \ker(N).$$

Let $F_{\lim}$ be the limit Hodge filtration. We have a bigrading of $V_{\mathbb{C}}$,

$$V_{\mathbb{C}} = I^{0,0} \oplus I^{0,1} \oplus I^{1,0} \oplus I^{1,1} \tag{7.6.7}$$

defined by the mixed Hodge structure $(W(N)[-1], F_{\lim})$. The nilpotent transformation N maps $I^{1,1}$ isomorphically onto $I^{0,0}$ and vanishes on the other summands. The form $Q(\cdot, N\cdot)$ polarizes the Hodge structure on Gr_2^W and hence defines a positive-definite Hermitian form on $I^{1,1}$. Similarly, we have that Q polarizes the Hodge decomposition on $V_1 := I^{1,0} \oplus I^{0,1}$. Thus, we may choose a basis of $V_{\mathbb{C}}$, adapted to the bigrading (7.6.7), and so that

$$N = \begin{pmatrix} 0 & 0 & I_\nu & 0 \\ 0 & 0 & 0 & 0 \\ 0 & 0 & 0 & 0 \\ 0 & 0 & 0 & 0 \end{pmatrix} \;;\quad Q = \begin{pmatrix} 0 & 0 & -I_\nu & 0 \\ 0 & 0 & 0 & -I_{n-\nu} \\ I_\nu & 0 & 0 & 0 \\ 0 & I_{n-\nu} & 0 & 0 \end{pmatrix},$$

where $\nu = \dim_{\mathbb{C}} I^{1,1}$ and the Hodge filtration $F_{\lim} = I^{1,0} \oplus I^{1,1}$ is the subspace spanned by the columns of the $2n \times n$ matrix

$$F_{\lim} = \begin{pmatrix} 0 & 0 \\ 0 & iI_{n-\nu} \\ I_\nu & 0 \\ 0 & I_{n-\nu} \end{pmatrix}.$$

The Lie algebra $\mathfrak{g}_-$ equals $\mathfrak{p}_{-1}$, and the period map can be written as

$$\Phi(t) = \exp\left(\frac{\log t}{2\pi i}\, N\right) \cdot \exp(\Gamma(t)) \cdot F_{\lim},$$

which takes the matrix form (cf. Example 1.1.23)

$$\Phi(t) = \begin{pmatrix} W(t) \\ I_n \end{pmatrix},$$

where

$$W(t) = \begin{pmatrix} \frac{\log t}{2\pi i} I_\nu + A_{11}(t) & A_{12}(t) \\ A_{12}^T(t) & A_{22}(t) \end{pmatrix},$$

with $A_{11}(t)$ and $A_{22}(t)$ symmetric and $A_{22}(0) = iI$; hence, $A_{22}(t)$ has positive-definite imaginary part for t near zero. This computation is carried out from scratch in [23, (13.3)].

EXAMPLE 7.6.3 This example arises in the study of mirror symmetry for quintic threefolds [20]. We will consider a polarized variation of Hodge structure $\mathbb{V} \to \Delta^*$ over the punctured disk Δ^*, of weight 3, and Hodge numbers $h^{3,0} = h^{2,1} = h^{1,2} = h^{0,3} = 1$. The classifying space for such Hodge structures is the homogeneous space $D = \mathrm{Sp}(2, \mathbb{R})/\mathrm{U}(1) \times \mathrm{U}(1)$. We will assume that the limit mixed Hodge structure $(W(N)[-3], F_0)$ is split over $\mathbb{R}$ and that the monodromy has maximal unipotency index, that is, $N^3 \neq 0$ while, of course, $N^4 = 0$. Hence, the bigrading defined by (W, F_0) is

$$V_{\mathbb{C}} = I^{0,0} \oplus I^{1,1} \oplus I^{2,2} \oplus I^{3,3},$$

where each $I^{p,q}$ is one-dimensional and defined over $\mathbb{R}$. We have $N(I^{p,p}) \subset I^{p-1,p-1}$ and therefore we may choose a basis e_p of $I^{p,p}$ such that $N(e_p) = e_{p-1}$. These elements may be chosen to be real and the polarization conditions mean that the skew-symmetric polarization form Q must satisfy

$$Q(e_3, e_0) = Q(e_2, e_1) = 1.$$

Choosing a coordinate t in Δ centered at 0, we can write the period map

$$\Phi(t) = \exp\left(\frac{\log t}{2\pi i}\right) \psi(t),$$

where $\psi(t) \colon \Delta \to \check{D}$ is holomorphic. Moreover, there exists a unique holomorphic map $\Gamma \colon \Delta \to \mathfrak{g}_-$, with $\Gamma(0) = 0$ and such that

$$\psi(t) = \exp(\Gamma(t)) \cdot F_0.$$

Recall also that Γ is completely determined by its (-1)-component which must be of the form

$$\Gamma_{-1}(t) = \begin{pmatrix} 0 & a(t) & 0 & 0 \\ 0 & 0 & b(t) & 0 \\ 0 & 0 & 0 & c(t) \\ 0 & 0 & 0 & 0 \end{pmatrix}, \tag{7.6.8}$$

and, since $\Gamma(t)$ is an infinitesimal automorphism of Q, then $c(t) = a(t)$.

The asymptotic description of the period map depend on the choice of local coordinates and we would like to understand this dependency. A change of coordinates fixing the origin must be of the form $\hat{t} = t \cdot f(t)$, with $f(0) = \lambda \neq 0$. Rescaling we may assume $\lambda = 1$ and, therefore, it follows from (7.5.8) that $\hat{F}_0 = F_0$. Hence the limit mixed Hodge structure is unchanged under the assumption $\lambda = 1$, and

$$\hat{\psi}(\hat{t}) = \exp\left(-\frac{\log f(t)}{2\pi i} N\right) \cdot \psi(t).$$

Consequently,

$$\begin{aligned} \hat{\psi}(\hat{t}) &= \exp\left(-\frac{\log f(t)}{2\pi i} N\right) \cdot \exp(\Gamma(t)) \cdot F_0 \\ &= \exp\left(-\frac{\log f(t)}{2\pi i} N\right) \cdot \exp(\Gamma(t)) \cdot \hat{F}_0. \end{aligned}$$

It then follows by uniqueness of the lifting that

$$\hat{\Gamma}_{-1}(\hat{t}) = -\frac{\log f(t)}{2\pi i} N + \Gamma_{-1}(t).$$

Hence, given (7.6.8) it follows that in the coordinate

$$\hat{t} := t \, \exp(2\pi i a(t)), \qquad (7.6.9)$$

the function $\hat{\Gamma}_{-1}(\hat{t})$ takes the form

$$\hat{\Gamma}_{-1}(\hat{t}) = \begin{pmatrix} 0 & 0 & 0 & 0 \\ 0 & 0 & \hat{b}(\hat{t}) & 0 \\ 0 & 0 & 0 & 0 \\ 0 & 0 & 0 & 0 \end{pmatrix},$$

and, consequently, the period mapping depends on the nilpotent orbit and just one analytic function $\hat{b}(\hat{t})$. The coordinate $\hat{t}$ defined by (7.6.9) is called the *canonical coordinate* and first appeared in the work on mirror symmetry (see [32]). We refer to [7] for a full discussion of the canonical coordinates. It was shown by Deligne that the holomorphic function $\hat{b}(\hat{t})$ is related to the so-called Yukawa coupling (see [20, 1]).

Bibliography

[1] José Bertin, Jean-Pierre Demailly, Luc Illusie, and Chris Peters. *Introduction to Hodge theory*, volume 8 of *SMF/AMS Texts and Monographs*. American Mathematical Society, Providence, RI, 2002. Translated from the 1996 French original by James Lewis and Peters.

[2] Jean-Luc Brylinski and Steven Zucker. An overview of recent advances in Hodge theory. In *Several complex variables, VI*, volume 69 of *Encyclopaedia Math. Sci.*, pages 39–142. Springer, Berlin, 1990.

[3] James Carlson, Stefan Müller-Stach, and Chris Peters. *Period mappings and period domains*, volume 85 of *Cambridge Studies in Advanced Mathematics*. Cambridge University Press, Cambridge, 2003.

[4] Eduardo Cattani. Mixed Hodge structures, compactifications and monodromy weight filtration. In *Topics in transcendental algebraic geometry (Princeton, NJ, 1981/1982)*, volume 106 of *Ann. of Math. Stud.*, pages 75–100. Princeton University Press, Princeton, NJ, 1984.

[5] Eduardo Cattani. Mixed Lefschetz theorems and Hodge–Riemann bilinear relations. *Int. Math. Res. Not.*, (10):Art. ID rnn025, 20, 2008.

[6] Eduardo Cattani, Pierre Deligne, and Aroldo Kaplan. On the locus of Hodge classes. *J. Amer. Math. Soc.*, 8(2):483–506, 1995.

[7] Eduardo Cattani and Javier Fernandez. Asymptotic Hodge theory and quantum products. In *Advances in algebraic geometry motivated by physics (Lowell, MA, 2000)*, volume 276 of *Contemp. Math.*, pages 115–136. Amer. Math. Soc., Providence, RI, 2001.

[8] Eduardo Cattani and Javier Fernandez. Frobenius modules and Hodge asymptotics. *Comm. Math. Phys.*, 238(3):489–504, 2003.

[9] Eduardo Cattani and Aroldo Kaplan. Polarized mixed Hodge structures and the local monodromy of a variation of Hodge structure. *Invent. Math.*, 67(1):101–115, 1982.

[10] Eduardo Cattani and Aroldo Kaplan. Degenerating variations of Hodge structure. *Astérisque* (179–180):9, 67–96, 1989. Actes du Colloque de Théorie de Hodge (Luminy, 1987).

[11] Eduardo Cattani, Aroldo Kaplan, and Wilfried Schmid. Degeneration of Hodge structures. *Ann. of Math. (2)*, 123(3):457–535, 1986.

[12] Eduardo Cattani, Aroldo Kaplan, and Wilfried Schmid. L^2 and intersection cohomologies for a polarizable variation of Hodge structure. *Invent. Math.*, 87(2):217–252, 1987.

[13] C. Herbert Clemens. Degeneration of Kähler manifolds. *Duke Math. J.*, 44(2):215–290, 1977.

[14] Pierre Deligne. Travaux de Griffiths. In *Séminaire Bourbaki, 23ème année (1969/70), Exp. No. 376*, pages 213–237.

[15] Pierre Deligne. *Équations différentielles à points singuliers réguliers*. Lecture Notes in Mathematics, Vol. 163. Springer, Berlin, 1970.

[16] Pierre Deligne. Théorie de Hodge. I. In *Actes du Congrès International des Mathématiciens (Nice, 1970), Tome 1*, pages 425–430. Gauthier-Villars, Paris, 1971.

[17] Pierre Deligne. Théorie de Hodge. II. *Inst. Hautes Études Sci. Publ. Math.*, (40):5–57, 1971.

[18] Pierre Deligne. Théorie de Hodge. III. *Inst. Hautes Études Sci. Publ. Math.*, (44):5–77, 1974.

[19] Pierre Deligne. Structures de Hodge mixtes réelles. In *Motives (Seattle, WA, 1991)*, volume 55 of *Proc. Sympos. Pure Math.*, pages 509–514. Amer. Math. Soc., Providence, RI, 1994.

[20] P. Deligne. Local behavior of Hodge structures at infinity. In *Mirror symmetry, II*, volume 1 of *AMS/IP Stud. Adv. Math.*, pages 683–699. Amer. Math. Soc., Providence, RI, 1997.

[21] Phillip A. Griffiths. Periods of integrals on algebraic manifolds. I. Construction and properties of the modular varieties. *Amer. J. Math.*, 90:568–626, 1968.

[22] Phillip A. Griffiths. Periods of integrals on algebraic manifolds. II. Local study of the period mapping. *Amer. J. Math.*, 90:805–865, 1968.

[23] Phillip A. Griffiths. Periods of integrals on algebraic manifolds: Summary of main results and discussion of open problems. *Bull. Amer. Math. Soc.*, 76:228–296, 1970.

[24] Phillip Griffiths. Topics in transcendental algebraic geometry. Ann. of Math. Stud., 106, Princeton University Press, Princeton, NJ, 1984.

[25] Phillip Griffiths and Wilfried Schmid. Locally homogeneous complex manifolds. *Acta Math.*, 123:253–302, 1969.

[26] Phillip Griffiths and Wilfried Schmid. Recent developments in Hodge theory: a discussion of techniques and results. In *Discrete subgroups of Lie groups and applicatons to moduli (Internat. Colloq., Bombay, 1973)*, pages 31–127. Oxford University Press, Bombay, 1975.

[27] Daniel Huybrechts. *Complex geometry*. Universitext. Springer, Berlin, 2005. An introduction.

[28] Matt Kerr and Gregory Pearlstein. An exponential history of functions with logarithmic growth. In *Topology of stratified spaces*, volume 58 of *Math. Sci. Res. Inst. Publ.*, pages 281–374. Cambridge University Press, Cambridge, 2011.

[29] Shoshichi Kobayashi. *Differential geometry of complex vector bundles*, volume 15 of *Publications of the Mathematical Society of Japan*. Princeton University Press, Princeton, NJ, 1987. Kanô Memorial Lectures, 5.

[30] Kunihiko Kodaira. *Complex manifolds and deformation of complex structures*. Classics in Mathematics. Springer, Berlin, English edition, 2005. Translated from the 1981 Japanese original by Kazuo Akao.

[31] Alan Landman. On the Picard–Lefschetz transformation for algebraic manifolds acquiring general singularities. *Trans. Amer. Math. Soc.*, 181:89–126, 1973.

[32] David R. Morrison. Mirror symmetry and rational curves on quintic threefolds: a guide for mathematicians. *J. Amer. Math. Soc.*, 6(1):223–247, 1993.

[33] Gregory Pearlstein. Variations of mixed Hodge structure, Higgs fields, and quantum cohomology. *Manuscripta Math.*, 102(3):269–310, 2000.

[34] Chris A. M. Peters and Joseph H. M. Steenbrink. *Mixed Hodge structures*, volume 52 of *Ergebnisse der Mathematik und ihrer Grenzgebiete. 3. Folge. A Series of Modern Surveys in Mathematics [Results in Mathematics and Related Areas. 3rd Series.]*. Springer, Berlin, 2008.

[35] Wilfried Schmid. Variation of Hodge structure: the singularities of the period mapping. *Invent. Math.*, 22:211–319, 1973.

[36] Wilfried Schmid. Abbildungen in arithmetische Quotienten hermitesch symmetrischer Räume. *Lecture Notes in Mathematics* **412**, 211–219. Springer, Berlin and New York, 1974.

[37] Joseph Steenbrink. Limits of Hodge structures. *Invent. Math.*, 31(3):229–257, 1975/76.

[38] Claire Voisin. *Hodge theory and complex algebraic geometry. I*, volume 76 of *Cambridge Studies in Advanced Mathematics*. Cambridge University Press, Cambridge, English edition, 2007. Translated from the French by Leila Schneps.

[39] Claire Voisin. *Hodge theory and complex algebraic geometry. II*, volume 77 of *Cambridge Studies in Advanced Mathematics*. Cambridge University Press, Cambridge, English edition, 2007. Translated from the French by Leila Schneps.

[40] Raymond O. Wells, Jr. *Differential analysis on complex manifolds*, volume 65 of *Graduate Texts in Mathematics*. Springer, New York, third edition, 2008. With a new appendix by Oscar Garcia-Prada.

Chapter Eight

Variations of Mixed Hodge Structure

by Patrick Brosnan and Fouad El Zein

INTRODUCTION

The object of the paper is to discuss the definition of admissible variations of mixed Hodge structure (VMHS), the results of Kashiwara in [22] and applications to the proof of algebraicity of the locus of Hodge cycles [4, 5]. Since we present an expository article, we explain the evolution of ideas starting from the geometric properties of algebraic families as they degenerate and acquire singularities.

The study of morphisms in algebraic geometry is at the origin of the theory of VMHS. Let $f : X \to V$ be a smooth proper morphism of complex algebraic varieties. By Ehresmann's theorem, the underlying differentiable structure of the various fibers does not vary: the fibers near a point of the parameter space V are diffeomorphic in this case but the algebraic or analytic structure on the fibers does vary.

From another point of view, locally near a point on the parameter space V, we may think of a morphism as being given by a fixed differentiable manifold and a family of analytic structures parametrized by the neighborhood of the point.

Hence the cohomology of the fibers does not vary, but in general the Hodge structure, which is sensitive to the analytic structure, does. In this case, the cohomology groups of the fibers form a local system $\mathcal{L}_{\mathbb{Z}}$.

So we start with the study of the structure of local systems and its relation to flat connections corresponding to the study of linear differential equations on manifolds. The local system of cohomology of the fibers define the Gauss–Manin connection.

The theory of variation of Hodge structure (VHS) (see Chapter 7) adds to the local system the Hodge structures on the cohomology of the fibers, and transforms geometric problems concerning smooth proper morphisms into problems in linear algebra involving the Hodge filtration by complex subspaces of cohomology vector spaces of the fibers. The data of a VHS:$(\mathcal{L}, F)$ consists of three objects:

(i) The local system of groups $\mathcal{L}_{\mathbb{Z}}$

(ii) The Hodge filtration F varying holomorphically with the fibers defined as a filtration by subbundles of $\mathcal{L}_{\mathcal{O}_V} := \mathcal{O}_V \otimes \mathcal{L}_{\mathbb{Z}}$ on the base V

(iii) The Hodge decomposition, which is a decomposition of the differentiable bundle
$$\mathcal{L}_{\mathcal{C}_V^\infty} := \mathcal{C}_V^\infty \otimes \mathcal{L}_{\mathbb{Z}} = \oplus_{p+q=i} \mathcal{L}^{p,q}$$

In general, a morphism onto the smooth variety V is smooth outside a divisor D called its discriminant, in which case the above description applies on the complement $V - D$. Then, the VHS is said to degenerate along D which means it acquires singularities.

The study of the singularities may be carried out in two ways, either by introducing the theory of mixed Hodge structure (MHS) on the singular fibers, or by the study of the asymptotic behavior of the VHS in the punctured neighborhood of a point in D. We may blow up closed subvarieties in D without modifying the family on $V - D$, hence in the study of the asymptotic behavior we may suppose, by Hironaka's results, D is a normal crossing divisor (NCD). Often, we may also suppose the parameter space reduced to a disc and D to a point, since many arguments are carried over an embedded disc in V with center a point of D.

In this setting, Grothendieck proved first in positive characteristic that the local monodromy around points in D of a local system of geometric origin is quasi-unipotent. In general, one can expect that geometric results over a field of positive characteristic correspond to results over a field of characteristic 0. Deligne gives in [2] a set of arguments to explain the correspondence of geometric results in such cases. Direct proofs of the quasi-unipotent monodromy exist using desingularization and spectral sequences [9, 29], or using the negativity of the holomorphic sectional curvature of period domains in the horizontal directions as in the proof due to Borel in [37].

Technically it is easier to write this expository article if we suppose D is a normal crossing divisor and the local monodromy is unipotent, although this does not change basically the results. For a local system with unipotent local monodromy, we need to introduce Deligne's canonical extension of the analytic flat vector bundle $\mathcal{L}_{V-D}$ into a bundle $\mathcal{L}_V$ on V characterized by the fact that the extended connection has logarithmic singularities with nilpotent residues. It is on $\mathcal{L}_V$ that the Hodge filtrations F will extend as a filtration by subbundles, but they no longer define a Hodge filtration on the fibers of the bundle over points of D.

Instead, combined with the local monodromy around the components of D near a point of D, a new structure called the limit mixed Hodge structure (MHS) and the companion results on the nilpotent orbit and SL(2)-orbit [37] describe in the best way the asymptotic behavior of the VHS near a point of D (see Chapter 7). We describe now the contents of this chapter.

Introduction. The above summary is the background needed to understand the motivations, the definitions, and the problems raised in the theory.

Section 8.1. We recall in the first section, the relations between local systems and linear differential equations as well as the Thom–Whitney results on the topological properties of morphisms of algebraic varieties. The definition of a VMHS on a smooth variety is given. The singularities of local systems are discussed in terms of the singularities of the corresponding flat connections.

Section 8.2. After introducing the theory of mixed Hodge structures (MHS) on the cohomology of algebraic varieties (see Chapter 3), Deligne proposed to study the variation of such linear MHS structures (VMHS) reflecting the variation of the geometry on cohomology of families of algebraic varieties ([13, Pb. 1.8.15]). We study the properties of degenerating geometric VMHS.

Section 8.3. The study of VMHS on the various strata of a stratified algebraic variety has developed in the last twenty years after the introduction of perverse sheaves. A remarkable decomposition of the higher direct image of a pure perverse sheaf by a projective morphism is proved in [2]. It follows in the transcendental case for a geometric VHS by subtle arguments of Deligne which deduce results from the case of positive characteristic. For abstract polarized VHS, the decomposition is stated and proved in terms of Hodge modules [35] following the work of Kashiwara [22]. In this section we give the definition and properties of admissible VMHS and review important local results of Kashiwara needed to understand the decomposition theory in the transcendental case and leading to a new direct proof without the use of the heavy machinery of Hodge modules.

Section 8.4. In the last section we recall the definition of normal functions and we explain recent results on the algebraicity of the zero set of normal functions to answer a question raised by Griffiths and Green.

Acknowledgments. P. Brosnan would like to thank G. Pearlstein for patiently teaching him Hodge theory.

8.1   VARIATION OF MIXED HODGE STRUCTURES

The classical theory of linear differential equations on an open subset of $\mathbb{C}$ has developed into the theory of connections on manifolds, while the monodromy of the solutions has developed into representation theory of the fundamental group of a space.

With the development of sheaf theory, a third definition of local systems as locally constant sheaves, appeared to be a powerful tool to study the cohomology of families of algebraic varieties. In his modern lecture notes [10] with a defiant classical title, "On linear differential equations with regular singular points," Deligne proved the equivalence between these three notions and studied their singularities. The study of singularities of morphisms lead to the problems on degeneration of VMHS.

8.1.1   Local Systems and Representations of the Fundamental Group

We refer to [10] for this section, in particular the generator of the fundamental group of $\mathbb{C}^*$ is defined by a clockwise loop γ; the notion of local system coincides with the theory of representations of the fundamental group of a topological space.

In this section, we suppose the topological space M *locally path connected and locally simply connected* (each point has a basis of connected neighborhoods $(U_i)_{i \in I}$ with trivial fundamental groups, i.e., $\pi_0(U_i) = e$ and $\pi_1(U_i) = e$). In particular, on complex algebraic varieties, we refer to the transcendental topology and not the Zariski topology to define local systems.

DEFINITION 8.1.1 (Local system)   *Let Λ be a ring and $\underline{\Lambda}$ be the constant sheaf defined by Λ on a topological space X. A local system or a locally constant sheaf of Λ-modules on X is a sheaf $\mathcal{L}$ of $\underline{\Lambda}$-modules such that, there is a nonnegative integer n, and for each $x \in X$, a neighborhood U such that $\mathcal{L}_{|U} \cong \underline{\Lambda}^n_{|U}$. A local system of Λ-modules is said to be* constant *if it is isomorphic on X to $\underline{\Lambda}^n$ for some fixed n.*

DEFINITION 8.1.2 (Representation of a group) *Let L be a finitely generated $\mathbb{Z}$-module. A representation of a group G is a homomorphism of groups*

$$G \xrightarrow{\rho} \mathrm{Aut}_{\mathbb{Z}}(L)$$

from G to the group of $\mathbb{Z}$-linear automorphisms of L, or equivalently a linear action of G on L.

Similarly, we define a representation by automorphisms of $\mathbb{Q}$-vector spaces instead of $\mathbb{Z}$-modules.

8.1.1.1   Monodromy

If $\mathcal{L}$ is a local system of Λ-modules on a topological space M and $f : N \to M$ is a continuous map, then the inverse image $f^{-1}(\mathcal{L})$ is a local system of Λ-modules on N.

LEMMA 8.1.3 *A local system $\mathcal{L}$ on the interval $[0, 1]$ is necessarily constant.*

PROOF. Let $\mathcal{L}^{\text{ét}}$ denote the étalé space of $\mathcal{L}$. Since $\mathcal{L}$ is locally constant, $\mathcal{L}^{\text{ét}} \to [0, 1]$ is a covering space. Since $[0, 1]$ is contractible, $\mathcal{L}^{\text{ét}}$ is a product. This implies that $\mathcal{L}$ is constant. □

Let $\gamma : [0, 1] \to M$ be a loop in M with origin a point v and let $\mathcal{L}$ be a $\mathbb{Z}$-local system on M with fiber L at v. The inverse image $\gamma^{-1}(\mathcal{L})$ of the local system is isomorphic to the constant sheaf defined by L on $[0, 1]$: $\gamma^{-1}\mathcal{L} \simeq L_{[0,1]}$, hence we deduce from this property the notion of monodromy.

DEFINITION 8.1.4 (Monodromy) *The composition of the linear isomorphisms*
$$L = \mathcal{L}_v = \mathcal{L}_{\gamma(0)} \simeq \Gamma([0, 1], \mathcal{L}) \simeq \mathcal{L}_{\gamma(1)} = \mathcal{L}_v = L$$
is denoted by T and called the monodromy along γ. It depends only on the homotopy class of γ.

The monodromy of a local system $\mathcal{L}$ defines a representation of the fundamental group $\pi_1(M, v)$ of a topological space M on the stalk at v, $\mathcal{L}_v = L$,

$$\pi_1(M, v) \xrightarrow{\rho} \mathrm{Aut}_{\mathbb{Z}}(\mathcal{L}_v),$$

which characterizes local systems on connected spaces in the following sense:

PROPOSITION 8.1.5 *Let M be a connected topological space. The above correspondence is an equivalence between the following categories:*

 (i) $\mathbb{Z}$-local systems with finitely generated $\mathbb{Z}$-modules L on M.

 (ii) Representations of the fundamental group $\pi_1(M, v)$ by linear automorphisms of finitely generated $\mathbb{Z}$-modules L.

8.1.2   Connections and Local Systems

The concept of connections on analytic manifolds (resp. smooth complex algebraic variety) is a generalization of the concept of a system of n-linear first-order differential equations.

DEFINITION 8.1.6 (Connection)   *Let $\mathcal{F}$ be a locally free holomorphic $\mathcal{O}_X$-module on a complex analytic manifold X (resp. smooth algebraic complex variety). A connection on $\mathcal{F}$ is a $\mathbb{C}_X$-linear map*

$$\nabla : \mathcal{F} \to \Omega^1_X \otimes_{\mathcal{O}_X} \mathcal{F}$$

satisfying the following condition for all sections f of $\mathcal{F}$ and φ of $\mathcal{O}_X$:

$$\nabla(\varphi f) = d\varphi \otimes f + \varphi \nabla f$$

known as the Leibniz condition.

We define a morphism of connections as a morphism of $\mathcal{O}_X$-modules which commutes with ∇.

8.1.2.1   *Flat Connections*

The definition of ∇ extends to differential forms in degree p as a $\mathbb{C}$-linear map

$$\nabla^p : \Omega^p_X \otimes_{\mathcal{O}_X} \mathcal{F} \to \Omega^{p+1}_X \otimes_{\mathcal{O}_X} \mathcal{F} \text{ such that } \nabla^p(\omega \otimes f) = d\omega \otimes f + (-1)^p \omega \wedge \nabla f.$$

The connection is said to be integrable if its curvature $\nabla^1 \circ \nabla^0 : F \to \Omega^2_X \otimes_{\mathcal{O}_X} \mathcal{F}$ vanishes ($\nabla = \nabla^0$, and the curvature is a linear morphism).

Then it follows that the composition of maps $\nabla^{i+1} \circ \nabla^i = 0$ vanishes for all $i \in \mathbb{N}$ for an integrable connection. In this case a de Rham complex is associated to ∇:

$$(\Omega^*_X \otimes_{\mathcal{O}_X} \mathcal{F}, \nabla): = \mathcal{F} \to \Omega^1_X \otimes_{\mathcal{O}_X} \mathcal{F} \cdots \Omega^p_X \otimes_{\mathcal{O}_X} \mathcal{F} \xrightarrow{\nabla^p} \cdots \Omega^n_X \otimes_{\mathcal{O}_X} \mathcal{F}.$$

PROPOSITION 8.1.7   *The horizontal sections $\mathcal{F}^\nabla$ of a connection ∇ on a module $\mathcal{F}$ on an analytic (resp. algebraic) smooth variety X, are defined as the solutions of the differential equation on X (resp. on the analytic associated manifold X^h)*

$$\mathcal{F}^\nabla = \{f : \nabla(f) = 0\}.$$

When the connection is integrable, $\mathcal{F}^\nabla$ is a local system of rank $\dim \mathcal{F}$.

PROOF.   This result is based on the relation between differential equations and connections. Locally, we consider a small open subset $U \subset X$ (isomorphic to an open subset of $\mathbb{C}^n$) such that $\mathcal{F}_{|U}$ is isomorphic to $\mathcal{O}^m_U$. This isomorphism is defined by the choice of a basis of sections $(e_i)_{i\in[1,m]}$ of $\mathcal{F}$ on U and extends to the tensor product of $\mathcal{F}$ with the module of differential forms $\Omega^1_U \otimes \mathcal{F} \simeq (\Omega^1_U)^m$.

In terms of the basis $e = (e_1, \ldots, e_m)$ of $F_{|U}$, a section s is written as

$$s = \textstyle\sum_{i\in[1,m]} y_i e_i \text{ and } \nabla s = \sum_{i\in[1,m]} dy_i \otimes e_i + \sum_{i\in[1,m]} y_i \nabla e_i,$$

where $\nabla e_i = \sum_{j\in[1,m]} \omega_{ij} \otimes e_j$.

The connection matrix Ω_U is the matrix of differential forms $(\omega_{ij})_{i,j\in[1,m]}$, sections of Ω^1_U; its ith column is the transpose of the line coordinates in $(\Omega^1_U)^m$ of the image of $\nabla(e_i)$.

Then the restriction of ∇ to U corresponds to a connection on $\mathcal{O}_U^m$ denoted ∇_U and defined on sections $y = (y_1, \ldots, y_m)$ of $\mathcal{O}_U^m$ on U, written in columns as $\nabla_U{}^t y = d({}^t y) + \Omega_U{}^t y$ or

$$
\nabla_U \begin{pmatrix} y_1 \\ \vdots \\ y_m \end{pmatrix} = \begin{pmatrix} dy_1 \\ \vdots \\ dy_m \end{pmatrix} + \Omega_U \begin{pmatrix} y_1 \\ \vdots \\ y_m \end{pmatrix}.
$$

The equation is in $\operatorname{End}(T, \mathcal{F})_{|U} \simeq (\Omega^1 \otimes \mathcal{F})_{|U}$, where T is the tangent bundle to X.

In terms of coordinates $(x_1, \ldots, x_n)$ on U, we write ω_{ij} as

$$
\omega_{ij} = \sum_{k \in [1,n]} \Gamma_{ij}^k(x)\, dx_k
$$

so that the equation of the coordinates of horizontal sections is given by linear partial differential equations for $i \in [1, m]$ and $k \in [1, n]$:

$$
\frac{\partial y_i}{\partial x_k} + \sum_{j \in [1,m]} \Gamma_{ij}^k(x)\, y_j = 0.
$$

The solutions form a local system of rank m, since the Frobenius condition is satisfied by the integrability hypothesis on ∇. $\qquad\square$

The connection appears as a global version of linear differential equations, independent of the choice of local coordinates on X.

REMARK 8.1.8 The natural morphism $\mathcal{L} \to (\Omega_X^* \otimes_{\mathbb{C}} \mathcal{L}, \nabla)$ defines a resolution of a complex local system $\mathcal{L}$ by coherent modules; hence it induces isomorphisms on cohomology

$$
H^i(X, \mathcal{L}) \simeq H^i(R\Gamma(X, (\Omega_X^* \otimes_{\mathbb{C}} \mathcal{L}, \nabla))),
$$

where we take hypercohomology on the right-hand side. On a smooth differentiable manifold X, the natural morphism $\mathcal{L} \to (\mathcal{E}_X^* \otimes_{\mathbb{C}} \mathcal{L}, \nabla)$ defines a soft resolution of $\mathcal{L}$ and induces isomorphisms on cohomology

$$
H^i(X, \mathcal{L}) \simeq H^i(\Gamma(X, (\mathcal{E}_X^* \otimes_{\mathbb{C}} \mathcal{L}, \nabla))).
$$

An algebraic flat connection on a complex algebraic variety X is defined on an algebraic bundle $\nabla : \mathcal{L}_{\mathcal{O}_X} \to \Omega_X^1 \otimes_{\mathcal{O}_X} \mathcal{L}_{\mathcal{O}_X}$; then, the whole complex $(\Omega_X^* \otimes_{\mathcal{O}_X} \mathcal{L}_{\mathcal{O}_X}, \nabla)$ is algebraic. However, the local system of flat sections is defined on the associated analytic manifold X^h. If X is proper, its hypercohomology for Zariski topology is isomorphic to the cohomology of the transcendental cohomology of the local system $\mathcal{L}$ on X^h by Grothendieck algebraic de Rham cohomology.

THEOREM 8.1.9 (Local systems and flat connections (Deligne [10])) *The functor $(\mathcal{F}, \nabla) \mapsto \mathcal{F}^\nabla$, defined by the flat sections, is an equivalence between the category of integrable connections on an analytic manifold X and the category of complex local systems on X with quasi-inverse defined by $\mathcal{L} \mapsto (\mathcal{L} \otimes_{\mathbb{C}} \mathcal{O}_X, \nabla)$.*

8.1.2.2   Local System of Geometric Origin

The structure of the local system appears naturally on the cohomology of a smooth and proper family of varieties.

THEOREM 8.1.10 (Differentiable fibrations)   *Let $f : M \to N$ be a proper differentiable submersive morphism of manifolds. For each point $v \in N$ there exists an open neighborhood U_v of v such that the differentiable structure of the inverse image $M_{U_v} = f^{-1}(U_v)$ decomposes as a product of a fiber at v with U_v:*

$$f^{-1}(U_v) \xrightarrow{\ \varphi\ } U_v \times M_v \quad \text{such that} \quad \mathrm{pr}_1 \circ \varphi = f_{|U_v}.$$

The proof follows from the existence of a tubular neighborhood of the submanifold M_v ([43, Thm. 9.3]).

COROLLARY 8.1.11 (Locally constant cohomology)   *In each degree i, the sheaf of cohomology of the fibers $R^i f_* \mathbb{Z}$ is constant on a small neighborhood U_v of any point v of fiber $H^i(M_v, \mathbb{Z})$, i.e., there exists an isomorphism between the restriction $(R^i f_* \mathbb{Z})_{|U_v}$ with the constant sheaf $\underline{H}^i_{|U_v}$ defined on U_v by the vector space $H^i = H^i(M_v, \mathbb{Z})$.*

PROOF.   Let U_v be isomorphic to a ball in $\mathbb{R}^n$ over which f is trivial; then for any small ball B_ρ included in U_v, the restriction $H^i(M_{U_v}, \mathbb{Z}) \to H^i(M_{B_\rho}, \mathbb{Z})$ is an isomorphism since M_{B_ρ} is a deformation retract of M_{U_v}. $\square$

REMARK 8.1.12 (Algebraic family of complex varieties)   Let $f : X \to V$ be a smooth proper morphism of complex varieties; then f defines a differentiable locally trivial fiber bundle on V. We still denote by f the differentiable morphism $X^{\mathrm{dif}} \to V^{\mathrm{dif}}$ associated to f; then the complex of real differential forms $\mathcal{E}^*_X$ is a fine resolution of the constant sheaf $\mathbb{R}$ and $R^i f_* \mathbb{R} \simeq \mathcal{H}^i(f_* \mathcal{E}^*_X)$ is a local system, which is said to be of geometric origin. Such local systems carry additional structures and have specific properties which are the subject of study in this article.

For example, in the geometric case of a smooth proper morphism $f : X \to (D^*)^p$, a loop γ_j defined by $z_j = e^{2\pi i t_j}$ defines an homotopy class of diffeomorphisms $[h_{\gamma_j} : X_{t_0 = \gamma_j(0)} \simeq X_{t_1 = \gamma_j(1)}]$ called the monodromy on the fiber at $t_0 = t_1$, and an isomorphism on the cohomology $(h^*_{\gamma_j})^{-1} : H^i(X_{t_0}, \mathbb{Z}) \simeq H^i(X_{t_0}, \mathbb{Z})$ (the monodromy T_j on the cohomology of the fiber). The representation defined by the local system $R^p f_* \mathbb{C}_X$ is related to the monodromy T_j

$$\rho : \pi_1(X, t_0) \to Aut\, H^p(X_{t_0}, \mathbb{C}), \quad [\gamma_j] \to (h^*_{\gamma_j})^{-1}.$$

Now we give the *abstract definition of variation of mixed Hodge structures* (VMHS) ([13, 1.8.14]) (see Chapter 3.2.2 for a summary of MHS), then we explain how the geometric situation leads to such structure.

DEFINITION 8.1.13 (Variation of mixed Hodge structure)   *A variation of mixed Hodge structure on an analytic manifold X consists of*

(i) a local system $\mathcal{L}_{\mathbb{Z}}$ of $\mathbb{Z}$-modules of finite type;

(ii) a finite increasing filtration $\mathcal{W}$ of $\mathcal{L}_{\mathbb{Q}} := \mathcal{L}_{\mathbb{Z}} \otimes \mathbb{Q}$ by sublocal systems of rational vector spaces;

(iii) a finite decreasing filtration $\mathcal{F}$ by locally free analytic subsheaves of $\mathcal{L}_{\mathcal{O}_X} := \mathcal{L}_{\mathbb{Z}} \otimes \mathcal{O}_X$, whose sections on X satisfy the infinitesimal Griffiths transversality relation with respect to the connection ∇ defined on $\mathcal{L}_{\mathcal{O}_X}$ by the local system $\mathcal{L}_{\mathbb{C}} := \mathcal{L}_{\mathbb{Z}} \otimes \mathbb{C}$,

$$\nabla(\mathcal{F}^p) \subset \Omega^1_X \otimes_{\mathcal{O}_X} \mathcal{F}^{p-1};$$

(iv) the filtrations $\mathcal{W}$ and $\mathcal{F}$ define an MHS on each fiber $(\mathcal{L}_{\mathcal{O}_X}(t), \mathcal{W}(t), \mathcal{F}(t))$ of the bundle $\mathcal{L}_{\mathcal{O}_X}$ at a point t.

The definition of VHS is obtained in the particular pure case when the weight filtration is trivial but for one index. The induced filtration by $\mathcal{F}$ on the graded objects $\mathrm{Gr}^{\mathcal{W}}_m \mathcal{L}$ forms a VHS. A morphism of VMHS is a morphism of local systems compatible with the filtrations.

DEFINITION 8.1.14 (Graded polarization)   *The variation of mixed Hodge structure is graded polarizable if the graded objects $\mathrm{Gr}^{\mathcal{W}}_m \mathcal{L}$ are polarizable variations of Hodge structure.*

8.1.3   Variation of Mixed Hodge Structure of Geometric Origin

Let $f : X \to V$ be a smooth proper morphism of complex algebraic varieties; then the cohomology of the fibers carry a Hodge structure (HS) which defines a variation of Hodge structure (VHS). On the other hand, if we do not assume the morphism $f : X \to V$ is smooth and proper, the sheaves $R^i f_* \mathbb{Z}$ are no longer locally constant.

There is, however, a stratification of V by locally closed subvarieties such that the restriction of the sheaves $R^i f_* \mathbb{Z}$ to the strata are locally constant and carry variations of mixed Hodge structure called of geometric origin. We will discuss this below, along with the asymptotic properties of a VMHS near the boundary of a stratum.

The study of the whole data, including VMHS over various strata, has developed in the last twenty years after the introduction of perverse sheaves in [2]. We discuss admissible VMHS and related results in Section 8.3.

8.1.3.1   *Background on Morphisms of Algebraic Varieties and Local Systems*

We describe here structural theorems of algebraic morphisms in order to deduce VMHS on various strata of the parameter space later.

Stratification theory on a variety consists of the decomposition of the variety into a disjoint union of smooth locally closed algebraic (or analytic) subvarieties called strata (a stratum is smooth but the variety may be singular along a stratum). By construction, the closure of a stratum is a union of additional strata of lower dimensions. A Whitney stratification satisfies two more conditions named after H. Whitney [30].

We are interested here in their consequence: the local topological trivial property at any point of a stratum proved by Mather [32] and useful in the study of local cohomology. Thom described in addition the topology of the singularities of algebraic morphisms. Next, we summarize these results.

8.1.3.2   Thom–Whitney Stratifications

Let $f : X \to V$ be an algebraic morphism. There exist finite Whitney stratifications $\mathfrak{X}$ of X and $\mathcal{S} = \{S_l\}_{l \leq d}$ of V by locally closed subsets S_l of dimension l $(d = \dim V)$, such that for each connected component S (a stratum) of S_l,

(i) $f^{-1}S$ is a topological fiber bundle over S, the union of connected components of strata of $\mathfrak{X}$, each mapped submersively to S;

(ii) local topological triviality: for all $v \in S$, there exist an open neighborhood $U(v)$ in S and a stratum-preserving homeomorphism $h : f^{-1}(U) \simeq f^{-1}(v) \times U$ such that $f_{|U} = p_U \circ h$, where p_U is the projection on U.

This statement can be found in the articles [30] by Lê and Teissier and [18] by Goresky and MacPherson. It follows easily from [41, Corollaire 5.1] by Verdier. Since the restriction f/S to a stratum S is a locally trivial topological bundle, we deduce the following corollary:

COROLLARY 8.1.15   *For $i \in \mathbb{Z}$, the higher direct cohomology sheaf $(R^i f_* \mathbb{Z}_X)/S$ is locally constant on each stratum S of V.*

We say that $R^i f_* \mathbb{Z}_X$ is constructible and $R f_* \mathbb{Z}_X$ is cohomologically constructible on V.

8.1.3.3   Geometric VMHS

We show that the abstract definition of VMHS above describes (or is determined by) the structure defined on the cohomology of the fibers of an algebraic morphism over the various strata of the parameter space.

The cohomology groups $H^i(X_t, \mathbb{Z})$ of the fibers X_t at points $t \in V$ underlie a mixed Hodge structure $(H^i(X_t, \mathbb{Z}), W, F)$. The following proposition describes the properties of the weight W and Hodge F filtrations:

PROPOSITION 8.1.16 (Geometric VMHS)   *Let $f : X \to V$ be an algebraic morphism and $S \subset V$ a Thom–Whitney smooth stratum over which the restriction of f is locally topologically trivial.*

(i) *For all integers $i \in \mathbb{N}$, the restriction to S of the higher direct image cohomology sheaves $(R^i f_* \mathbb{Z}_X)/S$ (resp. $(R^i f_* \mathbb{Z}_X/\text{Torsion})/S$) are local systems of $\mathbb{Z}_S$-modules of finite type (resp. free).*

(ii) *The weight filtration W on the cohomology $H^i(X_t, \mathbb{Q})$ of a fiber X_t at $t \in S$ defines a filtration $\mathcal{W}$ by sublocal systems of $(R^i f_* \mathbb{Q}_X)/S$.*

(iii) The Hodge filtration F on the cohomology $H^i(X_t, \mathbb{C})$ defines a filtration $\mathcal{F}$ by analytic subbundles of $(R^i f_* \mathbb{C}_X)/S \otimes \mathcal{O}_S$ whose locally free sheaf of sections on D^* satisfies the infinitesimal Griffiths transversality with respect to the Gauss–Manin connection ∇:

$$\nabla \mathcal{F}^p \subset \Omega^1_{D^*} \otimes_{\mathcal{O}_{D^*}} \mathcal{F}^{p-1}.$$

(iv) The graded objects $\mathrm{Gr}^{\mathcal{W}}_m (R^i f_* \mathbb{C}_X)/S$ with the induced filtration by $\mathcal{F}$ are polarizable variations of Hodge structure.

REMARK 8.1.17   (i) The category of geometric VMHS as defined by Deligne includes the VMHS defined by algebraic morphisms but it is bigger. See [2, 6.2.4].

(ii) *Introduction to degeneration of VMHS.* In the above proposition, the VMHS on the various strata are presented separately, while they are in fact related. The decomposition theorem [2] states, for example, in the case of a smooth X, that the complex defining the higher direct image decomposes in the derived category of constructible sheaves on V into a special kind of complex called perverse sheaves.

The first basic step is to understand the case when $V = \mathbb{P}$ is the projective space and the stratification is defined by a normal crossing divisor $Y \subset \mathbb{P}$. In this case the VMHS on the complement of Y is said to degenerate along Y.

Various properties of the degeneration have been stated by Deligne and proved by specialists in the theory. They will be introduced later:

(i) The canonical extension to a connection with logarithmic singularities along Y

(ii) Extensions of the Hodge filtration

(iii) Existence of a relative monodromy filtration along points of Y

The statement and the proof of Proposition 8.1.16 are a summary of the evolution of the theory by steps. In particular, the transversality condition has been a starting point. We give below various statements of the proposition according to the properties of the morphism f. We need first to introduce relative connections.

8.1.3.4   Relative Connections ([10, 2.20.3])

Let $f : X \to V$ denote a smooth morphism of analytic varieties and $\Omega^*_{X/V}$ the de Rham complex of relative differential forms. We denote by $f^{-1}\mathcal{F}$ the inverse image on X of a sheaf $\mathcal{F}$ on V.

DEFINITION 8.1.18 (Relative connection)   *(i) A relative connection on a coherent sheaf of modules $\mathcal{V}$ on X is an $f^{-1}\mathcal{O}_V$ linear map*

$$\nabla : \mathcal{V} \to \mathcal{V} \otimes \Omega^1_{X/V}$$

satisfying, for all local sections $\phi \in \mathcal{O}_X$ and $v \in \mathcal{V}$,

$$\nabla(\phi v) = f \cdot \nabla v + d\phi \cdot v.$$

(ii) A relative local system on X is a sheaf $\mathcal{L}$ with a structure of an $f^{-1}\mathcal{O}_V$-module, locally isomorphic on X to the inverse image of a coherent sheaf on an open subset of V, i.e., for all $x \in X$ there exists an open subset $W \subset S, f(x) \in W$:

$$\exists U \subset f^{-1}(W), x \in U, f_{|U} : U \to W, \exists \mathcal{F}_W \text{ coherent on } W : \mathcal{L}_{|U} \simeq (f_{|U}^{-1}\mathcal{F}_W).$$

PROPOSITION 8.1.19 (Deligne [10, Proposition 2.28])   *Let $f : X \to V$ be a separated smooth morphism of analytic varieties and $\mathcal{L}_{\mathbb{Z}}$ a local system on X with finite-dimensional cohomology on the fibers; then $\mathcal{L}_{\mathrm{rel}} := f^{-1}\mathcal{O}_V \otimes \mathcal{L}_{\mathbb{Z}}$ is a relative local system and $\Omega^*_{X/V} \otimes \mathcal{L}_{\mathbb{Z}}$ is its de Rham resolution. If we suppose f is locally topologically trivial over V, then we have an isomorphism:*

$$\mathcal{O}_V \otimes R^i f_* \mathcal{L}_{\mathbb{Z}} \simeq R^i f_* (\Omega^*_{X/V} \otimes \mathcal{L}_{\mathbb{Z}})$$

8.1.3.5   Variation of Hodge Structure Defined by a Smooth Algebraic Proper Morphism

We prove Proposition 8.1.16 in the case of a smooth proper morphism f with algebraic fibers over a nonsingular analytic variety. The main point is to prove that the variation of the Hodge filtration is analytic.

The original proof of transversality by Griffiths is based on the description of the Hodge filtration F as a map to the classifying space of all filtrations of the cohomology vector space of a fiber at a reference point, where the Hodge filtration of the cohomology of the fibers are transported horizontally to the reference point.

We summarize here a proof of the transversality by Katz–Oda [28, 43]. A decreasing filtration

$$L^r \Omega^i_X := f^* \Omega^r_V \otimes \Omega^{i-r}_X \subset \Omega^i_X$$

is defined on the complex Ω^*_X. The first term follows from the exact sequence of differential forms $0 \to f^*\Omega^1_V \to \Omega^1_X \to \Omega^1_{X/V} \to 0$, from which we deduce in general that $\mathrm{Gr}^r_L \Omega^i_X := f^*\Omega^r_V \otimes \Omega^{i-r}_{X/V}$.

Taking the first terms of the associated spectral sequence on the hypercohomology of the higher direct image, Katz–Oda prove that the associated connecting morphism

$$R^i f_* \Omega^*_{X/V} \to \Omega^r_V \otimes R^i f_* \Omega^*_{X/Y}$$

coincides with the terms of the de Rham complex defined by the Gauss–Manin connection on $\mathcal{O}_V \otimes R^i f_* \mathbb{C} \simeq R^i f_* \Omega^*_{X/V}$ with a sheaf of flat sections isomorphic to $R^i f_* \mathbb{C}$ (Proposition 8.1.19). Moreover, starting with the exact sequence of complexes

$$0 \to f^*\Omega^1_V \otimes F^{p-1}\Omega^{*-1}_X \to F^p\Omega^*_X \to F^p\Omega^*_{X/V} \to 0,$$

they prove the transversality condition: $R^i f_* F^p \Omega^*_{X/V} \to \Omega^1_V \otimes R^i f_* F^{p-1}\Omega^*_{X/Y}$.

8.1.3.6   Hints to the Proof of Proposition 8.1.16 in the Case of a Smooth Morphism

We prove Proposition 8.1.16 in the case of a smooth morphism f. We deduce from the fact that $f : X \to V$ is algebraic that it is the restriction to X of a proper morphism $X' \to V$. This is the case when f is quasi-projective for example. For an analytic map, it is an additional property.

DEFINITION 8.1.20 (Compactifiable morphism)   *An analytic morphism $f : X \to V$ is compactifiable if there exists a diagram $X \overset{i}{\to} X' \overset{h}{\to} D$ of analytic morphisms, where X is embedded as a dense open subset in X', h is proper, and $f = h \circ i$.*

We compactify f by adding a normal crossing divisor (NCD): there exists a proper morphism $h : X' \to V$ and an NCD, $Z \subset X'$ such that $X \simeq (X' - Z)$; moreover h induces f on X. However, X is not necessarily locally topologically trivial over V, hence we need to add this condition satisfied by definition over a stratum of the Thom–Whitney stratification:

We suppose the morphism $f : X \to V$ is smooth, compactifiable with algebraic fibers, locally topologically trivial over an analytic manifold V.

Even in this case, X' is not necessarily locally topologically trivial over V; instead there exists a big stratum S of the same dimension as V over which the subspace $Z'_{\text{rel}} := Z' \cap h^{-1}(S)$ is a relative NCD over S in $X'_S := X' \cap h^{-1}(S)$ so that we may consider the relative logarithmic complex $\Omega^*_{X'_S/S}(\text{Log } Z'_{\text{rel}})$ (see Section 3.4.1.1). The proof of Katz–Oda can be extended to this case. This is clear in the Poincaré dual situation of cohomology with relative proper support in $X \cap f^{-1}(S)$ over S, in which case we use the simplicial covering defined by the various intersections Z'_i of components of Z'_{rel} (see ([10, 2.20.3]) and Section 3.2.4.1). Then this case is deduced from the smooth proper case applied to X'_{rel} and the various Z'_i over S.

At points in $V - S$ the VMHS may degenerate in general, but since the local system extends under our hypothesis, the whole VMHS extends. Since we present a considerable amount of work on the degeneration of VMHS in the next section, we give a hint on the proof as an application (see also [2, Proposition 6.2.3]). First we reduce the study to the case of a morphism over a disc.

At a point $v \in V$, for any morphism $g : D \to V$ defined on a disc D and centered at v, the inverse image $g^{-1}X'$ and $g^{-1}X$ over D are necessarily topological fibrations over the punctured disc D^* for D small enough (since D^* is a Thom–Whitney stratum in this case). Moreover we can suppose $g^{-1}Z'$ is a relative NCD over D^* outside the central fiber. This technical reduction explains that important statements on degeneration are reduced to the case where V is a disc D.

PROPOSITION 8.1.21   *Let $f : X \to D$ be a compactifiable analytic morphism with algebraic fibers over a disc D. For D of radius small enough,*

  (i) *the restriction to D^* of the higher direct image cohomology sheaves $(R^i f_* \mathbb{Z}_X)/D^*$ for all integers $i \in \mathbb{N}$ are local systems of $\mathbb{Z}_{D^*}$-modules of finite type;*

  (ii) *the weight filtration W on the cohomology $H^i(X_t, \mathbb{Q})$ of a fiber X_t at $t \in D^*$ defines a filtration $\mathcal{W}$ by sublocal systems of $(R^i f_* \mathbb{Q}_X)/D^*$;*

(iii) *the Hodge filtration F on the cohomology $H^i(X_t, \mathbb{C})$ defines a filtration $\mathcal{F}$ by analytic subbundles of $(R^i f_* \mathbb{C}_X)/D^* \otimes \mathcal{O}_{D^*}$ whose locally free sheaf of sections on D^* satisfy the infinitesimal Griffiths transversality with respect to the Gauss–Manin connection ∇,*

$$\nabla \mathcal{F}^p \subset \Omega^1_{D^*} \otimes_{\mathcal{O}_{D^*}} \mathcal{F}^{p-1};$$

(iv) *if the higher direct image sheaf of cohomology of the fibers $(R^i f_* \mathbb{Z}_X)/D$ is a local system $\mathcal{L}_i$ on D, then it underlies a VMHS on D.*

PROOF OF THE EXTENSION. It remains to prove that the VMHS over D^* extends across v (iv). Since the local system extends to the whole disc, the local monodromy at v is trivial, but we do not know yet about the extension of the weight and the Hodge filtrations. Since the local monodromy at v of the weight local subsystems $\mathcal{W}_r$ is induced from $\mathcal{L}_i$, it is also trivial and $\mathcal{W}_r$ extend at v.

The extension of the Hodge filtration has been proved for the proper smooth case in [39, 37]. The proof of the extension of the Hodge filtration to an analytic bundle over D can be deduced from the study in Section 8.2 of the asymptotic behavior of the Hodge filtration (Theorem 8.2.1).

Over the center of D we obtain the limit MHS of the VMHS on D^* (Remark 8.2.2), and a comparison theorem via the natural morphism from the MHS on the fiber X_v to this limit MHS (see the basic local invariant cycle theorem in [19, VI] and theorem (5.3.7) in Chapter 5).

Since the local monodromy is trivial in our case, both MHS coincide above the point v, hence the Hodge filtration extends also by analytic bundles over D to define, with the weight at the center, an MHS isomorphic to the limit MHS.   □

8.1.3.7   *Hints to the Proof of Proposition 8.1.16 in the Case of a General Algebraic Morphism*

If X has singularities, we use the technique introduced by Deligne to cover X by simplicial smooth varieties [12], see Section 3.4: there exists a smooth compact simplicial variety $X'_\bullet$ defined by a family $\{X'_n\}_{n \in \mathbb{N}}$ containing a family $\{Z_n\}_{n \in \mathbb{N}}$ of normal crossing divisors (NCD) such that $\{X_n := X'_n - Z_n\}_{n \in \mathbb{N}}$ is a cohomological hyper-resolution of X, which gives, in particular, an isomorphism between the cohomology groups of the simplicial variety $X_\bullet$ and X.

A finite number of indices determine the cohomology, hence there exists a big stratum S in V such that we can suppose the NCD Z_n relative over S. Then, the proof is reduced to the above smooth case.

At a point v in $V - S$, we embed a disc D with center at v, then we can repeat the proof over the punctured disc above.

COROLLARY 8.1.22   *For each integer i and each stratum $S \subset V$ of a Thom–Whitney stratification of $f : X \to V$, the restriction of the higher direct cohomology sheaf $(R^i f_* \mathbb{Z}_X)/S$ underlies a VMHS on the stratum S.*

8.1.4   Singularities of Local Systems

In general the local system $(R^i f_* \mathbb{Z}_X)/S$ over a stratum S does not extend as a local system to the boundary of S.

The study of the singularities at the boundary may be carried out through the study of the singularities of the associated connection. It is important to distinguish in the geometric case between the data over the closure $\partial S := \overline{S} - S$ of S and the data that can be extracted from the asymptotic behavior on S. They are linked by the local invariant cycle theorem.

The degeneration can be studied locally at points of ∂S or globally, in which case we suppose the boundary of S to be an NCD, since we are often reduced to such case by the desingularization theorem of Hironaka.

We discuss in this section, the quasi-unipotent property of the monodromy and Deligne's canonical extension of the connection.

8.1.4.1   Local Monodromy

To study the local properties of $R^i f_* \mathbb{Z}_X$ at a point $v \in V$, we consider an embedding of a small disc D in V with center v. Thus, the study is reduced to the case of a proper analytic morphism $f : X \to D$ defined on an analytic space to a complex disc D. The inverse image $f^{-1}(D^*)$ of the punctured disc D^*, for D small enough, is a topological fiber bundle ([15, Exp. 14, (1.3.5)]). It follows that a monodromy homeomorphism is defined on the fiber X_t at a point $t \in D^*$ by restricting X to a closed path $\gamma : [0, 1] \to D^*, \gamma(u) = \exp(2\pi i u)t$. The inverse fiber bundle is trivial on the interval: $\gamma^* X \simeq [0, 1] \times X_t$ and the monodromy on X_t is defined by the path γ as

$$T : X_t \simeq (\gamma^* X)_0 \simeq (\gamma^* X)_1 \simeq X_t.$$

The homotopy class of the monodromy is independent of the choice of the trivialization, and can be achieved for singular X via the integration of a special type of vector field compatible with a Thom stratification of X [41].

REMARK 8.1.23   The following construction, suggested in the introduction of [15, Exp.13] shows how X can be recovered as a topological space from the monodromy. There exists a retraction $r_t : X_t \to X_0$ of the general fiber X_t to the special fiber X_0 at 0, satisfying $r_t \circ T = r_t$; then starting with the system (X_t, X_0, T, r_t),

(i) we define a topological space X' by gluing the boundaries of $X_t \times [0, 1]$, $X_t \times 0$, and $X_t \times 1$ via T; then a map $r' : X' \to X_0$ is deduced from r_t, and a map $f' : X' \to S^1$ is defined such that $f'(X'_t) := e^{2\pi i t}$;

(ii) then $f : X \to D$ is defined as the cone of f',

$$X \simeq X' \times [0, 1]/(X' \times 0 \xrightarrow{r'} X_0) \to S^1 \times [0, 1]/(S^1 \times 0 \to 0) \simeq D.$$

8.1.4.2   Quasi-unipotent Monodromy

If we suppose $D^* \subset S$, then the monodromy along a loop in D^* acts on the local system $(R^i f_* \mathbb{Z}_X)/D^*$ as a linear operator on the cohomology $H^i(X_t, \mathbb{Q})$ which may have only roots of unity as eigenvalues. This condition, discovered first by Grothendieck for algebraic varieties over a field of positive characteristic, is also true over $\mathbb{C}$ [20].

PROPOSITION 8.1.24   *The monodromy T of the local system $(R^i f_* \mathbb{Z}_X)/D^*$ defined by an algebraic (resp. proper analytic) morphism $f : X \to D$ is quasi-unipotent:*

$$\exists\, a, b \in \mathbb{N} \quad \text{such that} \quad (T^a - \mathrm{Id})^b = 0.$$

A proof in the analytic setting is explained in [9] and [29]. It is valid for an abstract VHS ([37], 4.5) where the definition of the local system over $\mathbb{Z}$ is used in the proof; hence it is true for VMHS. Finally, we remark that the theorem is true also for the local system defined by the Milnor fiber.

8.1.4.3   Universal Fiber

A canonical way to study the general fiber, independently of the choice of $t \in D^*$, is to introduce the universal covering $\widetilde{D}^*$ of D^* which can be defined by the Poincaré half plane $\mathbb{H} = \{u \in \mathbb{C} : \mathrm{Im}\, u > 0\}$, $\pi : \mathbb{H} \to D^* : u \mapsto \exp 2\pi i u$. The inverse image $\widetilde{X}^* := \mathbb{H} \times_D X$ is a topological fiber bundle trivial over $\mathbb{H}$ with fiber homeomorphic to X_t. Let $H : \mathbb{H} \times X_t \to \widetilde{X}^*$ denote a trivialization of the fiber bundle with fiber X_t at t; then the translation $u \to u + 1$ extends to $\mathbb{H} \times X_t$ and induces, via H, a transformation T of $\widetilde{X}^*$ such that the following diagram commutes:

$$
\begin{array}{ccc}
X_t & \xrightarrow{\;I_0\;} & \widetilde{X}^* \\
\downarrow T_t & & \downarrow T \\
X_t & \xrightarrow{\;I_1\;} & \widetilde{X}^*,
\end{array}
$$

where I_0 is defined by the choice of a point $u_0 \in \mathbb{H}$ satisfying $e^{2\pi i u_0} = t$ such that $I_0 : x \mapsto H(u_0, x)$, then $I_1 : x \mapsto H(u_0 + 1, x)$ and T_t is the monodromy on X_t. Hence T acts on $\widetilde{X}^*$ as a universal monodromy operator.

8.1.4.4   Canonical Extension with Logarithmic Singularities

For the global study of the asymptotic behavior of the local system on a stratum S near ∂S, we introduce the canonical extension by Deligne [10].

   Let Y be a normal crossing divisor (NCD) with smooth irreducible components $Y = \cup_{i \in I} Y_i$ in a smooth analytic variety X, $j : X^* := X - Y \to X$ the open embedding, and $\Omega_X^*(\log Y)$ the complex of sheaves of differential forms with logarithmic poles along Y. It is a complex of subsheaves of $j_* \Omega_{X-Y}^*$. There exists a global residue morphism $\mathrm{Res}_i : \Omega_X^1(\log Y) \to \mathcal{O}_{Y_i}$ defined locally as follows: given a point $y \in Y_i$ and z_i an analytic local coordinate equation of Y_i at y, a differential $\omega = \alpha \wedge dz_i/z_i + \omega'$ where ω' does not contain dz_i and α is regular along Y_i, then $\mathrm{Res}_i(\omega) = \alpha_{|Y_i}$.

DEFINITION 8.1.25 *Let F be a vector bundle on X. An analytic connection on F has logarithmic poles along Y if the entries of the connection matrix are 1-forms in $\Omega^1_X(\text{Log}\,Y)$; hence the connection is a $\mathbb{C}$- linear map satisfying the Leibniz condition*

$$\nabla : F \to \Omega^1_X(\text{Log}\,Y) \otimes_{\mathcal{O}_X} F.$$

Then the definition of ∇ extends to

$$\nabla^i : \Omega^i_X(\text{Log}\,Y) \otimes_{\mathcal{O}_X} F \to \Omega^{i+1}_X(\text{Log}\,Y) \otimes_{\mathcal{O}_X} F.$$

It is integrable if $\nabla^1 \circ \nabla = 0$, then a logarithmic complex is defined in this case:

$$(\Omega^*_X(\text{Log}\,Y) \otimes_{\mathcal{O}_X} F, \nabla).$$

The composition of ∇ with the residue map,

$$(\text{Res}_i \otimes \text{Id}) \circ \nabla : F \to \Omega^1_X(\text{Log}\,Y) \otimes_{\mathcal{O}_X} F \to \mathcal{O}_{Y_i} \otimes F,$$

vanishes on the product $\mathcal{I}_{Y_i} F$ of F with the ideal of definition $\mathcal{I}_{Y_i}$ of Y_i.

LEMMA 8.1.26 ([10, 3.8.3]) *The residue of the connection is a linear endomorphism of analytic bundles on Y_i:*

$$\text{Res}_i(\nabla) : F \otimes_{\mathcal{O}_X} \mathcal{O}_{Y_i} \to F \otimes_{\mathcal{O}_X} \mathcal{O}_{Y_i}.$$

THEOREM 8.1.27 (Logarithmic extension [10, 5.2]) *Let $\mathcal{L}$ be a complex local system on the complement of the NCD Y in X with locally unipotent monodromy along the components $Y_i, i \in I$ of Y. Then*

(i) *there exists a locally free module $\mathcal{L}_{\mathcal{O}_X}$ on X which extends $\mathcal{L}_{\mathcal{O}_{X^*}} := \mathcal{L} \otimes \mathcal{O}_{X^*}$;*

(ii) *moreover, the extension is unique if the connection ∇ has logarithmic poles with respect to $\mathcal{L}_{\mathcal{O}_X}$,*

$$\nabla : \mathcal{L}_{\mathcal{O}_X} \to \Omega^1_X(\text{Log}\,Y) \otimes \mathcal{L}_{\mathcal{O}_X},$$

with nilpotent residues along Y_i.

PROOF. The structure of a flat bundle is important here since it is not known how to extend any analytic bundle on X^* over X. Let y be a point in Y, and $U(y)$ a polydisc D^n with center y. If we fix a reference point $t_0 \in U(y)^* \simeq (D^*)^p \times D^{n-p}$, the restriction of the local system $\mathcal{L}$ to $U(y)^*$ is determined by a vector space L_{t_0} (the fiber at t_0), and the action on L_{t_0} of the various monodromy T_j for $j \le p$ defined by a set of generators γ_j around Y_j of the fundamental group of $U(y)^*$. The local system $\mathcal{L}$ is *locally unipotent along Y if at any point $y \in Y$ all T_j are unipotent*, in which case the extension we describe is *canonical*. We rely for the proof on a detailed exposition of Malgrange [31]. The construction has two steps: the first describes a local extension of the bundle, while the second consists in showing that local coordinates patching of the bundle over X^* extends to a local coordinates patching of the bundle over X. We describe explicitly the first step, since it will be useful in applications.

The universal covering $\widetilde{U}(y)^*$ of $(D^*)^p \times D^{n-p}$ is defined by

$$\mathbb{H}^p \times D^{n-p} = \{t = (t_1, \ldots, t_n) \in \mathbb{C}^n : \forall i \leq p, \operatorname{Im} t_i > 0 \text{ and } \forall i > p, |t_i| < \varepsilon\},$$

where $\operatorname{Im} t_i$ is the imaginary part of t_i and the covering map is

$$\pi : \mathbb{H}^p \times D^{n-p} \to (D^*)^p \times D^{n-p} : t \to (e^{2\pi i t_1}, \ldots, e^{2\pi i t_p}, t_{p+1}, \ldots, t_n).$$

The inverse image $\widetilde{\mathcal{L}} := \pi^*(\mathcal{L}_{|U(y)})$ on $\widetilde{U}(y)^*$ is a trivial local system with global sections a vector space L isomorphic to L_{t_0}:

$$L_{t_0} := \mathcal{L}(t_0) \xrightarrow{\sim} L = H^0(\widetilde{U}(y)^*, \widetilde{\mathcal{L}}) = H^0(\widetilde{U}(y)^*, \widetilde{\mathcal{L}}_{\mathcal{O}_{\widetilde{U}(y)^*}})^\nabla : v_0 \mapsto v$$

where v is defined by extension of v_0 along the loop γ_j defined by $z_j = e^{2\pi i t_j}$. We write $v(t)$ when the section is viewed as as a germ of $\widetilde{\mathcal{L}}_t = \mathcal{L}_{\pi(t)}$. The sections $v(t)$ may be viewed also as an horizontal analytic map $\widetilde{U}(y)^* \to L_{t_0}$ with value in the finite dimensional space L_{t_0}.

Let $\rho : \pi_1((D^*)^p, t_0) \to GL(L_{t_0}) \simeq GL(L)$ denote the representation defined by the local system $\mathcal{L}$, then the action of the monodromy $T_j := \rho(\gamma_j)$ on L for $j \leq p$ on a section $v \in L$ is given by the formula:

$$T_j v(t) = v(t_1, \ldots, t_j + 1, \ldots, t_n).$$

where $\widetilde{\mathcal{L}}_t = \mathcal{L}_{\pi(t)} = \widetilde{\mathcal{L}}_{t+e_j}$ and $e_j := (0, \cdots, 1_j, \cdots, 0)$.

We define $N_j = \operatorname{Log} T_j = -\sum_{k>0} (I - T_j)^k / k$. In particular, it is a finite sum as we suppose $I - T_j$ nilpotent, rationnally defined and N_j is also a nilpotent endomorphism of L. Then we construct an embedding of the vector space L of multivalued sections of $\mathcal{L}$ into the subspace of analytic sections of the sheaf $\mathcal{L}_{\mathcal{O}_{\widetilde{U}(y)^*}}$ on $\widetilde{U}(y)^*$ by the following formula:

$$L \to \mathcal{L}_{\mathcal{O}_{\widetilde{U}(y)^*}} : v \mapsto \widetilde{v} = (\exp(\ \Sigma_{i \leq p} t_i N_i)) \cdot v.$$

where the exponential is a linear sum of multiples of $\operatorname{Id} - T_j$ with analytic coefficients, hence its action defines an analytic section.

LEMMA 8.1.28 *Let $j_y : U(y)^* \to U(y)$; then the analytic section $\widetilde{v} \in \mathcal{L}_{\mathcal{O}_{\widetilde{U}(y)^*}}$ descends to an analytic section of $j_{y*} \mathcal{L}_{\mathcal{O}_{U(y)^*}}$.*

We show $\widetilde{v}(t + e_j) = \widetilde{v}(t)$ for all $t \in \widetilde{U}(y)^*$ and all vectors e_j:

$$\widetilde{v}(t + e_j) = ([\exp(-\Sigma_{i \leq p} t_i N_i] \exp(-N_j) \cdot v)(t + e_j),$$

where

$$(\exp(-N_j) \cdot v)(t + e_j) = (T_j^{-1} \cdot v)(t + e_j) = (T_j^{-1} T_j \cdot v)(t) = v(t)$$

since $(T_j \cdot v)(t) = v(t + e_j)$.

The bundle $\mathcal{L}_{\mathcal{O}_X}$ is defined by the locally free subsheaf of $j_*\mathcal{L}_{\mathcal{O}_{X^*}}$ with fiber $\mathcal{L}_{\mathcal{O}_{X,y}}$ generated as an $\mathcal{O}_{X,y}$-module by the sections $\tilde{v}$ for $v \in L$.

In terms of the local coordinates $z_j, j \in [1, n]$ of $U(y)$, $\log z_j = 2\pi i t_j$ is a determination of the logarithm map; then the analytic function on $U(y)^*$ defined by the horizontal section $v \in L$ is given by the formula

$$\tilde{v} = (exp(-\frac{1}{2\pi i}\Sigma_{i \le p}(\log z_i)N_i)) \cdot v, \quad \nabla\tilde{v} = -\frac{1}{2\pi i}\Sigma_{i \le p}\widetilde{(N_i \cdot v)} \otimes \frac{dz_i}{z_i}.$$

$\square$

REMARK 8.1.29   (i) The residue of ∇ along a component Y_i is a nilpotent endomorphism $\mathcal{N}_i$ of the analytic bundle $\mathcal{L}_{\mathcal{O}_{Y_i}} := \mathcal{L}_{\mathcal{O}_X} \otimes \mathcal{O}_{Y_i}$.

(ii) Let $Y_{i,j} := Y_i \cap Y_j$; then the restrictions of $\mathcal{N}_{Y_i}$ and $\mathcal{N}_{Y_j}$ commute ([10, 3.10]).

(iii) Let $Y_i^* := Y_i - \cup_{k \in I-i} Y_i \cap Y_k$. There is no global local system underlying $\mathcal{L}_{\mathcal{O}_{Y_i^*}}$. Locally, at any point $y \in Y_i$, a section $s_i : Y_i \cap U(y)^* \to U(y)^*$ may be defined by the hyperplane parallel to $Y_i \cap U(y)^*$ through the reference point t_0. Then $\mathcal{L}_{\mathcal{O}_{Y_i}}$ is isomorphic to the canonical extension of the inverse image local system $s_i^* \mathcal{L}$ ([10, 3.9.b]).

Equivalently, the formula $T_i = \exp(-2i\pi \operatorname{Res}_i(\nabla))$ ([10, 1.17, 3.11]) holds on the sheaf $\Psi_{z_i}\mathcal{L}$ of nearby cycles.

8.1.4.5   Relative Monodromy Filtration

In Proposition 8.1.24, the endomorphism $T^a - \operatorname{Id}$ is nilpotent. To study the degeneration, Deligne introduced a canonical monodromy filtration defined by a nilpotent endomorphism, satisfying some kind of degenerating Lefschetz formula and representing the Jordan form of the nilpotent endomorphism.

Let V be a vector space, W a finite increasing filtration of V by subspaces and N a nilpotent endomorphism of V compatible with W; then Deligne proves the following lemma (in [13, Proposition 1.6.13]):

LEMMA 8.1.30   *There exists at most a unique increasing filtration M of V satisfying $NM_i \subset M_{i-2}$ such that for all integers k and l, $k \ge 0$,*

$$N^k : \operatorname{Gr}_{l+k}^M V \operatorname{Gr}_l^W V \xrightarrow{\sim} \operatorname{Gr}_{l-k}^M \operatorname{Gr}_l^W V.$$

REMARK 8.1.31   The lemma is true for an object V of an abelian category $\mathbb{A}$, a finite increasing filtration W of V by subobjects of V in $\mathbb{A}$ and a nilpotent endomorphism N. Such generalization is particularly interesting when applied to the abelian category of perverse sheaves on a complex variety.

DEFINITION 8.1.32   *When it exists, such filtration is denoted by $M(N, W)$ and called the relative monodromy filtration of N with respect to W.*

LEMMA 8.1.33 *If the filtration W is trivial of weight a, the relative monodromy filtration M exists on the vector space V and satisfies*

$$NM_i \subset M_{i-2}; \quad N^k : \mathrm{Gr}^M_{a+k} V \xrightarrow{\sim} \mathrm{Gr}^M_{a-k} V.$$

The above definition has a striking similarity with the hard Lefschetz theorem.

EXAMPLE 8.1.34 Let $V = \oplus_i H^i(X, \mathbb{Q})$ denote the direct sum of cohomology spaces of a smooth projective complex variety X of dimension n and N the nilpotent endomorphism defined by the cup product with the cohomology class $c_1(H)$ of a hyperplane section H of X.

We consider the increasing filtration of V,

$$M_i(V) = \oplus_{j \geq n-i} H^j(X, \mathbb{Q}).$$

By the hard Lefschetz theorem, the repeated action $N^i : \mathrm{Gr}^M_i V \simeq H^{n-i}(X, \mathbb{Q}) \to \mathrm{Gr}^M_{-i} V \simeq H^{n+i}(X, \mathbb{Q})$ is an isomorphism. Hence M_i coincides with the monodromy filtration defined by N and centered at 0. Following this example, the property of the relative monodromy filtration appears as a degenerate form of the Lefschetz result on cohomology.

The monodromy filtration centered at 0. Let N be nilpotent on V such that $N^{l+1} = 0$. In the case of a trivial filtration W such that $W_0 = V$ and $W_{-1} = 0$, the filtration M exists and it is constructed by induction as follows. Let $M_l := V$, $M_{l-1} := \mathrm{Ker}\, N^l$ and $M_{-l} := N^l(V)$, $M_{-l-1} := 0$; then $N^l : \mathrm{Gr}^l V \simeq \mathrm{Gr}^{-l} V$ and the induced morphism N' by N on the quotient space V/M_{-l} satisfies $(N')^l = 0$ so that the definition of M is by induction on the index of nilpotency. The primitive part of $\mathrm{Gr}^M_i V$ is defined for $i \geq 0$ as

$$P_i := \mathrm{Ker}\, N^{i+1} : \mathrm{Gr}^M_i \to \mathrm{Gr}^M_{-i-2}, \quad \mathrm{Gr}^M_i V \simeq \oplus_{k \geq 0} N^k P_{i+2k}.$$

The decomposition on the right follows from this definition, and the proof is similar to the existence of a primitive decomposition following the hard Lefschetz theorem on compact Kähler manifolds. In this case, the filtration M gives a description of the Jordan form of the nilpotent endomorphism N, independent of the choice of a Jordan basis as follows. For all $i \geq 0$, we have the following properties:

- $N^i : M_i \to M_{-i}$ is surjective;

- $\mathrm{Ker}\, N^{i+1} \subset M_i$; and

- $\mathrm{Ker}\, N^{i+1}$ projects surjectively onto the primitive subspace $P_i \subset \mathrm{Gr}^M_i$.

Let $(e_i^j)_{j \in I_i}$ denote a subset of elements in $\mathrm{Ker}\, N^{i+1}$ which lift a basis of $P_i \subset \mathrm{Gr}^M_i V$; then the various elements $N^k(e_i^j)_{j \in I_i, i \geq 0, k \leq i}$ define a Jordan basis of V for N. In particular, each element $(e_i^j)_j$ for fixed i gives rise to a Jordan block of length $i+1$ in the matrix of N.

8.1.4.6   Limit Hodge Filtration

Let $(\mathcal{L}, F)$ be an abstract polarized VHS on a punctured disc D^*, where the local system is defined by a unipotent endomorphism T on a $\mathbb{Z}$-module L; then Schmid ([20]) showed that such a VHS is asymptotic to a "nilpotent orbit" defined by a filtration F called limit or asymptotic such that for $N = \operatorname{Log} T$ the nilpotent logarithm of T, and $W(N)$ the monodromy filtration, the data $(L, W(N), F)$ form an MHS.

This positive answer to a question of Deligne was one of the starting points of the linear aspect of degeneration theory developed here, but the main development occurred with the discovery of intersection cohomology.

There is no such limit filtration F in general for a VMHS.

8.2   DEGENERATION OF VARIATIONS OF MIXED HODGE STRUCTURES

Families of algebraic varieties parametrized by a nonsingular algebraic curve acquire, in general, singularities that change their topology at a finite number of points of the curve. If we center a disc D at one of these points, we are in the above case of a family over D^* which extends over the origin in an algebraic family over D. The fiber at the origin may be changed by modification along a subvariety, which does not change the family over D^*.

The study of the degeneration follows the same pattern as the definition of the VMHS. The main results have been established first for the degeneration of abstract VHS [37], then a geometric construction has been given in the case of degeneration of smooth families [39]. These results will be assumed since we concentrate our attention on the singular case here.

The degeneration of families of singular varieties is reduced to the case of smooth families by the technique of simplicial coverings already mentioned (see Section 3.4). Such coverings by simplicial smooth algebraic varieties with NCD above the origin and satisfying the descent cohomological property, induce a covering of the fibers over D^* which is fit to study the degeneration of the cohomology of the fibers.

In the case of open families, we use the fact that we can complete algebraic varieties with an NCD at infinity Z, which is moreover a relative NCD over the punctured disc of small radius.

8.2.1   Diagonal Degeneration of Geometric VMHS

The term diagonal refers to a type of construction of the weight as diagonal with respect to a simplicial covering. The next results describe the cohomological degeneration of the MHS of an algebraic family over a disc. Here $\mathbb{Z}(b)$ will denote the MHS on the group $(2\pi i)^b \mathbb{Z} \subset \mathbb{C}$ of type $(-b, -b)$ and its tensor product with an MHS on a group V is the twisted MHS on V denoted $V(b) := V \otimes \mathbb{Z}(b)$.

Hypothesis. Let $f : X \to D$ be a proper analytic morphism defined on an analytic manifold X to a complex disc D, let Z be a closed analytic subspace of X, and suppose the fibers of X and Z over D are algebraic; then for D small enough,

- the weight filtration on the family $H^n(X_t - Z_t, \mathbb{Q})$ defines a filtration by sublocal systems $\mathcal{W}$ of $R^n f_* \mathbb{Q}_{|D^*}$ on D^*;

- the graded objects $\mathrm{Gr}_i^{\mathcal{W}} R^n f_* \mathbb{Q}_{|D^*}$ underlie variations of Hodge structures (VHS) on D^* defining a limit MHS at the origin [37, 39].

The construction below gives back this limit MHS in the case of a VHS when $Z = \emptyset$. Moreover, the general case is deduced from this pure case on a simplicial family of varieties by a diagonalization process.

Let $\widetilde{X}^* := X \times_D \widetilde{D}^*$ (resp. $\widetilde{Z}^* := Z \times_D \widetilde{D}^*$) be the inverse image on the universal covering $\widetilde{D}^*$ of D^*; then the inverse image $\widetilde{\mathcal{W}}$ of $\mathcal{W}$ on $\widetilde{D}^*$ is trivial and defines a filtration W^f by subspaces of $H^n(\widetilde{X}^* - \widetilde{Z}^*, \mathbb{Q})$, called here the finite weight filtration.

THEOREM 8.2.1   *There exists an MHS on the cohomology $H^n(\widetilde{X}^* - \widetilde{Z}^*, \mathbb{Z})$ with weight filtration W defined over $\mathbb{Q}$ and Hodge filtration defined over $\mathbb{C}$ satisfying*

(i) *the finite filtration W^f is a filtration by sub-MHS of $H^n(\widetilde{X}^* - \widetilde{Z}^*, \mathbb{Q})$;*

(ii) *the induced MHS on $\mathrm{Gr}_i^{W^f} H^n(\widetilde{X}^* - \widetilde{Z}^*, \mathbb{Q})$ coincides with the limit MHS defined by the VHS on the family $\mathrm{Gr}_i^W H^n(X_t - Z_t, \mathbb{Q})$.*

(iii) *for a quasi-projective morphism and all integers a, b, the endomorphism $N = \mathrm{Log}\, T^u$, defined by the logarithm of the unipotent part T^u of the monodromy, induces an isomorphism for $b \geq 0$,*

$$\mathrm{Gr}\, N^b : \mathrm{Gr}_{a+b}^W \mathrm{Gr}_a^{W^f} H^n(\widetilde{X}^* - \widetilde{Z}^*, \mathbb{Q}) \simeq \mathrm{Gr}_{a-b}^W \mathrm{Gr}_a^{W^f} H^n(\widetilde{X}^* - \widetilde{Z}^*, \mathbb{Q})(-b).$$

REMARK 8.2.2 (Category of Limit MHS)   The assertion (iii) characterizes the weight W as the monodromy filtration of N relative to W^f, which proves its existence in the case of geometric variations.

The above MHS is called the limit MHS of the VMHS defined by the fibers of f. In general we define a structure called the limit MHS, by the data (V, W^f, W, F, N), where (V, W, F) form an MHS, W^f is an increasing filtration by sub-MHS, and N is a nilpotent endomorphism of MHS: $(V, W, F) \to (V, W, F)(-1)$ compatible with W^f such that W is the relative weight filtration of (V, W^f, N).

A morphism of limit MHS is compatible with the filtrations so that we have an additive category which is not abelian since W^f is not necessarily strict.

PLAN OF THE PROOF OF THEOREM 8.2.1.   The proof will occupy this section and is based on the reduction to the smooth case, via a simplicial hypercovering resolution of X, followed by a diagonalization process of the weight of each term of the hypercovering [12], as in the case of the weight in the MHS of a singular variety [11, 12].

To be precise, let $Y = f^{-1}(0)$ and consider a smooth hypercovering $\pi : X_\bullet \to X$ over X, where each term X_i is smooth and proper over X, such that $Z_\bullet := \pi^{-1}(Z)$, $Y_\bullet := \pi^{-1}(Y)$, and $Z_\bullet \cup Y_\bullet$ are NCD in $X_\bullet$ with no common irreducible component in $Y_\bullet$ and $Z_\bullet$. Let $V := X - Z$, $V_i := X_i - Z_i$; then $\pi|V_\bullet : V_\bullet \to V$ is a hypercovering.

$\square$

Notice that only a finite number of terms X_i (resp. V_i) of the hypercovering are needed to compute the cohomology of X (resp. V). Then by Thom–Whitney theorems, for D small enough, X_i and Z_i are topological fiber bundles over D^* and Z_i^* is a relative NCD in X_i^* for a large number of indices i, and for each $t \in D^*$, $(X_\bullet)_t$ (resp. $(V_\bullet)_t$) is a hypercovering of X_t (resp. V_t).

Then for each index i, X_i and the various intersections of the irreducible components of Z_i, are proper and smooth families over D^*, so that we can apply in this situation the results of Steenbrink on the degeneration process for a geometric family of HS [39]. The method consists first in computing the hypercohomology of the sheaf of the nearby cycles $\Psi_f(\mathbb{C}_V)$ on $Y - Y \cap Z$ as the hypercohomology of simplicial nearby cycles $\Psi_{f \circ \pi}(\mathbb{C}_{V_\bullet})$,

$$\mathbb{H}^n(Y - Y \cap Z, \Psi_f(\mathbb{C}_V)) \simeq H^n(\widetilde{V}^*, \mathbb{C}) \simeq H^n(\widetilde{V}_\bullet^*, \mathbb{C}) \simeq \mathbb{H}^n(Y_\bullet - Y_\bullet \cap Z_\bullet, \Psi_f(\mathbb{C}_{V_\bullet})),$$

where we denote by tilde the inverse image on the universal cover $\widetilde{D}^*$ of a space over D or D^*, and where the third term is the cohomology of the simplicial space $\widetilde{V}_\bullet^*$.

We describe the construction on the unipotent part of the cohomology $H^n(\widetilde{V}^*, \mathbb{C})^u$ (that is the subspace of $H^n(\widetilde{V}^*, \mathbb{C})$ where the action of the monodromy is unipotent) although the theorem is true without this condition.

The unipotent cohomology is computed as the hypercohomology of the simplicial variety $Y_\bullet$ with value in some sheaf denoted $\Psi_{f \circ \pi}(\mathrm{Log}\, Z_\bullet)$ that we *define here*. Such a complex is a logarithmic version of the nearby cycle complex of sheaves on $Y_\bullet$ endowed with the structure of a simplicial cohomological mixed Hodge complex, such that

$$H^n(\widetilde{V}^*, \mathbb{C})^u \simeq \mathbb{H}^n(Y_\bullet, \Psi_{f \circ \pi}(\mathrm{Log}\, Z_\bullet)).$$

Then Theorem 8.2.1(ii) on $\mathrm{Gr}_i^W H^n(X_t - Z_t, \mathbb{Q})$ refers to the case of geometric VHS. That is, the smooth proper case which is first generalized to the nonproper case on each X_i, then the simplicial case with all terms X_i is considered.

With this in mind, the method of proof is an application of general results on *simplicial trifiltered complexes* that we develop now; still later we describe explicitly the complexes involved.

8.2.2   Filtered Mixed Hodge Complex (FMHC)

The proof involves abstract results concerning FMHC, with three filtrations W^f, W, and F, where W and F define an MHS on cohomology and W^f induces a filtration by sub-MHS. We define first the category of complexes with three filtrations.

8.2.2.1   Three Filtered Complex

Let $\mathbb{A}$ be an abelian category, $F_3\mathbb{A}$ the category of three filtered objects of $\mathbb{A}$ with finite filtrations, and $K^+F_3\mathbb{A}$ the category of three filtered complexes of objects of $\mathbb{A}$ bounded on the left, with morphisms defined up to homotopy respecting the filtrations.

DEFINITION 8.2.3   *A morphism $f : (K, F_1, F_2, F_3) \to (K', F'_1, F'_2, F'_3)$ in $K^+ F_3 \mathbb{A}$, where F_1, F'_1 are increasing, is called a quasi-isomorphism if the following morphisms f^i_j are bifiltered quasi-isomorphisms for all $i \geq j$:*

$$f^i_j : (F_1^i K / F_1^j K, F_2, F_3) \to (F_1'^i K / F_1'^j K, F'_2, F'_3)$$

The category $D^+ F_3 \mathbb{A}$ is obtained from $K^+ F_3 \mathbb{A}$ by inverting the above quasi-isomorphisms; the objects in $D^+ F_3 \mathbb{A}$ are trifiltered complexes but the group of morphisms $\mathrm{Hom}(K, K')$ of complexes change, since we add to a quasi-isomorphism f in $\mathrm{Hom}(K, K')$ an inverse element in $\mathrm{Hom}(K', K)$ denoted $1/f$ such that $f \circ (1/f) = \mathrm{Id}$ (resp. $(1/f) \circ f = \mathrm{Id}$), where equal to the identity means homotopic to the identity of K' (resp. K). In fact this completely changes the category since different objects, not initially isomorphic, may become isomorphic in the new category.

DEFINITION 8.2.4 (Filtered mixed Hodge complex (FMHC))   *An FMHC is given by*

(i) *a complex $K_{\mathbb{Z}} \in \mathrm{Ob}\, D^+(\mathbb{Z})$ such that $H^k(K_{\mathbb{Z}})$ is a $\mathbb{Z}$-module of finite type for all k;*

(ii) *a bifiltered complex $(K_{\mathbb{Q}}, W^f, W) \in \mathrm{Ob}\, D^+ F_2(\mathbb{Q})$ and an isomorphism $K_{\mathbb{Q}} \simeq K_{\mathbb{Z}} \otimes \mathbb{Q}$ in $D^+(\mathbb{Q})$, where W (resp. W^f) is an increasing filtration by weight (resp. finite weight);*

(iii) *a trifiltered complex $(K_{\mathbb{C}}, W^f, W, F) \in \mathrm{Ob}\, D^+ F_3(\mathbb{C})$ and an isomorphism*

$$(K_{\mathbb{C}}, W^f, W) \simeq (K_{\mathbb{Q}}, W^f, W) \otimes \mathbb{C}$$

in $D^+ F_2(\mathbb{C})$ where F is a decreasing filtration called Hodge filtration.

The following axiom is satisfied: for all $j \leq i$, the following system is an MHC:

$$(W_i^f K_{\mathbb{Q}} / W_j^f K_{\mathbb{Q}}, W), \; (W_i^f K_{\mathbb{C}} / W_j^f K_{\mathbb{C}}, W, F),$$
$$\alpha_i^j : (W_i^f K_{\mathbb{Q}} / W_j^f K_{\mathbb{Q}}, W) \otimes \mathbb{C} \simeq (W_i^f K_{\mathbb{C}} / W_j^f K_{\mathbb{C}}, W).$$

8.2.2.2   Cohomological Filtered Mixed Hodge Complex

We define a cohomological FMHC on a topological space X, as a sheaf version of FMHC:

$$K_{\mathbb{Z}} \in \mathrm{Ob}\, D^+(X, \mathbb{Z}), \qquad (K_{\mathbb{Q}}, W^f, W) \in \mathrm{Ob}\, D^+ F_2(X, \mathbb{Q}),$$
$$(K_{\mathbb{C}}, W^f, W, F) \in \mathrm{Ob}\, D^+ F_3(X, \mathbb{C}); \; \alpha : (K_{\mathbb{Q}}, W^f, W) \otimes \mathbb{C} \simeq (K_{\mathbb{C}}, W^f, W)$$

such that W_i^f / W_j^f is a cohomological MHC on X.

The global section functor Γ on X can be filtered derived using acyclic trifiltered canonical resolutions such as Godement flabby resolutions.

LEMMA 8.2.5 *The derived global section functor $R\Gamma$ of a cohomological FMHC on X is an FMHC.*

PROPOSITION 8.2.6 *Let (K, W^f, W, F) denote an FMHC, then*

(i) *the filtrations $W[n]$ and F define an MHS on the cohomology $H^n(K)$;*

(ii) *the terms of the spectral sequence defined by the filtration W^f on K, with induced weight W and Hodge F filtrations*

$$({}_{W^f}E_r^{p,q}, W, F) = \mathrm{Gr}_{-p}^{W^f} H^{p+q}(W_{-p+r-1}^f K/W_{-p-r}^f K), W, F)$$

form an MHS and the differentials d_r are morphisms of MHS for $r \geq 1$;

(iii) *the filtration W^f is a filtration by sub-MHS on cohomology and we have*

$$(\mathrm{Gr}_{-p}^{W^f} H^{p+q}(K), W[p+q], F) \simeq ({}_{W^f}E_\infty^{p,q}, W, F).$$

This is a convenient generalization of Theorem 3.4.20 in order to prove the existence of the monodromy filtration in our case. The proof is in [16, Theorem 2.8]. The formula for ${}_{W^f}E_r^{p,q}$ above coincides with Deligne's definition of the spectral sequence (see Lemma 3.2.7). In this formula, the MHS on ${}_{W^f}E_r^{p,q}$ is defined as on the cohomology of any MHC. We prove that the differential

$$_{W^f}E_r^{p,q} \to {}_{W^f}E_r^{p+r,q-r+1} = \mathrm{Gr}_{-p-r}^{W^f} H^{p+q+1}(W_{-p-1}^f K/W_{-p-2r}^f K)$$

is compatible with MHS. It is deduced from the connection morphism ∂ defined by the exact sequence of complexes

$$0 \to W_{-p-r}^f K/W_{-p-2r}^f K \to W_{-p+r-1}^f K/W_{-p-2r}^f K$$
$$\to W_{-p+r-1}^f K/W_{-p-r}^f K \to 0.$$

Let

$$\varphi : H^{p+q+1}(W_{-p-r}^f K/W_{-p-2r}^f K) \to H^{p+q+1}(W_{-p-1}^f K/W_{-p-2r}^f K)$$

denotes the morphism induced by the natural embedding $W_{-p-r}^f K \to W_{-p-1}^f K$ and

$$\pi \circ (\varphi_|) : W_{-p}^f H^{p+q+1}(W_{-p-r}^f K/W_{-p-2r}^f K)$$
$$\to \mathrm{Gr}_{-p-r}^{W^f} H^{p+q+1}(W_{-p-1}^f K/W_{-p-2r}^f K)$$

the restriction $\varphi_|$ of φ to W_{-p}^f composed with the projection π onto ${}_{W^f}E_r^{p+r,q-r+1}$ and the restriction of ∂ to the subspace W_{-p}^f,

$$W_{-p}^f H^{p+q}(W_{-p+r-1}^f K/W_{-p-r}^f K) \xrightarrow{\partial_|} W_{-p}^f H^{p+q+1}(W_{-p-r}^f K/W_{-p-2r}^f K);$$

then $d_r = \pi \circ \varphi_| (\circ \partial_|)$. Since the connection ∂ is compatible with MHS, as is $\pi \circ (\varphi_|)$, so is d_r.

The isomorphism $H^{p,q}(_{W^f} E_r^{*,*}, d_r) \to {}_{W^f} E_{r+1}^{p,q}$ is induced by the embedding $W_{-p+r-1}^f K \to W_{-p+r}^f K$; hence it is also compatible with W and F. We deduce that the recurrent filtrations on $E_{r+1}^{p,q}$ induced by W and F on $E_r^{p,q}$ coincide with W and F on $_{W^f} E_{r+1}^{p,q}$.

DEFINITION 8.2.7 (Limit mixed Hodge complex (LMHC))

(i) *A limit mixed Hodge complex* (K, W^f, W, F) *in* $D^+ F_3 \mathbb{A}$ *is given by the above data in Proposition 8.2.6(i) to (iii) of an FMHC, satisfying the following:*

   (1) *The subcomplexes* $(W_i^f K, W, F)$ *are MHC for all indices* i.

   (2) *For all* $n \in \mathbb{Z}$, *we have the induced MHC*

$$(\mathrm{Gr}_n^{W^f} K_{\mathbb{Q}}, W), (\mathrm{Gr}_n^{W^f} K_{\mathbb{C}}, W, F),$$

$$\mathrm{Gr}_n \alpha : (\mathrm{Gr}_n^{W^f} K_{\mathbb{Q}}, W) \otimes \mathbb{C} \simeq (\mathrm{Gr}_n^{W^f} K_{\mathbb{C}}, W).$$

   (3) *The spectral sequence of* $(K_{\mathbb{Q}}, W^f)$ *degenerates at rank 2:*

$$E_2(K_{\mathbb{Q}}, W^f) \simeq E_\infty(K_{\mathbb{Q}}, W^f).$$

(ii) *A cohomological LMHC on a space* X *is given by a sheaf version of the data in Proposition 8.2.6(i) to (iii) such that* $R\Gamma(X, K, W^f, W, F)$ *is an LMHC.*

PROPOSITION 8.2.8   *Let* (K, W^f, W, F) *denote an LMHC; then*

(i) *the filtrations* $W[n]$ *and* F *define an MHS on the cohomology* $H^n(K)$, *and* W^f *induces a filtration by sub-MHS;*

(ii) *the MHS deduced from (i) on* $E_\infty^{p,q}(K, W^f) = \mathrm{Gr}_{-p}^{W^f} H^{p+q}(K)$ *coincides with the MHS on the terms of the spectral sequence* $_{W^f} E_2^{p,q}$ *deduced from the MHS on the terms* $_{W^f} E_1^{p,q} = H^{p+q}(\mathrm{Gr}_{-p}^{W^f} K)$.

The proof, similar to the above case of FMHC, is in [16, Theorem 2.13].

8.2.3   Diagonal Direct Image of a Simplicial Cohomological FMHC

We define the direct image $(R\pi_* K, W, W^f, F)$ of a simplicial cohomological FMHC K on a simplicial space $\pi : X_\bullet \to X$ over X [12]. The important point here is that the weight W is in fact a diagonal sum $\delta((\pi)_\bullet W, L)$ with respect to a filtration L. This operation is of the same nature as the mixed cone over a morphism of MHC where the sum of the weight in the cone is also diagonal.

DEFINITION 8.2.9   *A simplicial cohomological FMHC on a simplicial (resp. simplicial strict) space* $X_\bullet$ *is given by*

- *a complex $K_{\mathbb{Z}}$;*

- *a bifiltered complex $(K_{\mathbb{Q}}, W^f, W)$, an isomorphism $K_{\mathbb{Q}} \simeq K_{\mathbb{Z}} \otimes \mathbb{Q}$;*

- *a trifiltered complex $(K_{\mathbb{C}}, W^f, W, F)$ on $X_{\bullet}$, an isomorphism*

$$(K_{\mathbb{C}}, W^f, W) \simeq (K_{\mathbb{Q}}, W^f, W) \otimes \mathbb{C};$$

such that the restriction K_p of K to each X_p is a cohomological FMHC.

8.2.3.1 *Differential Graded Cohomological FMHC Defined by a Simplicial Cohomological FMHC*

Let $\pi : X_{\bullet} \to X$ be a simplicial space over X. We define $R\pi_* K$ as a cosimplicial cohomological FMHC on X by deriving first π_i on each space X_i, and then we deduce on X a structure of a complex called a differential graded cohomological FMHC as follows.

Let $I_{\bullet}^*$ be an injective (or in general a π-acyclic) resolution of K on $X_{\bullet}$, that is, a resolution I_i^* on X_i varying functorially with the index i; then $\pi_* I_{\bullet}^*$ is a cosimplicial complex of abelian sheaves on X with double indices where p on $\pi_* I_i^p$ is the complex degree and i the cosimplicial degree. It is endowed by the structure of a double complex with the differential $\sum_i (-1)^i \delta_i x^{pq}$ deduced from the cosimplicial structure, as in [12] (see Definition 3.4.18 to Theorem 3.4.20). Such a structure is known as a cohomological differential graded DG^+-complex. This operation is carried on the various levels, rational and complex. We obtain the following structure.

Differential Graded Cohomological FMHC.
A differential graded DG^+-complex $C_{\bullet}^*$ is a bounded below complex of graded objects, with two degrees, the first defined by the complex and the second by the gradings. It is endowed with two differentials and viewed as a double complex.

A differential graded cohomological MHC is defined by a system of a DG^+-complex, filtered and bifiltered with compatibility isomorphisms

$$C_{\bullet}^*, (C_{\bullet}^*, W), (C_{\bullet}^*, W, F)$$

such that for each degree n of the grading, the component (C_n^*, W, F) is a CMHC.

8.2.3.2 *The Higher Direct Image of a Simplicial Cohomological FMHC*

It is defined by the simple complex associated to the double complex $(\pi_* I_q^*)$ with total differential involving the face maps δ_i of the simplicial structure and the differentials d_q on I_q^*,

$$s(\pi_* I_q^*)_{q \in \mathbb{N}}^n := \oplus_{p+q=n} \pi_* I_q^p; \quad d(x_q^p) = d_q(x_q^p) + \Sigma_i (-1)^i \delta_i x_q^p.$$

The filtration L with respect to the second degree will be useful,

$$L^r(s(\pi_* I_q^*)_{q \in \mathbb{N}}) = s(\pi_* I_q^*)_{q \geq r},$$

so we can deduce a cohomological FMHC on X by summing into a simple complex

$$R\pi_*K := s(R\pi_*K|X_q)_{q\in\mathbb{N}} := s(\pi_*I_q^*)_{q\in\mathbb{N}}.$$

DEFINITION 8.2.10 (Diagonal filtration)   *The diagonal direct image of K on X_* is defined by the data $(R\pi_*K, W, W^f, F)$:*

(i) *The weight diagonal filtration W (resp. the finite weight diagonal filtration W^f) is*

$$\delta(W, L)_n R\pi_*K := \oplus_p W_{n+p}\pi_*I_p^*, \quad \delta(W^f, L)_n R\pi_*K := \oplus_p W_{n+p}^f\pi_*I_p^*.$$

(ii) *The simple Hodge filtration F is $F^n R\pi_*K := \oplus_p F^n\pi_*I_p^*$.*

LEMMA 8.2.11   *The complex $(R\pi_*K, W^f, W, F)$ is a cohomological FMHC and we have*

$$(W_i^f/W_j^f)(R\pi_*K, W, F) \simeq \oplus_p((W_{i+p}^f/W_{j+p}^f)\pi_*I_p^*, \delta(W, L), F),$$
$$(\mathrm{Gr}_i^W(W_i^f/W_j^f)R\pi_*K, F) \simeq \oplus_p(\mathrm{Gr}_{i+p}^W(W_{i+p}^f/W_{j+p}^f)\pi_*I_p^*[-p], F).$$

8.2.4   Construction of a Limit MHS on the Unipotent Nearby Cycles

We illustrate the above theory by an explicit construction of an LMHC on the nearby cycles that is applied to define the limit MHS of a geometric VMHS.

Let $f : V \to D$ be a quasi-projective morphism to a disc. If D is small enough, the morphism is a topological bundle on D^*, hence the higher direct images $R^i f_*\mathbb{C}$ underlie a variation of the MHS defined on the cohomology of the fibers.

In order to obtain at the limit a canonical structure not depending on the choice of the general fiber at a point $t \in D^*$, we introduce what we call here the universal fiber to define the *nearby cycle complex of sheaves* $\Psi_f(\mathbb{C})$, of which we recall the definition in the complex analytic setting.

Let $V_0 = f^{-1}(0)$, $V^* = V - V_0$, $p : \widetilde{D}^* \to D^*$ be a universal cover of the punctured unit disc D^*, and consider the diagram

$$
\begin{array}{ccccccc}
\widetilde{V}^* & \xrightarrow{\widetilde{p}} & V^* & \xrightarrow{j} & V & \xleftarrow{i} & V_0 \\
\downarrow & & \downarrow & & \downarrow & & \downarrow \\
\widetilde{D}^* & \xrightarrow{p} & D^* & \to & D & \hookleftarrow & \{0\},
\end{array}
$$

where $\widetilde{V}^* := V^* \times_D \widetilde{D}^*$. For each complex of sheaves $\mathcal{F}$ of abelian groups on V^*, the nearby cycle complex of sheaves $\Psi_f(\mathcal{F})$ is defined as

$$\Psi_f(\mathcal{F}) := i^* R j_* R\widetilde{p}_*\widetilde{p}^*(\mathcal{F}).$$

Let $\widetilde{D}^* = \{u = x + iy \in \mathbb{C} : x < 0\}$ and the exponential map $p(u) = \exp u$. The translation $u \to u + 2\pi i$ on $\widetilde{D}^*$ lifts to an action on $\widetilde{V}^*$, inducing an action on $R\widetilde{p}_*\widetilde{p}^*(\mathcal{F})$ and finally a monodromy action T on $\Psi_f(\mathcal{F})$.

To construct the limit MHS, we put an explicit structure of mixed Hodge complex on the nearby cycles $R\Gamma(Y, \Psi_f(\mathbb{C})) = R\Gamma(\tilde{V}^*, \mathbb{C})$ complex.

The technique used here puts such a structure on the complex of subsheaves $\Psi_f^u(\mathbb{C})$, where the action of T is unipotent.

In view of recent developments, the existence of the weight filtration with rational coefficients becomes clear in the frame of the abelian category of perverse sheaves since the weight filtration is exactly the monodromy filtration defined by the nilpotent action of the logarithm of T on the perverse sheaf $\Psi_f^u(\mathbb{Q})$, up to a shift in indices.

Hence we will concentrate here on the construction of the weight filtration on the complex $\Psi_f^u(\mathbb{C})$. The construction is carried first for a smooth morphism, then applied to each space of a smooth simplicial covering of V.

8.2.5   Case of a Smooth Morphism

For a smooth morphism $f : V \to D$, following [15] and [39] the limit MHS is constructed on the universal fiber $\tilde{V}^*$. Since the MHS depends on the properties at infinity of the fibers, we need to introduce a compactification of the morphism f.

We suppose there exists such a compactification: a proper morphism called also $f : X \to D$ with algebraic fibers with $V \subset X : \overline{V} = X$, which induces the given morphism on V. This will apply to a quasi-projective morphism in which case we may suppose the morphism $f : X \to D$ is projective. Moreover, we suppose the divisor at infinity $Z = X - V$, the special fiber $Y = f^{-1}(0)$, and their union $Z \cup Y$, are normal crossing divisors in X.

To study the case of the smooth morphism on $V = X - Z$, we still cannot use the logarithmic complex $\Omega^*_{\tilde{X}^*}(\mathrm{Log}\, \tilde{Z}^*)$ since $\tilde{X}^*$ is analytic in nature (it is defined via the exponential map). So we need to introduce a subcomplex of sheaves of $\Omega^*_{\tilde{X}^*}(\mathrm{Log}\, \tilde{Z}^*)$, which underlie the structure of cohomological FMHC.

Construction of an FMHC.
We may start with the following result in [15]. Let c be a generator of the cohomology $H^1(D^*, \mathbb{Q})$ and denote by $\eta_{\mathbb{Q}} = \smile f^*(c)$ the cup product with the inverse image $f^*(c) \in H^1(X^*, \mathbb{Q})$. Locally at a point $y \in Y$, a neighborhood X_y is a product of discs and $X_y^* := X_y - (Y \cap X_y)$ is homotopic to a product of n punctured discs; hence $H^i(X_y^*, \mathbb{Q}) \simeq \wedge^i H^1(X_y^*, \mathbb{Q}) \simeq \wedge^i(\mathbb{Q}^n)$. The morphism $\eta_{\mathbb{Q}}$

$$H^i(X_y^*, \mathbb{Q}) \xrightarrow{\eta_{\mathbb{Q}}} H^{i+1}(X_y^*, \mathbb{Q})$$

is the differential of an acyclic complex $(H^i(X_y^*, \mathbb{Q})_{i \geq 0}, d = \eta_{\mathbb{Q}})$. A truncation $H^{* \geq i+1}(X_y^*, \mathbb{Q})$ of this complex defines a resolution ([15, Lecture 14, Lemma 4.18.4])

$$H^i(\tilde{X}_y^*, \mathbb{Q})^u \to H^{i+1}(X_y^*, \mathbb{Q}) \to \cdots \to H^p(X_y^*, \mathbb{Q}) \xrightarrow{\eta_{\mathbb{Q}}} H^{p+1}(X_y^*, \mathbb{Q}) \to \cdots$$

of the cohomology in degree i of the space $\tilde{X}_y^* = X_y \times_{D^*} \tilde{D}^*$, which is homotopic to a Milnor fiber. Dually, we have an isomorphism $H^i(X_y{}^*, \mathbb{Q})/\eta_{\mathbb{Q}} H^{i-1}(X_y{}^*, \mathbb{Q}) \simeq H^i(\tilde{X}_y^*, \mathbb{Q})^u$. This construction can be lifted to the complex level to produce, in our case, a cohomological FMHC on Y computing the cohomology of the space $\widetilde{X}^* = X \times_{D^*} \widetilde{D}^*$ homotopic to a general fiber as follows.

The logarithmic complex $\Omega_X^*(Log(Y \cup Z))$ computes $\mathbf{R}\,j_* \mathbb{C}_{X-(Y \cup Z)}$ where $j : X - (Y \cup Z) \to X$ denotes the open embedding. On the level of differential forms, df/f represents the class $2i\pi f^*(c)$, since $\int_{|z|=1} dz/z = 2i\pi$, and $\wedge df/f$ realizes the cup product as a morphism (of degree 1)

$$i_Y^*(\Omega_X^*(Log\,Y \cup Z)) \xrightarrow{\eta := \wedge df/f} i_Y^*(\Omega_X^*(Log\,Y \cup Z))[1]$$

satisfying $\eta^2 = 0$ so as to get a double complex.

By the above local result, the simple associated complex is quasi-isomorphic to the subsheaf of unipotent nearby cycles Ψ_f^u ([15], [16, Theorem 2.6]).

Since Y and Z are NCD, we write the logarithmic complex as $\Omega_X^*(Log\,Y)(Log\,Z)$ so as to introduce *the weight filtration* W^Y *(resp.* W^Z*) with respect to Y (resp. Z) in addition to the total weight filtration* $W^{Y \cup Z}$.

Still, to get regular filtrations we need to work on a finite complex deduced as a quotient modulo an acyclic subcomplex, hence we construct the trifiltered complex on which the filtrations are regular,

$$(\Psi_Y^u(Log\,Z), W^f, W, F),$$

as follows:

$$(\Psi_Y^u)^r(Log\,Z) := \oplus_{p \geq 0, q \geq 0, p+q=r}\Omega_X^{p+q+1}(Log\,Y \cup Z)/W_p^Y,$$

$$W_i(\Psi_Y^u)^r(Log\,Z) := \oplus_{p+q=r}W_{i+2p+1}^{Y \cup Z}\Omega_X^{p+q+1}(Log\,Y \cup Z)/W_p^Y,$$

$$W_i^f = \oplus_{p+q=r}W_i^Z, W_i^Y = \oplus_{p+q=r}W_{i+p+1}^Y, F^i = \oplus_{p+q=r}F^{i+p+1}.$$

It is the simple complex associated to the double complex

$$(\Psi_Y^u)^{p,q}(Log\,Z) := \Omega_X^{p+q+1}(Log\,Y \cup Z)/W_p^Y, \quad p \geq 0, q \geq 0, \quad d, \eta,$$

with the usual differential d of forms for fixed p and the differential $\eta := \wedge df/f$ for fixed q; hence the total differential is $D\omega = d\omega + (df/f) \wedge \omega$.

The projection map $(\Psi_Y^u)^{p,q}(Log\,Z) \to (\Psi_Y^u)^{p+1,q-1}(Log\,Z)$ is the action of an endomorphism on the term of degree $(p+q)$ of the complex commuting with the differential, hence an endomorphism $\nu : \Psi_Y^u \to \Psi_Y^u$ of the complex. The study of such a complex is reduced to the smooth proper case applied to X and the intersections of the components of Z via the residue Res_Z on $\operatorname{Gr}_*^{W^f} \Psi_Y^u(Log\,Z)$.

REMARK 8.2.12 By construction, the differentials are compatible with the above embedding. We take the quotient by various submodules W_p^Y which form an acyclic subcomplex, hence we have an isomorphism $\mathbb{H}^*(Y, \Psi_Y^u(Log\,Z)) \simeq H^*(\widetilde{V}^*, \mathbb{C})^u$ such that the action of ν induces $-\frac{1}{2\pi i}LogT$.

THEOREM 8.2.13 *The trifiltered complex* $(\Psi_Y^u(Log\,Z), W^f, W, F)$ *is a cohomological limit mixed Hodge complex (LMHC) which endows the cohomology* $H^*(\widetilde{V}^*, \mathbb{C})^u$ *with a limit MHS such that the weight filtration W is equal to the monodromy weight filtration relative to* W^f.

The theorem results from the following proposition where $Z = \cup_{i \in I_1} Z_i$ denotes a decomposition into components of Z, $Z^J = Z_{i_1} \cap \cdots \cap Z_{i_r}$ for $J = \{i_1, \ldots, i_r\} \subset I_1$, $Z^r := \coprod_{J \subset I_1, |J| = r} Z^J$.

PROPOSITION 8.2.14   (i) *Let $k_Z : Y - (Z \cap Y) \to Y$ and $j_Z : (X^* - Z^*) \to X^*$. There exists a natural quasi-isomorphism*

$$\Psi^u_Y(\operatorname{Log} Z) \xrightarrow{\approx} Rk_{Z,*}(k_Z^* \Psi^u_f(\mathbb{C})) \xrightarrow{\approx} \Psi^u_f(Rj_{Z,*}\mathbb{C}_{X^* - Z^*}).$$

(ii) *The graded part for W^f is expressed with the LMHC for the various proper smooth maps $Z^{-p} \to D$ for $p < 0$, with singularities along the NCD $Z^{-p} \cap Y$,*

$$\operatorname{Res}_Z : (\operatorname{Gr}^{W^f}_{-p}(\Psi^u_Y)(\operatorname{Log} Z), W, F) \simeq (\Psi^u_{Z^{-p} \cap Y}[p], W[-p], F[p]),$$

and the spectral sequence with respect to W^f is given by the limit MHS of the unipotent cohomology of $(\widetilde{Z}^{-p})^* = Z^{-p} \times_D \widetilde{D}^*$ twisted by (p):

$$(_{W^f}E_1^{p,q}, W, F) \simeq \mathbb{H}^{p+q}(Z^{-p} \cap Y, (\Psi^u_{Z^{-p} \cap Y}[p], W[-p], F[p]))$$
$$\simeq (\mathbb{H}^{2p+q}((\widetilde{Z}^{-p})^*, \mathbb{C})^u, W[-2p], F[p])$$
$$\simeq (\mathbb{H}^{2p+q}((\widetilde{Z}^{-p})^*, \mathbb{C})^u, W, F)(p).$$

(iii) *The endomorphism ν shifts the weight by -2: $\nu(W_i \Psi^u_Y) \subset W_{i-2}\Psi^u_Y$, and preserves W^f. It induces an isomorphism*

$$\nu^i : \operatorname{Gr}^W_{a+i} \operatorname{Gr}^{W^f}_a \Psi^u_Y(\operatorname{Log} Z) \xrightarrow{\sim} \operatorname{Gr}^W_{a-i} \operatorname{Gr}^{W^f}_a \Psi^u_Y(\operatorname{Log} Z).$$

Moreover, the action of ν corresponds to the logarithm of the monodromy on the cohomology $H^*(\widetilde{V}^*, \mathbb{C})^u$.

(iv) *The induced monodromy action $\bar{\nu}$ defines an isomorphism*

$$\bar{\nu}^i : \operatorname{Gr}^W_{a+i} \operatorname{Gr}^{W^f}_a H^n(Y, \Psi^u_Y(\operatorname{Log} Z)) \xrightarrow{\sim} \operatorname{Gr}^W_{a-i} \operatorname{Gr}^{W^f}_a H^n(Y, \Psi^u_Y(\operatorname{Log} Z)).$$

COROLLARY 8.2.15 *The weight filtration induced by W on $H^*(\widetilde{V}^*, \mathbb{C})^u$ satisfies the characteristic property of the monodromy weight filtration relative to the weight filtration W^f.*

The main argument consists in deducing (iv) from the corresponding isomorphism on the complex level in (iii) after a reduction to the proper case. We remark also ([16, Proposition 3.5]) that the spectral sequence $_{W^f}E_r^{p,q}$ is isomorphic to the weight spectral sequence of any fiber $X_t - Z_t$ for $t \in D^*$ and degenerates at rank 2.

8.2.5.1   Proof of Proposition 8.2.14 in the Proper Smooth Case (VHS)

The complex $\Psi_Y^u(\text{Log}\,Z)$ for $Z = \emptyset$ coincides with Steenbrink's complex Ψ_Y^u on Y in $X = V$ [39]. In this case W^f is trivial, $W^{Y\cup Z} = W^Y$ on $\Omega_X^*(\text{Log}\,Y)$, and (Ψ_Y^u, W, F) is an MHC since its graded object is expressed in terms of the Hodge complexes defined by the embedding of the various s intersections of components Y_i of $Y = \cup_{i\in I} Y_i$ denoted as $a_s : Y^s := \coprod_{J\subset I, |J|=s} Y^J \to X$, where $Y^J = Y_{i_1} \cap \cdots \cap Y_{i_s}$ for $J = \{i_1, \ldots, i_s\}$.

The residue

$$(\text{Gr}_r^W \Psi_Y^u, F) \xrightarrow{\text{Res}} \oplus_{p\geq \sup(0,-r)} a_{r+2p+1,*}(\Omega_{Y^{r+2p+1}}^*[-r-2p], F[-p-r])$$

defines an isomorphism with the HC of weight r on the right, then Proposition 8.2.14(i) for $Z = \emptyset$ reduces to the quasi-isomorphism

$$\Psi_Y^u \xrightarrow{\approx} \Psi_f^u(\mathbb{C}).$$

Locally, the cohomology $H^i(\Psi_Y^u(\mathbb{C})_y)$ of the stalk at y is equal to the unipotent cohomology of the universal Milnor fiber $\tilde{X}_y^*$ of f at y; hence the quasi-isomorphism above follows by construction of $\Psi_{Y,y}^u$ since

$$H^i(\Psi_f^u(\mathbb{C})_y) \simeq H^i(\tilde{X}_y^*, \mathbb{C})^u \simeq H^i(\Psi_{Y,y}^u).$$

To prove this local isomorphism, we use the spectral sequence of Ψ_Y^u with respect to the columns of the underlying double complex. Since the pth column is isomorphic to $(Rj_*\mathbb{C}/W_p^Y)[p+1]$, the stalk at y is $E_{1,y}^{-p,q} \simeq H^{p+q+1}(X_y^*, \mathbb{C})$, for $p \geq 0, q \geq 0$, and 0 otherwise, with differential $d = \eta$; hence the term $E_{2,y}^{-p,q}$ is equal to 0 for $p > 0$ and equal to $H^{p+q}(\tilde{X}_y^*, \mathbb{C})$ for $p = 0$, where $(E_{1,y}^{-p,q}, \eta)$ is the resolution of the cohomology of the Milnor fiber mentioned earlier; then the global isomorphism follows:

$$\mathbb{H}^i(Y, \Psi_Y^u) \simeq \mathbb{H}^i(\tilde{X}^*, \mathbb{C})^u.$$

In Proposition 8.2.14(ii) we use the residue to define an isomorphism on the first terms of the spectral sequence with respect to W with the HS defined by Y^i after a twist:

$$_W E_1^{p,q} = \mathbb{H}^{p+q}(Y, \text{Gr}_{-p}^W \Psi_Y^u, F)$$
$$= \oplus_{q'\geq\sup(0,p)} H^{2p-2q'+q}(Y^{2q'-p+1}, \mathbb{C}), F)(p-q').$$

Proposition 8.2.14(iii) reduces to an isomorphism

$$\text{Gr}_i^W \Psi_Y^u \xrightarrow{\nu^i} \text{Gr}_{-i}^W \Psi_Y^u,$$

which can be checked easily since

$$W_i(\Psi_Y^u)^{p,q} := W_{i+2p+1}\Omega_X^{p+q+1}(\text{Log}\,Y)/W_p^Y = W_{i-2}(\Psi_Y^u)^{p+1,q-1}$$
$$= \cdots = W_{-i}(\Psi_Y^u)^{p+i,q-i},$$

while the two conditions $(p + i \geq 0, p \geq 0)$ for $W_i(\Psi_Y^u)^{p,q}$, become successively $(p+i) - i = p \geq 0, p+i \geq 0$ for $W_{-i}(\Psi_Y^u)^{p+i,q-i}$, hence they are interchanged. We end the proof in the next section.

8.2.6 Polarized Hodge–Lefschetz Structure

The first correct proof of Proposition 8.2.14(iv) is given in [35] in the more general setting of polarized Hodge–Lefschetz modules. We follow [33] for an easy exposition in our case.

8.2.6.1 Hodge–Lefschetz Structure

Two endomorphisms on a finite-dimensional bigraded real vector space $L = \oplus_{i,j \in \mathbb{Z}} L^{i,j}$, $l_1 : L^{i,j} \to L^{i+2,j}$ and $l_2 : L^{i,j} \to L^{i,j+2}$, define a Lefschetz structure if they commute and if moreover the morphisms obtained by composition,

$$l_1^i : L^{-i,j} \to L^{i,j}, i > 0 \text{ and } l_2^j : L^{i,-j} \to L^{i,j}, j > 0,$$

are isomorphisms.

It is classical to deduce from representation theory, as in the hard Lefschetz theorem, that such a structure corresponds to a finite-dimensional representation of the group $\mathrm{SL}(2, \mathbb{R}) \times \mathrm{SL}(2, \mathbb{R})$; then a primitive decomposition follows:

$$L^{i,j} = \oplus_{r,s \geq 0} l_1^r l_2^s L_0^{i-2r,j-2s},$$

where $L_0^{-i,-j} = L^{-i,-j} \cap \mathrm{Ker}\, l_1^{i+1} \cap \mathrm{Ker}\, l_2^{j+1}, i \geq 0, j \geq 0$.

A Lefschetz structure is called a *Hodge–Lefschetz structure* if in addition $L^{i,j}$ underlies real Hodge structures and l_1, l_2 are compatible with such structures.

A polarization of L is defined by a real bigraded bilinear form $S : L \otimes L \to \mathbb{R}$ compatible with HS, such that

$$S(l_i x, y) + S(x, l_i y) = 0, \ i = 1, 2.$$

It extends to a complex Hermitian form on $L \otimes \mathbb{C}$ such that the induced form $S(x, C l_1^i l_2^j y)$ is symmetric positive definite on $L_0^{-i,-j}$, where C is the Weil operator defined by the HS.

A differential $d : L \to L$ is a morphism compatible with HS satisfying

$$d : L^{i,j} \to L^{i+1,j+1}, i, j \in \mathbb{Z}, d^2 = 0, [d, l_i] = 0, i = 1, 2,$$
$$S(dx, y) = S(x, dy), x, y \in L.$$

THEOREM 8.2.16 ([35, 33]) *Let* (L, l_1, l_2, S, d) *be a bigraded Hodge–Lefschetz structure; then the cohomology* $(H^*(L, d), l_1, l_2, S)$ *is a polarized Hodge–Lefschetz structure.*

We assume the theorem, that we apply to the weight spectral sequence as follows.

Let $n = \dim X$ and let $K_{\mathbb{C}}^{i,j,k} = H^{i+j-2k+n}(Y^{2k-i+1}, \mathbb{C})(i-k)$, for $k \geq \sup(0, i)$, and $K_{\mathbb{C}}^{i,j,k} = 0$ otherwise. Then the residue induces an isomorphism of $K_{\mathbb{C}}^{i,j} = \oplus_{k \geq \sup(0,i)} K_{\mathbb{C}}^{i,j,k}$ with the terms of the spectral sequence above: $_W E_1^{r,q-r} \simeq K_{\mathbb{C}}^{r,q-n}$. Since the special fiber Y is projective, the cup product with a hyperplane section class defines a morphism $l_1 = \smile c$ satisfying the hard Lefschetz theorem on the various smooth proper intersections Y^i of the components of Y, while l_2 is defined by the action of $-\frac{1}{2\pi i}N$ on E_1 defined by the action of ν on the complex Ψ_Y^u. The differential d is defined on the terms of the spectral sequence which are naturally polarized as cohomology of smooth projective varieties. Then, all the conditions to apply the above result on differential polarized bigraded Hodge–Lefschetz structures are satisfied, so we can deduce the following result:

COROLLARY 8.2.17 *For all $q, r \geq 0$, the endomorphism N induces an isomorphism of HS,*

$$N^r : \mathrm{Gr}_{q+r}^W H^q(\widetilde{X}^*, \mathbb{Q})^u \to \mathrm{Gr}_{q-r}^W H^q(\widetilde{X}^*, \mathbb{Q})^u(-r).$$

This ends the proof in the smooth proper case.

REMARK 8.2.18 (Normal crossing divisor case) Let $X = \cup_{i \in I} X_i$ be embedded as an NCD with smooth irreducible components X_i in a smooth variety V projective over the disc D, such that the fiber Y at 0 and its union with X is an NCD in V. Then the restriction of f to the intersections $X_J = \cap_{i \in J} X_i$, $J \subset I$ is an NCD $Y_J \subset X_J$, and the limit MHC $\Psi_{Y_J}^u$ for various $J, \emptyset \neq J \subset I$ form a simplicial cohomological MHC on the semisimplicial variety X_* defined by X. In this case the finite filtration W^f on the direct image, coincides with the increasing filtration associated by change of indices to the canonical decreasing filtration L on the simplicial complex, that is, $W_i^f = L^{-i}$, so that we can apply the general theory to obtain a cohomological LMHC defining the LMHS on the cohomology $H^*(\widetilde{X}^*, \mathbb{Q})$. This is an example of the general singular case.

If we add $X_\emptyset = V$ to the simplicial variety X_* we obtain the cohomology with compact support of the general fiber of $V - X$, which is Poincaré dual to the cohomology of $V - X$, the complement in V of the NCD. This remark explains the parallel (in fact, dual) between the logarithmic complex case and the simplicial case.

8.2.6.2 *Proof of Proposition 8.2.14 in the Open Smooth Case*

We consider the maps $(\widetilde{X}^* - \widetilde{Z}^*) \xrightarrow{\widetilde{j}_Z} \widetilde{X}^* \xrightarrow{\widetilde{j}_Y} X$; then Proposition 8.2.14(i) follows from the isomorphisms

$$(\Psi_Y^u(\mathrm{Log}\,Z), W^f) \simeq i_Y^* R\widetilde{j}_{Y,*}(\Omega_{\widetilde{X}_*}^*(\mathrm{Log}\,\widetilde{Z}^*), W^{\widetilde{Z}^*})$$

$$\simeq i_Y^* R\widetilde{j}_{Y,*}(R\widetilde{j}_{Z,*}\mathbb{C}_{\widetilde{X}_* - \widetilde{Z}_*}, \tau) \simeq \Psi_f^u(R j_{Z,*}\mathbb{C}_{X^* - Z^*}, \tau).$$

Let $I_1 \subset I$ denote the set of indices of the components of Z, Z^i the union of the intersections Z^J for $J \subset I_1, |J| = i$ and $a_{Z^i} : Z^i \to X$. Proposition 8.2.14(ii) follows from the corresponding bifiltered residue isomorphism along Z:

$$(\mathrm{Gr}_i^{W^Z} \Omega_X^*(\mathrm{Log}\,Y)(\mathrm{Log}\,Z, W, F) \simeq a_{Z^i,*}(\Omega_{Z^i}^*(\mathrm{Log}\,Y \cap Z^i), W[i], F[-i]).$$

More generally, we have residue isomorphisms Res_Z and Res_Y ([16, 3.3.2])

$$\mathrm{Gr}_m^W (W_b^f/W_a^f)(\Psi_Y^u(\mathrm{Log}\,Z), F) \xrightarrow{\mathrm{Res}_Z} \mathrm{Gr}_{m-j}^W(\Psi_{Z^j \cap Y}^u[-j], F[-j])$$

$$\xrightarrow{\mathrm{Res}_Y} \oplus_{j \leq m+p, j \in [a,b], p \geq 0}(\Omega_{Z^j \cap Y}^{*\,m-j+2p+1}[-m-2p], F[-m-p]),$$

where $Z^i \cap Y$ is the union of $Z^J \cap Y$, so we can deduce the structure of LMHC in the open smooth case from the proper case.

The isomorphism of complexes in Proposition 8.2.14(iii) can be easily checked. While Proposition 8.2.14(iv) for a smooth proper $X \to D$ is deduced via the above Res_Z from the proper smooth projective case $Z^j \to D$ for various j as follows. The monodromy ν induces on the spectral sequence the isomorphism for $p \leq 0$,

$$(\mathrm{Gr}^W_{i+b}(_{W^f} E^{p,i}_1), d_1) \xrightarrow{\nu^b} (\mathrm{Gr}^W_{i-b}(_{W^f} E^{p,i}_1), d_1),$$

which commutes with the differential d_1 equal to a Gysin morphism alternating with respect to the embeddings of components of Z^{-p} into Z^{-p-1}. Since the isomorphism

$$\nu^b : \mathrm{Gr}^W_{2p+i+b}(H^{2p+i}(Z^{-p}, \mathbb{C})_{p \in \mathbb{Z}}, \mathrm{Gysin})$$
$$\xrightarrow{\sim} \mathrm{Gr}^W_{2p+i-b}(H^{2p+i}(Z^{-p}, \mathbb{C})_{p \in \mathbb{Z}}, \mathrm{Gysin})$$

has been checked in the proper case Z^{-p}, then we deduce Proposition 8.2.14(iv):

$$\mathrm{Gr}^W_{i+b} \mathrm{Gr}^{W^f}_i H^n(Y, \Psi^u_Y(\mathrm{Log}\, Z)) \xrightarrow{\nu^b} \mathrm{Gr}^W_{i-b} \mathrm{Gr}^{W^f}_i H^n(Y, \Psi^u_Y(\mathrm{Log}\, Z)).$$

8.2.7 Quasi-projective Case

Let $f : V \to D$ be quasi-projective. There exists a simplicial smooth hypercovering of V of the following type. First, we consider an extension into a projective morphism $f : X \to D$ by completing with $Z = X - V$, then we consider a simplicial smooth hypercovering $\pi : X_\bullet \to X$ with $\pi_i := \pi_{|X_i}$ such that $Z_i := \pi_i^{-1}(Z)$ consists of an NCD in X_i. Let $f_i := f \circ \pi_i : X_i \to D$; we may suppose Y_i and $Y_i \cup Z_i$ are NCD in X_i so to consider the simplicial cohomological limit MHC $(\Psi^u_{Y_\bullet}(\mathrm{Log}\, Z_\bullet), W^f, W, F)$ and its direct image $R\pi_\bullet(\Psi^u_{Y_\bullet}(\mathrm{Log}\, Z_\bullet), W^f, W, F)$ on X, then the theorem results from the following proposition ([16, 3.26, 3.29]):

PROPOSITION 8.2.19 *The trifiltered complex*

$$R\pi_*(\Psi^u_{Y_\bullet}(\mathrm{Log}\, Z_\bullet), W^f, W, F)$$

satisfies the following properties:

(i) *Let $k_Z : Y - (Z \cap Y) \to Y, j_Z : X^* - Z^* \to X^*$; then there exist natural quasi-isomorphisms*

$$R\pi_* \Psi^u_{Y_\bullet}(\mathrm{Log}\, Z_\bullet) \xrightarrow{\sim} Rk_{Z,*}(k_Z^* \Psi^u_f(\mathbb{C})) \xrightarrow{\sim} \Psi^u_f(Rj_{Z,*}\mathbb{C}_{X^*-Z^*}).$$

(ii) *The graded part for W^f is expressed in terms of the cohomological limit MHC for the various smooth maps $(X_i - Z_i) \to D$,*

$$\mathrm{Gr}^{W^f}_{-p} R\pi_*(\Psi^u_{Y_\bullet}(\mathrm{Log}\, Z_\bullet), W, F)$$
$$\simeq \oplus_i R\pi_{i,*}(\mathrm{Gr}^{W^{Z_i}}_{i-p} \Psi^u_{X_i}(\mathrm{Log}\, Z_i)[-i], W[-i], F)$$
$$\simeq \oplus_i R\pi_{i,*}(\Psi^u_{Z_i^{i-p} \cap Y_i}[p - 2i], W[-p], F[p - i]).$$

The spectral sequence with respect to W^f is given by the twisted limit MHS on the cohomology of $(\widetilde{Z}_i^{i-p})^* = Z_i^{i-p} \times_D \widetilde{D}^*$,

$$_{W^f} E_1^{p,q}(R\Gamma(Y, R\pi_*(\Psi_{Y_\bullet}^u(\operatorname{Log} Z_\bullet), W^f, W, F)))$$
$$\simeq \oplus_i \mathbb{H}^{p+q}(Z_i^{i-p} \cap Y_i, (\Psi_{Z_i^{i-p} \cap Y_i}^u[p-2i], W[-p], F[p-i]))$$
$$\simeq \oplus_i (H^{2p+q-2i}((\widetilde{Z}_i^{i-p})^*, \mathbb{C})^u, W, F)(p-i).$$

(iii) The monodromy ν shifts the weight by -2: $\nu(W_i \Psi_Y^u) \subset W_{i-2}\Psi_Y^u$ and preserves W^f. It induces an isomorphism

$$\nu^i : \operatorname{Gr}_{a+i}^W \operatorname{Gr}_a^{W^f} R\pi_* \Psi_{Y_\bullet}^u(\operatorname{Log} Z_\bullet) \xrightarrow{\sim} \operatorname{Gr}_{a-i}^W \operatorname{Gr}_a^{W^f} R\pi_* \Psi_{Y_\bullet}^u(\operatorname{Log} Z_\bullet).$$

(iv) The induced iterated monodromy action ν^i defines an isomorphism

$$\nu^i : \operatorname{Gr}_{a+i}^W \operatorname{Gr}_a^{W^f} H^n(Y, R\pi_* \Psi_{Y_\bullet}^u(\operatorname{Log} Z_\bullet))$$
$$\xrightarrow{\sim} \operatorname{Gr}_{a-i}^W \operatorname{Gr}_a^{W^f} H^n(Y, R\pi_* \Psi_{Y_\bullet}^u(\operatorname{Log} Z_\bullet)).$$

The proof is by reduction to the smooth proper case, namely, the various intersections Z_i^j of the components of the NCD Z_i in X_i as in the smooth open case. The spectral sequence is expressed as a double complex as in the case of the diagonal direct image in general. In particular, the differential $d_1 : {}_{W^f}E_1^{n-i-1,i} \to {}_{W^f}E_1^{n-i,i}$ is written in terms of alternating Gysin maps associated to $(\widetilde{Z}_j^{i+j-n})^* \to (\widetilde{Z}_j^{i+j-n-1})^*$ and simplicial maps d' associated to $(\widetilde{Z}_j^{i+j-n})^* \to (\widetilde{Z}_{j-1}^{i+j-n})^*$ in the double complex ([12, 8.1.19] and [16, 3.30.1]) written as

$$\begin{array}{ccc}
H^{2n-2j-i}((\widetilde{Z}_j^{i+j-n})^*, \mathbb{C})^u & \xrightarrow{\partial = \text{Gysin}} & H^{2n+2-2j-i}((\widetilde{Z}_j^{i+j-n-1})^*, \mathbb{C})^u \\
\uparrow d' & & \uparrow d' \\
H^{2n-2j-i}((\widetilde{Z}_{j-1}^{i+j-n+1})^*, \mathbb{C})^u & \xrightarrow{\partial = \text{Gysin}} & H^{2n+2-2j-i}((\widetilde{Z}_{j-1}^{i+j-n-1})^*, \mathbb{C})^u.
\end{array}$$

The isomorphism (iii) follows from the same property on each X_i while the isomorphism (iv) is deduced from the smooth case above.

8.2.8 Alternative Construction, Existence and Uniqueness

An easy example of simplicial variety is defined in the case of an open normal crossing divisor (NCD) (see Section 3.4.3) and reciprocally the MHS of an embedded variety in a smooth algebraic variety can be deduced from this case.

In parallel, we deduce the limit structure on cohomology of a quasi-projective family from the case of a relative open NCD in a projective smooth family.

8.2.8.1 Hypothesis

Let $f : X \to D$ be a projective family; then f may be written as the composition $X \xrightarrow{i_X} P_D \xrightarrow{h} D$ of the natural projection h of the relative projective space over a disc D with a closed embedding i_X. Let $i_Z : Z \to X$ be a closed embedding. By Hironaka's desingularization we construct diagrams

$$
\begin{array}{ccccc}
Z'' & \to & X'' & \to & P'' \\
\downarrow & & \downarrow & & \downarrow q \\
Z' & \to & X' & \to & P' \\
\downarrow & & \downarrow & & \downarrow p \\
Z & \xrightarrow{i_Z} & X & \xrightarrow{i_X} & P_D \\
& \searrow & \downarrow f & \swarrow h & \\
& & D & &
\end{array}
\qquad
\begin{array}{ccccc}
Z_0'' & \to & X_0'' & \to & P_0'' \\
\downarrow & & \downarrow & & \downarrow \\
Z_0' & \to & X_0' & \to & P_0' \\
\downarrow & & \downarrow & & \downarrow \\
Z_0 & \to & X_0 & \to & P_0 \\
& \searrow & \downarrow & \swarrow & \\
& & 0 & &
\end{array}
$$

first by blowing up centers over Z so as to obtain a smooth space P' with a projection p over P_D such that $P_0' := p^{-1}(P_0)$, $Z' := p^{-1}(Z)$, and $P_0' \cup Z'$ are all NCD. Set $X' := p^{-1}(X)$; then the restrictions of p are isomorphisms

$$
p_| : (X' - Z') \xrightarrow{\sim} (X - Z), \quad p_| : (P' - Z') \xrightarrow{\sim} (P_D - Z)
$$

since the modifications are all over Z. Next, by blowing up centers over X' we obtain a smooth space P'' with a projection q over P' such that $X'' := q^{-1}(X')$, $P_0'' := q^{-1}(P_0')$, $Z'' := q^{-1}(Z')$, and $P_0'' \cup X''$ are all NCD, and the restriction of q, $q_| : P'' - X'' \xrightarrow{\sim} P' - X'$ is an isomorphism. For D small enough, X'', Z'', and Z' are relative NCD over D^*. Hence we deduce the diagram

$$
\begin{array}{ccccc}
X'' - Z'' & \xrightarrow{i_X''} & P'' - Z'' & \xleftarrow{j''} & P'' - X'' \\
q_X \downarrow & & q \downarrow & & q_| \downarrow \wr \\
X' - Z' & \xrightarrow{i_X'} & P' - Z' & \xleftarrow{j'} & P' - X'.
\end{array}
$$

Since all modifications are above X', we still have an isomorphism $q_|$ induced by q on the right.

Since q is proper, for all integers i, the morphism $q^* : H_c^{2d-i}(P' - Z', \mathbb{Q}) \to H_c^{2d-i}(P'' - Z'', \mathbb{Q})$ is well defined on cohomology with compact support, where d is the dimension of P_D. Its Poincaré dual is called the trace morphism:

$$
\mathrm{Tr}\, q : H^i(P'' - Z'', \mathbb{Q}) \to H^i(P' - Z', \mathbb{Q}).
$$

It satisfies the relation $\mathrm{Tr}\, q \circ q^* = \mathrm{Id}$. The trace morphism is defined as a morphism of sheaves:

$$
q_* \mathbb{Z}_{P'' - Z''} \to \mathbb{Z}_{P' - Z'}
$$

in [42] and Section 3.4.3. Hence an induced morphism: $(\mathrm{Tr}\, q)_| : H^i(X'' - Z'', \mathbb{Q}) \to H^i(X' - Z', \mathbb{Q})$ is well defined. Taking the inverse image on a universal covering $\widetilde{D^*}$, we get a diagram of universal fibers

$$
\begin{array}{ccccc}
(\widetilde{X}'' - \widetilde{Z}'')^* & \xrightarrow{\tilde{i}_X''} & (\widetilde{P}'' - \widetilde{Z}'')^* & \xleftarrow{\tilde{j}''} & (\widetilde{P}'' - \widetilde{X}'')^* \\
\tilde{q}_X \downarrow & & \tilde{q} \downarrow & & \tilde{q}_| \downarrow \wr \\
(\widetilde{X}' - \widetilde{Z}')^* & \xrightarrow{\tilde{i}_X'} & (\widetilde{P}' - \widetilde{Z}')^* & \xleftarrow{\tilde{j}'} & (\widetilde{P}' - \widetilde{X}')^*.
\end{array}
$$

PROPOSITION 8.2.20   *With the notation of the above diagram, we have short exact sequences*

$$0 \to H^i((\widetilde{P}'' - \widetilde{Z}'')^*, \mathbb{Q}) \xrightarrow{(\widetilde{i}''_X)^* - \mathrm{Tr}\,\tilde{q}} H^i((\widetilde{X}'' - \widetilde{Z}'')^*, \mathbb{Q}) \oplus H^i((\widetilde{P}' - \widetilde{Z}')^*, \mathbb{Q})$$

$$\xrightarrow{(\widetilde{i}'_X)^* - (\mathrm{Tr}\,\tilde{q})_{|(\widetilde{X}'' - \widetilde{Z}'')}} H^i((\widetilde{X}' - \widetilde{Z}')^*, \mathbb{Q}) \to 0.$$

Since we have a vertical isomorphism $\tilde{q}_|$ on the right in the above diagram, we deduce a long exact sequence of cohomology spaces containing the sequences of the proposition; the injectivity of $(\widetilde{i}''_X)^* - \mathrm{Tr}\,\tilde{q}$ and the surjectivity of $(\widetilde{i}'_X)^* - (\mathrm{Tr}\,\tilde{q})_{|(\widetilde{X}'' - \widetilde{Z}'')}$ are deduced from the relation $\mathrm{Tr}\,\tilde{q} \circ \tilde{q}^* = \mathrm{Id}$ and $(\mathrm{Tr}\,\tilde{q})_{|(\widetilde{X}'' - \widetilde{Z}'')} \circ \tilde{q}_X^* = \mathrm{Id}$; hence the long exact sequence of cohomology deduced from the diagram splits into short exact sequences.

COROLLARY 8.2.21   *(i) We have an isomorphism*

$$H^i((\widetilde{X} - \widetilde{Z})^*, \mathbb{Z}) \xrightarrow{\tilde{p}^*_|} H^i((\widetilde{X}' - \widetilde{Z}')^*, \mathbb{Z}).$$

(ii) The above isomorphism defines on the cohomology $H^i((\widetilde{X} - \widetilde{Z})^*, \mathbb{Z})$ a limit MHS isomorphic to the cokernel of $(\widetilde{i}''_X)^* - \mathrm{Tr}\,\tilde{q}$ acting as a morphism of limit MHS.

The cohomology $H^i((\widetilde{X} - \widetilde{Z})^*, \mathbb{Z})$ is isomorphic to $H^i((\widetilde{X}' - \widetilde{Z}')^*, \mathbb{Z})$ since the morphism $\tilde{p}^*_|$ is induced by the isomorphism $p_| : X' - Z' \simeq X - Z$.

Although the cokernel is not defined in general in the additive category of limit MHS, we show that it is well defined in this case: in fact, we remark here that *the exact sequence is strict not only for the weight W, but also for W^f* since it is isomorphic to a similar exact sequence for each fiber at a point $t \in D^*$, where W^f is identified with the weight filtration on the respective cohomology groups over the fiber at t.

Hence the sequence remains exact after taking the graded part $\mathrm{Gr}^W_* \mathrm{Gr}^{W^f}_*$ of each term. The left-hand term carries a limit MHS as the special case of the complement of the relative NCD $Z'' \to D$ over D in the smooth proper variety $P'' \to D$, while the middle term is the complement of the intersection of the relative NCD $Z'' \to D$ over D with the relative NCD $X'' \to D$ over D. Both cases can be treated by the above special cases without the general theory of simplicial varieties.

Hence we deduce the limit MHS on the right as a quotient. This shows that the limit structure is uniquely defined by the construction in the NCD case and by duality in the logarithmic case for smooth families.

8.3   ADMISSIBLE VARIATION OF MIXED HODGE STRUCTURE

The degeneration properties of VMHS of geometric origin on a punctured unit disc are not necessarily satisfied for general VMHS as has been the case for VHS with the results of Schmid. The notion of admissible VMHS over a unit disc ([40]) extends the notion of good VMHS ([13]) and assumes by definition the degeneration properties of the geometric case. Such a definition has been extended in [22] to analytic spaces and is satisfactory for natural operations such as the direct image by a projective morphism of varieties [35]. We mention here the main local properties of admissible VMHS over the complement of a normal crossing divisor (DCN) proved by Kashiwara in [22].

As an application of this concept we describe a natural MHS on the cohomology of an admissible VMHS, and in the next section we recall the definition of normal functions and the proof of the algebraicity of the zero set of normal functions which answers a question raised by Griffiths and Green.

The results apply in general for a VMHS with quasi-unipotent local monodromy at the points of degeneration of the NCD; however we assume the local monodromy is unipotent, to simplify the exposition and the proofs.

8.3.1 Definition and Results

We consider a VMHS $(\mathcal{L}, W, F)$ (see Definition 8.1.13) on the complement X^* of an NCD in an analytic manifold X with unipotent local monodromy and we denote by $(\mathcal{L}_{\mathcal{O}_X}, \nabla)$ Deligne's canonical extension of $\mathcal{L} \otimes \mathcal{O}_{X^*}$ to an analytic vector bundle on X with a flat connection having logarithmic singularities [10].

To relate Deligne's previous notations with [22] and Chapter 7, we recall for each component Y_j of the NCD, the equality $N_j := Log\, T_j = -2\pi i Res \nabla_j$ (-2π the residue of the connection along ∂_j), where the monodromy T_j is defined by extension (or parallel transport) along a loop γ from $\gamma(0)$ to $\gamma(1)$.

The filtration by sublocal systems W defines a filtration by canonical extensions of $W \otimes \mathcal{O}_{X^*}$, subbundles of $\mathcal{L}_{\mathcal{O}_X}$, denoted $\mathcal{W}$. The graded object $Gr_k^{\mathcal{W}} \mathcal{L}_{\mathcal{O}_X}$ is the canonical extension of $Gr_k^W \mathcal{L}$ and we know that the Hodge filtration by subbundles extends on $Gr_k^{\mathcal{W}} \mathcal{L}_{\mathcal{O}_X}$ by Schmid's result [37].

Strengthening Deligne's good VMHS ([13, Problem (1.8.15)]), the following definition assumes by hypothesis the degeneration properties in the geometric case.

DEFINITION 8.3.1 (Preadmissible VMHS ([40, 3.13]))   *A variation of mixed Hodge structure $(\mathcal{L}, W, \mathcal{F})$ (see Definition 8.1.13) graded polarizable over the punctured unit disc D^* with local monodromy T, is called* preadmissible *if the following conditions are satisfied:*

(i) *The Hodge filtration $\mathcal{F} \subset \mathcal{L}_{\mathcal{O}_{D^*}}$ extends to a filtration $\mathcal{F}$ on Deligne's extension $\mathcal{L}_{\mathcal{O}_D}$ by subbundles inducing for each k on $Gr_k^{\mathcal{W}} \mathcal{L}_{\mathcal{O}_D}$, Schmid's extension of the Hodge filtration.*

(ii) *Let $W^0 := \mathcal{W}(0)$, $F_0 := \mathcal{F}(0)$ denote the filtrations on the fiber $L_0 := \mathcal{L}_{\mathcal{O}_D}(0)$ at $0 \in D$ and T the local monodromy at 0, $N = \log T$; then the following two conditions must be satisfied:*

- $N F_0^p \subset F_0^{p-1}$ *for all $p \in \mathbb{Z}$.*
- *The weight filtration $M(N, W^0)$ relative to W^0 exists.*

Notice that the extension of the filtration $\mathcal{F}$ to $\mathcal{L}_{\mathcal{O}_D}$ cannot be deduced in general from the various Schmid extensions to $Gr_k^{\mathcal{W}} \mathcal{L}_{\mathcal{O}_D}$.

We remark that the filtrations $M := M(N, W^0)$ and F_0 at the origin define an MHS:

LEMMA 8.3.2 (Deligne)   *Let $(\mathcal{L}, W, \mathcal{F})$ be a preadmissible VMHS. Then (L_0, M, F_0) is an MHS. The endomorphism N is compatible with the MHS of type $(-1, -1)$.*

The proof due to Deligne is stated in [40, Appendix]. The result follows from the following properties:

(i) (L_0, W^0, F_0, N) satisfies $NF_0^p \subset F_0^{p-1}$ and $NW_k^0 \subset W_k^0$.

(ii) The relative filtration $M(N, W^0)$ exists.

(iii) For each k, $(\mathrm{Gr}_k^{W^0} L_0, M, F_0)$ is an MHS.

The admissibility property in the next definition by Kashiwara coincides over D^* with the above definition in the unipotent case (but not the quasi-unipotent case) as proved in [22].

DEFINITION 8.3.3 (Admissible VMHS ([22, 1.9]))   *Let X be a complex analytic space and $U \subset X$ a nonsingular open subset, complement of a closed analytic subset. A graded polarizable variation of mixed Hodge structure $(\mathcal{L}, W, \mathcal{F})$ on U is called admissible if for every analytic morphism $f : D \to X$ on a unit disc which maps D^* to U, the inverse $(f_{|D^*})^*(\mathcal{L}, W, \mathcal{F})$ is a preadmissible variation on D^*.*

In the case of locally unipotent admissible VMHS, Kashiwara noticed that preadmissible VMHS in the disc are necessarily admissible.

The following criteria in [22] state that admissibility can be tested in codimension 1:

THEOREM 8.3.4 (Codimension 2 ([22, 4.5.2]))   *Let X be a complex manifold, $U \subset X$ the complement of an NCD and let Z be a closed analytic subset of codimension ≥ 2 in X. An admissible VMHS $(\mathcal{L}, W, \mathcal{F})$ on U whose restriction to $U - Z$ is admissible in $X - Z$, is necessarily admissible in X.*

In particular, the existence of the relative weight filtration at a point $y \in Z$ follows from its existence at the nearby points on $Y - Z$. Such a result is stated and checked locally in terms of nilpotent orbits localized at y.

Next, we cite the following fundamental result for admissible variations of MHS.

THEOREM 8.3.5 (Hypercohomology)   *Let $\mathcal{L}$ be an admissible graded polarized VMHS $(\mathcal{L}, W, \mathcal{F})$ on the complement $X - Y$ of an NCD Y in a complex compact smooth algebraic variety X, $j : X - Y \to X$, Z a sub-NCD of Y, $U := X - Z$; then for all degrees $k \in \mathbb{Z}$, the cohomology groups $\mathbb{H}^k(U, j_{!*}\mathcal{L})$ of the intermediate extension, carry a canonical mixed Hodge structure.*

This result follows from Saito's general theory of mixed Hodge modules [35] but it is obtained here directly via the logarithmic complex. In both cases it relies heavily on the local study of VMHS by Kashiwara in [22] that is highlighted in the next section. We use also the purity of the intersection cohomology in [24, 8]. The curve case is treated in [40].

8.3.1.1 Properties

(i) We describe below a logarithmic de Rham complex $\Omega_X^*(\mathrm{Log}\,Y) \otimes \mathcal{L}_{\mathcal{O}_X}$ with coefficients in $\mathcal{L}_{\mathcal{O}_X}$ on which the weight filtration is defined in terms of the local study in [22] while the Hodge filtration is easily defined and compatible with the result in [23].

(ii) We realize the cohomology of $U = X - Z$ as the cohomology of a cohomological mixed Hodge complex (MHC) $\Omega^*(\mathcal{L}, Z)$, a subcomplex of $\Omega_X^*(\mathrm{Log}\,Y) \otimes \mathcal{L}_{\mathcal{O}_X}$ containing the intermediate extension $j_{!*}\mathcal{L}$ of $\mathcal{L}$ as a sub-MHC $\mathrm{IC}(X, \mathcal{L})$ such that the quotient complex defines a structure of MHC on $i_Z^! j_{!*}\mathcal{L}[1]$.

(iii) If the weights of $j_{!*}\mathcal{L}$ are $\geq a$, then the weights on $\mathbb{H}^i(U, j_{!*}\mathcal{L})$ are $\geq a + i$.

(iv) The MHS is defined dually on $\mathbb{H}_c^i(U, j_{!*}\mathcal{L})$ (resp. $\mathbb{H}^i(Z, j_{!*}\mathcal{L})$) of weights $\leq a+i$ if the weights of $j_{!*}\mathcal{L}$ are $\leq a$.

(v) Let H be a smooth hypersurface intersecting transversally $Y \cup Z$ such that $H \cup Y \cup Z$ is an NCD; then the Gysin isomorphism $\Omega^*(\mathcal{L}_{|H}, Z \cap H) \simeq i_H^*\Omega^*(\mathcal{L}, Z) \simeq i_H^!\Omega^*(\mathcal{L}, Z)[2]$ is an isomorphism of cohomological MHC with a shift in degrees in the last isomorphism.

8.3.2   Local Study of Infinitesimal Mixed Hodge Structures After Kashiwara

The global results cited in the theorems above are determined by the study of the local properties of VMHS. We state here the local version of the definitions and results in [22], but we skip the proofs, as they are technically complex, although based on invariants in linear algebra, reflecting analysis and geometry.

Then, an extensive study of infinitesimal mixed Hodge structures (IMHS) below is needed to state and check locally the decomposition property of the graded complex for the weight filtration of the logarithmic complex.

8.3.2.1 Infinitesimal Mixed Hodge Structure (IMHS)

It is convenient in analysis to consider complex MHS $(L, W, F, \overline{F})$ (see 3.2.2.8) where we do not need W to be rational but we suppose the three filtrations $W, F, \overline{F}$ are opposed [11]. In particular, a complex HS of weight k is given as $(L, F, \overline{F})$ satisfying $L \simeq \oplus_{p+q=k} L^{pq}$, where $L^{pq} = F^p \cap \overline{F}^q$. In the case of an MHS with underlying rational structure on W and L, $\overline{F}$ on L is just the conjugate of F with respect to the rational structure.

To define polarization, we recall that the conjugate space $\overline{L}$ of a complex vector space L, is the same group L with a different complex structure, such that the identity map on the group L defines a real linear map $\sigma : L \to \overline{L}$ and the product by scalars satisfies the relation for all $\lambda \in \mathbb{C}, v \in L, \lambda \times_{\overline{L}} \sigma(v) := \sigma(\overline{\lambda} \times_L v)$, then the complex structure on $\overline{L}$ is unique. A morphism $f : V \to V'$ defines a morphism $\overline{f} : \overline{V} \to \overline{V}'$ satisfying $\overline{f}(\sigma(v)) = \sigma(f(v))$.

8.3.2.2   Hypothesis

We consider a complex vector space L of finite dimension, two filtrations $F, \overline{F}$ of L by complex subvector spaces, an integer k and a nondegenerate linear map $S : L \otimes \overline{L} \to \mathbb{C}$ satisfying

$$S(x, \sigma(y)) = (-1)^k \overline{S(y, \sigma(x))} \text{ for } x, y \in L \quad \text{and} \quad S(F^p, \sigma(\overline{F}^q)) = 0 \text{ for } p+q > k.$$

Let $(N_1, \ldots, N_l)$ be a set of mutually commuting nilpotent endomorphisms of L such that

$$S(N_j x, y) + S(x, N_j y) = 0 \quad \text{and} \quad N_j F^p \subset F^{p-1}, \ N_j \overline{F}^p \subset \overline{F}^{p-1}.$$

By definition, an MHS is of weight w if the HS on Gr_k^W is of weight $w + k$.

DEFINITION 8.3.6 (Nilpotent orbit)   *The above data $(L, F, \overline{F}, S)$ are called a (polarized) nilpotent orbit of weight w if the following equivalent conditions are satisfied [22]:*

(i) *There exists a real number c such that $(L, (e^{i \sum t_j N_j}) F, (e^{-i \sum t_j N_j}) \overline{F})$ is an HS of weight w polarized by S for all $t_j > c$.*

(ii) *The monodromy filtration W of the nilpotent endomorphism $N = \sum_j t_j N_j$ does not depend on the various t_j for $t_j > 0$ and all j. It defines, with the filtration F, an MHS on L of weight w, and the bilinear form S_k such that $S_k(x, y) = S(x, N^k y)$ polarizes the primitive subspace $P_k = \mathrm{Ker}(N^{k+1} : \mathrm{Gr}_k^W \to \mathrm{Gr}_{-k-2}^W)$ with its induced HS of weight $w + k$.*

Henceforth, all nilpotent orbits are polarized.

   We consider now a filtered version of the above data $(L; W; F; \overline{F}; N_1, \ldots, N_l)$ with an increasing filtration W such that $N_j W_k \subset W_k$ but without any given fixed bilinear product S.

DEFINITION 8.3.7 (Mixed nilpotent orbit)   *The above data $(L, W, F; \overline{F}; N_1, \ldots, N_l)$ are called a mixed nilpotent orbit (graded polarized) if for each integer i, $(\mathrm{Gr}_i^W L; F_|; \overline{F}_|; (N_1)_|, \ldots, (N_l)_|)$, with the restricted structures, is a nilpotent orbit of weight i for some polarization S_i.*

   This structure is called a preinfinitesimal mixed Hodge module in [22, 4.2]. A preadmissible VMHS $(f^*\mathcal{L}, W, \mathcal{F})_{|D^*}$ on D^* defines such a structure at $0 \in D$.

DEFINITION 8.3.8 (IMHS ([22, 4.3]))   *A mixed nilpotent orbit*

$$(L; W, F; \overline{F}; N_1, \ldots, N_l)$$

is called an infinitesimal mixed Hodge structure (IMHS) if the following conditions are satisfied:

(i) *For each $J \subset I = \{1, \ldots, l\}$, the monodromy filtration $M(J)$ of $\sum_{j \in J} N_j$ relative to W exists and satisfies $N_j M_i(J) \subset M_{i-2}(J)$ for all $j \in J$ and $i \in \mathbb{Z}$.*

(ii) The filtrations $M(I)$, F, $\overline{F}$ define a graded polarized MHS. The filtrations W and $M(J)$ are compatible with the MHS and the morphisms N_i are of type $(-1, -1)$.

In [22], IMHS are called IMHM; Deligne remarked that if the relative monodromy filtration $M(\sum_{i \in I} N_i, W)$ exists in the case of a mixed nilpotent orbit, then it is necessarily the weight filtration of an MHS.

The following criterion is the infinitesimal statement which corresponds to the result that admissibility may be checked in codimension 1.

THEOREM 8.3.9 ([22, 4.4.1]) *A mixed nilpotent orbit $(L; W, F; \overline{F}; N_1, \ldots, N_l)$ is an IMHS if the monodromy filtration of N_j relative to W exists for any $j = 1, \ldots, l$.*

This result is not used here and it is directly satisfied in most applications. Its proof is embedded in surprisingly important properties of IMHS valuable for their own sake needed here.

8.3.2.3 Properties of IMHS

We describe now fundamental properties frequently needed in various constructions in mixed Hodge theory with degenerating coefficients.

We start with an important property of a relative weight filtration, used in various proofs.

THEOREM 8.3.10 ([22, 3.2.9]) *Let (L, W, N) be a filtered space with a nilpotent endomorphism with a relative monodromy filtration $M(N, W)$; then for each l, there exists a canonical decomposition*

$$\mathrm{Gr}_l^M L \simeq \oplus_k \mathrm{Gr}_k^W \mathrm{Gr}_l^M L.$$

In the proof, Kashiwara describes a natural subspace of $\mathrm{Gr}_l^M L$ isomorphic to $\mathrm{Gr}_k^W \mathrm{Gr}_l^M L$ in terms of W and N.

In the case of an IMHS as above, $(\mathrm{Gr}_l^M \mathrm{Gr}_k^W L, F_|, \overline{F}_|)$ and $(\mathrm{Gr}_k^W \mathrm{Gr}_l^M L, F_|, \overline{F}_|)$ are endowed with an induced HS of weight l. The isomorphism in the Zassenhaus lemma between the two groups is compatible with HS in this case, and $(\mathrm{Gr}_l^M L, F_|, \overline{F}_|)$ is a direct sum of HS of weight l for various k. Deligne's remark that the relative weight filtration $M(\sum_{i \in I} N_i, W)$ is the weight filtration of an MHS, may be deduced from this result. Another application is the proof of the following proposition:

PROPOSITION 8.3.11 ([22, 5.2.5]) *Let $(L; W, F; \overline{F}; N_1, \ldots, N_l)$ be an IMHS and for $J \subset \{1, \ldots, l\}$ set $M(J)$ as the relative weight of $N \in C(J) = \{\Sigma_{j \in J} t_j N_j, t_j > 0\}$. Then, for $J_1, J_2 \subset \{1, \ldots, l\}$, $N_1 \in C(J_1)$, $M(J_1 \cup J_2)$ is the weight filtration of N_1 relative to $M(J_2)$.*

The geometric interpretation of this result in the case of a VMHS on the complement of an NCD $Y_1 \cup Y_2$ is that the degeneration to Y_1 followed by the degeneration along Y_1 to a point $x \in Y_1 \cap Y_2$ yields the same limit MHS as the degeneration to Y_2 then along Y_2 to x, and the direct degeneration to x along a curve in the complement of $Y_1 \cup Y_2$.

8.3.2.4   Abelian Category of IMHS

The morphisms of two mixed nilpotent orbits (resp. IMHS) $(L; W, F; \overline{F}; N_1, \ldots, N_l)$ and $(L'; W', F'; \overline{F}'; N'_1, \ldots, N'_l)$ are defined to be compatible with both the filtrations and the nilpotent endomorphisms.

PROPOSITION 8.3.12   *(i) The category of mixed nilpotent orbits is abelian.*

(ii) The category of IMHS is abelian.

(iii) For all $J \subset I$, $M(J)$ and W are filtrations by sub-MHS of the MHS defined by $M(I)$ and F and the functors defined by the filtrations $W_j, \mathrm{Gr}_j^W, M_j(J), \mathrm{Gr}_j^{M(J)}$ are exact functors.

We define a corresponding mixed nilpotent orbit structure Hom on the vector space $\mathrm{Hom}(L, L')$ with the following filtrations: $\mathrm{Hom}(W, W'), \mathrm{Hom}(F, F')$, and $\mathrm{Hom}(\overline{F}, \overline{F}')$ classically defined, and the natural endomorphisms denoted

$$- \mathrm{Hom}(N_1, N'_1), \ldots, - \mathrm{Hom}(N_l, N'_l)$$

on $\mathrm{Hom}(L, L')$. Similarly a structure called the tensor product is defined.

REMARK 8.3.13   Among the specific properties of the filtrations of IMHS, we mention the distributivity defined by Kashiwara in [25] and used in various proofs. In general three subspaces A, B, C of a vector space do not satisfy the following distributivity property: $(A+B) \cap C = (A \cap C) + (B \cap C)$. However, a distributive family of filtrations $F_1, \ldots, F_n$ of a vector space L satisfy for all $p, q, r \in \mathbb{Z}$ and indices i, j, k

$$(F_i^p + F_j^q) \cap F_k^r = (F_i^p \cap F_k^r) + (F_j^q \cap F_k^r)$$

In the case of an IMHS $(W, F, N_i, i \in I)$, for $J_1 \subset \cdots \subset J_k \subset I$, the family of filtrations $\{W, F, M(J_1), \ldots, M(J_k)\}$ is distributive ([22, Proposition 5.2.4]).

8.3.3   Deligne–Hodge Theory on the Cohomology of a Smooth Variety

We describe now a weight filtration on the logarithmic complex with coefficients in the canonical extension of an admissible VMHS on the complement of an NCD, based on the local study in [22].

Hypothesis
Let X be a smooth and compact complex algebraic variety, $Y = \cup_{i \in I} Y_i$ an NCD in X with smooth irreducible components Y_i, and $(\mathcal{L}, W, F)$ a graded polarized VMHS on $U := X - Y$ *admissible* on X with unipotent local monodromy along Y.

Notation
We denote by $(\mathcal{L}_{\mathcal{O}_X}, \nabla)$ Deligne's canonical extension of $\mathcal{L} \otimes \mathcal{O}_U$ into an analytic vector bundle on X with a flat connection having logarithmic singularities. The filtration by sublocal systems W of $\mathcal{L}$ define a filtration by canonical extensions denoted $\mathcal{W}$,

subbundles of $\mathcal{L}_{\mathcal{O}_X}$, while by definition of admissibility the filtration $\mathcal{F}$ over U extends by subbundles $\mathcal{F} \subset \mathcal{L}_{\mathcal{O}_X}$ over X.

The aim of this section is to deduce from the local study in [22] the following global result:

THEOREM 8.3.14 (Logarithmic mixed Hodge complex)   *There exists on the perverse sheaf, complex of logarithmic forms with coefficients in $\mathcal{L}_{\mathcal{O}_X}$ and $\mathbb{C}$-linear differentials a perverse weight filtration W and a filtration F by subcomplexes of analytic sheaves such that the bi-filtered complex*

$$(\Omega_X^*(\text{Log } Y) \otimes \mathcal{L}_{\mathcal{O}_X}, W, F)$$

(logarithmic complex) underlies a structure of mixed Hodge complex and induces a canonical MHS on the cohomology groups $H^i(U, \mathcal{L})$ of $U := X - Y$.

The filtration F is classically deduced on the logarithmic complex from the above bundles $\mathcal{F}$ on X:

$$F^p = 0 \to F^p \mathcal{L}_{\mathcal{O}_X} \to \cdots \to \Omega_X^i(\text{Log } Y) \otimes F^{p-i}\mathcal{L}_{\mathcal{O}_X} \to \cdots .$$

Before giving a proof, we need to describe the weight filtration W.

8.3.3.1   Local Definition of the Perverse Weight W on the Logarithmic Complex

For $J \subset I$, let $Y_J := \cap_{i \in J} Y_i$, $Y_J^* := Y_J - \cup_{i \notin J}(Y_i \cap Y_J)$ $(Y_\emptyset^* := X - Y)$. We denote uniformly by $j : Y_J^* \to X$ the various embeddings. Let $X(y) \simeq D^{m+l}$ be a neighborhood in X of a point y in Y, and $X^*(y) = X(y) \cap U \simeq (D^*)^m \times D^l$, where D is a complex disc, denoted with a star when the origin is deleted. The fundamental group $\pi_1(X^*(y))$ is a free abelian group generated by m elements representing classes of closed paths around the origin, one for each D^* with coordinate z_i (the hypersurface Y_i is defined locally by the equation $z_i = 0$). Then the local system $\mathcal{L}$ corresponds to a representation of $\pi_1(X^*(y))$ in a vector space L defined by the action of commuting unipotent automorphisms T_i for $i \in [1, m]$ indexed by the local components Y_i of Y and called monodromy action around Y_i.

We represent the fibre at y of Deligne's extended bundle $\mathcal{L}_{X,y} \otimes (\mathcal{O}_{X,y}/\mathcal{M}_y)$, by *the vector space L of multivalued sections of $\mathcal{L}$* isomorphic to L (that is the sections of the inverse of $\mathcal{L}$ on a universal covering of $X^*(y)$) via the embedding $L \to \mathcal{L}_{\mathcal{O}_X,y}$: $v \to \tilde{v}$ defined by

$$\tilde{v} = \left(\exp -\frac{1}{2\pi i}(\Sigma_{j \in J}(\log z_j)N_j) \right) \cdot v, \quad \nabla\tilde{v} = \Sigma_{j \in J} - \frac{1}{2\pi i}(\widetilde{N_j \cdot v}) \otimes \frac{dz_j}{z_j} \quad (8.3.1)$$

where a basis of L is sent on a basis of $\mathcal{L}_{X,y} \subset j_*\mathcal{L}_{\mathcal{O}_{X^*}}$.

8.3.3.2   Local Description of $Rj_*\mathcal{L}$: $\Omega(L, N.)$

The fiber at y of the complex $\Omega_X^*(\text{Log } Y) \otimes \mathcal{L}_{\mathcal{O}_X}$ is quasi-isomorphic to a Koszul complex as follows. We associate to (L, N_i), $i \in [1, m]$ a strict simplicial vector space such that for all sequences $(i.) = (i_1 < \cdots < i_p)$

$$L(i.) = L, \qquad N_{i_j} \colon L(i. - \{i_j\}) \to L(i.)$$

DEFINITION 8.3.15   *The simple complex defined by the simplicial vector space above is the Koszul complex defined by (L, N_i) and denoted by $\Omega(L, N.)$.*

Another item of notation is $s(L(J), N.)_{J \subset [1,m]}$, where J is identified with the strictly increasing sequence of its elements and where $L(J) = L$.

We remark that $\Omega(L, N.)$ is quasi-isomorphic to the Koszul complex $\Omega(L, \text{Id} - T.)$ defined by $(L, \text{Id} - T_i)$, $i \in [1, m]$.

LEMMA 8.3.16   *For $M \subset I$ and $y \in Y_M^*$, the above correspondence $v \mapsto \tilde{v}$ from L to $\mathcal{L}_{\mathcal{O}_X, y}$ extends to a quasi-isomorphism*

$$s(L(J), N.)_{J \subset [1,m]} \cong \Omega(L, N_j, j \in M)$$
$$\cong (\Omega_X^*(\text{Log } Y) \otimes \mathcal{L}_{\mathcal{O}_X})_y \cong (Rj_*\mathcal{L})_y \tag{8.3.2}$$

by the correspondence

$$v \mapsto \tilde{v} \frac{dz_{i_1}}{z_{i_1}} \wedge \cdots \wedge \frac{dz_{i_j}}{z_{i_j}} \text{ from } L(i_1, \ldots, i_j) \text{ to } (\Omega_X^*(\text{Log } Y) \otimes \mathcal{L}_{\mathcal{O}_X})_y.$$

This description of $(Rj_*\mathcal{L})_y$ is the model for the description of the next various perverse sheaves.

8.3.3.3   The Intermediate Extension $j_{!*}\mathcal{L}$: $IC(L)$

Let $N_J = \prod_{j \in J} N_j$ denote a composition of endomorphisms of L. We consider the strict simplicial subcomplex of the de Rham logarithmic complex (or Koszul complex) defined by $N_J L := \text{Im } N_J$ in $L(J) = L$.

DEFINITION 8.3.17   *The simple complex defined by the above simplicial subvector space is the intersection complex of L denoted by*

$$IC(L) \colon = s(N_J L, N.)_{J \subset M}, \quad N_J L \colon = N_{i_1} N_{i_2} \cdots N_{i_j} L, J = \{i_1, \ldots, i_j\}. \tag{8.3.3}$$

Locally, the germ of the intermediate extension $j_{!*}\mathcal{L}$ of $\mathcal{L}$ at a point $y \in Y_M^*$ is quasi-isomorphic to the above complex ([8, 24])

$$j_{!*}(\mathcal{L})_y \simeq IC(L) \simeq s(N_J L, N.)_{J \subset M}. \tag{8.3.4}$$

8.3.3.4   Definition of $N * W$

Let (L, W, N) denote an increasing filtration W on a vector space L with a nilpotent endomorphism N compatible with W such that the relative monodromy filtration $M(N, W)$ exists; then a new filtration $N * W$ of L is defined by the formula ([22, 3.4])

$$(N * W)_k := N W_{k+1} + M_k(N, W) \cap W_k = N W_{k+1} + M_k(N, W) \cap W_{k+1}, \tag{8.3.5}$$

where the last equality follows from ([22, Proposition 3.4.1]).

For each index k, the endomorphism $N : L \to L$ induces a morphism $N : W_k \to (N * W)_{k-1}$ and the identity on L induces a morphism $I : (N * W)_{k-1} \to W_k$. We remark upon two important properties of $N * W$ ([22, Lemma 3.4.2]):

(i) The relative weight filtration exists and $M(N, N * W) = M(N, W)$.

(ii) We have the decomposition property

$$\mathrm{Gr}_k^{N*W} \simeq \mathrm{Im}(N : \mathrm{Gr}_{k+1}^W L \to \mathrm{Gr}_k^{N*W} L) \oplus \mathrm{Ker}(I : \mathrm{Gr}_k^{N*W} L \to \mathrm{Gr}_{k+1}^W L),$$
$$\mathrm{Im}(N : \mathrm{Gr}_{k+1}^W L \to \mathrm{Gr}_k^{N*W} L) \simeq \mathrm{Im}(N : \mathrm{Gr}_{k+1}^W L \to \mathrm{Gr}_{k+1}^W L).$$
$$\tag{8.3.6}$$

In referring to (ii), we say that W and $N * W$ form a graded distinguished pair.

8.3.3.5   The Filtration W^J on an IMHS L

The fiber of Deligne's extension of an admissible VMHS on $X - Y$ at a point $y \in Y_M^*$ for $M \subset I$ a subset of the set of indices of Y, defines an IMHS $(L, W, F, N_i, i \in M)$; in particular, the relative weight filtration $M(J, W)$ of an element N in $C(J) = \{\Sigma_{j \in J} t_j N_j, t_j > 0\}$ exists for all $J \subset M$. A basic lemma [22, Corollary 5.5.4] asserts the following lemma:

LEMMA 8.3.18 $(L, N_j * W, F, N_i, i \in M)$ and $(L, M(N_j, W), F, N_i, i \in M - \{j\})$ are infinitesimal MHS.

In particular, an increasing filtration W^J of L may be defined recursively by the star operation

$$W^J := N_{i_1} * (\cdots (N_{i_j} * W) \cdots) \text{ for } J = \{i_1, \ldots, i_j\} \tag{8.3.7}$$

(denoted $\Psi_J * W$ in [22, (5.8.2)]; see also [1]). It describes the fiber of the proposed weight filtration on $\mathcal{L}(y)$ for $y \in Y_J^*, J \subset M$. The filtration W^J does not depend on the order of composition of the respective transformations $N_{i_k} *$ since in the case of an IMHS $N_{i_p} * (N_{i_q} * W) = N_{i_q} * (N_{i_p} * W)$ for all $i_p, i_q \in J$ according to [22, Proposition 5.5.5].

DEFINITION 8.3.19   *The filtration W of the de Rham complex, associated to an IMHS $(L, W, F, N_i, i \in M)$, is defined as the Koszul complex*

$$W_k(\Omega(L, N.)) := s(W_{k-|J|}^J, N.)_{J \subset M},$$

where for each index $i \in M - J$, the endomorphism $N_i : L \to L$ induces a morphism $N_i : W_k^J \to (N_i * W^J)_{k-1}$ (the same notation is used for W on L and W on $\Omega(L, N.)$). It is important to add the canonical inclusion $I : (N_i * W^J)_{k-1} \to W_k^J$ to the data defining the filtration W of the de Rham complex. For example, for $|M| = 2$, the data with alternating differentials N_i is written as follows:

$$
\begin{array}{ccccccc}
L & \xrightarrow[I]{N_1} & L & & W_k & \xrightarrow[I]{N_1} & (N_1 * W)_{k-1} \\
I \uparrow\downarrow N_2 & & I \uparrow\downarrow N_2 & & I \uparrow\downarrow N_2 & & I \uparrow\downarrow N_2 \\
L & \xrightarrow[I]{N_1} & L & & (N_2 * W)_{k-1} & \xrightarrow[I]{N_1} & (N_1 * N_2 * W)_{k-2}.
\end{array}
$$

8.3.3.6 Local Decomposition of W

The proof of the decomposition involves a general description of perverse sheaves in the normal crossing divisor case ([22, Section 2]) in terms of the following de Rham data $\mathrm{DR}(L)$ deduced from the IMHS:

$$\mathrm{DR}(L) := \{L_J, W^J, F_J, N_{K,J} = L_J \to L_K, I_{J,K} = L_K \to L_J\}_{K \subset J \subset M},$$

where for all $J \subset M$, $L_J := L$, $F_J := F$, W^J is the filtration (8.3.7), $N_{K,J} := \prod_{i \in J-K} N_i$, and $I_{J,K} := \mathrm{Id} : L \to L$. The correspondence $L \to \mathcal{L}_{X,y}, v \mapsto \tilde{v}$, extends to $L_J \to (\Omega_X^{|J|}(LogY) \otimes \mathcal{L}_X)_y, v \mapsto \tilde{v} \wedge_{j \in J} dz_j/z_j$.

The object of this study is to show that the local filtrations W^J and $F_J := F$ are induced at a point $y \in X$ by global weight and Hodge filtrations which define the structure of a global mixed Hodge complex on X. To this end, a set of properties stated by Kashiwara ([22, 5.6]) are satisfied. In particular, we mention the following two basic results, where N_A denotes the composition of the linear endomorphisms $N_j, j \in A$ for $A \subset M$ and we put: $M(A, W) := M(\sum_{i \in A} N_i, W)$.

LEMMA 8.3.20 *Let $(L, W, F, N_i, i \in M)$ be an IMHS, and for $K \subset J \subset M$, let $A := J - K$, $N_A = W_i^K L \to W_{i-|A|}^J L$, $I_A : W_{i-|A|}^J L \hookrightarrow W_i^K L$ be the inclusion, then for each $J \subset M$:*

i) the data $(L_J, W^J, F_J, \{N_j, j \in (M - J)\})$ is an IMHS, and

ii) we have a local decomposition:

$$\mathrm{Gr}_{i-|A|}^{W^J} L \simeq \mathrm{Im}(\mathrm{Gr}_i N_A : \mathrm{Gr}_i^{W^K} L \to \mathrm{Gr}_{i-|A|}^{W^J} L)$$

$$\oplus \mathrm{Ker}(\mathrm{Gr}_i I_A : \mathrm{Gr}_{i-|A|}^{W^J} L \to \mathrm{Gr}_i^{W^K} L).$$

For each $L_J := L$, the filtration $F_J := F$ is defined by the given IMHS, and $(M(M, W), F)$ defines a MHS on $L_J = L$.

To prove i) we must show that $M(K, W^J)$ exists for all $K \subset M - J$, which is a property of the star operation since for all $J_1, J_2 \subset M$ and $J \subset K \subset M$:

$$M(J_1, W^{J_2}) = M(J_1, W)^{J_2}, \quad M(K, W^J) = M(K, W)$$

see ([22, theorem 5.5.1 and formulas 5.8.5, 5.8.6]).

ii) The proof of the local decomposition is by induction on the length $|A|$ of A ([22, 5.6.7 and Lemma 5.6.5]), and is based at each step for $a \in J-K$, on the decomposition:

$$\mathrm{Gr}_{i-1}^{W^{K \cup a}} L \simeq \mathrm{Im}(\mathrm{Gr}_i N_a : \mathrm{Gr}_i^{W^K} L \to \mathrm{Gr}_{i-1}^{W^{K \cup a}})L$$
$$\oplus \mathrm{Ker}(\mathrm{Gr}_i I_a : \mathrm{Gr}_{i-1}^{W^{K \cup a}} L \to \mathrm{Gr}_i^{W^K} L).$$

COROLLARY 8.3.21 ([22, 5.6.10])   *(i) Set*

$$P_k^J(L) := \cap_{K \subset J, K \neq J} \mathrm{Ker}(I_{J,K} : \mathrm{Gr}_k^{W^J} \to \mathrm{Gr}_{k+|J-K|}^{W^K}) \subset \mathrm{Gr}_k^{W^J};$$

then $P_k^J(L)$ has pure weight k with respect to the filtration $M(\sum_{j \in J} N_j, L)$.

(ii) $(P_k^J(L), N_j, j \in M - J, F)$ is an infinitesimal VHS.

(iii) We have $\mathrm{Gr}_k^{W^J} L \simeq \oplus_{K \subset J} N_{J-K} P_{k-|J-K|}^K(L)$.

(iv) For $J = \emptyset$, $P_k^\emptyset(L) = \mathrm{Gr}_k^W L$.

The assertion (ii) means that $P_k^J(L)$ underlies the structure of a polarised VHS with Hodge filtrations defined by $e^{i \sum_{j \in M-J} t_j N_j} F$ for $t_j > c$ where c is a positive constant. Equivalently, $P_k^J(L)$ defines a local system on the restriction of $\mathcal{L}_{\mathcal{O}_X}$ to Y_J^* underlying a VHS induced by F on Y_J^*.

The proof is based on the remark that the filtration induced by $M(\sum_{j \in J} N_j, L)$ on $\mathrm{Gr}_k^{W^J} L$ coincides with the filtration induced by $M(\sum_{j \in J} N_j, \mathrm{Gr}_k^{W^J} L)$ shifted by k, while the action of $\sum_{j \in J} N_j$ on the subset $P_k^J(L) \subset \mathrm{Gr}_k^{W^J} L$ is zero by construction; the statement (iii) is proved in [22, 2.3.1].

LEMMA 8.3.22 ([22, Proposition 2.3.1])   *The complex of graded vector spaces of the filtration $W_k, k \in \mathbb{Z}$ on $\Omega(L, N.)$ satisfies the decomposition property into a direct sum of intersection complexes*

$$\mathrm{Gr}_k^W(\Omega(L, N.)) \simeq \oplus \mathrm{IC}(P_{k-|J|}^J(L)[-|J|])_{J \subset M}. \tag{8.3.8}$$

The lemma follows from the corollary. It is the local statement of the structure of MHC on the logarithmic de Rham complex.

8.3.3.7 Global Definition and Properties of the Weight W

The local study ended with the local decomposition into intersection complexes. We develop now the corresponding global results. Taking the residue of the connection, we define nilpotent analytic linear endomorphisms of $\mathcal{L}_{\mathcal{O}_{Y_i}}$ compatible with the filtration $\mathcal{W}$ by subanalytic bundles:

$$\mathcal{N}_i := -2\pi i \, \mathrm{Res}_i(\nabla) : \mathcal{L}_{\mathcal{O}_{Y_i}} \to \mathcal{L}_{\mathcal{O}_{Y_i}}, \quad \mathcal{N}_J = \mathcal{N}_{i_1} \cdots \mathcal{N}_{i_j} : \mathcal{L}_{\mathcal{O}_{Y_J}} \to \mathcal{L}_{\mathcal{O}_{Y_J}}.$$

The Intersection Complex

We introduce the global definition of the intersection complex $\mathrm{IC}(X, \mathcal{L})$ as the subcomplex of $\Omega_X^*(\mathrm{Log}\, Y) \otimes \mathcal{L}_{\mathcal{O}_X}$ whose terms in each degree are $\mathcal{O}_X$-modules with singularities along the strata Y_J of Y, defined in terms of the analytic nilpotent endomorphisms $\mathcal{N}_i$ and $\mathcal{N}_J$ for subsets $J \subset I$ of the set I of indices of the components of Y:

DEFINITION 8.3.23 (Intersection complex) *The intersection complex is the subanalytic complex* $\mathrm{IC}(X, \mathcal{L}) \subset \Omega_X^*(\mathrm{Log}\, Y) \otimes \mathcal{L}_{\mathcal{O}_X}$ *whose fiber at a point* $y \in Y_M^*$ *is defined, in terms of a set of coordinates* $z_i, i \in M$ *defining equations of* Y_M, *as an* $\Omega_{X,y}^*$ *submodule, generated by the sections* $\tilde{v} \wedge_{j \in J} \frac{dz_j}{z_j}$ *for* $v \in N_J \mathcal{L}(y)$ *and* $J \subset M$ *($N_\emptyset = \mathrm{Id}$).*

This definition is independent of the choice of coordinates; moreover, the restriction of the section is still defined in the subcomplex since at each point y, $N_J L \subset N_{J-i} L$ for all $i \in J$.

For example, for $M = \{1, 2\}$ at $y \in Y_M^*$, the sections in $(\Omega_X^2(\mathrm{Log}\, Y) \otimes \mathcal{L}_X)_y \cap \mathrm{IC}(X, \mathcal{L})_y$ are generated by $\tilde{v}\frac{dz_1}{z_1} \wedge \frac{dz_2}{z_2}$ for $v \in N_J \mathcal{L}(y)$, $\tilde{v}\frac{dz_1}{z_1} \wedge dz_2$ for $v \in N_1 \mathcal{L}(y)$, $\tilde{v} \in N_1 \mathcal{L}(y)$, $\tilde{v} dz_1 \wedge \frac{dz_2}{z_2}$ for $v \in N_2 \mathcal{L}(y)$, and $\tilde{v} dz_1 \wedge dz_2$ for $v \in \mathcal{L}(y)$. We deduce the following lemma from the local result:

LEMMA 8.3.24 *Let $\mathcal{L}$ be locally unipotent and a polarized VHS on $X - Y$. The intersection complex* $\mathrm{IC}(X, \mathcal{L})[n]$ *shifted by* $n := \dim X$ *is quasi-isomorphic to the unique autodual complex on X, intermediate extension* $j_{!*}\mathcal{L}[n]$ *satisfying*

$$\mathrm{RHom}(j_{!*}\mathcal{L}[n], \mathbb{Q}_X[2n]) \simeq j_{!*}\mathcal{L}[n].$$

The shift by n is needed for the compatibility with the definitions in [2]. The next theorem is proved in [24] and [8]; see also [23].

THEOREM 8.3.25 *Let $(\mathcal{L}, F)$ be a polarized variation of Hodge structure of weight a. The subcomplex* $(\mathrm{IC}(X, \mathcal{L}), F)$ *of the logarithmic complex with induced filtration F is a Hodge complex which defines a pure HS of weight $a + i$ on the intersection cohomology* $\mathrm{IH}^i(X, \mathcal{L})$.

The proof is in terms of L^2-cohomology defined by square-integrable forms with coefficients in Deligne's extension $\mathcal{L}_{\mathcal{O}_X}$ for an adequate metric. The filtration F on $\mathrm{IC}(X, \mathcal{L})$ defined in an algebraic geometric way yields the same Hodge filtration as in L^2-cohomology as proved elegantly in [23] using the autoduality of the intersection cohomology.

8.3.3.8 The Global Filtration by Analytic Sheaves $\mathcal{W}^J$ and Perverse Sheaves W

The relative monodromy weight filtrations $\mathcal{M}(J, \mathcal{W}_{\mathcal{O}_{Y_J}}) := \mathcal{M}(\sum_{i \in J} \mathcal{N}_i, \mathcal{W}_{\mathcal{O}_{Y_J}})$ of $\sum_{i \in J} \mathcal{N}_i$ with respect to the restriction $\mathcal{W}_{\mathcal{O}_{Y_J}}$ on Y_J of $\mathcal{W}$ on $\mathcal{L}_{\mathcal{O}_X}$, exist for all $J \subset I$, so that we can define the global filtrations

$$(\mathcal{N}_i * \mathcal{W}_{\mathcal{O}_{Y_i}})_k := \mathcal{N}_i(\mathcal{W}_{\mathcal{O}_{Y_i}})_{k+1} + \mathcal{M}_k(\mathcal{N}_i, \mathcal{W}_{\mathcal{O}_{Y_i}}) \cap (\mathcal{W}_{\mathcal{O}_{Y_i}})_k$$
$$= \mathcal{N}_i(\mathcal{W}_{\mathcal{O}_{Y_i}})_{k+1} + \mathcal{M}_k(\mathcal{N}_i, \mathcal{W}_{\mathcal{O}_{Y_i}}) \cap (\mathcal{W}_{\mathcal{O}_{Y_i}})_{k+1},$$

and for all $J \subset I$ an increasing filtration $\mathcal{W}^J$ of $\mathcal{L}_{\mathcal{O}_{Y_J}}$ is defined recursively by the star operation

$$\mathcal{W}^J_{\mathcal{O}_{Y_J}} := \mathcal{N}_{i_1} * (\cdots (\mathcal{N}_{i_j} * \mathcal{W}_{\mathcal{O}_{Y_J}}) \cdots) \text{ for } J = \{i_1, \ldots, i_j\}.$$

DEFINITION 8.3.26 (Weight W)   *The filtration W by perverse sheaves on the logarithmic de Rham complex with coefficients in the canonical extension $\mathcal{L}_{\mathcal{O}_X}$ defined by $\mathcal{L}$ is constructed by induction on the decreasing dimension of the strata Y_J^* as follows:*

  (i) *On $U := X - Y$, the subcomplex $(W_r)_{|U}$ coincides with $\Omega_U^* \otimes (W_r)_{|U} \subset \Omega_U^* \otimes \mathcal{L}_{\mathcal{O}_U}$.*

  (ii) *We suppose W_r defined on the complement of the closure of the union of strata $\cup_{|M|=m} Y_M$; then for each point $y \in Y_M^*$ we define W_r locally in a neighborhood of $y \in Y_M^*$, on $(\Omega_X^*(\text{Log } Y) \otimes \mathcal{L}_X)_y$, in terms of the IMHS $(L, W, F, N_i, i \in M)$ at y and a set of coordinates z_i for $i \in M$, defining a set of local equations of Y_M at y, as follows:*

  - *W_r is generated as an $\Omega_{X,y}^*$-submodule by the germs of the sections $\wedge_{j \in J} \frac{dz_j}{z_j} \otimes \tilde{v}$ for $v \in W_{r-|J|}^J L$, where $\tilde{v}$ is the corresponding germ of $\mathcal{L}_{\mathcal{O}_{X,y}}$.*

The definition of W above is independent of the choice of coordinates on a neighborhood $X(y)$, since if we choose a different coordinate $z_i' = f z_i$ instead of z_i, with f invertible holomorphic at y, we check first that the submodule $\mathcal{W}_{r-|J|}^J(\mathcal{L}_{\mathcal{O}_X})_y$ of $\mathcal{L}_{\mathcal{O}_{X,y}}$ defined by the image of $W_{r-|J|}^J L$ is independent of the coordinates as in the construction of the canonical extension. Then we check that for a fixed $\alpha \in \mathcal{W}_{r-|J|}^J(\mathcal{L}_{\mathcal{O}_X})_y$, since the difference $\frac{dz_i'}{z_i'} - \frac{dz_i}{z_i} = \frac{df}{f}$ is holomorphic at y, the difference of the sections $\wedge_{j \in J} \frac{dz_j'}{z_j'} \otimes \alpha - \wedge_{j \in J} \frac{dz_j}{z_j} \otimes \alpha$ is still a section of the $\Omega_{X,y}^*$-submodule generated by the germs of the sections $\wedge_{j \in (J-i)} \frac{dz_j}{z_j} \otimes \mathcal{W}_{r-|J|}^J(\mathcal{L}_{\mathcal{O}_X})_y$.

Finally, we remark that the sections defined by induction at y restrict to sections already defined by induction on $X(y) - (Y_M^* \cap X(y))$.

8.3.3.9   The Bundles $\mathcal{P}_k^J(\mathcal{L}_{\mathcal{O}_{Y_J}})$

Given a subset $J \subset I$, the filtration $\mathcal{W}^J$ defines a filtration by subanalytic bundles of $\mathcal{L}_{\mathcal{O}_{Y_J}}$; then we introduce the analytic bundles

$$\mathcal{P}_k^J(\mathcal{L}_{\mathcal{O}_{Y_J}}) := \cap_{K \subset J, K \neq J} \text{Ker}(I_{J,K} : \text{Gr}_k^{\mathcal{W}^J} \mathcal{L}_{\mathcal{O}_{Y_J}} \to \text{Gr}_{k+|J-K|}^{\mathcal{W}^K} \mathcal{L}_{\mathcal{O}_{Y_J}})$$
$$\subset \text{Gr}_k^{\mathcal{W}^J} \mathcal{L}_{\mathcal{O}_{Y_J}},$$

where $I_{J,K}$ is induced by the natural inclusion $\mathcal{W}_k^J \mathcal{L}_{\mathcal{O}_{Y_J}} \subset \mathcal{W}_{k+|J-K|}^K \mathcal{L}_{\mathcal{O}_{Y_J}}$. In particular, $\mathcal{P}_k^\emptyset(\mathcal{L}_{\mathcal{O}_X}) = \text{Gr}_k^{\mathcal{W}} \mathcal{L}_{\mathcal{O}_X}$ and $\mathcal{P}_k^J(\mathcal{L}_{\mathcal{O}_{Y_J}}) = 0$ if $Y_J^* = \emptyset$.

PROPOSITION 8.3.27   *(i) The weight $W[n]$ shifted by $n := \dim X$ is a filtration by perverse sheaves defined over $\mathbb{Q}$, subcomplexes of the logarithmic complex quasi-isomorphic to $Rj_*\mathcal{L}[n]$.*

(ii) The bundles $\mathcal{P}_k^J(\mathcal{L}_{\mathcal{O}_{Y_J}})$ are Deligne's extensions of local systems $\mathcal{P}_k^J(\mathcal{L})$ on Y_J^*.

(iii) The graded perverse sheaves for the weight filtration, satisfy the decomposition property into intermediate extensions for all k

$$\operatorname{Gr}_k^W(\Omega_X^*(\operatorname{Log} Y)\otimes\mathcal{L}_{\mathcal{O}_X})[n]\simeq\oplus_{J\subset I}(i_{Y_J})_*j_{!*}\mathcal{P}_{k-|J|}^J(\mathcal{L})[n-|J|],$$

where j denotes uniformly the inclusion of Y_J^* into Y_J for each $J\subset I$, and $\mathcal{P}_p^J(\mathcal{L})$ on Y_J^* is a polarized VHS with respect to the weight induced by

$$\mathcal{M}(\textstyle\sum_{j\in J}\mathcal{N}_j,\mathcal{L}_{Y_J^*}).$$

The proof is essentially based on the previous local study which makes sense over $\mathbb{Q}$ as $\mathcal{L}$ and W are defined over $\mathbb{Q}$. In particular, we deduce that the graded complexes $\operatorname{Gr}_k^W(\Omega_X^*(\operatorname{Log} Y)\otimes\mathcal{L}_{\mathcal{O}_X})[n]$ are direct sums of intersection complexes over $\mathbb{C}$, from which we deduce that the extended filtration W_k on the de Rham complex satisfies the condition of support of perverse sheaves with respect to the stratification defined by Y. Similarly, the proof applies to the Verdier dual of W_k.

We need to prove that the local rational structure of the complexes W_k glues into a global rational structure, as perverse sheaves may be glued as the usual sheaves, although they are not concentrated in a unique degree. Since the total complex $Rj_*\mathcal{L}$ is defined over $\mathbb{Q}$, the gluing isomorphisms induced on the various extended W_k are also defined over $\mathbb{Q}$. Another proof of the existence of the rational structure is based on Verdier's specialization [17]. The next result is compatible with [13, Corollary 3.3.5].

COROLLARY 8.3.28   *The de Rham logarithmic mixed Hodge complex of an admissible VMHS of weight $\omega\geq a$ induces on the cohomology $\mathbb{H}^i(X-Y,\mathcal{L})$ an MHS of weight $\omega\geq a+i$.*

Indeed, the weight $W_k=0$ for $k\leq a$.

COROLLARY 8.3.29   *The intersection complex $(\operatorname{IC}(X,\mathcal{L})[n],W,F)$ of an admissible VMHS, with induced filtration as an embedded subcomplex of the de Rham logarithmic mixed Hodge complex, is a mixed Hodge complex satisfying for all k,*

$$\operatorname{Gr}_k^W\operatorname{IC}(X,\mathcal{L})=\operatorname{IC}(X,\operatorname{Gr}_k^W\mathcal{L}).$$

In general the intersection complex of an extension of two local systems, is not the extension of their intersection complex. We use here the existence of relative filtrations to check, for each $J\subset I$ of length j, the following property of the induced filtration $\mathcal{W}^J\cap\mathcal{N}_J\mathcal{L}_{\mathcal{O}_{Y_J}}$ on $\mathcal{N}_J\mathcal{L}_{\mathcal{O}_{Y_J}}$:

$$\operatorname{Gr}_{k-j}^{\mathcal{W}^J}(\mathcal{N}_J\mathcal{L}_{\mathcal{O}_{Y_J}})\simeq\operatorname{Im}(\mathcal{N}_J:\operatorname{Gr}_k^W\mathcal{L}_{\mathcal{O}_{Y_J}}\to\operatorname{Gr}_k^W\mathcal{L}_{\mathcal{O}_{Y_J}}).$$

The problem is local and the solution is based on the graded distinguished pair decomposition of $Gr_*^{N*W}L$ (formula 8.3.6), and the corollary ([22], cor 3.4.3).

We prove by induction on the length j of J the following lemma:

LEMMA 8.3.30 (Graded split sequence)   *Let (L, W, N_i) be an IMHS. The natural exact sequence defined, for each J of length $j > 0$, by the induced filtrations on each term by W^J on L*

$$0 \to \mathrm{Gr}^{W^J}_{k-j}(N_J L) \to \mathrm{Gr}^{W^J}_{k-j} L \to \mathrm{Gr}^{W^J}_{k-j}(L/N_J L) \to 0$$

splits, and we have : $\mathrm{Gr}^{W^J}_{k-j}(N_J L) \simeq (\mathrm{Im}\, N_J : \mathrm{Gr}^W_k L \to \mathrm{Gr}^W_k L)$.

We start with the following assertion:

Let (L, W, N) be a an IMHS and let $M := M(N, W)$. The natural exact sequence defined by the induced filtrations on each term by $N * W$ on L

$$0 \to \mathrm{Gr}^{N*W}_k(NL) \to \mathrm{Gr}^{N*W}_k L \to \mathrm{Gr}^{N*W}_k(L/NL) \to 0$$

is split, with the following properties:

$$\mathrm{Gr}^{N*W}_k NL \simeq \mathrm{Im}(N : \mathrm{Gr}^W_{k+1} L \to \mathrm{Gr}^{N*W}_k L) \simeq \mathrm{Im}(N : \mathrm{Gr}^W_{k+1} L \to \mathrm{Gr}^W_{k+1} L),$$

$$\mathrm{Gr}^{N*W}_k(L/NL) \simeq \mathrm{Ker}(I : \mathrm{Gr}^{N*W}_k L \to \mathrm{Gr}^W_{k+1} L) \simeq \mathrm{Gr}^M_k(L/NL)$$

PROOF. The exact sequence : $0 \to NL \to L \to L/NL \to 0$ with induced filtrations on each term gives naturally the graded exact sequence. The splitting is deduced from the graded distinguished pair decomposition of $\mathrm{Gr}^{N*W}_* L$ (formula 8.3.6). The natural injective embedding of $\mathrm{Gr}^{N*W}_k(NL)$ is into $\mathrm{Im}(N : \mathrm{Gr}^W_{k+1} L \to \mathrm{Gr}^W_{k+1} L)$ and it is surjective since $(N * W)_k(NL)$ contains $N(W_{k+1} L)$.

The full assertion follows from the isomorphisms proved in ([22], lemma 3.4.2 and corollary 3.4.3):

$\mathrm{Gr}^W_a \mathrm{Gr}^{N*W}_k L \simeq 0$ for $a > k + 1$, $\mathrm{Gr}^W_{k+1} \mathrm{Gr}^{N*W}_k L \simeq \mathrm{Im}(N : \mathrm{Gr}^W_{k+1} L \to \mathrm{Gr}^W_{k+1} L)$,

and for $a \le k$:

$$\mathrm{Gr}^W_a \mathrm{Gr}^{N*W}_k L \simeq \mathrm{Gr}^W_a \mathrm{Ker}(I : \mathrm{Gr}^{N*W}_k L \to \mathrm{Gr}^W_{k+1} L)$$
$$\simeq \mathrm{Coker}(N : \mathrm{Gr}^W_a \mathrm{Gr}^M_{k+2} L \to \mathrm{Gr}^W_a \mathrm{Gr}^M_k L)$$

from which we deduce the isomorphism:

$$\mathrm{Ker}(I : \mathrm{Gr}^{N*W}_k L \to \mathrm{Gr}^W_{k+1} L) \simeq \mathrm{Coker}(N : W_k \mathrm{Gr}^M_{k+2} L \to W_k \mathrm{Gr}^M_k L)$$
$$\simeq W_k \mathrm{Gr}^M_k L/NL$$

The isomorphism $W_k \mathrm{Gr}^M_k(L/NL) \simeq \mathrm{Gr}^M_k(L/NL)$, follows since the morphism $\mathrm{Gr}^M_{k+2} \mathrm{Gr}^W_a L \xrightarrow{N} \mathrm{Gr}^M_k \mathrm{Gr}^W_a L$ is surjective for $a > k$.   □

We apply the above assertion to the IMHS defined by L, W^J, N_i for all $i \notin J$, and the filtration $N_i * W^J$ on L, to deduce an isomorphism:

$$\mathrm{Gr}^{W^{J \cup i}}_{k-j-1}(N_i L) \simeq \mathrm{Im}(N_i : \mathrm{Gr}^{W^J}_{k-j} L \to \mathrm{Gr}^{W^J}_{k-j} L)$$

inducing on $N_J L \subset L$, an isomorphism $\mathrm{Gr}_{k-j-1}^{W^{J \cup i}}(N_i N_J L) \simeq N_i(\mathrm{Gr}_{k-j}^{W^J}(N_J L))$. Then, we deduce from the inductive hypothesis on $\mathrm{Gr}_{k-j}^{W^J}(N_J L)$:

$$\mathrm{Gr}_{k-j-1}^{W^{J \cup i}}(N_i N_J L) \simeq \mathrm{Im}(N_i : \mathrm{Gr}_{k-j}^{W^J}(N_J L) \to \mathrm{Gr}_{k-j}^{W^J}(N_J L)) \simeq$$
$$\mathrm{Im}(N_i N_J : \mathrm{Gr}_k^W L \to \mathrm{Gr}_k^W L).$$

8.3.3.10   MHS on Hypercohomology Groups of the Intersection Complex
$\mathbb{H}^*(X - Z, j_{!*}\mathcal{L})$

Let I_1 be a finite subset of I and let $Z := \cup_{i \in I_1} Y_i$ be a sub-NCD of Y. We describe an MHS on the hypercohomology $\mathbb{H}^*(X - Z, j_{!*}\mathcal{L})$.

Let $j' : (X - Y) \to (X - Z)$, $j'' : (X - Z) \to X$ such that $j = j'' \circ j'$. The fiber at a point $y \in Z$ of the logarithmic de Rham complex is isomorphic to the Koszul complex $\Omega(L, N_i, i \in M)$ for some subset M of I. To describe the fiber of the complex $Rj''_*(j'_{!*}\mathcal{L})_y$ as a subcomplex, we consider $M_1 := M \cap I_1$ and $M_2 := M - M_1$, and for $J \subset M$: $J_1 = J \cap M_1$ and $J_2 = J \cap M_2$.

DEFINITION 8.3.31 ($\Omega^*(\mathcal{L}, Z)$ for $Z \subset Y$)   *With the above notation, the subcomplex of analytic sheaves $\Omega^*(\mathcal{L}, Z)$ of the logarithmic de Rham complex $\Omega_X^*(\mathrm{Log}\, Y) \otimes \mathcal{L}_X$ is defined locally at a point $y \in Y_M^*$ in terms of a set of coordinates $z_i, i \subset M$, equations of Y_M, as follows:*

The fiber $\Omega^*(\mathcal{L}, Z)$ is generated as an $\Omega_{X,y}^*$ submodule, by the sections $\tilde{v} \wedge_{j \in J} \frac{dz_j}{z_j}$ for each $J \subset M$ and $v \in N_{J_2} L$.

LEMMA 8.3.32   *We have $(Rj''_*(j'_{!*}\mathcal{L}))_y \simeq \Omega^*(\mathcal{L}, Z)_y$.*

The intersection of a neighborhood of y with $X - Z$ is homeomorphic to $U = U_1^* \times U_2 := (D^*)^{n_1} \times D^{n_2}$ with coordinates $z_i, i \in M_1$ on U_1^* (defining local equations of Z at y) and $z_j, j \in M_2$ on U_2. At a point $a = (a_1, a_2) \in U$, the fiber of the Intersection complex is isomorphic to a Koszul complex: $(j'_{!*}\mathcal{L})_a \simeq IC(U_2, L) := s(N_{J_2}L, N_i, i \in M_2)_{J_2 \subset M_2}$ on which $N_i, i \in M_1$ acts. By comparison with the logarithmic de Rham complex along Z, the double complex $s(IC(U_2, L), N_i)_{i \in M_1}$ is quasi-isomorphic to the fiber $(Rj''_*(j'_{!*}\mathcal{L}))_y$, since the terms of its classical spectral sequence are $E_2^{p,q} \simeq R^p j''_*(\mathcal{H}^q(j'_{!*}\mathcal{L}))_y$.

EXAMPLE 8.3.33   In three-dimensional space, let $Y = Y_1 \cup Y_2 \cup Y_3$ be the union of the coordinates hyperplanes, $Z = Y_1 \cup Y_2$, with Y_3 defined by $z_3 = 0$, D a small disc at the origin of $\mathbb{C}$, then $(Rj''_*(j'_{!*}\mathcal{L}))_y = R\Gamma(D^3 - (D^3 \cap Z), j'_{!*}\mathcal{L})$ is defined by the following diagram with differentials defined by N_i with $+$ or $-$ sign:

$$
\begin{array}{ccccc}
L & \xrightarrow{N_1, N_2} & L \oplus L & \xrightarrow{N_1, N_2} & L \\
\downarrow N_3 & & \downarrow N_3 & & \downarrow N_3 \\
N_3 L & \xrightarrow{N_1, N_2} & N_3 L \oplus N_3 L & \xrightarrow{N_1, N_2} & N_3 L
\end{array}
$$

PROPOSITION 8.3.34   *The filtrations W and F of the logarithmic de Rham complex induce on $\Omega^*(\mathcal{L}, Z)$ a structure of MHC, defining an MHS on the hypercohomology*

$\mathbb{H}^*(X - Z, j_{!*}\mathcal{L})$, *such that the graded perverse sheaves for the weight filtration satisfy for all k the decomposition property into intermediate extensions*

$$\mathrm{Gr}_k^W \Omega^*(\mathcal{L}, Z)[n] \simeq \oplus_{J_1 \subset I_1}(i_{Y_{J_1}})_* j_{!*} P_{k-|J_1|}^{J_1}(\mathcal{L})[n - |J_1|],$$

where j denotes uniformly the inclusion of $Y_{J_1}^*$ into Y_{J_1} for each $J_1 \subset I_1 \subset I$, $Y := \cup_{i \in I} Y_i$ and $Z := \cup_{i \in I_1} Y_i$. In particular, for $J_1 = \emptyset$, we have $j_{!*} \mathrm{Gr}_k^W(\mathcal{L})[n]$; otherwise the summands are supported by Z.

The proof is local and based on the properties of relative filtrations. If the complex is written at a point $y \in Y_M^*$ as a double complex

$$s(s(N_{J_2}L, N_i, i \in M_1)_{J_1 \subset M_1})_{J_2 \subset M_2} = s(\Omega(N_{J_2}L, N_i, i \in M_1))_{J_2 \subset M_2}$$

for each term of index $J = (J_1 \cup J_2)$, the filtration on $N_{J_2}L$ is induced by W^J on L; hence,

$$\mathrm{Gr}_{k-|J_1|-|J_2|}^{W^{(J_1,J_2)}} N_{J_2}L \simeq N_{J_2} \mathrm{Gr}_{k-|J_1|}^{W^{J_1}} L \simeq \oplus_{K \supset J_1} N_{J_2} N_{K-J_1} P_{k-|K|}^K(L),$$

where the first isomorphism follows from the lemma on the graded split sequence (Lemma 8.3.30), and the second isomorphism follows from the the decomposition of the second term (Corollary 8.3.21,(iii)). Then, we can write Gr_k^W of the double complex as $\oplus_{J_1 \subset M_1} \mathrm{IC}(P_{k-|K|}^K(L))[-|J_1|]$.

REMARK 8.3.35   (i) If Z is not in Y but $Z \cup Y$ is an NCD, we may always suppose that $\mathcal{L}$ is a VMHS on $X - (Y \cup Z)$ (that is, to enlarge Y) and consider Z as a subspace of Y equal to a union of components of Y.

(ii) If Z is a union of intersections of components of Y, these techniques should apply to construct a subcomplex of the logarithmic de Rham complex endowed with the structure of MHC with the induced filtrations and hypercohomology $\mathbb{H}^*(X - Z, j_{!*}\mathcal{L})$; for example, we give below the fiber of the complex at the intersection of two lines in the plane, first when $Z = Y_1$, then for $Z = Y_1 \cap Y_2$:

$$
\begin{array}{ccc}
L & \xrightarrow{N_1} & L \\
\downarrow N_2 & & \downarrow N_2 \\
N_2 L & \xrightarrow{N_1} & N_2 L
\end{array}
\qquad\qquad
\begin{array}{ccc}
L & \xrightarrow{N_1} & N_1 L \\
\downarrow N_2 & & \downarrow N_2 \\
N_2 L & \xrightarrow{N_1} & N_1 L \cap N_2 L.
\end{array}
$$

8.3.3.11   Thom–Gysin Isomorphism

Let H be a smooth hypersurface transversally intersecting Y such that $H \cup Y$ is an NCD; then $i_H^* j_{!*}\mathcal{L}$ is isomorphic to the intermediate extension $(j_{Y \cap H})_{!*}(i_H^*\mathcal{L})$ of the restriction of $\mathcal{L}$ to H and the residue with respect to H induces an isomorphism $R_H : i_H^*(\Omega^*(\mathcal{L}, H)/j_{!*}\mathcal{L}) \simeq i_H^* j_{!*}\mathcal{L}[-1]$ inverse to the Thom–Gysin isomorphism $i_H^! j_{!*}\mathcal{L}[-1] \simeq i_H^! j_{!*}\mathcal{L}[1] \simeq i_H^*(\Omega^*(\mathcal{L}, H)/j_{!*}\mathcal{L})$ where $j_{!*}\mathcal{L}$ is a logarithmic subcomplex.

Moreover, if H transversally intersects $Y \cup Z$ such that $H \cup Y \cup Z$ is an NCD, then we have a triangle

$$(i_H)_* i_H^! \Omega^*(\mathcal{L}, Z) \to \Omega^*(\mathcal{L}, Z) \to \Omega^*(\mathcal{L}, Z \cup H) \xrightarrow{[1]}.$$

The following isomorphisms are compatible with the filtrations up to a shift in degrees:

- $(\Omega^*(\mathcal{L}, Z \cup H)/\Omega^*(\mathcal{L}, Z)) \simeq i_H^! \Omega^*(\mathcal{L}, Z)[1]$ (the quotient complex computes the cohomology with support).

- $i_H^* \Omega^*(\mathcal{L}, Z) \simeq \Omega^*(i_H^* \mathcal{L}, Z \cap H)$ (the restriction to H coincides with the complex constructed directly on H).

- The inverse to the Thom–Gysin isomorphism $i_H^* \Omega^*(\mathcal{L}, Z) \simeq i_H^! \Omega^*(\mathcal{L}, Z)[2]$ is induced by the residue with respect to H vanishing on $\Omega^*(\mathcal{L}, Z)$:
  $\Omega^*(\mathcal{L}, Z \cup H) \to i_{H,*} \Omega^*(i_H^* \mathcal{L}, Z \cap H)[1].$

8.3.3.12   Duality and Cohomology with Compact Support

First we recall, Verdier's dual of a bifiltered complex. Let (K, W, F) be a complex with two filtrations on a smooth compact Kähler or complex algebraic variety X, and $\omega_X := \mathbb{Q}_X[2n](n)$ a dualizing complex with the trivial filtration and a Tate twist of the filtration by $\dim X = n$ and a degree shift by $2n$ (so that the weight remains 0). We denote by $\mathcal{D}(K)$ the complex dual to K with filtrations

$$\begin{aligned}
\mathcal{D}(K) &:= \mathrm{RHom}(K, \mathbb{Q}_X[2n](n)), \\
W_{-i}\mathcal{D}(K) &:= \mathcal{D}(K/W_{i-1}), \\
F^{-i}\mathcal{D}(K) &:= \mathcal{D}(K/F^{i+1});
\end{aligned}$$

then we have $\mathcal{D} \mathrm{Gr}_i^W K \simeq \mathrm{Gr}_{-i}^W \mathcal{D}K$ and $\mathcal{D} \mathrm{Gr}_F^i K \simeq \mathrm{Gr}_F^{-i} \mathcal{D}K$. The dual of a mixed Hodge complex is an MHC.

In the case of $K = \Omega_X^*(\mathrm{Log}\, Y) \otimes \mathcal{L}_{\mathcal{O}_X}[n] = Rj_*\mathcal{L}[n]$, the dual $\mathcal{D}K - j_!\mathcal{L}[n]$ with the dual structure of an MHC defines the MHS on cohomology with compact support.

COROLLARY 8.3.36   *(i) An admissible VMHS $\mathcal{L}$ of weight $\omega \le a$ induces on the cohomology with compact support $\mathbb{H}_c^i(X - Y, \mathcal{L})$ an MHS of weight $\omega \le a + i$.*

(ii) The cohomology $\mathbb{H}^i(Y, j_!\mathcal{L})$ carries an MHS of weight $\omega \le a + i$.

This result is compatible with [13, Theorem 3.3.1].

(i) The dual admissible VMHS $\mathcal{L}^*$ is of weight $\omega \ge -a$. Its de Rham logarithmic mixed Hodge complex is of weight $\omega \ge -a$, whose dual is quasi-isomorphic to $j_!\mathcal{L}[2n](n)$ of weight $\omega \le a$.

By duality, the MHS on the cohomology $H_c^i(X - Y, \mathcal{L})$ is of weight $\omega \le a + i$.

(ii) The weights on $\mathbb{H}_Y^i(X, j_{!*}\mathcal{L})$ satisfy $\omega \ge a + i$ and by duality, $\mathbb{H}^i(Y, j_{!*}\mathcal{L})$, has weights $\omega \le a + i$.

8.3.3.13   Complementary Results

We consider from now on a pure Hodge complex of weight $\omega(K) = a$; its dual $\mathcal{D}K$ is pure of weight $\omega(\mathcal{D}K) = -a$. For a pure polarized VHS $\mathcal{L}$ on $X - Y$ of weight $\omega = b$, hence $\mathcal{L}[n]$ is of weight $\omega = a = b + n$, and the polarization defines an isomorphism $\mathcal{L}(a)[n] \simeq \mathbb{R}\operatorname{Hom}(\mathcal{L}[n], \mathbb{Q}_{X-Y}(n)[2n]) \simeq \mathcal{L}^*[n]$; then Verdier's autoduality of the intersection complex is written as follows:

$$j_{!*}\mathcal{L}[n] \simeq (\mathcal{D}(j_{!*}\mathcal{L}[n]))(-a),$$

$$\mathbb{H}^{-i}(X, j_{!*}\mathcal{L}[n])^* = H^i(\operatorname{Hom}(\mathbb{R}\Gamma(X, j_{!*}\mathcal{L}[n]), \mathbb{C}))$$
$$\simeq \mathbb{H}^i(X, \mathcal{D}(j_{!*}\mathcal{L}[n])) \simeq \mathbb{H}^i(X, j_{!*}\mathcal{D}(\mathcal{L}[n]))$$
$$\simeq \mathbb{H}^i(X, j_{!*}(\mathcal{L}^*[n])) \simeq \mathbb{H}^i(X, j_{!*}(\mathcal{L}[n]))(a).$$

For $K = \mathcal{L}[n]$, the exact sequence $i^! j_{!*}K \to j_{!*}K \to j_*K$ yields an isomorphism $i^! j_{!*}K[1] \simeq j_*K/j_{!*}K$; hence

$$\mathcal{D}(i^* j_*K/i^* j_{!*}K) \simeq \mathcal{D}(i^! j_{!*}K[1]) \simeq i^* j_{!*}\mathcal{D}K[-1].$$

For K pure of weight $\omega(K) = a$, $W_a j_*K = j_{!*}K$, $\omega(\mathcal{D}K) = -a$, and for $r > a$,

$$W_{-r}i^* j_{!*}\mathcal{D}K[-1] \simeq \mathcal{D}(i^* j_*K/W_{r-1}).$$

We deduce from the polarization $\mathcal{D}K \simeq K(a)$, where (a) drops the weight by $-2a$, a definition of the weight on $i^* j_{!*}K$ for $r > a$,

$$W_{-r+2a}i^* j_{!*}K \simeq \mathcal{D}(i^* j_*K/W_{r-1})[1], \quad \operatorname{Gr}^W_{-r+2a} i^* j_{!*}K \simeq \mathcal{D}\operatorname{Gr}^W_r(i^* j_*K)[1].$$

COROLLARY 8.3.37   *The complex* $(i^* j_{!*}K[-1], F)$ *with induced Hodge filtration and weight filtration for $k > 0$,*

$$W_{a-k}i^* j_{!*}K[-1] \simeq \mathcal{D}(i^* j_*K/W_{a+k-1}),$$

has the structure of an MHC satisfying $\operatorname{Gr}^W_{a-k} i^* j_{!*}K[-1] \simeq \mathcal{D}\operatorname{Gr}^W_{a+k}(i^* j_*K/i^* j_{!*}K)$. *(In particular, the cohomology* $\mathbb{H}^i(Y, j_{!*}K)$ *carries an MHS of weight $\omega \leq a + i$.)*

We remark that since $W_{a-1}i^* j_{!*}K[-1] \simeq i^* j_{!*}K[-1] = \mathcal{D}(i^* j_*K/j_{!*}K)$, it follows that $\mathbb{H}^i(Y, j_{!*}K) = \mathbb{H}^{i+1}(Y, j_{!*}K[-1])$ has weight $\omega \leq a - 1 + i + 1$.

By the above corollary, we have

$$\operatorname{Gr}^W_{-r+2a} i^* j_{!*}K \simeq \mathcal{D}(\operatorname{Gr}^W_r(\Omega^*_X(\operatorname{Log} Y) \otimes \mathcal{L}_{\mathcal{O}_X})[n])[1]$$
$$\simeq \oplus_{J \subset I} j_{!*}((\mathcal{P}^J_{r-|J|}(\mathcal{L}))^*)[n + 1 - |J|].$$

EXAMPLE 8.3.38   For a polarized unipotent VHS on a punctured disc D^*, $i^* j_{!*}K[-1]$ is the complex $(L \xrightarrow{N} NL)$ in degree 0 and 1, quasi-isomorphic to $\operatorname{Ker} N$, while

$(i^*j_*K/j_{!*}K) \simeq L/NL$ and $\mathcal{D}(i^*j_*K/j_{!*}K) \simeq (L/NL)^*$. The isomorphism is induced by the polarization Q as follows:

$$
\begin{array}{ccc}
(L/NL)^* & \simeq & (L^* \xrightarrow{N^*} (NL)^*) \\
\uparrow\simeq & & \uparrow\simeq \qquad \uparrow\simeq \\
\operatorname{Ker} N & \simeq & (L \xrightarrow{-N} NL),
\end{array}
$$

where $-N$ corresponds to N^* since $Q(Na, b) + Q(a, Nb) = 0$. The isomorphism we use is $\operatorname{Ker} N \simeq (L/NL)^*$ in degree 0. We set, for $k > 0$,

$$
M(N)_{a-k+1}(\operatorname{Ker} N) := W_{a-k}(L \to NL)
$$
$$
\simeq \mathcal{D}(i^*j_*K/W_{a+k-1}) = (L/(NL + M_{a+k-2}))^*.
$$

8.3.3.14 *The Dual Filtration $N!W$*

We introduce the filtration ([22, 3.4.2])

$$
(N!W)_k := W_{k-1} + M_k(N, W) \cap N^{-1}W_{k-1}.
$$

The morphisms induced by Id (resp. N) on L are $I : W_{k-1} \to (N!W)_k$ and $N : (N!W)_k \to W_{k-1}$ satisfying $N \circ I = N$ and $I \circ N = N$ on L. We prove the duality with $N * W$.

LEMMA 8.3.39 *Let W^* denote the filtration on the vector space $L^* := \operatorname{Hom}(L, \mathbb{Q})$ dual to a filtration W on L; then for all a,*

$$
(N^*!W^*)_a = (N * W)_a^* \subset L^*.
$$

Let $M_i := M_i(N, W)$; it is autodual as a filtration of the vector space L: $M_i^* = M_i(N^*, W^*)$. To prove the inclusion of the left-hand term into the right-hand term for $a = -k$, we write an element $\varphi \in L^*$ vanishing on $NW_k + M_{k-1} \cap W_{k-1} = NW_k + M_{k-1} \cap W_k$ as a sum of elements $\varphi = \gamma + \delta$ where $\delta \in W^*_{-k-1}$ vanishes on W_k and $\gamma \in M^*_{-k} \cap (N^*)^{-1}W^*_{-k-1}$ vanishes on $NW_k + M_{k-1}$. We construct γ such that $\gamma(NW_k+M_{k-1}) = 0$ and $\gamma_{|W_k} = \varphi_{|W_k}$, which is possible as $\varphi(M_{k-1}\cap W_k) = 0$; then we put $\delta = \varphi - \gamma$. The opposite inclusion is clear.

Hence, we deduce the following corollary from the decomposition in the case of $N * W$:

COROLLARY 8.3.40 *We have the decomposition*

$$
\operatorname{Gr}_k^{N!W} \simeq \operatorname{Im}(I : \operatorname{Gr}_{k-1}^W \to \operatorname{Gr}_k^{N!W}) \oplus \operatorname{Ker}(N : \operatorname{Gr}_k^{N!W} \to \operatorname{Gr}_{k-1}^W).
$$

The Filtration $\overline{W}$

We associate to each IMHS with nilpotent endomorphisms N_i for $i \in M$ and to each subset $J \subset M$, an increasing filtration $\overline{W}^J$ of L recursively by the ! operation,

$$
\overline{W}^J := N_{i_1}!(\cdots (N_{i_j}!W) \cdots) \text{ for } J = \{i_1, \ldots, i_j\}
$$

and for all i, j, $N_i! N_j! W = N_j! N_i! W$. This family satisfies the data in [22, Proposition 2.3.1] determined by the following morphisms, defined for $K \subset J$:

$$I_{K,J} : \overline{W}_k^K \to \overline{W}_{k+|J-K|}^J \text{ and } N_{J,K} : \overline{W}_k^J \to \overline{W}_{k-|J-K|}^K.$$

So we deduce the following lemma:

LEMMA 8.3.41   (i) *Set for each $J \subset M$,*

$$Q_k^J(L) := \cap_{K \subset J, K \neq J} \operatorname{Ker}\left(N_{J,K} : \operatorname{Gr}_k^{\overline{W}^J} \to \operatorname{Gr}_{k-|J-K|}^{\overline{W}^K} \right);$$

then $Q_k^J(L)$ has pure weight with respect to the weight $M(\sum_{j \in J} N_j, L)$.

(ii) *We have $\operatorname{Gr}_k^{\overline{W}^J} \simeq \oplus_{K \subset J} I_{K,J}(Q_{k-|J-K|}^K(L))$.*

(iii) *Let (L^*, W^*) be dual to (L, W); then $Q_k^J(L^*) \simeq (P_{-k}^J(L))^*$.*

8.3.3.15   Final Remark on the Decomposition Theorem

We do not cover here the ultimate results in Hodge theory with coefficients in an admissible VMHS (see [2, 35]). Instead we explain basic points in the theory. Given a projective morphism $f : X \to V$ and a polarized VHS of weight a on a smooth open subset $X - Y$ with embedding $j : X - Y \to X$, the intermediate extension $j_{!*}\mathcal{L}$ is defined on X and a theory of perverse filtration ${}^p\tau_i(Rf_* j_{!*}\mathcal{L})$ on the higher direct image on V has been developed in [2]. It defines on the cohomology of the inverse image of an open algebraic set $U \subset V$, a perverse filtration ${}^p\tau_i \mathbb{H}^r(f^{-1}(U), j_{!*}\mathcal{L})$.

The study in this section can be used for a new proof of the decomposition theorem as follows.

By desingularizing X, we can suppose X smooth and Y an NCD. Then, the MHS on the cohomology of the open sets complement of normal crossing subdivisors Z of Y with coefficients in the intersection complex has been defined above.

(i) For any open subset $U \subset V$ such that $X_U := f^{-1}(U)$ is the complement of a sub-NCD of Y, the theory asserts that the subspaces ${}^p\tau_i$ are sub-MHS of $\mathbb{H}^r(f^{-1}(U), j_{!*}\mathcal{L})$.

(ii) Let $v \in V$ be a point in the zero-dimensional strata of a Thom–Whitney stratification compatible with f and $Rf_* j_{!*}\mathcal{L}$. Let $X_v := f^{-1}(v)$ be a sub-NCD of Y, B_v a neighborhood of v in V, $X_{B_v} := f^{-1}(B_v)$ a tubular neighborhood of X_v. An MHS can be deduced on the cohomology of $X_{B_v} - X_v$. Then the local-purity theorem in [14] corresponds to the following semipurity property of the weights ω of this MHS:

(1) $\omega > a + r$ on ${}^p\tau_{\leq r} \mathbb{H}^r(X_{B_v} - X_v, j_{!*}\mathcal{L})$, and dually

(2) $\omega \leq a + r$ on $\mathbb{H}^r(X_{B_v} - X_v, j_{!*}\mathcal{L})/{}^p\tau_{\leq r}\mathbb{H}^r(X_{B_v} - X_v, j_{!*}\mathcal{L})$.

(iii) The corresponding decomposition theorem on $B_v - v$ states the isomorphism with the cohomology of a perverse cohomology:

$$\mathrm{Gr}_i^{p\tau} \, \mathbb{H}^r(X_{B_v} - X_v, j_{!*}\mathcal{L}) \simeq \mathbb{H}^{r-i}(B_v - v, {}^p\mathcal{H}^i(Rf_* j_{!*}\mathcal{L})).$$

(iv) By iterating the cup product with the class η of a relative hyperplane section, we have Lefschetz-type isomorphisms of perverse cohomology sheaves

$$ {}^p\mathcal{H}^{-i}(Rf_* j_{!*}\mathcal{L}[n]) \xrightarrow{\eta^i} {}^p\mathcal{H}^i(Rf_* j_{!*}\mathcal{L}[n]).$$

In [35], the proof is carried out via the techniques of differential modules and is based on extensive use of Hodge theory on the perverse sheaves of nearby and vanishing cycles. A direct proof may be obtained by induction on the strata on V. If we suppose the decomposition theorem and Lefschetz-type isomorphisms on $V - v$, one may prove directly that the perverse filtration on the cohomology $\mathbb{H}^r(X_{B_v} - X_v, j_{!*}\mathcal{L})$ is compatible with MHS. Then it makes sense to prove the semipurity property, from which we deduce the extension of the decomposition over the point v and check the Lefschetz isomorphism, which completes the inductive step (for a general stratum we intersect with a transversal section so as to reduce to the case of a zero-dimensional stratum).

8.4   ADMISSIBLE NORMAL FUNCTIONS

The purpose of this section is to explain the result on the algebraicity of the zero locus of an admissible normal function, which has been proved in three different ways by Brosnan and Pearlstein, by Kato, Nakayama, and Usui, and by Schnell. For complete proofs we refer the reader to the original papers [4, 27, 38]. Our goal here is to explain what the result says and to give some indication of the methods used in [4]. In particular, we sketch the proof in the one-dimensional case following the argument given in [3].

Suppose $j : S \to \bar{S}$ is an embedding of a complex manifold S as a Zariski open subset of the complex manifold $\bar{S}$. In other words, suppose that S is the complement of a closed analytic subset of $\bar{S}$. Recall that a variation of mixed Hodge structure $\mathcal{L}$ on S is *graded-polarizable* if, for each $k \in \mathbb{Z}$, the graded $\mathrm{Gr}_k^W \mathcal{L}$ admits a polarization. (For many purposes, which polarization we use is unimportant, but being polarizable is essential.) Write VMHS(S) for the (abelian) category of graded-polarizable variations of mixed Hodge structure on S and VMHS(S)$_{\bar{S}}^{\mathrm{ad}}$ for the full abelian subcategory consisting of all variations which are admissible with respect to $\bar{S}$.

The main case of interest is when S is a smooth, quasi-projective, complex variety and $\bar{S}$ is a smooth, projective compactification of S. However, it is also interesting and technically very useful to consider the case when $\bar{S}$ is a polydisk and S is the complement of the coordinate hyperplanes. To be absolutely explicit about this and to fix notation, we recall that we write $D := \{z \in \mathbb{C} : |z| < 1\}$ and $D^* := D \setminus \{0\}$. If r is a nonnegative integer, then the r-dimensional polydisk is D^r and the r-dimensional complement of the coordinate hyperplanes is $(D^*)^r$.

We now follow Saito's paper [36] in defining normal functions and admissible normal functions. Suppose $\mathcal{L}$ is a graded-polarizable variation of mixed Hodge structure on S with negative weights. That is, assume that $\mathrm{Gr}_n^W \mathcal{L} = 0$ for $n \geq 0$. In addition, assume that the underlying local system $\mathcal{L}_{\mathbb{Z}}$ of $\mathcal{L}$ is torsion-free.

DEFINITION 8.4.1 *With $\mathcal{L}$ as above, the group of normal functions on S is the group*

$$\mathrm{NF}(S, \mathcal{L}) := \mathrm{Ext}^1_{\mathrm{VMHS}(S)}(\mathbb{Z}, \mathcal{L})$$

of Yoneda extensions of $\mathbb{Z}$ by $\mathcal{L}$. If $\mathcal{L}$ is admissible relative to $\bar{S}$, then the group of admissible normal functions on S is the subgroup

$$\mathrm{NF}(S, \mathcal{L})_{\bar{S}}^{\mathrm{ad}} := \mathrm{Ext}^1_{\mathrm{VMHS}(S)_{\bar{S}}^{\mathrm{ad}}}(\mathbb{Z}, \mathcal{L})$$

consisting of Yoneda extensions which are admissible relative to $\bar{S}$.

The definition of $\mathrm{NF}(S, \mathcal{L})$ is motivated by a theorem of Carlson which computes the extension group in the case that S is a point. If L is a torsion-free mixed Hodge structure with negative weights, then the *intermediate Jacobian* of L is the torus $J(L) := L_{\mathbb{C}}/(F^0 L_{\mathbb{C}} + L_{\mathbb{Z}})$. In [6], Carlson proved the following.

THEOREM 8.4.2 (Carlson) *Suppose L is a graded-polarizable, torsion-free, mixed Hodge structure with negative weights. Then, there is a natural isomorphism of abelian groups*

$$\mathrm{Ext}^1_{\mathrm{MHS}}(\mathbb{Z}, L) = J(L).$$

Remark. To get the map $\mathrm{Ext}^1_{\mathrm{MHS}}(\mathbb{Z}, L) \to J(L)$ of Theorem 8.4.2, consider an exact sequence

$$0 \to L \xrightarrow{i} V \xrightarrow{p} \mathbb{Z} \to 0 \tag{8.4.1}$$

in the category of mixed Hodge structures. Pick $v_{\mathbb{Z}} \in V_{\mathbb{Z}}$ such that $p(v_{\mathbb{Z}}) = 1$. Since morphisms of mixed Hodge structures are strict with respect to the Hodge filtration, we can also find $v_F \in F^0 V_{\mathbb{C}}$ such that $p(v_F) = 1$. Set $\sigma(V) := v_{\mathbb{Z}} - v_F \in \mathrm{Ker}\, p_{\mathbb{C}} = L_{\mathbb{C}}$. It is well defined modulo the choices of $v_{\mathbb{Z}}$ and the choice of v_F. In other words, it is well defined modulo $L_{\mathbb{Z}} + F^0 L_{\mathbb{C}}$. Thus the class of σ in $J(L)$ is well defined.

A variation $\mathcal{L}$ as in Definition 8.4.1 gives rise to a morphism $\pi : J(\mathcal{L}) \to S$ of complex manifolds where the fiber of $J(\mathcal{L})$ over a point $s \in S$ is the intermediate Jacobian $J(\mathcal{L}_s)$. Thus, by restriction and Carlson's theorem, a normal function ν gives rise to a section $\sigma_\nu : S \to J(\mathcal{L})$ of π. In [36], Saito showed that, if $\mathcal{L}$ is torsion-free and concentrated in negative weights, then the resulting map from $\mathrm{Ext}^1_{\mathrm{VMHS}(S)}(\mathbb{Z}, \mathcal{L})$ to the space of holomorphic sections of $J(\mathcal{L})$ over S is injective. Moreover, the flat connection on the vector bundle $\mathcal{L}_{\mathcal{O}_S}$ induces a map ∇ taking section of $J(\mathcal{L})$ over S to sections of the vector bundle $(\mathcal{L}_{\mathcal{O}_S}/F^{-1}\mathcal{L}) \otimes \Omega^1_S$. A holomorphic section of $J(\mathcal{L})$ is *horizontal* if it is in the kernel of ∇. As explained in [36], the normal functions coincide exactly with the horizontal sections of $J(\mathcal{L})$. This explains why elements of the group $\mathrm{NF}(S, \mathcal{L})$ are called "functions": they are holomorphic functions $\sigma : S \to J(\mathcal{L})$ such that $\pi \circ \sigma$ is the identity and σ is horizontal.

DEFINITION 8.4.3 *Suppose* $\nu \in \mathrm{NF}(S, \mathcal{L})$ *is a normal function. The zero locus of* ν
is the subset $Z(\nu) := \{s \in S : \nu(s) = 0\}$.

LEMMA 8.4.4 *Suppose* $0 \to L \to V \xrightarrow{p} \mathbb{Z} \to 0$ *is an extension of the pure Hodge
structure* $\mathbb{Z}$ *by a mixed Hodge structure* L *with negative weights as in* (8.4.1). *Then the
extension class is 0 in* $\mathrm{Ext}^1_{\mathrm{MHS}}(\mathbb{Z}, L)$ *if and only if* $V \cong \mathbb{Z} \oplus L'$ *for some mixed Hodge
structure* L'.

PROOF. Obviously, if the extension class is 0, then $V \cong \mathbb{Z} \oplus L$. Conversely,
suppose $V \cong \mathbb{Z} \oplus L'$. We then have a short-exact sequence of mixed Hodge structures

$$0 \to L \to \mathbb{Z} \oplus L' \xrightarrow{p} \mathbb{Z} \to 0.$$

Since $W_{-1}L = L$, any morphism from L to $\mathbb{Z}$ is 0. Thus L maps injectively into the L'
factor of $\mathbb{Z} \oplus L'$. Since the cokernel of the map $L' \to \mathbb{Z} \oplus L'$ is $\mathbb{Z}$, the map $L \to L'$ must
be an isomorphism. But then $p(L') = 0$ and $p(\mathbb{Z}) = \mathbb{Z}$. It follows that the extension
class is trivial.   □

COROLLARY 8.4.5 *If* ν *is a normal function corresponding to an extension* $\mathcal{V}$ *of* $\mathbb{Z}$
by $\mathcal{L}$ *in* $\mathrm{VMHS}(S)$, *then* $Z(\nu)$ *is the locus of points* $s \in S$ *where* $\mathcal{V}_s$ *contains* $\mathbb{Z}$ *as a
direct summand.*

PROOF. This follows directly from Lemma 8.4.4.   □

The following theorem was proved in [5, Proposition 6].

THEOREM 8.4.6 *Suppose* $j : S \to \bar{S}$ *is a Zariski open embedding of complex manifolds and* $\mathcal{L}$ *is a graded-polarizable, torsion-free, admissible variation of mixed Hodge
structure of negative weights on* S. *Let* $\nu \in \mathrm{NF}(S, \mathcal{L})^{\mathrm{ad}}_{\bar{S}}$ *be an admissible normal function. Then the closure* $\mathrm{cl}\, Z(\nu)$ *of* $Z(\nu)$ *in the usual (classical) topology is an analytic
subvariety of* $\bar{S}$.

Our goal here is not to give a complete proof of the theorem, but rather to put it in
context and to explain some of the ideas of the proof in a way that is less technical and,
hopefully, easier to follow than the original papers. After defining admissible normal
functions, Saito proved a special case of Theorem 8.4.6 in [36, Corollary 2.8]. (Saito
assumes that $\mathcal{L}$ is induced via pull-back from a pure variation of weight -1 on a curve,
and he makes a restrictive assumption concerning the monodromy of the variation on
the curve.) For variations of pure Hodge structure of weight -1 on curves, the result
was proved by Brosnan and Pearlstein in [3]. For variations of pure Hodge structure
of weight -1 with S of arbitrary dimension the result was proved independently by
Schnell in [38] using his Néron models, which are partial compactifications of the
family $J(\mathcal{L}) \to S$ of intermediate Jacobians. In [4], the theorem is proved for pure
variations of Hodge structure of negative weight. The full theorem follows from this
by induction using the exact sequence

$$0 \to \mathrm{NF}(S, W_{n-1}\mathcal{L})^{\mathrm{ad}}_{\bar{S}} \to \mathrm{NF}(S, W_n\mathcal{L})^{\mathrm{ad}}_{\bar{S}} \to \mathrm{NF}(S, \mathrm{Gr}^W_n \mathcal{L})^{\mathrm{ad}}_{\bar{S}}. \qquad (8.4.2)$$

See [5] for more details. As mentioned above, there is also a proof by Kato, Nakayama, and Usui [27].

Remark. The theorem is a natural extension of the theorem of Cattani, Deligne, and Kaplan on the locus of Hodge classes [7]. The proofs are also analogous. The main technical tool used to prove the theorem in [7] is the SL_2-orbit theorem of Cattani, Kaplan, and Schmid [8]. The main technical tool used in all known proofs of Theorem 8.4.6 is the mixed SL_2-orbit theorem of Kato, Nakayama, and Usui [26]. (The paper [3] uses an SL_2-orbit theorem proved by Pearlstein [34] to handle the case of pure variations of weight -1 on curves.)

8.4.1   Reducing Theorem 8.4.6 to a Special Case

The first thing to notice in the proof of Theorem 8.4.6 is that the statement is local in the analytic topology on $\bar{S}$ (because being a closed analytic subvariety is a local property). Therefore, we may assume that $\bar{S}$ is the polydisk D^r. Similarly, by arguments using resolution of singularities, we may assume that S is the punctured polydisk D^{*r}. As explained in [5], by the induction using (8.4.2) mentioned above, we may assume that $\mathcal{L}$ is concentrated in one negative weight. By Borel's theorem, the monodromy of $\mathcal{L}$ around the coordinate hyperplanes is quasi-unipotent. So, by pulling back to a cover of S ramified along the coordinate hyperplanes, we can assume that the monodromy is, in fact, unipotent. So from now on we might as well assume that $\bar{S}$ and S are r-dimensional polydisks and punctured polydisks respectively, and $\mathcal{L}$ is a pure variation of negative weight in $\mathrm{VMHS}(S)_{\bar{S}}^{\mathrm{ad}}$ with unipotent monodromy.

8.4.2   Examples

We want to point out that admissibility is an absolutely crucial hypothesis in Theorem 8.4.6. For this, it seems instructive to look at a couple of simple examples of Theorem 8.4.6 where the variation $\mathcal{L}$ is constant. We will see that $\mathrm{cl}\, Z(\nu)$ is analytic for $\nu \in \mathrm{NF}(S, \mathcal{L})_{\bar{S}}^{\mathrm{ad}}$, but definitely not analytic for arbitrary $\nu \in \mathrm{NF}(S, \mathcal{L})$.

For the simple examples we want to study, it will be useful to have a simple lemma.

LEMMA 8.4.7   *Suppose $(V, W, F, \bar{F}, N)$ is an IMHS with $N = 0$ on $\mathrm{Gr}_*^W V$. Then the relative weight filtration $M = M(N, W)$ coincides with the weight filtration W.*

PROOF.   The condition that $N^j : \mathrm{Gr}_{k+j}^M \mathrm{Gr}_k^W V \to \mathrm{Gr}_{k-j}^M \mathrm{Gr}_k^W V$ is an isomorphism for $j \geq 0$ and $k, j \in \mathbb{Z}$ immediately implies that $\mathrm{Gr}_a^M \mathrm{Gr}_b^W V = 0$ for $a \neq b$. By a simple induction, this in turn implies that $M = W$. $\square$

8.4.2.1 $\mathcal{L} = \mathbb{Z}(1)$

In the first example, we consider the constant variation $\mathbb{Z}(1)$ on $S = D^{*r}$. It follows directly from Carlson's theorem that $J(\mathbb{Z}(1)) = \mathbb{C}/2\pi i\mathbb{Z} = \mathbb{C}^\times$. Thus, $\mathrm{NF}(S, \mathbb{Z}(1))$ is a subgroup of $\mathcal{O}_S^\times(S)$. The condition of horizontality on a variation of mixed Hodge structure $\mathcal{V} \in \mathrm{Ext}^1_{\mathrm{VMHS}(S)}(\mathbb{Z}, \mathbb{Z}(1))$ is vacuous because $F^{-1}\mathcal{V}_{\mathcal{O}_S} = \mathcal{V}_{\mathcal{O}_S}$. From this, it is easy to see that $\mathrm{NF}(S, \mathbb{Z}(1)) = \mathcal{O}_S^\times(S)$.

On the other hand, with $\bar{S} = D^r$, $\mathrm{NF}(S, \mathbb{Z}(1))^{\mathrm{ad}}_{\bar{S}}$ is the subset $\mathcal{O}_S^{\times,\mathrm{mer}}$ of $\mathcal{O}_S^\times(S)$ consisting of nonvanishing holomorphic functions with meromorphic extension to $\bar{S}$.

To see this, let $\mathcal{V}$ be an admissible variation of mixed Hodge structure on S representing a class in $\mathrm{NF}(S, \mathbb{Z}(1))^{\mathrm{ad}}_{\bar{S}}$. The local system $\mathcal{V}_{\mathbb{Z}}$ underlying $\mathcal{V}$ is determined by the action of the monodromy operators T_i on a reference fiber, which we may take to be the free abelian group on generators e and f. We may assume that the generator f corresponds to the trivial sublocal system $\mathbb{Z}(1)$ of $\mathcal{V}_{\mathbb{Z}}$, and then the generator e maps onto the generator of the quotient $\mathcal{V}_{\mathbb{Z}}/\mathbb{Z}(1)$. Then $T_i f = f$ for $i = 1, \ldots, r$ and there are integers a_i such that $T_i e = e + a_i f$.

Set $N_j = -\frac{1}{2\pi i}\log T_j = \frac{1}{2\pi i}(I - T_j)$ for $j = 1, \ldots, r$, and let $\mathcal{V}_{\mathrm{can}}$ denote the canonical extension of $\mathcal{V}_{\mathcal{O}_S}$ to $\bar{S}$. Then $\mathcal{V}_{\mathrm{can}}$ is freely generated as an $\mathcal{O}_{\bar{S}}$-module by sections $\tilde{e}$ and $\tilde{f}$ as in Theorem 8.1.27. In terms of the multivalued sections e and f of $\mathcal{V}_{\mathcal{O}_S}$, the restrictions of $\tilde{e}$ and $\tilde{f}$ to S are given by

$$\tilde{e} = e - \frac{1}{2\pi i}\sum_{i=1}^r a_i \log z_i f, \quad \tilde{f} = f.$$

Since $\mathbb{Z}(1)$ has weight -2, the subbundle of $\mathcal{V}_{\mathcal{O}_S}$ corresponding to $\mathbb{Z}(1)$ does not intersect $F^0\mathcal{V}_{\mathcal{O}_S}$. Moreover, both $F^0\mathcal{V}_{\mathcal{O}_S}$ and $\mathrm{Gr}_0^F\mathcal{V}_{\mathcal{O}_S}$ are trivial line bundles, since $\bar{S}$ is Stein and we have $H^2(\bar{S}, \mathbb{Z}) = 0$. It follows that $F^0\mathcal{V}_{\mathrm{can}}$ is freely generated as an $\mathcal{O}_{\bar{S}}$ module by a section of the form $g := \alpha(z)\tilde{e} + \beta(z)\tilde{f}$ with α, β holomorphic functions on $\bar{S}$. Let $W_{k,\mathrm{can}}$ denote the canonical extension of the vector bundle $W_k \otimes_{\mathbb{Q}} \mathcal{O}_S$ to $\bar{S}$. It is a subbundle of $\mathcal{V}_{\mathrm{can}}$. Since $\mathcal{V}$ is admissible, $\mathrm{Gr}_F^p \mathrm{Gr}_0^{W_{\mathrm{can}}} \mathcal{V}_{\mathrm{can}} = 0$ for $p \neq 0$. So F^0 surjects onto $\mathrm{Gr}_0^{W_{\mathrm{can}}}$. It follows that α is a nonvanishing holomorphic function on $\bar{S}$. Thus, dividing out by α, we can assume $\alpha = 1$. So, on S, F^0 is generated by the section

$$g(z) = \tilde{e} + \beta(z)\tilde{f} = e + \left(\beta(z) - \frac{1}{2\pi i}\sum_{i=1}^r a_i \log z_i\right)f.$$

Using Carlson's formula, we find that the class $\sigma(\mathcal{V}_z)$ of $\mathcal{V}_z$ in $J(\mathbb{Z}(1))$ is given by $h(z) := \frac{1}{2\pi i}\sum_{i=1}^r a_i \log z_i - \beta(z)$ modulo $\mathbb{Z}$. Exponentiating, we find that the class of $\mathcal{V}_z$ in $\mathbb{C}^\times$ is

$$e^{2\pi i h(z)} = e^{-2\pi i \beta(z)}\prod_{i=1}^r z_i^{a_i}. \tag{8.4.3}$$

This is a nonvanishing holomorphic function on S with meromorphic extension to $\bar{S}$. So we have shown that $\mathrm{NF}(S, \mathbb{Z}(1))^{\mathrm{ad}}_{\bar{S}} \subset \mathcal{O}_S^{\times,\mathrm{mer}}$.

To show the reverse inclusion, note that any $\varphi \in \mathcal{O}_S^{\times,\mathrm{mer}}$ can be written in the form of the function on the right-hand side of (8.4.3) by choosing suitable $a_i \in \mathbb{Z}$ and $\beta \in \mathcal{O}_{\bar{S}}$. Suppose then that we are given such $(a_i)_{i=1}^r$ and β. Define a variation $\mathcal{V}$, where a fiber of the underlying local system $\mathcal{V}_{\mathbb{Z}}$ is the free abelian group generated by e, f as above with $T_i f = f, T_i e = e + a_i f$. Let $\mathcal{V}_{\mathrm{can}}$, and $\tilde{e}, \tilde{f}$ be as above. And, using β as above, set $F^0\mathcal{V}_{\mathrm{can}} = \mathcal{O}_{\bar{S}}(\tilde{e} + \beta(z)\tilde{f})$. Define $F^p\mathcal{V}_{\mathrm{can}} = 0$ for $p > 0$ and

$F^p \mathcal{V}_{\text{can}} = \mathcal{V}_{\text{can}}$ for $p < 0$. Let the weight filtration of $\mathcal{V}$ be the one where, on the reference fiber, W_{-2} is generated by f and $\text{Gr}_p^W = 0$ for $p \neq -2$ or 0. Then it is easy to see that $\mathcal{V}$ is a variation of mixed Hodge structure (e.g., horizontality is trivial). Using the map from $\mathbb{Z}(1) \to \mathcal{V}$ sending the generator to f and the map from $\mathcal{V}$ to $\mathbb{Z}$ sending e to 1 and f to 0, we see that $\mathcal{V}$ is an extension of $\mathbb{Z}$ by $\mathbb{Z}(1)$. Moreover, since $N_i = 0$ on Gr_*^W, it follows easily that $M(J) = W$ for any $J \subset \{1, \dots, r\}$. Thus $\mathcal{V}$ is admissible. So $\text{NF}(S, \mathbb{Z}(1))_{\bar{S}}^{\text{ad}} = \mathcal{O}_S^{\times, \text{mer}}$.

Note that it is easy to find $\nu \in \text{NF}(S, \mathbb{Z}(1))$ such that $\text{cl}\, Z(\nu)$ fails to be an analytic subvariety of S. For example, take $r = 1$ and consider the nonvanishing function $e^{1/s}$ on $S = D^*$. Let ν be the corresponding normal function. Then

$$Z(\nu) = \{z \in D : e^{1/s} = 1\} = \left\{ \frac{1}{2\pi i n} : n \in \mathbb{Z} \right\}.$$

The closure of $Z(\nu)$ is simply $Z(\nu) \cup \{0\}$. Obviously this set is not analytic at the origin.

On the other hand, for $\nu \in \text{NF}(S, \mathbb{Z}(1))_{\bar{S}}^{\text{ad}}$, it is easy to see that $\text{cl}\, Z(\nu)$ is always analytic.

8.4.2.2 $\mathcal{L} = H_1(E)$ for E an Elliptic Curve

In the previous example, the existence of the relative weight filtration was automatic. Now, we look at another simple example with a constant variation. Let E denote the elliptic curve $\mathbb{C}/(\mathbb{Z} + i\mathbb{Z})$ and let $\mathcal{H}$ denote the constant variation of Hodge structure $H_1(E)$ on $S = D^*$. Because horizontality is automatic, it is not hard to see that $\text{NF}(S, \mathcal{H})$ is the group of all analytic maps from S to E.

On the other hand, $\text{NF}(S, \mathcal{H})_{\bar{S}}^{\text{ad}}$ consists of the analytic maps from $\bar{S}$ into E. To see this, suppose $\mathcal{V}$ is a variation in $\text{Ext}^1_{\text{VMHS}(S)_{\bar{S}}^{\text{ad}}}(\mathbb{Z}, \mathcal{H})$. Since $\mathcal{H}$ is a constant variation, the monodromy logarithm $N = \log T$ is trivial on $\mathcal{H}_{\mathbb{Z}}$. Since $\mathcal{V}$ is admissible, the relative weight filtration $M = M(N, W)$ exists. From the fact that N is trivial on $\text{Gr}^W \mathcal{V}$, we know that $M = W$. But then, since $N : M_k \subset M_{k-2}$, it follows that $N = 0$. Thus, the local system underlying $\mathcal{V}$ is trivial. From the fact that the Hodge filtration extends in $\mathcal{V}_{\text{can}}$, it then follows that any normal function in $\text{NF}(S, \mathcal{L})_{\bar{S}}^{\text{ad}}$ is the restriction of an element of $\text{NF}(\bar{S}, \mathcal{L})$. In other words, $\text{NF}(S, \mathcal{L})_{\bar{S}}^{\text{ad}} \subset \text{NF}(\bar{S}, \mathcal{L})$. The reverse inclusion is obvious.

Clearly $\text{cl}\, Z(\nu)$ is analytic for any $\nu \in \text{NF}(S, \mathcal{L})_{\bar{S}}^{\text{ad}} = \text{NF}(\bar{S}, \mathcal{L})$, because $\text{NF}(\bar{S}, \mathcal{L})$ is simply the set of analytic maps from $\bar{S}$ to E. On the other hand, there are many $\nu \in \text{NF}(S, \mathcal{L})$ for which $\text{cl}\, Z(\nu)$ is not an analytic subset of S. For example, consider the map $\pi : S \to E = \mathbb{C}/(\mathbb{Z} + i\mathbb{Z})$ given by $s \mapsto \frac{\log s}{2\pi}$. Clearly, this map is well defined and analytic. Thus it corresponds to a normal function $\nu \in \text{NF}(S, \mathcal{L})$. We have

$$Z(\nu) = \left\{ s \in D^* : \frac{\log s}{2\pi} \in \mathbb{Z} + i\mathbb{Z} \right\}$$
$$= \{e^{2\pi n} : n \in \mathbb{Z}\} \cap D^*.$$

Since $Z(\nu)$ is a countable subset of S with a limit point in $\bar{S}$, $\text{cl}\, Z(\nu)$ is not analytic.

8.4.3 Classifying Spaces

In order to go further toward the proof of Theorem 8.4.6, we want to be able to get a hold of variations of mixed Hodge structures on products of punctured disks in a way that is as concrete as possible. For this it helps to talk a little bit about classifying spaces in the mixed setting. We are going to do this following Cattani's lectures in Chapter 7.

Remark.   We change the notation here slightly in that Chapter 7 fixes the dimensions of the flags in a classifying space, while we consider possibly disconnected classifying spaces where the dimensions of the flags may vary from component to component. This is in line with the notation of [26].

8.4.4 Pure Classifying Spaces

Suppose V is a finite-dimensional real vector space, k is an integer and $Q : V \otimes V \to \mathbb{R}$ is a nondegenerate bilinear form. Assume that Q is $(-1)^k$-symmetric, that is, assume that Q is symmetric if k is even and skew symmetric otherwise. Then the classifying space $D = D(V, Q)$ of the pair (V, Q) is the set of all decreasing filtrations F of $V_\mathbb{C}$ such that (V, F) is a pure real Hodge structure of weight k polarized by Q. We write $\hat{D}$ for the set of all finite decreasing filtrations F of V such that $Q(F^p, F^{k-p+1}) = 0$ for all $p \in \mathbb{Z}$. Clearly $D \subset \hat{D}$. In fact, D is included as an (analytic) open subset of $\hat{D}$, because the condition for (V, F) to be a Hodge structure of weight k is an open one, and so is the condition for (V, F) to be polarized by Q. The compact dual classifying space is the union $D^\vee$ of the components of $\hat{D}$ meeting D.

Write G for the real algebraic group consisting of all automorphisms of V preserving Q. Then $G(\mathbb{R})$ acts on D. Similarly, the group $G(\mathbb{C})$ of complex points of G acts on $D^\vee$. Clearly, the dimensions $f^p := \dim F^p$ of the spaces in the flags are fixed on each $G(\mathbb{C})$-orbit. Theorem 7.4.3 tells us that, for each sequence $\{f_p\}_{p \in \mathbb{Z}}$ of dimensions, there is at most one component of D. More importantly, Theorem 7.4.3 tells us that both D and $D^\vee$ are smooth complex manifolds, and that the connected components of D (resp. $D^\vee$) are $G(\mathbb{R})$- (resp. $G(\mathbb{C})$-) orbits.

8.4.5 Mixed Classifying Spaces

Suppose now that V is a real vector space equipped with a finite, increasing filtration W. (I.e., there exists $a, b \in \mathbb{Z}$ such that $W_a = 0, W_b = V$.) Suppose that for each $k \in \mathbb{Z}$ we are given a nondegenerate, $(-1)^k$-symmetric bilinear form Q_k on $\mathrm{Gr}_k^W V$. Write D_k for the classifying space of the pair $(\mathrm{Gr}_k^W V, Q_k)$. Similarly, write $D_k^\vee$ for the compact dual classifying space and G_k for the group of automorphisms of Q_k.

The classifying space $D = D(V, W, Q)$ of the triple (V, W, Q) is the set of all decreasing filtrations F of $V_\mathbb{C}$, such that (V, W, F) is a real mixed Hodge structure and, for each $k \in \mathbb{Z}$, $\mathrm{Gr}_k^W V$ is polarized by Q_k. In other words, D is the set of all filtrations F on V such that, for each k, the filtration $F \, \mathrm{Gr}_k^W V$ induced by F on $\mathrm{Gr}_k^W V$ lies in D_k. Similarly, the compact dual classifying space $D^\vee$ is the set of all finite, decreasing filtrations F of V such that $F \, \mathrm{Gr}_k^W V$ lies in $D_k^\vee$.

Write G for the real algebraic group of linear automorphisms of the pair (V, W) preserving the forms Q_k. Then $G(\mathbb{R})$ acts on D and $G(\mathbb{C})$ acts on $D^\vee$. We have an obvious homomorphism $G \to \prod G_k$ making the obvious map $D \to \prod D_k$ (resp. $D^\vee \to \prod D_k^\vee$) into a map of $G(\mathbb{R})$ (resp. $G(\mathbb{C})$) spaces. By choosing a splitting of the filtration W, it is easy to see that the homomorphism $G \to \prod G_k$ is surjective with kernel the subgroup U consisting of all $g \in G$ which act trivially on $\mathrm{Gr}^W V$.

LEMMA 8.4.8 *The group $G(\mathbb{C})$ acts transitively on every component of $D^\vee$.*

SKETCH. Using elementary linear algebra, it is not hard to see that $U(\mathbb{C})$ acts transitively on the fibers of the map $D^\vee \to \prod D_k^\vee$. Since G acts transitively on $\prod D_k^\vee$, the result follows. □

Now suppose that V, W, and Q are as above, and that $F \in D^\vee$. Suppose M is another filtration of V. Furthermore, suppose that, for each integer k, F induces a mixed Hodge structure on the pair $(W_k, M \cap W_k)$. In particular, F induces a mixed Hodge structure on the pair (V, M). This mixed Hodge structure also induces a mixed Hodge structure on the Lie algebra $\mathfrak{g}$ of the real algebraic group G. So, by Theorem 7.5.6, we obtain a decomposition $\mathfrak{g}_\mathbb{C} = \oplus \mathfrak{g}_{(F,M)}^{p,q}$, where we write $\mathfrak{g}_{(F,M)}^{p,q}$ for $I_{(F,M)}^{p,q} \mathfrak{g}$. As in equation (7.6.3), we write $\mathfrak{g}_- := \oplus_{p<0,q} \mathfrak{g}_{(F,M)}^{p,q}$. We write $\mathfrak{g}^F := \oplus_{p>0,q} \mathfrak{g}_{(F,M)}^{p,q}$. Then $\mathfrak{g}^F$ is the subspace of $\mathfrak{g}$ stabilizing F, and $\mathfrak{g}_-$ is complementary. It follows that $\mathfrak{g}_-$ is isomorphic to the tangent space of $D^\vee$ at F.

Remark. It is not true that $G(\mathbb{R})$ acts transitively on D. To see this, we work out an example. Let V denote the real vector space $\mathbb{R}^2$ with basis e and f. Let W be the filtration on V with $W_{-3} = 0, W_{-2} = \langle f \rangle = W_{-1}$, and $W_0 V = V$. Then $\mathrm{Gr}_{-2}^W V = f$, $\mathrm{Gr}_0^W V$ is spanned by the image of e, and all other graded pieces of V are trivial. Let $Q_{-2}(f, f) = Q_0(e, e) = 1$. Then, with respect to the ordered basis e, f,

$$G = \left\{ \pm \begin{pmatrix} 1 & 0 \\ z & 1 \end{pmatrix} \in \mathrm{GL}_2 \right\}.$$

The spaces D_0 and D_{-2} are both trivial, i.e., one-point sets. To give a flag in D amounts to giving the subspace F^0, of the form $\langle e + zf \rangle$ for some (unique) $z \in \mathbb{C}$. So $D \cong \mathbb{C}$. Thus, we see that $G(\mathbb{C})$ acts simply transitively on the space D, which, in this case, coincides with $D^\vee$. On the other hand, clearly $G(\mathbb{R})$ does not act transitively on D.

8.4.6   Local Normal Form

Now suppose $\mathcal{V}$ is a variation of mixed Hodge structure on $S = (D^*)^r$ which is admissible relative to $\bar{S} = D^r$. Assume that the monodromy operators $T_1, \dots, T_r$ are unipotent with monodromy logarithms $N_1, \dots, N_r$. Let $V_\mathbb{Z}$ denote the space of global sections of the pull-back of the local system $\mathcal{V}$ on S to the product U^r of upper half planes via the map $(z_1, \dots, z_r) \mapsto (s_1, \dots, s_r)$ where $s_i = e^{2\pi i z_i}$. Write V for the real vector space $V_\mathbb{Z} \otimes \mathbb{R}$, and $W_k V$ for the filtration of V induced by $W_k V$.

The pull-back of $\mathcal{V}$ to U^r gives a variation of mixed Hodge structure on U^r whose underlying local system is the constant local system with fiber $V_{\mathbb{Z}}$. Thus we obtain a map $\Phi : U^r \to D$ sending a point $z \in U^r$ to the Hodge filtration $F(z)$ on the trivial vector bundle $V \otimes \mathcal{O}_{U^r}$. For each $z \in U^r$, set $N(z) := \sum z_i N_i$. Note that $N(z) \in \mathfrak{g}_{\mathbb{C}}$, so $e^{N(z)} \in G(\mathbb{C})$. Owing to the way the monodromy acts on the pull-back $\mathcal{V}_{U^r}$, the map $\tilde{\Psi} : U^r \to D$ given by $z \mapsto e^{-N(z)}\Phi$ is invariant under the transformation $z \mapsto z + n$, where $n \in \mathbb{Z}^r$. Thus $\tilde{\Psi}$ descends to a map from S to $D^\vee$ which we will denote by Ψ.

LEMMA 8.4.9 *The map $\Psi : S \to D^\vee$ extends to a neighborhood of 0 in $\bar{S}$.*

SKETCH. This follows from the assumption of admissibility. The crucial fact is that the Hodge filtration extends to a filtration of Deligne's canonical extension in such a way that the F^p are vector subbundles of the weight-graded subquotients of the canonical extension.   □

Now, suppose we shrink $\bar{S}$ to a neighborhood of 0 to which the map Ψ extends. Write $F = \Psi(0)$, and let M denote the relative weight filtration $M = M(N_1 + \cdots + N_r, W)$ which exists by the assumption of admissibility. From Schmid's SL_2-orbit theorem [37], it follows that F induces a mixed Hodge structure on the pair (V, M). Moreover, since for each k, $W_k V$ is a subvariation of $\mathcal{V}$, the subspace W_k of V is a mixed Hodge substructure. Thus (F, M) induces a mixed Hodge structure on the Lie algebra $\mathfrak{g}$ of the group G. As we pointed out above, $\mathfrak{g}_-$ is isomorphic to the tangent space of $D^\vee$ at F. Since $\mathfrak{g}_-$ is a nilpotent Lie subalgebra of $\mathfrak{g}$, the exponential map is a polynomial map on $\mathfrak{g}_-$ sending a neighborhood of 0 biholomorphically onto a neighborhood of F in $D^\vee$. It follows that, in a neighborhood of F, there is a uniquely determined, holomorphic function $\Gamma : \bar{S} \to \mathfrak{g}_-$ such that $\Gamma(0) = 0$ and

$$\Psi(s) = e^{\Gamma(s)} F.$$

Unraveling the definitions of $\Psi(s)$, we find that

$$F(z) = \Phi(z) = e^{N(z)} e^{\Gamma(s)} F \qquad (8.4.4)$$

for all z in the inverse image of S in U^r. This is the *local normal form* of the admissible variation $\mathcal{V}$. It is the mixed version of the local form of the period map described for pure orbits in Section 7.6. By shrinking $\bar{S}$ around 0, we can assume that equation (8.4.4) holds for all $z \in U$.

8.4.7 Splittings

Suppose (V, W, F) is a $\mathbb{Z}$-mixed Hodge structure. For a linear transformation Y of $V_{\mathbb{C}}$ and a number $\lambda \in \mathbb{C}$, let $E_\lambda(Y)$ denote the λ-eigenspace of Y. A *complex splitting* or *grading* of W is a semisimple endomorphism Y of $V_{\mathbb{C}}$ with integral eigenvalues such that $W_k = \oplus_{j \le k} E_j(Y)$.

Suppose $V_{\mathbb{C}} = \oplus I^{p,q}$ is the decomposition of $V_{\mathbb{C}}$ from Chapter 7. Define an endomorphism $Y_{(F,W)}$ of $V_{\mathbb{C}}$ by setting $Y_{(F,W)}v = (p + q)v$ for $v \in I^{p,q}$. Then, clearly

$Y_{(F,W)}$ is a complex splitting of W. We call it the *canonical grading*. Note that $Y_{(F,W)}$ preserves the spaces $I^{p,q}$.

LEMMA 8.4.10   *Suppose H is a pure Hodge structure of weight -1 and $V = (V, F, W)$ is an extension of $\mathbb{Z}(0)$ by H. Then $Y_{(F,W)}$ is a real endomorphism of $V_\mathbb{C}$ which is integral if and only if V is a trivial extension.*

PROOF.   If V is the trivial extension, then $Y_{(F,W)}$ is obviously integral. Conversely, suppose $Y_{(F,W)}$ is integral. Since $F^p = \oplus_{p' \geq p} I^{p',q}$ and $W_k \otimes \mathbb{C} = \oplus_{p+q \leq k} I^{p,q}$, $Y_{(F,W)}$ preserves both F and $W_k \otimes \mathbb{C}$. If $Y_{(F,W)}$ is integral, then it is, by definition, an endomorphism of mixed Hodge structures. But then $-Y_{(F,W)}$ is an idempotent morphism of mixed Hodge structures, whose kernel is isomorphic to $\mathbb{Z}$. Thus $\mathbb{Z}$ is a direct factor of V. So, by Lemma 8.4.4, the extension is trivial.

From the formula (7.5.10) for the subspaces $I^{p,q}$ and the fact that $\mathrm{Gr}_k^W V = 0$ for $k \notin [-1, 0]$, it is easy to see that $Y_{(F,W)}$ is real.   □

8.4.8   A Formula for the Zero Locus of a Normal Function

We want to introduce the main tool for proving Theorem 8.4.6 in the one-dimensional case. So, from now on, suppose $\mathcal{H}$ is a variation of pure Hodge structure of weight -1 with unipotent monodromy on the punctured disk $S = D^*$ and $\nu \in \mathrm{NF}(S, \mathcal{H})^{\mathrm{ad}}_{\bar{S}}$ is an admissible normal function where $\bar{S} = D$. Let $\mathcal{V}$ denote the corresponding admissible variation of mixed Hodge structure. Pulling back $\mathcal{V}$ to the upper half plane and shrinking $\bar{S}$ around 0 if necessary, we can write $\mathcal{V}$ in terms of its local normal form (8.4.4). As above, the $F(z)$ are a holomorphically varying family of filtrations of the fixed vector space $V_\mathbb{C} = V_\mathbb{Z} \otimes \mathbb{C}$. Write $Y(z) := Y_{(F(z),W)}$. By Lemma 8.4.10, the inverse image of $Z(\nu)$ in U^r is the set of all $z \in U^r$ such that $Y(z)$ is integral.

The following theorem is the main tool that is used in the proof of Theorem 8.4.6 in the case that $\mathcal{H}$ has weight -1 and S is one-dimensional. The proof is an application of Pearlstein's SL_2-orbit theorem [34].

THEOREM 8.4.11 ([3])   *The limit $Y^\ddagger$ of $Y(z)$ as z tends to infinity along a vertical strip in the upper half plane exists.*

Remark.   The limit is real because all of the $Y(z)$ are real by Lemma 8.4.10. In [3], a formula is given for $Y^\ddagger$. It depends only on the nilpotent orbit associated to $\mathcal{V}$.

LEMMA 8.4.12   *The limit $Y^\ddagger$ does not depend on the vertical strip used in the limit. Moreover $Y^\ddagger$ commutes with N.*

PROOF.   Since the limit exists along any vertical strip, the limit cannot depend on the vertical strip the limit is taken over. On the other hand, if $u \in \mathbb{R}$, then e^{uN} is real. It follows (from formula (7.5.10) for the spaces $I^{p,q}$) that

$$Y(z + u) = Y_{(e^{uN}e^{zN}F,W)} = \mathrm{Ad}(e^{uN})Y_{(e^{zN}F,W)}$$
$$= \mathrm{Ad}(e^{uN})Y(z).$$

By taking the limit as z tends to ∞ along the vertical strip, it follows that $Y^{\ddagger}$ commutes with N. $\square$

One obvious corollary of Theorem 8.4.11 is that Theorem 8.4.6 must hold unless $Y^{\ddagger}$ is integral. Therefore we can restrict our attention to the integral case. To handle this case we need to know a little bit more about $Y^{\ddagger}$. In particular, we need to know about the relationship between $Y^{\ddagger}$ and the limit Hodge structure on V induced by F and the relative weight filtration M. We have the Deligne decomposition $V = \oplus I^{p,q}_{(F,M)}$ of V with respect to this filtration, and we write $\mathfrak{gl}_{\mathbb{C}} = \oplus \mathfrak{gl}^{p,q}_{(F,M)}$ for the corresponding decomposition of the Lie algebra $\mathfrak{gl} := \mathfrak{gl}(V)$. Explicitly,

$$\mathfrak{gl}^{p,q} = \{X \in \mathfrak{gl}_{\mathbb{C}} : X(I^{r,s}_{(F,M)}) \subset I^{r+p,s+q}_{(F,M)}\}.$$

This gives us a splitting $\mathfrak{gl}_{\mathbb{C}} = \mathfrak{gl}_{-} \oplus \mathfrak{gl}^{F}$ where $\mathfrak{gl}^{F}$ is the subset of $\mathfrak{gl}_{\mathbb{C}}$ preserving F and $\mathfrak{gl}_{-} = \oplus_{p<0} \mathfrak{gl}^{p,q}$. Finally, write $\Lambda^{-1,-1}$ for the subset $\oplus_{p,q<0} \mathfrak{gl}^{p,q}$. The subalgebra $\Lambda^{-1,-1}$ of $\mathfrak{gl}$ is important in Hodge theory because of its role in the SL_2-orbit theorem of Cattani, Kaplan, and Schmid (see [8, Equation 2.18] where the notation $L^{-1,-1}$ is used). For us, the most important thing about $\Lambda^{-1,-1}$ is that it is (by definition) contained in $\mathfrak{gl}_{-}$.

THEOREM 8.4.13 ([3])   *We have* $Y^{\ddagger} = \mathrm{Ad}(e^{-\xi})Y_{\infty}$, *where* $\xi \in \Lambda^{-1,-1}$ *and* Y_{∞} *is an endomorphism of* $V_{\mathbb{C}}$ *preserving the subspaces* $I^{(p,q)}_{(F,M)}$. *Both* ξ *and* Y_{∞} *commute with* N.

We refer the reader to [3] for the proof of Theorem 8.4.13.

LEMMA 8.4.14   *Suppose* X *is an operator in* $\mathfrak{gl}_{-}$. *Then* $\mathrm{Ad}(e^{X})Y_{\infty} - Y_{\infty} \in \mathfrak{gl}_{-}$.

PROOF.   Since Y_{∞} preserves the subspace $I^{(p,q)}_{(F,M)}$, then XY_{∞} and $Y_{\infty}X$ are both in $\mathfrak{g}_{-}$. The result follows by taking the Taylor expansion of e^{X}. $\square$

COROLLARY 8.4.15   *If* $X_1, X_2 \in \mathfrak{gl}_{-}$, *then* $\mathrm{Ad}(e^{X_1}e^{X_2})Y_{\infty} \equiv Y_{\infty}$ mod $\mathfrak{gl}_{-}$.

PROOF.   We have $\mathrm{Ad}(e^{X_1})\mathfrak{gl}_{-} \subset \mathfrak{gl}_{-}$. So $\mathrm{Ad}(e^{X_1}e^{X_2})Y_{\infty} \equiv \mathrm{Ad}(e^{X_1})Y_{\infty} \equiv Y_{\infty}$ mod $\mathfrak{gl}_{-}$. $\square$

8.4.9   Proof of Theorem 8.4.6 for Curves

Suppose $\mathcal{V}$ is a variation as in Section 8.4.8, where the Hodge filtration has local normal form $e^{zN}e^{\Gamma(s)}F$, and where the limit $Y^{\ddagger}$ along a vertical strip in the upper half plane is integral. Without loss of generality, we can assume that the local normal form holds for all z in the upper half plane. So, for $z \in U$, set $Y_{\infty}(z) := \mathrm{Ad}(e^{zN}e^{\Gamma(s)})Y_{\infty}$. Since Y_{∞} preserves F, then $Y_{\infty}(z)$ preserves $F(z)$. So, since $Y(z)$ also preserves $F(z)$, then $Y(z) - Y_{\infty}(z)$ lies in $\mathfrak{gl}^{F(z)}$. It follows that

$$h(z) := \mathrm{Ad}(e^{-\Gamma(s)}e^{-zN})Y(z) - Y_{\infty} \in \mathfrak{gl}^{F}. \tag{8.4.5}$$

Write $\exp : U \to D^*$ for the map $z \mapsto e^{2\pi i z}$. The zero locus $Z := Z(\nu)$ of the normal function ν corresponding to $\mathcal{V}$ is the locus of points $s \in D^*$ such that $\mathcal{V}_z$ is split. To show that $\mathrm{cl}\,\nu$ is analytic in D, it suffices to show

$$Z(\nu) = D^* \text{ if } 0 \in \mathrm{cl}\, Z(\nu). \tag{8.4.6}$$

If the imaginary parts of the points in $\exp^{-1}(Z)$ are bounded, then we have nothing to prove. Otherwise, there is a sequence of points $z_k \in \exp^{-1}(Z)$ such that $\mathrm{Im}\, z_k \to \infty$. Without loss of generality, we can assume that the sequence $\mathrm{Re}\, z_k$ is bounded. Then, since $Y(z_k) \to Y^{\ddagger}$ and the $Y(z_k)$ are all integral, we must have $Y(z_k) = Y^{\ddagger}$ for $k \gg 0$. So, by throwing away finitely many elements of the sequence, we can assume that $Y(z_k) = Y^{\ddagger}$ for all k.

Now, for each k, let $s_k = \exp(z_k)$. Using Lemma 8.4.12, we see that

$$
\begin{aligned}
h(z_k) &= \mathrm{Ad}(e^{-\Gamma(s_k)}e^{-z_k N})Y(z_k) - Y_\infty \\
&= \mathrm{Ad}(e^{-\Gamma(s_k)}e^{-z_k N})Y^{\ddagger} - Y_\infty \\
&= \mathrm{Ad}(e^{-\Gamma(s_k)})Y^{\ddagger} - Y_\infty \\
&= \mathrm{Ad}(e^{-\Gamma(s_k)}e^{-\xi})Y_\infty - Y_\infty.
\end{aligned}
$$

Since ξ and Γ are both in $\mathfrak{gl}_-$, Corollary 8.4.15 shows that $h(z_k)$ lies in $\mathfrak{gl}_-$. But $h(z) \in \mathfrak{gl}^F$ for all z, and $\mathfrak{gl}_-$ is a vector space complement to $\mathfrak{gl}^F$. So, $h(z_k) = 0$ for all k. On the other hand, $H(s) := \mathrm{Ad}(e^{-\Gamma(s)}e^{-\xi})Y_\infty - Y_\infty$ is clearly a (matrix-valued) holomorphic function of s on D. So, since $s_k \to 0$ and $H(s_k) = 0$ for all k, it follows that $H(s)$ is identically 0. Examining the limit of $H(s)$ as s tends to 0, using the fact that $\Gamma(0) = 0$ shows that $Y^{\ddagger} = Y_\infty$. So, $\mathrm{Ad}(e^{-\Gamma(s)})Y^{\ddagger} = Y^{\ddagger}$ for all $s \in D$. It follows that $Y_\infty(z) = \mathrm{Ad}(e^{zN}e^{\Gamma(s)})Y^{\ddagger} = \mathrm{Ad}(e^{zN})Y^{\ddagger} = Y^{\ddagger}$ for all $z \in U$. In particular, $Y_\infty(z)$ is integral for all $z \in U$. Since $Y_\infty(z)$ preserves $F(z)$ and grades W, the integrality of $Y_\infty(z)$ implies that $Y_\infty(z)$ is a morphism of mixed Hodge structures. Therefore, $\mathcal{V}_z$ is a split mixed Hodge structure for all z. So the zero locus of the corresponding admissible normal function is all of D^*.

Remark. The proof of Theorem 8.4.6 in the general case is analogous to the above proof for weight -1 variations on a punctured disk, but there are several subtle complications. First, in the higher-dimensional case, the limit $Y(z)$ of Theorem 8.4.11 does not exist except along certain sequences, which are well adapted to the SL_2-orbit theorem of [26]. Second, if the weight of $\mathcal{H}$ is less than -1, the grading $Y(z)$ is not, in general, real. However, there is a related grading of W which is real, and this is the one that plays a role in the proof in [4].

8.4.10 An Example

We want to illustrate the objects appearing in the above proof with a simple example. So let $V := \mathbb{Z}^3$ and write $e_1 := (1,0,0), e_2 := (0,1,0), e_3 := (0,0,1)$ for the standard generators of V. Set $H = \langle e_2, e_3 \rangle$ and define a filtration W on $V_{\mathbb{Q}}$ by setting $W_0 V_{\mathbb{Q}} = V_{\mathbb{Q}}, W_{-1} V_{\mathbb{Q}} = H_{\mathbb{Q}}$, and $W_0 V_{\mathbb{Q}} = 0$. Let N denote the nilpotent matrix $N e_1 = N e_2 = e_3, N e_3 = 0$.

LEMMA 8.4.16   *The filtration* $M = M(N, W)$ *exists and is given by* $M_0 V_{\mathbb{Q}} = V_{\mathbb{Q}}, M_{-1} V_{\mathbb{Q}} = M_{-2} V_{\mathbb{Q}} = \langle e_3 \rangle, M_{-3} V_{\mathbb{Q}} = 0.$

PROOF. With this definition of M, we clearly have $N M_k \subset M_{k-2}$. We have $\mathrm{Gr}_0^M \mathrm{Gr}_0^W V_{\mathbb{Q}} = \langle e_1 \rangle, \mathrm{Gr}_0^M \mathrm{Gr}_{-1}^W V_{\mathbb{Q}} = \langle e_2 \rangle$ and $\mathrm{Gr}_{-2}^M \mathrm{Gr}_{-1}^W V_{\mathbb{Q}} = \langle e_3 \rangle$. Now it is immediate to check that M satisfies the conditions to be the relative weight filtration. □

Remark.   In the above proof, we abused notation. For example, we wrote e_1 for the image of e_1 in $\mathrm{Gr}_0^M \mathrm{Gr}_0^W V_{\mathbb{Q}}$. We will repeat this abuse in the future.

Now define $Q_k : \mathrm{Gr}_k^W V_{\mathbb{Q}} \otimes \mathrm{Gr}_k^W V_{\mathbb{Q}} \to \mathbb{Q}$ to be the $(-1)^k$-symmetric nondegenerate bilinear form such that

$$Q_0(e_1, e_1) = Q_{-1}(e_2, e_3) = 1. \tag{8.4.7}$$

We have

$$G = G(V, W, Q) = \left\{ \begin{pmatrix} \pm 1 & 0 & 0 \\ * & a & b \\ * & c & d \end{pmatrix} : ad - bc = 1 \right\}.$$

Pick $\alpha \in \mathbb{C}$ and define a decreasing filtration F on $V_{\mathbb{C}}$ by setting

$$F^{-1} V_{\mathbb{C}} = V_{\mathbb{C}}, \quad F^0 = \langle e_1 + \alpha e_3, e_2 \rangle, \quad F^1 = 0. \tag{8.4.8}$$

LEMMA 8.4.17   *We have $F \in D^{\vee}$ and, for $z \in U$, $F(z) := e^{zN} F \in D$.*

PROOF. On $\mathrm{Gr}_0^W V$, F induces the Hodge structure $\mathbb{Z}(0)$, so there is really nothing to prove as far as $\mathrm{Gr}_0^W V$ goes. We have $F \mathrm{Gr}_{-1}^W = \langle e_2 \rangle$. So $Q_{-1}(F^p, F^{-1-p+1}) = Q_{-1}(F^p, F^{-p}) = 0$ because $Q_{-1}(F^0, F^0) = 0$ and, for $p \neq 0$, either F^p or F^{-p} is 0. This shows that F is in $\hat{D}$.

Suppose we take $z \in U$. Then $e^{zN} F \cap W_{-1} = \langle e_2 + ze_3 \rangle$. It is easy to see that $e^{zN} F$ induces a Hodge structure of weight -1 on W_{-2}. It is polarized, because $Q_{-1}(C(e_2 + ze_3), (e_2 + \bar{z}e_3)) = i(\bar{z} - z) > 0$. □

COROLLARY 8.4.18   *The map $z \mapsto e^{zN} F$ with the filtration W defines an admissible nilpotent orbit $\mathcal{V} = (V, W, F, N)$. The restriction of this nilpotent orbit to H is a pure nilpotent orbit $\mathcal{H}$ of weight -1. The map $\pi : \mathcal{V} \to \mathbb{Z}$ given by $\pi(e_1) = 1, \pi(e_2) = \pi(e_3) = 0$ makes (V, W, F, N) into an extension of the trivial variation $\mathbb{Z}$ by $\mathcal{H}$. In other words, we have an exact sequence of nilpotent orbits*

$$0 \to \mathcal{H} \to \mathcal{V} \to \mathbb{Z}(0) \to 0. \tag{8.4.9}$$

PROOF. This is really just rephrasing what we have done. Of course, we should check that transversality is satisfied, i.e., that $N F^p \subset F^{p-1}$. But that is obvious here. □

8.4.10.1   Zero Locus

The sequence (8.4.9) is the local normal form of an admissible normal function ν on D^*. We want to calculate the zero locus of ν, or, equivalently, the inverse image of the zero locus on the upper half plane.

We can always lift the generator 1 of $\mathbb{Z}$ in (8.4.9) to $v_{\mathbb{Z}} := e_1 \in V$. On the other hand, we have

$$F^0(z) = \langle e_1 + (\alpha + z)e_3, e_2 + ze_3 \rangle.$$

So, we can lift the generator 1 in $\mathbb{Z}$ to the section $v_F := e_1 + (\alpha + z)e_3 \in F(z)^0$. The difference $\sigma := v_F - v_{\mathbb{Z}}$ is $(\alpha + z)e_3$. Now $F(z)^0 H = \langle e_2 + ze_3 \rangle$. So

$$
\begin{aligned}
\nu(z) = 0 &\Leftrightarrow (\alpha + z)e_3 \in \mathbb{C}(e_2 + ze_3) + H \\
&\Leftrightarrow \gamma(e_2 + ze_3) + me_2 + ne_3 = (\alpha + z)e_3 \quad \text{for some } \gamma \in \mathbb{C}, m, n \in \mathbb{Z} \\
&\Leftrightarrow mz + n = \alpha + z \quad\quad\quad\quad\quad\quad\quad\;\; \text{for some } m, n \in \mathbb{Z} \\
&\Leftrightarrow (m - 1)z = n + \alpha \quad\quad\quad\quad\quad\quad\;\; \text{for some } m, n \in \mathbb{Z}.
\end{aligned}
$$

If $\alpha \in \mathbb{Z}$, then the above equation can be satisfied for any z by taking $m = 1$. In that case, $\nu(z) = 0$ for all z. So $Z(\nu) = D^*$.

On the other hand, if $\alpha \notin \mathbb{Z}$, then the above equation holds if and only if $z = (n + \alpha)/m$ for some integers n, m with $m \neq 0$. It follows that $\operatorname{Im} z$ is bounded from above by $\operatorname{Im} \alpha$. Thus the zero locus of ν is bounded away from 0 in D. It follows that the closure of $Z(\nu)$ is analytic in D.

Now we want to look at $Y(z) = Y_{(F(z),W)}$. To compute this we need to compute the decomposition $V = \oplus I^{p,q}_{(F(z),W)}$. It is easy to see that $H^{-1,0} = \langle e_2 + ze_3 \rangle$, $H^{0,-1} = \langle e_2 + \bar{z}e_3 \rangle$. Using the formula (7.5.10) for the spaces $I^{p,q}$, we see that $I^{0,0} = F(z)^0 \cap \bar{F}(z)^0$ in this case. Setting $w = z + \alpha$, and using a little Gaussian elimination to compute the intersection of two vector spaces, we see that

$$I^{0,0} = \left\langle e_1 - \frac{\operatorname{Im} w}{\operatorname{Im} z}e_2 + \frac{\operatorname{Im} z\bar{w}}{\operatorname{Im} z}e_3 \right\rangle.$$

Now, note that

$$\frac{\operatorname{Im} z\bar{w}}{\operatorname{Im} z} = \frac{\operatorname{Im} z\bar{\alpha}}{\operatorname{Im} z}.$$

So set $z = x + iy$ and $\alpha = u + iv$ with x, y, u, v real. Then, since $Y(z)e_2 = -e_2$ and $Y(z)e_3 = -e_3$, it follows that

$$
Y(z) = \begin{pmatrix} 0 & 0 & 0 \\ -\frac{\operatorname{Im} w}{\operatorname{Im} z} & -1 & 0 \\ \frac{\operatorname{Im} z\bar{w}}{\operatorname{Im} z} & 0 & -1 \end{pmatrix}
$$

$$
= \begin{pmatrix} 0 & 0 & 0 \\ -1 - \frac{v}{y} & -1 & 0 \\ u - \frac{xv}{y} & 0 & -1 \end{pmatrix}.
$$

Taking the limit as $\operatorname{Im} z \to \infty$, we get

$$
Y^{\ddagger} = \begin{pmatrix} 0 & 0 & 0 \\ -1 & -1 & 0 \\ \operatorname{Re}\alpha & 0 & -1 \end{pmatrix}. \tag{8.4.10}
$$

Thus, we see again that, if $\operatorname{Re}\alpha$ is nonintegral, then the zero locus of ν in D^* is bounded away from 0.

Now we want to compute Y_∞ as in Theorem 8.4.13. In fact, the paper [3] gives a formula for a specific matrix Y_∞ satisfying the conditions of Theorem 8.4.13. We will write down this matrix and check that it satisfies the conditions of the theorem. To do this, we need to compute the decomposition of V induced by the limit Hodge structure (F, M). In other words, we need to compute $I^{p,q}_{(F,M)}$.

LEMMA 8.4.19   *We have*

$$
\begin{aligned}
I^{0,0}_{(F,M)} &= \langle e_1 + \alpha e_3, e_2 \rangle, \\
I^{-1,-1}_{(F,M)} &= \langle e_3 \rangle.
\end{aligned}
$$

PROOF. The second line is obvious. For the first, note that by (7.5.10), in this case,

$$
\begin{aligned}
I^{0,0}_{(F,M)} &= F^0 \cap (\bar{F}^0 + F^{-1} \cap M_{-2}) \\
&= F^0 \cap (\bar{F}^0 + M_{-2}) \\
&= F^0.
\end{aligned}
$$

$\square$

COROLLARY 8.4.20   *Set*

$$
Y_\infty := \begin{pmatrix} 0 & 0 & 0 \\ -1 & -1 & 0 \\ \alpha & 0 & -1 \end{pmatrix},
$$

and set

$$
\xi := \begin{pmatrix} 0 & 0 & 0 \\ 0 & 0 & 0 \\ -i\operatorname{Im}\alpha & 0 & 0 \end{pmatrix}.
$$

Then ξ and Y_∞ satisfy the conditions of Theorem 8.4.13.

PROOF. It is straightforward to check that $\operatorname{Ad}(e^\xi)Y_\infty = Y^{\ddagger}$ by multiplying out the matrices. Similarly, it is straightforward to check that $\xi \in \Lambda^{-1,-1}_{F,M}$, because we know the decomposition $V_{\mathbb{C}} = \oplus I^{p,q}_{(F,M)}$ from Lemma 8.4.19. It is clear on inspection that ξ and Y_∞ commute with N.

To check that Y_∞ preserves $I^{p,q}_{(F,M)}$, note that $Y_\infty(e_1 + \alpha e_3) = -e_2 + \alpha e_3 - \alpha e_3 = -e_2$, which is in $I^{0,0}_{(F,M)}$. We have $Y_\infty e_2 = -e_2$ and $Y_\infty e_3 = -e_3$, so the rest is obvious.

$\square$

Remark.   As mentioned above, in [3], the matrix Y_∞ is given in terms of the admissible nilpotent orbit (V, W, F, N) by means of a formula. In fact, this formula originally appeared in an (unpublished) letter of Deligne to Cattani and Kaplan. If we restrict what Deligne says in general to the specific case of this example, it is that there is a unique splitting Y_∞ of W commuting with the splitting $Y_{(F,M)}$ of M and with N. In this case, using Lemma 8.4.19, we have

$$Y_{(F,M)} = \begin{pmatrix} 0 & 0 & 0 \\ 0 & 0 & 0 \\ 2\alpha & 0 & -2 \end{pmatrix},$$

and the fact that Y_∞ is the unique splitting commuting with N and $Y_{(F,M)}$ can be checked by hand. The contents of Deligne's letter are mentioned in [3], and [4] contains an appendix devoted to recovering some of the results of the letter. The appendix to [21] also contains an exposition of some of the results from Deligne's letter.

Bibliography

[1] D. Arapura, Mixed Hodge structures associated to geometric variations. ArXiv:math/0611837v3, Apr. 2008.

[2] A. A. Beilinson, J. Bernstein, P. Deligne, *Analyse et topologie sur les espaces singuliers, Vol.I.* Astérisque **100**, France, 1982.

[3] P. Brosnan, G. Pearlstein, The zero locus of an admissible normal function. *Ann. of Math.* 170(2):883–897, 2009.

[4] P. Brosnan, G. Pearlstein, On the algebraicity of the zero locus of an admissible normal function. ArXiv:0910.0628v1, [math.AG], 4 Oct. 2009. To appear in *Compositio Mathematica*.

[5] A. Brosnan, G. Pearlstein, C. Schnell, *On the algebraicity of the locus of Hodge classes in an admissible variation of mixed Hodge structure.* ArXiv:1002.4422v1, [math.AG], 23 Feb. 2010.

[6] J. A. Carlson, The geometry of the extension class of a mixed Hodge structure. In *Algebraic geometry, Bowdoin, 1985 (Brunswick, ME, 1985)*, volume 46 of *Proc. Sympos. Pure Math.*, pages 199–222. Amer. Math. Soc., Providence, RI, 1987.

[7] E. Cattani, P. Deligne, A. Kaplan, On the locus of Hodge classes. *J. Amer. Math. Soc.* 8(2):483–506, 1995.

[8] E. Cattani, A. Kaplan, W. Schmid, L^2 and intersection cohomologies for a polarizable variation of Hodge structure. *Invent. Math.* 87(2):217–252, 1987.

[9] C. H. Clemens, Degeneration of Kähler manifolds. *Duke Math. J.* 44(2):215–290, 1977.

[10] P. Deligne, *Equations différentielles à points singuliers réguliers.* Lecture Notes, 163, Springer, Berlin, 1970.

[11] P. Deligne, Théorie de Hodge II. *Publ. Math. IHES* 40:5–57, 1972.

[12] P. Deligne, Théorie de Hodge III. *Publ. Math. IHES* 44:6–77, 1975.

[13] P. Deligne, Conjecture de Weil II. *Publ. Math. IHES* 52, 1980.

[14] P. Deligne, O. Gabber, *Théorème de pureté d'après Gabber.* Note written by Deligne and distributed at IHES, 1981.

[15] P. Deligne, N. Katz, *Groupes de monodromie en géométrie algébrique*. Lecture Notes in Math. 340, Springer, 1973.

[16] F. El Zein, Théorie de Hodge des cycles évanescents. *Ann. Scient. Ec. Norm. Sup.* 19:107–184, 1986. *Notes C.R. Acad. Sc., Paris, Série I* 295:669-672, 1982; 292:51–54, 1983; 296:199–202, 1983.

[17] F. El Zein, Le D. T., L. Migliorini, A topological construction of the weight filtration. *Manuscripta Mathematica* 133(1–2):173–182, 2010.

[18] M. Goresky, R. MacPherson, Stratified Morse theory. Ergebnisse der Mathematik, 3.folge. Band 2, Springer, Berlin, Heidelberg, 1988.

[19] P. Griffiths, *Topics in transcendental algebraic geometry*. Ann. of Math. Studies 106, Princeton University Press, 1984.

[20] P. Griffiths, W. Schmid, *Recent developments in Hodge theory*. In Discrete subgroups of Lie groups, Oxford University Press, 1973.

[21] A. Kaplan, G. Pearlstein, Singularities of variations of mixed Hodge structure. *Asian J. Math.*, 7(3):307–336, 2003.

[22] M. Kashiwara, A study of variation of mixed Hodge structure. *Publ. RIMS, Kyoto University* 22:991–1024, 1986.

[23] M. Kashiwara, T. Kawai, Hodge structure and holonomic systems. *Proc. Japan Acad. Ser. A* 62:1–4, 1986.

[24] M. Kashiwara, T. Kawai, Poincaré lemma for a variation of Hodge structure. *Publ. RIMS, Kyoto University* 23:345–407, 1987.

[25] M. Kashiwara, The asymptotic behavior of a variation of polarized Hodge structure. *Publ. RIMS, Kyoto University* 21:353–875, 1985.

[26] K. Kato, C. Nakayama, S. Usui, SL(2)-orbit theorem for degeneration of mixed Hodge structure. *J. Algebraic Geom.* 17(3):401–479, 2008.

[27] K. Kato, C. Nakayama, S. Usui, Moduli of log mixed Hodge structures. *Proc. Japan Acad. Ser. A Math. Sci.* 86(7):107–112, 2010.

[28] N. M. Katz, T. Oda, On the differentiation of de Rham cohomology classes with respect to parameters. *J. Math. Kyoto University* 8:199–213, 1968.

[29] A. Landman, On the Picard Lefschetz transformation acquiring general singularities. *Trans. Amer. Math. Soc.* 181:89–126, 1973.

[30] D.T. Lê, B. Teissier, *Cycles évanescents et conditions de Whitney II*. Proc. Symp. Pure Math. 40, part 2, Amer. Math. Soc. Providence, RI, 1983, 65–103.

[31] B. Malgrange, *Regular connections, after Deligne, in algebraic D-modules, by Borel A*. Perspective in Math. 2, Academic Press, Boston, 1987.

[32] J. Mather, Notes on topological stability. *Bull. Amer. Math. Soc.* 49(4):475–506, 2012.

[33] A. V. Navarro, F. Guillén, Sur le théorème local des cycles invariants. *Duke Math. J.* 61:133–155, 1990.

[34] G. Pearlstein, SL_2-orbits and degenerations of mixed Hodge structure. *J. Differential Geom.* 74(1):1–67, 2006.

[35] M. Saito, (1) Modules de Hodge polarisables. *Publ. RIMS, Kyoto University* 24:849–995, 1988. (2) Mixed Hodge modules. *Publ. RIMS, Kyoto University* 26:221–333, 1990.

[36] M. Saito, Admissible normal functions. *J. Algebraic Geom.* 5(2):235–276, 1996.

[37] W. Schmid, Variation of Hodge structure: the singularities of the period mapping. *Invent. Math.* 22:211–319, 1973.

[38] C. Schnell, Complex analytic Néron models for arbitrary families of intermediate Jacobians. *Invent. Math.* 188(1):1–81, 2012.

[39] J. H. M. Steenbrink, Limits of Hodge structures. *Invent. Math.* 31:229–257, 1976.

[40] J. H. M. Steenbrink, S. Zucker, Variation of mixed Hodge structures I. *Invent. Math.* 80:489–542, 1985.

[41] J. L. Verdier, Stratifications de Whitney et théorème de Bertini–Sard. *Invent. Math.* 36:295–312, 1976.

[42] J. L. Verdier, *Dualité dans la cohomologie des espaces localement compacts.* Séminaire Bourbaki 300, 1965.

[43] C. Voisin, *Hodge theory and complex algebraic geometry I, II.* Volumes 76, 77, Cambridge University Press, 2007.

Chapter Nine

Lectures on Algebraic Cycles and Chow Groups
by Jacob Murre

These are the notes of my lectures in the ICTP Summer School and conference on "Hodge Theory and Related Topics" in 2010.

The notes are informal and close to the lectures themselves. As much as possible I have concentrated on the main results. Especially in the proofs I have tried to outline the main ideas and mostly omitted the technical details. In order not to "wave hands" I have often written "outline or indication of proof" instead of "proof"; on the other hand when possible I have given references where the interested reader can find the details for a full proof.

The first two lectures are over an arbitrary field (for simplicity always assumed to be algebraically closed), Lectures III and IV are over the complex numbers and in Lecture V we return to an arbitrary field.

I have tried to stress the difference between the theory of divisors and the theory of algebraic cycles of codimension larger than 1. In Lectures IV and V, I have discussed results of Griffiths and Mumford which—in my opinion—are the two most striking facts which make this difference clear.

Acknowledgments. First of all I want to thank the organizers of the summer school for inviting me to, and giving me the opportunity to lecture in, this very interesting and inspiring meeting. Next I would like to thank the audience for their patience and their critical remarks. Finally I thank Javier Fresán, first for valuable suggestions on my notes and second, last but not least, for LaTeXing my notes; he did a splendid job.

9.1 LECTURE I: ALGEBRAIC CYCLES. CHOW GROUPS

9.1.1 Assumptions and Conventions

In the first two lectures, k is an *algebraically closed field*. We work with *algebraic varieties* defined over k (i.e., k-schemes which are reduced, i.e., there are no nilpotent elements in the structure sheaves). We *assume moreover* (unless otherwise stated) that our varieties are *smooth, quasi-projective, and irreducible*. We denote the category of such varieties by $\mathrm{Var}(k)$ (the morphisms are the usual morphisms, i.e., rational maps which are everywhere regular). (See [14, Ch. 1].) If X is such a variety, let $d = \dim X$; in the following we often denote this in short by X_d.

9.1.2 Algebraic Cycles

Let $X_d \in \mathrm{Var}(k)$; let $0 \le i \le d$ and $q = d - i$. Let $\mathcal{Z}_q(X) = \mathcal{Z}^i(X)$ be the group of *algebraic cycles of dimension q (i.e., codimension i) on* X,[1] i.e., the free abelian group generated by the k-irreducible subvarieties W on X of dimension q, but W *not* necessarily smooth. Therefore such an algebraic cycle $Z \in \mathcal{Z}_q(X) = \mathcal{Z}^i(X)$ can be written as $Z = \sum_\alpha n_\alpha W_\alpha$, a finite sum with $n_\alpha \in \mathbb{Z}$ and $W_\alpha \subset X$ q-dimensional subvarieties of X defined over k and irreducible but not necessarily smooth.

EXAMPLE 9.1.1

(a) $\mathcal{Z}^1(X) = \mathrm{Div}(X)$ is the group of (Weil) *divisors* on X.

(b) $\mathcal{Z}_0(X) = \mathcal{Z}^d(X)$ is the group of 0-cycles on X, so $Z \in \mathcal{Z}_0(X)$ is a formal sum $Z = \sum n_\alpha P_\alpha$ with $P_\alpha \in X$ points. Put $\deg(Z) = \sum n_\alpha$.

(c) $\mathcal{Z}_1(X) = \mathcal{Z}^{d-1}(X)$ is the group of *curves* on X, i.e., $Z = \sum n_\alpha C_\alpha$ with $C_\alpha \subset X$ curves.

9.1.2.1 Operations on Algebraic Cycles

There are three *basic* operations and a number of other operations which are built from these basic operations:

1. **Cartesian product**. If $W \subset X_1$ (resp. $V \subset X_2$) is a subvariety of dimension q_1 (resp. q_2) then $W \times V \subset X_1 \times X_2$ is a subvariety of dimension $q_1 + q_2$. Proceeding by linearity we get

$$\mathcal{Z}_{q_1}(X_1) \times \mathcal{Z}_{q_2}(X_2) \longrightarrow \mathcal{Z}_{q_1+q_2}(X_1 \times X_2).$$

2. **Push-forward**. (See [8, p. 11]) Given a morphism $f : X \to Y$ we get a homomorphism $f_* : \mathcal{Z}_q(X) \to \mathcal{Z}_q(Y)$. By linearity it suffices to define this only for a subvariety $W \subset X$. Now consider the *set-theoretical* image $f(W) \subset Y$; its Zariski closure[2] $\overline{f(W)}$ is an algebraic subvariety of Y, irreducible if W itself is irreducible and $\dim \overline{f(W)} \le \dim W = q$. Now define

$$f_*(W) = \begin{cases} 0 & \text{if } \dim \overline{f(W)} < \dim W, \\ [k(W) : k(\overline{f(W)})] \cdot \overline{f(W)} & \text{if } \dim \overline{f(W)} = \dim W, \end{cases}$$

where $k(W)$ is the function field of W (i.e., the field of rational functions on W) and $k(\overline{f(W)})$ is the function field of $\overline{f(W)}$ (note that we have a finite extension of fields in the case $\dim \overline{f(W)} = \dim W$).

3. **Intersection product** (only defined under a *restriction*!). Let $V \subset X$ (resp. $W \subset X$) be an irreducible subvariety of codimension i (resp. j). Then $V \cap W$ is a finite union $\bigcup A_l$ of irreducible subvarieties $A_l \subset X$. Since X is smooth all A_l have codimension $\le i + j$ ([14, p. 48], [8, p. 120]).

---

[1] Usually we prefer to work with the codimension i, but sometimes it is more convenient to work with the dimension q.

[2] Or $f(W)$ itself if f is proper.

DEFINITION 9.1.2 *The intersection of V and W at A_l is called proper (or good) if the codimension of A_l in X is $i + j$.*

In that case we define the *intersection multiplicity* $i(V \cdot W; A_l)$ of V and W at A_l as follows:

DEFINITION 9.1.3 (See [14, p. 427] and/or [25, p. 144])

$$i(V \cdot W; A_l) := \sum_{r=0}^{\dim X} (-1)^r \mathrm{length}_{\mathcal{O}} \{ \mathrm{Tor}_r^{\mathcal{O}}(\mathcal{O}/J(V), \mathcal{O}/J(W)) \}.$$

Here $\mathcal{O} = \mathcal{O}_{A_l, X}$ is the local ring of A_l in X and $J(V)$ (resp. $J(W)$) is the ideal defining V (resp. W) in $\mathcal{O}$.

If the intersection is proper at every A_l, then one defines the *intersection product as a cycle* by

$$V \cdot W := \sum_l i(V \cdot W; A_l) A_l.$$

This is an algebraic cycle in $Z^{i+j}(X)$.

By linearity one defines now in an obvious way the intersection product of two cycles $Z_1 = \sum n_\alpha V_\alpha \in \mathcal{Z}^i(X)$ and $Z_2 = \sum m_\beta W_\beta \in \mathcal{Z}^j(X)$ as

$$Z_1 \cdot Z_2 := \sum n_\alpha m_\beta (V_\alpha \cdot W_\beta) \in \mathcal{Z}^{i+j}(X).$$

REMARK 9.1.4

(a) For the notion of length of a module see—for instance—[8, p. 406].

(b) For the intersection multiplicity $i(V \cdot W; A_l)$ one could try—more naively— to work *only* with the tensor product of $\mathcal{O}/J(V)$ and $\mathcal{O}/J(W)$ but this is not correct (see [14, p. 428, Ex. 1.1.1]). One needs for correction the terms with the Tor's. The Tor-functors are the so-called higher derived functors for the tensor product functor. See, for instance, Eisenbud [5, p. 159] or Hilton and Stammbach [15, Chs. III and IV].

(c) The above definition of the intersection multiplicity of Serre coincides with the older and more geometric definitions of Weil, Chevalley, and Samuel (see [25, p. 144]).

Now we discuss *further operations* on algebraic cycles built via the basic operations.

4. **Pull-back of cycles** (not always defined!). Given a morphism $f : X \to Y$ we want to define a homomorphism $f^* : \mathcal{Z}^i(Y) \to \mathcal{Z}^i(X)$. So let $Z \in \mathcal{Z}^i(Y)$.

DEFINITION 9.1.5 $f^*(Z) := (\mathrm{pr}_X)_*(\Gamma_f \cdot (X \times Z))$, *where Γ_f is the graph of f.*

But this is only defined if the intersection $\Gamma_f \cdot (X \times Z)$ is defined.[3]

REMARK 9.1.6 This is defined if $f : X \to Y$ is *flat* (see [8, p. 18]). This happens, in particular, if $X = Y \times Y'$ and f is the projection on Y.

5. **Correspondences and operations of correspondences on algebraic cycles.** Let $X_d, Y_e \in \mathrm{Var}(k)$. A *correspondence* $T \in \mathrm{Cor}(X, Y)$ from X to Y is an element $T \in \mathcal{Z}^n(X \times Y)$ for a certain $n \geq 0$, i.e., $\mathrm{Cor}(X, Y)$ equals $\mathcal{Z}(X \times Y)$. We denote the transpose by ${}^t T \in \mathcal{Z}^n(X \times Y)$, so ${}^t T \in \mathrm{Cor}(Y, X)$. Given $T \in \mathcal{Z}^n(X \times Y)$ then we define the homomorphism

$$T : \mathcal{Z}^{\bullet i}(X_d) \longrightarrow \mathcal{Z}^{i+n-d}(Y_e)$$

by the formula

$$T(Z) := (\mathrm{pr}_Y)_* \{ T \cdot (Z \times Y) \},$$

but this is *only defined* on a *subgroup* $\mathcal{Z}^{\bullet i} \subset \mathcal{Z}^i(X)$, namely, on those Z for which the intersection product $T \cdot (Z \times Y)$ is defined (on $X \times Y$).

REMARK 9.1.7 If we have a morphism $f : X \to Y$, then for $T = \Gamma_f$ we get back f_* and when T is ${}^t\Gamma_f$ we get f^*.

9.1.3   Adequate Equivalence Relations

It will be clear from the above that one wants to introduce on the group of algebraic cycles a "good" equivalence relation in such a way that—in particular—the above operations are *always defined* on the corresponding *cycle classes*.

Samuel introduced in 1958 the notion of *adequate* (or "good") *equivalence relations* ([24, p. 470]). Roughly speaking, an equivalence relation is adequate if it is compatible with addition and intersection and if it is functorial. The precise conditions are as follows.

An *equivalence relation* "$\sim$" given on the groups of algebraic cycles $Z(X)$ of all varieties $X \in \mathrm{Var}(k)$ is *adequate* if it satisfies the following conditions:

(R1) $\mathcal{Z}^i_\sim(X) := \{ Z \in \mathcal{Z}^i(X) : Z \sim 0 \} \subset \mathcal{Z}^i(X)$ is a subgroup.

(R2) If $Z \in \mathcal{Z}^i(X)$, $Z' \in \mathcal{Z}^i(X)$, $W \in \mathcal{Z}^j(X)$ are such that $Z \cdot W$ and $Z' \cdot W$ are defined and $Z \sim Z'$, then also $Z' \cdot W \sim Z \cdot W$.

(R3) Given $Z \in \mathcal{Z}^i(X)$ and a finite number of subvarieties $W_\alpha \subset X$, then there exists $Z' \in \mathcal{Z}^i(X)$ such that $Z' \sim Z$ and such that all $Z' \cdot W_\alpha$ are defined.

(R4) Let $Z \in \mathcal{Z}(X)$ and $T \in Z(X \times Y)$ be such that $T \cdot (Z \times Y)$ is defined. Assume that Y is proper (for instance, projective) and that $Z \sim 0$. Then also $T(Z) \sim 0$ in $\mathcal{Z}(Y)$. Recall that $T(Z) = (\mathrm{pr}_Y)_*(T \cdot (Z \times Y))$.

---

[3] The intersection is of course on $X \times Y$.

Now let $\sim$ be an *adequate* equivalence relation for algebraic cycles. Put

$$C_\sim^i(X) := \mathcal{Z}^i(X)/\mathcal{Z}_\sim^i(X)$$

(and similarly $C_q^\sim(X)$ if $q = d - i$ with $d = \dim X$ if we want to work with dimension instead of codimension). Then we have the following result:

PROPOSITION 9.1.8

(a) $C_\sim(X) = \bigoplus_{i=0}^d C_\sim^i(X)$ is a commutative ring with respect to the intersection product.

(b) If $f : X \to Y$ is proper, then $f_* : C_q^\sim(X) \to C_q^\sim(Y)$ is an additive homomorphism.

(c) If $f : X \to Y$ (arbitrary!), then $f^* : C_\sim(X) \to C_\sim(X)$ is a ring homomorphism.

PROOF. Left to the reader (or see [24]). Hint: (a) and (b) are straightforward. For (c) one can use the "reduction to the diagonal." Namely, if Z_1 and Z_2 are algebraic cycles on X such that $Z_1 \cdot Z_2$ is defined then $Z_1 \cdot Z_2 = \Delta_*(Z_1 \times Z_2)$, where $\Delta : X \hookrightarrow X \times X$ is the diagonal (see [25, V-25]). □

PROPOSITION 9.1.9 (Supplement) *Let $T \in \mathcal{Z}(X \times Y)$. Then T defines an additive homomorphism $T : C_\sim(X) \to C_\sim(Y)$ and this homomorphism depends only on the class of T in $\mathcal{Z}(X \times Y)$.*

PROOF. For the definition of T as an operator on the cycles see Section 9.1.2. For the proof see [24, Prop. 7, p. 472]. □

We shall discuss in the remaining part of this Lecture I and in Lecture II the following adequate equivalence relations:

(a) Rational equivalence (Samuel and Chow independently, 1956)

(b) Algebraic equivalence (Weil, 1952)

(c) Smash-nilpotent equivalence (Voevodsky, 1995)

(d) Homological equivalence

(e) Numerical equivalence

Homological (at least if $k = \mathbb{C}$) and numerical equivalence are kind of classical and the origin is difficult to trace.

9.1.4   Rational Equivalence. Chow Groups

Rational equivalence, defined and studied independently in 1956 by Samuel and Chow, is a *generalization* of the classical concept of *linear equivalence for divisors*.

9.1.4.1   Linear Equivalence for Divisors

Let $X = X_d$ be an irreducible variety but for the moment (for technical reasons, see Section 9.1.4.2) *not* necessarily smooth. Let $\varphi \in k(X)^*$ be a rational function on X. Recall [8, p. 8] that

$$\mathrm{div}(\varphi) := \sum_{\substack{Y \subset X \\ \mathrm{codim}\ 1}} \mathrm{ord}_Y(\varphi) \cdot Y.$$

Here Y "runs" through the irreducible subvarieties of codimension 1 and $\mathrm{ord}_Y(\varphi)$ is defined as follows:

(a) If $\varphi \in \mathcal{O}_{Y,X}$, then $\mathrm{ord}_Y(\varphi) := \mathrm{length}_{\mathcal{O}_{Y,X}}(\mathcal{O}_{Y,X}/(\varphi))$.

(b) Otherwise write $\varphi = \varphi_1/\varphi_2$ with $\varphi_1, \varphi_2 \in \mathcal{O}_{Y,X}$ and

$$\mathrm{ord}_Y(\varphi) := \mathrm{ord}_Y(\varphi_1) - \mathrm{ord}_Y(\varphi_2)$$

(this is well defined!).

REMARK 9.1.10   If X is smooth at Y then $\mathcal{O}_{Y,X}$ is a discrete valuation ring and $\mathrm{ord}_Y(\varphi) = \mathrm{val}_Y(\varphi)$.

So (always) $\mathrm{div}(\varphi)$ is a Weil divisor and put $\mathrm{Div}_l(X) \subset \mathrm{Div}(X)$ for the subgroup generated by such divisors; in fact,

$$\mathrm{Div}_l(X) = \{D = \mathrm{div}(\varphi);\ \varphi \in k(X)^*\}$$

and $\mathrm{CH}^1(X) := \mathrm{Div}(X)/\mathrm{Div}_l(X)$ is the (Chow) group of the *divisor classes* with respect to linear equivalence.

9.1.4.2   Rational Equivalence. Definition

Let $X = X_d \in \mathrm{Var}(k)$, i.e., smooth, quasi-projective, and irreducible of dimension d. Let $0 \leq i \leq d$ and put $q = d - i$.

DEFINITION 9.1.11 $\mathcal{Z}_q^{\mathrm{rat}}(X) = \mathcal{Z}_{\mathrm{rat}}^i(X) \subset \mathcal{Z}^i(X)$ *is the subgroup generated by the algebraic cycles of type* $Z = \mathrm{div}(\varphi)$ *with* $\varphi \in k(Y)^*$ *and* $Y \subset X$ *an irreducible subvariety of codimension* $(i-1)$ *(i.e., of dimension* $(q+1)$*) (see [8, Ch. 1], in particular, page 10). Note that we do not require* Y *to be smooth.*

REMARK 9.1.12

(a) Equivalently, let $Z \in \mathcal{Z}_q(X)$. Then $Z \sim_{\mathrm{rat}} 0$ if and only if there exists a finite collection $\{Y_\alpha, \varphi_\alpha\}$ with $Y_\alpha \subset X$ irreducible and of dimension $(q+1)$ and $\varphi_\alpha \in k(Y_\alpha)^*$ such that $Z = \sum_\alpha \mathrm{div}(\varphi_\alpha)$.

(b) We do *not* assume that the Y_α are smooth, therefore it is important that $\mathrm{div}(\varphi)$ is defined for nonzero rational functions on *arbitrary* varieties.

(c) Clearly $Z^1_{\text{rat}}(X) = Z^1_{\text{lin}}(X) = \text{Div}_l(X)$, i.e., for divisors rational equivalence is linear equivalence.

There is another equivalent formulation [8, p. 15] for rational equivalence (which was in fact used in the original definition by Samuel and by Chow):

PROPOSITION 9.1.13 *Let $Z \in Z^i(X)$. The following conditions are equivalent:*

(a) *Z is rationally equivalent to zero.*

(b) *There exists a correspondence $T \in Z^i(\mathbb{P}^1 \times X)$ and two points $a, b \in \mathbb{P}^1$ such that $Z = T(b) - T(a)$.[4]*

PROOF. The first implication is easy. Assume for simplicity $Z = \text{div}(\varphi)$ with $\varphi \in k(Y)^*$ and $Y \subset X$ an irreducible subvariety of dimension $(q + 1)$; then take $T = {}^t\Gamma_\varphi$ on $\mathbb{P}^1 \times Y$ and consider it as a cycle on $\mathbb{P}^1 \times Y$ via $\iota : Y \hookrightarrow X$ (so strictly speaking $T = (\text{id} \times \iota)_*({}^t\Gamma_\varphi)$). If $Z = \sum \text{div}(\varphi_\alpha)$ do this for every φ_α.

The second implication is less easy and depends on the following theorem (see [8, Prop. 1.4], also for the proof):

THEOREM 9.1.14 *Let $f : V \to W$ be a proper, surjective morphism of irreducible varieties and $\varphi \in k(V)^*$. Then*

(a) *$f_*(\text{div}(\varphi)) = 0$ if $\dim V > \dim W$;*

(b) *$f_*(\text{div}(\varphi)) = \text{div}(N(\varphi))$ if $\dim V = \dim W$ where $N = \text{Norm}_{k(V)/k(W)}$.*

Now for (b) implies (a) in Proposition 9.1.13: we can assume that T is irreducible, $b = 0$ and $a = \infty$ on $\mathbb{P}^1$. We have on T the function φ induced by the "canonical function" t on $\mathbb{P}^1$ (i.e., $\varphi = \text{pr}^{-1}_{\mathbb{P}^1}(t)$). Now apply the theorem with $V = T$ and $W = \text{pr}_X(T) \subset X$ (the set-theoretic projection). $\square$

9.1.4.3   Properties of Rational Equivalence (see [8, Ch. 1])

PROPOSITION 9.1.15   *Rational equivalence is an adequate equivalence relation.*

PROOF. (Indications only!) (R1) is immediate from the definition. (R2) is also easy if we use the alternative definition from Proposition 9.1.13. Indeed, with some easy modifications we can get a $T \in Z^i(\mathbb{P}^1 \times X)$ such that $T(a) = Z$, $T(b) = Z'$ and then from the assumptions we get that $T \cdot (\mathbb{P}^1 \times W) = T_1 \in Z^{i+j}(\mathbb{P}^1 \times X)$ is defined, $T_1(a) = Z \cdot W$ and $T_1(b) = Z' \cdot W$. The proof of (R4) is left to the reader; see [8, Th. 1.4, p. 11]. The crucial property is (R3). This is the so-called *Chow's moving lemma*; the origin of the idea of the proof is classical and goes back to Severi who used it for his so-called dynamical theory of intersection numbers. We outline the main idea; for details see [23].

---

[4] Recall that for $t \in \mathbb{P}^1$ we have $T(t) = (\text{pr}_X)_*(T \cdot (t \times X))$.

We can assume that $X_d \subset \mathbb{P}^N$, that $Z \in \mathcal{Z}^i(X_d)$ is itself an irreducible subvariety and that we have only one (irreducible) $W \subset X_d$ of codimension j in X. The intersection $Z \cap W$ would be proper ("good") if all the components have codimension $i + j$ in X, so let us assume that there is a component of codimension $(i + j - e)$ with $e > 0$, e is called the *excess*, denoted by $e(Z, W)$.

Assume first that $X = \mathbb{P}^N$ itself. Let $\tau : \mathbb{P}^N \to \mathbb{P}^N$ be a projective transformation. Consider the transform $Z' = \tau(Z)$ of Z, then $Z' \sim Z$ is rationally equivalent. To make this more explicit, remember that such a projective transformation is given by linear equations in the coordinates of $\mathbb{P}^N$. Now take the coefficients occurring in these equations; they determine a point in an affine space $\mathbb{A}^M$, where $M = (N + 1)^2$. So both the transformation τ and the identity transformation, say τ_0, can be considered as points in this $\mathbb{A}^M$; connect them by a line L (itself a space $\mathbb{A}^1$) and consider in $\mathbb{P}^N$ the transformations corresponding with the point $t \in L$; then the cycles $t(Z)$ give a family of cycles which determine a cycle $T \in \mathcal{Z}^i(L \times X)$ as in Proposition 9.1.13. Taking on L the points $a = \tau$ and $b = \tau_0$ we get by Proposition 9.1.13 that $T(\tau) = \tau(Z) = Z'$ is rationally equivalent to $T(\tau_0) = Z$. Now taking τ "sufficiently general" we can show that $\tau(Z) \cap W$ intersects properly.[5]

Next consider the general case $X_d \subset \mathbb{P}^N$. Choose a linear space $L \subset \mathbb{P}^N$ of codimension $(d + 1)$ and such that $X \cap L = \emptyset$. Now consider the cone $C_L(Z)$ on Z with "vertex" L. One can show (see [23]) that if we take L "sufficiently general" then $C_L(Z) \cdot X = 1 \cdot Z + Z_1$ with $Z_1 \in \mathcal{Z}^i(X)$ and where, moreover, the excess $e(Z_1, W) < e$. Next take again a "sufficiently general" projective transformation $\tau : \mathbb{P}^N \to \mathbb{P}^N$; then we have $Z \sim \tau(C_L(Z)) \cdot X - Z_1 =: Z_2$ rationally equivalent and moreover, $e(Z_2, W) = e(Z_1, W) < e = e(Z, W)$. Hence proceeding by induction on the excess we are done. $\square$

9.1.4.4 Chow Groups

As before, let $X \in \mathrm{Var}(k)$. Define

$$\mathrm{CH}^i(X) := \mathcal{Z}^i(X)/\mathcal{Z}^i_{\mathrm{rat}}(X), \quad \mathrm{CH}(X) := \bigoplus_{i=0}^{\dim X} \mathrm{CH}^i(X).$$

$\mathrm{CH}^i(X)$ is called the ith *Chow group* of X and $\mathrm{CH}(X)$ the total Chow group.

REMARK 9.1.16

(a) So $\mathrm{CH}^i(X) = C^i_\sim(X)$ if "$\sim$" is rational equivalence.

(b) If $d = \dim X$ and $q = d - i$ we put also $\mathrm{CH}_q(X) = \mathrm{CH}^i(X)$.

(c) The Chow groups are in fact also defined in a completely similar way if X is an arbitrary variety; see [8, Ch. 1].

---

[5]Think for instance of the simple case in which Z and W are surfaces in $\mathbb{P}^4$, then $e > 0$ if and only if Z and W have a curve (or curves) in common. We have to move Z such that $\tau(Z) \cap W$ consists only of points.

Since rational equivalence is an adequate equivalence relation we get the following theorem (see Propositions 9.1.8 and 9.1.9 in Section 9.1.3):

THEOREM 9.1.17 (Chow, Samuel, 1956)   *Let $X_d, Y_n \in \mathrm{Var}(k)$, i.e., smooth, projective, irreducible varieties. Then*

(a) *$\mathrm{CH}(X)$ is a commutative ring (Chow ring) with respect to the intersection product;*

(b) *for a proper morphism $f : X \to Y$ we have additive homomorphisms $f_* : \mathrm{CH}_q(X) \to \mathrm{CH}_q(Y)$;*

(c) *for an arbitrary morphism $f : X \to Y$ we have additive homomorphisms $f^* : \mathrm{CH}^i(Y) \to \mathrm{CH}^i(X)$ and in fact a ring homomorphism $f^* : \mathrm{CH}(Y) \to \mathrm{CH}(X)$;*

(d) *let $T \in \mathrm{CH}^n(X_d \times Y_n)$; then $T_* : \mathrm{CH}^i(X) \to \mathrm{CH}^{i+n-d}(Y)$ is an additive homomorphism (depending only on the class of T).*

We mention two other important properties of Chow groups (for easy proofs we refer to [8, Ch. 1]).

THEOREM 9.1.18 (Homotopy property)   *Let $\mathbb{A}^n$ be affine n-space. Consider the projection $p : X \times \mathbb{A}^n \to X$. Then*

$$p^* : \mathrm{CH}^i(X) \to \mathrm{CH}^i(X \times \mathbb{A}^n)$$

is an isomorphism ($0 \le i \le \dim X$).

REMARK 9.1.19   In [8, p. 22] it is stated only that p is surjective; however by taking a point $P \in \mathbb{A}^n$ we get a section $i_P : X \to X \times \mathbb{A}^n$ of p which gives the injectivity.

THEOREM 9.1.20 (Localization sequence; see [8, p. 21])   *Let $\iota : Y \hookrightarrow X$ be a closed subvariety of X, let $U = X - Y$ and let $j : U \hookrightarrow X$ be the inclusion. Then the following sequence is exact:*

$$\mathrm{CH}_q(Y) \xrightarrow{\iota_*} \mathrm{CH}_q(X) \xrightarrow{j^*} \mathrm{CH}_q(U) \longrightarrow 0.$$

REMARK 9.1.21   This holds for arbitrary X and Y (not necessarily smooth or projective). Recall from Remark 9.1.16 that the definition of $\mathrm{CH}_q(X)$ for X an arbitrary variety is entirely similar to the case when X is smooth and projective (see [8, p. 10, Sec. 1.3]

REMARK 9.1.22   (On the coefficients) If we want to work with $\mathbb{Q}$-coefficients we write $\mathrm{CH}_{\mathbb{Q}}(X) := \mathrm{CH}(X) \otimes_{\mathbb{Z}} \mathbb{Q}$. Of course then we lose the torsion aspects!

9.2   LECTURE II: EQUIVALENCE RELATIONS. SHORT SURVEY ON THE RESULTS FOR DIVISORS

As in Lecture I we assume X_d, Y_n, etc. to be smooth, irreducible, projective varieties defined over an algebraically closed field k.

9.2.1   Algebraic Equivalence (Weil, 1952)

DEFINITION 9.2.1 $Z \in \mathcal{Z}^i(X)$ *is algebraically equivalent to* 0 *if there exists a smooth curve* C, *a cycle* $T \in \mathcal{Z}^i(C \times X)$ *and two points* $a, b \in C$ *such that* $Z = T(a) - T(b)$ *(equivalently, replacing* T *by* $T - C \times T(b)$ *we could say* $T(a) = Z$, $T(b) = 0$*).*

Put $Z^i_{\text{alg}}(X) = \{Z \in \mathcal{Z}^i(X) : Z \sim 0$ algebraically. Clearly from the "alternative" formulation of rational equivalence given in Proposition 9.1.13, we see that there is an inclusion $Z^i_{\text{rat}}(X) \subset Z^i_{\text{alg}}(X)$; but in general $Z^i_{\text{rat}}(X) \neq Z^i_{\text{alg}}(X)$. For instance, take $X = E$ an elliptic curve, and $Z = P_1 - P_2$ with $P_i \in E$ two distinct points.

Algebraic equivalence is an adequate equivalence relation (see [24, p. 474]). Put $\text{CH}^i_{\text{alg}}(X) := \mathcal{Z}^i_{\text{alg}}(X)/\mathcal{Z}^i_{\text{rat}}(X) \subset \text{CH}^i(X)$.[6]

REMARK 9.2.2

(a) We may replace C by any algebraic variety V and $a, b \in V$ smooth points.

(b) By the theory of *Hilbert schemes* (or more elementary *Chow varieties*) we know that the group $\mathcal{Z}^i(X)/\mathcal{Z}^i_{\text{alg}}(X)$ is *discrete*.

9.2.2   Smash-Nilpotent Equivalence

Around 1995 Voevodsky introduced the notion of *smash nilpotence*, also denoted by $\otimes$-*nilpotence* (see [1, p. 21]).

DEFINITION 9.2.3 $Z \in \mathcal{Z}^i(X)$ *is called smash nilpotent to* 0 *on* X *if there exists an integer* $N > 0$ *such that the product of* N *copies of* Z *is rationally equivalent to* 0 *on* X^N.

Let $\mathcal{Z}^i_{\otimes}(X) \subset \mathcal{Z}^i(X)$ be the subgroup generated by the cycles smash nilpotent to 0. It can be proved that this is an adequate equivalence relation.

There is the following important theorem:

THEOREM 9.2.4 (Voisin, Voevodsky independently)

$$\mathcal{Z}^i_{\text{alg}}(X) \otimes \mathbb{Q} \subset \mathcal{Z}^i_{\otimes}(X) \otimes \mathbb{Q}.$$

PROOF.   It goes beyond the scope of these lectures. See [28, Ch. 11].   □

REMARK 9.2.5   Recently, Kahn and Sebastian have shown that in the above theorem inclusion is strict. For instance, on the Jacobian variety $X = J(C)$ of a general curve of genus 3, the so-called Ceresa cycle $Z = C - C^-$ is $\otimes$-nilpotent to 0 but not algebraically equivalent to 0 (see Lecture IV, after Theorem 9.4.14).

---

[6]Not to be confused with $C^i_{\text{alg}}(X) = \mathcal{Z}^i(X)/\mathcal{Z}^i_{\text{alg}}(X)$!

9.2.3 Homological Equivalence

Let $H(X)$ be a "good" (so-called Weil) cohomology theory. Without going into details, let us say that this means that the $H^i(X)$ are F-vector spaces with F a field of characteristic 0 and that all the "classical" properties for cohomology hold, so, in particular, there are cup products, Poincaré duality and the Künneth formula both hold, and there is a cycle map (see below). If $\mathrm{char}(k) = 0$ then one can assume that $k \subset \mathbb{C}$ and one can take $H(X) = H_B(X_{\mathrm{an}}, \mathbb{Q})$ or $H_B(X_{\mathrm{an}}, \mathbb{C})$, i.e., the classical Betti cohomology on the underlying analytic manifold X_{an}; instead of the Betti cohomology one can also take the classical de Rham cohomology on X_{an} or the algebraic de Rham cohomology with respect to the Zariski topology on X. In the case of a general field k one can take the étale cohomology $H(X) = H_{\mathrm{et}}(X_{\bar{k}}, \mathbb{Q}_\ell)$ ($l \neq \mathrm{char}(k)$). Note that for general k one has to make the base change from k to $\bar{k}$ (in our case we assume that $k = \bar{k}$ already); for arbitrary k the $H_{\mathrm{et}}(X, \mathbb{Q}_\ell)$ on X itself (i.e., without base change) is certainly an interesting cohomology but it is not a Weil cohomology in general.

For a Weil cohomology one has a *cycle map*

$$\gamma_X : \mathrm{CH}^i(X) \longrightarrow H^{2i}(X)$$

having "nice" properties, in particular, the intersection product is compatible with the cup product, i.e., if $\alpha, \beta \in \mathrm{CH}(X)$, then $\gamma_X(\alpha \cdot \beta) = \gamma_X(\alpha) \cup \gamma_X(\beta)$.

DEFINITION 9.2.6 *A cycle* $Z \in \mathcal{Z}^i(X)$ *is homologically equivalent to* 0 *if*

$$\gamma_X(Z) = 0.$$

This turns out (using the "nice" properties of the cycle map) to be again an adequate equivalence relation. Put $\mathcal{Z}^i_{\mathrm{hom}}(X) \subset \mathcal{Z}^i(X)$ for the subgroup of cycles homologically equivalent to 0.

REMARK 9.2.7

(a) This $\mathcal{Z}^i_{\mathrm{hom}}(X)$ depends—at least a priori—on the choice of the cohomology theory. If $\mathrm{char}(k) = 0$ then Artin's comparison theorem implies that $H_{\mathrm{et}}(X_{\bar{k}}, \mathbb{Q}_\ell) \cong H_B(X_{\mathrm{an}}, \mathbb{Q}) \otimes \mathbb{Q}_\ell$ and one gets from the classical theory and the étale theory (for all l) the same homological equivalence. In the general case this would follow from the—still wide open—standard conjectures of Grothendieck and also from the conjecture of Voevodsky (see (c) below).

(b) We have $\mathcal{Z}^i_{\mathrm{alg}}(X) \subset \mathcal{Z}^i_{\mathrm{hom}}(X)$ as follows from the fact that two points a and b on a curve C are homologically equivalent and from the functoriality properties of the equivalence relations. For divisors we have by a *theorem of Matsusaka* that $\mathcal{Z}^1_{\mathrm{alg}}(X) \otimes \mathbb{Q} = \mathcal{Z}^1_{\mathrm{hom}}(X) \otimes \mathbb{Q}$; however, for $1 < i < d$ "algebraic" and "homological equivalence" are, in general, *different* by a famous *theorem of Griffiths* (see Lecture IV).

(c) *Smash-nilpotent equivalence* versus *homological equivalence*. The Künneth formula shows that

$$\mathcal{Z}^i_\otimes(X) \subset \mathcal{Z}^i_{\mathrm{hom}}(X).$$

Voevodsky *conjectures* that in fact we have equality (and this would imply—in particular—that homological equivalence would be independent of the choice of the cohomology theory). Indeed he makes the even stronger conjecture that it coincides with numerical equivalence; see Section 9.2.4.

9.2.4 Numerical Equivalence

Let $X = X_d \in \mathrm{Var}(k)$. If $Z \in \mathcal{Z}^i(X)$ and $W \in \mathcal{Z}^{d-i}(X)$ then their intersection product is a zero-cycle $Z \cdot W = \sum n_\alpha P_\alpha \in \mathcal{Z}_0(X)$, where $P_\alpha \in X$, and hence it has a *degree* $\sum n_\alpha$. (We can assume that $Z \cdot W$ is defined because replacing Z by Z' rationally equivalent to Z we have that $Z \cdot W$ and $Z' \cdot W$ have the same degree.)

DEFINITION 9.2.8 $Z \in \mathcal{Z}^i(X)$ *is numerically equivalent to* 0 *if* $\deg(Z \cdot W) = 0$ *for all* $W \in \mathcal{Z}^{d-i}(X)$. *Let* $\mathcal{Z}^i_{\mathrm{num}}(X) \subset \mathcal{Z}^i(X)$ *be the subgroup of the cycles numerically equivalent to* 0.

Numerical equivalence is an adequate equivalence relation ([24, p. 474]).

REMARK 9.2.9

(a) $\deg(Z \cdot W)$ is called the intersection number of Z and W and is sometimes denoted by $\sharp(Z \cdot W)$.

(b) Because of the compatibility of intersection with the cup product of the corresponding cohomology classes we have

$$\mathcal{Z}^i_{\mathrm{hom}}(X) \subseteq \mathcal{Z}^i_{\mathrm{num}}(X).$$

For divisors we have $\mathrm{Div}_{\mathrm{hom}}(X) = \mathrm{Div}_{\mathrm{num}}(X)$ (theorem of Matsusaka). It is a *fundamental conjecture* that the equality $\mathcal{Z}^i_{\mathrm{hom}}(X) = \mathcal{Z}^i_{\mathrm{num}}(X)$ should hold for all i. This is part of the standard conjectures of Grothendieck (and it is usually denoted as conjecture $D(X)$).

REMARK 9.2.10

(a) For $k = \mathbb{C}$, conjecture $D(X)$ would follow from the famous Hodge conjecture (see Lecture III).

(b) For arbitrary $k = \bar{k}$, Voevodsky *conjectures* that

$$\mathcal{Z}^i_\otimes(X) \otimes \mathbb{Q} = \mathcal{Z}^i_{\mathrm{num}}(X) \otimes \mathbb{Q},$$

i.e., nilpotent equivalence, homological equivalence, and numerical equivalence should coincide (at least up to torsion). Of course this would imply conjecture $D(X)$.

9.2.5   Final Remarks and Résumé of Relations and Notation

There are also other interesting equivalence relations for algebraic cycles (see, for instance, [24, 17]); however in these lectures we restrict to those above. One can show (see [24, p. 473]) that rational equivalence is the most fine adequate equivalence, i.e., for any adequate relation $\sim$ there is an inclusion $\mathcal{Z}^i_{\mathrm{rat}}(X) \subset \mathcal{Z}^i_\sim(X)$.

Résumé of the relations:

$$\mathcal{Z}^i_{\mathrm{rat}}(X) \subsetneqq \mathcal{Z}^i_{\mathrm{alg}}(X) \subsetneqq \mathcal{Z}^i_\otimes(X) \subseteq \mathcal{Z}^i_{\mathrm{hom}}(X) \subseteq \mathcal{Z}^i_{\mathrm{num}}(X) \subset \mathcal{Z}^i(X).$$

Dividing out by rational equivalence we get the following subgroups in the Chow groups:

$$\mathrm{CH}^i_{\mathrm{alg}}(X) \subsetneqq \mathrm{CH}^i_\otimes(X) \subseteq \mathrm{CH}^i_{\mathrm{hom}}(X) \subseteq \mathrm{CH}^i_{\mathrm{num}}(X) \subset \mathrm{CH}^i(X).$$

9.2.6   Cartier Divisors and the Picard Group

$\mathrm{Div}(X) = \mathcal{Z}^i(X)$ is the group of *Weil divisors*. There are also *Cartier divisors* which are more suited if one works with an arbitrary variety (see [14, pp. 140–145]). So let X be an arbitrary variety, but irreducible (and always defined over k). Let $K = k(X)$ be the function field of X and K^*_X the constant sheaf K^* on X. Let $\mathcal{O}^*_X \subset K^*_X$ be the sheaf of units in $\mathcal{O}_X$ and define the quotient sheaf $\underline{\mathrm{Div}}_X := K^*_X / \mathcal{O}^*_X$ (always in the Zariski topology). So we have an exact sequence:

$$1 \to \mathcal{O}^*_X \to K^*_X \to \underline{\mathrm{Div}}_X \to 1.$$

$\underline{\mathrm{Div}}_X$ is called the *sheaf of Cartier divisors* and the global sections $\Gamma(X, \underline{\mathrm{Div}}_X)$ are the *Cartier divisors*; we denote the corresponding group by $\mathrm{CaDiv}(X)$. So concretely, a Cartier divisor D is given via a collection $\{U_\alpha, f_\alpha\}$ with $\{U_\alpha\}$ an open Zariski covering of X and $f_\alpha \in K^*$ rational functions such that f_α / f_β is in $\Gamma(U_\alpha \cap U_\beta, \mathcal{O}^*_X)$.

A Cartier divisor D is called *linear equivalent to* 0 (or *principal*) if there exists $f \in K^*$ such that $D = \mathrm{div}(f)$, i.e., if D is in the image of

$$K^* = \Gamma(X, K^*_X) \to \Gamma(X, \underline{\mathrm{Div}}_X)$$

and the quotient group is denoted by $\mathrm{CaCl}(X) := \Gamma(X, \underline{\mathrm{Div}}_X)/\sim$.

Since X is irreducible we have $H^1(X, K^*_X) = H^1(X, K^*) = 1$ and the above exact sequence gives the following isomorphism (see [14, p. 145]):

$$\mathrm{CaCl}(X) \cong H^1(X, \mathcal{O}^*_X).$$

9.2.6.1   Picard Group

On a ringed space, in particular, on an algebraic variety, the isomorphism classes of invertible sheaves form an *abelian group* under the tensor product, the so-called *Picard* *group* $\text{Pic}(X)$. From the definition of invertible sheaves, i.e., from the fact that such an invertible sheaf is Zariski-locally isomorphic to $\mathcal{O}_X^*$, we get

$$\text{Pic}(X) \cong H^1(X, \mathcal{O}_X^*)$$

(see [14, p. 224, Ex. 4.5]). Combining with the above we have

$$\text{CaCl}(X) \cong H^1(X, \mathcal{O}_X^*) \cong \text{Pic}(X).$$

9.2.6.2 *Weil Divisors and the Picard Group*

Let us now assume again that X is a *smooth* irreducible variety. Then for every point $P \in X$ the local ring $\mathcal{O}_{P,X}$ is a unique factorization domain and every Weil divisor is given in $\mathcal{O}_{P,X}$ by an equation f_P unique up to a unit in $\mathcal{O}_{P,X}$. From this it follows easily that, if X is smooth, Weil divisors and Cartier divisors coincide, i.e., $\text{Div}(X) \xrightarrow{\sim} \text{CaDiv}(X)$ (see [14, p. 141, Prop. 6.11]). Moreover, $D \in \text{Div}(X)$ is of the form $\text{div}(f)$, $f \in K^*$ if and only if D is principal. Therefore we get if X is smooth and projective

$$\text{CH}^1(X) = \text{CaCl}(X) = H^1(X, \mathcal{O}_X^*) = \text{Pic}(X).$$

REMARK 9.2.11 See [14, p. 129, Ex. 5.18(d)]. There is also a one-to-one correspondence between isomorphism classes of invertible sheaves on X and isomorphism classes of line bundles on X.

9.2.7 **Résumé of the Main Facts for Divisors**

Let X be a smooth, irreducible, projective variety. In $\text{CH}^1(X)$ we have the following subgroups:

$$\text{CH}^1_{\text{alg}}(X) \subset \text{CH}^1_\tau(X) \subset \text{CH}^1_{\text{hom}}(X) \subset \text{CH}^1_{\text{num}}(X) \subset \text{CH}^1(X).$$

The following facts are known (see Mumford [29, app. to Ch. V]):

(a) $\text{CH}^1_{\text{alg}}(X)$ has the structure of an *abelian variety*, the so-called *Picard variety* $\text{Pic}^0_{red}(X)$ (classically for $k = \mathbb{C}$ this goes back to Italian algebraic geometry, see [29, p. 104], in char $= p > 0$ to Matsusaka, Weil, and Chow. Moreover $\text{Pic}^0_{red}(X)$ is the reduced scheme of the component of the identity of the Picard scheme of Grothendieck).

(b) $\text{CH}^1_\tau(X)$ is by definition the set of divisors classes D such that $nD \sim 0$ algebraically equivalent to 0 for some $n \neq 0$. By a theorem of Matsusaka (1956), $\text{CH}^1_\tau(X) = \text{CH}^1_{\text{hom}}(X) \subset \text{CH}^1(X)$.

(c) $\text{NS}(X) := \text{CH}^1(X)/\text{CH}^1_{\text{alg}}(X)$ is a finitely generated (abelian) group, the so-called *Néron–Severi* group (classically Severi around 1908, in general Néron 1952).

REMARK 9.2.12 For cycles of codim $i > 1$ almost all of the above facts *fail* as we shall see later (Lectures IV and V). However $\mathrm{CH}^i(X)/\mathrm{CH}^i_{\mathrm{num}}(X)$ is still, for all i and all $k = \bar{k}$, a finitely generated abelian group, as Kleiman proved in 1968 [19, Thm. 3.5, p. 379]; this follows from the existence of the Weil cohomology theory $H_{\mathrm{et}}(X_{\bar{k}}, \mathbb{Q}_\ell)$.

REMARK 9.2.13 (Comparison between the algebraic and the analytic theory) [14, App. B] If $k = \mathbb{C}$ we have for X a smooth projective variety, several topologies, namely, algebraically the Zariski topology and the étale topology, as well as the classical topology on the underlying analytic space X_{an} which is a complex manifold, compact and connected. Now there are the following comparison theorems:

(a) For the étale topology (Artin, 1965),

$$H_{\mathrm{et}}^i(X, \mathbb{Q}_\ell) \cong H^i(X_{\mathrm{an}}, \mathbb{Q}) \otimes_{\mathbb{Q}} \mathbb{Q}_\ell.$$

(b) For the Zariski topology the famous theorem of Serre (GAGA, 1956) saying that for coherent sheaves $\mathcal{F}$ the functor $\mathcal{F} \mapsto \mathcal{F} \otimes_{\mathcal{O}_{X_{\mathrm{alg}}}} \mathcal{O}_{X_{\mathrm{an}}} =: \mathcal{F}_{\mathrm{an}}$ is an equivalence of categories between algebraic and analytic coherent sheaves and moreover,

$$H_{\mathrm{Zar}}^i(X, \mathcal{F}) \cong H^i(X_{\mathrm{an}}, \mathcal{F}_{\mathrm{an}})$$
$$H_{\mathrm{Zar}}^1(X, \mathcal{O}_X^*) \cong H^1(X_{\mathrm{an}}, \mathcal{O}_{X_{\mathrm{an}}}^*)$$

(the latter via interpretation as invertible sheaves), so, in particular, $\mathrm{Pic}(X_{\mathrm{alg}}) = \mathrm{Pic}(X_{\mathrm{an}})$ (see [14, App. B]).

Now using this GAGA theorem and the exponential exact sequence

$$0 \longrightarrow \mathbb{Z} \longrightarrow \mathcal{O}_{X_{\mathrm{an}}} \overset{\exp}{\longrightarrow} \mathcal{O}_{X_{\mathrm{an}}}^* \longrightarrow 1,$$

most of the above *facts for divisors* become at *least plausible*. Namely, from the exact exponential sequence we get the following exact sequence

$$H^1(X_{\mathrm{an}}, \mathbb{Z}) \overset{\alpha}{\longrightarrow} H^1(X_{\mathrm{an}}, \mathcal{O}_{X_{\mathrm{an}}}) \overset{\beta}{\longrightarrow}$$
$$H^1(X_{\mathrm{an}}, \mathcal{O}_{X_{\mathrm{an}}}^*) \cong \mathrm{Pic}(X) \overset{\gamma}{\longrightarrow} H^2(X_{\mathrm{an}}, \mathbb{Z}).$$

Now γ is the cycle map (see later), hence $\mathrm{CH}^1(X)/\mathrm{CH}^1_{\mathrm{hom}}(X) = \mathrm{Im}(\gamma)$ is finitely generated as a subgroup of $H^2(X_{\mathrm{an}}, \mathbb{Z})$, which is itself a finitely generated group. Next,

$$\mathrm{CH}^1_{\mathrm{hom}}(X) \cong \mathrm{Im}(\beta) \cong H^1(X_{\mathrm{an}}, \mathcal{O}_{X_{\mathrm{an}}})/\mathrm{Im}(\alpha)$$

is a complex torus since $H^1(X_{\mathrm{an}}, \mathcal{O}_{X_{\mathrm{an}}})$ is a finite-dimensional $\mathbb{C}$-vector space and $\mathrm{Im}(\alpha)$ is a lattice in this vector space (see Lecture III).

9.2.8   References for Lectures I and II

In these lectures we assume knowledge of the "basic material" of algebraic geometry which can be found amply (for instance) in the book of Hartshorne [14, Ch. 1 and parts of Chs. 2 and 3]. A nice introduction to algebraic cycles and Chow groups is given in [14, App. A]. The basic standard book for algebraic cycles and Chow groups is the book of Fulton [8], but here we have mostly only needed Chapter 1. Fulton's theory of intersection theory is much more advanced and precise; his main tool is not the moving lemma but the so-called deformation of the normal cone, but this is much more technical. For the theory of Chow groups (over $\mathbb{C}$) one can also look at the book by Voisin [28, Ch. 9]. For the definition of the intersection multiplicities we have used Serre's approach [25, Ch. 5C].

9.3   LECTURE III: CYCLE MAP. INTERMEDIATE JACOBIAN. DELIGNE COHOMOLOGY

In this lecture we assume that $k = \mathbb{C}$ is the field of complex numbers. As before, let X be a smooth, irreducible, projective variety now defined over $\mathbb{C}$. Then we have (see, for instance, [14, App. B]) the underlying complex analytic space which is a complex manifold X_{an} compact and connected, on which we have the classical "usual" topology. If there is no danger of confusion we shall sometimes, by abuse of notation, use the same letter for X and X_{an}.

9.3.1   The Cycle Map

PROPOSITION 9.3.1   *Let $X = X_d$ be a smooth, projective, irreducible variety defined over $\mathbb{C}$. Let $0 \le p \le d$, and put $q = d - p$. Then there exists a homomorphism, the cycle map $\gamma_{X,\mathbb{Z}}$ (in short, $\gamma_{\mathbb{Z}}$), as follows:*

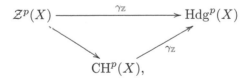

where $\mathrm{Hdg}^p(X) \subset H^{2p}(X_{\mathrm{an}}, \mathbb{Z})$ is the subgroup defined as follows. Let j denote the natural map $j : H^{2p}(X_{\mathrm{an}}, \mathbb{Z}) \to H^{2p}(X_{\mathrm{an}}, \mathbb{C})$ and

$$H^{2p}(X_{\mathrm{an}}, \mathbb{C}) = \bigoplus_{r+s=2p} H^{r,s}(X_{\mathrm{an}})$$

the Hodge decomposition, then

$$\mathrm{Hdg}^p(X) := H^{2p}(X_{\mathrm{an}}, \mathbb{Z}) \cap j^{-1}(H^{p,p}(X_{\mathrm{an}})).$$

REMARK 9.3.2   By abuse of notation we have denoted the factorization of the map $\gamma_{\mathbb{Z}} : \mathcal{Z}^p(X) \to \mathrm{Hdg}^p(X)$ through the Chow group by the same symbol $\gamma_{\mathbb{Z}}$. Also it would be more correct to write $\mathrm{Hdg}^p(X_{\mathrm{an}})$.

REMARK 9.3.3 We shall only construct the map for $\mathcal{Z}^p(X)$ itself. The factorization exists because $\mathcal{Z}_{\mathrm{rat}}(\cdot) \subseteq \mathcal{Z}_{\mathrm{hom}}(\cdot)$ and this is true since it is trivially true for $X = \mathbb{P}^1$ and true in general via functoriality.[7]

9.3.1.1 Outline of the Construction of the $\gamma_{\mathbb{Z}}$

(For details see [27, 11.1.2].) Let $Z_q \subset X_d$ be a closed subvariety. There is (in the analytic topology) the following exact sequence with $U = X - Z$:

$$\cdots \to H^{2p-1}(U;\mathbb{Z}) \to H^{2p}(X,U;\mathbb{Z}) \xrightarrow{\rho} H^{2p}(X;\mathbb{Z}) \to H^{2p}(U;\mathbb{Z}) \to \cdots$$

Assume for simplicity that Z is also smooth. By a theorem of Thom (see [27, 11.1.2]) we have an isomorphism

$$T : H^{2p}(X,U;\mathbb{Z}) \xrightarrow{\cong} H^0(Z;\mathbb{Z}) = \mathbb{Z}.$$

Now take $\gamma_{\mathbb{Z}}(Z) = \rho \circ T^{-1}(1_{\mathbb{Z}})$. If Z is not smooth we replace it by $Z - Z_{\mathrm{sing}}$ in the above sequence (for details see [27, 11.1.2]).

REMARK 9.3.4 If the variety X is defined over an algebraically closed field k but is otherwise of arbitrary characteristic, we have essentially the same construction (see [20, p. 268]) working with $H_{\mathrm{et}}(X,\mathbb{Z}_\ell)$, where $\ell \neq \mathrm{char}(k)$, and using instead of the above sequence, the sequence

$$\cdots \to H^{2p}_Z(X,\mathbb{Z}_\ell) \to H^{2p}(X,\mathbb{Z}_\ell) \to H^{2p}(U,\mathbb{Z}_\ell) \to \cdots.$$

9.3.1.2 Position of $\gamma_{\mathbb{Z}}(Z)$ in the Hodge Decomposition

For this we must use the *de Rham interpretation* of the cohomology. Recall (see [13, p. 44]) that there exists an isomorphism

$$H^i_{dR}(X_{\mathrm{an}},\mathbb{C}) \xrightarrow{\cong} H^i_{\mathrm{sing}}(X_{\mathrm{an}},\mathbb{C}) \cong H^{\mathrm{sing}}_i(X_{\mathrm{an}},\mathbb{C})^*,$$

where $H_{\mathrm{sing}}(\cdot)$ is the singular cohomology and $*$ is the dual, given by

$$\varphi \longmapsto \langle \sigma, \varphi \rangle := \int_\sigma \varphi,$$

where φ is a closed $\mathcal{C}^\infty$-differential form of degree i and σ is a differentiable i-chain. Moreover, in terms of the de Rham cohomology the Poincaré duality is given by the pairing

$$H^i_{dR}(X_{\mathrm{an}},\mathbb{C}) \times H^{2d-i}_{dR}(X_{\mathrm{an}},\mathbb{C}) \longrightarrow \mathbb{C},$$

$$(\alpha,\beta) \longmapsto \langle \alpha,\beta \rangle := \int_X \alpha \wedge \beta,$$

and by the pairing

$$H^{r,s}(X_{\mathrm{an}}) \xrightarrow{\sim} H^{d-r,d-s}(X_{\mathrm{an}})^*.$$

Returning to $\gamma_{\mathbb{Z}}(Z)$ we have the following lemma which proves Proposition 9.3.1:

---

[7]Recall that we have in fact the stronger inclusion $\mathcal{Z}_{\mathrm{alg}}(\cdot) \subset \mathcal{Z}_{\mathrm{hom}}(\cdot)$; see Remark 9.2.7(b).

LEMMA 9.3.5 $j \circ \gamma_{\mathbb{Z}}(Z) \in H^{p,p}(X_{\mathrm{an}}) \subset H^{2p}(X_{\mathrm{an}}, \mathbb{C})$.

INDICATION OF THE PROOF. See [27, Secs. 11.1.2 and 11.1.3]. The cycle class $j \circ \gamma_{\mathbb{Z}}(Z)$ is, by Poincaré duality, a functional on $H^{2d-2p}(X_{\mathrm{an}}, \mathbb{C})$. Working with de Rham cohomology, let $\beta \in H^{2d-2p}(X_{\mathrm{an}}, \mathbb{C})$. Then the class is determined by the relation

$$\int_X j \circ \gamma_{\mathbb{Z}}(Z) \wedge \beta = \int_Z \beta_{|Z} = \int_Z i^*(\beta),$$

where $i : Z \hookrightarrow X$. Since Z is a complex manifold of complex dimension $d - p$, this is 0 unless β is of type $(d - p, d - p)$. Hence the class $j \circ \gamma_{\mathbb{Z}}(Z)$ is orthogonal to all $H^{r,s}(X_{\mathrm{an}})$ unless $(r, s) = (d - p, d - p)$. Hence $j \circ \gamma_{\mathbb{Z}}(Z) \in H^{p,p}(X_{\mathrm{an}})$.   □

9.3.2   Hodge Classes. Hodge Conjecture

Recall (see Proposition 9.3.1)

$$\mathrm{Hdg}^p(X) := \{\eta \in H^{2p}(X_{\mathrm{an}}, \mathbb{Z}) : j(\eta) \in H^{p,p}(X_{\mathrm{an}})\},$$

where $j : H^{2p}(X_{\mathrm{an}}, \mathbb{Z}) \to H^{2p}(X_{\mathrm{an}}, \mathbb{C})$ is the natural map. The elements of $\mathrm{Hdg}^p(X)$ are called *Hodge classes* or "*Hodge cycles*" of type (p, p). So we have by Proposition 9.3.1 from Section 9.3.1:

THEOREM 9.3.6   *The cohomology classes $\gamma_{\mathbb{Z}}(Z)$ of the algebraic cycles $\mathcal{Z}^p(X)$ are Hodge classes of type (p, p), i.e.,*

$$\gamma_{\mathbb{Z},X}^p : \mathcal{Z}^p(X) \longrightarrow \mathrm{Hdg}^p(X) \subset H^{2p}(X_{\mathrm{an}}, \mathbb{Z}).$$

Of course, immediately there comes up the question, what is the image?

9.3.2.1   Lefschetz $(1, 1)$ Theorem

For *divisors* there is a famous theorem:

THEOREM 9.3.7 (Lefschetz $(1, 1)$, 1924)   *Let X be a smooth, irreducible, projective variety defined over $\mathbb{C}$. Then*

$$\gamma_{\mathbb{Z},X}^1 : \mathrm{Div}(X) \longrightarrow \mathrm{Hdg}^1(X) \subset H^2(X_{\mathrm{an}}, \mathbb{Z})$$

is onto, i.e., every Hodge class of type $(1, 1)$ is "algebraic" (i.e., is the cohomology class of a divisor).

INDICATION OF THE PROOF. See [13, p. 163] for details. Using the GAGA theorems one goes from the algebraic to the analytic theory, namely, $H^i(X, \mathcal{O}_X) \cong H^i(X_{\mathrm{an}}, \mathcal{O}_{X_{\mathrm{an}}})$ and $\mathrm{CH}^1(X) \cong H^1(X, \mathcal{O}_X^*) \cong H^1(X_{\mathrm{an}}, \mathcal{O}_{X_{\mathrm{an}}}^*)$. Therefore we have to show that "the cycle map" $H^1(X_{\mathrm{an}}, \mathcal{O}_{X_{\mathrm{an}}}^*) \to H^2(X_{\mathrm{an}}, \mathbb{Z})$ maps *onto* $\mathrm{Hdg}^1(X) \subset H^2(X_{\mathrm{an}}, \mathbb{Z})$.

For this one uses the exponential sequence

$$0 \to \mathbb{Z} \to \mathcal{O}_{X_{\mathrm{an}}} \xrightarrow{\exp} \mathcal{O}_{X_{\mathrm{an}}}^* \longrightarrow 1$$

from which one gets an exact sequence

$$H^1(X_{\mathrm{an}}, \mathcal{O}^*_{X_{\mathrm{an}}}) \xrightarrow{\alpha} H^2(X_{\mathrm{an}}, \mathbb{Z}) \xrightarrow{\beta} H^2(X_{\mathrm{an}}, \mathcal{O}_{X_{\mathrm{an}}}).$$

Now one shows that under the identification, the boundary map α corresponds to the cycle map $\gamma_{\mathbb{Z}} : \mathrm{CH}^1(X) \to H^2(X_{\mathrm{an}}, \mathbb{Z})$ and the map β is the "projection" of the image of $j : H^2(X_{\mathrm{an}}, \mathbb{Z}) \to H^2(X_{\mathrm{an}}, \mathbb{C})$ to $H^{0,2}(X_{\mathrm{an}})$. From these identifications *the* theorem follows from the fact that $\mathrm{Im}(\alpha) = \ker(\beta)$ and because $\overline{H^{2,0}} = H^{0,2}$ and therefore $\ker(\beta) = \mathrm{Im}(j) \cap H^{1,1}$, where $j : H^1(X_{\mathrm{an}}, \mathbb{Z}) \to H^2(X_{\mathrm{an}}, \mathbb{C})$.   □

REMARK 9.3.8 This is the modern proof due to Kodaira–Spencer (1953). For a discussion of the ideas of the original proof of Lefschetz see the very interesting paper of Griffiths [12].

9.3.2.2 Hodge Conjecture

Motivated by the Lefschetz $(1,1)$ theorem for divisors, Hodge conjectured that, or at least raised the question of whether $\gamma_{\mathbb{Z}}$ is onto always for all p ("integral Hodge conjecture"). However Atiyah–Hirzebruch discovered that this integral form is *not* true (1962); later other counterexamples were given by Kollár (1992) and Totaro (1997). Therefore the question has to be modified to rational coefficients.

CONJECTURE 9.3.9 (Hodge) $\gamma_{\mathbb{Q}} : \mathcal{Z}^p(X) \otimes \mathbb{Q} \longrightarrow \mathrm{Hdg}^p(X) \otimes \mathbb{Q}$ *is onto.*

This fundamental conjecture is wide open and only known for special cases (see, for instance, lectures by Murre and van Geemen in [9]).

REMARK 9.3.10 In fact Hodge raised an even more general question (see [16, p. 214]) known under the name "generalized Hodge conjecture." However, in 1969 Grothendieck pointed out that this generalized Hodge conjecture *as stated by Hodge* is not true and he corrected the statement. For this GHC (Grothendieck–Hodge conjecture) see [27, 11.3.2] or [22, p. 164].

9.3.3   Intermediate Jacobian and Abel–Jacobi Map

9.3.3.1   Intermediate Jacobian (of Griffiths)

Let X be a smooth, irreducible, projective variety defined over $\mathbb{C}$. Recall the *Hodge decomposition*

$$H^i(X, \mathbb{C}) = \bigoplus_{r+s=i} H^{r,s}(X), \quad H^{s,r}(X) = \overline{H^{r,s}(X)}$$

(now write, by abuse of notation, $X = X_{\mathrm{an}}$) and the corresponding descending *Hodge filtration*

$$F^j H^i(X, \mathbb{C}) = \bigoplus_{r \geq j} H^{r,i-r} = H^{i,0} + H^{i-1,1} + \cdots + H^{j,i-j}.$$

DEFINITION 9.3.11 *The p-th* intermediate Jacobian *of X is*

$$J^p(X) = H^{2p-1}(X, \mathbb{C})/F^p H^{2p-1}(X, \mathbb{C}) + H^{2p-1}(X, \mathbb{Z}).$$

So writing $V = H^{p-1,p} + \cdots + H^{0,2p-1}$ we have that $J^p(X) = V/H^{2p-1}(X, \mathbb{Z})$ (where—of course—we mean the image of $H^{2p-1}(X, \mathbb{Z})$ in V).

LEMMA 9.3.12 $J^p(X)$ *is a complex torus of dimension half the* $(2p-1)$th Betti number of X:

$$\dim J^p(X) = \frac{1}{2} B_{2p-1}(X),$$

and hence for the conjugate $\overline{V}$ *of V we have* $\overline{V} = H^{p,p-1} + \cdots + H^{2p-1,0}$ *and also that the Betti number is even.*

PROOF. First note that due to $H^{s,r}(X) = \overline{H^{r,s}(X)}$, the Betti number is even, so let $B_{2p-1}(X) = 2m$. We have to show that the image of $H^{2p-1}(X, \mathbb{Z})$ is a *lattice* in the complex vector space V. Therefore if $\alpha_1, \ldots, \alpha_{2m}$ is a $\mathbb{Q}$-basis of $H^{2p-1}(X, \mathbb{Q})$, we must show that if $\omega = \sum r_i \alpha_i \in F^p H^{2p-1}(X, \mathbb{C})$ with $r_i \in \mathbb{Q}$ then $r_i = 0$ for all i. But $F^p H^{2p-1}(X, \mathbb{C}) = V$ and $\omega = \overline{\omega}$ so $\omega \in V \cap \overline{V} = (0)$. Therefore $\omega = 0$, but $\{\alpha_i\}$ is a $\mathbb{Q}$-basis for $H^{2p-1}(X, \mathbb{Q})$; hence all $r_i = 0$. $\square$

REMARK 9.3.13 The complex torus $J^p(X)$ is *in general not* an *abelian variety*, i.e., cannot be embedded in projective space. For a torus $T = V/L$ to be an abelian variety it is necessary and sufficient that there exists a so-called *Riemann form*. This is an $\mathbb{R}$-bilinear alternating form $E : V \times V \to \mathbb{R}$ satisfying

(a) $E(iv, iw) = E(v, w)$;

(b) $E(v, w) \in \mathbb{Z}$ whenever $v, w \in L$;

(c) $E(v, iw)$ symmetric and *positive definite*.

In our case there is a nondegenerate form on V given by $E(v, w) = v \cup w \cup h^{d+2-2p}$, where h is the hyperplane class in $H^2(X, \mathbb{Z})$. However, this form is in general not positive definite because it changes sign on the different $H^{r,s}(X)$, but it is if only one $H^{r,s}$ occurs in V, for instance only $H^{p-1,p}$.

Special Cases

(a) $p = 1$. Then $J^1(X) = H^1(X, \mathbb{C})/H^{1,0} + H^1(X, \mathbb{Z})$. This is the *Picard variety* of X, which is an abelian variety.

(b) $p = d$. Then $J^d(X) = H^{2d-1}(X, \mathbb{C})/H^{d,d-1} + H^{2d-1}(X, \mathbb{Z})$. This is the *Albanese variety* of X, which is an abelian variety.

(c) If $X = C$ is a curve then $J^1(X)$ is the so-called *Jacobian variety* of C, an abelian variety which is at the same time the Picard variety and the Albanese variety of X.

9.3.3.2 Abel–Jacobi map

Recall

$$\mathcal{Z}^p_{\mathrm{hom}}(X) = \{Z \in \mathcal{Z}^p(X) : \gamma_{\mathbb{Z}}(Z) = 0\},$$

i.e., the algebraic cycles which are *homologically equivalent* to zero.

THEOREM 9.3.14 *There exists a homomorphism* $\mathrm{AJ}^p : \mathcal{Z}^p_{\mathrm{hom}} \to J^p(X)$ *which factors through* $\mathrm{CH}^p_{\mathrm{hom}}(X)$. AJ *is called the Abel–Jacobi map.*

OUTLINE OF THE PROOF. Recall that by the Poincaré duality,

$$H^{2p-1}(X) \overset{\mathrm{dual}}{\longleftrightarrow} H^{2d-2p+1}(X), \quad H^{r,s} \overset{\mathrm{dual}}{\longleftrightarrow} H^{d-r,d-s},$$

where $d = \dim X$. Hence we have that the $\mathbb{C}$-vector space V which occurs in the description of the intermediate Jacobian $J^p(X) = V/H^{2p-1}(X,\mathbb{Z})$ is the dual of $F^{d-p+1}H^{2d-2p+1}(X,\mathbb{C})$, because

$$V = H^{p-1,p} + \cdots + H^{0,2p-1} \overset{\mathrm{dual}}{\longleftrightarrow} H^{d-p+1,d-p} + \cdots + H^{d,d-2p+1}$$
$$= F^{d-p+1}H^{2d-2p+1}(X,\mathbb{C}).$$

Therefore an element $v \in V$ is a *functional* on $F^{d-p+1}H^{2d-2p+1}(X,\mathbb{C})$. (Note that for $H^{r,s}$ we have $0 \le r, s \le d$, so the above expressions may stop "earlier.")

Now let $Z \in \mathcal{Z}^p_{\mathrm{hom}}(X)$. Since $\dim Z = d - p$, there exists a topological $(2d - 2p+1)$-chain Γ such that $Z = \partial\Gamma$. Now Γ is a functional on $F^{d-p+1}H^{2d-2p+1}(X,\mathbb{C})$, because we can represent $\omega \in F^{d-p+1}H^{2d-2p+1}(X,\mathbb{C})$ by a closed $\mathcal{C}^\infty$-differential form φ of degree $2d - 2p + 1$, and we get the functional

$$\omega \mapsto \int_\Gamma \varphi.$$

Of course we must check that this does not depend on the choice of φ in the cohomology class ω. If φ' is another choice then $\varphi' = \varphi + d\psi$, but now one can show that one can take ψ such that in ψ there occur at least $(d - p + 1)$ $dz's$ ([11, p. 188]); therefore we get by the Stokes theorem, $\int_\Gamma d\psi = \int_Z \psi = 0$ because Z is a complex manifold of dimension $(d - p)$.

Hence the choice of Γ determines an element of V and hence also an element of the intermediate Jacobian $J^p(X) = V/H^{2p-1}(X,\mathbb{Z})$. Now if we have another Γ' such that $\partial\Gamma' = Z$, then $\Gamma' - \Gamma \in H_{2d-2p+1}(X,\mathbb{Z})$, i.e., $\Gamma' - \Gamma$ is an *integral* cycle and therefore they give the same element in $J^p(X)$: this is the element $\mathrm{AJ}(Z)$.

For the fact that this factors through $\mathrm{CH}^p_{\mathrm{hom}}(X)$ see [9, p. 85] $\square$

EXAMPLE 9.3.15 Let $X = C$ be a smooth projective curve defined over $\mathbb{C}$ of genus g. Now $p = 1$ and $\mathcal{Z}^1_{\mathrm{hom}}(C) = \mathrm{Div}^{(0)}(C)$ are the divisors of degree 0, so $D = \sum P_i - \sum Q_j$ ($P_i, Q_j \in C, 1 \le i, j \le m$); so we can arrange things so that $D = \sum_i (P_i - Q_i)$. Let Γ_i be a path from Q_i to P_i, so $D = \partial\Gamma$ with $\Gamma = \sum \Gamma_i$. Now

$$J(C) = H^1(C,\mathbb{C})/(F^1H^1 + H^1(C,\mathbb{Z})) = H^{01}(C)/H^1(C,\mathbb{Z})$$

and $F^1 H^1(C, \mathbb{C}) = H^{1,0}(\mathbb{C}) = H^0(C, \Omega^1_C)$, i.e., the space of holomorphic differentials, so $\dim H^{1,0} = \dim H^{0,1} = g$. Now Γ defines a functional on $H^0(C, \Omega^1_C)$, namely, if $\omega \in H^0(C, \Omega^1_C)$ consider $\int_\Gamma \omega \in \mathbb{C}$. If we choose other paths (or another ordering of the points Q_j) then we get a 1-chain Γ' and $\Gamma' - \Gamma \in H_1(C, \mathbb{Z})$ and the functionals Γ and Γ' seen as elements in the g-dimensional vector space $H^{01}(C)$ give the same element in $J(C) = H^{01}(C)/H^1(C, \mathbb{Z})$.

9.3.3.3   The Image of $\mathcal{Z}^p_{\mathrm{alg}}(X)$ under the Abel–Jacobi Map

The intermediate Jacobian behaves *functorially* under correspondences. Namely, if $T \in \mathcal{Z}^p(Y_e \times X_d)$ then we get a homomorphism $T : J^r(Y) \to J^{p+r-e}(X)$. So, in particular, if $Y = C$ is a curve and $T \in \mathcal{Z}^p(C \times X)$, then we get a homomorphism $T : J(C) \to J^p(X)$.

The tangent space at the origin of $J^p(X)$ is the vector space used in the construction (see Section 9.3.3.1), i.e., $V = H^{p-1,p} + \cdots + H^{0,2p-1}$. From the Künneth decomposition of the cycle class of T in $H^{2p}(C \times X)$, we see that the tangent space $H^{01}(C)$ to $J(C)$ is mapped into a subspace of $H^{p-1,p}(X) \subset V$ of the tangent space to $J^p(X)$.

Let $J^p(X)_{\mathrm{alg}} \subseteq J^p(X)$ be the *largest subtorus* of $J^p(X)$ for which the tangent space is contained in $H^{p-1,p}(X)$. This subtorus $J^p(X)_{\mathrm{alg}}$ is in fact an abelian variety (see Remark 9.3.13). Of course it may happen that $J^p(X)_{\mathrm{alg}} = 0$ (see, in particular, the next lecture, Lecture IV).

From the above it will be clear that we have the following lemma:

LEMMA 9.3.16 $\mathrm{AJ}(\mathcal{Z}^p_{\mathrm{alg}}(X)) \subseteq J^p(X)_{\mathrm{alg}}$.

REMARK 9.3.17   To $J^p(X)_{\mathrm{alg}}$ corresponds a subgroup $W \subset H^{2p-1}(X, \mathbb{Z})$ which comes from the lattice which we have in the tangent space to $J^p(X)_{\mathrm{alg}}$; in fact, it is the counterimage in $H^{2p-1}(X, \mathbb{Z})$ of this lattice. This $W \subset H^{2p-1}(X, \mathbb{Z})$ is a so-called *sub-Hodge structure* of the Hodge structure

$$(H^{2p-1}(X, \mathbb{Z}), H^{2p-1}(X, \mathbb{C}) = \bigoplus H^{r,s}),$$

i.e., the Hodge structure of $H^{2p-1}(X, \mathbb{C})$ induces a Hodge structure on $W \otimes_{\mathbb{Z}} \mathbb{C}$.

9.3.4   Deligne Cohomology. Deligne Cycle Map

9.3.4.1   Deligne Cohomology

In this section, $X = X_d$ is a smooth, irreducible, quasi-projective variety defined over $\mathbb{C}$. We denote the associated analytic space X_{an} by the same letter (X_{an} is now a complex manifold, connected but not necessarily compact). Let Ω^i_X denote the *holomorphic* differential forms of degree i (so $\Omega^0_X = \mathcal{O}_{X_{\mathrm{an}}}$) and

$$\Omega^\bullet_X := 0 \to \Omega^0_X \xrightarrow{d} \Omega^1_X \xrightarrow{d} \Omega^2_X \to \cdots$$

is the *holomorphic de Rham complex*.

Recall that by the classical *holomorphic Poincaré lemma* ([13, p. 448]), $0 \to \mathbb{C} \to \Omega_X^\bullet$ is a resolution for $\mathbb{C}$.

Let $\Omega_X^{\bullet < n} := 0 \to \Omega_X^0 \to \Omega_X^1 \to \cdots \to \Omega_X^{n-1} \to 0$ be the truncated complex and $\Omega_X^{\le n}[-1]$ the complex shifted one place to the right. Furthermore, let $A \subset \mathbb{C}$ be a subring (usually $A = \mathbb{Z}, \mathbb{Q}$, or $\mathbb{R}$). Then Deligne considered the complex (with $A(n) = (2\pi i)^n A$)

$$A(n)_\mathcal{D}^\bullet : 0 \to A(n) \to \Omega_X^0 \to \Omega_X^1 \to \cdots \to \Omega_X^{n-1} \to 0 \to \cdots$$

in degrees 0 up to n; hence

$$A(n)_\mathcal{D}^\bullet : 0 \to A(n) \to \Omega_X^{\le n}[-1].$$

DEFINITION 9.3.18 (Deligne–Beilinson cohomology with coefficients in $A(n)$)

$$H_\mathcal{D}^i(X, A(n)) := \mathbb{H}^i(X, A(n)_\mathcal{D}),$$

where the $\mathbb{H}^i(X, \cdot)$ are the hypercohomology groups ([13, p. 445]).

DEFINITION 9.3.19 (Deligne–Beilinson cohomology with support in Y)   *More generally, if $Y \hookrightarrow X$ is a closed immersion of analytic manifolds,*

$$H_{Y,\mathcal{D}}^i(X, A(n)) := \mathbb{H}_Y^i(X, A(n)_\mathcal{D}).$$

EXAMPLE 9.3.20   For $n = 0$, $H_\mathcal{D}^i(X, \mathbb{Z}(0)) = H^i(X, \mathbb{Z})$. Let $n = 1$. The complex $\mathbb{Z}(1)_\mathcal{D}$ is *quasi-isomorphic* to the complex $\mathcal{O}_X^*[-1]$ via the map $z \mapsto \exp(z)$, where $(\mathcal{O}_X^*)^\bullet$ is the complex $(\mathcal{O}_X^*)^\bullet := 1 \to \mathcal{O}_X^* \to 1$. Indeed this follows from the commutative diagram below and the exactness of the following exponential sequence:

$$
\begin{array}{ccccccccc}
0 & \longrightarrow & (2\pi i)\mathbb{Z} & \longrightarrow & \mathcal{O}_X & \longrightarrow & 0 \\
& & \downarrow{\scriptstyle \exp} & & \downarrow{\scriptstyle \exp} & & \downarrow \\
1 & \longrightarrow & 1 & \longrightarrow & \mathcal{O}_X^* & \longrightarrow & 1
\end{array}
$$

Hence we get $H_\mathcal{D}^2(X, \mathbb{Z}(1)) \cong H^1(X_{\mathrm{an}}, \mathcal{O}_{X_{\mathrm{an}}}^*) = \mathrm{Pic}(X_{\mathrm{an}}) = \mathrm{Pic}(X_{\mathrm{alg}})$. Now recall the exact sequence

$$0 \to \mathrm{Pic}^0(X) \to \mathrm{Pic}(X) \to \mathrm{NS}(X) \to 0 \tag{9.3.1}$$

and $\mathrm{Pic}^0(X) = J^1(X)$.

EXAMPLE 9.3.21 ($n = p$, the general case! ($1 \le p \le d = \dim X$))   The following theorem shows that the Deligne cohomology gives the following beautiful generalization of the sequence (9.3.1):

THEOREM 9.3.22 (Deligne)   *There is an exact sequence*

$$0 \to J^p(X) \to H_\mathcal{D}^{2p}(X, \mathbb{Z}(p)) \to \mathrm{Hdg}^p(X) \to 0. \tag{9.3.2}$$

INDICATION OF THE PROOF. (See [27, p. 304].) The *exact sequence of complexes*

$$0 \to \Omega_X^{\bullet < p}[-1] \to \mathbb{Z}(p)_{\mathcal{D}} \to \mathbb{Z}(p) \to 0$$

gives a long exact sequence of (hyper)cohomology groups:

$$\cdots \to H^{2p-1}(X, \mathbb{Z}(p)) \xrightarrow{\alpha} \mathbb{H}^{2p}(X, \Omega_X^{\bullet < p}[-1]) \xrightarrow{\beta} H_{\mathcal{D}}^{2p}(X, \mathbb{Z}(p)) \xrightarrow{\lambda}$$

$$\xrightarrow{\lambda} H^{2p}(X, \mathbb{Z}(p)) \xrightarrow{\mu} \mathbb{H}^{2p+1}(X, \Omega_X^{\leq p}[-1]) \to \cdots.$$

So we want to see, first, that $\mathrm{Im}(\beta) \cong J^p(X)$, but this amounts essentially to seeing that

$$\mathbb{H}^{2p-1}(X, \Omega_X^{\bullet < p}) \cong H^{2p-1}(X, \mathbb{C})/F^p H(X, \mathbb{C}).$$

However this follows from the short exact sequence of complexes

$$0 \to \Omega_X^{\bullet \geq p} \to \Omega_X^{\bullet} \to \Omega_X^{\bullet < p} \to 0$$

together with the corresponding long exact sequence of hypercohomology groups plus the fact that $\mathbb{H}^{2p-1}(X, \Omega_X^{\bullet \geq p}) = F^p H^{2p-1}(X, \mathbb{C})$ (see [27, 12.3]).

Second, we need to see that $\ker(\mu) - \mathrm{Hdg}^p(X)$, but using the same facts as above we get that

$$\mu : H^{2p}(X, \mathbb{Z}(p)) \to H^{2p}(X, \mathbb{C})/F^p H^{2p}(X, \mathbb{C})$$

and from this we get $\ker(\mu) \cong \mathrm{Hdg}^p(X)$. $\qquad \square$

9.3.4.2 *Deligne Cycle Map*

We make assumptions as before, but now assume, moreover, that again X is *projective*, hence X_{an} is *compact*.

THEOREM 9.3.23 (Deligne) *There is a cycle map*

$$\gamma_{\mathcal{D}} : \mathcal{Z}^p(X) \longrightarrow H_{\mathcal{D}}^{2p}(X, \mathbb{Z}(p))$$

such that the following diagram is commutative:

$$
\begin{array}{ccccccccc}
0 & \longrightarrow & \mathcal{Z}_{\mathrm{hom}}^p(X) & \longrightarrow & \mathcal{Z}^p(X) & \longrightarrow & \mathcal{Z}^p(X)/\mathcal{Z}_{\mathrm{hom}}^p(X) & \longrightarrow & 0 \\
& & \downarrow{\scriptstyle \mathrm{AJ}} & & \downarrow{\scriptstyle \gamma_{\mathcal{D}}} & & \downarrow{\scriptstyle \gamma_{\mathbb{Z}}} & & \\
0 & \longrightarrow & J^p(X) & \longrightarrow & H_{\mathcal{D}}^{2p}(X, \mathbb{Z}(p)) & \longrightarrow & \mathrm{Hdg}^p(X) & \longrightarrow & 0.
\end{array}
$$

Moreover these vertical maps factor through $\mathrm{CH}^p(X)$.

ABOUT THE CONSTRUCTION AND PROOF. The construction and proof is very involved and goes beyond the scope of this lecture. We refer to [27, 12.3.3] or to the lectures of Green and Murre in [9]. One can also consult [7] or [6, Lect. 15]. $\qquad \square$

9.3.5   References for Lecture III

For the basics of complex algebraic geometry see the book of Griffiths and Harris [13]. The topics discussed in this lecture are all thoroughly treated in the books [27] and [28] of Voisin; these books are the English translation of the original French book [26]. For some of the topics one can also consult the relevant lectures by Green, Voisin, and the author in [9] which are CIME lectures held in Torino in 1993. The Deligne–Beilinson cohomology is treated in greater detail in [7].

9.4   LECTURE IV: ALGEBRAIC VERSUS HOMOLOGICAL EQUIVALENCE. GRIFFITHS GROUP

Recall that we have $\mathcal{Z}^i_{\mathrm{alg}}(X) \subseteq \mathcal{Z}^i_{\mathrm{hom}}(X)$. For divisors $(i = 1)$ Matsusaka proved that $\mathrm{Div}_{\mathrm{alg}}(X) \otimes \mathbb{Q} = \mathrm{Div}_{\mathrm{hom}}(X) \otimes \mathbb{Q}$ (see Remark 9.2.7(b)). Also for 0-cycles $(i = d = \dim X)$ we have $Z \in \mathcal{Z}^{\mathrm{alg}}_0(X)$ if and only if $\deg Z = 0$ if and only if $Z \in \mathcal{Z}^{\mathrm{hom}}_0(X)$.

However in 1969 Griffiths proved that there exist varieties X and $i > 1$ such that $\mathcal{Z}^i_{\mathrm{alg}}(X) \otimes \mathbb{Q} \neq \mathcal{Z}^i_{\mathrm{hom}}(X) \otimes \mathbb{Q}$, i.e., *for $i > 1$ essential new features* happen for $\mathrm{CH}^i(X)$: the theory of algebraic cycles of codimension greater than 1 is *very different* from the theory of divisors!

THEOREM 9.4.1 (Griffiths, 1969)   *There exist smooth, irreducible, projective varieties of dimension 3 (defined over $\mathbb{C}$) such that $\mathcal{Z}^2_{\mathrm{alg}}(X) \otimes \mathbb{Q} \neq \mathcal{Z}^2_{\mathrm{hom}}(X) \otimes \mathbb{Q}$.*

Therefore it is interesting to introduce the following group, nowadays called the Griffiths group:

DEFINITION 9.4.2 $\mathrm{Gr}^i(X) := \mathcal{Z}^i_{\mathrm{hom}}(X)/\mathcal{Z}^i_{\mathrm{alg}}(X)$.

So the above theorem implies the following corollary:

COROLLARY 9.4.3   *There exist smooth, irreducible, projective varieties X (defined over $\mathbb{C}$) and codimensions $i > 1$ such that the Griffiths group $\mathrm{Gr}^i(X)$ is not 0 and in fact $\mathrm{Gr}^i(X) \otimes \mathbb{Q} \neq 0$.*

Griffiths heavily uses results and methods of Lefschetz, so in order to discuss this theorem we have to make some preparations.

9.4.1   Lefschetz Theory

For simplicity and for the application made by Griffiths, we assume that *the base field is* $\mathbb{C}$ (although most of the results are also true for étale cohomology with $\mathbb{Q}_\ell$-coefficients, $\ell \neq \mathrm{char}(k)$, if $k = \bar{k}$).

THEOREM 9.4.4 (Lefschetz hyperplane section theorem)   *Let $V_{d+1} \subset \mathbb{P}^N$ be a smooth, irreducible variety and $W = V \cap H$ a smooth hyperplane section. Then the restriction map*

$$H^j(V, \mathbb{Z}) \longrightarrow H^j(W, \mathbb{Z})$$

is an isomorphism for $j < d = \dim W$ and injective for $j = d$.

PROOF. See [13, p. 156] or [28, 1.2.2]. □

REMARK 9.4.5

(a) This holds also if W is a *hypersurface* section of V (use the Veronese embedding).

(b) Special case: take $V = \mathbb{P}^{d+1}$ itself and $W \subset \mathbb{P}^{d+1}$ hypersurface. Then we get $H^j(W, \mathbb{Z}) = 0$ for odd $j < \dim W$ and $H^{2j}(W, \mathbb{Z}) = \mathbb{Z} \cdot h^j$ if $2j < \dim W$, where $h = \gamma_{\mathbb{Z}}(W \cap H)$, i.e., the class of the hyperplane section on W. Using Poincaré duality we get also $H^j(W, \mathbb{Z}) = 0$ for odd $j > \dim W$ and $H^{2j}(W, \mathbb{Z}) = \mathbb{Z}$ if $2j > \dim W$. So the only "interesting cohomology" is in $H^d(W, \mathbb{Z})$, i.e., the "middle cohomology."

(c) The same results are true, using Theorem 9.4.4, for the cohomology of *smooth, complete intersections* $W \subset \mathbb{P}^N$.

9.4.1.1 Hard (Strong) Lefschetz Theorem

Let $V_{d+1} \subset \mathbb{P}^N$ be smooth, irreducible. Let $W = V \cap H$ be a smooth hyperplane section. Let $h = \gamma_{\mathbb{Z},V}(W) \in H^2(V, \mathbb{Z})$. Then there is the so-called *Lefschetz operator*

$$L_V : H^j(V, \mathbb{Z}) \longrightarrow H^{j+2}(V, \mathbb{Z}),$$
$$\alpha \longmapsto h \cup \alpha.$$

By iterating it we get (writing $n = d + 1 = \dim V$)

$$L^r : H^{n-r}(V, \mathbb{Z}) \longrightarrow H^{n+r}(V, \mathbb{Z}) \quad (0 \leq r \leq n).$$

THEOREM 9.4.6 (Hard Lefschetz theorem) $L^r : H^{n-r}(V, \mathbb{Q}) \to H^{n+r}(V, \mathbb{Q})$ *is an isomorphism for all $r \leq n$ (note: $\mathbb{Q}$-coefficients!).*

For V defined over $\mathbb{C}$ this is proved by Hodge theory (see [27, 6.2.3] and [5.2] in Chapter 1). In arbitrary characteristic for $k = \bar{k}$ this holds also for $H^{\bullet}_{\mathrm{et}}(X, \mathbb{Q}_\ell)$ and it was proved in 1973 by Deligne at the same time as the Weil conjectures.

We mention also the following

DEFINITION 9.4.7 (Primitive cohomology)

$$H^{n-r}_{\mathrm{prim}}(V, \mathbb{Q}) := \ker(L^{r+1} : H^{n-r}(V, \mathbb{Q}) \to H^{n-r+2}(V, \mathbb{Q})).$$

Using the Lefschetz operator there is the so-called Lefschetz decomposition of the cohomology into primitive cohomology. Since we are not going to use this we refer only to [27, 6.2.3].

9.4.1.2 Pencils and Lefschetz Pencil

Let $V_n \subset \mathbb{P}^N$ be smooth, irreducible. Take two hyperplanes H_0 and H_1 in $\mathbb{P}^N$ and consider the *pencil* $H_t := H_0 + tH_1$ ($t \in \mathbb{C}$) or better $H_\lambda = \lambda_0 H_0 + \lambda_1 H_1$ ($\lambda = [\lambda_0 : \lambda_1] \in \mathbb{P}^1$), then by intersecting with V we get a *pencil of hyperplane sections* $\{W_\lambda = V \cap H_\lambda\}$ on V.

Now take H_0 and H_1 "sufficiently general." Then this pencil has the following properties (see [13, p. 509] or [28, Ch. 2]):

(a) There is a finite set S of points $t \in \mathbb{P}^1$ such that W_t is *smooth* outside S. Put $U := \mathbb{P}^1 - S$.

(b) For $s \in S$, the W_s has only *one singular point* x and this is an "ordinary double point" (that means that in a sufficiently small analytic neighborhood of x, the W_s is given analytically by a set of equations starting with transversal linear forms plus one more equation starting with a nondegenerate quadratic form; see [28, 2.1.1]).

Such a family of hyperplane sections is called a *Lefschetz pencil*. The *axis* of the pencil is $A := V \cap H_0 \cap H_1$. Consider the blow-up $\tilde{V}$ of V along the axis A, given by $\tilde{V} = \{(x,t) \in X \times \mathbb{P}^1; x \in W_t\}$ and let $f : \tilde{V} \to \mathbb{P}^1$ with $f(x,t) = t$. Then we have the following diagram with $W_t = f^{-1}(t)$:

$$
\begin{array}{ccccccc}
W := W_t & \overset{\iota_t}{\hookrightarrow} & \tilde{V}_{|U} & \hookrightarrow & \tilde{V} \cong B_A(V) & \longrightarrow & V \\
\downarrow & & \downarrow & & \downarrow & & \vdots \\
t & & U & \longrightarrow & \mathbb{P}^1 & = = & \mathbb{P}^1.
\end{array}
$$

Now let $t \in U$; then we have the Lefschetz theorem

$$\iota_t^* : H^j(V, \mathbb{Z}) \longrightarrow H^j(W_t, \mathbb{Z}),$$

which is an *isomorphism* for $j < \dim W_t$ and injective for $j = \dim W_t$.

9.4.1.3 Monodromy of Lefschetz Pencils

(See [28, Ch. 3].) Recall from above that $U = \{t \in \mathbb{P}^1 : W_t \subset V \text{smooth} \}$, $S = \mathbb{P}^1 - U$, and $S = \{s_1, \ldots, s_l\}$ finite set of points. Fix $t_0 \in U$ and take $W = W_{t_0}$. Consider $\pi_1(U) = \pi_1(U, t_0)$ the fundamental group of U with base point t_0. The $\pi_1(U)$ is generated by loops $\sigma_1, \ldots, \sigma_l$, where σ_i is a loop with origin t_0 and winding one time around s_i and the σ_i are not crossing with each other (there is one relation $\sigma_l \sigma_{l-1} \cdots \sigma_2 \sigma_1 = 1$). The group $\pi_1(U)$ operates on the $H^j(W, \mathbb{Q})$, but due to the Lefschetz theorems it acts trivially if $j \neq d = \dim W$ (because the $H^j(W)$ comes for $j \neq d$ from (and via) the cohomology of V). Consider the action

$$\rho : \pi_1(U) \twoheadrightarrow \Gamma := \mathrm{Im}(\rho) \subset \mathrm{Aut}(H^d(W_d, \mathbb{Q})).$$

Γ is called the *monodromy group*.

Let $\iota : W \hookrightarrow V$ and $\iota_* : H^j(W, \mathbb{Q}) \to H^{j+2}(V, \mathbb{Q})$, the induced morphism in co-homology. One defines $H^j(W, \mathbb{Q})_{\mathrm{van}} := \ker \iota_*$. Again due to the Lefschetz theorems, $H^j(W, \mathbb{Q})_{\mathrm{van}} = 0$ for $j \neq d$.

DEFINITION 9.4.8 $H^d(W_d, \mathbb{Q})_{\mathrm{van}} = \ker(\iota_* : H^d(W, \mathbb{Q}) \to H^{d+2}(V, \mathbb{Q}))$ *is called the vanishing cohomology.*[8]

We have for the Lefschetz operator on W that $L_W = \iota^* \circ \iota_*$ and therefore

$$H^d(W, \mathbb{Q})_{\mathrm{van}} \subseteq H^d(W, \mathbb{Q})_{\mathrm{prim}};$$

moreover there is the orthogonal decomposition (see [28, 2.3.3]):

$$H^d(W, \mathbb{Q}) = H^d(W, \mathbb{Q})_{\mathrm{van}} \oplus \iota^* H^d(V, \mathbb{Q}).$$

Now the following fact is fundamental (see [28, 3.2.3]):

THEOREM 9.4.9 (Lefschetz) *Let $\{W_t\} \subset V_{d+1}$ be a Lefschetz pencil. Then the vanishing cohomology $E = H^d(W, \mathbb{Q})_{\mathrm{van}}$ is an irreducible Γ-module (i.e., E is irreducible under the monodromy action. So, in particular, the action of Γ on $H^d(W, \mathbb{Q})$ is completely reducible, i.e., $H^d(W, \mathbb{Q})$ is a direct sum of irreducible Γ-modules and E is one of them).*

9.4.2 Return to the Griffiths Theorem

The key point in the Griffiths theorem that $\mathcal{Z}^i_{\mathrm{alg}}$ can be different from $\mathcal{Z}^i_{\mathrm{hom}}$ for $i > 1$ is the following:

THEOREM 9.4.10 (Griffiths, 1969) *Let $Y \subset \mathbb{P}^N$ be smooth, irreducible, and defined over $\mathbb{C}$. Let $\dim Y = 2m$, and assume $H^{2m-1}(Y, \mathbb{C}) = 0$. Let $\{X_t\}_{t \in \mathbb{P}^1}$ be a Lefschetz pencil on Y. Let $t \in \mathbb{P}^1$ be very general, i.e., $t \in \mathbb{P}^1 - B$, where B is a countable set of points on $\mathbb{P}^1$ containing, in particular, the points $s \in \mathbb{P}^1$ for which X_s is singular. Assume $H^{2m-1}(X_t) \neq H^{m,m-1}(X_t) \oplus H^{m,m-1}(X_t)$. Finally let $Z \in \mathcal{Z}^m(Y)$ and assume that for such t as above, $Z_t = Z \cdot X_t \in \mathcal{Z}^m_{\mathrm{alg}}(X_t)$. Then Z is homologically equivalent to 0 on Y, i.e., $Z \in \mathcal{Z}^m_{\mathrm{hom}}(Y)$.*

INDICATION OF THE MAIN POINTS IN THE PROOF. (See [21, Section 7.2].) Write in short $X = X_t$.

Step 1. $\mathrm{Im}\ \mathrm{AJ}(\mathcal{Z}^m_{\mathrm{alg}}(X)) = 0$ in $J^m(X)$.

PROOF. Consider the action of the monodromy group Γ on $H^{2m-1}(X, \mathbb{Q})$ (note $\dim X = 2m - 1$). Because of our assumptions,

$$H^{2m-1}(X, \mathbb{Q}) = H^{2m-1}(X, \mathbb{Q})_{\mathrm{van}} + \iota^* H^{2m-1}(Y, \mathbb{Q}) = E \oplus 0 = E$$

---

[8]Because it is the subvector space of $H^d(W, \mathbb{Q})$ generated by the so-called *vanishing cycles* ([28, 2.3.3]), but we do not need to discuss this.

with $E = H^{2m-1}(X, \mathbb{Q})_{\text{van}}$. Therefore by the fundamental Theorem 9.4.9, then $H^{2m-1}(X, \mathbb{Q}) = E$ is an irreducible Γ-module. On the other hand, $\text{AJ}(\mathscr{Z}^m_{\text{alg}}(X))$ determines, via its tangent space, also an irreducible Γ-module, say $H' \subset H^{2m-1}(X, \mathbb{Q}) = E$. However $H' \subset H^{m,m-1} + H^{m-1,m}$ as we have seen above (because $\text{AJ}(\mathscr{Z}^m_{\text{alg}}(X)) \subset J^m_{\text{alg}}(X)$), and since by assumption $H^{m,m-1} + H^{m-1,m} \neq H^{2m-1}(X, \mathbb{Q})$, we have $H' = 0$; hence the image of $\text{AJ}(\mathscr{Z}^m_{\text{alg}}(X))$ equals 0. $\square$

Step 2. Put $U_1 = \mathbb{P}^1 - S$, where S is the finite set of points s, where X_s is singular, and consider the family $J^m(X_t), t \in U_1$. These intermediate Jacobians fit together to give a fiber space

$$J^m(X/U_1) := \bigcup_{t \in U_1} J^m(X_t)$$

of complex analytic tori. For each $t \in U_1$ we have an element $\text{AJ}(Z_t) \in J^m(X_t)$, where $Z_t = Z \cdot X_t \in \mathscr{Z}^m_{\text{alg}}(X_t)$. These elements fit together to give a holomorphic function

$$\nu_Z : U_1 \longrightarrow J^m(X/U_1)$$

(see [28, Thm. 7.9]). This function is a so-called *normal function*. However in our case $\nu_Z(t) = 0$ for $t \in U \subset U_1$; U is dense in U_1, hence $\nu_Z = 0$.

Step 3. $\nu_Z(t) = 0$ implies $Z \in \mathscr{Z}^m_{\text{hom}}(Y)$. The proof depends on an infinitesimal study of normal functions which goes beyond the scope of our lectures. We refer therefore (for instance) to [21, Nagel lecture, Sec. 7.2] or to [9, Voisin, Lect. 7]. $\square$

9.4.2.1   Application (Griffiths)

Let us prove the Griffiths theorem, that there exist varieties X such that for certain $i > 1$, $\mathscr{Z}^i_{\text{alg}}(X) \neq \mathscr{Z}^i_{\text{hom}}(X)$.

First of all we are going to construct a Lefschetz pencil that satisfies the conditions of the above Theorem 9.4.10.

Let $V_4 = V(2) \subset \mathbb{P}^5$ be a smooth *quadric* hypersurface in $\mathbb{P}^5$. On such a quadric we have two families of planes $\{P\}$ and $\{P'\}$ and $H^4(V(2), \mathbb{Q}) = \mathbb{Q} \cdot P \oplus \mathbb{Q} \cdot P'$ with the linear section $V \cdot H_1 \cdot H_2 = P + P'$ ($H_i \subset \mathbb{P}^5$ hyperplanes) and $Z = P - P'$. Now note that the intersection number $\sharp(Z \cdot Z) = -2$, hence $Z \in \mathscr{Z}^2(V)$ is *not* homologically equivalent to 0 (see for these facts, for instance, [28, Ch. 2, Exercices]). Take this $V(2)$ as Y in the theorem. Note $H^3(V, \mathbb{Q}) = 0$ by the Lefschetz theorem on hypersurface sections.

Next take the Lefschetz pencil $\{X_t = F_t \cdot V\}$, where F_t are hypersurfaces of a certain degree r to be specified later (use the Veronese embedding of $\mathbb{P}^5$ to make them hyperplanes). Now take $X = X_t$ *very general* (in the sense explained in the Theorem 9.4.10) and consider in $H^3(X, \mathbb{C})$ the subspace $H^{3,0}(X) = H^0(X, \Omega^3_X)$. Now Ω^3_X corresponds to the linear system $\mathcal{O}_X(K_X)$ with K_X the canonical class and $K_X = \mathcal{O}_X(-4 + r)$, because by the adjunction formula we have $K_V = (K_{\mathbb{P}^5} + V) \cdot V = \mathcal{O}_V(-4)$ and $K_X = (K_V + X) \cdot X = \mathcal{O}_X(-4 + r)$. Therefore, for $r \geq 5$, then $H^{3,0}(X) \neq 0$ and $H^3(X, \mathbb{C}) \neq H^{2,1} + H^{1,2}$, as required in the theorem.

Now consider $Z_t = Z \cdot X_t = (P - P') \cdot X_t = C_t - C'_t$ in $\mathcal{Z}^2(X_t)$, where C_t and C'_t are curves on X_t. We have $Z_t = C_t - C'_t$ homologically equivalent to 0 on X_t because $\sharp H \cdot (C_t - C'_t) = 0$ for a hyperplane H in $\mathbb{P}^5$, and so the cohomology class $\gamma_{Z,X_t}(Z_t) = 0$ in $H^4(X_t, \mathbb{Q}) = H^2(X_t^*, \mathbb{Q})^*$ since $H^2(X_t, \mathbb{Q}) = \mathbb{Q}(H \cdot X_t) \cong \mathbb{Q}$, so $Z_t \in \mathcal{Z}^2_{\mathrm{hom}}(X_t)$. However it follows from the theorem that $Z_t \notin \mathcal{Z}^2_{\mathrm{alg}}(X_t)$ because that would imply $Z \in \mathcal{Z}^2_{\mathrm{hom}}(Y) = \mathcal{Z}^2_{\mathrm{hom}}(V(2))$, which is not the case as we have seen.

In fact the above argument works for nZ for all $n > 0$. This proves the Griffiths theorem and we have for such $X = X_t$ an element in $\mathrm{Gr}^2(X_t)$ which is nonzero in $\mathrm{Gr}(X) \otimes \mathbb{Q}$.

REMARK 9.4.11   In fact this gives a nontorsion element $\mathrm{AJ}(Z_t)$ in $J^2(X)$. We have seen that the proof heavily uses the fact that $J^2(X) \neq 0$ and the theory of normal functions.

9.4.2.2 Further Facts

$\mathrm{Gr}^i(X)$ is, for all i and all X, a *countable group*. This follows from the existence of the so-called *Hilbert schemes* or, less technically, from the existence of the so-called *Chow varieties*.

"Recall" that there is the following fact (see [4]): given $X \subset \mathbb{P}^N$ the algebraic cycles on X of a fixed dimension q and a fixed degree r are *parametrized* by an *algebraic variety* $\mathrm{Ch}(X, q, r)$, a so called Chow variety (not necessarily connected!).

Since two such cycles Z and Z', which are in the same connected component of $\mathrm{Ch}(X, q, r)$, are algebraically equivalent, the $\mathrm{Ch}(X, q, r)$ itself gives only a finite number of generators to $\mathrm{Gr}^i(X)$.[9]

Further Examples (mentioned only, without proofs).

(a)

> **THEOREM 9.4.12** (Griffiths, 1969)   *Consider a very general $X_3 = V(5) \subset \mathbb{P}^4$, i.e., a very general quintic hypersurface of dimension 3. Such a quintic threefold contains a finite number of lines $\{l\}$. Let $Z = l_1 - l_2$; then $\mathrm{AJ}(Z)$ is not a torsion point in $J^2(X)$, Z is homologically equivalent to 0 but not algebraically equivalent to 0. This type of Z therefore gives nonzero elements in $\mathrm{Gr}^2(X) \otimes \mathbb{Q}$ (see [27, 12.2.2]).*

(b) The above result was improved in 1983 by Clemens:

> **THEOREM 9.4.13** (Clemens, 1983)   *For such a very general quintic hypersurface in $\mathbb{P}^4$ we have $\dim_{\mathbb{Q}} \mathrm{Gr}^2(X) \otimes \mathbb{Q} = \infty$.*

This result was extended in 2000 by Voisin to very general Calabi–Yau threefolds.

---

[9]But, of course, we must take into account all r's.

(c) *Ceresa cycle*

> **THEOREM 9.4.14** (Ceresa, 1983)   *For a very general curve C of genus $g \geq 3$, the cycle $Z = C - C^-$ in $J(C)$ is not algebraically equivalent to 0 (but it is homologically equivalent to 0). Here C^- is the image of C under the map $x \mapsto -x$ on $J(C)$.*

This also gives an example of a cycle Z such that $Z \in \mathcal{Z}_1^{\otimes}(J(C))$, but $Z \notin \mathcal{Z}_1^{\mathrm{alg}}(J(C))$ because Kahn and Sebastian proved in 2009 that on abelian varieties A of dimension 3, then $\mathcal{Z}_{\otimes}^2(A) = \mathcal{Z}_{\mathrm{hom}}^2(A)$ (Voevodsky's conjecture!).

(d) All the previous examples were on "very general varieties," i.e., varieties defined over "large" fields. However there are also examples over number fields. To my knowledge the first such example is due to Harris. He proved in 1983 that the Ceresa cycle $C - C^-$ is not algebraically equivalent to 0 on the Jacobian $J(C)$ where C is the *Fermat curve* $X^4 + Y^4 = 1$; hence an example over $\mathbb{Q}$!

(e) All the above examples are in general "in the spirit of Griffiths," i.e., they use the (intermediate) Jacobian. However Nori proved in 1993 that there exist varieties X such that $\mathrm{Gr}^i(X) \otimes \mathbb{Q} \neq 0$ for $i > 2$ but for which $J^i(X) = 0$. See [28, Ch. 8] or [21].

9.4.3   References for Lecture IV

For the results of Griffiths and related issues see Voisin's books [27, 12.2.2] and [28, Ch. 3] or also [9, Voisin, Lect. 7]. See also [21, Nagel lecture]. For the original paper of Griffiths see [10].

9.5   LECTURE V: THE ALBANESE KERNEL. RESULTS OF MUMFORD, BLOCH, AND BLOCH–SRINIVAS

In 1969 Mumford proved a theorem which shows another important difference between the theory of divisors and the theory of algebraic cycles of larger codimension; this time it concerns 0-cycles.

   In this lecture, k is (again) an algebraically closed field of arbitrary characteristic (unless stated explicitly otherwise) and the varieties are smooth, irreducible, projective and defined over k.

9.5.1   The Result of Mumford

Let $X = X_d$ be such a variety. Recall from Lecture II that

$$\mathrm{CH}_{\mathrm{alg}}^1(X) \xrightarrow{\sim} (\mathrm{Pic}^0(X))_{red}$$

with $(\mathrm{Pic}^0(X))_{red}$ an abelian variety, the *Picard variety* of X. If X is a curve C, then this is the Jacobian variety $J(C)$ of C.

For 0-cycles $\mathrm{CH}^d_{\mathrm{alg}}(X) = \mathrm{CH}^{\mathrm{alg}}_0(X)$ is the group of rational equivalence classes of 0-cycles of degree 0 and we have a homomorphism (coming from the Albanese map)

$$\alpha_X : \mathrm{CH}^d_{\mathrm{alg}}(X) \twoheadrightarrow \mathrm{Alb}(X),$$

where $\mathrm{Alb}(X)$ is an abelian variety, the *Albanese variety* of X. If X is a curve C, then $\mathrm{Alb}(X) = J(X)$ and α_X is the Abel–Jacobi map. The map α_X is surjective but in general *not injective* (on the contrary, see further on)! Put

$$T(X) := \mathrm{CH}^d_{alb}(X) := \ker(\alpha_X),$$

the *Albanese kernel*.

THEOREM 9.5.1 (Mumford, 1969) *Let S be an algebraic surface (smooth, projective, irreducible) defined over $\mathbb{C}$. Let $p_g(S) := \dim H^0(S, \Omega^2_S)$ (geometric genus of S). If $p_g(S) \neq 0$, then $T(S) \neq 0$ and in fact $T(S)$ is "infinite-dimensional."*

Below, we shall make this notion more precise but it implies that the "size" of $T(S)$ is so large that it can *not* be parametrized by an algebraic variety.

Write, in short, $\mathrm{CH}_0(S)_0 := \mathrm{CH}^{\mathrm{alg}}_0(S)$.

It is somewhat more convenient to formulate things in terms of $\mathrm{CH}_0(S)_0$ itself but remember that the "sizes" of $T(S)$ and $\mathrm{CH}_0(S)_0$ only differ by a finite number, namely, the dimension of $\mathrm{Alb}(S)$ (which itself is half the dimension of $H^{2d-1}(S)$).

Consider more generally $X = X_d$ and its n-fold *symmetric product* $X^{(n)}$ (the quotient of the n-fold product $X \times X \times \cdots \times X$ by the symmetric group; $X^{(n)}$ has mild singularities!). Clearly $X^{(n)}$ parametrizes the 0-cycles of degree n and we have a map

$$\varphi_n : X^{(n)} \times X^{(n)} \longrightarrow \mathrm{CH}_0(X)_0,$$
$$(Z_1, Z_2) \longmapsto \text{class of } (Z_1 - Z_2).$$

Mumford calls $\mathrm{CH}_0(X)_0$ *finite-dimensional* if there exists an n such that φ_n is surjective, otherwise *infinite-dimensional*.

REMARK 9.5.2 One can elaborate further on this concept as follows. There is the following fact (see for instance [28, 10.1]): the fibers of φ_n consist of a countable number of algebraic varieties (this is proved via the existence of the Chow varieties or the Hilbert schemes). Since each of these subvarieties has bounded dimension (at most $2dn$) it makes sense to take the maximum and call that the dimension of the fiber. Let r_n be the dimension of the generic fiber and consider $2dn - r_n$; this can be considered as the "dimension" of $\mathrm{Im}(\varphi_n)$. So intuitively "$\dim \mathrm{CH}_0(X)_0 = \lim_{\longrightarrow_n} (2dn - r_n)$." Now there is the following fact (see [28, 10.10]): $\mathrm{CH}_0(X)_0$ is finite-dimensional if and only if $\lim_{\longrightarrow_n} (2dn - r_n)$ is finite.

We mention (in passing) the following beautiful *theorem of Roitman* (see [28, Prop. 10.11]):

THEOREM 9.5.3 (Roitman, 1972)   *If* $CH_0(X)_0$ *is finite-dimensional then the Albanese morphism* $\alpha_X : CH_0(X)_0 \twoheadrightarrow \mathrm{Alb}(X)$ *is an isomorphism.*

PROOF. We refer to [28, 10.1.2]. $\square$

In this lecture we want to reduce the proof of the Mumford's theorem to the method used by Bloch (see below). For that we need one further result for which we must refer to Voisin's book ([28, 10.12]).

LEMMA 9.5.4   *The following properties are equivalent:*

(a) $CH_0(X)_0$ *is finite-dimensional.*

(b) *If* $C = X \cap F_1 \cap \cdots \cap F_{d-1} \overset{j}{\hookrightarrow} X$ *is a smooth curve cut out on X by hypersurfaces* F_i *then the induced homomorphism* $j_* : J(C) = CH^0(C)_0 \to CH^0(X)_0$ *is surjective.*

REMARK 9.5.5   For a very precise list of properties equivalent to finite dimensionality one can look to Jannsen [18, Prop. 1.6].

9.5.2   Reformulation and Generalization by Bloch

9.5.2.1   Reformulation of Finite Dimensionality

Bloch introduced in 1976 the notion of "*weak representability.*"

Let $\Omega \supset k$ be a so-called "*universal domain*," i.e., Ω is an algebraically closed field of infinite transcendence degree over k; hence every $L \supset k$ of finite transcendence degree over k can be embedded in Ω, i.e., $k \subset L \subset \Omega$ (for example, if $k = \bar{\mathbb{Q}}$ then one can take $\Omega = \mathbb{C}$).

DEFINITION 9.5.6   *Let X be defined over k.* $CH^j_{\mathrm{alg}}(X)$ *is weakly representable if there exists a curve C smooth, but not necessarily irreducible, and a cycle class* $T \in CH^j(C \times X)$ *such that the corresponding homomorphism* $T_* : CH_0^{\mathrm{alg}}(C_L) \to CH^j_{\mathrm{alg}}(X_L)$ *is surjective for all L with* $k \subset L \subset \Omega$ *and* $L = \bar{L}$.

REMARK 9.5.7

(a) If $T(S) = 0$ then it is easy to see that $CH_0(S)_0$ is weakly representable. Namely, $T(S) = 0$ gives $CH_0(S)_0 \overset{\sim}{\to} \mathrm{Alb}(S)$ and by taking a sufficiently general curve C on S we get a surjective map $J(C) \twoheadrightarrow \mathrm{Alb}(S)$.

(b) Assume tacitly that we have chosen on each connected component of C a "base point" e such that T_* is defined on this component as $T(x) - T(e)$ for $x \in C$.

(c) If follows (easily) from Lemma 9.5.4 that $CH_0(X)_0$ is finite-dimensional if and only if it is weakly representable.

(d) Often this notion is denoted by "representability," however it seems better to use the name "weak representability" to distinguish it from the much stronger concept of representability in the sense of Grothendieck.

9.5.2.2   Transcendental Cohomology

Let again $X = X_d$ be as usual and assume we have chosen a Weil cohomology theory with coefficient field $F \supset \mathbb{Q}$. Consider the cycle map

$$\mathrm{NS}(X) \otimes_{\mathbb{Q}} F \longrightarrow H^2(X)$$

(recall the definition $\mathrm{NS}(X) = \mathrm{CH}^1(X)/\mathrm{CH}^1_{\mathrm{alg}}(X)$). Put $H^2(X)_{\mathrm{alg}}$ for the image and $H^2(X)_{tr} := H^2(X)/H^2(X)_{\mathrm{alg}}$; $H^2(X)_{tr}$ is called the group of *transcendental* (cohomology) cycles of degree 2.

So if $X = S$ a surface, then we have via Poincaré duality an orthogonal decomposition

$$H^2(S) = H^2(S)_{\mathrm{alg}} \oplus H^2(S)_{tr}.$$

9.5.2.3   A Theorem by Bloch

THEOREM 9.5.8 (Bloch, 1979) *(See [2, p. I.24].) Let S be an algebraic surface defined over $k = \bar{k}$. Assume that $H^2(S)_{tr} \neq 0$; then $\mathrm{CH}^2_{\mathrm{alg}}(S)$ is not weakly representable.*

The proof will be given in Section 9.5.3. Bloch's theorem implies Mumford's theorem. Namely, let $k = \mathbb{C}$ and $p_g(S) \neq 0$. Now $p_g(S) = \dim H^0(S, \Omega^2) = \dim H^{2,0}(S)$ and since $H^2(S)_{\mathrm{alg}} \subset H^{1,1}(S)$ we have $H^2(S)_{tr} \supset H^{2,0}(S)$. Therefore $p_g(S) \neq 0$ implies $H^2(S)_{tr} \neq 0$, hence by Theorem 9.5.8, $\mathrm{CH}^2_{\mathrm{alg}}(S)$ is not weakly representable, hence $\mathrm{CH}^2_{\mathrm{alg}}(S)$ is not finite-dimensional by (c) in Remark 9.5.7.

9.5.3   A Result on the Diagonal

First, as a matter of notation, if $Z \in Z^i(X)$ let us write $|Z| \subset X$ for the support of Z.

THEOREM 9.5.9 (Bloch, 1979; Bloch–Srinivas, 1983) *Let $X = X_d$ be smooth, irreducible, projective, and defined over k. Let Ω be a "universal domain" as before. Assume that there exists (over k) a closed algebraic subset $Y \subsetneq X$ such that for $U = X - Y$ we have $\mathrm{CH}_0(U_\Omega) = 0$. Then there exist d-dimensional cycles Γ_1 and Γ_2 with supports $|\Gamma_1| \subset X \times Y$ and $|\Gamma_2| \subset W \times X$, where $W \subsetneq X$ is a closed algebraic subset (defined over k) and an integer $N > 0$ such that*

$$N \cdot \Delta(X) = \Gamma_1 + \Gamma_2,$$

where $\Delta(X) \subset X \times X$ is the diagonal.

PROOF. Take the generic point η of X. Consider the 0-cycle (η) in $\mathrm{CH}_0(X_L)$, where $L = k(\eta)$; in fact, $(\eta) \in \mathrm{CH}_0(U_L)$.

LEMMA 9.5.10 $\ker(\mathrm{CH}_0(U_L) \to \mathrm{CH}_0(U_\Omega))$ *is torsion.*

PROOF. We proceed in three steps: $L \subset L' \subset \bar{L} \subset \Omega$ with $[L' : L]$ finite. By the first step, $L' \supset L$ the kernel is finite because of the existence of a norm map $\mathrm{CH}_0(U_{L'}) \to \mathrm{CH}_0(U_L)$, and the composition $\mathrm{CH}(U_L) \to \mathrm{CH}(U_{L'}) \to \mathrm{CH}(U_L)$ is $[L' : L]\,\mathrm{id}$. Next, the step $L \subset \bar{L}$ is also torsion since $\mathrm{CH}_0(U_{\bar{L}}) = \varinjlim \mathrm{CH}_0(U_{L'})$. Finally, the step $\mathrm{CH}_0(U_{\bar{L}}) \to \mathrm{CH}_0(U_\Omega)$ is an isomorphism because $\Omega = \varinjlim R$, where R are finitely generated $\bar{L}$-algebras and $\mathrm{CH}(U_{\bar{L}}) \to \mathrm{CH}(U \times_{\bar{L}} \mathrm{Spec}\, R)$ is an isomorphism because we get a section by taking an $\bar{L}$-rational point in $\mathrm{Spec}(R)$. This proves the lemma. $\square$

Returning to the 0-cycle $(\eta) \in \mathrm{CH}_0(U_L)$, our assumption gives that $\mathrm{CH}_0(U_\Omega) = 0$ and from the lemma it follows that there exists $N > 0$ such that $N \cdot (\eta) = 0$ in $\mathrm{CH}_0(U_L)$. Now we apply to X_L the localization Theorem 9.1.20 and we see that in $\mathrm{CH}_0(Y_L)$ there exists a 0-cycle A such that in $\mathrm{CH}_0(X_L)$, the 0-cycle

$$N \cdot (\eta) - A = 0.$$

REMARK 9.5.11 Strictly speaking we assumed in Lecture I that the base field is algebraically closed; however the localization sequence holds for any base field ([8, Prop. 1.8]) so we can apply it also in our case to $L = k(\eta)$.

Next we consider X_L as the fiber in $X \times X$ over the point η (η in the first factor, X_η in the second) and take in $X \times X$ the k-Zariski closure of $N \cdot (\eta) - A$. Then we get a d-dimensional cycle $(N \cdot \Delta(X) - \Gamma_1)$ on $X \times X$, where $\Gamma_1 \in \mathcal{Z}_d(X \times X)$ and $|\Gamma_1| \subset X \times Y$, Γ_1 restricted to $X_{k(\eta)}$ is A and the restriction of the cycle $N\Delta(X) - A$ to $\mathrm{CH}_0(X_{k(\eta)})$ is 0. However,

$$\mathrm{CH}_0(X_{k(\eta)}) = \varinjlim_{D} \mathrm{CH}_0((X - D) \times X),$$

where the limit runs over divisors $D \subset X$. Therefore there exists a divisor, say $D = W$, and a cycle $\Gamma_2 \in \mathcal{Z}_d(X \times X)$ with $|\Gamma_2| \subset W \times X$ such that $N \cdot \Delta(X) = \Gamma_1 + \Gamma_2$. This completes the proof. $\square$

9.5.3.1 End of the Proof of Mumford's Theorem

Claim. Theorem 9.5.9 implies Theorem 9.5.8 (and we have already seen in Section 9.5.2.3 that Theorem 9.5.8 implies Mumford's theorem).

PROOF OF THE CLAIM. Let $X = S$ be a surface such that $H^2(S)_{tr} \neq 0$. We proceed by contradiction. If $\mathrm{CH}^2_{\mathrm{alg}}(S) = \mathrm{CH}_0(S)_0$ is weakly representable, then there exists a curve C and $T \in \mathrm{CH}^2(C \times S)$ such that $\mathrm{CH}^{\mathrm{alg}}_0(C_L) \to \mathrm{CH}^{\mathrm{alg}}_0(S_L)$ is surjective for all $k \subset L \subset \Omega$, $L = \bar{L}$. Therefore if we take $Y = \mathrm{pr}_2 T$ then the condition of the theorem of Bloch–Srinivas is satisfied. Now we have Lemma 9.5.12 below and clearly this proves the claim because $N \cdot \Delta(X)$ operates nontrivially on $H^2(S)_{tr}$, since clearly it operates as multiplication by N. $\square$

LEMMA 9.5.12 Both correspondences Γ_1 and Γ_2 operate trivially on $H^2(S)_{tr}$.

PROOF. Let us start with Γ_2. Now $|\Gamma_2| \subset C \times S$, where $\iota : C \hookrightarrow S$ is a curve. We have a commutative diagram

$$
\begin{array}{ccc}
C \times S & \xrightarrow{\iota_1} & S \times S \\
{\scriptstyle q = \mathrm{pr}_1}\downarrow & & \downarrow{\scriptstyle \mathrm{pr}_1 = p} \\
C & \xrightarrow{\iota} & S,
\end{array}
$$

where $\iota_1 = \iota \times \mathrm{id}_S$. Now let $\alpha \in H^2(S)_{tr}$. We have $\Gamma_2(\alpha) = (\mathrm{pr}_2)_* \{p^*(\alpha) \cup (\iota_1)_*(\Gamma_2)\}$. By the projection formula, $p^*(\alpha) \cup (\iota_1)_*(\Gamma_2) = (\iota_1)_* \{\iota_1^* p^*(\alpha) \cup \Gamma_2\}$; now $\iota_1^* p^*(\alpha) = q^* \iota^*(\alpha)$, but $\iota^*(\alpha) = 0$ because $\alpha \in H^2(S)_{tr}$ is orthogonal to the class of C in $H^2(S)$. Therefore $\Gamma_2(\alpha) = 0$.

Next Γ_1: we have $|\Gamma_1| \subset S \times C'$, where again $\iota' : C' \to S$ is a curve. We have a commutative diagram

$$
\begin{array}{ccc}
S \times C' & \xrightarrow{\iota_2} & S \times S \\
{\scriptstyle q' = \mathrm{pr}_2}\downarrow & & \downarrow{\scriptstyle \mathrm{pr}_2 = p'} \\
C' & \xrightarrow{\iota'} & S,
\end{array}
$$

where $\iota_2 = \mathrm{id}_S \times \iota'$. For $\alpha \in H^2(S)_{tr}$ we get $\Gamma_1(\alpha) = (\mathrm{pr}_2)_* \{(\iota_2)_*(\Gamma_1) \cup \mathrm{pr}_1^*(\alpha)\}$, by the projection formula $(\iota_2)_*(\Gamma_1) \cup \mathrm{pr}_1^*(\alpha) = (\iota_2)_* \{\Gamma_1 \cup \iota_2^* \mathrm{pr}_1^*(\alpha)\}$. Writing $\beta = \Gamma_1 \cup \iota_2^* \mathrm{pr}_1^*(\alpha)$, we get $\Gamma_1(\alpha) = (\mathrm{pr}_2)_*(\iota_2)_*(\beta) = (\iota')_* q_*'(\beta) \in H^2(S)_{\mathrm{alg}} = \mathrm{NS}(S)$, therefore the image in $H^2(S)_{tr}$ is 0. $\square$

9.5.3.2 Bloch's Conjecture

Let S be a surface as before defined over $k = \bar{k}$. Assume $H^2(S)_{tr} = 0$. Then Bloch conjectures that the Albanese kernel $T(S) = 0$, i.e., that the Albanese map $\alpha_S : \mathrm{CH}^2_{\mathrm{alg}}(S) \to \mathrm{Alb}(S)$ is an isomorphism, which implies that $\mathrm{CH}^2_{\mathrm{alg}}(S)$ is "finite-dimensional."

Bloch's conjecture has been proved for all surfaces which are *not* of "general type" [28, 11.10], i.e., surfaces for which the so-called Kodaira dimension is less than 2. For surfaces of general type it has only been proved in special cases, for instance for the so-called Godeaux surfaces.

Bloch's conjecture is one of the most important conjectures in the theory of algebraic cycles.

9.5.4 References for Lecture V

For Mumford's theorem, see [28, Ch. 10]. For Bloch's theorem see Bloch's book [2] on his Duke lectures held in 1979; there is now a second edition which appeared with Cambridge University Press. For the Bloch–Srinivas theorem and its consequences see the original paper [3]. One can find this material also in of [28, Chs. 10 and 11].

Bibliography

[1] Yves André. *Une introduction aux motifs (motifs purs, motifs mixtes, périodes)*, volume 17 of *Panoramas et Synthèses*. Société Mathématique de France, Paris, 2004.

[2] Spencer Bloch. *Lectures on algebraic cycles*, volume IV of Duke Univ. Math. Series. Cambridge University Press, 2010. Second edition.

[3] Spencer Bloch and Vasudevan Srinivas. Remarks on correspondences and algebraic cycles. *Amer. J. of Math*, 105:1235–1253, 1983.

[4] Wei-Liang Chow and Bartel van der Waerden. *Math. Ann.*, 113:692–704, 1937.

[5] David Eisenbud. *Commutative algebra*, volume 150 of *Graduate Text in Mathematics*. Springer, 1993.

[6] Fouad El Zein and Steven Zucker. *Topics in transcendental algebraic geometry* (ed. Griffiths). Ann. of Math. Studies **106**.

[7] Hélène Esnault and Eckart Viehweg. *Deligne–Beilinson cohomology*. In "Beilinson's Conjectures on Special Values of L-Functions" (eds. Rapoport, Schappacher, and Schneider). Perspectives in Math. Volume 4, Academic Press (1988), p. 43–91.

[8] William Fulton. *Intersection theory*. Ergebnisse der Mathematik und ihrer Grenzgebiete, 3 Folge, Band 2, Springer, 1984.

[9] Mark Green, Jacob P. Murre, and Claire Voisin. *Algebraic cycles and Hodge theory*. Lecture Notes in Mathematics. **1594**, Springer, 1994.

[10] Phillip Griffiths. On the periods of certain rational integrals, I and II. *Ann. of Math.*, 90:460–541, 1969.

[11] Phillip Griffiths. Some results on algebraic cycles on algebraic manifolds. In *Algebraic Geometry* (Internat. Colloq., Bombay, 1968), 93–191. Oxford University Press, Bombay, 1969.

[12] Phillip Griffiths. A theorem concerning the differential equations satisfied by normal functions associated to algebraic cycles. *Amer. J. of Math*, 101:94–131, 1979.

[13] Phillip Griffiths and Joseph Harris. *Principles of algebraic geometry*. J. Wiley and Sons, New York, 1978.

[14] Robin Hartshorne. *Algebraic geometry*, volume 52 of *Graduate Texts in Mathematics*. Springer, 1977.

[15] Peter J. Hilton and Urs Stammbach. *A course in homological algebra*, volume 4 of *Graduate Texts in Mathematics*. Springer, 1971.

[16] William Hodge. *Harmonic integrals*. Cambridge University Press, 1941.

[17] Uwe Jannsen. Equivalence relations on algebraic cycles. "The arithmetic and geometry of algebraic cycles", (eds. B. Gordon et al.). Banff Conference 1998, Kluwer Academic.

[18] Uwe Jannsen. *Motive sheaves and fibration on Chow groups*. In "Motives" (eds. S. Kleiman et al.). Proc. of Symp. in Pure Math, volume 55, part I, AMS 1994, p. 245–302.

[19] Steven L. Kleiman. *Algebraic cycles and the Weil conjectures*. In "Dix exposés sur la cohomologie des schémas" (J. Giraud et al.). North-Holland, 1968.

[20] James S. Milne. *Étale cohomology*. Princeton Math. Series 33, 1980.

[21] Jan Nagel. *Lectures on Nori's connectivity theorem*. In "Transcendental Aspects of Algebraic Cycles", Proc. Grenoble Summer School 2001, (eds. S. Müller-Stach and C. Peters). London Mathematical Society Lecture Notes 313, Cambridge University Press, 2004.

[22] Chris A. M. Peters and Joseph H. M. Steenbrink. *Mixed Hodge structures*, volume 52 of *Ergeb. der Mathematik und ihrer Grenzgebiete*. Springer, 2008.

[23] Joel Roberts. *Appendix 2*. In "Algebraic Geometry", Oslo 1970 (eds. F. Oort). Groningen: Wolters-Noordhoff, 1972, p. 89–96.

[24] Pierre Samuel. Rélations d'équivalence en géométrie algébrique. In *Proceedings of the International Congress of Mathematicians 1958*, Edinburgh, p. 470–487.

[25] Jean-Pierre Serre. *Algèbre locale. Multiplicités*. Cours Collège de France, 1957–1958, rédigé par Pierre Gabriel. Seconde édition, Lecture Notes in Mathematics 11, Springer, 1965.

[26] Claire Voisin. *Théorie de Hodge et géométrie algébrique complexe*, volume 710 of *Cours spécialisés*. Société Mathématique de France, Paris, 2002.

[27] Claire Voisin. *Hodge theory and complex algebraic geometry I*, volume 76 of *Cambridge Studies in Advanced Mathematics*. Cambridge University Press, Cambridge, 2007. Translated from the French by Leila Schneps.

[28] Claire Voisin. *Hodge theory and complex algebraic geometry II*, volume 77 of *Cambridge Studies in Advanced Mathematics*. Cambridge University Press, Cambridge, 2007. Translated from the French by Leila Schneps.

[29] Oscar Zariski. *Algebraic surfaces*, volume 61 of *Ergebnisse der Mathematik und ihrer Grenzgebiete*. Springer, 1971. Second supplemented edition.

Chapter Ten

The Spread Philosophy in the Study of Algebraic Cycles

by Mark L. Green

10.1 INTRODUCTION TO SPREADS

There are two ways of looking at a smooth projective variety in characteristic 0:

Geometric: X is a compact Kähler manifold plus a Hodge class, embedded in $\mathbb{CP}^N$ so that the hyperplane bundle H pulls back to a multiple of the Hodge class.

Algebraic: X is defined by homogeneous polynomials in $k[x_0, \ldots, x_N]$ for a field k of characteristic 0. We may take k to be the field generated by ratios of coefficients of the defining equations of X, and hence we may take k to be finitely generated over $\mathbb{Q}$. There are algebraic equations with coefficients in $\mathbb{Q}$ that, applied to the coefficients of the defining equations, tell us when X is (not) smooth, irreducible, of dimension n. The crucial additional ingredient to do Hodge theory is an embedding $k \hookrightarrow \mathbb{C}$.

To get a Hodge structure associated to X, we need $k \hookrightarrow \mathbb{C}$. The cohomology groups of X can be computed purely in terms of k, but the integral lattice requires us to have an embedding of k in $\mathbb{C}$.

A field k that is finitely generated/$\mathbb{Q}$ is of the form

$$k = \mathbb{Q}(\alpha_1, \ldots, \alpha_T, \beta_1, \ldots, \beta_A),$$

where $\alpha_1, \ldots, \alpha_T$ are algebraically independent over $\mathbb{Q}$ and $[k : \mathbb{Q}(\alpha_1, \ldots, \alpha_T)] < \infty$. Note $T = \operatorname{tr} \deg(k)$, the transcendence degree of k. Alternatively,

$$k \cong \mathbb{Q}(x_1, \ldots, x_T)[y_1, \ldots, y_A]/(p_1, \ldots, p_B),$$

where

$$p_1, \ldots, p_B \in \mathbb{Q}(x_1, \ldots, x_T)[y_1, \ldots, y_A].$$

IMPORTANT IDEA We can make k geometric.

The idea here is to find a variety S, defined /$\mathbb{Q}$, such that

$$k \cong \mathbb{Q}(S) = \text{field of rational functions of } S.$$

Further, by definition,

$$\mathbb{Q}(S_1) \cong \mathbb{Q}(S_2) \Longleftrightarrow S_1 \text{ birationally equivalent to } S_2.$$

If

$$k \cong \mathbb{Q}(x_1, \ldots, x_T)[y_1, \ldots, y_A]/(p_1, \ldots, p_B),$$

then we may take

$$S = \text{projectivization of affine variety in } \mathbb{Q}^{T+A} \text{ defined by } p_1, \ldots, p_B.$$

EXAMPLE 10.1.1 (Elliptic curve $X = \{y^2 = x(x-1)(x-\alpha)\}$)   If $\alpha \notin \bar{\mathbb{Q}}$, i.e., α is transcendental, then $k = \mathbb{Q}(\alpha)$ and $S \cong \mathbb{P}^1$. Note S is defined $/\mathbb{Q}$. We can think of X as giving a family

$$\pi \colon \mathcal{X} \to S(\mathbb{C})$$

where

$$X_s = \pi^{-1}(s).$$

Note that

$$X_s \text{ is singular} \Longleftrightarrow s \in \{0, 1, \infty\}.$$

Note that the latter is a subvariety of S defined $/\mathbb{Q}$. We have

$$\text{field of definition of } X_s \cong k \text{ for } s \notin \bar{\mathbb{Q}}.$$

The varieties X_{s_1} and X_{s_2} are indistinguishable algebraically if $s_1, s_2 \notin \bar{\mathbb{Q}}$, but we can tell them apart analytically and in general they will have different Hodge structures.

DEFINITION 10.1.2   *For S defined $/\mathbb{Q}$, we will say that $s \in S(\mathbb{C})$ is a **very general point** if the Zariski closure over $\mathbb{Q}$ of s is S, i.e., s does not belong to any proper subvariety of S defined $/\mathbb{Q}$.*

We have

$$\{\text{very general points of } S\} \Longleftrightarrow \{ \text{embeddings } k \xrightarrow{i_s} \mathbb{C}\}.$$

We get from a very general point of S and a variety X defined $/k$ a variety X_s defined $/\mathbb{C}$. We piece these together to get a family of complex varieties

$$\pi \colon \mathcal{X} \to S,$$

which we call the **spread of X over k**. If X is defined over k by homogeneous polynomials $f_1, \ldots, f_r$ in $k[z_0, \ldots, z_N]$, then expanding out the coefficients of the f_i's in terms of x's and y's, we may take $\mathcal{X}$ to be the projectivization of the variety $/\mathbb{Q}$ defined by $f_1, \ldots f_r, p_1, \ldots, p_B$ in the variables $x_1, \ldots, x_T, y_1, \ldots, y_B, z_0, \ldots, z_N$. We thus have

$$\pi \colon \mathcal{X} \to S,$$

where $\mathcal{X}$, S are defined $/\mathbb{Q}$ and the map π is a map defined $/\mathbb{Q}$. Then π will be smooth and of maximal rank outside a proper subvariety $\Sigma \subset S$ defined $/\mathbb{Q}$.

The spread of X over k is not unique, but the nonuniqueness can be understood and kept under control.

EXAMPLE 10.1.3 (k a number field)   Here, S consists of a finite set of $[k : \mathbb{Q}]$ points. As a variety, S is defined $/\mathbb{Q}$, although the individual points are defined over a splitting field of k. We get a finite number of complex varieties X_s, corresponding to the $[k : \mathbb{Q}]$ embeddings $k \overset{i_s}{\hookrightarrow} \mathbb{C}$.

EXAMPLE 10.1.4 (Very general points)   Let Y be a projective algebraic variety, irreducible, defined (say) over $\mathbb{Q}$. Let $y \in Y$ be a very general point. Take

$$k = \mathbb{Q}(\text{ ratios of coordinates of } y).$$

Then $k \cong \mathbb{Q}(Y)$, so we may take $S = Y$.

EXAMPLE 10.1.5 (Ordered pairs of very general points)   Let X be a variety defined $/\mathbb{Q}, X \subseteq \mathbb{P}^N$, $\dim(X) = n$. Let (p, q) be a very general point of $X \times X$. Let

$$k = \mathbb{Q}(\text{ ratios of coordinates of } p, \text{ ratios of coordinates of } q).$$

Now

$$\operatorname{tr} \deg(k) = 2n$$

and we can take $S = X \times X$. The moral of this story is that complicated 0-cycles on X potentially require ever more complicated fields of definition.

EXAMPLE 10.1.6 (Hypersurfaces in $\mathbb{P}^{n+1}$)   Take $F \in \mathbb{C}[z_0, \dots, z_{n+1}]$, homogeneous of degree d, $X = \{F = 0\}$. Let

$$F = \sum_{|I|=d} a_I z^I$$

using multiindex notation. Assume that $a = (a_I)_{|I|=d}$ is chosen to be a very general point of $\mathbb{P}^{\binom{n+1+d}{d}-1}$. Then the field of definition of X is $\mathbb{Q}(\text{ ratios of the } a_I)$, and $S = \mathbb{P}^{\binom{n+1+d}{d}-1}$. Now

$$\pi : \mathcal{X} \to S$$

is the universal family of hypersurfaces of degree d.

ESSENTIAL OBSERVATION   It is usually productive to make use of this geometry.

To do this, we need constructions that are robust to birational changes.

One natural idea is to look at the associated **variation of Hodge structure**

$$\begin{array}{ccc} S - \Sigma & \overset{P}{\longrightarrow} & \Gamma/D, \\ s & \mapsto & H^r(X_s, \mathbb{C}), \end{array}$$

where D is the Hodge domain (or the appropriate Mumford–Tate domain) and Γ is the group of automorphisms of the integral lattice preserving the intersection pairing. If we have an **algebraic cycle** Z on X, taking spreads yields a cycle $\mathcal{Z}$ on $\mathcal{X}$. Applying Hodge theory to $\mathcal{Z}$ on $\mathcal{X}$ gives invariants of the cycle. Another related situation is **algebraic K-theory**. For example, to study $K_p^{\text{Milnor}}(k)$, the geometry of S can be used to construct invariants.

This is, overall, the **spread philosophy**.

10.2 CYCLE CLASS AND SPREADS

Let X be a smooth projective variety $/k$. By $\Omega^\bullet_{X(k)/k}$ we denote the differentials on $X(k)$ over k. The sheaf $\Omega^1_{X(k)/k}$ is defined to be objects of the form $\sum_i f_i dg_i$, where $f_i, g_i \in \mathcal{O}_{X(k)}$ and subject to the rules

(i) $d(f + g) = df + dg$;

(ii) $d(fg) = fdg + gdf$;

(iii) $d^2 = 0$;

(iv) $dc = 0$ for $c \in k$.

We note that a consequence of (iv) is that $dc = 0$ if $c \in \bar{k}$. We set $\Omega^p_{X(k)/k} = \wedge^p \Omega^1_{X(k)/k}$ and these are made into a complex using d,

$$\mathcal{O}_{X(k)} \overset{d}{\to} \Omega^1_{X(k)/k} \overset{d}{\to} \Omega^2_{X(k)/k} \overset{d}{\to} \cdots$$

which we denote $\Omega^\bullet_{X(k)/k}$. By $\Omega^{\geq p}_{X(k)/k}$ we denote the complex

$$\Omega^p_{X(k)/k} \overset{d}{\to} \Omega^{p+1}_{X(k)/k} \overset{d}{\to} \cdots,$$

but indexed to be a subcomplex of $\Omega^\bullet_{X(k)/k}$.

COMPARISON THEOREMS OF GROTHENDIECK ([6])   If $k \overset{i_s}{\hookrightarrow} \mathbb{C}$ is a complex embedding of k then

(i) $\mathbf{H}^r(\Omega^\bullet_{X(k)/k}) \otimes_{i_s} \mathbb{C} \cong H^r(X_s, \mathbb{C})$;

(ii) $\mathbf{H}^r(\Omega^{\geq p}_{X(k)/k}) \otimes_{i_s} \mathbb{C} \cong F^p H^r(X_s, \mathbb{C})$, where $F^\bullet H^r(X_s, \mathbb{C})$ denotes the Hodge filtration;

(iii) $H^q(\Omega^p_{X(k)/k}) \otimes_{i_s} \mathbb{C} \cong H^{p,q}(X_s)$.

If in the definition of differentials we instead only require
(iv') $dc = 0$ for $c \in \mathbb{Q}$;
then we get $\Omega^\bullet_{X(k)/\mathbb{Q}}$ etc. By $\Omega^1_{k/\mathbb{Q}}$ we denote expressions $\sum_i a_i db_i$, where $a_i, b_i \in k$, subject to the rules (i)–(iii) and (iv'). Note that $\Omega^1_{k/\mathbb{Q}}$ is a k-vector space of dimension $\operatorname{tr}\deg(k)$. If

$$k \cong \mathbb{Q}(x_1, \ldots, x_T)[y_1, \ldots, y_A]/(p_1, \ldots, p_B),$$

then $dx_1, \ldots, dx_T$ give a k-basis for $\Omega^1_{k/\mathbb{Q}}$.

There is a natural filtration

$$F^m \Omega^\bullet_{X(k)/\mathbb{Q}} = \operatorname{Im}(\Omega^m_{k/\mathbb{Q}} \otimes \Omega^{\bullet-m}_{X(k)/\mathbb{Q}} \to \Omega^\bullet_{X(k)/\mathbb{Q}}).$$

The associated graded is

$$\mathrm{Gr}^m \, \Omega^\bullet_{X(k)/\mathbb{Q}} = \frac{F^m \Omega^\bullet_{X(k)/\mathbb{Q}}}{F^{m+1}\Omega^\bullet_{X(k)/\mathbb{Q}}} \cong \Omega^m_{k/\mathbb{Q}} \otimes \Omega^{\bullet-m}_{X(k)/k}.$$

We thus have an exact sequence

$$0 \to \mathrm{Gr}^1 \, \Omega^\bullet_{X(k)/\mathbb{Q}} \to \frac{\Omega^\bullet_{X(k)/\mathbb{Q}}}{F^2\Omega^\bullet_{X(k)/\mathbb{Q}}} \to \mathrm{Gr}^0 \, \Omega^\bullet_{X(k)/\mathbb{Q}} \to 0.$$

This can be rewritten:

$$0 \to \Omega^1_{k/\mathbb{Q}} \otimes \Omega^{\bullet-1}_{X(k)/k} \to \frac{\Omega^\bullet_{X(k)/\mathbb{Q}}}{F^2\Omega^\bullet_{X(k)/\mathbb{Q}}} \to \Omega^\bullet_{X(k)/k} \to 0.$$

From the long exact sequence for (hyper)cohomology, we obtain a map

$$\mathbf{H}^r(\Omega^\bullet_{X(k)/k}) \xrightarrow{\nabla} \Omega^1_{k/\mathbb{Q}} \otimes \mathbf{H}^r(\Omega^\bullet_{X(k)/k});$$

this is the **Gauss–Manin connection**. Similarly, using $F^m\Omega^\bullet_{X(k)/\mathbb{Q}}/F^{m+2}\Omega^\bullet_{X(k)/\mathbb{Q}}$, we get

$$\Omega^m_{k/\mathbb{Q}} \otimes \mathbf{H}^r(\Omega^\bullet_{X(k)/k}) \xrightarrow{\nabla} \Omega^{m+1}_{k/\mathbb{Q}} \otimes \mathbf{H}^r(\Omega^\bullet_{X(k)/k}).$$

We have

$$\nabla^2 = 0 \qquad \text{(integrability of the Gauss–Manin connection).}$$

Finally, we note that

$$0 \to \Omega^1_{k/\mathbb{Q}} \otimes \Omega^{\geq p-1}_{X(k)/k} \to \frac{\Omega^{\geq p}_{X(k)/\mathbb{Q}}}{F^2\Omega^{\geq p}_{X(k)/\mathbb{Q}}} \to \Omega^{\geq p}_{X(k)/k} \to 0$$

gives

$$F^p \mathbf{H}^r(\Omega^\bullet_{X(k)/k}) \xrightarrow{\nabla} \Omega^1_{k/\mathbb{Q}} \otimes F^{p-1}\mathbf{H}^r(\Omega^\bullet_{X(k)/k});$$

i.e.,

$$\nabla(F^p) \subseteq \Omega^1_{k/\mathbb{Q}} \otimes F^{p-1},$$

which is known as the **infinitesimal period relation** or **Griffiths transversality** (see [3]).

   Note that all of this takes place in the abstract world of k, without the need to choose an embedding $k \xrightarrow{i_s} \mathbb{C}$. The essential new feature, once we pick i_s, is the **integral lattice**

$$H^r(X_s, \mathbb{Z}) \to \mathbf{H}^r(\Omega^\bullet_{X(k)/k}) \otimes_{i_s} \mathbb{C}.$$

ESSENTIAL OBSERVATION Expressing this map involves transcendentals not already in k. The fields $\mathbb{Q}(\pi) \cong \mathbb{Q}(e)$, where π, e are transcendentals, and thus there is no algebraic construction over the field $k = \mathbb{Q}(x)$ that distinguishes the cohomology of varieties defined over k when we take different embeddings $k \to \mathbb{C}$ taking x to π or e. This is why the integral lattice involves maps transcendental in the elements of k.

EXAMPLE 10.2.1 (Elliptic curves)   We take X to be the projectivization of $y^2 = f(x)$, where $f(x) = x(x - 1)(x - \alpha)$, and α is transcendental; $k = \mathbb{Q}(\alpha)$. Differentiating,

$$2y dy = f'(x) dx$$

in $\Omega^1_{X(k)/k}$. If $U_1 = \{y \neq 0\}$ and $U_2 = \{f'(x) \neq 0\}$, then

$$\frac{2 dy}{f'(x)} = \frac{dx}{y}$$

and thus we get an element ω of $\mathbf{H}^0(\Omega^{\geq 1}_{X(k)/k})$ that is dx/y in U_1 and $2dy/f'(x)$ in U_2. However, in $\Omega^1_{X(k)/\mathbb{Q}}$, we have

$$2y dy = f'(x) dx - x(x - 1) d\alpha$$

on $U_1 \cap U_2$, and thus ω does not lift to $\mathbf{H}^0(\Omega^{\geq 1}_{X(k)/\mathbb{Q}})$. Thus

$$\nabla(\omega) = \frac{x(x - 1)}{f'(x)y} d\alpha \in \Omega^1_{k/\mathbb{Q}} \otimes H^1(\mathcal{O}_{X(k)}),$$

where $x(x - 1)/f'(x)y$ on $U_1 \cap U_2$ represents a class in $H^1(\mathcal{O}_{X(k)})$.

There is another construction in terms of k that is significant.

BLOCH–QUILLEN THEOREM

$$H^p(\mathcal{K}_p(\mathcal{O}_{X(k)})) \cong \mathrm{CH}^p(X(k)),$$

where $\mathcal{K}_p$ denotes the sheaf of K_p's from algebraic K-theory and $\mathrm{CH}^p(X(k))$ is cycles on X defined over k modulo rational equivalences defined over k (see [9]).

If we are willing to neglect torsion, we can replace $\mathcal{K}_p$ with the more intuitive $\mathcal{K}_p^{\mathrm{Milnor}}$.

SOULÉ'S BLOCH–QUILLEN THEOREM ([11])

$$H^p(\mathcal{K}_p^{\mathrm{Milnor}}(\mathcal{O}_{X(k)})) \otimes_{\mathbb{Z}} \mathbb{Q} \cong \mathrm{CH}^p(X(k)) \otimes_{\mathbb{Z}} \mathbb{Q}.$$

The description of $\mathcal{K}_p^{\mathrm{Milnor}}(\mathcal{O}_{X(k)})$ proceeds as follows: Regard $\mathcal{O}^*_{X(k)}$ as a $\mathbb{Z}$-module under exponentiation. One takes (locally) a quotient of $\otimes^p_{\mathbb{Z}} \mathcal{O}^*_{X(k)}$, representing $f_1 \otimes \cdots \otimes f_p$ by the symbol $\{f_1, \ldots, f_p\}$. The quotient is defined by the relations generated by
(Steinberg relations): $\{f_1, \ldots, f_p\} = 1$ if $f_i = 1 - f_j$ for some $i \neq j$.
There is now a map

$$
\begin{aligned}
\mathcal{K}_p^{\mathrm{Milnor}}(\mathcal{O}_{X(k)}) &\longrightarrow \Omega^p_{X(k)/\mathbb{Q}}, \\
\{f_1, \ldots, f_p\} &\longmapsto \frac{df_1 \wedge df_2 \wedge \cdots \wedge df_p}{f_1 f_2 \cdots f_p}.
\end{aligned}
$$

We may also regard this as a map:

$$\mathcal{K}_p^{\mathrm{Milnor}}(\mathcal{O}_{X(k)}) \longrightarrow \Omega_{X(k)/\mathbb{Q}}^{\geq p}.$$

We thus get maps

$$H^p(\mathcal{K}_p^{\mathrm{Milnor}}(\mathcal{O}_{X(k)})) \to H^p(\Omega_{X(k)/\mathbb{Q}}^p)$$

and

$$H^p(\mathcal{K}_p^{\mathrm{Milnor}}(\mathcal{O}_{X(k)})) \to \mathbf{H}^{2p}(\Omega_{X(k)/\mathbb{Q}}^{\geq p}).$$

The shift in index in the last cohomology group is to align it with the indexing for $\Omega_{X(k)/\mathbb{Q}}^\bullet$. These are called the **arithmetic cycle class** and were studied by Grothendieck, Srinivas [12] and Esnault–Paranjape [1].

If we move from differentials over $\mathbb{Q}$ to differentials over k, we obtain the **cycle class map**

$$H^p(\mathcal{K}_p^{\mathrm{Milnor}}(\mathcal{O}_{X(k)})) \overset{\psi_{X(k)}}{\longrightarrow} F^p\mathbf{H}^{2p}(\Omega_{X(k)/k}^\bullet).$$

Note that this is constant under the Gauss–Manin connection, i.e.,

$$\nabla \circ \psi_{X(k)} = 0.$$

If we choose a complex embedding of k, we then have a map $\psi_{X(k)} \otimes_{i_s} \mathbb{C}$, which we will denote as ψ_{X_s} which maps

$$H^p(\mathcal{K}_p^{\mathrm{Milnor}}(\mathcal{O}_{X(k)})) \overset{\psi_{X_s}}{\longrightarrow} F^p H^{2p}(X_s, \mathbb{C}) \cap \mathrm{Im}(H^{2p}(X_s, \mathbb{Q})).$$

Since integral classes are flat under ∇, this is consistent with $\nabla \circ \psi_{X(k)} = 0$. We denote

$$\mathrm{Hg}^p(X_s) = F^p H^{2p}(X_s, \mathbb{C}) \cap \mathrm{Im}(H^{2p}(X_s, \mathbb{Z})).$$

HODGE CONJECTURE

$$\mathrm{CH}^p(X_s(\mathbb{C})) \otimes_{\mathbb{Z}} \mathbb{Q} \overset{\psi_{X_s}}{\longrightarrow} \mathrm{Hg}^p(X_s) \otimes_z \mathbb{Q}$$

is surjective.

ABSOLUTE HODGE CONJECTURE ([7])   Given a class $\xi \in F^p\mathbf{H}^{2p}(\Omega_{X(k)/k}^\bullet)$, the set of $s \in S$ such that $i_{s*}(\xi) \in \mathrm{Hg}^p(X_s)$ is a subvariety of S defined $/\mathbb{Q}$.

Note that the absolute Hodge conjecture is weaker than the Hodge conjecture.

10.3   THE CONJECTURAL FILTRATION ON CHOW GROUPS FROM A SPREAD PERSPECTIVE

- X is a smooth projective variety defined $/k$.

- $Z^p(X(k)) = $ codimension-p cycles defined over k.

- $\mathrm{CH}^p(X(k)) = Z^p(X(k))/$ rational equivalences defined over k.

- $\mathrm{CH}^p(X(k))_{\mathbb{Q}} = \mathrm{CH}^p(X(k)) \otimes_{\mathbb{Z}} \mathbb{Q}$.

- $\mathrm{CH}^p(X(k))_{\mathbb{Q},\mathrm{Hom}} = \{Z \in \mathrm{CH}^p(X(k))_{\mathbb{Q}} \mid NZ \cong_{\mathrm{Hom}} 0 \text{ for some } N \neq 0\}$.

We tensor the Chow group with $\mathbb{Q}$ in order to eliminate torsion phenomena, which tend to be especially difficult—for example, the fact that the Hodge conjecture is not true over $\mathbb{Z}$.

CONJECTURAL FILTRATION ([9])   There is a decreasing filtration

$$\mathrm{CH}^p(X(k))_{\mathbb{Q}} = F^0\mathrm{CH}^p(X(k))_{\mathbb{Q}} \supseteq F^1\mathrm{CH}^p(X(k))_{\mathbb{Q}} \supseteq F^2\mathrm{CH}^p(X(k))_{\mathbb{Q}} \supseteq \cdots$$

with the following properties:

(i) $F^{m_1}\mathrm{CH}^{p_1}(X(k))_{\mathbb{Q}} \otimes F^{m_2}\mathrm{CH}^{p_2}(X \times Y(k))_{\mathbb{Q}} \to F^{m_1+m_2}\mathrm{CH}^{p_1+p_2-\dim(X)}(Y(k))_{\mathbb{Q}}$.

(ii) $F^1\mathrm{CH}^p(X(k)) = \mathrm{CH}^p(X(k))_{\mathbb{Q},\mathrm{Hom}}$.

(iii) $F^{p+1}\mathrm{CH}^p(X(k))_{\mathbb{Q}} = 0$.

An essential feature of this conjectural filtration is that it should be defined in terms of k and not depend on a choice of complex embedding $k \to \mathbb{C}$.

EXAMPLE 10.3.1 ($F^1\mathrm{CH}^p(X(k))$)   Because Grothendieck identifies

$$\mathbf{H}^r(\Omega^\bullet_{X(k)/k}) \otimes_{i_s} \mathbb{C} \cong H^r(X_s, \mathbb{C}),$$

the condition that $[Z] \in \mathbf{H}^{2p}(\Omega^\bullet_{X(k)/k})$ is 0 is equivalent to $i_{s*}[Z] = 0$ in $H^{2p}(X_s, \mathbb{C})$ for any complex embedding $k \overset{i_s}{\to} \mathbb{C}$.

EXAMPLE 10.3.2 ($F^2\mathrm{CH}^p(X(k))$)   The expectation is that if

$$\mathrm{AJ}^p_{X_s,\mathbb{Q}} : \mathrm{CH}^p(X(k))_{\mathbb{Q},\mathrm{Hom}} \to J^p(X_s) \otimes_{\mathbb{Z}} \mathbb{Q}$$

is the Abel–Jacobi map for X_s tensored with $\mathbb{Q}$, then

$$F^2\mathrm{CH}^p(X(k)) \cong \ker(\mathrm{AJ}^p_{X_s,\mathbb{Q}}).$$

The Abel–Jacobi map is highly transcendental, and it is not known that the kernel of the Abel–Jacobi map tensored with $\mathbb{Q}$ is independent of the complex embedding of k.

Let $\mathrm{Gr}^m \mathrm{CH}^p(X(k))_{\mathbb{Q}} = F^m \mathrm{CH}^p(X(k))_{\mathbb{Q}}/F^{m+1} \mathrm{CH}^p(X(k))_{\mathbb{Q}}$.

EXAMPLE 10.3.3 (Cycle classes)   We have the cycle class map

$$\psi_{X_s} : \mathrm{Gr}^0 \mathrm{CH}^p(X(k)) \hookrightarrow H^{2p}(X_s, \mathbb{C}).$$

The Hodge conjecture says that

$$\mathrm{Im}(\psi_{X_s}) = \mathrm{Hg}^p(X_s),$$

the Hodge classes of X_s. Note that the set of Hodge classes is thus conjecturally isomorphically the same for any very general $s \in S$.

EXAMPLE 10.3.4 (Image of the Abel–Jacobi map)   We have

$$\mathrm{AJ}^p_{X_s,\mathbb{Q}} : \mathrm{Gr}^1 \mathrm{CH}^p(X(k)) \hookrightarrow J^p(X_s) \otimes_{\mathbb{Z}} \mathbb{Q}.$$

Thus conjecturally, $\mathrm{Im}(\mathrm{AJ}^p_{X_s,\mathbb{Q}})$ is isomorphically the same for any very general $s \in S$.

It is thus expected that there should be something nice happening on the Hodge theory side.

BEILINSON'S CONJECTURAL FORMULA

$$\mathrm{Gr}^m \mathrm{CH}^p(X(k))_{\mathbb{Q}} \cong \mathrm{Ext}^m_{\mathcal{M}_k}(\mathbb{Q}, H^{2p-m}(X)(p)),$$

where $\mathcal{M}_k$ means that the extensions are in the category of mixed motives over k (see [10]).

Unfortunately, we do not have an explicit description of what these Ext groups should look like. One explicit consequence of this conjecture is that a cycle $\Gamma \in \mathrm{CH}^r(X \times Y(k))$ induces the zero map

$$\mathrm{CH}^p(X(k))_{\mathbb{Q}} \xrightarrow{\Gamma_*} \mathrm{Gr}^m \mathrm{CH}^{p+r-\dim(X)}(Y(k))_{\mathbb{Q}}$$

if the $H^{2\dim(X)-2p+m}(X) \otimes H^{2r+2p-2\dim(X)-m}(Y)$ component of $[\Gamma]$ is 0.

A very different aspect of the conjectures is that the arithmetic properties of k limit the possible graded pieces of the filtration on Chow groups.

CONJECTURE (DELIGNE–BLOCH–BEILINSON)

(i) $\mathrm{Gr}^m \mathrm{CH}^p(X(k))_{\mathbb{Q}} = 0$ for $m > \mathrm{tr}\deg(k) + 1$.

(ii) In particular, for X defined $/\mathbb{Q}$, $\mathrm{AJ}^p_{X,\mathbb{Q}} : \mathrm{CH}^p(X(\mathbb{Q}))_{\mathbb{Q},\mathrm{Hom}} \to J^p(X) \otimes_{\mathbb{Z}} \mathbb{Q}$ is injective (see [10]).

It is thus very natural to try to express the conjectural filtration of Chow groups in terms of spreads.

A first step is to have as good an understanding of the Abel–Jacobi map as possible. There is a nice interpretation of the Abel–Jacobi map as the extension class of an **extension of mixed Hodge structures**. To phrase this in our context, if $Z \in Z^p(X(k))$, we may represent Z by taking smooth varieties Z_i defined over k, maps $f_i \colon Z_i \to X$ defined over k, with

$$Z = \sum_i n_i f_{i*} Z_i.$$

We may construct a complex of differentials $\Omega^\bullet_{(X,|Z|)(k)/k}$, where

$$\Omega^m_{(X,|Z|)(k)/k} = \Omega^m_{X(k)/k} \oplus \bigoplus_i \Omega^{m-1}_{Z_i(k)/k}$$

with the differential

$$d(\omega, \oplus_i \phi_i) = (d\omega, \oplus_i d\phi_i - f_i^* \omega).$$

If $\dim(X) = n$, and thus $\dim(Z_i) = n - p$, we have an exact sequence

$$0 \to \mathrm{coker}(\mathbf{H}^{2n-2p}(\Omega^\bullet_{X(k)/k}) \to \oplus_i \mathbf{H}^{2n-2p}(Z_i)) \to \mathbf{H}^{2n-2p+1}(\Omega^\bullet_{(X,|Z|)(k)/k})$$
$$\to \mathbf{H}^{2n-2p+1}(\Omega^\bullet_{X(k)/k}) \to 0.$$

If $Z \equiv_{\mathrm{Hom}} 0$ on X, we can derive from this an exact sequence

$$0 \to k(-(n - p)) \to V \to \mathbf{H}^{2n-2p+1}(\Omega^\bullet_{X(k)/k}) \to 0,$$

where V is a k-vector space with a Hodge filtration and a Gauss–Manin connection ∇_V. If we choose a complex embedding of k, and tensor the sequence above $\otimes_{i_s} \mathbb{C}$, we obtain an extension of mixed Hodge structures

$$0 \to \mathbb{Z}(-(n - p)) \xrightarrow{f_s} V_s \xrightarrow{g_s} H^{2n-2p+1}(X_s, \mathbb{C}) \to 0,$$

where the new element added by the complex embedding is the integral lattice. If we pick an integral lifting $\phi_{\mathbb{Z}} \colon H^{2n-2p+1}(X_s, \mathbb{Z}) \to V_{s,\mathbb{Z}}$ and a complex lifting preserving the Hodge filtration, $\phi_{\mathrm{Hodge}} \colon H^{2n-2p+1}(X_s, \mathbb{C}) \to V_s$, then the extension class

$$e_s = f_s^{-1}(\phi_{\mathbb{Z}} - \phi)$$

and e_s lies in

$$\frac{\mathrm{Hom}_{\mathbb{C}}(H^{2n-2p+1}(X_s, \mathbb{C}), \mathbb{C}(-n'))}{F^0 \mathrm{Hom}_{\mathbb{C}}(H^{2n-2p+1}(X_s, \mathbb{C}), \mathbb{C}(-n')) + \mathrm{Hom}_{\mathbb{Z}}(H^{2n-2p+1}(X_s, \mathbb{Z}), \mathbb{Z}(-n'))},$$

where $n' = n - p$.

We may rewrite this using Poincaré duality as

$$e_s \in \frac{H^{2p-1}(X_s, \mathbb{C})}{F^p H^{2p-1}(X_s, \mathbb{C}) + H^{2p-1}(X_s, \mathbb{Z})} = J^p(X_s),$$

the pth **intermediate Jacobian** of X_s. The intermediate Jacobians fit together to give a family $\mathcal{J} \to S$, and $s \mapsto e_s$ gives a section $\nu_Z \colon S \to \mathcal{J}$ of this family, which is called the **normal function associated to the cycle** Z. By an argument analogous to the one used to define the Gauss–Manin connection by looking at the obstruction to lifting to differentials $/\mathbb{Q}$, we get for a local lifting $\tilde{\nu}_Z$ of ν_Z to the variation of Hodge structures $H^{2p-1}(X_s, \mathbb{C})$ that

$$\nabla \tilde{\nu}_Z \in \Omega^1_{S/\mathbb{C}} \otimes F^{p-1} H^{2p-1}(X_s, \mathbb{C})$$

for all s; the fact that we land in $F^{p-1} H^{2p-1}(X_s, \mathbb{C})$ is known as the **infinitesimal relation on normal functions**. The actual value we get depends on how we lift ν_Z, but $\nabla \nu_Z$ gives a well-defined element

$$\delta \nu_Z \in \mathbf{H}^1(\Omega^\bullet_S \otimes F^{p-\bullet} H^{2p-1}(X_s, \mathbb{C}))$$

which is **Griffiths' infinitesimal invariant of the normal function** ν_Z (see [3, 2]).

We may also encapsulate the information in the construction above as an extension involving the variation of Hodge structure $\mathcal{H}^{2p-1}_X \to S$ of the form

$$0 \to \mathcal{H}^{2p-1}_X(p) \to \mathcal{V}^* \to \mathbb{Z} \to 0.$$

We may regard such extensions as elements of the group $\mathrm{Ext}^1_S(\mathbb{Z}, \mathcal{H}^{2p-1}_X(p))$, where we must make some technical assumptions about how the families behave over the subvariety Σ, where the map $\pi \colon \mathcal{X} \to S$ is not of maximal rank, and thus X_s is singular. Conjecturally, one would expect that

$$\mathrm{Gr}^1 \mathrm{CH}^p(X(k)) \otimes_{\mathbb{Z}} \mathbb{Q} \hookrightarrow \mathrm{Ext}^1_S(\mathbb{Q}, \mathcal{H}^{2p-1}_X(p))$$

is well defined and injective.

  * There are a number of different ways that, conjecturally, produce the conjectural filtration. Two of my favorites are those of Murre and of Saito. Let $\dim(X) = n$. The Hodge conjecture says that the **Künneth decomposition of the diagonal** $\Delta \in X \times X$ as $\Delta = \sum_i \pi_i$, where $\pi_i \in Z^n(X \times X)$ and $[\pi_i] \in H^{2n-i}(X, \mathbb{Q}) \otimes H^i(X, \mathbb{Q})$, represents the identity map under Poincaré duality. These induce maps

$$\mathrm{CH}^p(X)_{\mathbb{Q}} \xrightarrow{\pi^p_{i*}} \mathrm{CH}^p(X)_{\mathbb{Q}}.$$

Now we want for $m \geq 1$,

$$F^m \mathrm{CH}^p(X)_{\mathbb{Q}} = \cap^{2p}_{i=2p-m+1} \ker(\pi^p_{i*}).$$

This is Murre's definition [9].

The definition of Saito (see [8]) generates $F^m \mathrm{CH}^p(X(k))_{\mathbb{Q}}$ by taking auxiliary varieties T defined $/k$ and cycles $Z_1 \in \mathrm{CH}^{r_1}(X \times T(k))_{\mathrm{hom}}$ and for $i = 2, \ldots, m$ cycles $Z_i \in \mathrm{CH}^{r_i}(T(k))_{\mathrm{hom}}$ and looking at

$$p_{X*}(Z_1 \cdot p^*_T Z_2 \cdot p^*_T Z_3 \cdots p^*_T Z_m),$$

where $\sum_i r_i - \dim(T) = p$. Clearly all such elements must lie in $F^m \mathrm{CH}^p(X(k))_{\mathbb{Q}}$ for any definition satisfying the conditions of the conjectural filtration. However, it is not clear that $F^{p+1} \mathrm{CH}^p(X(k))_{\mathbb{Q}} = 0$.

10.4   THE CASE OF X DEFINED OVER $\mathbb{Q}$

This lecture is based on joint work with Phillip Griffiths [4].

We now look at the case X a smooth projective connected variety defined $/\mathbb{Q}$, as discussed in a joint paper with Phillip Griffiths. We consider cycles defined over a finitely generated extension k of the rationals. Thus

$$\mathcal{X} \cong X \times S.$$

If $Z \in Z^p(X(k))$, then its spread

$$\mathcal{Z} \in Z^p(X \times S(\mathbb{Q}))$$

is well defined up to an ambiguity in the form of a cycle

$$\mathcal{W} \in Z^{p-\mathrm{codim}(W)}(X \times W(\mathbb{Q})),$$

where W is a lower-dimensional subvariety of S defined $/\mathbb{Q}$. Cycles rationally equivalent to 0 over k are generated by taking a codimension $p - 1$ subvariety $Y \overset{i}{\hookrightarrow} X$ defined $/k$ and $f \in k(Y)$ and taking $i_* \mathrm{div}(f)$. Taking spreads over k, we have $\mathcal{Y} \subset X \times S$ of codimension $p - 1$ defined $/\mathbb{Q}$ and $F \in \mathbb{Q}(Y)$. Once again, the ambiguities in this process are supported on a variety of the form $X \times W$, where W is a lower-dimensional subvariety of S defined over $\mathbb{Q}$.

At this point, we invoke the conjecture of Deligne–Bloch–Beilinson mentioned in Lecture 3 that for cycles and varieties over $\mathbb{Q}$, the cycle class and the Abel–Jacobi map are a complete set of invariants for cycles modulo rational equivalence, tensored with $\mathbb{Q}$. Thus, if $\mathcal{Z} \in Z^p(X \times S(\mathbb{Q}))$, the invariants are

(i) $[\mathcal{Z}] \in H^{2p}(X \times S, \mathbb{C})$; and

(ii) if $[\mathcal{Z}] = 0$, then $\mathrm{AJ}^p_{X \times S}(\mathcal{Z}) \otimes_{\mathbb{Z}} \mathbb{Q}$.

It follows from this conjecture that $Z = 0$ in $\mathrm{CH}^p(X(k))_{\mathbb{Q}}$ if and only if there exists a cycle $\mathcal{W} \in Z^{p-\mathrm{codim}(W)}(X \times W)$ for some lower-dimensional subvariety $W \subset S$ such that

(i) $[\mathcal{Z} + \mathcal{W}] = 0$ in $H^{2p}(X \times S)$; and

(ii) $\mathrm{AJ}_{X \times S, \mathbb{Q}}(\mathcal{Z} + \mathcal{W}) = 0$ in $J^p(X \times S) \otimes_{\mathbb{Z}} \mathbb{Q}$.

The Künneth decomposition of $X \times S$ allows us to write over $\mathbb{Q}$:

$$H^{2p}(X \times S) \cong \bigoplus_m H^{2p-m}(X) \otimes H^m(S)$$

and

$$J^p(X \times S) \otimes_{\mathbb{Z}} \mathbb{Q} \cong \bigoplus_m J^p(X \times S)_m,$$

where

$$J^p(X \times S)_m = \frac{H^{2p-1-m}(X, \mathbb{C}) \otimes H^m(S, \mathbb{C})}{F^p(H^{2p-1-m}(X, \mathbb{C}) \otimes H^m(S, \mathbb{C})) + H^{2p-1-m}(X, \mathbb{Q}) \otimes H^m(S, \mathbb{Q})}.$$

We will denote the Künneth components of $[\mathcal{Z}]$ as $[\mathcal{Z}]_m$ and the $J^p(X \times S)_m$ component of $AJ_{X \times S}(\mathcal{Z})$ as $AJ_{X \times S}(\mathcal{Z})_m$.

It is important to note that while we need the vanishing of the cycle class in order to define the Abel–Jacobi map, we only need the vanishing of $[\mathcal{Z}]_i$ for $i \leq m + 1$ in order to define $AJ_{X \times S}(\mathcal{Z})_m$.

DEFINITION OF FILTRATION ON CHOW GROUPS $Z \in F^m CH^p(X(k))$ if and only if for some $\mathcal{W}$ as above, $[\mathcal{Z} + \mathcal{W}]_i = 0$ for all $i < m$ and (this is now defined) $AJ_{X \times S}(\mathcal{Z} + \mathcal{W})_i = 0$ for all $i < m - 1$.

In order to understand this definition, it is essential to understand what happens for cycles $\mathcal{W} \in Z^{p - \text{codim}(W)}(X \times W)$. If $r = \text{codim}(W)$, then we have the Gysin map

$$\text{Gy}_W^m : H^m(W) \to H^{m+2r}(S).$$

This induces a map

$$H^{2p-m}(X) \otimes H^{m-2r}(W) \to H^{2p-m}(X) \otimes H^m(S).$$

We see that

$$[\mathcal{W}]_m \in H^{2p-m}(X) \otimes \text{Im}(\text{Gy}_W^{m-2r}).$$

Now $\text{Im}(\text{Gy}_W^{m-2r})$ is contained in the largest weight-$(m - 2r)$ sub-Hodge structure of $H^m(S)$.

Let H denote the largest weight-$(m - 2r)$ sub-Hodge structure of $H^m(S)$. The generalized Hodge conjecture implies that there is a dimension $m - 2r$ subvariety V of $S(\mathbb{C})$ such that $\text{Im}(\text{Gy}_V^{m-2r}) = H$. Now V might in principle require a finitely generated field of definition L, with $L = \mathbb{Q}(T)$ for some variety T defined $/\mathbb{Q}$. Taking the spread of V over T, with the complex embedding of V in S represented by $t_0 \in T$, we know that there is a lower-dimensional subvariety of T such that away from it, for t in the same connected component of T, $H^{m-2r}(V_t)$ and $H^{m-2r}(V_{t_0})$ have the same image in $H^m(S)$. We may therefore find a point $t_1 \in T(\bar{\mathbb{Q}})$ in the same connected component of $T(\mathbb{C})$ as t_0, such that V_{t_1} is defined over $\bar{\mathbb{Q}}$ and $\text{Gy}_{V_{t_1}}^{m-2r}$ has the same image as for V_{t_0}. We take W to be the union of the Galois conjugates of V_{t_1}. Then $H \in \text{Im}(\text{Gy}_W^{m-2r})$. It follows that any Hodge class in $H^{2p-m}(X) \otimes H$ is, by the Hodge conjecture, a $\mathbb{Q}$-multiple of the Hodge class of a cycle $\mathcal{W} \in Z^{p-\text{codim}(W)}(X \times W)$, i.e., an ambiguity. It follows that

$$CH^p(X(k))_{\mathbb{Q}} \to H^{2p-m}(X, \mathbb{Q}) \otimes \frac{H^m(S, \mathbb{Q})}{H^m(S, \mathbb{Q})_{m-2}},$$

where $H^m(S)_{m-2}$ is the largest weight-$(m-2)$ sub-Hodge structure of $H^m(S)$, is well defined and captures all of the information in the invariant $[\mathcal{Z}]_m$ modulo ambiguities.

We note that if $\dim(S) < m$, then by the Lefschetz theorem $H^m(S) = H^m(S)_{m-2}$, so this invariant vanishes.

We note that a Hodge class in $H^{2p-m}(X) \otimes H^m(S)$ gives us a map of Hodge structures, with a shift, $H^{2n-2p+m}(X) \to H^m(S)$. However, $H^{2n-2p+m}(X) \cong H^{2p-m}(X)(-(n-2p+m))$. If $m > 2p - m$, this implies that the Hodge class actually lies in $H^{2p-m}(X) \otimes H^m(S)_{m-2}$ and thus is an ambiguity. This happens precisely when $m > p$. We thus have that the invariant $[\mathcal{Z}]_m = 0$ modulo ambiguities if $m > p$. An alternative proof is that if X_P is a general $\mathbb{Q}$-linear section of X of dimension $2p - m$, then let $\mathcal{Z}_P = \mathcal{Z} \cap (X_P \times S)$. By the Lefschetz theorem, we have

$$r_P^{2p-m} : H^{2p-m}(X) \hookrightarrow H^{2p-m}(X_P)$$

and

$$r_P^{2p-m} \otimes \mathrm{id}_{H^m(S)}([\mathcal{Z}]_m) = [\mathcal{Z}_P]_m.$$

However, if $p > 2p - m$, then of necessity $\mathcal{Z}_P$ projects to a proper $\mathbb{Q}$-subvariety of S, and hence is an ambiguity.

The second argument also shows that taking P so that X_P has dimension $2p - m - 1$, then

$$r_P^{2p-m-1} \otimes \mathrm{id}_{H^m(S)}(\mathrm{AJ}_{X \times S}^p(\mathcal{Z})_m) = \mathrm{AJ}_{X_P \times S}^p(\mathcal{Z}_P)_m,$$

and thus for $p > 2p - m - 1$, $\mathcal{Z}_P$ must project to a proper $\mathbb{Q}$-subvariety of S and hence involves only $H^m(S)_{m-2}$. One can also use the linear section argument to use cycles defined on $X \times W$ for proper $\mathbb{Q}$-subvarieties W of S to kill off portions of the Abel–Jacobi map that involve $H^m(S)_{m-2}$.

We define $[\mathcal{Z}]_m^{\mathrm{red}}$ to be the image of $[\mathcal{Z}]_m$ in $H^{2p-m}(X) \otimes H^m(S)/H^m(S)_{m-2}$ and we define $\mathrm{AJ}_{X \times S}^p(\mathcal{Z})_m^{\mathrm{red}}$ to be the image of $\mathrm{AJ}_{X \times S}^p(\mathcal{Z})_m$ in the intermediate Jacobian constructed from $H^{2p-m-1}(X) \otimes H^m(S)/H^m(S)_{m-2}$.

INVARIANTS OF CYCLES For X defined over $\mathbb{Q}$, a complete set of invariants of $\mathrm{CH}^p(X(k))_{\mathbb{Q}}$ are $[\mathcal{Z}]_m^{\mathrm{red}}$ for $0 \le m \le p$ and $\mathrm{AJ}_{X \times S}^p(\mathcal{Z})_m^{\mathrm{red}}$ for $0 \le m \le p - 1$. Note that $[\mathcal{Z}]_m^{\mathrm{red}}$ and $\mathrm{AJ}_{X \times S}^p(\mathcal{Z})_m^{\mathrm{red}}$ both vanish if $m > \dim(S)$, i.e., $m > \mathrm{tr\,deg}(k)$. We then get that $F^m\mathrm{CH}^p(X(k))_{\mathbb{Q}}$ is defined by the vanishing (tensored with $\mathbb{Q}$) of $[\mathcal{Z}]_i^{\mathrm{red}}$ for $i < m$ and of $\mathrm{AJ}_{X \times S}^p(\mathcal{Z})_m^{\mathrm{red}}$ for $i < m - 1$. This forces $F^m\mathrm{CH}^p(X(k))_{\mathbb{Q}} = 0$ if $m > p$ or if $m > \mathrm{tr\,deg}(k) + 1$.

EXAMPLE 10.4.1 ($p = 1$)   Here $[\mathcal{Z}]_0^{\mathrm{red}}$ is just the cycle class of Z if S is connected, or the various cycle classes coming from different complex embeddings if S is not connected. Next, $[\mathcal{Z}]_1^{\mathrm{red}}$ is in $H^1(X, \mathbb{Q}) \otimes H^1(S, \mathbb{Q})$ and is equivalent to the induced map on cohomology coming from $\mathrm{Alb}(S) \to J^1(X) \otimes_{\mathbb{Z}} \mathbb{Q}$ coming from $s \mapsto \mathrm{AJ}_X^1(Z_s)$. If this is zero, then this map is constant on connected components of S, and this allows us to define $\mathrm{AJ}_{X \times S}^1(\mathcal{Z})_0^{\mathrm{red}} \in J^1(X) \otimes_{\mathbb{Z}} H^0(S, \mathbb{Q})$. Note that we do not need to reduce modulo lower-weight sub-Hodge structures on S for either of these.

EXAMPLE 10.4.2 ($p = 2$)   In this case, the only really new invariants are $[\mathcal{Z}]_2^{\mathrm{red}}$ and $\mathrm{AJ}_{X \times S}^p([\mathcal{Z}])_1^{\mathrm{red}}$, which are the invariants of $\mathrm{Gr}^2\,\mathrm{CH}^2(X(k))_{\mathbb{Q}}$. The former is an element of $H^2(X)/\mathrm{Hg}^1(X) \otimes H^2(S)/\mathrm{Hg}^1(S)$. This invariant was discussed by

Voisin, and comes by integrating $(2n-2)$-forms on X over two-dimensional families of cycles Z_s. The latter invariant is in

$$\frac{H^2(X,\mathbb{C}) \otimes H^1(S,\mathbb{C})}{F^2(H^2(X,\mathbb{C}) \otimes H^1(S,\mathbb{C})) + H^2(X,\mathbb{Q}) \otimes H^1(S,\mathbb{Q})}.$$

We note that the portion of this coming from $\mathrm{Hg}^1(X) \otimes H^1(S)$ can be realized by taking a divisor Y on X representing the Hodge class in $\mathrm{Hg}^1(X)$ cross a codimension-1 cycle on S, and hence comes from an ambiguity. Thus $H^2(X)/\mathrm{Hg}^1(X) = 0$ implies both invariants of $\mathrm{Gr}^2\,\mathrm{CH}^2(X(k))_\mathbb{Q}$ are 0. It is worth noting that the geometry of S comes in—if S has no H^1 and no transcendental part of H^2; then $\mathrm{Gr}^2\,\mathrm{CH}^2(X(k))_\mathbb{Q} = 0$.

10.5   THE TANGENT SPACE TO ALGEBRAIC CYCLES

This lecture is based on joint work with Phillip Griffiths [5].

It was noted by Van der Kallen that there is a natural tangent space to algebraic K-theory, and that

$$T\mathcal{K}_p^{\mathrm{Milnor}}(\mathcal{O}_{X(k)}) \cong \Omega_{X(k)/\mathbb{Q}}^{p-1}.$$

The map is

$$\{f_1,\dots,f_p\} \mapsto \frac{\dot{f}_1 df_2 \wedge \cdots \wedge df_p + \cdots + (-1)^{p-1}\dot{f}_p df_1 \wedge \cdots \wedge df_{p-1}}{f_1 f_2 \cdots f_p}.$$

This generalizes the statement for $p = 1$ that

$$T\mathcal{O}_{X(k)}^* \cong \mathcal{O}_{X(k)},$$

but the more exotic differentials over $\mathbb{Q}$ only manifest themselves once we reach $p \geq 2$. Bloch then derived from this the natural formula that

$$TH^p(\mathcal{K}_p(\mathcal{O}_{X(k)})) \cong H^p(\Omega_{X(k)/\mathbb{Q}}^{p-1}),$$

which thus is a formula for $TCH^p(X(k))$.

EXAMPLE 10.5.1 $(p = 1)$ The formula above reduces to the classical formula

$$TCH^1(X(k)) \cong TH^1(\mathcal{O}_{X(k)}^*) \cong H^1(\mathcal{O}_{X(k)}),$$

where the second map is induced by

$$(f_{\alpha\beta}) \mapsto \left(\frac{\dot{f}_{\alpha\beta}}{f_{\alpha\beta}}\right).$$

We have a filtration on differentials

$$F^m\Omega_{X(k)/\mathbb{Q}}^r = \mathrm{Im}(\Omega_{k/\mathbb{Q}}^m \otimes_k \Omega_{X(k)/\mathbb{Q}}^{r-m}) \to \Omega_{X(k)/\mathbb{Q}}^r.$$

We thus have

$$\mathrm{Gr}^m \, \Omega^r_{X(k)/\mathbb{Q}} \cong \Omega^m_{k/\mathbb{Q}} \otimes_k \Omega^{r-m}_{X(k)/k}.$$

When X is smooth, there is a spectral sequence that computes the $H^*(\Omega^r_{X(k)/\mathbb{Q}})$, which degenerates at the E_2 term and has

$$E_2^{p,q} = H^p(\Omega^p_{k/\mathbb{Q}} \otimes_k H^{q-p}(\Omega^{r-p}_{X(k)/k}), \nabla).$$

This gives a natural filtration $F^m H^p(\Omega^{p-1}_{X(k)/\mathbb{Q}})$ with

$$\mathrm{Gr}^m \, H^p(\Omega^{p-1}_{X(k)/\mathbb{Q}}) \cong H^m(\Omega^\bullet_{k/\mathbb{Q}} \otimes_k H^p(\Omega^{r-\bullet}_{X(k)/k}), \nabla).$$

EXAMPLE 10.5.2 ($p = 2$)   There are two graded pieces to $H^2(\Omega^1_{X(k)/\mathbb{Q}})$:

- $\mathrm{Gr}^0 \, H^2(\Omega^1_{X(k)/\mathbb{Q}}) = \ker(H^2(\Omega^1_{X(k)/k}) \xrightarrow{\nabla} \Omega^1_{k/\mathbb{Q}} \otimes_k H^3(\mathcal{O}_{X(k)}))$.

- $\mathrm{Gr}^1 \, H^2(\Omega^1_{X(k)/\mathbb{Q}}) = \mathrm{coker}(H^1(\Omega^1_{X(k)/k}) \xrightarrow{\nabla} \Omega^1_{k/\mathbb{Q}} \otimes H^2(\mathcal{O}_{X(k)}))$.

We know that geometrically $/\mathbb{C}$, the generalized Hodge conjecture predicts that the image of AJ^2_X is precisely the part of $J^2(X)$ constructed from $H^3(X, \mathbb{C})_1$, i.e., the maximal weight-1 sub-Hodge structure of $H^3(X, \mathbb{C})$. The absolute Hodge conjecture implies that this is contained in the image of

$$\ker(H^2(\Omega^1_{X(k)/k}) \xrightarrow{\nabla} \Omega^1_{k/\mathbb{Q}} \otimes_k H^3(\mathcal{O}_{X(k)})) \otimes_{i_s} \mathbb{C}.$$

However, one does not expect that the two coincide, and indeed the tangent space to Chow groups is correct formally but not geometrically. Note that $\mathrm{Gr}^1 \, H^2(\Omega^1_{X(k)/\mathbb{Q}})$ has dimension over k that grows linearly with the transcendence degree of k once $H^{2,0}(X) \neq 0$, which is in line with our expectations from Roitman's theorem. This example was discussed by [1].

Geometrically, there are two problems with the tangent space formula for Chow groups. First-order tangent vectors may fail to be part of an actual geometric family, and first-order tangents to rational equivalences may fail to be part of an actual geometric family of rational equivalences. In order to understand this phenomenon better, we need to lift from tangent spaces to Chow groups to obtain a formula for tangent spaces to algebraic cycles. The strategy adopted for doing this is to look at a kind of Zariski tangent space.

EXAMPLE 10.5.3 (0-cycles on an algebraic curve)

Let X be an irreducible algebraic curve defined $/k$. If $\{U_\alpha\}$ is a k-Zariski cover of X, then a 0-cycle is defined by giving nonzero k-rational functions r_α on U_α whose ratios on overlaps $U_\alpha \cap U_\beta$ belong to $\mathcal{O}^*_{X(k)}$. Now $\dot{r}_\alpha/r_\alpha$ describes a tangent vector. We may define the sheaf of **principal parts** $\mathcal{PP}_{X(k)}$ to be given as the additive sheaf $\mathcal{M}_{X(k)}/\mathcal{O}_{X(k)}$, where $\mathcal{M}_{X(k)}$ is the sheaf of germs of k-rational functions. Then

$$TZ^1(X(k)) \cong H^0(X, \mathcal{PP}_{X(k)}).$$

From the exact sequence

$$0 \to \mathcal{O}_{X(k)} \to \mathcal{M}_{X(k)} \to \mathcal{PP}_{X(k)} \to 0$$

we get a natural map

$$H^0(X, \mathcal{PP}_{X(k)}) \to H^1(\mathcal{O}_{X(k)})$$

which we may think of as a map

$$TZ^1(X(k)) \to TCH^1(X(k)).$$

Because $\mathcal{M}_{X(k)}$ is flasque, this map is surjective. We may think of the exact sequence above as the tangent exact sequence to

$$0 \to \mathcal{O}^*_{X(k)} \to \mathcal{M}^*_{X(k)} \to \mathcal{D}_{X(k)} \to 0,$$

where $\mathcal{D}_{X(k)}$ is the sheaf of k-divisors on X.

Of course, what is missing in this discussion is the **exponential sheaf sequence**

$$0 \to \mathbb{Z} \to \mathcal{O}_{X(\mathbb{C})} \to \mathcal{O}^*_{X(\mathbb{C})} \to 0.$$

This leaves the algebraic category in two ways—we need to use the classical topology rather than the Zariski topology, and we need to use the transcendental function $f \mapsto \exp(2\pi i f)$. Once we have this, we get the exact sequence

$$0 \to J^1(X(\mathbb{C})) \to \mathrm{CH}^1(X(\mathbb{C})) \to \mathrm{Hg}^1(X(\mathbb{C})) \to 0,$$

which completely solves the problem of describing $\mathrm{CH}^1(X(\mathbb{C}))$ in the analytic category. One of the enduringly nice features of taking derivatives is that transcendental maps in algebraic geometry frequently have algebraic derivatives.

Once we pass to higher codimension, we no longer have the exponential sheaf sequence, and we do not really know what the right transcendental functions to invoke are—in some cases, these turn out to involve **polylogarithms**.

EXAMPLE 10.5.4 $(K_2^{\mathrm{Milnor}}(k))$

Here we have

$$TK_2^{\mathrm{Milnor}}(k) \cong \Omega^1_{k/\mathbb{Q}}.$$

Unfortunately, this is deceptively simple. Consider, for example, the family of elements

$$t \mapsto \{a, t\},$$

where $a \in k$. The derivative is $da/a \in \Omega^1_{k/\mathbb{Q}}$. If $a \in \bar{\mathbb{Q}}$, then $da = 0$ and thus the derivative of this map vanishes identically. However, it is known that this map is actually constant if $\mathrm{tr} \deg(k) > 0$ only when a is a root of unity. This is an example of what we call a **null curve**, one whose formal derivative is identically 0 but which is not constant. The problem is that $TK_2^{\mathrm{Milnor}}(k)$ is really a quotient of two tangent spaces—to the space of possible products of Steinberg symbols, and to the space of Steinberg relations. If we are unable to integrate a tangent vector in the space of Steinberg relations up to a geometric family of Steinberg relations, we might expect this to produce a null curve.

EXAMPLE 10.5.5 (0-cycles on a surface)   This turns out to already embody many of the complexities of tangent spaces to cycles. The answer is

$$TZ^2(X(k)) \cong \oplus_{|Z|} H^2_{|Z|}(\Omega^1_{X(k)/\mathbb{Q}}),$$

where the sum is over supports of irreducible codimension-2 k-subvarieties of X. This involves local cohomology. There is a natural map

$$H^2_{|Z|}(\Omega^1_{X(k)/\mathbb{Q}}) \to H^2(\Omega^1_{X(k)/\mathbb{Q}})$$

which we may interpret as giving us a map

$$TZ^2(X(k)) \to TCH^2(X(k)).$$

These are quite complicated objects. For example, if we work over $\mathbb{C}$, then the tangent space to 0-cycles supported at a point $x \in X$ is

$$TZ^2(X(\mathbb{C}))_x \cong \lim_{|Z|=x} \operatorname{Ext}^2_{\mathcal{O}_X(\mathbb{C})}(\mathcal{O}_Z, \Omega^1_{X(\mathbb{C})/\mathbb{Q}}).$$

An example of the simplest family where the distinction between differentials over k and differentials over $\mathbb{Q}$ comes in is the family

$$Z(t) = \operatorname{Var}(x^2 - \alpha y^2, xy - t),$$

where $\alpha \in \mathbb{C}^*$ is transcendental.

Bibliography

[1] Hélène Esnault and Kapil H. Paranjape. Remarks on absolute de Rham and absolute Hodge cycles. *C. R. Acad. Sci. Paris Sér. I Math.*, 319(1):67–72, 1994.

[2] Mark L. Green. Griffiths' infinitesimal invariant and the Abel–Jacobi map. *J. Differential Geom.*, 29(3):545–555, 1989.

[3] Mark L. Green. Infinitesimal methods in Hodge theory. In *Algebraic cycles and Hodge theory (Torino, 1993)*, volume 1594 of *Lecture Notes in Math.*, pages 1–92. Springer, Berlin, 1994.

[4] Mark L. Green and Phillip A. Griffiths. Hodge-theoretic invariants for algebraic cycles. *Int. Math. Res. Not.*, 9:477–510, 2003.

[5] Mark L. Green and Phillip A. Griffiths. *On the tangent space to the space of algebraic cycles on a smooth algebraic variety*, volume 157 of *Annals of Mathematics Studies*. Princeton University Press, Princeton, NJ, 2005.

[6] A. Grothendieck. On the de Rham cohomology of algebraic varieties. *Inst. Hautes Études Sci. Publ. Math.*, 29:95–103, 1966.

[7] A. Grothendieck. Hodge's general conjecture is false for trivial reasons. *Topology*, 8:299–303, 1969.

[8] Uwe Jannsen. Equivalence relations on algebraic cycles. In *The arithmetic and geometry of algebraic cycles (Banff, AB, 1998)*, volume 548 of *NATO Sci. Ser. C Math. Phys. Sci.*, pages 225–260. Kluwer Academic, Dordrecht, 2000.

[9] J. P. Murre. Algebraic cycles and algebraic aspects of cohomology and K-theory. In *Algebraic cycles and Hodge theory (Torino, 1993)*, volume 1594 of *Lecture Notes in Math.*, pages 93–152. Springer, Berlin, 1994.

[10] Dinakar Ramakrishnan. Regulators, algebraic cycles, and values of L-functions. In *Algebraic K-theory and algebraic number theory (Honolulu, HI, 1987)*, volume 83 of *Contemp. Math.*, pages 183–310. Amer. Math. Soc., Providence, RI, 1989.

[11] Christophe Soulé. Opérations en K-théorie algébrique. *Canad. J. Math.*, 37(3):488–550, 1985.

[12] V. Srinivas. Decomposition of the de Rham complex. *Proc. Indian Acad. Sci. Math. Sci.*, 100(2):103–106, 1990.

Chapter Eleven

Notes on Absolute Hodge Classes
by François Charles and Christian Schnell

INTRODUCTION

Absolute Hodge classes first appear in Deligne's proof of the Weil conjectures for K3 surfaces in [14] and are explicitly introduced in [16]. The notion of absolute Hodge classes in the singular cohomology of a smooth projective variety stands between that of Hodge classes and classes of algebraic cycles. While it is not known whether absolute Hodge classes are algebraic, their definition is both of an analytic and arithmetic nature.

The paper [14] contains one of the first appearances of the notion of motives, and is among the first unconditional applications of motivic ideas. Part of the importance of the notion of absolute Hodge classes is indeed to provide an unconditional setting for the application of motivic ideas. The papers [14], [17] and [1], among others, give examples of this train of thought. The book [23] develops a theory of mixed motives based on absolute Hodge classes.

In these notes, we survey the theory of absolute Hodge classes. The first section of these notes recalls the construction of cycle maps in de Rham cohomology. As proved by Grothendieck, the singular cohomology groups of a complex algebraic variety can be computed using suitable algebraic de Rham complexes. This provides an algebraic device for computing topological invariants of complex algebraic varieties.

The preceding construction is the main tool behind the definition of absolute Hodge classes, the object of Section 11.2. Indeed, comparison with algebraic de Rham cohomology makes it possible to conjugate singular cohomology with complex coefficients by automorphisms of $\mathbb{C}$. In Section 11.2, we discuss the definition of absolute Hodge classes. We try to investigate two aspects of this subject. The first one pertains to the Hodge conjecture. Absolute Hodge classes shed some light on the problem of the algebraicity of Hodge classes, and make it possible to isolate the number-theoretic content of the Hodge conjecture. The second aspect we hint at is the motivic meaning of absolute Hodge classes. While we do not discuss the construction of motives for absolute Hodge classes as in [17], we show various functoriality and semisimplicity properties of absolute Hodge classes which lie behind the more general motivic constructions cited above. We try to phrase our results so as to get results and proofs which are valid for André's theory of motivated cycles as in [1]. We do not define motivated cycles, but some of our proofs are very much inspired by that paper.

The third section deals with variational properties of absolute Hodge classes. After stating the variational Hodge conjecture, we prove Deligne's principle B as in [16] which is one of the main technical tools of the paper. In the remainder of the section, we discuss consequences of the algebraicity of Hodge bundles and of the Galois action on relative de Rham cohomology. Following [38], we investigate the meaning of the theorem of Deligne–Cattani–Kaplan on the algebraicity of Hodge loci (see [10]), and discuss the link between Hodge classes being absolute and the field of definition of Hodge loci.

The last two sections are devoted to important examples of absolute Hodge classes. Section 11.4 discusses the Kuga–Satake correspondence following Deligne in [14]. In Section 11.5, we give a full proof of Deligne's theorem which states that Hodge classes on abelian varieties are absolute [16].

In writing these notes, we did not strive for concision. Indeed, we did not necessarily prove properties of absolute Hodge cycles in the shortest way possible, but we rather chose to emphasize a variety of techniques and ideas.

Acknowledgment

This text is an expanded version of five lectures given at the ICTP Summer School on Hodge Theory in Trieste in June 2010. The first two lectures were devoted to absolute Hodge cycles and arithmetic aspects of the Hodge conjecture. The remaining three lectures outlined Deligne's proof that every Hodge class on an abelian variety is an absolute Hodge class. We would like to thank the organizers for this very nice and fruitful summer school.

We thank Daniel Huybrechts, Chris Lyons, and Tamás Szamuely for pointing out several small mistakes in an earlier version of these notes.

Claire Voisin was supposed to give these lectures in Trieste, but she could not attend. It would be hard to acknowledge enough the influence of her work on these notes and the lectures we gave. We are grateful to her for allowing us to use the beautiful survey [40] for our lectures and thank her sincerely. We also want to thank Matt Kerr for giving one of the lectures.

11.1   ALGEBRAIC DE RHAM COHOMOLOGY

Shortly after Hironaka's paper on resolutions of singularities had appeared, it was observed by Grothendieck that the cohomology groups of a complex algebraic variety could be computed algebraically. More precisely, he showed in [20] that on a nonsingular n-dimensional algebraic variety X (of finite type over the field of complex numbers $\mathbb{C}$), the hypercohomology of the algebraic de Rham complex

$$\mathscr{O}_X \to \Omega^1_{X/\mathbb{C}} \to \cdots \to \Omega^n_{X/\mathbb{C}}$$

is isomorphic to the singular cohomology $H^*(X^{\mathrm{an}}, \mathbb{C})$ of the complex manifold corresponding to X. Grothendieck's theorem makes it possible to ask arithmetic questions in Hodge theory, and is the founding stone for the theory of absolute Hodge classes. In this lecture, we briefly review Grothendieck's theorem, as well as the construction of cycle classes in algebraic de Rham cohomology.

11.1.1 Algebraic de Rham Cohomology

We begin by describing algebraic de Rham cohomology in a more general setting. Let X be a nonsingular quasi-projective variety, defined over a field K of characteristic 0. This means that we have a morphism $X \to \operatorname{Spec} K$, and we let $\Omega^1_{X/K}$ denote the sheaf of Kähler differentials on X. We also define $\Omega^i_{X/K} = \bigwedge^i \Omega^1_{X/K}$.

DEFINITION 11.1.1 *The algebraic de Rham cohomology of $X \to K$ consists of the K-vector spaces*

$$H^i(X/K) = \mathbb{H}^i\big(\mathscr{O}_X \to \Omega^1_{X/K} \to \cdots \to \Omega^n_{X/K}\big),$$

where $n = \dim X$.

This definition is compatible with field extensions, for the following reason. Given a field extension $K \subseteq L$, we let $X_L = X \times_{\operatorname{Spec} K} \operatorname{Spec} L$ denote the variety obtained from X by extension of scalars. Since $\Omega^1_{X_L/L} \simeq \Omega^1_{X/K} \otimes_K L$, we obtain $H^i(X_L/L) \simeq H^i(X/K) \otimes_K L$.

The algebraic de Rham complex $\Omega^\bullet_{X/K}$ is naturally filtered by the subcomplexes $\Omega^{\bullet \geq p}_{X/K}$. Let $\phi_p : \Omega^{\bullet \geq p}_{X/K} \to \Omega^\bullet_{X/K}$ be the canonical inclusion. It induces a filtration on algebraic de Rham cohomology which we will denote by

$$F^p H^i(X/K) = \operatorname{Im}(\phi_p)$$

and refer to it as the Hodge filtration. We can now state Grothendieck's comparison theorem.

THEOREM 11.1.2 (Grothendieck [20]) *Let X be a nonsingular projective variety over $\mathbb{C}$, and let X^{an} denote the associated complex manifold. Then there is a canonical isomorphism*

$$H^i(X/\mathbb{C}) \simeq H^i(X^{\mathrm{an}}, \mathbb{C}),$$

and under this isomorphism, $F^p H^i(X/\mathbb{C}) \simeq F^p H^i(X^{\mathrm{an}}, \mathbb{C})$ gives the Hodge filtration on singular cohomology.

PROOF. The theorem is a consequence of the GAGA theorem of Serre [33]. Let $\mathscr{O}_{X^{\mathrm{an}}}$ denote the sheaf of holomorphic functions on the complex manifold X^{an}. We then have a morphism $\pi : (X^{\mathrm{an}}, \mathscr{O}_{X^{\mathrm{an}}}) \to (X, \mathscr{O}_X)$ of locally ringed spaces. For any coherent sheaf $\mathscr{F}$ on X, the associated coherent analytic sheaf on X^{an} is given by $\mathscr{F}^{\mathrm{an}} = \pi^* \mathscr{F}$, and according to Serre's theorem, $H^i(X, \mathscr{F}) \simeq H^i(X^{\mathrm{an}}, \mathscr{F}^{\mathrm{an}})$.

It is easy to see from the local description of the sheaf of Kähler differentials that $(\Omega^1_{X/\mathbb{C}})^{\mathrm{an}} = \Omega^1_{X^{\mathrm{an}}}$. This implies that $H^q(X, \Omega^p_{X/\mathbb{C}}) \simeq H^q(X^{\mathrm{an}}, \Omega^p_{X^{\mathrm{an}}})$ for all

$p, q \geq 0$. Now pull-back via π induces homomorphisms $\mathbb{H}^i(\Omega^\bullet_{X/\mathbb{C}}) \to \mathbb{H}^i(\Omega^\bullet_{X^{\mathrm{an}}})$, which are isomorphisms by Serre's theorem. Indeed, the groups on the left are computed by a spectral sequence with $E_2^{p,q}(X) = H^q(X, \Omega^p_{X/\mathbb{C}})$, and the groups on the right by a spectral sequence with terms $E_2^{p,q}(X^{\mathrm{an}}) = H^q(X^{\mathrm{an}}, \Omega^p_{X^{\mathrm{an}}})$, and the two spectral sequences are isomorphic starting from the E_2-page. By the Poincaré lemma, the holomorphic de Rham complex $\Omega^\bullet_{X^{\mathrm{an}}}$ is a resolution of the constant sheaf $\mathbb{C}$, and therefore $H^i(X^{\mathrm{an}}, \mathbb{C}) \simeq \mathbb{H}^i(\Omega^\bullet_{X^{\mathrm{an}}})$. Putting everything together, we obtain a canonical isomorphism

$$H^i(X/\mathbb{C}) \simeq H^i(X^{\mathrm{an}}, \mathbb{C}).$$

Since the Hodge filtration on $H^i(X^{\mathrm{an}}, \mathbb{C})$ is induced by the naive filtration on the complex $\Omega^\bullet_{X^{\mathrm{an}}}$, the second assertion follows by the same argument. $\square$

Remark. A similar result holds when X is nonsingular and quasi-projective. Using resolution of singularities, one can find a nonsingular projective variety $\overline{X}$ and a divisor D with normal crossing singularities, such that $X = \overline{X} - D$. Using differential forms with at worst logarithmic poles along D, one still has

$$H^i(X^{\mathrm{an}}, \mathbb{C}) \simeq \mathbb{H}^i(\Omega^\bullet_{\overline{X}^{\mathrm{an}}}(\log D^{\mathrm{an}})) \simeq \mathbb{H}^i(\Omega^\bullet_{\overline{X}/\mathbb{C}}(\log D));$$

under this isomorphism, the Hodge filtration is again induced by the naive filtration on the logarithmic de Rham complex $\Omega^\bullet_{\overline{X}^{\mathrm{an}}}(\log D^{\mathrm{an}})$. Since algebraic differential forms on X have at worst poles along D, it can further be shown that those groups are still isomorphic to $H^i(X/\mathbb{C})$.

The general case of a possibly singular quasi-projective variety is dealt with in [15]. It involves the previous construction together with simplicial techniques.

Now suppose that X is defined over a subfield $K \subseteq \mathbb{C}$. Then the complex vector space $H^i(X^{\mathrm{an}}, \mathbb{C})$ has two additional structures: a $\mathbb{Q}$-structure, coming from the universal coefficients theorem,

$$H^i(X^{\mathrm{an}}, \mathbb{C}) \simeq H^i(X^{\mathrm{an}}, \mathbb{Q}) \otimes_{\mathbb{Q}} \mathbb{C},$$

and a K-structure, coming from Grothendieck's theorem,

$$H^i(X^{\mathrm{an}}, \mathbb{C}) \simeq H^i(X/K) \otimes_K \mathbb{C}.$$

In general, these two structures are not compatible with each other. It should be noted that the Hodge filtration is defined over K.

The same construction works in families to show that Hodge bundles and the Gauss–Manin connection are algebraic. Let $f \colon X \to B$ be a smooth projective morphism of varieties over $\mathbb{C}$. For each i, it determines a variation of Hodge structure on B whose underlying vector bundle is

$$\mathcal{H}^i = R^i f_* \mathbb{Q} \otimes_{\mathbb{Q}} \mathscr{O}_{B^{\mathrm{an}}} \simeq R^i f_*^{\mathrm{an}} \Omega^\bullet_{X^{\mathrm{an}}/B^{\mathrm{an}}} \simeq \left(R^i f_* \Omega^\bullet_{X/B}\right)^{\mathrm{an}}.$$

By the relative version of Grothendieck's theorem, the Hodge bundles are given by

$$F^p \mathcal{H}^i \simeq \left(R^i f_* \Omega^{\bullet \geq p}_{X/B}\right)^{\mathrm{an}}.$$

Katz and Oda have shown that the Gauss–Manin connection $\nabla \colon \mathcal{H}^i \to \Omega^1_{B^{\mathrm{an}}} \otimes \mathcal{H}^i$ can also be constructed algebraically [24]. Starting from the exact sequence

$$0 \to f^*\Omega^1_{B/\mathbb{C}} \to \Omega^1_{X/\mathbb{C}} \to \Omega^1_{X/B} \to 0,$$

let $L^r\Omega^i_{X/\mathbb{C}} = f^*\Omega^r_{B/\mathbb{C}} \wedge \Omega^{i-r}_{X/\mathbb{C}}$. We get a short exact sequence of complexes

$$0 \to f^*\Omega^1_{B/\mathbb{C}} \otimes \Omega^{\bullet-1}_{X/B} \to \Omega^\bullet_{X/\mathbb{C}}/L^2\Omega^\bullet_{X/\mathbb{C}} \to \Omega^\bullet_{X/B} \to 0,$$

and hence a connecting morphism

$$R^i f_*\Omega^\bullet_{X/B} \to R^{i+1}f_*\bigl(f^*\Omega^1_{B/\mathbb{C}} \otimes \Omega^{\bullet-1}_{X/B}\bigr) \simeq \Omega^1_{B/\mathbb{C}} \otimes R^i f_*\Omega^\bullet_{X/B}.$$

The theorem of Katz–Oda is that the associated morphism between analytic vector bundles is precisely the Gauss–Manin connection ∇.

For our purposes, the most interesting conclusion is the following: if f, X, and B are all defined over a subfield $K \subseteq \mathbb{C}$, then the same is true for the Hodge bundles $F^p\mathcal{H}^i$ and the Gauss–Manin connection ∇. We shall make use of this fact later when discussing absolute Hodge classes and Deligne's principle B.

11.1.2   Cycle Classes

Let X be a nonsingular projective variety over $\mathbb{C}$ of dimension n. Integration of differential forms gives an isomorphism

$$H^{2n}\bigl(X^{\mathrm{an}}, \mathbb{Q}(n)\bigr) \to \mathbb{Q}, \qquad \alpha \mapsto \frac{1}{(2\pi i)^n} \int_{X^{\mathrm{an}}} \alpha.$$

The reason for including the factor of $(2\pi i)^n$ is that this functional is actually the Grothendieck trace map (up to a sign factor that depends on the exact set of conventions used); see [30]. This is important when considering the comparison with algebraic de Rham cohomology below.

Remark.   Let us recall that $\mathbb{Z}(p)$ is defined to be the Hodge structure of type $(-p, -p)$ and weight $(-2p)$ on the lattice $(2i\pi)^p\mathbb{Z} \subset \mathbb{C}$, and that $\mathbb{Q}(p) = \mathbb{Z}(p) \otimes_\mathbb{Z} \mathbb{Q}$. If H is any integral Hodge structure, we denote by $H(p)$ the Hodge structure $H \otimes \mathbb{Z}(p)$; when H is a rational Hodge structure, $H(p) = H \otimes \mathbb{Q}(p)$. If X is a variety over a field K of characteristic 0, the de Rham cohomology group is filtered K-vector space $H^i_{dR}(X/K)$. We will denote by $H^i_{dR}(X/K)(p)$ the K-vector space $H^i_{dR}(X/K)$ with the filtration $F^j H^i_{dR}(X/K)(p) = F^{j+p}H^i_{dR}(X/K)$. Tensor products with $\mathbb{Z}(p)$ or $\mathbb{Q}(p)$ are called Tate twists.

Now let $Z \subseteq X$ be an algebraic subvariety of codimension p, and hence of dimension $n - p$. It determines a cycle class

$$[Z^{\mathrm{an}}] \in H^{2p}\bigl(X^{\mathrm{an}}, \mathbb{Q}(p)\bigr)$$

in Betti cohomology, as follows. Let $\widetilde{Z}$ be a resolution of singularities of Z, and let $\mu: \widetilde{Z} \to X$ denote the induced morphism. By Poincaré duality, the linear functional

$$H^{2n-2p}\big(X^{\mathrm{an}}, \mathbb{Q}(n-p)\big) \to \mathbb{Q}, \qquad \alpha \mapsto \frac{1}{(2\pi i)^{n-p}} \int_{\widetilde{Z}^{\mathrm{an}}} \mu^*(\alpha)$$

is represented by a unique class $\zeta \in H^{2p}\big(X^{\mathrm{an}}, \mathbb{Q}(p)\big)$, with the property that

$$\frac{1}{(2\pi i)^{n-p}} \int_{\widetilde{Z}^{\mathrm{an}}} \mu^*(\alpha) = \frac{1}{(2\pi i)^n} \int_{X^{\mathrm{an}}} \zeta \cup \alpha.$$

This class belongs to the group $H^{2p}\big(X^{\mathrm{an}}, \mathbb{Q}(p)\big)$ which is endowed with a weight-0 Hodge structure. In fact, one can prove, using triangulations and simplicial cohomology groups, that it actually comes from a class in $H^{2p}\big(X^{\mathrm{an}}, \mathbb{Z}(p)\big)$.

The class ζ is a Hodge class. Indeed, if $\alpha \in H^{2n-2p}\big(X^{\mathrm{an}}, \mathbb{Q}(n-p)\big)$ is of type $(n-i, n-j)$ with $i \neq j$, then either i or j is strictly greater than p, and $\int_{\widetilde{Z}^{\mathrm{an}}} \mu^*(\alpha) = 0$. This implies that $\int_{X^{\mathrm{an}}} \zeta \cup \alpha = 0$ and that ζ is of type $(0,0)$ in $H^{2p}\big(X^{\mathrm{an}}, \mathbb{Q}(p)\big)$.

An important fact is that one can also define a cycle class

$$[Z] \in F^p H^{2p}(X/\mathbb{C})$$

in algebraic de Rham cohomology such that the following comparison theorem holds.

THEOREM 11.1.3 *Under the isomorphism $H^{2p}(X/\mathbb{C}) \simeq H^{2p}(X^{\mathrm{an}}, \mathbb{C})$, we have*

$$[Z] = [Z^{\mathrm{an}}].$$

Consequently, if Z and X are both defined over a subfield $K \subseteq \mathbb{C}$, then the cycle class $[Z^{\mathrm{an}}]$ is actually defined over the algebraic closure $\bar{K}$.

In the remainder of this section, our goal is to understand the construction of the algebraic cycle class. This will also give a second explanation for the factor $(2\pi i)^p$ in the definition of the cycle class. We shall first look at a nice special case, due to Grothendieck in [22]; see also [5]. Assume for now that Z is a local complete intersection of codimension p. This means that X can be covered by open sets U, with the property that $Z \cap U = V(f_1, \ldots, f_p)$ is the zero scheme of p regular functions $f_1, \ldots, f_p$. Then $U - (Z \cap U)$ is covered by the open sets $D(f_1), \ldots, D(f_p)$, and

$$\frac{df_1}{f_1} \wedge \cdots \wedge \frac{df_p}{f_p} \tag{11.1.1}$$

is a closed p-form on $D(f_1) \cap \cdots \cap D(f_p)$. Using Čech cohomology, it determines a class in

$$H^{p-1}\big(U - (Z \cap U), \Omega^{p,\mathrm{cl}}_{X/\mathbb{C}}\big),$$

where $\Omega^{p,\mathrm{cl}}_{X/\mathbb{C}}$ is the subsheaf of $\Omega^p_{X/\mathbb{C}}$ consisting of closed p-forms. Since we have a map of complexes $\Omega^{p,\mathrm{cl}}_{X/\mathbb{C}}[-p] \to \Omega^{\bullet \geq p}_{X/\mathbb{C}}$, we get

$$H^{p-1}\big(U - (Z \cap U), \Omega^{p,\mathrm{cl}}_{X/\mathbb{C}}\big) \to \mathbb{H}^{2p-1}\big(U - (Z \cap U), \Omega^{\bullet \geq p}_{X/\mathbb{C}}\big) \to \mathbb{H}^{2p}_{Z \cap U}\big(\Omega^{\bullet \geq p}_{X/\mathbb{C}}\big).$$

One can show that the image of (11.1.1) in the cohomology group with supports on the right does not depend on the choice of local equations $f_1, \ldots, f_p$. (A good exercise is to prove this for $p = 1$ and $p = 2$.) It therefore defines a global section of the sheaf $\mathcal{H}_Z^{2p}(\Omega_{X/\mathbb{C}}^{\bullet \geq p})$. Using that $\mathcal{H}_Z^i(\Omega_{X/\mathbb{C}}^{\bullet \geq p}) = 0$ for $i \leq 2p-1$, we get from the local-to-global spectral sequence that

$$\mathbb{H}_Z^{2p}(\Omega_{X/\mathbb{C}}^{\bullet \geq p}) \simeq H^0(X, \mathcal{H}_Z^{2p}(\Omega_{X/\mathbb{C}}^{\bullet \geq p})).$$

In this way, we obtain a well-defined class in $\mathbb{H}_Z^{2p}(\Omega_{X/\mathbb{C}}^{\bullet \geq p})$, and hence in the algebraic de Rham cohomology $\mathbb{H}^{2p}(\Omega_{X/\mathbb{C}}^{\bullet \geq p}) = F^p H^{2p}(X/\mathbb{C})$.

For the general case, one uses the theory of Chern classes, which associates to a locally free sheaf $\mathcal{E}$ of rank r a sequence of Chern classes $c_1(\mathcal{E}), \ldots, c_r(\mathcal{E})$. We recall their construction in Betti cohomology and in algebraic de Rham cohomology, referring to [35, 11.2] for details and references.

First, consider the case of an algebraic line bundle $\mathcal{L}$; we denote the associated holomorphic line bundle by $\mathcal{L}^{\mathrm{an}}$. The first Chern class $c_1(\mathcal{L}^{\mathrm{an}}) \in H^2(X^{\mathrm{an}}, \mathbb{Z}(1))$ can be defined using the exponential sequence

$$0 \to \mathbb{Z}(1) \to \mathcal{O}_{X^{\mathrm{an}}} \xrightarrow{\exp} \mathcal{O}_{X^{\mathrm{an}}}^* \to 0.$$

The isomorphism class of $\mathcal{L}^{\mathrm{an}}$ belongs to $H^1(X^{\mathrm{an}}, \mathcal{O}_{X^{\mathrm{an}}}^*)$, and $c_1(\mathcal{L}^{\mathrm{an}})$ is the image of this class under the connecting homomorphism.

To relate this to differential forms, cover X by open subsets U_i on which $\mathcal{L}^{\mathrm{an}}$ is trivial, and let $g_{ij} \in \mathcal{O}_{X^{\mathrm{an}}}^*(U_i \cap U_j)$ denote the holomorphic transition functions for this cover. If each U_i is simply connected, say, then we can write $g_{ij} = e^{f_{ij}}$, and then

$$f_{jk} - f_{ik} + f_{ij} \in \mathbb{Z}(1)$$

form a 2-cocycle that represents $c_1(\mathcal{L}^{\mathrm{an}})$. Its image in $H^2(X^{\mathrm{an}}, \mathbb{C}) \simeq \mathbb{H}^2(\Omega_{X^{\mathrm{an}}}^\bullet)$ is cohomologous to the class of the 1-cocycle df_{ij} in $H^1(X^{\mathrm{an}}, \Omega_{X^{\mathrm{an}}}^1)$. But $df_{ij} = dg_{ij}/g_{ij}$, and so $c_1(\mathcal{L}^{\mathrm{an}})$ is also represented by the cocycle dg_{ij}/g_{ij}. This explains the special case $p = 1$ in Bloch's construction.

To define the first Chern class of $\mathcal{L}$ in algebraic de Rham cohomology, we use the fact that a line bundle is also locally trivial in the Zariski topology. If U_i are Zariski-open sets on which $\mathcal{L}$ is trivial, and $g_{ij} \in \mathcal{O}_X^*(U_i \cap U_j)$ denotes the corresponding transition functions, we can define $c_1(\mathcal{L}) \in F^1 H^2(X/\mathbb{C})$ as the hypercohomology class determined by the cocycle dg_{ij}/g_{ij}. In conclusion, we then have $c_1(\mathcal{L}) = c_1(\mathcal{L}^{\mathrm{an}})$ under the isomorphism in Grothendieck's theorem.

Now suppose that $\mathcal{E}$ is a locally free sheaf of rank r on X. On the associated projective bundle $\pi: \mathbb{P}(\mathcal{E}) \to X$, we have a universal line bundle $\mathcal{O}_\mathcal{E}(1)$, together with a surjection from $\pi^* \mathcal{E}$. In Betti cohomology, we have

$$H^{2r}(\mathbb{P}(\mathcal{E}^{\mathrm{an}}), \mathbb{Z}(r)) = \bigoplus_{i=0}^{r-1} \xi^i \cdot \pi^* H^{2r-2i}(X^{\mathrm{an}}, \mathbb{Z}(r-i)),$$

where $\xi \in H^2(\mathbb{P}(\mathscr{E}^{\mathrm{an}}), \mathbb{Z}(1))$ denotes the first Chern class of $\mathscr{O}_{\mathscr{E}}(1)$. Consequently, there are unique classes $c_k \in H^{2k}(X^{\mathrm{an}}, \mathbb{Z}(k))$ that satisfy the relation

$$\xi^r - \pi^*(c_1) \cdot \xi^{r-1} + \pi^*(c_2) \cdot \xi^{r-2} + \cdots + (-1)^r \pi^*(c_r) = 0,$$

and the kth Chern class of $\mathscr{E}^{\mathrm{an}}$ is defined to be $c_k(\mathscr{E}^{\mathrm{an}}) = c_k$. The same construction can be carried out in algebraic de Rham cohomology, producing Chern classes $c_k(\mathscr{E}) \in F^k H^{2k}(X/\mathbb{C})$. It follows easily from the case of line bundles that we have

$$c_k(\mathscr{E}) = c_k(\mathscr{E}^{\mathrm{an}})$$

under the isomorphism in Grothendieck's theorem.

Since coherent sheaves on regular schemes admit finite resolutions by locally free sheaves, it is possible to define Chern classes for arbitrary coherent sheaves. One consequence of the Riemann–Roch theorem is the equality

$$[Z^{\mathrm{an}}] = \frac{(-1)^{p-1}}{(p-1)!} c_p(\mathscr{O}_{Z^{\mathrm{an}}}) \in H^{2p}(X^{\mathrm{an}}, \mathbb{Q}(p)).$$

Thus it makes sense to define the cycle class of Z in algebraic de Rham cohomology by the formula

$$[Z] = \frac{(-1)^{p-1}}{(p-1)!} c_p(\mathscr{O}_Z) \in F^p H^{2p}(X/\mathbb{C}).$$

It follows that $[Z] = [Z^{\mathrm{an}}]$, and so the cycle class of Z^{an} can indeed be constructed algebraically, as claimed.

EXERCISE 11.1.4   Let X be a nonsingular projective variety defined over $\mathbb{C}$, let $D \subseteq X$ be a nonsingular hypersurface, and set $U = X - D$. One can show that $H^i(U/\mathbb{C})$ is isomorphic to the hypercohomology of the log complex $\Omega^\bullet_{X/\mathbb{C}}(\log D)$. Use this to construct a long exact sequence

$$\cdots \to H^{i-2}(D) \to H^i(X) \to H^i(U) \to H^{i-1}(D) \to \cdots$$

for the algebraic de Rham cohomology groups. Conclude by induction on the dimension of X that the restriction map

$$H^i(X/\mathbb{C}) \to H^i(U/\mathbb{C})$$

is injective for $i \le 2 \operatorname{codim} Z - 1$, and an isomorphism for $i \le 2 \operatorname{codim} Z - 2$.

11.2   ABSOLUTE HODGE CLASSES

In this section, we introduce the notion of absolute Hodge classes in the cohomology of a complex algebraic variety. While Hodge theory applies to general compact Kähler manifolds, absolute Hodge classes are brought in as a way to deal with cohomological properties of a variety coming from its algebraic structure.

This circle of ideas is closely connected to the motivic philosophy as envisioned by Grothendieck. One of the goals of this text is to give a hint of how absolute Hodge classes can allow one to give unconditional proofs for results of a motivic flavor.

11.2.1   Algebraic Cycles and the Hodge Conjecture

As an example of the need for a suitable structure on the cohomology of a complex algebraic variety that uses more than the usual Hodge theory, let us first discuss some aspects of the Hodge conjecture.

Let X be a smooth projective variety over $\mathbb{C}$. The singular cohomology groups of X are endowed with pure Hodge structures such that $H^{2p}(X, \mathbb{Z}(p))$ has weight 0 for any integer $p \in \mathbb{Z}$. We denote by $\mathrm{Hdg}^p(X)$ the group of Hodge classes in $H^{2p}(X, \mathbb{Z}(p))$.

As we showed earlier, if Z is a subvariety of X of codimension p, its cohomology class $[Z]$ in $H^{2p}(X, \mathbb{Q}(p))$ is a Hodge class. The Hodge conjecture states that the cohomology classes of subvarieties of X span the $\mathbb{Q}$-vector space generated by Hodge classes.

CONJECTURE 11.2.1   *Let X be a smooth projective variety over $\mathbb{C}$. For any nonnegative integer p, the subspace of degree-p rational Hodge classes*

$$\mathrm{Hdg}^p(X) \otimes \mathbb{Q} \subset H^{2p}(X, \mathbb{Q}(p))$$

is generated over $\mathbb{Q}$ by the cohomology classes of codimension-p subvarieties of X.

If X is only assumed to be a compact Kähler manifold, the cohomology groups $H^{2p}(X, \mathbb{Z}(p))$ still carry Hodge structures, and analytic subvarieties of X still give rise to Hodge classes. While a general compact Kähler manifold can have very few analytic subvarieties, Chern classes of coherent sheaves also are Hodge classes on the cohomology of X.

Note that on a smooth projective complex variety, analytic subvarieties are algebraic by the GAGA principle of Serre [33], and that Chern classes of coherent sheaves are linear combinations of cohomology classes of algebraic subvarieties of X. Indeed, this is true for locally free sheaves and coherent sheaves on a smooth variety have finite free resolutions. This latter result is no longer true for general compact Kähler manifolds, and indeed Chern classes of coherent sheaves can generate a strictly larger subspace than that generated by the cohomology classes of analytic subvarieties.

These remarks show that the Hodge conjecture could be generalized to the Kähler setting by asking whether Chern classes of coherent sheaves on a compact Kähler manifold generate the space of Hodge classes. This would be the natural Hodge-theoretic framework for this question. However, the answer to this question is negative, as proved by Voisin in [36].

THEOREM 11.2.2   *There exists a compact Kähler manifold X such that $\mathrm{Hdg}^2(X)$ is nontorsion while for any coherent sheaf $\mathcal{F}$ on X, the second Chern class $c_2(\mathcal{F}) = 0$.*

The proof of the preceding theorem takes X to be a general Weil torus. Weil tori are complex tori with a specific linear algebra condition which endows them with a nonzero space of Hodge classes. Note that Weil tori will be instrumental, in the projective case, in proving Deligne's theorem on absolute Hodge classes.

To our knowledge, there is no tentative formulation of a Hodge conjecture for compact Kähler manifolds. This makes it important to use those circumstances that are specific to algebraic geometry, such as the field of definition of algebraic de Rham cohomology, to deal with the Hodge conjecture for projective varieties.

11.2.2 Galois Action, Algebraic de Rham Cohomology, and Absolute Hodge Classes

The preceding subsection suggests that the cohomology of projective complex varieties has a richer underlying structure than that of a general Kähler manifold.

This brings us very close to the theory of motives, which Grothendieck envisioned in the sixties as a way to encompass cohomological properties of algebraic varieties. Even though these notes will not use the language of motives, the motivic philosophy is pervasive to all the results we will state.

Historically, absolute Hodge classes were introduced by Deligne in [16] as a way to make an unconditional use of motivic ideas. We will review his results in the next sections. The main starting point is, as we showed earlier, that the singular cohomology of a smooth proper complex algebraic variety with complex coefficients can be computed algebraically, using algebraic de Rham cohomology.

Indeed, let X be a smooth proper complex algebraic variety defined over $\mathbb{C}$. As proved in Theorem 11.1.2, we have a canonical isomorphism

$$H^*(X^{\mathrm{an}}, \mathbb{C}) \simeq \mathbb{H}^*(\Omega^\bullet_{X/\mathbb{C}}),$$

where $\Omega^\bullet_{X/\mathbb{C}}$ is the algebraic de Rham complex of the variety X over $\mathbb{C}$. A striking consequence of this isomorphism is that the singular cohomology of the manifold X^{an} with complex coefficients can be computed algebraically. Note that the topology of the field of complex numbers does not come into play in the definition of algebraic de Rham cohomology. More generally, if X is a smooth proper variety defined over any field k of characteristic 0, the hypercohomology of the de Rham complex of X over $\mathrm{Spec}\, k$ gives a k-algebra which by definition is the algebraic de Rham cohomology of X over k.

Now let Z be an algebraic cycle of codimension p in X. As we showed earlier, Z has a cohomology class

$$[Z] \in H^{2p}(X^{\mathrm{an}}, \mathbb{Q}(p))$$

which is a Hodge class, that is, the image of $[Z]$ in $H^{2p}(X^{\mathrm{an}}, \mathbb{C}(p)) \simeq H^{2p}(X/\mathbb{C})(p)$ lies in

$$F^0 H^{2p}(X/\mathbb{C})(p) = F^p H^{2p}(X/\mathbb{C}).$$

Given any automorphism σ of the field $\mathbb{C}$, we can form the conjugate variety X^σ defined as the complex variety $X \times_\sigma \operatorname{Spec} \mathbb{C}$, that is, by the Cartesian diagram

$$
\begin{array}{ccc}
X^\sigma & \xrightarrow{\ \sigma^{-1}\ } & X \\
\downarrow & & \downarrow \\
\operatorname{Spec} \mathbb{C} & \xrightarrow{\ \sigma^*\ } & \operatorname{Spec} \mathbb{C}.
\end{array}
\tag{11.2.1}
$$

It is another smooth projective variety. When X is defined by homogeneous polynomials $P_1, \ldots, P_r$ in some projective space, then X^σ is defined by the conjugates of the P_i by σ. In this case, the morphism from X^σ to X in the Cartesian diagram sends the closed point with coordinates $(x_0 : \cdots : x_n)$ to the closed point with homogeneous coordinates $(\sigma^{-1}(x_0) : \cdots : \sigma^{-1}(x_n))$, which allows us to denote it by σ^{-1}.

The morphism $\sigma^{-1} : X^\sigma \to X$ is an isomorphism of abstract schemes, but it is not a morphism of complex varieties. Pull-back of Kähler forms still induces an isomorphism between the de Rham complexes of X and X^σ,

$$
(\sigma^{-1})^* \Omega^\bullet_{X/\mathbb{C}} \xrightarrow{\sim} \Omega^\bullet_{X^\sigma/\mathbb{C}}.
\tag{11.2.2}
$$

Taking hypercohomology, we get an isomorphism

$$
(\sigma^{-1})^* : H^*(X/\mathbb{C}) \xrightarrow{\sim} H^*(X^\sigma/\mathbb{C}), \alpha \mapsto \alpha^\sigma.
$$

Note however that this isomorphism is not $\mathbb{C}$-linear, but σ-linear, that is, if $\lambda \in \mathbb{C}$, we have $(\lambda \alpha)^\sigma = \sigma(\lambda) \alpha^\sigma$. We thus get an isomorphism of complex vector spaces

$$
H^*(X/\mathbb{C}) \otimes_\sigma \mathbb{C} \xrightarrow{\sim} H^*(X^\sigma/\mathbb{C})
\tag{11.2.3}
$$

between the de Rham cohomology of X and that of X^σ. Here the notation $\otimes_\sigma$ means that we are taking the tensor product with $\mathbb{C}$ mapping to $\mathbb{C}$ via the morphism σ. Since this isomorphism comes from an isomorphism of the de Rham complexes, it preserves the Hodge filtration.

The preceding construction is compatible with the cycle map. Indeed, Z being as before a codimension-p cycle in X, we can form its conjugate Z^σ by σ. It is a codimension-p cycle in X^σ. The construction of the cycle class map in de Rham cohomology shows that we have

$$
[Z^\sigma] = [Z]^\sigma
$$

in $H^{2p}(X^\sigma/\mathbb{C})(p)$. It lies in $F^0 H^{2p}(X^\sigma/\mathbb{C})(p)$.

Now as before X^σ is a smooth projective complex variety, and its de Rham cohomology group $H^{2p}(X^\sigma/\mathbb{C})(p)$ is canonically isomorphic to the singular cohomology group $H^{2p}((X^\sigma)^{\mathrm{an}}, \mathbb{C}(p))$. The cohomology class $[Z^\sigma]$ in $H^{2p}((X^\sigma)^{\mathrm{an}}, \mathbb{C}(p)) \simeq H^{2p}(X^\sigma/\mathbb{C})(p)$ is a Hodge class. This leads to the following definition.

DEFINITION 11.2.3 *Let X be a smooth complex projective variety. Let p be a nonnegative integer, and let α be an element of $H^{2p}(X/\mathbb{C})(p)$. The cohomology class α*

is an absolute Hodge class if for every automorphism σ of $\mathbb{C}$, the cohomology class $\alpha^\sigma \in H^{2p}((X^\sigma)^{\mathrm{an}}, \mathbb{C}(p)) \simeq H^{2p}(X^\sigma/\mathbb{C}(p))$ is a Hodge class.[1]

The preceding discussion shows that the cohomology class of an algebraic cycle is an absolute Hodge class. Taking $\sigma = \mathrm{id}_\mathbb{C}$, we see that absolute Hodge classes are Hodge classes.

Using the canonical isomorphism $H^{2p}(X^{\mathrm{an}}, \mathbb{C}(p)) \simeq H^{2p}(X/\mathbb{C})(p)$, we will say that a class in $H^{2p}(X^{\mathrm{an}}, \mathbb{C})$ is absolute Hodge if its image in $H^{2p}(X/\mathbb{C})(p)$ is.

We can rephrase the definition of absolute Hodge cycles in a slightly more intrinsic way. Let k be a field of characteristic 0, and let X be a smooth projective variety defined over k. Assume that there exist embeddings of k into $\mathbb{C}$. Note that any variety defined over a field of characteristic 0 is defined over such a field as it is defined over a field generated over $\mathbb{Q}$ by a finite number of elements.

DEFINITION 11.2.4   *Let p be an integer, and let α be an element of the de Rham cohomology space $H^{2p}(X/k)$. Let τ be an embedding of k into $\mathbb{C}$, and let τX be the complex variety obtained from X by base change to $\mathbb{C}$. We say that α is a Hodge class relative to τ if the image of α in*

$$H^{2p}(\tau X/\mathbb{C}) = H^{2p}(X/k) \otimes_\tau \mathbb{C}$$

is a Hodge class. We say that α is absolute Hodge if it is a Hodge class relative to every embedding of k into $\mathbb{C}$.

Let τ be any embedding of k into $\mathbb{C}$. Since by standard field theory, any two embeddings of k into $\mathbb{C}$ are conjugated by an automorphism of $\mathbb{C}$, it is straightforward to check that such a cohomology class α is absolute Hodge if and only if its image in $H^{2p}(\tau X/\mathbb{C})$ is. Definition 11.2.4 has the advantage of not making use of automorphisms of $\mathbb{C}$.

This definition allows us to work with absolute Hodge classes in a wider setting by using other cohomology theories.

DEFINITION 11.2.5   *Let $\overline{k}$ be an algebraic closure of k. Let p be an integer, ℓ a prime number, and let α be an element of the étale cohomology space $H^{2p}(X_{\overline{k}}, \mathbb{Q}_\ell(p))$. Let τ be an embedding of $\overline{k}$ into $\mathbb{C}$, and let τX be the complex variety obtained from $X_{\overline{k}}$ by base change to $\mathbb{C}$. We say that α is a Hodge class relative to τ if the image of α in*

$$H^{2p}((\tau X)^{\mathrm{an}}, \mathbb{Q}_\ell(p)) \simeq H^{2p}(X_{\overline{k}}, \mathbb{Q}_\ell(p))$$

is a Hodge class, that is, if it is a Hodge class and lies in the rational subspace $H^{2p}((\tau X)^{\mathrm{an}}, \mathbb{Q}(p))$ of $H^{2p}((\tau X)^{\mathrm{an}}, \mathbb{Q}_\ell(p))$. We say that α is absolute Hodge if it is a Hodge class relative to every embedding of $\overline{k}$ into $\mathbb{C}$.

---

[1] Since $H^{2p}((X^\sigma)^{\mathrm{an}}, \mathbb{C})$ is only considered as a vector space here, the Tate twist might seem superfluous. We put it here to emphasize that the comparison isomorphism with de Rham cohomology contains a factor $(2\pi i)^{-p}$.

Remark. The original definition of absolute Hodge classes in [16] covers both Betti and étale cohomology. It is not clear whether absolute Hodge classes in the sense of Definitions 11.2.4 and 11.2.5 are the same; see [16, Question 2.4].

Remark. It is possible to encompass crystalline cohomology in a similar framework; see [4, 28].

Remark. It is possible to work with absolute Hodge classes on more general varieties. Indeed, while the definitions we gave above only deal with the smooth projective case, the fact that the singular cohomology of any quasi-projective variety can be computed using suitable versions of algebraic de Rham cohomology—whether through logarithmic de Rham cohomology, algebraic de Rham cohomology on simplicial schemes, or a combination of the two—makes it possible to consider absolute Hodge classes in the singular cohomology groups of a general complex variety.

Note here that if H is a mixed Hodge structure defined over $\mathbb{Z}$ with weight filtration $W_\bullet$ and Hodge filtration $F^\bullet$, a Hodge class in H is an element of $H_{\mathbb{Z}} \cap F^0 H_{\mathbb{C}} \cap W_0 H_{\mathbb{C}}$. One of the specific features of absolute Hodge classes on quasi-projective varieties is that they can be found in the odd singular cohomology groups. Let us consider the one-dimensional case as an example. Let C be a smooth complex projective curve, and let D be a divisor of degree 0 on C. Let Z be the support of D, and let C' be the complement of Z in C. It is a smooth quasi-projective curve.

As in Exercise 11.1.4, we have an exact sequence

$$0 \to H^1(C, \mathbb{Q}(1)) \to H^1(C', \mathbb{Q}(1)) \to H^0(Z, \mathbb{Q}) \to H^2(C, \mathbb{Q}(1)).$$

The divisor D has a cohomology class $d \in H^0(Z, \mathbb{Q})$. Since the degree of D is 0, d maps to 0 in $H^2(C, \mathbb{Q}(1))$. As a consequence, it comes from an element in $H^1(C', \mathbb{Q}(1))$. Now it can be proved that there exists a Hodge class in $H^1(C', \mathbb{Q}(1))$ mapping to d if and only if some multiple of the divisor D is rationally equivalent to 0.

In general, the existence of Hodge classes in extensions of mixed Hodge structures is related to Griffiths' Abel–Jacobi map; see [9]. The problem of whether these are absolute Hodge classes is linked with problems pertaining to the Bloch–Beilinson filtration and comparison results with regulators in étale cohomology; see [23].

While we will not discuss here specific features of this problem, most of the results we will state in the pure case have extensions to the mixed case; see, for instance, [12].

11.2.3 Variations on the Definition and Some Functoriality Properties

While the goal of these notes is neither to construct nor to discuss the category of motives for absolute Hodge classes, we will need to use functoriality properties of absolute Hodge classes that are very close to those motivic constructions. In this subsection, we extend the definition of absolute Hodge classes to encompass morphisms, multilinear forms, etc. This almost amounts to defining motives for absolute Hodge classes as in [17]. The next subsection will be devoted to semisimplicity results through the use of polarized Hodge structures.

The following generalizes Definition 11.2.4.

DEFINITION 11.2.6 *Let k be a field of characteristic 0 with cardinality less than or equal to the cardinality of $\mathbb{C}$. Let $(X_i)_{i \in I}$ and $(X_j)_{j \in J}$ be smooth projective varieties over $\mathbb{C}$, and let $(p_i)_{i \in I}$, $(q_j)_{j \in J}$, n be integers. Let α be an element of the tensor product*

$$\left(\bigotimes_{i \in I} H^{p_i}(X_i/k) \right) \otimes \left(\bigotimes_{j \in J} H^{q_j}(X_j/k)^* \right)(n).$$

Let τ be an embedding of k into $\mathbb{C}$. We say that α is a Hodge class relative to τ if the image of α in

$$\left(\bigotimes_{i \in I} H^{p_i}(X_i/k) \right) \otimes \left(\bigotimes_{j \in J} H^{q_j}(X_j/k)^* \right)(n) \otimes_\tau \mathbb{C}$$

$$= \left(\bigotimes_{i \in I} H^{p_i}(\tau X_i/\mathbb{C}) \right) \otimes \left(\bigotimes_{j \in J} H^{q_j}(\tau X_j/\mathbb{C})^* \right)(n)$$

is a Hodge class. We say that α is absolute Hodge if it is a Hodge class relative to every embedding of k into $\mathbb{C}$.

As before, if $k = \mathbb{C}$, we can speak of absolute Hodge classes in the group

$$\left(\bigotimes_{i \in I} H^{p_i}(X_i, \mathbb{Q}) \right) \otimes \left(\bigotimes_{j \in J} H^{q_j}(X_j, \mathbb{Q})^* \right)(n).$$

If X and Y are two smooth projective complex varieties, and if

$$f : H^p(X, \mathbb{Q}(i)) \to H^q(Y, \mathbb{Q}(j))$$

is a morphism of Hodge structures, we will say that f is absolute Hodge, or is given by an absolute Hodge class, if the element corresponding to f in

$$H^q(Y, \mathbb{Q}) \otimes H^p(X, \mathbb{Q})^*(j - i)$$

is an absolute Hodge class. Similarly, we can define what it means for a multilinear form, e.g., a polarization, to be absolute Hodge.

This definition allows us to exhibit elementary examples of absolute Hodge classes as follows.

PROPOSITION 11.2.7 *Let X be a smooth projective complex variety.*

1. Cup product defines a map

$$H^p(X, \mathbb{Q}) \otimes H^q(X, \mathbb{Q}) \to H^{p+q}(X, \mathbb{Q}).$$

This map is given by an absolute Hodge class.

2. *Poincaré duality defines an isomorphism*

$$H^p(X, \mathbb{Q}) \to H^{2d-p}(X, \mathbb{Q}(d))^*,$$

where d is the dimension of X. This map is given by an absolute Hodge class.

PROOF. This is formal. Let us write down the computations involved. Assume X is defined over k (which might be $\mathbb{C}$). We have a cup-product map

$$H^p(X/k) \otimes H^q(X/k) \to H^{p+q}(X/k).$$

Let τ be an embedding of k into $\mathbb{C}$. The induced map

$$H^p(\tau X/\mathbb{C}) \otimes H^q(\tau X/\mathbb{C}) \to H^{p+q}(\tau X/\mathbb{C})$$

is a cup product on the de Rham cohomology of τX. We know that cup product on a smooth complex projective variety is compatible with Hodge structures, which shows that it is given by a Hodge class. The conclusion follows, and a very similar argument proves the result regarding Poincaré duality. □

Morphisms given by absolute Hodge classes behave in a functorial way. The following properties are easy to prove, working as in the preceding example to track down compatibilities.

PROPOSITION 11.2.8 *Let X, Y, and Z be smooth projective complex varieties, and let*

$$f : H^p(X, \mathbb{Q}(i)) \to H^q(Y, \mathbb{Q}(j)), \; g : H^q(Y, \mathbb{Q}(j)) \to H^r(Y, \mathbb{Q}(k))$$

be morphisms of Hodge structures.

1. *If f is induced by an algebraic correspondence, then f is absolute Hodge.*

2. *If f and g are absolute Hodge, then g ∘ f is absolute Hodge.*

3. *Let*
$$f^\dagger : H^{2d'-q}(Y, \mathbb{Q}(d'-j)) \to H^{2d-p}(X, \mathbb{Q}(d-i))$$

be the adjoint of f with respect to Poincaré duality. Then f is absolute Hodge if and only if $f^\dagger$ is absolute Hodge.

4. *If f is an isomorphism, then f is absolute Hodge if and only if f^{-1} is absolute Hodge.*

Note that the last property is not known to be true for algebraic correspondences. For these, it is equivalent to the Lefschetz standard conjecture; see the next subsection. We will need a refinement of this property as follows.

PROPOSITION 11.2.9 *Let X and Y be smooth projective complex varieties, and let*

$$p: H^p(X, \mathbb{Q}(i)) \to H^p(X, \mathbb{Q}(i)) \text{ and } q: H^q(Y, \mathbb{Q}(j)) \to H^q(Y, \mathbb{Q}(j))$$

be projectors. Assume that p and q are absolute Hodge. Let V (resp. W) be the image of p (resp. q), and let

$$f: H^p(X, \mathbb{Q}(i)) \to H^q(Y, \mathbb{Q}(j))$$

be absolute Hodge. Assume that qfp induces an isomorphism from V to W. Then the composition

$$H^q(Y, \mathbb{Q}(j)) \longrightarrow W \xrightarrow{(qfp)^{-1}} V \lhook\joinrel\longrightarrow H^p(X, \mathbb{Q}(i))$$

is absolute Hodge.

PROOF. We need to check that after conjugating by any automorphism of $\mathbb{C}$, the above composition is given by a Hodge class. Since q, f, and p are absolute Hodge, we only have to check that this is true for the identity automorphism, which is the case. $\square$

This is to compare with Grothendieck's construction of the category of pure motives as a pseudo-abelian category; see for instance [3].

11.2.4   Classes Coming from the Standard Conjectures and Polarizations

Let X be a smooth projective complex variety of dimension d. The cohomology of $X \times X$ carries a number of Hodge classes which are not known to be algebraic. The standard conjectures, as stated in [21], predict that the Künneth components of the diagonal and the inverse of the Lefschetz isomorphism are algebraic. A proof of these would have a lot of consequences in the theory of pure motives. Let us prove that they are absolute Hodge classes. More generally, any cohomology class obtained from absolute Hodge classes by canonical (rational) constructions can be proved to be absolute Hodge.

First, let Δ be the diagonal of $X \times X$. It is an algebraic cycle of codimension d in $X \times X$, hence it has a cohomology class $[\Delta]$ in $H^{2d}(X \times X, \mathbb{Q}(d))$. By the Künneth formula, we have a canonical isomorphism of Hodge structures

$$H^{2d}(X \times X, \mathbb{Q}) \simeq \bigoplus_{i=0}^{2d} H^i(X, \mathbb{Q}) \otimes H^{2d-i}(X, \mathbb{Q}),$$

hence projections $H^{2d}(X \times X, \mathbb{Q}) \to H^i(X, \mathbb{Q}) \otimes H^{2d-i}(X, \mathbb{Q})$. Let π_i be the component of $[\Delta]$ in $H^i(X, \mathbb{Q}) \otimes H^{2d-i}(X, \mathbb{Q})(d) \subset H^{2d}(X \times X, \mathbb{Q})(d)$. The cohomology classes π_i are the called the Künneth components of the diagonal.

PROPOSITION 11.2.10 *The Künneth components of the diagonal are absolute Hodge cycles.*

PROOF. Clearly the π_i are Hodge classes. Let σ be an automorphism of $\mathbb{C}$. Denote by Δ^σ the diagonal of $X^\sigma \times X^\sigma = (X \times X)^\sigma$, and by π_i^σ the Künneth components of Δ^σ. These are also Hodge classes.

Let $\pi_{i,dR}$ (resp. $(\pi_i^\sigma)_{dR}$) denote the images of the π_i (resp. π_i^σ) in the de Rham cohomology of $X \times X$ (resp. $X^\sigma \times X^\sigma$). The Künneth formula holds for de Rham cohomology and is compatible with the comparison isomorphism between de Rham and singular cohomology. It follows that

$$(\pi_i^\sigma)_{dR} = (\pi_{i,dR})^\sigma.$$

Since the $(\pi_i^\sigma)_{dR}$ are Hodge classes, the conjugates of $\pi_{i,dR}$ are, which concludes the proof. $\qquad\square$

Fix an embedding of X into a projective space, and let $h \in H^2(X, \mathbb{Q}(1))$ be the cohomology class of a hyperplane section. The hard Lefschetz theorem states that for all $i \leq d$, the morphism

$$L^{d-i} = \cup h^{d-i} : H^i(X, \mathbb{Q}) \to H^{2d-i}(X, \mathbb{Q}(d-i)), \quad x \mapsto x \cup \xi^{d-i}$$

is an isomorphism.

PROPOSITION 11.2.11 *The inverse $f_i : H^{2d-i}(X, \mathbb{Q}(d-i)) \to H^i(X, \mathbb{Q})$ of the Lefschetz isomorphism is absolute Hodge.*

PROOF. This an immediate consequence of Proposition 11.2.8. $\qquad\square$

As an immediate corollary, we get the following result.

COROLLARY 11.2.12 *Let i be an integer such that $2i \leq d$. An element $x \in H^{2i}(X, \mathbb{Q})$ is an absolute Hodge class if and only if $x \cup \xi^{d-2i} \in H^{2d-2i}(X, \mathbb{Q}(d-2i))$ is an absolute Hodge class.*

Using the preceding results, one introduces polarized Hodge structures in the setting of absolute Hodge classes. Let us start with an easy lemma.

LEMMA 11.2.13 *Let X be a smooth projective complex variety of dimension d, and let $h \in H^2(X, \mathbb{Q}(1))$ be the cohomology class of a hyperplane section. Let L denote the operator given by cup product with ξ. Let i be an integer. Consider the Lefschetz decomposition*

$$H^i(X, \mathbb{Q}) = \bigoplus_{j \geq 0} L^j H^{i-2j}(X, \mathbb{Q})_{\mathrm{prim}}$$

of the cohomology of X into primitive parts. Then the projection of $H^i(X, \mathbb{Q})$ onto the component $L^j H^{i-2j}(X, \mathbb{Q})_{\mathrm{prim}}$ with respect to the Lefschetz decomposition is given by an absolute Hodge class.

PROOF. By induction, it is enough to prove that the projection of $H^i(X, \mathbb{Q})$ onto $LH^{i-2}(X, \mathbb{Q})$ is given by an absolute Hodge class. While this could be proved by an argument of Galois equivariance as before, consider the composition

$$L \circ f_i \circ L^{d-i+1} : H^i(X, \mathbb{Q}) \to H^i(X, \mathbb{Q})$$

where $f_i : H^{2d-i}(X, \mathbb{Q}) \to H^i(X, \mathbb{Q})$ is the inverse of the Lefschetz operator. It is the desired projection since $H^k(S, \mathbb{Q})_{\text{prim}}$ is the kernel of L^{d-i+1} in $H^i(S, \mathbb{Q})$. $\square$

This allows for the following result, which shows that the Hodge structures on the cohomology of smooth projective varieties can be polarized by absolute Hodge classes.

PROPOSITION 11.2.14 *Let X be a smooth projective complex variety and k be an integer. There exists an absolute Hodge class giving a pairing*

$$Q : H^k(X, \mathbb{Q}) \otimes H^k(X, \mathbb{Q}) \to \mathbb{Q}(-k)$$

which turns $H^k(X, \mathbb{Q})$ into a polarized Hodge structure.

PROOF. Let d be the dimension of X. By the hard Lefschetz theorem, we can assume $k \leq d$. Let H be an ample line bundle on X with first Chern class $h \in H^2(X, \mathbb{Q}(1))$, and let L be the endomorphism of the cohomology of X given by cup product with h. Consider the Lefschetz decomposition

$$H^k(X, \mathbb{Q}) = \bigoplus_{i \geq 0} L^i H^{k-2i}(X, \mathbb{Q})_{\text{prim}}$$

of $H^k(X, \mathbb{Q})$ into primitive parts. Let s be the linear automorphism of $H^k(X, \mathbb{Q})$ which is given by multiplication by $(-1)^i$ on $L^i H^{k-2i}(X, \mathbb{Q})_{\text{prim}}$.

By the Hodge index theorem, the pairing

$$H^k(X, \mathbb{Q}) \otimes H^k(X, \mathbb{Q}) \to \mathbb{Q}(-k), \quad \alpha \otimes \beta \mapsto \int_X \alpha \cup L^{d-k}(s(\beta))$$

turns $H^k(X, \mathbb{Q})$ into a polarized Hodge structure.

It follows from Lemma 11.2.13 that the projections of $H^k(X, \mathbb{Q})$ onto the factors $L^i H^{k-2i}(X, \mathbb{Q})_{\text{prim}}$ are given by absolute Hodge classes. Hence, the morphism s is given by an absolute Hodge class.

Since cup product is given by an absolute Hodge class, see Section 11.2.3, and L is induced by an algebraic correspondence, it follows that the pairing Q is given by an absolute Hodge class, which concludes the proof of the proposition. $\square$

PROPOSITION 11.2.15 *Let X and Y be smooth projective complex varieties, and let*

$$f : H^p(X, \mathbb{Q}(i)) \to H^q(Y, \mathbb{Q}(j))$$

be a morphism of Hodge structures. Fix polarizations on the cohomology groups of X and Y given by absolute Hodge classes. Then the orthogonal projection of $H^p(X, \mathbb{Q}(i))$ onto $\operatorname{Ker} f$ and the orthogonal projection of $H^q(Y, \mathbb{Q}(j))$ onto $\operatorname{Im} f$ are given by absolute Hodge classes.

PROOF. The proof of this result is a formal consequence of the existence of polarizations by absolute Hodge classes. It is easy to prove that the projections we consider are absolute using an argument of Galois equivariance as in the preceding subsection. Let us however give an alternate proof from linear algebra. The abstract argument corresponding to this proof can be found in [1, Section 3]. We will only prove that the orthogonal projection of $H^p(X, \mathbb{Q}(i))$ onto Ker f is absolute Hodge, the other statement being a consequence via Poincaré duality.

For ease of notation, we will not write down Tate twists. They can be recovered by weight considerations. By Poincaré duality, the polarization on $H^p(X, \mathbb{Q})$ induces an isomorphism

$$\phi : H^p(X, \mathbb{Q}) \to H^{2d-p}(X, \mathbb{Q}),$$

where d is the dimension of X, which is absolute Hodge since the polarization is. Similarly, the polarization on $H^q(Y, \mathbb{Q})$ induces a morphism

$$\psi : H^q(Y, \mathbb{Q}) \to H^{2d'-q}(Y, \mathbb{Q})$$

where d' is the dimension of Y, which is given by an absolute Hodge class.

Consider the following diagram, which does not commute

$$
\begin{array}{ccc}
H^p(X, \mathbb{Q}) & \xrightarrow{\phi} & H^{2d-p}(X, \mathbb{Q}) \\
\downarrow{\scriptstyle f} & & \uparrow{\scriptstyle f^\dagger} \\
H^q(Y, \mathbb{Q}) & \xrightarrow{\psi} & H^{2d'-q}(Y, \mathbb{Q}),
\end{array}
$$

and consider the morphism

$$h : H^p(X, \mathbb{Q}) \to H^p(X, \mathbb{Q}), \quad x \mapsto (\phi^{-1} \circ f^\dagger \circ \psi \circ f)(x).$$

Since all the morphisms in the diagram above are given by absolute Hodge classes, then h is. Let us compute the kernel and the image of h.

Let $x \in H^p(X, \mathbb{Q})$. We have $h(x) = 0$ if and only if $f^\dagger \psi f(x) = 0$, which means that for all y in $H^p(X, \mathbb{Q})$,

$$f^\dagger \psi f(x) \cup y = 0;$$

that is, since f and $f^\dagger$ are the transpose of each other,

$$\psi f(x) \cup f(y) = 0,$$

which exactly means that $f(x)$ is orthogonal to $f(H^p(X, \mathbb{Q}))$ with respect to the polarization of $H^q(Y, \mathbb{Q})$. Now the space $f(H^p(X, \mathbb{Q}))$ is a Hodge substructure of the polarized Hodge structure $H^q(Y, \mathbb{Q})$. As such, it does not contain any nonzero totally isotropic element. This implies that $f(x) = 0$ and shows that

$$\text{Ker } h = \text{Ker } f.$$

Since f and $f^\dagger$ are the transpose of each other, the image of h is clearly contained in $(\operatorname{Ker} f)^\perp$. Considering the rank of h, this readily shows that

$$\operatorname{Im} h = (\operatorname{Ker} f)^\perp.$$

The two subspaces $\operatorname{Ker} h = \operatorname{Ker} f$ and $\operatorname{Im} h = (\operatorname{Ker} f)^\perp$ of $H^p(X, \mathbb{Q})$ are in direct sum. By standard linear algebra, it follows that the orthogonal projection p of $H^p(X, \mathbb{Q})$ onto $(\operatorname{Ker} f)^\perp$ is a polynomial in h with rational coefficients. Since h is given by an absolute Hodge class, so is p, as well as $\operatorname{id} - p$, which is the orthogonal projection onto $\operatorname{Ker} f$.   □

COROLLARY 11.2.16 *Let X and Y be two smooth projective complex varieties, and let*

$$f : H^p(X, \mathbb{Q}(i)) \to H^q(Y, \mathbb{Q}(j))$$

be a morphism given by an absolute Hodge class. Let α be an absolute Hodge class in the image of f. Then there exists an absolute Hodge class $\beta \in H^p(X, \mathbb{Q}(i))$ such that $f(\beta) = \alpha$.

PROOF. By Proposition 11.2.15, the orthogonal projection of $H^p(X, \mathbb{Q}(i))$ to the subspace $(\operatorname{Ker} f)^\perp$ and the orthogonal projection of $H^q(Y, \mathbb{Q}(j))$ to $\operatorname{Im} f$ are given by absolute Hodge classes. Now Proposition 11.2.9 shows that the composition

$$H^q(Y, \mathbb{Q}(j)) \longrightarrow \operatorname{Im} f \xrightarrow{(qfp)^{-1}} (\operatorname{Ker} f)^\perp \hookrightarrow H^p(X, \mathbb{Q}(i))$$

is absolute Hodge. As such, it sends α to an absolute Hodge class β. Since α belongs to the image of f, we have $f(\beta) = \alpha$.   □

The results we proved in this subsection and the preceding one are the ones needed to construct a category of motives for absolute Hodge cycles and prove it is a semisimple abelian category. This is done in [17]. In that sense, absolute Hodge classes provide a way to work with an unconditional theory of motives, to quote André.

We actually proved more. Indeed, while the explicit proofs for Lemma 11.2.13 and Proposition 11.2.15 might seem a little longer than what would be needed, they provide the cohomology classes we need using only classes coming from the standard conjectures. This is the basis for André's notion of motivated cycles described in [1]. This paper shows that a lot of the results we obtain here about the existence of some absolute Hodge classes can actually be strengthened to motivated cycles. In particular, the algebraicity of the absolute Hodge classes we consider, which is a consequence of the Hodge conjecture, is most of the time implied by the standard conjectures.

11.2.5 Absolute Hodge Classes and the Hodge Conjecture

Let X be a smooth projective complex variety. We proved earlier that the cohomology class of an algebraic cycle in X is absolute Hodge. This remark allows us to split the Hodge conjecture into the two following conjectures.

CONJECTURE 11.2.17 *Let X be a smooth projective complex variety. Let p be a nonnegative integer, and let α be an element of $H^{2p}(X, \mathbb{Q}(p))$. Then α is a Hodge class if and only if it is an absolute Hodge class.*

CONJECTURE 11.2.18 *Let X be a smooth projective complex variety. For any non-negative integer p, the subspace of degree-p absolute Hodge classes is generated over $\mathbb{Q}$ by the cohomology classes of codimension-p subvarieties of X.*

These statements do address the problem we raised in Section 11.2.1. Indeed, while these two conjectures together imply the Hodge conjecture, neither of them makes sense in the setting of Kähler manifolds. Indeed, automorphisms of $\mathbb{C}$ other than the identity and complex conjugation are very discontinuous—e.g., they are not measurable. This makes it impossible to give a meaning to the conjugate of a complex manifold by an automorphism of $\mathbb{C}$.

Even for algebraic varieties, the fact that automorphisms of $\mathbb{C}$ are highly discontinuous appears. Let σ be an automorphism of $\mathbb{C}$, and let X be a smooth projective complex variety. Equation (11.2.3) induces a σ-linear isomorphism

$$(\sigma^{-1})^* : H^*(X^{\mathrm{an}}, \mathbb{C}) \to H^*((X^\sigma)^{\mathrm{an}}, \mathbb{C})$$

between the singular cohomology with complex coefficients of the complex manifolds underlying X and X^σ. Conjecture 11.2.17 means that Hodge classes in $H^*(X^{\mathrm{an}}, \mathbb{C})$ should map to Hodge classes in $H^*((X^\sigma)^{\mathrm{an}}, \mathbb{C})$. In particular, they should map to elements of the rational subspace $H^*((X^\sigma)^{\mathrm{an}}, \mathbb{Q})$.

However, it is not to be expected that $(\sigma^{-1})^*$ maps $H^*(X^{\mathrm{an}}, \mathbb{Q})$ to $H^*((X^\sigma)^{\mathrm{an}}, \mathbb{Q})$. It can even happen that the two algebras $H^*(X^{\mathrm{an}}, \mathbb{Q})$ and $H^*((X^\sigma)^{\mathrm{an}}, \mathbb{Q})$ are not isomorphic; see [11]. This implies, in particular, that the complex varieties X^{an} and $(X^\sigma)^{\mathrm{an}}$ need not be homeomorphic, as was first shown by Serre in [34], while the schemes X and X^σ are isomorphic. This also shows that singular cohomology with rational algebraic coefficients cannot be defined algebraically.[2]

The main goal of these notes is to discuss Conjecture 11.2.17. We will give a number of examples of absolute Hodge classes which are not known to be algebraic, and describe some applications. While Conjecture 11.2.18 seemed to be completely open at the time, we can make two remarks about it.

Let us first state a result which might stand as a motivation for the statement of this conjecture. We mentioned above that conjugation by an automorphism of $\mathbb{C}$ does not in general preserve singular cohomology with rational coefficients, but it does preserve absolute Hodge classes by definition.

Let X be a smooth projective complex variety. The singular cohomology with rational coefficients of the underlying complex manifold X^{an} is spanned by the cohomology classes of images of real submanifolds of X^{an}. The next result (see [39, Lemma 28]

---

[2]While the isomorphism we gave between the algebras $H^*(X^{\mathrm{an}}, \mathbb{C})$ and $H^*((X^\sigma)^{\mathrm{an}}, \mathbb{C})$ is not $\mathbb{C}$-linear, it is possible to show using étale cohomology that there exists a $\mathbb{C}$-linear isomorphism between these two algebras, depending on an embedding of $\mathbb{Q}_l$ into $\mathbb{C}$.

for a related statement) shows that among closed subsets of X^{an} for the usual topology, algebraic subvarieties are the only ones that remain closed after conjugation by an automorphism of $\mathbb{C}$.

Recall that if σ is an automorphism of $\mathbb{C}$, we have an isomorphism of schemes

$$\sigma : X \to X^\sigma.$$

It sends complex points of X to complex points of X^σ.

PROPOSITION 11.2.19 *Let X be a complex variety, and let F be a closed subset of X^{an}. Assume that for any automorphism σ of $\mathbb{C}$, the subset*

$$\sigma(F) \subset X^\sigma(\mathbb{C})$$

is closed in $(X^\sigma)^{\mathrm{an}}$. Then F is a countable union of algebraic subvarieties of X. If furthermore X is proper, then F is an algebraic subvariety of X.

Note that we consider closed subsets for the usual topology of X^{an}, not only for the analytic one.

PROOF. Using induction on the dimension of X, we can assume that F is not contained in a countable union of proper subvarieties of X. We want to prove that $F = X$. Using a finite map from X to a projective space, we can assume that $X = \mathbb{A}^n_{\mathbb{C}}$. Our hypothesis is thus that F is a closed subset of $\mathbb{C}^n$ which is not contained in a countable union of proper subvarieties of $\mathbb{C}^n$, such that for any automorphism of $\mathbb{C}$, $\sigma(F) = \{(\sigma(x_1), \ldots, \sigma(x_n)), (x_1, \ldots, x_n) \in \mathbb{C}^n\}$ is closed in $\mathbb{C}^n$. We will use an elementary lemma.

LEMMA 11.2.20 *Let k be a countable subfield of $\mathbb{C}$. There exists a point $(x_1, \ldots, x_n)$ in F such that the complex numbers $(x_1, \ldots, x_n)$ are algebraically independent over k.*

PROOF. Since k is countable, there exists only a countable number of algebraic subvarieties of $\mathbb{C}^n$ defined over k. By our assumption on F, there exists a point of F which does not lie in any proper algebraic variety defined over k. Such a point has coordinates which are algebraically independent over k. □

Using the preceding lemma and induction, we can find a sequence of points

$$p_i = (x_1^i, \ldots, x_n^i) \in F$$

such that the $(x_j^i)_{i \in \mathbb{N}, j \le n}$ are algebraically independent over $\mathbb{Q}$. Now let $(y_j^i)_{i \in \mathbb{N}, j \le n}$ be a sequence of algebraically independent points in $\mathbb{C}^n$ such that $\{(y_1^i, \ldots, y_n^i), i \in \mathbb{N}\}$ is dense in $\mathbb{C}^n$. We can find an automorphism σ of $\mathbb{C}$ mapping x_j^i to y_j^i for all i, j. The closed subset $\sigma(F) \subseteq \mathbb{C}^n$ contains a dense subset of $\mathbb{C}^n$; hence $\sigma(F) = \mathbb{C}^n$. This shows that $F = \mathbb{C}^n$ and concludes the proof of the first part. The proper case follows using a standard compactness argument. □

Given that absolute Hodge classes are classes in the singular cohomology groups that are, in some sense, preserved by automorphisms of $\mathbb{C}$,[3] and that by the preceding result, algebraic subvarieties are the only closed subsets with good behavior with respect to the Galois action, this might serve as a motivation for Conjecture 11.2.18.

Another, more precise, reason that explains why Conjecture 11.2.18 might be more tractable than the Hodge conjecture is given by the work of André around motivated cycles in [1]. Through motivic considerations, André does indeed show that for most of the absolute Hodge classes we know, Conjecture 11.2.18 is actually a consequence of the standard conjectures, which, at least in characteristic 0, seem considerably weaker than the Hodge conjecture.

While we will not prove such results, it is to be noted that the proofs we gave in Sections 11.2.3 and 11.2.4 were given so as to imply André's results for the absolute Hodge classes we will consider. The interested reader should have no problem filling in the details.

In the following sections, we will not use the notation X^{an} for the complex manifold underlying a complex variety X anymore, but rather, by an abuse of notation, use X to refer to both objects. The context will hopefully help the reader avoid any confusion.

11.3   ABSOLUTE HODGE CLASSES IN FAMILIES

This section deals with the behavior of absolute Hodge classes under deformations. We will focus on consequences of the algebraicity of Hodge bundles. We prove Deligne's principle B, which states that absolute Hodge classes are preserved by parallel transport, and discuss the link between Hodge loci and absolute Hodge classes as in [38]. The survey [40] contains a beautiful account of similar results.

We only work here with projective families. Some aspects of the quasi-projective case are treated in [12].

11.3.1   The Variational Hodge Conjecture and the Global Invariant Cycle Theorem

Before stating Deligne's principle B of [16], let us explain a variant of the Hodge conjecture.

Let S be a smooth connected complex quasi-projective variety, and let $\pi : \mathcal{X} \to S$ be a smooth projective morphism. Let 0 be a complex point of S, and, for some integer p let α be a cohomology class in $H^{2p}(\mathcal{X}_0, \mathbb{Q}(p))$. Assume that α is the cohomology class of some codimension p algebraic cycle Z_0, and that α extends as a section $\widetilde{\alpha}$ of the local system $R^{2p}\pi_*\mathbb{Q}(p)$ on S.

In [20, footnote 13], Grothendieck makes the following conjecture.

---

[3] See [16, Question 2.4], where the questions of whether these are the only ones is raised.

CONJECTURE 11.3.1 (Variational Hodge conjecture)   *For any complex point s of S, the class $\widetilde{\alpha}_s$ is the cohomology class of an algebraic cycle.*

Using the Gauss–Manin connection and the isomorphism between de Rham and singular cohomology, we can formulate an alternative version of the variational Hodge conjecture in de Rham cohomology. For this, keeping the notation as above, we have a coherent sheaf $\mathcal{H}^{2p} = \mathbb{R}^{2p}\pi_*\Omega^\bullet_{\mathcal{X}/S}$ which computes the relative de Rham cohomology of $\mathcal{X}$ over S. As we saw earlier, it is endowed with a canonical connection: the Gauss–Manin connection ∇.

CONJECTURE 11.3.2 (Variational Hodge conjecture for de Rham cohomology)   *Let β be a cohomology class in $H^{2p}(\mathcal{X}_0/\mathbb{C})$. Assume that β is the cohomology class of some codimension-p algebraic cycle Z_0, and that β extends as a section $\widetilde{\beta}$ of the coherent sheaf $\mathcal{H}^{2p} = \mathbb{R}^{2p}\pi_*\Omega^\bullet_{\mathcal{X}/S}$ such that $\widetilde{\beta}$ is flat for the Gauss–Manin connection.*

The variational Hodge conjecture states that for any complex point s of S, the class $\widetilde{\beta}_s$ is the cohomology class of an algebraic cycle.

Remark.   Note that both these conjectures are clearly false in the analytic setting. Indeed, if one takes S to be a simply connected subset of $\mathbb{C}^n$, the hypothesis that α extends to a global section of $R^{2p}\pi_*\mathbb{Q}(p)$ over S is automatically satisfied since the latter local system is trivial. This easily gives rise to counterexamples even in degree 2.

PROPOSITION 11.3.3   *Conjectures 11.3.1 and 11.3.2 are equivalent.*

PROOF.   The de Rham comparison isomorphism between singular and de Rham cohomology in a relative context takes the form of a canonical isomorphism

$$\mathbb{R}^{2p}\pi_*\Omega^\bullet_{\mathcal{X}/S} \simeq R^{2p}\pi_*\mathbb{Q}(p) \otimes_\mathbb{Q} \mathcal{O}_S. \tag{11.3.1}$$

Note that this formula is not one from algebraic geometry. Indeed, the sheaf $\mathcal{O}_S$ denotes here the sheaf of holomorphic functions on the complex manifold S. The derived functor $\mathbb{R}^{2p}\pi_*$ on the left is a functor between categories of complexes of holomorphic coherent sheaves, while the one on the right is computed for sheaves with the usual complex topology. The Gauss–Manin connection is the connection on $\mathbb{R}^{2p}\pi_*\Omega^\bullet_{\mathcal{X}/S}$ for which the local system $R^{2p}\pi_*\mathbb{Q}(p)$ is constant. As we saw earlier, the locally free sheaf $R^{2p}\pi_*\Omega^\bullet_{\mathcal{X}/S}$ is algebraic, i.e., is induced by a locally free sheaf on the algebraic variety S, as well as the Gauss–Manin connection.

Given β a cohomology class in the de Rham cohomology group $H^{2p}(\mathcal{X}_0/\mathbb{C})$ as above, we know that β belongs to the rational subspace $H^{2p}(\mathcal{X}_0, \mathbb{Q}(p))$ because it is the cohomology class of an algebraic cycle. Furthermore, since $\widetilde{\beta}$ is flat for the Gauss–Manin connection and is rational at one point, it corresponds to a section of the local system $R^{2p}\pi_*\mathbb{Q}(p)$ under the comparison isomorphism above. This shows that Conjecture 11.3.1 implies Conjecture 11.3.2.

On the other hand, sections of the local system $R^{2p}\pi_*\mathbb{Q}(p)$ induce flat holomorphic sections of the coherent sheaf $\mathbb{R}^{2p}\pi_*\Omega^\bullet_{\mathcal{X}/S}$. We have to show that they are algebraic. This is a consequence of the following important result, which is due to Deligne.

THEOREM 11.3.4 (Global invariant cycle theorem) *Let $\pi : \mathcal{X} \to S$ be a smooth projective morphism of quasi-projective complex varieties, and let $i : \mathcal{X} \hookrightarrow \overline{\mathcal{X}}$ be a smooth compactification of $\mathcal{X}$. Let 0 be complex point of S, and let $\pi_1(S, 0)$ be the fundamental group of S. For any integer k, the space of monodromy-invariant classes of degree k*

$$H^k(\mathcal{X}_0, \mathbb{Q})^{\pi_1(S,0)}$$

is equal to the image of the restriction map

$$i_0^* : H^k(\overline{\mathcal{X}}, \mathbb{Q}) \to H^k(\mathcal{X}_0, \mathbb{Q}),$$

where i_0 is the inclusion of $\mathcal{X}_0$ in $\overline{\mathcal{X}}$.

In the theorem, the monodromy action is the action of the fundamental group $\pi_1(S, 0)$ on the cohomology groups of the fiber $\mathcal{X}_0$. Note that the theorem implies that the space $H^k(\mathcal{X}_0, \mathbb{Q})^{\pi_1(S,0)}$ is a sub-Hodge structure of $H^k(\mathcal{X}_0, \mathbb{Q})$. However, the fundamental group of S does not in general act by automorphisms of Hodge structures.

The global invariant cycle theorem implies the algebraicity of flat holomorphic sections of the vector bundle $R^{2p}\pi_*\Omega^\bullet_{\mathcal{X}/S}$ as follows. Let $\widetilde{\beta}$ be such a section, and keep the notation of the theorem. By definition of the Gauss–Manin connection, $\widetilde{\beta}$ corresponds to a section of the local system $R^{2p}\pi_*\mathbb{C}$ under the isomorphism (11.3.1), that is, to a monodromy-invariant class in $H^{2p}(\mathcal{X}_0, \mathbb{C})$. The global invariant cycle theorem shows, using the comparison theorem between singular and de Rham cohomology on $\overline{\mathcal{X}}$, that $\widetilde{\beta}$ comes from a de Rham cohomology class b in $H^{2p}(\overline{\mathcal{X}}/\mathbb{C})$. As such, it is algebraic.

The preceding remarks readily show the equivalence of the two versions of the variational Hodge conjecture. $\qquad\square$

The next proposition shows that the variational Hodge conjecture is actually a part of the Hodge conjecture. This fact is a consequence of the global invariant cycle theorem. The following proof will be rewritten in the next subsection to give results on absolute Hodge cycles.

PROPOSITION 11.3.5 *Let S be a smooth connected quasi-projective variety, and let $\pi : \mathcal{X} \to S$ be a smooth projective morphism. Let 0 be a complex point of S, and let p be an integer.*

1. *Let α be a cohomology class in $H^{2p}(\mathcal{X}_0, \mathbb{Q}(p))$. Assume that α is a Hodge class and that α extends as a section $\widetilde{\alpha}$ of the local system $R^{2p}\pi_*\mathbb{Q}(p)$ on S. Then for any complex point s of S, the class $\widetilde{\alpha}_s$ is a Hodge class.*

2. *Let β be a cohomology class in $H^{2p}(\mathcal{X}_0/\mathbb{C})$. Assume that β is a Hodge class and that β extends as a section $\widetilde{\beta}$ of the coherent sheaf $R^{2p}\pi_*\Omega^\bullet_{\mathcal{X}/S}$ such that $\widetilde{\beta}$ is flat for the Gauss–Manin connection. Then for any complex point s of S, the class $\widetilde{\beta}_s$ is a Hodge class.*

As an immediate corollary, we get the following.

COROLLARY 11.3.6 *The Hodge conjecture implies the variational Hodge conjecture.*

PROOF OF PROPOSITION 11.3.5. The two statements are equivalent by the arguments of Proposition 11.3.3. Let us keep the notation as above. We want to prove that for any complex point s of S, the class $\widetilde{\alpha}_s$ is a Hodge class. Let us show how this is a consequence of the global invariant cycle theorem. This is a simple consequence of Corollary 11.2.16 in the—easier—context of Hodge classes. Let us prove the result from scratch.

As in Proposition 11.2.14, we can find a pairing

$$H^{2p}(\overline{\mathcal{X}}, \mathbb{Q}) \otimes H^{2p}(\overline{\mathcal{X}}, \mathbb{Q}) \to \mathbb{Q}(-p)$$

which turns $H^{2p}(\overline{\mathcal{X}}, \mathbb{Q})$ into a polarized Hodge structure.

Let $i : \mathcal{X} \hookrightarrow \overline{\mathcal{X}}$ be a smooth compactification of $\mathcal{X}$, and let i_0 be the inclusion of $\mathcal{X}_0$ in $\overline{\mathcal{X}}$.

By the global invariant cycle theorem, the morphism

$$i_0^* : H^{2p}(\overline{\mathcal{X}}, \mathbb{Q}) \to H^{2p}(\mathcal{X}_0, \mathbb{Q})^{\pi_1(S,0)}$$

is surjective. It restricts to an isomorphism of Hodge structures

$$i_0^* : (\operatorname{Ker} i_0^*)^{\perp} \to H^{2p}(\mathcal{X}_0, \mathbb{Q})^{\pi_1(S,0)},$$

hence a Hodge class $a \in (\operatorname{Ker} i_0^*)^{\perp} \subset H^{2p}(\overline{\mathcal{X}}, \mathbb{Q})$ mapping to α. Indeed, saying that α extends to a global section of the local system $R^{2p}\pi_*\mathbb{Q}(p)$ exactly means that α is monodromy-invariant.

Now let i_s be the inclusion of $\mathcal{X}_s$ in $\overline{\mathcal{X}}$. Since S is connected, we have $\widetilde{\alpha}_s = i_s^*(a)$, which shows that $\widetilde{\alpha}_s$ is a Hodge class. $\qquad\square$

It is an important fact that the variational Hodge conjecture is a purely algebraic statement. Indeed, we saw earlier that both relative de Rham cohomology and the Gauss–Manin connection can be defined algebraically. This is to be compared to the above discussion of the transcendental aspect of the Hodge conjecture, where one cannot avoid using singular cohomology, which cannot be defined in a purely algebraic fashion as it does depend on the topology of $\mathbb{C}$.

Very little seems to be known about the variational Hodge conjecture; see however [5].

11.3.2   Deligne's Principle B

In this subsection, we state and prove the so-called principle B for absolute Hodge cycles, which is due to Deligne. It shows that the variational Hodge conjecture is true if one replaces algebraic cohomology classes by absolute Hodge classes.

THEOREM 11.3.7 (Principle B, [16, Theorem 2.12]) *Let S be a smooth connected complex quasi-projective variety, and let $\pi : \mathcal{X} \to S$ be a smooth projective morphism.*

Let 0 be a complex point of S, and, for some integer p let α be a cohomology class in $H^{2p}(\mathcal{X}_0, \mathbb{Q}(p))$. Assume that α is an absolute Hodge class and that α extends as a section $\widetilde{\alpha}$ of the local system $R^{2p}\pi_*\mathbb{Q}(p)$ on S. Then for any complex point s of S, the class $\widetilde{\alpha}_s$ is absolute Hodge.

As in Proposition 11.3.3, this is equivalent to the following rephrasing.

THEOREM 11.3.8 (Principle B for de Rham cohomology)   *Let S be a smooth connected quasi-projective variety, and let $\pi : \mathcal{X} \to S$ be a smooth projective morphism. Let 0 be a complex point of S, and, for some integer p let β be a cohomology class in $H^{2p}(\mathcal{X}_0/\mathbb{C})$. Assume that β is an absolute Hodge class and that β extends as a flat section $\widetilde{\beta}$ of the locally free sheaf $\mathcal{H}^{2p} = \mathbb{R}^{2p}\pi_*\Omega^\bullet_{\mathcal{X}/S}$ endowed with the Gauss–Manin connection. Then for any complex point s of S, the class $\widetilde{\beta}_s$ is absolute Hodge.*

We will give two different proofs of this result to illustrate the techniques we introduced earlier. Both rely on Proposition 11.3.5, and on the global invariant cycle theorem. The first one proves the result as a consequence of the algebraicity of the Hodge bundles and of the Gauss–Manin connection. It is essentially Deligne's proof in [16]. The second proof elaborates on polarized Hodge structures and is inspired by André's approach in [1].

PROOF. We work with de Rham cohomology. Let σ be an automorphism of $\mathbb{C}$. Since $\widetilde{\beta}$ is a global section of the locally free sheaf $\mathcal{H}^{2p}$, we can form the conjugate section $\widetilde{\beta}^\sigma$ of the conjugate sheaf $(\mathcal{H}^{2p})^\sigma$ on S^σ. Now as in Section 11.2.2, this sheaf identifies with the relative de Rham cohomology of $\mathcal{X}^\sigma$ over S^σ.

Fix a complex point s in S. We want to show that the class $\widetilde{\beta}_s$ is absolute Hodge. This means that for any automorphism σ of $\mathbb{C}$, the class $\widetilde{\beta}^\sigma_{\sigma(s)}$ is a Hodge class in the cohomology of $\mathcal{X}^\sigma_{\sigma(s)}$. Now since $\beta = \widetilde{\beta}_0$ is an absolute Hodge class by assumption, $\widetilde{\beta}^\sigma_{\sigma(0)}$ is a Hodge class.

Since the construction of the Gauss–Manin connection commutes with base change, the Gauss–Manin connection ∇^σ on the relative de Rham cohomology of $\mathcal{X}^\sigma$ over S^σ is the conjugate by σ of the Gauss–Manin connection on $\mathcal{H}^{2p}$.

These remarks allow us to write

$$\nabla^\sigma\widetilde{\beta}^\sigma = (\nabla\widetilde{\beta})^\sigma = 0$$

since $\widetilde{\beta}$ is flat. This shows that $\widetilde{\beta}^\sigma$ is a flat section of the relative de Rham cohomology of $\mathcal{X}^\sigma$ over S^σ. Since $\widetilde{\beta}^\sigma_{\sigma(0)}$ is a Hodge class, then Proposition 11.3.5 shows that $\widetilde{\beta}^\sigma_{\sigma(s)}$ is a Hodge class, which is what we needed to prove. □

Note that while the above proof may seem just a formal computation, it actually uses in an essential way the important fact that both relative de Rham cohomology and the Gauss–Manin connection are algebraic objects, which makes it possible to conjugate them by field automorphisms.

Let us give a second proof of principle B.

PROOF. This is a consequence of Corollary 11.2.16. Indeed, let $i : \mathcal{X} \hookrightarrow \overline{\mathcal{X}}$ be a smooth compactification of $\mathcal{X}$, and let i_0 be the inclusion of $\mathcal{X}_0$ in $\overline{\mathcal{X}}$.

By the global invariant cycle theorem, the morphism

$$i_0^* : H^{2p}(\overline{\mathcal{X}}, \mathbb{Q}) \to H^{2p}(\mathcal{X}_0, \mathbb{Q})^{\pi_1(S,0)}$$

is surjective. As a consequence, since α is monodromy invariant, it belongs to the image of i_0^*. By Corollary 11.2.16, we can find an absolute Hodge class $a \in H^{2p}(\overline{\mathcal{X}}, \mathbb{Q})$ mapping to α. Now let i_s be the inclusion of $\mathcal{X}_s$ in $\overline{\mathcal{X}}$. Since S is connected, we have

$$\widetilde{\alpha}_s = i_s^*(a),$$

which shows that $\widetilde{\alpha}_s$ is an absolute Hodge class, and concludes the proof. □

Note that following the remarks we made around the notion of motivated cycles, this argument could be used to prove that the standard conjectures imply the variational Hodge conjecture; see [1].

Principle B will be one of our main tools in proving that some Hodge classes are absolute. When working with families of varieties, it allows us to work with specific members of the family where algebraicity results might be known. When proving that the Kuga–Satake correspondence between a projective K3 surface and its Kuga–Satake abelian variety is absolute Hodge, it will make it possible to reduce to the case of Kummer surfaces, while in the proof of Deligne's theorem that Hodge classes on abelian varieties are absolute, it allows for a reduction to the case of abelian varieties with complex multiplication. Its mixed case version is instrumental to the results of [12].

11.3.3 The Locus of Hodge Classes

In this subsection, we recall the definitions of the Hodge locus and the locus of Hodge classes associated to a variation of Hodge structures and discuss their relation to the Hodge conjecture . The study of those was started by Griffiths in [21]. References on this subject include [35, Chapter 17] and [40]. To simplify matters, we will only deal with variations of Hodge structures coming from geometry, that is, coming from the cohomology of a family of smooth projective varieties. We will point out statements that generalize to the quasi-projective case.

Let S be a smooth complex quasi-projective variety, and let $\pi : \mathcal{X} \to S$ be a smooth projective morphism. Let p be an integer. As earlier, consider the Hodge bundles

$$\mathcal{H}^{2p} = \mathbb{R}^{2p} \pi_* \Omega^\bullet_{\mathcal{X}/S}$$

together with the Hodge filtration

$$F^k \mathcal{H}^{2p} = \mathbb{R}^{2p} \pi_* \Omega^{\bullet \geq k}_{\mathcal{X}/S}.$$

These are algebraic vector bundles over S, as we saw before. They are endowed with the Gauss–Manin connection

$$\nabla : \mathcal{H}^{2p} \to \mathcal{H}^{2p} \otimes \Omega^1_{\mathcal{X}/S}.$$

Furthermore, the local system

$$H^{2p}_{\mathbb{Q}} = R^{2p}\pi_* \mathbb{Q}(p)$$

injects into $\mathcal{H}^{2p}$ and is flat with respect to the Gauss–Manin connection.

Let us start with a set-theoretic definition of the locus of Hodge classes.

DEFINITION 11.3.9 *The locus of Hodge classes in $\mathcal{H}^{2p}$ is the set of pairs (α, s), $s \in S(\mathbb{C})$, $\alpha \in \mathcal{H}^{2p}_s$, such that α is a Hodge class, that is, $\alpha \in F^p \mathcal{H}^{2p}_s$ and $\alpha \in H^{2p}_{\mathbb{Q}, s}$.*

It turns out that the locus of Hodge classes is the set of complex points of a countable union of analytic subvarieties of $\mathcal{H}^{2p}$. This can be seen as follows—see the above references for a thorough description. Let (α, s) be in the locus of Hodge classes of $\mathcal{H}^{2p}$. We want to describe the component of the locus of Hodge classes passing through (α, s) as an analytic variety in a neighborhood of (α, s).

On a neighborhood of s, the class $\widetilde{\alpha}$ extends to a flat holomorphic section of $\mathcal{H}^{2p}$. Now the points $(\widetilde{\alpha}_t, t)$, for t in the neighborhood of s, which belong to the locus of Hodge classes are the points of an analytic variety, namely, the variety defined by the $(\widetilde{\alpha}_t, t)$ such that $\widetilde{\alpha}_t$ vanishes in the holomorphic (and even algebraic) vector bundle $\mathcal{H}^{2p}/F^p\mathcal{H}^{2p}$.

It follows from this remark that the locus of Hodge classes is a countable union of analytic subvarieties of $\mathcal{H}^{2p}$. Note that if we were to consider only integer cohomology classes to define the locus of Hodge classes, we would actually get an analytic subvariety. The locus of Hodge classes was introduced in [10]. It is of course very much related to the more classical Hodge locus.

DEFINITION 11.3.10 *The Hodge locus associated to $\mathcal{H}^{2p}$ is the projection on S of the locus of Hodge classes. It is a countable union of analytic subvarieties of S.*

Note that the Hodge locus is interesting only when $\mathcal{H}^{2p}$ has no flat global section of type (p, p). Indeed, if it has, the Hodge locus is S itself. However, in this case, one can always split off any constant variation of Hodge structures for $\mathcal{H}^{2p}$ and consider the Hodge locus for the remaining variation of Hodge structures.

The reason why we are interested in these loci is the way they are related to the Hodge conjecture. Indeed, one has the following.

PROPOSITION 11.3.11 *If the Hodge conjecture is true, then the locus of Hodge classes and the Hodge locus for $\mathcal{H}^{2p} \to S$ are countable unions of closed algebraic subsets of $\mathcal{H}^{2p}$ and S respectively.*

PROOF. We only have to prove the proposition for the locus of Hodge classes. If the Hodge conjecture is true, the locus of Hodge classes is the locus of cohomology classes of algebraic cycles with rational coefficients. These algebraic cycles are parametrized by Hilbert schemes for the family $\mathcal{X}/B$. Since these are proper and have countably many connected components, the Hodge locus is a countable union of closed algebraic subsets of $\mathcal{H}^{2p}$. □

This consequence of the Hodge conjecture is a theorem proved in [10].

THEOREM 11.3.12 *(Cattani–Deligne–Kaplan) With the notation above, the locus of Hodge classes and the Hodge locus for $\mathcal{H}^{2p} \rightarrow S$ are countable unions of closed algebraic subsets of $\mathcal{H}^{2p}$ and S, respectively.*

As before, the preceding constructions can be carried out in the quasi-projective case. Generalized versions of the Hodge conjecture lead to similar algebraicity predictions, and indeed the corresponding algebraicity result for variations of mixed Hodge structures is proved in [8], after the work of Brosnan–Pearlstein on the zero locus of normal functions in [7].

11.3.4   Galois Action on Relative de Rham Cohomology

Let S be a smooth irreducible quasi-projective variety over a field k, and let $\pi : \mathcal{X} \rightarrow S$ be a smooth projective morphism. Let p be an integer. Consider again the Hodge bundles $\mathcal{H}^{2p}$ together with the Hodge filtration $F^k \mathcal{H}^{2p} = R^{2p} \pi_* \Omega_{\mathcal{X}/S}^{\bullet \geq k}$. They are defined over k.

Let α be a section of $\mathcal{H}^{2p}$ over S. Let η be the generic point of S. The class α induces a class α_η in the de Rham cohomology of the generic fiber $\mathcal{X}_\eta$ of π.

Let σ be any embedding of $k(S)$ in $\mathbb{C}$ over k. The morphism σ corresponds to a morphism $\operatorname{Spec}(\mathbb{C}) \rightarrow \eta \rightarrow S$, hence it induces a complex point s of $S_\mathbb{C}$. We have an isomorphism

$$\mathcal{X}_\eta \times_{k(S)} \mathbb{C} \simeq \mathcal{X}_{\mathbb{C},s}$$

and the cohomology class α_η pulls back to a class α_s in the cohomology of $\mathcal{X}_{\mathbb{C},s}$.

The class α_s only depends on the complex point s. Indeed, it can be obtained the following way. The class α pulls back as a section $\alpha_\mathbb{C}$ of $\mathcal{H}_\mathbb{C}^{2p}$ over $S_\mathbb{C}$. The class α_s is the value of $\alpha_\mathbb{C}$ at the point $s \in S(\mathbb{C})$.

The following rephrases the definition of an absolute Hodge class.

PROPOSITION 11.3.13 *Assume that α_η is an absolute Hodge class. If α_η is absolute, then α_s is a Hodge class. Furthermore, in the case $k = \mathbb{Q}$, α_η is absolute if and only if α_s is a Hodge class for all s induced by embeddings $\sigma : \mathbb{Q}(S) \rightarrow \mathbb{C}$.*

We now investigate the implications of the previous rephrasing.

LEMMA 11.3.14 *Assume the field k is countable. Then the set of points $s \in S_\mathbb{C}(\mathbb{C})$ induced by embeddings of $k(S)$ in $\mathbb{C}$ over k is dense in $S_\mathbb{C}(\mathbb{C})$ for the usual topology.*

PROOF. Say that a complex point of $S_{\mathbb{C}}$ is very general if it does not lie in any proper algebraic subset of $S_{\mathbb{C}}$ defined over k. Since k is countable, the Baire theorem shows that the set of general points is dense in $S_{\mathbb{C}}(\mathbb{C})$ for the usual topology.

Now consider a very general point s. There exists an embedding of $k(S)$ into $\mathbb{C}$ such that the associated complex point of $S_{\mathbb{C}}$ is s. Indeed, s being very general exactly means that the image of the morphism

$$\operatorname{Spec}(\mathbb{C}) \xrightarrow{s} S_{\mathbb{C}} \longrightarrow S$$

is η, the generic point of S, hence a morphism $\operatorname{Spec}(\mathbb{C}) \to \eta$ giving rise to s. This concludes the proof of the lemma. □

We say that a complex point of $S_{\mathbb{C}}$ is very general if it lies in the aforementioned subset.

THEOREM 11.3.15 *Let S be a smooth irreducible quasi-projective variety over a subfield k of $\mathbb{C}$ with generic point η, and let $\pi : \mathcal{X} \to S$ be a smooth projective morphism. Let p be an integer, and let α be a section of $\mathcal{H}^{2p}$ over S.*

1. *Assume the class $\alpha_\eta \in H^{2p}(\mathcal{X}_\eta/k(S))$ is absolute Hodge. Then α is flat for the Gauss–Manin connection and $\alpha_{\mathbb{C}}$ is a Hodge class at every complex point of $S_{\mathbb{C}}$.*

2. *Assume that $k = \mathbb{Q}$. Then the class $\alpha_\eta \in H^{2p}(\mathcal{X}_\eta/\mathbb{Q}(S))$ is absolute Hodge if and only if α is flat for the Gauss–Manin connection and for any connected component S' of $S_{\mathbb{C}}$, there exists a complex point s of S' such that α_s is a Hodge class.*

PROOF. All the objects we are considering are defined over a subfield of k that is finitely generated over $\mathbb{Q}$, so we can assume that k is finitely generated over $\mathbb{Q}$, hence countable. Let $\alpha_{\mathbb{C}}$ be the section of $\mathcal{H}_{\mathbb{C}}^{2p}$ over $S_{\mathbb{C}}$ obtained by pulling back α. The value of the class $\alpha_{\mathbb{C}}$ at any general point is a Hodge class. Locally on $S_{\mathbb{C}}$, the bundle $\mathcal{H}_{\mathbb{C}}^{2p}$ with the Gauss–Manin connection is biholomorphic to the flat bundle $S \times \mathbb{C}^n$, n being the rank of $\mathcal{H}_{\mathbb{C}}^{2p}$, and we can assume such a trivialization respects the rational subspaces.

Under such trivializations, the section $\alpha_{\mathbb{C}}$ is given locally on $S_{\mathbb{C}}$ by n holomorphic functions which take rational values on a dense subset. It follows that $\alpha_{\mathbb{C}}$ is locally constant, that is, that $\alpha_{\mathbb{C}}$, hence α, is flat for the Gauss–Manin connection. Since α is absolute Hodge, $\alpha_{\mathbb{C}}$ is a Hodge class at any very general point of $S_{\mathbb{C}}$. Since these are dense in $S_{\mathbb{C}}(\mathbb{C})$, Proposition 11.3.5 shows that $\alpha_{\mathbb{C}}$ is a Hodge class at every complex point of $S_{\mathbb{C}}$. This proves the first part of the theorem.

For the second part, assuming α is flat for the Gauss–Manin connection and α_s is Hodge for points s in all the connected components of $S_{\mathbb{C}}$, Proposition 11.3.5 shows that α_s is a Hodge class at all the complex points s of $S_{\mathbb{C}}$. In particular, this true for the general points of $S_{\mathbb{C}}$, which proves that α_η is an absolute Hodge class by Proposition 11.3.13. □

As a corollary, we get the following important result.

COROLLARY 11.3.16 *Let k be an algebraically closed subfield of $\mathbb{C}$, and let X be a smooth projective variety over k. Let α be an absolute Hodge class of degree $2p$ in $X_{\mathbb{C}}$. Then α is defined over k, that is, α is the pull-back of an absolute Hodge class in X.*

PROOF. The cohomology class α belongs to $H^{2p}(X_{\mathbb{C}}/\mathbb{C}) = H^{2p}(X/k) \otimes \mathbb{C}$. We need to show that it lies in $H^{2p}(X/k) \subset H^{2p}(X_{\mathbb{C}}/\mathbb{C})$, that is, that it is defined over k.

The class α is defined over a field K finitely generated over k. Since K is generated by a finite number of elements over k, we can find a smooth irreducible quasi-projective variety S defined over k such that K is isomorphic to $k(S)$. Let $\mathcal{X} = X \times S$, and let π be the projection of $\mathcal{X}$ onto S. Saying that α is defined over $k(S)$ means that α is a class defined at the generic fiber of π. Up to replacing S by a Zariski-open subset, we can assume that α extends to a section $\widetilde{\alpha}$ of the relative de Rham cohomology group $\mathcal{H}^{2p}$ of $\mathcal{X}$ over S. Since α is an absolute Hodge class, Theorem 11.3.15 shows that $\widetilde{\alpha}$ is flat with respect to the Gauss–Manin connection on $\mathcal{H}^{2p}$.

Since $\mathcal{X} = X \times S$, relative de Rham cohomology is trivial, that is, the flat bundle $\mathcal{H}^{2p}$ is isomorphic to $H^{2p}(X/k) \otimes \mathcal{O}_S$ with the canonical connection. Since $\widetilde{\alpha}$ is a flat section over S which is irreducible over the algebraically closed field k, it corresponds to the constant section with value some α_0 in $H^{2p}(X/k)$. Then α is the image of α_0 in $H^{2p}(X_{\mathbb{C}}/\mathbb{C}) = H^{2p}(X/k) \otimes \mathbb{C}$, which concludes the proof. $\square$

Remark. In the case α is the cohomology class of an algebraic cycle, the preceding result is a consequence of the existence of Hilbert schemes. If Z is an algebraic cycle in $X_{\mathbb{C}}$, Z is algebraically equivalent to an algebraic cycle defined over k. Indeed, Z corresponds to a point in some product of Hilbert schemes parametrizing subschemes of X. These Hilbert schemes are defined over k, so their points with value in k are dense. This shows the result. Of course, classes of algebraic cycles are absolute Hodge, so this is a special case of the previous result.

11.3.5 The Field of Definition of the Locus of Hodge Classes

In this subsection, we present some of the results of Voisin in [38]. While they could be proved using principle B and the global invariant cycle theorem along a line of arguments we used earlier, we focus on deducing the theorems as consequences of statements from the previous subsection. The reader can consult [40] for the former approach.

Let S be a smooth complex quasi-projective variety, and let $\pi : \mathcal{X} \to S$ be a smooth projective morphism. Let p be an integer, and let $\mathcal{H}^{2p} = \mathbb{R}^{2p}\pi_*\Omega^{\bullet}_{\mathcal{X}/S}$ together with the Hodge filtration $F^k\mathcal{H}^{2p} = \mathbb{R}^{2p}\pi_*\Omega^{\bullet \geq k}_{\mathcal{X}/S}$. Assume π is defined over $\mathbb{Q}$. Then $\mathcal{H}^{2p}$ is defined over $\mathbb{Q}$, as well as the Hodge filtration. Inside $\mathcal{H}^{2p}$, we have the locus of Hodge classes as before. It is an algebraic subset of $\mathcal{H}^{2p}$.

Note that any smooth projective complex variety is isomorphic to the fiber of such a morphism π over a complex point. Indeed, if X is a smooth projective complex variety, it is defined over a field finitely generated over $\mathbb{Q}$. Noticing that such a field is the function field of a smooth quasi-projective variety S defined over $\mathbb{Q}$ allows us to find $\mathcal{X} \to S$ as before. Of course, S might not be geometrically irreducible.

THEOREM 11.3.17 *Let s be a complex point of S, and let α be a Hodge class in $H^{2p}(\mathcal{X}_s/\mathbb{C})$. Then α is an absolute Hodge class if and only if the connected component Z_α of the locus of Hodge classes passing through α is defined over $\overline{\mathbb{Q}}$ and the conjugates of Z_α by $\mathrm{Gal}(\overline{\mathbb{Q}}/\mathbb{Q})$ are contained in the locus of Hodge classes.*

PROOF. Let Z' be the smallest algebraic subset defined over $\mathbb{Q}$ containing Z_α. It is the $\mathbb{Q}$-Zariski closure of Z_α. We want to show that Z' is contained in the locus of Hodge classes if and only if α is absolute Hodge.

Pulling back to the image of Z' and spreading the base scheme S if necessary, we can reduce to the situation where Z' dominates S, and there exists a smooth projective morphism

$$\pi_{\mathbb{Q}} : \mathcal{X}_{\mathbb{Q}} \to S_{\mathbb{Q}}$$

defined over $\mathbb{Q}$, such that π is the pull-back of $\pi_{\mathbb{Q}}$ to $\mathbb{C}$, a class $\alpha_{\mathbb{Q}}$ in $H^{2p}(\mathcal{X}_{\mathbb{Q}}/S)$, and an embedding of $\mathbb{Q}(S_{\mathbb{Q}})$ into $\mathbb{C}$ corresponding to the complex point $s \in S(\mathbb{C})$ such that $\mathcal{X}_s$ and α are the pull-backs of $\mathcal{X}_{\mathbb{Q},\eta}$ and α_η respectively, where η is the generic point of S.

In this situation, by the definition of absolute Hodge classes, α is an absolute Hodge class if and only if α_η is. Also, since Z' dominates S, Z' is contained in the locus of Hodge classes if and only if α extends as a flat section of $\mathcal{H}^{2p}$ over S which is a Hodge class at every complex point. Such a section is automatically defined over $\mathbb{Q}$ since the Gauss–Manin connection is. Theorem 11.3.15, item 2 allows us to conclude the proof.   □

Remark. It is to be noted that the proof uses in an essential way the theorem of Cattani–Deligne–Kaplan on the algebraicity of Hodge loci.

Recall that Conjecture 11.2.17 predicts that Hodge classes are absolute. As an immediate consequence, we get the following reformulation.

COROLLARY 11.3.18 *Conjecture 11.2.17 is equivalent to the following:*
Let S be a smooth complex quasi-projective variety, and let $\pi : \mathcal{X} \to S$ be a smooth projective morphism. Assume π is defined over $\mathbb{Q}$. Then the locus of Hodge classes for π is a countable union of algebraic subsets of the Hodge bundles defined over $\mathbb{Q}$.

By using Proposition 11.2.19, it is possible to prove the preceding corollary without resorting to the Cattani–Deligne–Kaplan theorem.

In the light of this result, the study of whether Hodge classes are absolute can be seen as a study of the field of definition of the locus of Hodge classes. An intermediate property is to ask for the component of the locus of Hodge classes passing through a class α to be defined over $\overline{\mathbb{Q}}$. In [38], Voisin shows how one can use arguments from the theory of variations of Hodge structures to give infinitesimal criteria for this to happen.

This is closely related to the rigidity result of Theorem 11.3.16. Indeed, using the fact that the Gauss–Manin connection is defined over $\mathbb{Q}$, it is easy to show that the component of the locus of Hodge classes passing through a class α in the cohomology of a complex variety defined over $\overline{\mathbb{Q}}$ is defined over $\overline{\mathbb{Q}}$ if and only if α is defined over $\overline{\mathbb{Q}}$ as a class in algebraic de Rham cohomology.

Let us conclude this section by showing how the study of fields of definition for Hodge loci is related to the Hodge conjecture. The following is due to Voisin in [38].

THEOREM 11.3.19 *Let S be a smooth complex quasi-projective variety, and let π : $\mathcal{X} \to S$ be a smooth projective morphism. Assume π is defined over $\mathbb{Q}$. Let s be a complex point of S and let α be a Hodge class in $H^{2p}(\mathcal{X}_s, \mathbb{Q}(p))$. If the image in S of the component of the locus of Hodge classes passing through α is defined over $\overline{\mathbb{Q}}$, then the Hodge conjecture for α can be reduced to the Hodge conjecture for varieties defined over number fields.*

PROOF. This is a consequence of the global invariant cycle theorem. Indeed, with the notation of Theorem 11.3.4, one can choose the compactification $\overline{\mathcal{X}}$ to be defined over $\overline{\mathbb{Q}}$. The desired result follows easily. $\qquad\square$

11.4 THE KUGA–SATAKE CONSTRUCTION

In this section, we give our first nontrivial example of absolute Hodge classes. It is due to Deligne in [14].

Let S be a complex projective K3 surface. We want to construct an abelian variety A and an embedding of Hodge structures

$$H^2(S, \mathbb{Q}) \hookrightarrow H^1(A, \mathbb{Q}) \otimes H^1(A, \mathbb{Q})$$

which is absolute Hodge. This is the Kuga–Satake correspondence; see [25, 14].

We will take a representation-theoretic approach to this problem. This subsection merely outlines the construction of the Kuga–Satake correspondence, leaving aside part of the proofs. We refer to the survey [19] for more details. Properties of spin groups and their representations can be found in [18, Chapter 20] or [6, Paragraph 9].

11.4.1 Recollection on Spin Groups

We follow Deligne's approach in [14]. Let us start with some linear algebra. Let V be a finite-dimensional vector space over a field k of characteristic 0 with a nondegenerate quadratic form Q. Recall that the Clifford algebra $C(V)$ over V is the algebra defined as the quotient of the tensor algebra $\bigoplus_{i \leq 0} V^{\otimes i}$ by the relation $v \otimes v = Q(v), v \in V$. Even though the natural grading of the tensor algebra does not descend to the Clifford algebra, there is a well-defined subalgebra $C^+(V)$ of $C(V)$ which is the image of $\bigoplus_{i \leq 0} V^{\otimes 2i}$ in $C(V)$. The algebra $C^+(V)$ is the even Clifford algebra over V.

The Clifford algebra is endowed with an antiautomorphism $x \mapsto x^*$ such that $(v_1 \cdots v_i)^* = v_i \cdots v_1$ if $v_1, \ldots, v_i \in V$. The Clifford group of V is the algebraic group defined by

$$\mathrm{CSpin}(V) = \{x \in C^+(V)^*, x \cdot V \cdot x^{-1} \subset V\}.$$

It can be proved that $\mathrm{CSpin}(V)$ is a connected algebraic group. By definition, it acts on V. Let $x \in \mathrm{CSpin}(V), v \in V$. We have $Q(xvx^{-1}) = xvx^{-1}xvx^{-1} = xQ(v)x^{-1} = Q(v)$, which shows that $\mathrm{CSpin}(V)$ acts on V through the orthogonal group $O(V)$, hence a map from $\mathrm{CSpin}(V)$ to $O(V)$. Since $\mathrm{CSpin}(V)$ is connected, this map factors through $\tau : \mathrm{CSpin}(V) \to \mathrm{SO}(V)$. We have an exact sequence

$$1 \longrightarrow \mathbb{G}_m \xrightarrow{\ w\ } \mathrm{CSpin}(V) \xrightarrow{\ \tau\ } \mathrm{SO}(V) \longrightarrow 1 .$$

The spinor norm is the morphism of algebraic groups

$$N : \mathrm{CSpin}(V) \to \mathbb{G}_m, x \mapsto xx^*.$$

It is well defined. Let t be the inverse of N. The composite map

$$t \circ w : \mathbb{G}_m \to \mathbb{G}_m$$

is the map $x \mapsto x^{-2}$. The spin group $\mathrm{Spin}(V)$ is the algebraic group defined as the kernel of N. The Clifford group is generated by homotheties and elements of the spin group.

The spin group is connected and simply connected. The exact sequence

$$1 \to \pm 1 \to \mathrm{Spin}(V) \to \mathrm{SO}(V) \to 1$$

realizes the spin group as the universal covering of $\mathrm{SO}(V)$.

11.4.2 Spin Representations

The Clifford group has two different representations on $C^+(V)$. The first one is the adjoint representation $C^+(V)_{\mathrm{ad}}$. The adjoint action of $\mathrm{CSpin}(V)$ is defined as

$$x \cdot_{\mathrm{ad}} v = xvx^{-1},$$

where $x \in \mathrm{CSpin}(V), v \in C^+(V)$. It factors through $\mathrm{SO}(V)$ and is isomorphic to $\bigoplus_i \bigwedge^{2i} V$ as a representation of $\mathrm{CSpin}(V)$.

The group $\mathrm{CSpin}(V)$ acts on $C^+(V)$ by multiplication on the left, hence a representation $C^+(V)_s$, with

$$x \cdot_s v = xv,$$

where $x \in \mathrm{CSpin}(V), v \in C^+(V)$. It is compatible with the structure of a right $C^+(V)$-module on $C^+(V)$, and we have

$$\mathrm{End}_{C^+(V)}(C^+(V)_s) = C^+(V)_{\mathrm{ad}}.$$

Assume k is algebraically closed. We can describe these representations explicitly. In the case that the dimension of V is odd, let W be a simple $C^+(V)$-module. The Clifford group $\mathrm{CSpin}(V)$ acts on W. This is the spin representation of $\mathrm{CSpin}(V)$. Then $C^+(V)_s$ is isomorphic to a sum of a copy of W, and $C^+(V)_{\mathrm{ad}}$ is isomorphic to $\mathrm{End}_k(W)$ as representations of $\mathrm{CSpin}(V)$.

In the case that the dimension of V is even, let W_1 and W_2 be nonisomorphic simple $C^+(V)$-modules. These are the half-spin representations of $\mathrm{CSpin}(V)$. Their sum W is called the spin representation. Then $C^+(V)_s$ is isomorphic to a sum of a copy of W, and $C^+(V)_{\mathrm{ad}}$ is isomorphic to $\mathrm{End}_k(W_1) \times \mathrm{End}_k(W_2)$ as representations of $\mathrm{CSpin}(V)$.

11.4.3 Hodge Structures and the Deligne Torus

Recall the definition of Hodge structures à la Deligne; see [13]. Let S be the Deligne torus; that is, the real algebraic group of invertible elements of $\mathbb{C}$. It can be defined as the Weil restriction of $\mathbb{G}_m$ from $\mathbb{C}$ to $\mathbb{R}$. We have morphisms of real algebraic groups

$$\mathbb{G}_m \xrightarrow{\;w\;} S \xrightarrow{\;t\;} \mathbb{G}_m \;,$$

where w is the inclusion of $\mathbb{R}^*$ into $\mathbb{C}^*$ and t maps a complex number z to $|z|^{-2}$. The composite map

$$t \circ w : \mathbb{G}_m \to \mathbb{G}_m$$

is the map $x \mapsto x^{-2}$.

Let $V_{\mathbb{Z}}$ be a free $\mathbb{Z}$-module of finite rank, and let $V = V_{\mathbb{Q}}$. The datum of a Hodge structure of weight k on V (or $V_{\mathbb{Z}}$) is the same as the datum of a representation $\rho : S \to \mathrm{GL}(V_{\mathbb{R}})$ such that $\rho w(x) = x^k \,\mathrm{Id}_{V_{\mathbb{R}}}$ for all $x \in \mathbb{R}^*$. Given a Hodge structure of weight n, then $z \in \mathbb{C}^*$ acts on $V_{\mathbb{R}}$ by $z \cdot v = z^p \bar{z}^q v$ if $v \in V^{p,q}$.

11.4.4 From Weight 2 to Weight 1

Now assume V is polarized of weight 0 with Hodge numbers $V^{-1,1} = V^{1,-1} = 1$, $V^{0,0} \neq 0$. We say that V (or $V_{\mathbb{Z}}$) is of K3 type. We get a quadratic form Q on $V_{\mathbb{R}}$, and the representation of S on $V_{\mathbb{R}}$ factors through the special orthogonal group of V as $h : S \to \mathrm{SO}(V_{\mathbb{R}})$.

LEMMA 11.4.1   *There exists a unique lifting of h to a morphism $\widetilde{h} : S \to \mathrm{CSpin}(V_{\mathbb{R}})$ such that the following diagram commutes:*

$$
\begin{array}{ccccc}
\mathbb{G}_m & \xrightarrow{\;w\;} & S & \xrightarrow{\;t\;} & \mathbb{G}_m \\
\Big\| & & \Big\downarrow{\scriptstyle \widetilde{h}} & & \Big\| \\
\mathbb{G}_m & \xrightarrow{\;w\;} & \mathrm{CSpin}(V_{\mathbb{R}}) & \xrightarrow{\;t\;} & \mathbb{G}_m.
\end{array}
$$

PROOF.   It is easy to prove that such a lifting is unique if it exists. The restriction of Q to $P = V_{\mathbb{R}} \cap (V^{-1,1} \oplus V^{1,-1})$ is positive definite. Furthermore, P has a canonical orientation. Let e_1, e_2 be a direct orthonormal basis of P. We have $e_1 e_2 = -e_2 e_1$ and $e_1^2 = e_2^2 = 1$. As a consequence, $(e_2 e_1)^2 = -1$. An easy computation shows that the morphism $a + ib \mapsto a + be_2 e_1$ defines a suitable lifting of h. $\square$

Using the preceding lemma, consider such a lifting $\tilde{h} : S \to \mathrm{CSpin}(V)$ of h. Any representation of $\mathrm{CSpin}(V_{\mathbb{R}})$ thus gives rise to a Hodge structure. Let us first consider the adjoint representation. We know that $C^+(V)_{\mathrm{ad}}$ is isomorphic to $\bigoplus_i \wedge^{2i} V$, where $\mathrm{CSpin}(V)$ acts on V through $\mathrm{SO}(V)$. It follows that $\tilde{h}$ endows $C^+(V)_{\mathrm{ad}}$ with a weight-0 Hodge structure. Since $V^{-1,1} = 1$, the type of the Hodge structure $C^+(V)_{\mathrm{ad}}$ is $\{(-1, 1), (0, 0), (1, -1)\}$.

Now assume the dimension of V is odd, and consider the spin representation W. It is a weight-1 representation. Indeed, Lemma 11.4.1 shows that $C^+(V)_s$ is of weight 1, and it is isomorphic to a sum of copies of W. Since $C^+(V)_{\mathrm{ad}}$ is isomorphic to $\mathrm{End}_k(W)$ as representations of $\mathrm{CSpin}(V)$, the type of W is $\{(1, 0), (0, 1)\}$.

It follows that $\tilde{h}$ endows $C^+(V)_s$ with an effective Hodge structure of weight 1. It is possible to show that this Hodge structure is polarizable, see [14]. The underlying vector space has $C^+(V_{\mathbb{Z}})$ as a natural lattice. This construction thus defines an abelian variety. Similar computations show that the same result holds if the dimension of V is even.

DEFINITION 11.4.2   *The abelian variety defined by the Hodge structure on $C^+(V)_s$ with its natural lattice $C^+(V_{\mathbb{Z}})$ is called the Kuga–Satake variety associated to $V_{\mathbb{Z}}$. We denote it by $\mathrm{KS}(V_{\mathbb{Z}})$.*

THEOREM 11.4.3   *Let $V_{\mathbb{Z}}$ be a polarized Hodge structure of K3 type. There exists a natural injective morphism of Hodge structures*

$$V_{\mathbb{Q}}(-1) \hookrightarrow H^1(\mathrm{KS}(V_{\mathbb{Z}}), \mathbb{Q}) \otimes H^1(\mathrm{KS}(V_{\mathbb{Z}}), \mathbb{Q}).$$

This morphism is called the Kuga–Satake correspondence.

PROOF. Let $V = V_{\mathbb{Q}}$. Fix an element $v_0 \in V$ that is invertible in $C(V)$ and consider the vector space $M = C^+(V)$. It is endowed with a left action of V by the formula

$$v \cdot x = vxv_0$$

for $v \in V$, $x \in C^+(V)$. This action induces an embedding

$$V \hookrightarrow \mathrm{End}_{\mathbb{Q}}(C^+(V)_s)$$

which is equivariant with respect to the action of $\mathrm{CSpin}(V)$.

Now we consider $\mathrm{End}_{\mathbb{Q}}(C^+(V)_s)(-1)$ as a subspace of $C^+(V) \otimes C^+(V) = H^1(\mathrm{KS}(V_{\mathbb{Z}}), \mathbb{Q}) \otimes H^1(\mathrm{KS}(V_{\mathbb{Z}}), \mathbb{Q})$ via a polarization of $\mathrm{KS}(V_{\mathbb{Z}})$, and V as a subspace of $C^+(V)$. This gives an injection

$$V(-1) \hookrightarrow H^1(\mathrm{KS}(V_{\mathbb{Z}}), \mathbb{Q}) \otimes H^1(\mathrm{KS}(V_{\mathbb{Z}}), \mathbb{Q})$$

as desired. The equivariance property stated above shows that this is a morphism of Hodge structures. $\square$

Remark. Let V be a Hodge structure of K3 type. In order to construct the Kuga–Satake correspondence associated to V, we can slightly relax the assumption that V is polarized. Indeed, it is enough to assume that V is endowed with a quadratic form that is positive definite on $(V^{-1,1} \oplus V^{1,-1}) \cap V_{\mathbb{R}}$ and such that $V^{1,-1}$ and $V^{-1,1}$ are totally isotropic subspaces of V.

11.4.5   The Kuga–Satake Correspondence Is Absolute

Let X be a polarized complex $K3$ surface. Denote by $\mathrm{KS}(X)$ the Kuga–Satake variety associated to $H^2(X, \mathbb{Z}(1))$ endowed with the intersection pairing. Even though this pairing only gives a polarization on the primitive part of cohomology, the construction is possible by the preceding remark. Theorem 11.4.3 gives us a correspondence between the cohomology groups of X and its Kuga–Satake variety. This is the Kuga–Satake correspondence for X. We can now state and prove the main theorem of this section. It is proved by Deligne in [14].

THEOREM 11.4.4 *Let X be a polarized complex $K3$ surface. The Kuga–Satake correspondence*

$$H^2(X, \mathbb{Q}(1)) \hookrightarrow H^1(\mathrm{KS}(X), \mathbb{Q}) \otimes H^1(\mathrm{KS}(X), \mathbb{Q})$$

is absolute Hodge.

PROOF. Any polarized complex $K3$ surface deforms to a polarized Kummer surface in a polarized family. Now the Kuga–Satake construction works in families. As a consequence, by principle B (see Theorem 11.3.7), it is enough to prove that the Kuga–Satake correspondence is absolute Hodge for a variety X which is the Kummer variety associated to an abelian surface A. In this case, we can even prove the Kuga–Satake correspondence is algebraic. Let us outline the proof of this result, which was proved first by Morrison in [27]. We follow a slightly different path.

First, note that the canonical correspondence between A and X identifies the transcendental part of the Hodge structure $H^2(X, \mathbb{Z}(1))$ with the transcendental part of $H^2(A, \mathbb{Z}(1))$. Note that the latter Hodge structure is of K3 type. Since this isomorphism is induced by an algebraic correspondence between X and A, standard reductions show that it is enough to show that the Kuga–Satake correspondence between A and the Kuga–Satake abelian variety associated to $H^2(A, \mathbb{Z}(1))$ is algebraic. Let us write $U = H^1(A, \mathbb{Q})$ and $V = H^2(A, \mathbb{Q})$, considered as vector spaces.

We have $V = \bigwedge^2 U$. The vector space U is of dimension 2, and the weight-1 Hodge structure on U induces a canonical isomorphism $\bigwedge^2 V = \bigwedge^4 U \simeq \mathbb{Q}$. The intersection pairing Q on V satisfies

$$\forall x, y \in V, \quad Q(x, y) = x \wedge y.$$

Let $g \in \mathrm{SL}(U)$. The determinant of g being 1, g acts trivially on $\bigwedge^2 V = \bigwedge^4 U$. As a consequence, $g \wedge g$ preserves the intersection form on V. This gives a morphism

$\mathrm{SL}(U) \to \mathrm{SO}(V)$. The kernel of this morphism is $\pm \mathrm{Id}_U$, and it is surjective by dimension counting. Since $\mathrm{SL}(U)$ is a connected algebraic group, this gives a canonical isomorphism $\mathrm{SL}(U) \simeq \mathrm{Spin}(V)$.

The group $\mathrm{SL}(U)$ acts on U by the standard action and on its dual U^* by $g \mapsto {}^tg^{-1}$. These representations are irreducible, and they are not isomorphic since no nontrivial bilinear form on U is preserved by $\mathrm{SL}(U)$. By standard representation theory, these are the two half-spin representations of $\mathrm{SL}(U) \simeq \mathrm{Spin}(V)$. As a consequence, the Clifford algebra of V is canonically isomorphic to $\mathrm{End}(U) \times \mathrm{End}(U^*)$, and we have a canonical identification

$$\mathrm{CSpin}(V) = \{(\lambda g, \lambda^t g^{-1}), g \in \mathrm{SL}(U), \lambda \in \mathbb{G}_m\}.$$

An element $(\lambda g, \lambda^t g^{-1})$ of the Clifford group acts on the half-spin representations U and U^* through its first and second component respectively.

The preceding identifications allow us to conclude the proof. Let $h' : S \to \mathrm{GL}(U)$ be the morphism that defines the weight-1 Hodge structure on U, and let $h : S \to \mathrm{SO}(V)$ endow V with its Hodge structure of K3 type. Note that if $s \in \mathbb{C}^*$, the determinant of $h'(s)$ is $|s|^4$ since U is of dimension 4 and weight 1. Since $V = \bigwedge^2 U(1)$ as Hodge structures, we get that h is the morphism

$$h : s \mapsto |s|^{-2} h'(s) \wedge h'(s).$$

It follows that the morphism

$$\widetilde{h} : S \to \mathrm{CSpin}(V),$$
$$s \mapsto (h'(s), |s|^2 {}^t h'(s)^{-1}) = (|s||s|^{-1} h'(s)s, |s| {}^t(|s|^{-1} h'(s))^{-1})$$

is a lifting of h to $\mathrm{CSpin}(V)$.

Following the previous identifications shows that the Hodge structure induced by $\widetilde{h}$ on U and U^* are the ones induced by the identifications $U = H^1(A, \mathbb{Q})$ and $U^* = H^1(\hat{A}, \mathbb{Q})$, where $\hat{A}$ is the dual abelian variety. Since the representation $C^+(V)_s$ is a sum of four copies of $U \oplus U^*$, this gives an isogeny between $\mathrm{KS}(A)$ and $(A \times \hat{A})^4$ and shows that the Kuga–Satake correspondence is algebraic, using the identity correspondence between A and itself and the correspondence between A and its dual induced by the polarization. This concludes the proof. $\square$

Remark. Since the cohomology of a Kummer variety is a direct factor of the cohomology of an abelian variety, it is an immediate consequence of Deligne's theorem on absolute cycles on abelian varieties that the Kuga–Satake correspondence for Kummer surfaces is absolute Hodge. However, our proof is more direct and also gives the algebraicity of the correspondence in the Kummer case. Few algebraicity results are known for the Kuga–Satake correspondence, but see [29] for the case of K3 surfaces which are a double cover of $\mathbb{P}^2$ ramified over six lines. See also [19, 37, 32] for further discussion of this problem.

Remark. In Definition 11.2.5, we extended the notion of absolute Hodge classes to the setting of étale cohomology. While we did not use this notion, most results we stated, for instance principle B, can be generalized in this setting with little additional work. This makes it possible to show that the Kuga–Satake correspondence is absolute Hodge in the sense of Definition 11.2.5. In the paper [14], Deligne uses this to deduce the Weil conjectures for K3 surfaces from the Weil conjectures for abelian varieties.

11.5 DELIGNE'S THEOREM ON HODGE CLASSES ON ABELIAN VARIETIES

Having introduced the notion of absolute Hodge classes, Deligne went on to prove the following remarkable theorem, which has already been mentioned several times in these notes.

THEOREM 11.5.1 (Deligne [16])   *On an abelian variety, all Hodge classes are absolute.*

The purpose of the remaining lectures is to explain the proof of Deligne's theorem. We follow Milne's account of the proof [16], with some simplifications due to André in [2] and Voisin in [40].

11.5.1   Overview

In the lectures of Griffiths and Kerr, we have already seen that rational Hodge structures whose endomorphism algebra contains a CM-field are very special. Since abelian varieties of CM-type also play a crucial role in the proof of Deligne's theorem, we shall begin by recalling two basic definitions.

DEFINITION 11.5.2   *A CM-field is a number field E, such that for every embedding $s \colon E \hookrightarrow \mathbb{C}$, complex conjugation induces an automorphism of E that is independent of the embedding. In other words, E admits an involution $\iota \in \mathrm{Aut}(E/\mathbb{Q})$, such that for every embedding $s \colon E \hookrightarrow \mathbb{C}$, one has $\bar{s} = s \circ \iota$.*

The fixed field of the involution is a totally real field F; concretely, this means that $F = \mathbb{Q}(\alpha)$, where α and all of its conjugates are real numbers. The field E is then of the form $F[x]/(x^2 - f)$, for some element $f \in F$ that is mapped to a negative number under all embeddings of F into $\mathbb{R}$.

DEFINITION 11.5.3   *An abelian variety A is said to be of CM-type if a CM-field E is contained in $\mathrm{End}(A) \otimes \mathbb{Q}$, and if $H^1(A, \mathbb{Q})$ is one-dimensional as an E-vector space. In that case, we clearly have $2 \dim A = \dim_{\mathbb{Q}} H^1(A, \mathbb{Q}) = [E : \mathbb{Q}]$.*

We will carry out a more careful analysis of abelian varieties and Hodge structures of CM-type below. However, to motivate what follows, let us briefly look at a criterion for a simple abelian variety A to be of CM-type that involves the (special) Mumford–Tate group $\mathrm{MT}(A) = \mathrm{MT}\big(H^1(A)\big)$.

Recall that the Hodge structure on $H^1(A, \mathbb{Q})$ can be described by a morphism of $\mathbb{R}$-algebraic groups $h \colon U(1) \to \mathrm{GL}\big(H^1(A, \mathbb{R})\big)$; the weight being fixed, $h(z)$ acts as multiplication by z^{p-q} on the space $H^{p,q}(A)$. Recalling Section 11.4.3, the group $U(1)$ is the kernel of the weight $w \colon S \to \mathbb{G}_m$. Representations of $\mathrm{Ker}(w)$ correspond to Hodge structures of fixed weight.

We can define $\mathrm{MT}(A)$ as the smallest $\mathbb{Q}$-algebraic subgroup of $\mathrm{GL}\big(H^1(A, \mathbb{Q})\big)$ whose set of real points contains the image of h. Equivalently, it is the subgroup fixing every Hodge class in every tensor product

$$T^{p,q}(A) = H^1(A)^{\otimes p} \otimes H_1(A)^{\otimes q}.$$

We have the following criterion.

PROPOSITION 11.5.4   *A simple abelian variety is of CM-type if and only if its Mumford–Tate group* $\mathrm{MT}(A)$ *is an abelian group.*

Here is a quick outline of the proof of the fact that the Mumford–Tate group of a simple abelian variety of CM-type is abelian; a more general discussion can be found in Section 11.5.2.

PROOF. Let $H = H^1(A, \mathbb{Q})$. The abelian variety A is simple, which implies that $E = \mathrm{End}(A) \otimes \mathbb{Q}$ is a division algebra. It is also the space of Hodge classes in $\mathrm{End}_\mathbb{Q}(H)$, and therefore consists exactly of those endomorphisms that commute with $\mathrm{MT}(A)$. Because the Mumford–Tate group is abelian, its action splits $H^1(A, \mathbb{C})$ into a direct sum of character spaces

$$H \otimes_\mathbb{Q} \mathbb{C} = \bigoplus_\chi H_\chi,$$

where $m \cdot h = \chi(m)h$ for $h \in H_\chi$ and $m \in \mathrm{MT}(A)$. Now any endomorphism of H_χ obviously commutes with $\mathrm{MT}(A)$, and is therefore contained in $E \otimes_\mathbb{Q} \mathbb{C}$. By counting dimensions, we find that

$$\dim_\mathbb{Q} E \geq \sum_\chi \big(\dim_\mathbb{C} H_\chi\big)^2 \geq \sum_\chi \dim_\mathbb{C} H_\chi = \dim_\mathbb{Q} H.$$

On the other hand, we have $\dim_\mathbb{Q} E \leq \dim_\mathbb{Q} H$; indeed, since E is a division algebra, the map $E \to H$, $e \mapsto e \cdot h$ is injective for every nonzero $h \in H$. Therefore $[E : \mathbb{Q}] = \dim_\mathbb{Q} H = 2 \dim A$; moreover, each character space H_χ is one-dimensional, and this implies that E is commutative, hence a field. To construct the involution $\iota \colon E \to E$ that makes E into a CM-field, choose a polarization $\psi \colon H \times H \to \mathbb{Q}$, and define ι by the condition that, for every $h, h' \in H$,

$$\psi(e \cdot h, h') = \psi\big(h, \iota(e) \cdot h'\big).$$

The fact that $-i\psi$ is positive definite on the subspace $H^{1,0}(A)$ can then be used to show that ι is nontrivial, and that $\bar{s} = s \circ \iota$ for any embedding of E into the complex numbers. $\square$

After this preliminary discussion of abelian varieties of CM-type, we return to Deligne's theorem on an arbitrary abelian variety A. The proof consists of the following three steps.

1. The first step is to reduce the problem to abelian varieties of CM-type. This is done by constructing an algebraic family of abelian varieties that links a given A and a Hodge class in $H^{2p}(A, \mathbb{Q})$ to an abelian variety of CM-type and a Hodge class on it, and then applying principle B.

2. The second step is to show that every Hodge class on an abelian variety of CM-type can be expressed as a sum of pull-backs of so-called split Weil classes. The latter are Hodge classes on certain special abelian varieties, constructed by linear algebra from the CM-field E and its embeddings into $\mathbb{C}$. This part of the proof is due to André [2].

3. The last step is to show that all split Weil classes are absolute. For a fixed CM-type, all abelian varieties of split Weil type are naturally parametrized by a certain Hermitian symmetric domain; by principle B, this allows the problem to be reduced to split Weil classes on abelian varieties of a very specific form, for which the proof of the result is straightforward.

The original proof by Deligne uses Baily–Borel theory to show that certain families of abelian varieties are algebraic. Following a suggestion by Voisin, we have chosen to replace this by the following two results: the existence of a quasi-projective moduli space for polarized abelian varieties with level structure and the theorem of Cattani–Deligne–Kaplan in [10] concerning the algebraicity of Hodge loci.

11.5.2   Hodge Structures of CM-Type

When A is an abelian variety of CM-type, $H^1(A, \mathbb{Q})$ is an example of a Hodge structure of CM-type. We now undertake a more careful study of this class of Hodge structures. Let V be a rational Hodge structure of weight n, with Hodge decomposition

$$V \otimes_{\mathbb{Q}} \mathbb{C} = \bigoplus_{p+q=n} V^{p,q}.$$

Once we fix the weight n, there is a one-to-one correspondence between such decompositions and group homomorphisms $h \colon U(1) \to \mathrm{GL}(V \otimes_{\mathbb{Q}} \mathbb{R})$. Namely, $h(z)$ acts as multiplication by $z^{p-q} = z^{2p-n}$ on the subspace $V^{p,q}$. We define the (special) Mumford–Tate group $\mathrm{MT}(V)$ as the smallest $\mathbb{Q}$-algebraic subgroup of $\mathrm{GL}(V)$ whose set of real points contains the image of h.

DEFINITION 11.5.5   *We say that V is a Hodge structure of CM-type if the following two equivalent conditions are satisfied:*

(a) *The group of real points of $\mathrm{MT}(V)$ is a compact torus.*

(b) *$\mathrm{MT}(V)$ is abelian and V is polarizable.*

A proof of the equivalence may be found in Schappacher's book [31, Section 1.6.1].

It is not hard to see that any Hodge structure of CM-type is a direct sum of irreducible Hodge structures of CM-type. Indeed, since V is polarizable, it admits a finite decomposition $V = V_1 \oplus \cdots \oplus V_r$, with each V_i irreducible. As subgroups of $\mathrm{GL}(V) = \mathrm{GL}(V_1) \times \cdots \times \mathrm{GL}(V_r)$, we then have $\mathrm{MT}(V) \subseteq \mathrm{MT}(V_1) \times \cdots \times \mathrm{MT}(V_r)$, and since the projection to each factor is surjective, it follows that $\mathrm{MT}(V_i)$ is abelian. But this means that each V_i is again of CM-type. It is therefore sufficient to concentrate on irreducible Hodge structures of CM-type. For those, there is a nice structure theorem that we shall now explain.

Let V be an irreducible Hodge structure of weight n that is of CM-type, and as above, denote by $\mathrm{MT}(V)$ its special Mumford–Tate group. Because V is irreducible, its algebra of endomorphisms

$$E = \mathrm{End}_{\mathbb{Q}\text{-HS}}(V)$$

must be a division algebra. In fact, since the endomorphisms of V as a Hodge structure are exactly the Hodge classes in $\mathrm{End}_{\mathbb{Q}}(V)$, we see that E consists of all rational endomorphisms of V that commute with $\mathrm{MT}(V)$. If $T_E = E^\times$ denotes the algebraic torus in $\mathrm{GL}(V)$ determined by E, then we get $\mathrm{MT}(V) \subseteq T_E$ because $\mathrm{MT}(V)$ is commutative by assumption.

Since $\mathrm{MT}(V)$ is commutative, it acts on $V \otimes_{\mathbb{Q}} \mathbb{C}$ by characters, and so we get a decomposition

$$V \otimes_{\mathbb{Q}} \mathbb{C} = \bigoplus_\chi V_\chi,$$

where $m \in \mathrm{MT}(V)$ acts on $v \in V_\chi$ by the rule $m \cdot v = \chi(m)v$. Any endomorphism of V_χ therefore commutes with $\mathrm{MT}(V)$, and so $E \otimes_{\mathbb{Q}} \mathbb{C}$ contains the spaces $\mathrm{End}_{\mathbb{C}}(V_\chi)$. This leads to the inequality

$$\dim_{\mathbb{Q}} E \geq \sum_\chi \left(\dim_{\mathbb{C}} V_\chi\right)^2 \geq \sum_\chi \dim_{\mathbb{C}} V_\chi = \dim_{\mathbb{Q}} V.$$

On the other hand, we have $\dim_{\mathbb{Q}} V \leq \dim_{\mathbb{Q}} E$ because every nonzero element in E is invertible. It follows that each V_χ is one-dimensional, that E is commutative, and therefore that E is a field of degree $[E : \mathbb{Q}] = \dim_{\mathbb{Q}} V$. In particular, V is one-dimensional as an E-vector space.

The decomposition into character spaces can be made more canonical in the following way. Let $S = \mathrm{Hom}(E, \mathbb{C})$ denote the set of all complex embeddings of E; its cardinality is $[E : \mathbb{Q}]$. Then

$$E \otimes_{\mathbb{Q}} \mathbb{C} \xrightarrow{\sim} \bigoplus_{s \in S} \mathbb{C}, \quad e \otimes z \mapsto \sum_{s \in S} s(e)z$$

is an isomorphism of E-vector spaces; E acts on each summand on the right through the corresponding embedding s. This decomposition induces an isomorphism

$$V \otimes_{\mathbb{Q}} \mathbb{C} \xrightarrow{\sim} \bigoplus_{s \in S} V_s,$$

where $V_s = V \otimes_{E,s} \mathbb{C}$ is a one-dimensional complex vector space on which E acts via s. The induced homomorphism $U(1) \to \mathrm{MT}(V) \to E^\times \to \mathrm{End}_{\mathbb{C}}(V_s)$ is a character of $U(1)$, hence of the form $z \mapsto z^k$ for some integer k. Solving $k = p - q$ and $n = p + q$, we find that $k = 2p - n$, which means that V_s is of type $(p, n - p)$ in the Hodge decomposition of V. Now define a function $\varphi\colon S \to \mathbb{Z}$ by setting $\varphi(s) = p$; then any choice of isomorphism $V \simeq E$ puts a Hodge structure of weight n on E, whose Hodge decomposition is given by

$$E \otimes_{\mathbb{Q}} \mathbb{C} \simeq \bigoplus_{s \in S} \mathbb{C}^{\varphi(s), n - \varphi(s)}.$$

From the fact that $\overline{e \otimes z} = e \otimes \bar{z}$, we deduce that

$$\overline{\sum_{s \in S} z_s} = \sum_{s \in S} \overline{z_{\bar{s}}}.$$

Since complex conjugation has to interchange $\mathbb{C}^{p,q}$ and $\mathbb{C}^{q,p}$, this implies that $\varphi(\bar{s}) = n - \varphi(s)$, and hence that $\varphi(s) + \varphi(\bar{s}) = n$ for every $s \in S$.

DEFINITION 11.5.6 *Let E be a number field, and $S = \mathrm{Hom}(E, \mathbb{C})$ the set of its complex embeddings. Any function $\varphi\colon S \to \mathbb{Z}$ with the property that $\varphi(s) + \varphi(\bar{s}) = n$ defines a* Hodge structure l_φ *of weight n on the $\mathbb{Q}$-vector space E, whose Hodge decomposition is given by*

$$E_\varphi \otimes_{\mathbb{Q}} \mathbb{C} \simeq \bigoplus_{s \in S} \mathbb{C}^{\varphi(s), \varphi(\bar{s})}.$$

By construction, the action of E on itself respects this decomposition.

In summary, we have $V \simeq E_\varphi$, which is an isomorphism both of E-modules and of Hodge structures of weight n. Next, we would like to prove that in all interesting cases, E must be a CM-field. Recall from Definition 11.5.2 that a field E is called a *CM-field* if there exists a nontrivial involution $\iota\colon E \to E$, such that complex conjugation induces ι under any embedding of E into the complex numbers. In other words, we must have $s(\iota e) = \bar{s}(e)$ for any $s \in S$ and any $e \in E$. We usually write $\bar{e}$ in place of ιe, and refer to it as complex conjugation on E. The fixed field of E is then a totally real subfield F, and E is a purely imaginary quadratic extension of F.

To prove that E is either a CM-field or $\mathbb{Q}$, we choose a polarization ψ on E_φ. We then define the so-called *Rosati involution* $\iota\colon E \to E$ by the condition that

$$\psi(e \cdot x, y) = \psi(x, \iota e \cdot y)$$

for every $x, y, e \in E$. Denoting the image of $1 \in E$ by $\sum_{s \in S} 1_s$, we have

$$\sum_{s \in S} \psi(1_s, 1_{\bar{s}}) s(e \cdot x) \bar{s}(y) = \sum_{s \in S} \psi(1_s, 1_{\bar{s}}) s(x) \bar{s}(\iota e \cdot y),$$

which implies that $s(e) = \bar{s}(\iota e)$. Now there are two cases: either ι is nontrivial, in which case E is a CM-field and the Rosati involution is complex conjugation, or ι is trivial, which means that $\bar{s} = s$ for every complex embedding. In the second case, we see that $\varphi(s) = n/2$ for every s, and so the Hodge structure must be $\mathbb{Q}(-n/2)$, being irreducible and of type $(n/2, n/2)$. This implies that $E = \mathbb{Q}$.

From now on, we exclude the trivial case $V = \mathbb{Q}(-n/2)$ and assume that E is a CM-field.

DEFINITION 11.5.7 *A* CM-type *of E is a mapping $\varphi \colon S \to \{0,1\}$ with the property that $\varphi(s) + \varphi(\bar{s}) = 1$ for every $s \in S$.*

When φ is a CM-type, E_φ is a polarizable rational Hodge structure of weight 1. As such, it is the rational Hodge structure of an abelian variety with complex multiplication by E. This variety is unique up to isogeny. In general, we have the following structure theorem.

PROPOSITION 11.5.8 *Any Hodge structure V of CM-type and of even weight $2k$ with $V^{p,q} = 0$ for $p < 0$ or $q < 0$ occurs as a direct factor of $H^{2k}(A, \mathbb{Q})$, where A is a finite product of simple abelian varieties of CM-type.*

PROOF. In our classification of irreducible Hodge structures of CM-type above, there were two cases: $\mathbb{Q}(-n/2)$ and Hodge structures of the form E_φ, where E is a CM-field and $\varphi \colon S \to \mathbb{Z}$ is a function satisfying $\varphi(s) + \varphi(\bar{s}) = n$. Clearly φ can be written as a linear combination (with integer coefficients) of CM-types for E. Because of the relations

$$E_{\varphi + \psi} \simeq E_\varphi \otimes_E E_\psi \quad \text{and} \quad E_{-\varphi} \simeq E_\varphi^\vee,$$

every irreducible Hodge structure of CM-type can thus be obtained from Hodge structures corresponding to CM-types by tensor products, duals, and Tate twists.

As we have seen, every Hodge structure of CM-type is a direct sum of irreducible Hodge structures of CM-type. The assertion follows from this by simple linear algebra. $\square$

To conclude our discussion of Hodge structures of CM-type, we will consider the case when the CM-field E is a Galois extension of $\mathbb{Q}$. In that case, the Galois group $G = \mathrm{Gal}(E/\mathbb{Q})$ acts on the set of complex embeddings of E by the rule

$$(g \cdot s)(e) = s(g^{-1}e).$$

This action is simply transitive. Recall that we have an isomorphism

$$E \otimes_\mathbb{Q} E \xrightarrow{\sim} \bigoplus_{g \in G} E, \quad x \otimes e \mapsto g(e)x.$$

For any E-vector space V, this isomorphism induces a decomposition

$$V \otimes_\mathbb{Q} E \xrightarrow{\sim} \bigoplus_{g \in G} V, \quad v \otimes e \mapsto g(e)v.$$

When V is an irreducible Hodge structure of CM-type, a natural question is whether this decomposition is compatible with the Hodge decomposition. The following lemma shows that the answer to this question is yes.

LEMMA 11.5.9 *Let E be a CM-field that is a Galois extension of $\mathbb{Q}$, with Galois group $G = \mathrm{Gal}(E/\mathbb{Q})$. Then for any $\varphi \colon S \to \mathbb{Z}$ with $\varphi(s) + \varphi(\bar{s}) = n$, we have*

$$E_\varphi \otimes_\mathbb{Q} E \simeq \bigoplus_{g \in G} E_{g\varphi}.$$

PROOF. We chase the Hodge decompositions through the various isomorphisms that are involved in the statement. To begin with, we have

$$\left(E_\varphi \otimes_\mathbb{Q} E\right) \otimes_\mathbb{Q} \mathbb{C} \simeq \left(E_\varphi \otimes_\mathbb{Q} \mathbb{C}\right) \otimes_\mathbb{Q} E \simeq \bigoplus_{s \in S} \mathbb{C}^{\varphi(s), n - \varphi(s)} \otimes_\mathbb{Q} E \simeq \bigoplus_{s,t \in S} \mathbb{C}^{\varphi(s), n - \varphi(s)},$$

and the isomorphism takes $(v \otimes e) \otimes z$ to the element

$$\sum_{s,t \in S} t(e) \cdot z \cdot s(v).$$

On the other hand,

$$\left(E_\varphi \otimes_\mathbb{Q} E\right) \otimes_\mathbb{Q} \mathbb{C} \simeq \bigoplus_{g \in G} E \otimes_\mathbb{Q} \mathbb{C} \simeq \bigoplus_{g \in G} \bigoplus_{s \in S} \mathbb{C}^{\varphi(s), n - \varphi(s)},$$

and under this isomorphism, $(v \otimes e) \otimes z$ is sent to the element

$$\sum_{g \in G} \sum_{s \in S} s(ge) \cdot s(v) \cdot z.$$

If we fix $g \in G$ and compare the two expressions, we see that $t = sg$, and hence

$$E \otimes_\mathbb{Q} \mathbb{C} \simeq \bigoplus_{t \in S} \mathbb{C}^{\varphi(s), n - \varphi(s)} \simeq \bigoplus_{t \in S} \mathbb{C}^{\varphi(tg^{-1}), n - \varphi(tg^{-1})}.$$

But since $(g\varphi)(t) = \varphi(tg^{-1})$, this is exactly the Hodge decomposition of $E_{g\varphi}$. □

11.5.3 Reduction to Abelian Varieties of CM-Type

The proof of Deligne's theorem involves the construction of algebraic families of abelian varieties, in order to apply principle B. For this, we shall use the existence of a fine moduli space for polarized abelian varieties with level structure. Recall that if A is an abelian variety of dimension g, the subgroup $A[N]$ of its N-torsion points is isomorphic to $(\mathbb{Z}/N\mathbb{Z})^{\oplus 2g}$. A *level N-structure* is a choice of symplectic isomorphism $A[N] \simeq (\mathbb{Z}/N\mathbb{Z})^{\oplus 2g}$. Also recall that a *polarization of degree d* on an abelian variety A is a finite morphism $\theta \colon A \to \hat{A}$ of degree d.

THEOREM 11.5.10 *Fix integers $g, d \geq 1$. Then for any $N \geq 3$, there is a smooth quasi-projective variety $\mathcal{M}_{g,d,N}$ that is a fine moduli space for g-dimensional abelian varieties with polarization of degree d and level N-structure. In particular, we have a universal family of abelian varieties over $\mathcal{M}_{g,d,N}$.*

The relationship of this result with Hodge theory is the following. Fix an abelian variety A of dimension g, with level N-structure and polarization $\theta\colon A \to \hat{A}$ of degree d. The polarization corresponds to an antisymmetric bilinear form $\psi\colon H^1(A,\mathbb{Z}) \times H^1(A,\mathbb{Z}) \to \mathbb{Z}$ that polarizes the Hodge structure; we shall refer to ψ as a *Riemann form*. Define $V = H^1(A,\mathbb{Q})$, and let D be the corresponding period domain; D parametrizes all possible Hodge structures of type $\{(1,0),(0,1)\}$ on V that are polarized by the form ψ. Then D is isomorphic to the universal covering space of the quasi-projective complex manifold $\mathcal{M}_{g,d,N}$.

We now turn to the first step in the proof of Deligne's theorem, namely, the reduction of the general problem to abelian varieties of CM-type. This is accomplished by the following theorem and principle B; see Theorem 11.3.7.

THEOREM 11.5.11 *Let A be an abelian variety, and let $\alpha \in H^{2p}(A,\mathbb{Q}(p))$ be a Hodge class on A. Then there exists a family $\pi\colon \mathcal{A} \to B$ of abelian varieties, with B nonsingular, irreducible, and quasi-projective, such that the following three things are true:*

(a) $\mathcal{A}_0 = A$ for some point $0 \in B$.

(b) There is a Hodge class $\tilde{\alpha} \in H^{2p}(\mathcal{A},\mathbb{Q}(p))$ whose restriction to A equals α.

(c) For a dense set of $t \in B$, the abelian variety $\mathcal{A}_t = \pi^{-1}(t)$ is of CM-type.

Before giving the proof, let us briefly recall the following useful interpretation of period domains. Say D parametrizes all Hodge structures of weight n on a fixed rational vector space V that are polarized by a given bilinear form ψ. The set of real points of the group $G = \mathrm{Aut}(V,\psi)$ then acts transitively on D by the rule $(gH)^{p,q} = g \cdot H^{p,q}$, and so $D \simeq G(\mathbb{R})/K$.

Now points of D are in one-to-one correspondence with homomorphisms of real algebraic groups $h\colon U(1) \to G_{\mathbb{R}}$, and we denote the Hodge structure corresponding to h by V_h. Then $V_h^{p,q}$ is exactly the subspace of $V \otimes_{\mathbb{Q}} \mathbb{C}$ on which $h(z)$ acts as multiplication by z^{p-q}, and from this, it is easy to verify that $gV_h = V_{ghg^{-1}}$. In other words, the points of D can be thought of as conjugacy classes of a fixed h under the action of $G(\mathbb{R})$.

PROOF OF THEOREM 11.5.11. After choosing a polarization $\theta\colon A \to \hat{A}$, we may assume that the Hodge structure on $V = H^1(A,\mathbb{Q})$ is polarized by a Riemann form ψ. Let $G = \mathrm{Aut}(V,\psi)$, and recall that $M = \mathrm{MT}(A)$ is the smallest $\mathbb{Q}$-algebraic subgroup of G whose set of real points $M(\mathbb{R})$ contains the image of the homomorphism $h\colon U(1) \to G(\mathbb{R})$. Let D be the period domain whose points parametrize all possible Hodge structures of type $\{(1,0),(0,1)\}$ on V that are polarized by the form ψ. With $V_h = H^1(A)$ as the base point, we then have $D \simeq G(\mathbb{R})/K$; the points of D are thus exactly the Hodge structures $V_{ghg^{-1}}$, for $g \in G(\mathbb{R})$ arbitrary.

The main idea of the proof is to consider the Mumford–Tate domain

$$D_h = M(\mathbb{R})/K \cap M(\mathbb{R}) \hookrightarrow D.$$

By definition, D_h consists of all Hodge structures of the form $V_{ghg^{-1}}$, for $g \in M(\mathbb{R})$. As explained in Griffiths' lecture (see Chapter 1, Appendix B), these are precisely the Hodge structures whose Mumford–Tate group is contained in M.

To find Hodge structures of CM-type in D_h, we appeal to a result by Borel. Since the image of h is abelian, it is contained in a maximal torus T of the real Lie group $M(\mathbb{R})$. One can show that, for a generic element ξ in the Lie algebra $\mathfrak{m}_{\mathbb{R}}$, this torus is the stabilizer of ξ under the adjoint action by $M(\mathbb{R})$. Now $\mathfrak{m}$ is defined over $\mathbb{Q}$, and so there exist arbitrarily small elements $g \in M(\mathbb{R})$ for which $\mathrm{Ad}(g)\xi = g\xi g^{-1}$ is rational. The stabilizer gTg^{-1} of such a rational point is then a maximal torus in M that is defined over $\mathbb{Q}$. The Hodge structure $V_{ghg^{-1}}$ is a point of the Mumford–Tate domain D_h, and by definition of the Mumford–Tate group, we have $\mathrm{MT}(V_{ghg^{-1}}) \subseteq T$. In particular, $V_{ghg^{-1}}$ is of CM-type, because its Mumford–Tate group is abelian. This reasoning shows that D_h contains a dense set of points of CM-type.

To obtain an algebraic family of abelian varieties with the desired properties, we can now argue as follows. Let $\mathcal{M}$ be the moduli space of abelian varieties of dimension $\dim A$, with polarization of the same type as θ, and level 3-structure. Then $\mathcal{M}$ is a smooth quasi-projective variety, and since it is a fine moduli space, it carries a universal family $\pi : \mathcal{A} \to \mathcal{M}$.

By general properties of reductive algebraic groups (see [16, Proposition 3.1] or Griffiths' lecture in Chapter 1, Appendix B), we can find finitely many Hodge tensors $\tau_1, \ldots, \tau_r$ for $H^1(A)$—that is, the elements τ_i are Hodge classes in spaces of the form $H^1(A)^{\otimes a} \otimes (H^1(A)^*)^{\otimes b} \otimes \mathbb{Q}(c)$—such that $M = \mathrm{MT}(A)$ is exactly the subgroup of G fixing every τ_i. Given τ_i, we can consider the irreducible component B_i of the Hodge locus of τ_i in $\mathcal{M}$ passing through the point A. These Hodge loci are associated to the local systems of the form $(R^1\pi_*\mathbb{Q})^{\otimes a} \otimes ((R^1\pi_*\mathbb{Q})^*)^{\otimes b} \otimes \mathbb{Q}(c)$ corresponding to the τ_i.

Let $B \subseteq \mathcal{M}$ be the intersection of the B_i. By the theorem of Cattani–Deligne–Kaplan, B is again a quasi-projective variety. Let $\pi \colon \mathcal{A} \to B$ be the restriction of the universal family to B. Then (a) is clearly satisfied for this family.

Now D is the universal covering space of $\mathcal{M}$, with the point $V_h = H^1(A)$ mapping to A. By construction, the preimage of B in D is exactly the Mumford–Tate domain D_h. Indeed, consider a Hodge structure $V_{ghg^{-1}}$ in the preimage of B. By construction, every τ_i is a Hodge tensor for this Hodge structure, which shows that $\mathrm{MT}(V_{ghg^{-1}})$ is contained in M. As explained above, this implies that $V_{ghg^{-1}}$ belongs to D_h. Since D_h contains a dense set of Hodge structures of CM-type, (c) follows. Since B is also contained in the Hodge locus of α, and since the monodromy action of $\pi_1(B,0)$ on the space of Hodge classes has finite orbits, we may pass to a finite étale cover of B and assume that the local system $R^{2p}\pi_*\mathbb{Q}(p)$ has a section that is a Hodge class at every point of B. We now obtain (b) from the global invariant cycle theorem (see Theorem 11.3.4). $\qquad\square$

11.5.4   Background on Hermitian Forms

The second step in the proof of Deligne's theorem involves the construction of special Hodge classes on abelian varieties of CM-type, the so-called split Weil classes. This requires some background on Hermitian forms, which we now provide. Throughout, E is a CM-field, with totally real subfield F and complex conjugation $e \mapsto \bar{e}$, and $S = \mathrm{Hom}(E, \mathbb{C})$ denotes the set of complex embeddings of E. An element $\zeta \in E^{\times}$ is called *totally imaginary* if $\bar{\zeta} = -\zeta$; concretely, this means that $\bar{s}(\zeta) = -s(\zeta)$ for every complex embedding s. Likewise, an element $f \in F^{\times}$ is said to be *totally positive* if $s(f) > 0$ for every $s \in S$.

DEFINITION 11.5.12   *Let V be an E-vector space. A $\mathbb{Q}$-bilinear form $\phi\colon V \times V \to E$ is said to be E-Hermitian if $\phi(e \cdot v, w) = e \cdot \phi(v, w)$ and $\phi(v, w) = \overline{\phi(w, v)}$ for every $v, w \in V$ and every $e \in E$.*

Now suppose that V is an E-vector space of dimension $d = \dim_E V$, and that ϕ is an E-Hermitian form on V. We begin by describing the numerical invariants of the pair (V, ϕ). For any embedding $s\colon E \hookrightarrow \mathbb{C}$, we obtain a Hermitian form ϕ_s (in the usual sense) on the complex vector space $V_s = V \otimes_{E,s} \mathbb{C}$. We let a_s and b_s be the dimensions of the maximal subspaces where ϕ_s is, respectively, positive definite and negative definite.

A second invariant of ϕ is its discriminant. To define it, note that ϕ induces an E-Hermitian form on the one-dimensional E-vector space $\bigwedge_E^d V$, which up to a choice of basis vector, is of the form $(x, y) \mapsto f x \bar{y}$. The element f belongs to the totally real subfield F, and a different choice of basis vector only changes f by elements of the form $\mathrm{Nm}_{E/F}(e) = e \cdot \bar{e}$. Consequently, the class of f in $F^{\times} / \mathrm{Nm}_{E/F}(E^{\times})$ is well defined, and is called the *discriminant* of (V, ϕ). We denote it by the symbol $\mathrm{disc}\, \phi$.

Now suppose that ϕ is nondegenerate. Let $v_1, \ldots, v_d$ be an orthogonal basis for V, and set $c_i = \phi(v_i, v_i)$. Then we have $c_i \in F^{\times}$, and

$$a_s = \#\{\, i \mid s(c_i) > 0 \,\} \quad \text{and} \quad b_s = \#\{\, i \mid s(c_i) < 0 \,\}$$

satisfy $a_s + b_s = d$. Moreover, we have

$$f = \prod_{i=1}^{d} c_i \quad \mathrm{mod}\ \mathrm{Nm}_{E/F}(E^{\times});$$

this implies that $\mathrm{sgn}\big(s(f)\big) = (-1)^{b_s}$ for every $s \in S$. The following theorem by Landherr [26] shows that the discriminant and the integers a_s and b_s are a complete set of invariants for E-Hermitian forms.

THEOREM 11.5.13 (Landherr)   *Let $a_s, b_s \geq 0$ be a collection of integers, indexed by the set S, and let $f \in F^{\times} / \mathrm{Nm}_{E/F}(E^{\times})$ be an arbitrary element. Suppose that they satisfy $a_s + b_s = d$ and $\mathrm{sgn}\big(s(f)\big) = (-1)^{b_s}$ for every $s \in S$. Then there exists a nondegenerate E-Hermitian form ϕ on an E-vector space V of dimension d with these invariants; moreover, (V, ϕ) is unique up to isomorphism.*

This classical result has the following useful consequence.

COROLLARY 11.5.14   *If (V, ϕ) is nondegenerate, then the following two conditions are equivalent:*

(a) $a_s = b_s = d/2$ for every $s \in S$, and disc $\phi = (-1)^{d/2}$.

(b) There is a totally isotropic subspace of V of dimension $d/2$.

PROOF. If $W \subseteq V$ is a totally isotropic subspace of dimension $d/2$, then $v \mapsto \phi(-, v)$ induces an antilinear isomorphism $V/W \xrightarrow{\sim} W^\vee$. Thus we can extend a basis $v_1, \dots, v_{d/2}$ of W to a basis $v_1, \dots, v_d$ of V, with the property that

$$\phi(v_i, v_{i+d/2}) = 1 \qquad \text{for } 1 \le i \le d/2,$$
$$\phi(v_i, v_j) = 0 \qquad \text{for } |i - j| \ne d/2.$$

We can use this basis to check that (a) is satisfied. For the converse, consider the Hermitian space $(E^{\oplus d}, \phi)$, where

$$\phi(x, y) = \sum_{1 \le i \le d/2} \left(x_i \bar{y}_{i+d/2} + x_{i+d/2} \bar{y}_i \right)$$

for every $x, y \in E^{\oplus d}$. By Landherr's theorem, this space is (up to isomorphism) the unique Hermitian space satisfying (a), and it is easy to see that it satisfies (b) too. □

DEFINITION 11.5.15 An E-Hermitian form ϕ that satisfies the two equivalent conditions in Corollary 11.5.14 is said to be split.

We shall see below that E-Hermitian forms are related to polarizations on Hodge structures of CM-type. We now describe one additional technical result that will be useful in that context. Suppose that V is a Hodge structure of type $\{(1,0), (0,1)\}$ that is of CM-type and whose endomorphism ring contains E; let $h: U(1) \to E^\times$ be the corresponding homomorphism. Recall that a *Riemann form* for V is a $\mathbb{Q}$-bilinear antisymmetric form $\psi: V \times V \to \mathbb{Q}$, with the property that

$$(x, y) \mapsto \psi\big(x, h(i) \cdot \bar{y}\big)$$

is Hermitian and positive definite on $V \otimes_{\mathbb{Q}} \mathbb{C}$. We only consider Riemann forms whose Rosati involution induces complex conjugation on E; that is, which satisfy

$$\psi(ev, w) = \psi(v, \bar{e}w).$$

LEMMA 11.5.16 Let $\zeta \in E^\times$ be a totally imaginary element ($\bar{\zeta} = -\zeta$), and let ψ be a Riemann form for V as above. Then there exists a unique E-Hermitian form ϕ with the property that $\psi = \mathrm{Tr}_{E/\mathbb{Q}}(\zeta\phi)$.

We begin with a simpler statement.

LEMMA 11.5.17 Let V and W be finite-dimensional vector spaces over E, and let $\psi: V \times W \to \mathbb{Q}$ be a $\mathbb{Q}$-bilinear form such that $\psi(ev, w) = \psi(v, ew)$ for every $e \in E$. Then there exists a unique E-bilinear form ϕ such that $\psi(v, w) = \mathrm{Tr}_{E/\mathbb{Q}} \phi(v, w)$.

PROOF. The trace pairing $E \times E \to \mathbb{Q}$, $(x, y) \mapsto \mathrm{Tr}_{E/\mathbb{Q}}(xy)$, is nondegenerate. Consequently, composition with $\mathrm{Tr}_{E/\mathbb{Q}}$ induces an injective homomorphism

$$\mathrm{Hom}_E(V \otimes_E W, E) \to \mathrm{Hom}_{\mathbb{Q}}(V \otimes_E W, \mathbb{Q}),$$

which has to be an isomorphism because both vector spaces have the same dimension over $\mathbb{Q}$. By assumption, ψ defines a $\mathbb{Q}$-linear map $V \otimes_E W \to \mathbb{Q}$, and we let ϕ be the element of $\mathrm{Hom}_E(V \otimes_E W, E)$ corresponding to ψ under the above isomorphism. $\square$

PROOF OF LEMMA 11.5.16. We apply the preceding lemma with $W = V$, but with E acting on W through complex conjugation. This gives a sesquilinear form ϕ_1 such that $\psi(x, y) = \mathrm{Tr}_{E/\mathbb{Q}}\, \phi_1(x, y)$. Now define $\phi = \zeta^{-1}\phi_1$, so that we have $\psi(x, y) = \mathrm{Tr}_{E/\mathbb{Q}}(\zeta\phi(x, y))$. The uniqueness of ϕ is obvious from the preceding lemma.

It remains to show that we have $\phi(y, x) = \overline{\phi(x, y)}$. Because ψ is antisymmetric, $\psi(y, x) = -\psi(x, y)$, which implies that

$$\mathrm{Tr}_{E/\mathbb{Q}}(\zeta\phi(y, x)) = -\mathrm{Tr}_{E/\mathbb{Q}}(\zeta\phi(x, y)) = \mathrm{Tr}_{E/\mathbb{Q}}(\bar{\zeta}\phi(x, y)).$$

On replacing y by ey, for arbitrary $e \in E$, we obtain

$$\mathrm{Tr}_{E/\mathbb{Q}}(\zeta e \cdot \phi(y, x)) = \mathrm{Tr}_{E/\mathbb{Q}}(\bar{\zeta}e \cdot \phi(x, y)).$$

On the other hand, we have

$$\mathrm{Tr}_{E/\mathbb{Q}}(\zeta e \cdot \phi(y, x)) = \mathrm{Tr}_{E/\mathbb{Q}}(\overline{\bar{\zeta}e \cdot \phi(y, x)}) = \mathrm{Tr}_{E/\mathbb{Q}}(\bar{\zeta}\bar{e} \cdot \overline{\phi(y, x)}).$$

Since $\bar{\zeta}e$ can be an arbitrary element of E, the nondegeneracy of the trace pairing implies that $\phi(x, y) = \overline{\phi(y, x)}$. $\square$

11.5.5 Construction of Split Weil Classes

Let E be a CM-field; as usual, we let $S = \mathrm{Hom}(E, \mathbb{C})$ be the set of complex embeddings; it has $[E : \mathbb{Q}]$ elements.

Let V be a rational Hodge structure of type $\{(1, 0), (0, 1)\}$ whose endomorphism algebra contains E. We shall assume that $\dim_E V = d$ is an even number. Let $V_s = V \otimes_{E,s} \mathbb{C}$. Corresponding to the decomposition

$$E \otimes_{\mathbb{Q}} \mathbb{C} \xrightarrow{\sim} \bigoplus_{s \in S} \mathbb{C}, \quad e \otimes z \mapsto \sum_{s \in S} s(e)z,$$

we get a decomposition

$$V \otimes_{\mathbb{Q}} \mathbb{C} \simeq \bigoplus_{s \in S} V_s.$$

The isomorphism is E-linear, where $e \in E$ acts on the complex vector space V_s as multiplication by $s(e)$. Since $\dim_{\mathbb{Q}} V = [E : \mathbb{Q}] \cdot \dim_E V$, each V_s has dimension d

over $\mathbb{C}$. By assumption, E respects the Hodge decomposition on V, and so we get an induced decomposition

$$V_s = V_s^{1,0} \oplus V_s^{0,1}.$$

Note that $\dim_{\mathbb{C}} V_s^{1,0} + \dim_{\mathbb{C}} V_s^{0,1} = d$.

LEMMA 11.5.18 *The rational subspace $\bigwedge_E^d V \subseteq \bigwedge_{\mathbb{Q}}^d V$ is purely of type $(d/2, d/2)$ if and only if $\dim_{\mathbb{C}} V_s^{1,0} = \dim_{\mathbb{C}} V_s^{0,1} = d/2$ for every $s \in S$.*

PROOF. We have

$$\left(\bigwedge_E^d V\right) \otimes_{\mathbb{Q}} \mathbb{C} \simeq \bigwedge_{E \otimes_{\mathbb{Q}} \mathbb{C}}^d (V \otimes_{\mathbb{Q}} \mathbb{C}) \simeq \bigoplus_{s \in S} \bigwedge_{\mathbb{C}}^d V_s \simeq \bigoplus_{s \in S} \left(\bigwedge_{\mathbb{C}}^{p_s} V_s^{1,0}\right) \otimes \left(\bigwedge_{\mathbb{C}}^{q_s} V_s^{0,1}\right),$$

where $p_s = \dim_{\mathbb{C}} V_s^{1,0}$ and $q_s = \dim_{\mathbb{C}} V_s^{0,1}$. The assertion follows because the Hodge type of each summand is evidently (p_s, q_s). $\square$

We will now describe a condition on V that guarantees that the space $\bigwedge_E^d V$ consists entirely of Hodge cycles.

DEFINITION 11.5.19 *Let V be a rational Hodge structure of type $\{(1,0), (0,1)\}$ with $E \hookrightarrow \mathrm{End}_{\mathbb{Q}\text{-}HS}(V)$ and $\dim_E V = d$. We say that V is of split Weil type relative to E if there exists an E-Hermitian form ϕ on V with a totally isotropic subspace of dimension $d/2$, and a totally imaginary element $\zeta \in E$, such that $\mathrm{Tr}_{E/\mathbb{Q}}(\zeta\phi)$ defines a polarization on V.*

According to Corollary 11.5.14, the condition on the E-Hermitian form ϕ is the same as saying that the pair (V, ϕ) is split.

PROPOSITION 11.5.20 *If V is of split Weil type relative to E, and $\dim_E V = d$ is even, then the space*

$$\bigwedge_E^d V \subseteq \bigwedge_{\mathbb{Q}}^d V$$

consists of Hodge classes of type $(d/2, d/2)$.

PROOF. Since $\psi = \mathrm{Tr}_{E/\mathbb{Q}}(\zeta\phi)$ defines a polarization, ϕ is nondegenerate; by Corollary 11.5.14, it follows that (V, ϕ) is split. Thus for any complex embedding $s \colon E \hookrightarrow \mathbb{C}$, we have $a_s = b_s = d/2$. Let ϕ_s be the induced Hermitian form on $V_s = V \otimes_{E,s} \mathbb{C}$. By Lemma 11.5.18, it suffices to show that $\dim_{\mathbb{C}} V_s^{1,0} = \dim_{\mathbb{C}} V_s^{0,1} = d/2$. By construction, the isomorphism

$$\alpha \colon V \otimes_{\mathbb{Q}} \mathbb{C} \xrightarrow{\sim} \bigoplus_{s \in S} V_s$$

respects the Hodge decompositions on both sides. For any $v \in V$, we have

$$\psi(v, v) = \mathrm{Tr}_{E/\mathbb{Q}}(\zeta\phi(v, v)) = \sum_{s \in S} s(\zeta) \cdot s(\phi(v, v)) = \sum_{s \in S} s(\zeta) \cdot \phi_s(v \otimes 1, v \otimes 1).$$

Now if we choose a nonzero element $x \in V_s^{1,0}$, then under the above isomorphism,

$$-s(\zeta)i \cdot \phi_s(x, \bar{x}) = \psi(\alpha^{-1}(x), h(i) \cdot \overline{\alpha^{-1}(x)}) > 0.$$

Likewise, we have $s(\zeta)i \cdot \phi_s(x, \bar{x}) > 0$ for $x \in V_s^{0,1}$ nonzero. Consequently, $\dim_{\mathbb{C}} V_s^{1,0}$ and $\dim_{\mathbb{C}} V_s^{0,1}$ must both be less than or equal to $d/2 = a_s = b_s$; since their dimensions add up to d, we get the desired result. □

11.5.6 André's Theorem and Reduction to Split Weil Classes

The second step in the proof of Deligne's theorem is to reduce the problem from arbitrary Hodge classes on abelian varieties of CM-type to Hodge classes of split Weil type. This is accomplished by the following pretty theorem due to Yves André in [2].

THEOREM 11.5.21 (André) *Let V be a rational Hodge structure of type $\{(1, 0), (0, 1)\}$, which is of CM-type. Then there exist a CM-field E, rational Hodge structures V_α of split Weil type (relative to E), and morphisms of Hodge structure $V_\alpha \to V$, such that every Hodge cycle $\xi \in \bigwedge_{\mathbb{Q}}^{2k} V$ is a sum of images of Hodge cycles $\xi_\alpha \in \bigwedge_{\mathbb{Q}}^{2k} V_\alpha$ of split Weil type.*

PROOF. Let $V = V_1 \oplus \cdots \oplus V_r$, with V_i irreducible; then each $E_i = \mathrm{End}_{\mathbb{Q}\text{-HS}}(V_i)$ is a CM-field. Define E to be the Galois closure of the compositum of the fields $E_1, \ldots, E_r$. Since V is of CM-type, E is a CM-field which is Galois over $\mathbb{Q}$. Let G be its Galois group over $\mathbb{Q}$. After replacing V by $V \otimes_{\mathbb{Q}} E$ (of which V is a direct factor), we may assume without loss of generality that $E_i = E$ for all i.

As before, let $S = \mathrm{Hom}(E, \mathbb{C})$ be the set of complex embeddings of E; we then have a decomposition

$$V \simeq \bigoplus_{i \in I} E_{\varphi_i}$$

for some collection of CM-types φ_i. Applying Lemma 11.5.9, we get

$$V \otimes_{\mathbb{Q}} E \simeq \bigoplus_{i \in I} \bigoplus_{g \in G} E_{g\varphi_i}.$$

Since each $E_{g\varphi_i}$ is one-dimensional over E, we get

$$\left(\bigwedge_{\mathbb{Q}}^{2k} V\right) \otimes_{\mathbb{Q}} E \simeq \bigwedge_{E}^{2k}(V \otimes_{\mathbb{Q}} E) \simeq \bigwedge_{E}^{2k} \bigoplus_{(i,g) \in I \times G} E_{g\varphi_i} \simeq \bigoplus_{\substack{\alpha \subseteq I \times G \\ |\alpha| = 2k}} \bigotimes_{(i,g) \in \alpha} E_{g\varphi_i},$$

where the tensor product is over E. If we now define Hodge structures of CM-type

$$V_\alpha = \bigoplus_{(i,g) \in \alpha} E_{g\varphi_i}$$

for any subset $\alpha \subseteq I \times G$ of size $2k$, then V_α has dimension $2k$ over E. The above calculation shows that

$$\left(\bigwedge_{\mathbb{Q}}^{2k} V\right) \otimes_{\mathbb{Q}} E \simeq \bigoplus_{\alpha} \bigwedge_{E}^{2k} V_\alpha,$$

which is an isomorphism both as Hodge structures and as E-vector spaces. Moreover, since V_α is a sub-Hodge structure of $V \otimes_{\mathbb{Q}} E$, we clearly have morphisms $V_\alpha \to V$, and any Hodge cycle $\xi \in \bigwedge_{\mathbb{Q}}^{2k} V$ is a sum of Hodge cycles $\xi_\alpha \in \bigwedge_E^{2k} V_\alpha$.

It remains to see that V_α is of split Weil type whenever ξ_α is nonzero. Fix a subset $\alpha \subseteq I \times G$ of size $2k$, with the property that $\xi_\alpha \neq 0$. Note that we have

$$\bigwedge_E^{2k} V_\alpha \simeq \bigotimes_{(i,g)\in\alpha} E_{g\varphi_i} \simeq E_\varphi,$$

where $\varphi \colon S \to \mathbb{Z}$ is the function

$$\varphi = \sum_{(i,g)\in\alpha} g\varphi_i.$$

The Hodge decomposition of E_φ is given by

$$E_\varphi \otimes_{\mathbb{Q}} \mathbb{C} \simeq \bigoplus_{s\in S} \mathbb{C}^{\varphi(s),\varphi(\bar{s})}.$$

The image of the Hodge cycle ξ_α in E_φ must be purely of type (k,k) with respect to this decomposition. But

$$\xi_\alpha \otimes 1 \mapsto \sum_{s\in S} s(\xi_\alpha),$$

and since each $s(\xi_\alpha)$ is nonzero, we conclude that $\varphi(s) = k$ for every $s \in S$. This means that the sum of the $2k$ CM-types $g\varphi_i$, indexed by $(i,g) \in \alpha$, is constant on S. We conclude by the criterion in Proposition 11.5.22 that V_α is of split Weil type. $\square$

The proof makes use of the following criterion for a Hodge structure to be of split Weil type. Let $\varphi_1, \ldots, \varphi_d$ be CM-types attached to E. Let $V_i = E_{\varphi_i}$ be the Hodge structure of CM-type corresponding to φ_i, and define

$$V = \bigoplus_{i=1}^d V_i.$$

Then V is a Hodge structure of CM-type with $\dim_E V = d$.

PROPOSITION 11.5.22 *If $\sum \varphi_i$ is constant on S, then V is of split Weil type.*

PROOF. To begin with, it is necessarily the case that $\sum \varphi_i = d/2$; indeed,

$$\sum_{i=1}^d \varphi_i(s) + \sum_{i=1}^d \varphi(\bar{s}) = \sum_{i=1}^d \big(\varphi_i(s) + \varphi_i(\bar{s})\big) = d,$$

and the two sums are equal by assumption. By construction, we have

$$V \otimes_{\mathbb{Q}} \mathbb{C} \simeq \bigoplus_{i=1}^d (E_{\varphi_i} \otimes_{\mathbb{Q}} \mathbb{C}) \simeq \bigoplus_{i=1}^d \bigoplus_{s\in S} \mathbb{C}^{\varphi_i(s),\varphi_i(\bar{s})}.$$

This shows that

$$V_s = V \otimes_{E,s} \mathbb{C} \simeq \bigoplus_{i=1}^{d} \mathbb{C}^{\varphi_i(s), \varphi_i(\bar{s})}.$$

Therefore $\dim_{\mathbb{C}} V_s^{1,0} = \sum \varphi_i(s) = d/2$, and likewise $\dim_{\mathbb{C}} V_s^{0,1} = \sum \varphi_i(\bar{s}) = d/2$.

Next, we construct the required E-Hermitian form on V. For each i, choose a Riemann form ψ_i on V_i, whose Rosati involution acts as complex conjugation on E. Since $V_i = E_{\varphi_i}$, there exist totally imaginary elements $\zeta_i \in E^\times$ such that

$$\psi_i(x, y) = \mathrm{Tr}_{E/\mathbb{Q}}(\zeta_i x \bar{y})$$

for every $x, y \in E$. Set $\zeta = \zeta_d$, and define $\phi_i(x, y) = \zeta_i \zeta^{-1} x \bar{y}$, which is an E-Hermitian form on V_i with the property that $\psi_i = \mathrm{Tr}_{E/\mathbb{Q}}(\zeta \phi_i)$.

For any collection of totally positive elements $f_i \in F$,

$$\psi = \sum_{i=1}^{d} f_i \psi_i$$

is a Riemann form for V. As E-vector spaces, we have $V = E^{\oplus d}$, and so we can define a nondegenerate E-Hermitian form on V by the rule

$$\phi(v, w) = \sum_{i=1}^{d} f_i \phi_i(v_i, w_i).$$

We then have $\psi = \mathrm{Tr}_{E/\mathbb{Q}}(\zeta \phi)$. By the same argument as before, $a_s = b_s = d/2$, since $\dim_{\mathbb{C}} V_s^{1,0} = \dim_{\mathbb{C}} V_s^{0,1} = d/2$. By construction, the form ϕ is diagonalized, and so its discriminant is easily found to be

$$\mathrm{disc}\, \phi = \zeta^{-d} \prod_{i=1}^{d} f_i \zeta_i \mod \mathrm{Nm}_{E/F}(E^\times).$$

On the other hand, we know from general principles that, for any $s \in S$,

$$\mathrm{sgn}(s(\mathrm{disc}\, \phi)) = (-1)^{b_s} = (-1)^{d/2}.$$

This means that $\mathrm{disc}\, \phi = (-1)^{d/2} f$ for some totally positive element $f \in F^\times$. Upon replacing f_d by $f_d f^{-1}$, we get $\mathrm{disc}\, \phi = (-1)^{d/2}$, which proves that (V, ϕ) is split. $\square$

11.5.7 Split Weil Classes are Absolute

The third step in the proof of Deligne's theorem is to show that split Weil classes are absolute. We begin by describing a special class of abelian varieties of split Weil type where this can be proved directly.

Let V_0 be a rational Hodge structure of even rank d and type $\{(1, 0), (0, 1)\}$. Let ψ_0 be a Riemann form that polarizes V_0, and W_0 a maximal isotropic subspace of dimension $d/2$. Also fix an element $\zeta \in E^\times$ with $\bar{\zeta} = -\zeta$.

Now set $V = V_0 \otimes_{\mathbb{Q}} E$, with Hodge structure induced by the isomorphism

$$V \otimes_{\mathbb{Q}} \mathbb{C} \simeq V_0 \otimes_{\mathbb{Q}} (E \otimes_{\mathbb{Q}} \mathbb{C}) \simeq \bigoplus_{s \in S} V_0 \otimes_{\mathbb{Q}} \mathbb{C}.$$

Define a $\mathbb{Q}$-bilinear form $\psi \colon V \times V \to \mathbb{Q}$ by the formula

$$\psi(v_0 \otimes e, v_0' \otimes e') = \mathrm{Tr}_{E/\mathbb{Q}}(e\overline{e'}) \cdot \psi_0(v_0, v_0').$$

This is a Riemann form on V, for which $W = W_0 \otimes_{\mathbb{Q}} E$ is an isotropic subspace of dimension $d/2$. By Lemma 11.5.16, there is a unique E-Hermitian form $\phi \colon V \times V \to E$ such that $\psi = \mathrm{Tr}_{E/\mathbb{Q}}(\zeta\phi)$. By Corollary 11.5.14, (V, ϕ) is split, and V is therefore of split Weil type. Let A_0 be an abelian variety with $H^1(A_0, \mathbb{Q}) = V_0$. The integral lattice of V_0 induces an integral lattice in $V = V_0 \otimes_{\mathbb{Q}} E$. We denote by $A_0 \otimes_{\mathbb{Q}} E$ the corresponding abelian variety. It is of split Weil type since V is.

The next result, albeit elementary, is the key to proving that split Weil classes are absolute.

PROPOSITION 11.5.23 *Let A_0 be an abelian variety with $H^1(A_0, \mathbb{Q}) = V_0$ as above, and define $A = A_0 \otimes_{\mathbb{Q}} E$. Then the subspace $\bigwedge_E^d H^1(A, \mathbb{Q})$ of $H^d(A, \mathbb{Q})$ consists entirely of absolute Hodge classes.*

PROOF. We have $H^d(A, \mathbb{Q}) \simeq \bigwedge_{\mathbb{Q}}^d H^1(A, \mathbb{Q})$, and the subspace

$$\bigwedge_E^d H^1(A, \mathbb{Q}) \simeq \bigwedge_E^d V_0 \otimes_{\mathbb{Q}} E \simeq \left(\bigwedge_{\mathbb{Q}}^d V_0 \right) \otimes_{\mathbb{Q}} E \simeq H^d(A_0, \mathbb{Q}) \otimes_{\mathbb{Q}} E$$

consists entirely of Hodge classes by Proposition 11.5.20. But since $\dim A_0 = d/2$, the space $H^d(A_0, \mathbb{Q})$ is generated by the fundamental class of a point, which is clearly absolute. This implies that every class in $\bigwedge_E^d H^1(A, \mathbb{Q})$ is absolute. $\square$

The following theorem, together with principle B as in Theorem 11.3.7, completes the proof of Deligne's theorem.

THEOREM 11.5.24 *Let E be a CM-field, and let A be an abelian variety of split Weil type (relative to E). Then there exists a family $\pi \colon \mathcal{A} \to B$ of abelian varieties, with B irreducible and quasi-projective, such that the following three statements are true:*

(a) $\mathcal{A}_0 = A$ for some point $0 \in B$.

(b) For every $t \in B$, the abelian variety $\mathcal{A}_t = \pi^{-1}(t)$ is of split Weil type (relative to E).

(c) The family contains an abelian variety of the form $A_0 \otimes_{\mathbb{Q}} E$.

The proof of Theorem 11.5.24 takes up the remainder of this section. Throughout, we let $V = H^1(A, \mathbb{Q})$, which is an E-vector space of some even dimension d. The polarization on A corresponds to a Riemann form $\psi \colon V \times V \to \mathbb{Q}$, with the property that the Rosati involution acts as complex conjugation on E. Fix a totally imaginary element

$\zeta \in E^\times$; then $\psi = \mathrm{Tr}_{E/\mathbb{Q}}(\zeta\phi)$ for a unique E-Hermitian form ϕ by Lemma 11.5.16. Since A is of split Weil type, the pair (V, ϕ) is split.

As before, let D be the period domain whose points parametrize Hodge structures of type $\{(1,0), (0,1)\}$ on V that are polarized by the form ψ. Let $D^{\mathrm{sp}} \subseteq D$ be the subset of those Hodge structures that are of split Weil type (relative to E, and with polarization given by ψ). We shall show that D^{sp} is a certain Hermitian symmetric domain.

We begin by observing that there are essentially $(2^{[E:\mathbb{Q}]}/2)$ different choices for the totally imaginary element ζ, up to multiplication by totally positive elements in $F^\times$. Indeed, if we fix a choice of $i = \sqrt{-1}$, and define $\varphi_\zeta \colon S \to \{0, 1\}$ by the rule

$$\varphi_\zeta(s) = \begin{cases} 1 & \text{if } s(\zeta)i > 0, \\ 0 & \text{if } s(\zeta)i < 0, \end{cases} \tag{11.5.1}$$

then $\varphi_\zeta(s) + \varphi_\zeta(\bar{s}) = 1$ because $\bar{s}(\zeta) = -s(\zeta)$, and so φ_ζ is a CM-type for E. Conversely, one can show that any CM-type is obtained in this manner.

LEMMA 11.5.25 *The subset D^{sp} of the period domain D is a Hermitian symmetric domain; in fact, it is isomorphic to the product of $|S| = ([E : \mathbb{Q}])$ copies of Siegel upper half-space.*

PROOF. Recall that V is an E-vector space of even dimension d, and that the Riemann form $\psi = \mathrm{Tr}_{E/\mathbb{Q}}(\zeta\phi)$ for a split E-Hermitian form $\phi\colon V \times V \to E$ and a totally imaginary $\zeta \in E^\times$. The Rosati involution corresponding to ψ induces complex conjugation on E; this means that $\psi(ev, w) = \psi(v, \bar{e}w)$ for every $e \in E$.

By definition, D^{sp} parametrizes all Hodge structures of type $\{(1,0),(0,1)\}$ on V that admit ψ as a Riemann form and are of split Weil type (relative to the CM-field E). Such a Hodge structure amounts to a decomposition

$$V \otimes_\mathbb{Q} \mathbb{C} = V^{1,0} \oplus V^{0,1}$$

with $V^{0,1} = \overline{V^{1,0}}$, with the following two properties:

(a) The action by E preserves $V^{1,0}$ and $V^{0,1}$.

(b) The form $-i\psi(x, \bar{y}) = \psi\big(x, h(i)\bar{y}\big)$ is positive definite on $V^{1,0}$.

Let $S = \mathrm{Hom}(E, \mathbb{C})$, and consider the isomorphism

$$V \otimes_\mathbb{Q} \mathbb{C} \xrightarrow{\sim} \bigoplus_{s \in S} V_s, \quad v \otimes z \mapsto \sum_{s \in S} v \otimes z,$$

where $V_s = V \otimes_{E,s} \mathbb{C}$. Since V_s is exactly the subspace on which $e \in E$ acts as multiplication by $s(e)$, the condition in (a) is equivalent to demanding that each complex vector space V_s decomposes as $V_s = V_s^{1,0} \oplus V_s^{0,1}$.

On the other hand, ϕ induces a Hermitian form ϕ_s on each V_s, and we have

$$\psi(v, w) = \mathrm{Tr}_{E/\mathbb{Q}}\big(\zeta\phi(v, w)\big) = \sum_{s \in S} s(\zeta)\phi_s(v \otimes 1, w \otimes 1).$$

Therefore ψ polarizes the Hodge structure $V^{1,0} \oplus V^{0,1}$ if and only if the form $x \mapsto -s(\zeta)i \cdot \phi_s(x, \bar{x})$ is positive definite on the subspace $V_s^{1,0}$. Referring to the definition of φ_ζ in (11.5.1), this is equivalent to demanding that $x \mapsto (-1)^{\varphi_\zeta(s)}\phi_s(x, \bar{x})$ be positive definite on $V_s^{1,0}$.

In summary, Hodge structures of split Weil type on V for which ψ is a Riemann form are parametrized by a choice of $d/2$-dimensional complex subspaces $V_s^{1,0} \subseteq V_s$, one for each $s \in S$, with the property that

$$V_s^{1,0} \cap \overline{V_s^{1,0}} = \{0\},$$

and such that $x \mapsto (-1)^{\varphi_\zeta(s)}\phi_s(x, \bar{x})$ is positive definite on $V_s^{1,0}$. Since for each $s \in S$, we have $a_s = b_s = d/2$, then the Hermitian form ϕ_s has signature $(d/2, d/2)$; this implies that the space

$$D_s = \left\{ W \in \mathrm{Grass}_{d/2}(V_s) \mid W \cap \overline{W} = \{0\} \text{ and } (-1)^{\varphi_\zeta(s)}\phi_s(x, \bar{x}) > 0 \text{ for } 0 \neq x \in W \right\}$$

is isomorphic to Siegel upper half-space. The parameter space D^{sp} for our Hodge structures is therefore the Hermitian symmetric domain

$$D^{\mathrm{sp}} \simeq \prod_{s \in S} D_s.$$

In particular, it is a connected complex manifold. $\square$

To be able to satisfy the final condition in Theorem 11.5.24, we need to know that D^{sp} contains Hodge structures of the form $V_0 \otimes_{\mathbb{Q}} E$. This is the content of the following lemma.

LEMMA 11.5.26 *With notation as above, there is a rational Hodge structure V_0 of weight 1, such that $V_0 \otimes_{\mathbb{Q}} E$ belongs to D^{sp}.*

PROOF. Since the pair (V, ϕ) is split, there is a totally isotropic subspace $W \subseteq V$ of dimension $\dim_E W = d/2$. Arguing as in the proof of Corollary 11.5.14, we can therefore find a basis $v_1, \ldots, v_d$ for the E-vector space V, with the property that

$$\phi(v_i, v_{i+d/2}) = \zeta^{-1} \quad \text{for } 1 \leq i \leq d/2,$$
$$\phi(v_i, v_j) = 0 \quad \text{for } |i - j| \neq d/2.$$

Let V_0 be the $\mathbb{Q}$-linear span of $v_1, \ldots, v_d$; then we have $V = V_0 \otimes_{\mathbb{Q}} E$. Now define $V_0^{1,0} \subseteq V_0 \otimes_{\mathbb{Q}} \mathbb{C}$ as the $\mathbb{C}$-linear span of the vectors $h_k = v_k + iv_{k+d/2}$ for $k = 1, \ldots, d/2$. Evidently, this gives a Hodge structure of weight 1 on V_0, hence a Hodge structure on $V = V_0 \otimes_{\mathbb{Q}} E$. It remains to show that ψ polarizes this Hodge structure.

But we compute that

$$\psi\left(\sum_{j=1}^{d/2} a_j h_j, i \sum_{k=1}^{d/2} \overline{a_k h_k}\right) = \sum_{k=1}^{d/2} |a_k|^2 \psi\bigl(v_k + i v_{k+d/2}, i(v_k - i v_{k+d/2})\bigr)$$

$$= 2 \sum_{k=1}^{d/2} |a_k|^2 \psi(v_k, v_{k+d/2})$$

$$= 2 \sum_{k=1}^{d/2} |a_k|^2 \, \mathrm{Tr}_{E/\mathbb{Q}}\bigl(\zeta\phi(v_k, v_{k+d/2})\bigr) = 2[E:\mathbb{Q}] \sum_{k=1}^{d/2} |a_k|^2,$$

which proves that $x \mapsto \psi(x, i\bar{x})$ is positive definite on the subspace $V_0^{1,0}$. The Hodge structure $V_0 \otimes_{\mathbb{Q}} E$ therefore belongs to D^{sp} as desired. $\qquad\square$

PROOF OF THEOREM 11.5.24. Let $\theta\colon A \to \hat{A}$ be the polarization on A. As before, let $\mathcal{M}$ be the moduli space of abelian varieties of dimension $d/2$, with polarization of the same type as θ, and level 3-structure. Then $\mathcal{M}$ is a quasi-projective complex manifold, and the period domain D is its universal covering space (with the Hodge structure $II^1(\Lambda)$ mapping to the point A). Let $B \subseteq \mathcal{M}$ be the locus of those abelian varieties whose endomorphism algebra contains E. Note that the original abelian variety A is contained in B. Since every element $e \in E$ is a Hodge class in $\mathrm{End}(A) \otimes \mathbb{Q}$, it is clear that B is a Hodge locus; in particular, B is a quasi-projective variety by the theorem of Cattani–Deligne–Kaplan. As before, we let $\pi\colon \mathcal{A} \to B$ be the restriction of the universal family of abelian varieties to B.

Now we claim that the preimage of B in D is precisely the set D^{sp} of Hodge structures of split Weil type. Indeed, the endomorphism ring of any Hodge structure in the preimage of B contains E by construction; since it is also polarized by the form ψ, all the conditions in Definition 11.5.19 are satisfied, and so the Hodge structure in question belongs to D^{sp}. Because D is the universal covering space of $\mathcal{M}$, this implies, in particular, that B is connected and smooth, hence a quasi-projective complex manifold.

The first two assertions are obvious from the construction, whereas the third follows from Lemma 11.5.26. This concludes the proof. $\qquad\square$

To complete the proof of Deligne's theorem, we have to show that every split Weil class is an absolute Hodge class. For this, we argue as follows: Consider the family of abelian varieties $\pi\colon \mathcal{A} \to B$ from Theorem 11.5.24. By Proposition 11.5.20, the space of split Weil classes $\bigwedge_E^d H^1(\mathcal{A}_t, \mathbb{Q})$ consists of Hodge classes for every $t \in B$. The family also contains an abelian variety of the form $A_0 \otimes_{\mathbb{Q}} E$, and according to Proposition 11.5.23, all split Weil classes on this particular abelian variety are absolute. But now B is irreducible, and so principle B applies and shows that for every $t \in B$, all split Weil classes on $\mathcal{A}_t$ are absolute. This finishes the third step of the proof, and finally establishes Deligne's theorem.

Bibliography

[1] André, Y.: *Pour une théorie inconditionnelle des motifs*, Pub. Math. IHÉS **83**, pp. 5–49 (1996).

[2] André, Y.: *Une remarque à propos des cycles de Hodge de type CM*, in Séminaire de Théorie des Nombres, Paris, 1989–1990. Progress in Mathematics **102**, pp. 1–7, Birkhäuser, Boston, MA (1992).

[3] André, Y.: *Une introduction aux motifs (motifs purs, motifs mixtes, périodes)*, Panoramas et Synthèses **17**, Société Mathématique de France (2004).

[4] Blasius, D.: *A p-adic property of Hodge classes on abelian varieties*, in Motives (Seattle, WA, 1991), pp. 293–308, Proc. Sympos. Pure Math. **55**, Part 2, Amer. Math. Soc., Providence, RI (1994).

[5] Bloch, S.: *Semi-regularity and de Rham cohomology*, Invent. Math. **17**, pp. 51–66 (1972).

[6] Bourbaki, N.: *Éléments de mathématique. Première partie: les structures fondamentales de l'analyse. Livre II: Algèbre. Chapitre 9: Formes sesquilinéaires et formes quadratiques*, Actualités Sci. Ind. **1272** Hermann, Paris (1959).

[7] Brosnan, P., Pearlstein, G.: *On the algebraicity of the zero locus of an admissible normal function*, preprint, arxiv.org/abs/0910.0628.

[8] Brosnan, P., Pearlstein, G., Schnell, C.: *The locus of Hodge classes in an admissible variation of mixed Hodge structure*, C. R. Acad. Sci. Paris, Ser. I **348**, pp. 657–660 (2010).

[9] Carlson, J.: *Extensions of mixed Hodge structures*, in Journées de Géometrie Algébrique d'Angers, juillet 1979, pp. 107–127, Sijthoff & Noordhoff, Alphen aan den Rijn–Germantown, Md. (1980).

[10] Cattani, E., Deligne, P., Kaplan, A.: *On the locus of Hodge classes*, J. Amer. Math. Soc. **8**(2), pp. 483–506 (1995).

[11] Charles, F.: *Conjugate varieties with distinct real cohomology algebras*, J. reine angew. Math. **630**, pp. 125–139 (2009).

[12] Charles, F.: *On the zero locus of normal functions and the étale Abel–Jacobi map*, Int. Math. Res. Not. IMRN **12**, pp. 2283–2304 (2010).

[13] Deligne, P.: *Théorie de Hodge, II*, Inst. Hautes Études Sci. Publ. Math. **40**, pp. 5–57 (1971).

[14] Deligne, P.: *La conjecture de Weil pour les surfaces K3* Invent. Math. **15**, pp. 206–226 (1972).

[15] Deligne, P.: *Théorie de Hodge, III*, Inst. Hautes Études Sci. Publ. Math. **44**, pp. 5–77 (1974).

[16] Deligne, P.: *Hodge cycles on abelian varieties* (notes by J. S. Milne), in Lecture Notes in Mathematics **900**, pp. 9–100, Springer (1982).

[17] Deligne, P., Milne, J.: *Tannakian categories*, in Lecture Notes in Mathematics **900**, pp. 101–220, Springer (1982).

[18] Fulton, W., Harris, J.: *Representation theory. A first course*, Graduate Texts in Mathematics **129**, Readings in Mathematics, Springer, New York (1991).

[19] van Geemen, B.: *Kuga–Satake varieties and the Hodge conjecture*, in The Arithmetic and Geometry of Algebraic Cycles (Banff, AB, 1998), pp. 51–82, NATO Sci. Ser. C Math. Phys. Sci. **548**, Kluwer Academic, Dordrecht, (2000).

[20] Grothendieck, A.: *On the de Rham cohomology of algebraic varieties*, Pub. Math. IHÉS **29**, pp. 95–103 (1966).

[21] Grothendieck, A.: *Standard conjectures on algebraic cycles*, Algebraic Geometry (Internat. Colloq., Tata Inst. Fund. Res., Bombay, 1968) pp. 193–199, Oxford University Press, London (1969).

[22] Hartshorne, R.: *Residues and duality*, Lecture Notes in Mathematics **20**, Springer (1966).

[23] Jannsen, U.: *Mixed motives and algebraic K-theory*, Lecture Notes in Mathematics **1400**, Springer (1990).

[24] Katz, N. M., Oda, T.: *On the differentiation of de Rham cohomology classes with respect to parameters*, J. Math. Kyoto Univ. **8**, pp. 199–213 (1968).

[25] Kuga, M., Satake, I.: *Abelian varieties attached to polarized K3 surfaces*, Math. Ann. **169**, pp. 239–242 (1967).

[26] Landherr, W.: *Äquivalenz Hermitescher Formen über einem beliebigen algebraischen Zahlkörper*, Abhandlungen des Mathematischen Seminars der Universität Hamburg **11**, pp. 245–248 (1936).

[27] Morrison, D.: *The Kuga–Satake variety of an abelian surface*, J. Algebra **92**, pp. 454–476 (1985).

[28] Ogus, A.: *Hodge cycles and crystalline cohomology*, in Lecture Notes in Mathematics **900**, pp. 357–414, Springer (1982).

[29] Paranjape, K.: *Abelian varieties associated to certain $K3$ surfaces*, Compos. Math. **68**(1), pp. 11–22 (1988).

[30] Sastry, P., Tong, Y.: *The Grothendieck trace and the de Rham integral*, Canad. Math. Bull **46**(3), pp. 429–440 (2003).

[31] Schappacher, N.: *Periods of Hecke characters*, Lecture Notes in Mathematics **1301**, Springer (1988).

[32] Schlickewei, U.: *The Hodge conjecture for self-products of certain K3 surfaces*, J. Algebra **324**(3), pp. 507–529.

[33] Serre, J.: *Géométrie algébrique et géométrie analytique*, Ann. Inst. Fourier **6**, pp. 1–42 (1955–1956).

[34] Serre, J.-P.: *Exemples de variétés projectives conjuguées non homéomorphes*, C. R. Acad. Sci. Paris **258**, pp. 4194–4196 (1964).

[35] Voisin, C.: *Théorie de Hodge et géométrie algébrique complexe*, Cours Spécialisés **10**, Société Mathématique de France, Paris (2002).

[36] Voisin, C.: *A counterexample to the Hodge conjecture extended to Kähler varieties*, Int. Math. Res. Not. **20**, pp. 1057–1075 (2002).

[37] Voisin, C.: *A generalization of the Kuga–Satake construction*, Pure Appl. Math. Q. 1 **3**, part 2, pp. 415–439 (2005).

[38] Voisin, C.: *Hodge loci and absolute Hodge classes*, Compos. Math. **143**(4), pp. 945–958 (2007).

[39] Voisin, C.: *Some aspects of the Hodge conjecture*, Jpn. J. Math. **2**(2), pp. 261–296 (2007).

[40] Voisin, C.: *Hodge loci*, in Handbook of moduli, Advanced Lectures in Mathematics **25**, Volume III, International Press, pp. 507–546.

Chapter Twelve

Shimura Varieties: A Hodge-Theoretic Perspective by Matt Kerr

INTRODUCTION

In algebraic geometry there is a plethora of objects which *turn out* by big theorems to be algebraic, but which are *defined* analytically:

- projective varieties, as well as functions and forms on them (by Chow's theorem or GAGA [Serre 1956]);

- Hodge loci, and zero loci of normal functions (work of Cattani–Deligne–Kaplan [CDK1995] and Brosnan–Pearlstein [BP2009]);

- complex tori with a polarization (using theta functions, or using the embedding theorem [Kodaira 1954]);

- Hodge classes (if a certain $1,000,000 problem could be solved)

and of concern to us presently,

- modular (locally symmetric) varieties

which can be thought of as the $\Gamma \backslash D$'s for the period maps of certain special VHS's. The fact that they are algebraic is the Baily–Borel theorem [BB1966].

What one does *not* know in the Hodge/zero locus setting above is the field of definition—a question related to the existence of Bloch–Beilinson filtrations, which are discussed in Chapter 10 of this volume. For certain cleverly constructed *unions* of modular varieties, called **Shimura varieties**, one actually knows the minimal (i.e., reflex) field of definition, and also quite a bit about the interplay between "upstairs" and "downstairs" (in $\check{D}$ and $\Gamma \backslash D$, respectively) fields of definition of subvarieties. My interest in the subject stems from investigating Mumford–Tate domains of Hodge structures, where for example the reflex fields can still be defined even though the $\Gamma \backslash D$'s are not algebraic varieties in general (see Section 5 and [GGK2010]). Accordingly, I have tried to pack as many Hodge-theoretic punchlines into the exposition below as possible.

Of course, Shimura varieties are of central importance from another point of view, that of the Langlands program. For instance, they provide a major test case for the conjecture, generalizing Shimura–Taniyama, that all motivic L-functions (arising from

Galois representations on étale cohomology of varieties over number fields) are automorphic, i.e., arise from automorphic forms (or more precisely, from Hecke eigenforms of adélic algebraic groups). The modern theory is largely due to Deligne, Langlands, and Shimura (with crucial details by Shih, Milne, and Borovoi), though many others are implicated in the huge amount of underlying mathematics, e.g.,

- complex multiplication for abelian varieties (Shimura, Taniyama, Weil);

- algebraic groups (Borel, Chevalley, Harish-Chandra);

- class field theory (Artin, Chevalley, Weil; Hensel for p-adics);

- modular varieties (Hilbert, Hecke, Siegel) and their compactifications (Baily, Borel, Satake, Serre, Mumford).

It seems that much of the impetus, historically, for the study of locally symmetric varieties can be credited to Hilbert's twelfth problem generalizing Kronecker's Jugendtraum. Its goal was the construction of abelian extensions (i.e., algebraic extensions with abelian Galois groups) of certain number fields by means of special values of abelian functions in several variables, and it directly underlay the work of Hilbert and his students on modular varieties and the theory of CM.

What follows is based on the course I gave in Trieste, and (though otherwise self-contained) makes free use of Chapters 7 and 4 of this volume. It was a pleasure to speak at such a large and successful summer school, and I heartily thank the organizers for the invitation to lecture there. A brief outline follows:

1. Hermitian symmetric domains: D

2. Locally symmetric varieties: $\Gamma \backslash D$

3. The theory of complex multiplication

4. Shimura varieties: $\amalg_i(\Gamma_i \backslash D)$

5. The field of definition

I acknowledge partial support under the aegis of NSF Standard Grant DMS-1068974 during the preparation of this chapter.

12.1 HERMITIAN SYMMETRIC DOMAINS

A. Algebraic Groups and Their Properties

DEFINITION 12.1.1 *An **algebraic group** G over a field k (of characteristic 0) is a smooth algebraic variety G together with morphisms*

$$\bullet : \ G \times G \to G \quad \textit{(multiplication)},$$
$$(\cdot)^{-1} : \ G \to G \quad \textit{(inversion)},$$

defined over k and an element

$$\mathbf{e} \in G(k) \qquad (identity),$$

subject to rules which make $G(L)$ into a group for each L/k. Here $G(L)$ denotes the L-rational points of G, i.e., the morphisms $\operatorname{Spec} L \rightarrow G$. In particular, $G(\mathbb{R})$ and $G(\mathbb{C})$ have the structure of real and complex Lie groups, respectively.

EXERCISE 12.1.2   Write out these rules as commutative diagrams.

EXAMPLE 12.1.3   As an algebraic variety, the multiplicative group

$$(\mathrm{GL}_1 \cong) \, \mathbb{G}_m := \{XY = 1\} \subset \mathbb{A}^2,$$

with $\mathbb{G}_m(k) = k^*$.

We review the definitions of some basic properties,[1] starting with

- G **connected** $\iff G_{\bar{k}}$ irreducible,

where the subscript denotes the extension of scalars: for L/k, $G_L := G \times_{\operatorname{Spec} k} \operatorname{Spec} L$. There are two fundamental building blocks for algebraic groups: simple groups and tori:

- G **simple** $\iff G$ nonabelian, with no normal connected subgroups $\neq \{\mathbf{e}\}$, G.

EXAMPLE 12.1.4   (a) $k = \mathbb{C}$: SL_n (Cartan type A), SO_n (types B, D), Sp_n (type C), and the exceptional groups of types E_6, E_7, E_8, F_4, G_2.

(b) $k = \mathbb{R}$: have to worry about real forms (of these groups) which can be isomorphic over $\mathbb{C}$ but not over $\mathbb{R}$.

(c) $k = \mathbb{Q}$: $\mathbb{Q}$-simple does not imply $\mathbb{R}$-simple; in other words, all hell breaks loose.

- G **(algebraic) torus** $\iff G_{\bar{k}} \cong \mathbb{G}_m \times \cdots \times \mathbb{G}_m$

EXAMPLE 12.1.5   (a) $k = \mathbb{C}$: the algebraic tori are all of the form $(\mathbb{C}^*)^{\times n}$.

(b) $k = \mathbb{Q}$: given a number field E, the "Weil restriction" or "restriction of scalars" $G = \operatorname{Res}_{E/\mathbb{Q}} \mathbb{G}_m$ is a torus of dimension $[E : \mathbb{Q}]$ with the property that $G(\mathbb{Q}) \cong E^*$, and more generally $G(k) \cong E^* \otimes_{\mathbb{Q}} k$. If $k \supseteq E$ then G splits, i.e.,

$$G(k) \cong (k^*)^{[E:\mathbb{Q}]},$$

with the factors corresponding to the distinct embeddings of E into k.

---

[1]For the analogous definitions for real and complex Lie groups, see [Rotger2005].

(c) $k \subset \mathbb{C}$ arbitrary: inside GL_2, one has tori $\mathbb{U} \subset \mathbb{S} \supset \mathbb{G}_m$ with k-rational points

$$\mathbb{U}(k) = \left\{ \left(\begin{array}{cc} a & b \\ -b & a \end{array} \right) \middle| \begin{array}{c} a^2 + b^2 = 1, \\ a, b \in k \end{array} \right\},$$

$$\mathbb{S}(k) = \left\{ \left(\begin{array}{cc} a & b \\ -b & a \end{array} \right) \middle| \begin{array}{c} a^2 + b^2 \neq 0, \\ a, b \in k \end{array} \right\},$$

$$\mathbb{G}_m(k) = \left\{ \left(\begin{array}{cc} \alpha & 0 \\ 0 & \alpha \end{array} \right) \middle| a \in k^* \right\}.$$

In particular, their complex points take the form

$$\mathbb{C}^* \lhook\joinrel\longrightarrow \mathbb{C}^* \times \mathbb{C}^* \longleftarrow\!\supset \mathbb{C}^*$$

$$z \longmapsto (z, \bar{z}) \; ; \; (\alpha, \alpha) \longleftarrow\!\mid \alpha$$

by considering the eigenvalues of matrices. Writing S^1 or U_1 for the unit circle in $\mathbb{C}^*$, the real points of these groups are $U_1 \subset \mathbb{C}^* \supset \mathbb{R}^*$, the smaller two of which exhibit distinct real forms of $\mathbb{G}_{m,\mathbb{C}}$. The map

$$\jmath : \mathbb{U} \to \mathbb{G}_m$$

sending

$$\left(\begin{array}{cc} a & b \\ -b & a \end{array} \right) \mapsto \left(\begin{array}{cc} a + bi & 0 \\ 0 & a + bi \end{array} \right)$$

is an isomorphism on the complex points but does not respect real points (hence is not defined over $\mathbb{R}$).

Next we put the building blocks together:

- G **semisimple** $\iff$ G an *almost-direct* product of simple subgroups,

i.e., the morphism from the direct product to G is an isogeny (has zero-dimensional kernel). More generally,

- G **reductive** $\iff$ G an almost-direct product of simple groups and tori,

which turns out to be equivalent to the complete reducibility of G's finite-dimensional linear representations.

EXAMPLE 12.1.6 One finite-dimensional representation is the **adjoint map**

$$\begin{array}{ccc} G & \xrightarrow{\mathrm{Ad}} & \mathrm{GL}(\mathfrak{g}), \\ g & \longmapsto & \{X \mapsto gXg^{-1}\}, \end{array}$$

where $\mathfrak{g} = \mathrm{Lie}(G) = T_e G$ and we are taking the differential (at **e**) of $\Psi_g \in \mathrm{Aut}(G)$, i.e., conjugation by g.

We conclude the review with a bit of structure theory. For semisimple groups,

- G **adjoint** $\iff$ Ad is injective; and

- G **simply connected** $\iff$ any isogeny $G' \twoheadrightarrow G$ with G' connected is an isomorphism.

Basically, the center $Z := Z(G)$ is zero-dimensional, hence has finitely many points over $\bar{k}$; we can also say that Z is *finite*. Adjointness is equivalent to its triviality; whereas in the simply connected case Z is as large as possible (with the given Lie algebra).

For reductive groups, we have short-exact sequences

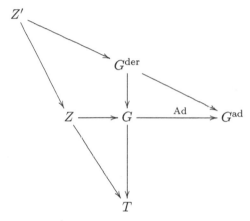

with $G^{\mathrm{der}} := [G, G]$ and $G^{\mathrm{ad}} := \mathrm{Ad}(G)$ both semisimple (and the latter adjoint), $Z' := Z \cap G^{\mathrm{der}}$ finite, and T a torus. Clearly, G is semisimple if and only if the "maximal abelian quotient" T is trivial.

Finally, let G be a reductive real algebraic group,

$$\theta : G \to G,$$

an involution.

DEFINITION 12.1.7 θ is **Cartan** $\iff$

$$\{g \in G(\mathbb{C}) \mid g = \theta(\bar{g})\} =: G^{(\theta)}(\mathbb{R}) \text{ is compact.}$$

Equivalently, $\theta = \Psi_C$ *for* $C \in G(\mathbb{R})$ *with*

- $C^2 \in Z(\mathbb{R})$; *and*

- $G \hookrightarrow \mathrm{Aut}(V, Q)$ *for some symmetric bilinear form* Q *satisfying* $Q(\cdot, C(\bar{\cdot})) > 0$ *on* $V_{\mathbb{C}}$.

Of course, to a Hodge theorist this last condition suggests polarizations and the Weil operator.

Cartan involutions always exist, and

$$G(\mathbb{R}) \text{ compact} \iff \theta = \mathrm{id}.$$

B. Three Characterizations of Hermitian Symmetric Domains

I. Hermitian Symmetric Space of Noncompact Type

This is the "intrinsic analytic" characterization. The basic object is (X, g), a connected complex manifold with Hermitian metric, or equivalently a Riemannian manifold with integrable almost-complex structure such that J acts by isometries. The real Lie group $\mathrm{Is}(X, g)$ of holomorphic isometries must

- act **transitively** on X; and

- contain (for each $p \in X$) **symmetries** $s_p : X \to X$ with $s_p^2 = \mathrm{id}_X$ and p as isolated fixed point; moreover,

- the identity connected component $\mathrm{Is}(X, g)^+$ must be a semisimple adjoint **noncompact** (real Lie) group.

The noncompactness means that (a) the Cartan involution projects to the identity in *no* factor and (b) X has negative sectional curvatures.

II. Bounded Symmetric Domain

Next we come to the "extrinsic analytic" approach, with X a connected open subset of $\mathbb{C}^n$ with compact closure such that the (real Lie) group $\mathrm{Hol}(X)$ of holomorphic automorphisms

- acts **transitively**; and

- contains **symmetries** s_p as above.

The Bergman metric makes X into a noncompact Hermitian space, and the Satake (or Harish-Chandra) embedding does the converse job. For further discussion of the equivalence between I and II; see [Milne2005, Sec. 1].

III. Circle Conjugacy Class

Finally we have the "algebraic" version, which will be crucial for any field of definition questions. Moreover, up to a square root, this is the definition that Mumford–Tate domains generalize as we shall see later.

Let G be a real adjoint (semisimple) algebraic group. We take X to be the orbit, under conjugation by $G(\mathbb{R})^+$, of a homomorphism

$$\phi : \mathbb{U} \to G$$

of algebraic groups defined over $\mathbb{R}$ subject to the constraints

- only z, 1, and z^{-1} appear as eigenvalues in the representation $\mathrm{Ad} \circ \phi$ on $\mathrm{Lie}(G)_{\mathbb{C}}$;

- $\theta := \Psi_{\phi(-1)}$ is Cartan; and

- $\phi(-1)$ does not project to the identity in any simple factor of G.

The points of X are of the form $g\phi g^{-1}$ for $g \in G(\mathbb{R})^+$, and we shall think of it as "a connected component of the conjugacy class of a circle in $G(\mathbb{R})$." Obviously the definition is independent of the choice of ϕ in a fixed conjugacy class.

Under the equivalence of the three characterizations,

$$\mathrm{Is}(X, g)^+ = \mathrm{Hol}(X)^+ = G(\mathbb{R})^+.$$

If, in any of these groups, K_p denotes the stabilizer of a point $p \in X$, then

$$G(\mathbb{R})^+ / K_p \xrightarrow{\cong} X.$$

We now sketch the proof of the equivalence of the algebraic and (intrinsic) analytic versions.

FROM III TO I. Let p denote a point of X given by a circle homomorphism ϕ. Since the centralizer

$$K := Z_{G(\mathbb{R})^+}(\phi)$$

belongs to $G^{(\theta)}(\mathbb{R})$, then

(a) $\mathfrak{k}_{\mathbb{C}} := \mathrm{Lie}(K_{\mathbb{C}})$ is the 1-eigenspace of $\mathrm{Ad}\,\phi(z)$ in $\mathfrak{g}_{\mathbb{C}}$; and

(b) K is compact (in fact, maximally so).

By (a), we have a decomposition

$$\mathfrak{g}_{\mathbb{C}} = \mathfrak{k}_{\mathbb{C}} \oplus \mathfrak{p}^- \oplus \mathfrak{p}^+$$

into $\mathrm{Ad}\,\phi(z)$-eigenspaces with eigenvalues 1, z, z^{-1}. Identifying $\mathfrak{p}^- \cong \mathfrak{g}_{\mathbb{R}}/\mathfrak{k}$ puts a complex structure on $T_\phi X$ for which $d(\Psi_{\phi(z)})$ is multiplication by z. Using $G(\mathbb{R})^+$ to translate $J := d(\Psi_{\phi(i)})$ to all of TX yields an almost-complex structure.

One way to see that this is integrable, making X into a (connected) complex manifold, is as follows: define the **compact dual** $\check{X}$ to be the $\mathrm{Ad}\,G(\mathbb{C})$-translates of the flag

$$\begin{cases} F^1 &= \mathfrak{p}^+, \\ F^0 &= \mathfrak{p}^+ \oplus \mathfrak{k}_{\mathbb{C}}, \end{cases}$$

on $\mathfrak{g}_{\mathbb{C}}$. We have

$$\check{X} \cong G(\mathbb{C})/P \cong G^{(\theta)}(\mathbb{R})\big/ K$$

for P a parabolic subgroup with $\mathrm{Lie}(P) = \mathfrak{k}_{\mathbb{C}} \oplus \mathfrak{p}^+$, which exhibits $\check{X}$ as a $\mathbb{C}$-manifold and as compact. (It is in fact projective.) The obvious map from decompositions to flags yields an injection $X \hookrightarrow \check{X}$ which is an isomorphism on tangent spaces, exhibiting X as an (analytic) open subset of $\check{X}$.

Now by (b), there exists a K-invariant symmetric and positive-definite bilinear form on $T_\phi X$. Translating this around yields a $G(\mathbb{R})^+$-invariant Riemannian metric g on X. Since $J \in K$, then g commutes with (translates of) J and is thereby Hermitian. The symmetry at ϕ is given by $s_\phi := \Psi_{\phi(-1)}$ (acting on X). Since $\Psi_{\phi(-1)}$ (as an element of $\mathrm{Aut}(G)$) is Cartan and does not project to $\mathbf{e}$ in any factor, G is noncompact. $\square$

FROM I TO III. It follows from [Borel1991, Thm. 7.9] that $\mathrm{Is}(X, g)^+$ is adjoint and semisimple. Hence, it is $G(\mathbb{R})^+$ for an algebraic group $G \subset \mathrm{GL}(\mathrm{Lie}(\mathrm{Is}(X, g)^+))$. (Note that this can *only* make sense if $\mathrm{Is}(X, g)^+$ is adjoint and, consequently, embeds in $\mathrm{GL}(\mathrm{Lie}(\mathrm{Is}(X, g)^+))$. Further, the "plus" on $G(\mathbb{R})^+$ is necessary: if $\mathrm{Is}(X, g)^+ = \mathrm{SO}(p, q)^+$, this is *not* $G(\mathbb{R})$ for algebraic G.)

Any $p \in X$ is an isolated fixed point of the associated symmetry $s_p \in \mathrm{Aut}(X)$ with $s_p^2 = \mathrm{id}_X$, and so ds_p is multiplication by (-1) on T_pX. In fact, by a delicate argument involving sectional curvature (cf. [Milne2005, Sec. 1]), for any $z = a + bi$ with $|z| = 1$, there exists a unique isometry $u_p(z)$ of (X, g) such that on T_pX, $du_p(z)$ is multiplication by z (i.e., $a + bJ$). Since du_p yields a homomorphism $U_1 \to \mathrm{GL}(T_pX)$, the uniqueness means that

$$u_p : U_1 \to \mathrm{Is}(X, g)^+$$

is also a homomorphism. It algebraizes to the homomorphism

$$\phi_p : \mathbb{U} \to G$$

of real algebraic groups.

To view X as a conjugacy class, recall that $G(\mathbb{R})^+$ acts transitively. For $g \in G(\mathbb{R})^+$ sending $p \mapsto q$, the uniqueness of u_q means that

$$\phi_q(z) = g \circ \phi_p(z) \circ g^{-1} = (\Psi_g \circ \phi_p)(z).$$

We must show that ϕ_p satisfies the three constraints. In the decomposition

$$\mathfrak{g}_\mathbb{C} = \mathfrak{k}_\mathbb{C} \oplus T_p^{1,0}X \oplus T_p^{0,1}X,$$

$d(\Psi_{\phi_p(z)})$ has eigenvalue z on $T^{1,0}$; hence $\bar{z} = z^{-1}$ on $T^{0,1}$, since ϕ_p is real and $z \in U_1$. Using the uniqueness once more, $\Psi_k \circ \phi_p = \phi_p$ for any $k \in K_p$, and so $\Psi_{\phi_p(z)}$ acts by the identity on $\mathfrak{k} = \mathrm{Lie}(K_p)$. Therefore $1, z, z^{-1}$ are the eigenvalues of $\mathrm{Ad}\,\phi_p$. Finally, from the fact that X has negative sectional curvatures we deduce that Ψ_{s_p} is Cartan, which together with the noncompactness of X implies that s_p projects to $\mathbf{e}$ in no factor of G. $\qquad\qquad\square$

C. Cartan's Classification of Irreducible Hermitian Symmetric Domains

Let X be an irreducible Hermitian symmetric domain (HSD), G the corresponding simple $\mathbb{R}$-algebraic group, and $T \subset G_\mathbb{C}$ a maximal algebraic torus. The restriction to T

$$T \hookrightarrow G_\mathbb{C} \xrightarrow[\mathrm{Ad}]{} \mathrm{GL}(\mathfrak{g}_\mathbb{C})$$

of the adjoint representation breaks into one-dimensional eigenspaces on which T acts through characters:

$$\mathfrak{g}_\mathbb{C} = \mathfrak{t} \oplus \left(\bigoplus_{\alpha \in R} \mathfrak{g}_\alpha \right),$$

where

$$R^+ \amalg R^- = R \subset \mathrm{Hom}(T, \mathbb{G}_m) \cong \mathbb{Z}^n$$

are the roots (with $R^- = -R^+$). With this choice, one has uniquely the

- **simple roots**: $\{\alpha_1, \ldots, \alpha_n\}$ such that each $\alpha \in R^+$ is of the form $\sum m_i \alpha_i$ where $m_i \geq 0$; and the

- **highest root**: $\hat{\alpha} = \sum \hat{m}_i \alpha_i \in R^+$ such that $\hat{m}_i \geq m_i$ for any other $\alpha \in R^+$.

The α_i give the nodes on the Dynkin diagram of G, in which α_i and α_j are connected if they pair nontrivially under a standard inner product, the **Killing form**

$$B(X, Y) := \mathrm{Tr}(\mathrm{ad}\, X \,\mathrm{ad}\, Y).$$

EXAMPLE 12.1.8

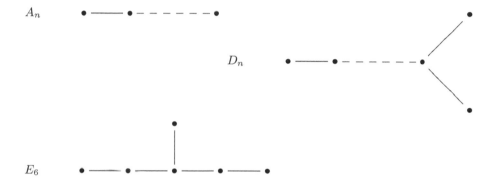

Over $\mathbb{C}$, our circle map ϕ defines a cocharacter

$$\mathbb{G}_m \overset{\mu}{\hookrightarrow} \mathbb{U} \xrightarrow{\phi_{\mathbb{C}}} G_{\mathbb{C}}.$$

This has a *unique* conjugate factoring through T in such a way that, under the pairing of characters and cocharacters given by

$$\mathbb{G}_m \underset{\mu}{\to} T \underset{\alpha}{\to} \mathbb{G}_m,$$

$$z \longmapsto z^{\langle \mu, \alpha \rangle},$$

one has $\langle \mu, \alpha \rangle \geq 0$ for all $\alpha \in R^+$. Since μ must act through the eigenvalues $z, 1, z^{-1}$, we know

$$\begin{aligned} \langle \mu, \alpha \rangle &= \quad 0 \text{ or } 1 \quad &&\text{for all } \alpha \in R^+ \\ \text{and} \quad &\neq \quad 0 \quad &&\text{for some } \alpha \in R^+ \end{aligned}.$$

By considering $\langle \mu, \hat{\alpha} \rangle$, we deduce from this that $\langle \mu, \alpha_i \rangle = 1$ for a unique i, and that the corresponding α_i is **special**: i.e., $\hat{m}_i = 1$. So we have a 1-to-1 correspondence

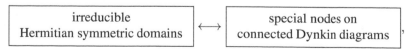

,

and hence a list of the number of distinct isomorphism classes of irreducible HSDs corresponding to each simple complex Lie algebra:

A_n	B_n	C_n	D_n	E_6	E_7	E_8	F_4	G_2
n	1	1	3	2	1	0	0	0

EXAMPLE 12.1.9

(a) A_n: $X \cong \mathrm{SU}(p,q)/S(U_p \times U_q)$ with $p + q = n + 1$ (n possibilities).

(b) B_n: $X \cong \mathrm{SO}(n,2)^+/\mathrm{SO}(n) \times \mathrm{SO}(2)$.

(c) C_n: $X \cong \mathrm{Sp}_{2n}(\mathbb{R})/U(n) \cong \{Z \in M_n(\mathbb{C}) \mid Z = {}^t Z, \mathrm{Im}(Z) > 0\}$.

From Chapter 4, we know that in (b) X is a period domain for H^2 of $K3$ surfaces; in (c), X is the Siegel upper half-space $\mathfrak{H}^n$, and parametrizes weight/level-1 Hodge structures.

D. Hodge-Theoretic Interpretation

Let V be a $\mathbb{Q}$-vector space.

DEFINITION 12.1.10   *A Hodge structure on V is a homomorphism*

$$\tilde{\varphi} : \mathbb{S} \to \mathrm{GL}(V)$$

defined over $\mathbb{R}$, such that the weight homomorphism

$$w_{\tilde{\varphi}} : \mathbb{G}_m \hookrightarrow \mathbb{S} \xrightarrow{\tilde{\varphi}} \mathrm{GL}(V)$$

is defined over $\mathbb{Q}$.[2]

Associated to $\tilde{\varphi}$ is

$$\begin{aligned} \mu_{\tilde{\varphi}} : \quad \mathbb{G}_m &\to \quad \mathrm{GL}(V), \\ z &\mapsto \quad \tilde{\varphi}_{\mathbb{C}}(z,1). \end{aligned}$$

REMARK 12.1.11   Recalling that $\mathbb{S}(\mathbb{C}) \cong \mathbb{C}^* \times \mathbb{C}^*$, $V^{p,q} \subset V_{\mathbb{C}}$ is the

$$\begin{cases} z^p w^q\text{-eigenspace of } \tilde{\varphi}_{\mathbb{C}}(z,w), \\ z^p\text{-eigenspace of } \mu(z), \end{cases}$$

and $w_{\tilde{\varphi}}(r) = \tilde{\varphi}(r,r)$ acts on it by r^{p+q}.

Fix a weight n, Hodge numbers $\{h^{p,q}\}_{p+q=n}$, and polarization $Q : V \times V \to \mathbb{Q}$. Let

---

[2]These are precisely the $\mathbb{Q}$-split mixed Hodge structures (no nontrivial extensions). More generally, mixed Hodge structures have weight *filtration* $W_\bullet$ defined over $\mathbb{Q}$ but with the canonical splitting of $W_\bullet$ defined over $\mathbb{C}$, so that the weight *homomorphism* is only defined over $\mathbb{C}$.

- D be the period domain parametrizing Hodge structures of this type, polarized by Q, on V;

- $\mathbf{t} \in \oplus_i V^{\otimes k_i} \otimes \check{V}^{\otimes \ell_i}$ be a finite sum of Hodge tensors;

- $D_{\mathbf{t}}^+ \subset D$ be a connected component of the subset of HS for which these tensors are **Hodge**: $\mathbf{t}_i \in F^{n(k_i - \ell_i)/2}$ for each i; and

- $M_{\mathbf{t}} \subset \mathrm{GL}(V)$ be the smallest $\mathbb{Q}$-algebraic subgroup with $M_{\mathbf{t}}(\mathbb{R}) \supset \tilde{\varphi}(\mathbb{S}(\mathbb{R}))$ for all $\tilde{\varphi} \in D_{\mathbf{t}}^+$. (This is reductive.)

Then given any $\tilde{\varphi} \in D_{\mathbf{t}}^+$, the orbit

$$D_{\mathbf{t}}^+ = M_{\mathbf{t}}(\mathbb{R})^+ . \tilde{\varphi} \cong M_{\mathbf{t}}(\mathbb{R})^+ / H_{\tilde{\varphi}},$$

under action by conjugation, is called a **Mumford–Tate domain**. This is a connected component of the **full Mumford–Tate domain** $D_{\mathbf{t}} := M_{\mathbf{t}}(\mathbb{R}).\tilde{\varphi}$, which will become relevant later (see Section 12.4.B). In either case, $M_{\mathbf{t}}$ is the **Mumford–Tate group** of $D_{\mathbf{t}}^{(+)}$.

EXERCISE 12.1.12   Check that $\Psi_{\mu_{\tilde{\varphi}}(-1)}$ is a Cartan involution.

Now, consider the condition that the tautological family $\mathcal{V} \to D_{\mathbf{t}}^+$ be a variation of Hodge structure, i.e., that Griffiths's infinitesimal period relation (IPR) $\mathfrak{I} \subset \Omega^\bullet(D_{\mathbf{t}}^+)$ be trivial. This is equivalent to the statement that the HS induced on $\mathrm{Lie}(M_{\mathbf{t}}) \subset \mathrm{End}(V)$ "at $\tilde{\varphi}$" (by $\mathrm{Ad} \circ \tilde{\varphi}$) be of type $(-1, 1) + (0, 0) + (1, -1)$, since terms in the Hodge decomposition of type $(-2, 2)$ or worse would violate Griffiths transversality. Another way of stating this is that

$$\mathrm{Ad} \circ \mu_{\tilde{\varphi}}(z) \text{ has only the eigenvalues } z, 1, z^{-1}, \qquad (12.1.1)$$

and so we have proved part (a) of the following proposition:

PROPOSITION 12.1.13   *(a) A Mumford–Tate domain with trivial IPR (and $M_{\mathbf{t}}$ adjoint) admits the structure of a Hermitian symmetric domain with G defined over $\mathbb{Q}$, and*

(b) the converse—that is, such Hermitian symmetric domains parametrize VHS.

REMARK 12.1.14   (i) Condition (12.1.1) implies that $\Psi_{\mu_{\tilde{\varphi}}(-1)}$ gives a symmetry of $D_{\mathbf{t}}^+$ at $\tilde{\varphi}$, but not conversely; e.g., an example of a Hermitian symmetric MT domain with nontrivial IPR is the period domain for HS of weight 6 and type $(1, 0, 1, h, 1, 0, 1)$.

(ii) This does not contradict (b), because the same Hermitian symmetric domain can have different MT domain structures.

(iii) Strictly speaking, to get the HSD structure in (a) one must put $\phi := \mu_{\tilde{\varphi}} \circ \jmath$, which unlike $\mu_{\tilde{\varphi}}$ is actually defined over $\mathbb{R}$.

(iv) The isotropy group $H_{\tilde{\varphi}}$ above is a maximal compact subgroup of $M_{\mathfrak{t}}(\mathbb{R})^+$ if and only if $D_{\mathfrak{t}}^+$ is Hermitian symmetric.

PROOF OF PROPOSITION 12.1.13. To show (b), let X be an HSD with real circle $\mathbb{U} \xrightarrow{\phi} G$. Since a product of MT domains is an MT domain, we may assume G is $\mathbb{Q}$-simple. The composition

$$(z, w) \longmapsto z/w,$$

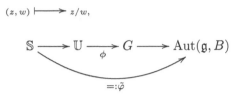

is a Hodge structure on $V = \mathfrak{g}$, polarized by $-B$ since $\Psi_{\phi(-1)}$ is Cartan. The $\mathbb{Q}$-closure of a generic $G(\mathbb{R})$-conjugate is G by nontriviality of $\phi(-1)$, and G is of the form $M_{\mathfrak{t}}$ by Chevalley's theorem (cf. [GGK2010], section I.B). Consequently, $X = G(\mathbb{R})^+.\tilde{\varphi}$ is an MT domain. The IPR vanishes because $\mathrm{Ad} \circ \phi$ has eigenvalues $z, 1, z^{-1}$. $\square$

The proposition is essentially a theorem of [Deligne1979].

EXAMPLE 12.1.15 (Due to Mark Green)   Applied to one of the HSDs for E_6, this procedure yields an MT domain parametrizing certain HS of type $(h^{2,0}, h^{1,1}, h^{0,2}) = (16, 46, 16)$. This is a submanifold $D_{\mathfrak{t}}^+$ of the period domain D for such HS. The IPR $\mathfrak{I} \subset \Omega^\bullet(D)$ is nontrivial but *pulls back to zero* on $D_{\mathfrak{t}}^+$.

PROBLEM 12.1.16   Find a family of varieties over $D_{\mathfrak{t}}^+$ with this family of HS. More generally, it is conjectured that the tautological VHS over every MT domain with trivial IPR, is *motivic*; i.e., comes from algebraic geometry.

The proof of Proposition 12.1.13(b) always produces HS of even weight. Sometimes, by replacing the adjoint representation by a "standard" representation, we can parametrize HS of odd weight: for instance, $\mathfrak{H}^4 \cong \mathrm{Sp}_8(\mathbb{R})/U(4)$ parametrizes HS of weight/level 1 and rank 8, or equivalently abelian varieties of dimension 4. There are "two" types of MT subdomains in $\mathfrak{H}^4$:

(a) Those corresponding to $\mathrm{End}_{\mathrm{HS}}(V) (= \mathrm{End}(A)_{\mathbb{Q}})$ containing a nontrivial fixed subalgebra $\mathcal{E}$ isomorphic to a product of matrix algebras over $\mathbb{Q}$-division algebras. These must be of the four types occurring in the Albert classification:

  (I) totally real field

  (II) indefinite quaternion algebra over a totally real field

  (III) definite quaternion algebra over a totally real field

  (IV) division algebra over a CM-field

  For example, an imaginary quadratic field is a type (IV) division algebra. All four types do occur in $\mathfrak{H}^4$.

(b) Those corresponding to fixed endomorphisms $\mathcal{E}$ together with *higher* Hodge tensors. We regard $\mathrm{End}_{\mathrm{HS}}(V)$ as a subspace of $T^{1,1}V$ and the polarization Q as an element of $T^{0,2}V$; "higher" means (in a $T^{k,\ell}V$) of degree $k + \ell > 2$.

Here are two such examples:

EXAMPLE 12.1.17 (Type (a))   Fix an embedding

$$\mathbb{Q}(i) \overset{\beta}{\hookrightarrow} \mathrm{End}(V),$$

and write $\mathrm{Hom}(\mathbb{Q}(i), \mathbb{C}) = \{\eta, \bar{\eta}\}$. We consider Hodge structures on V such that

$$V^{1,0} = V_\eta^{1,0} \oplus V_{\bar{\eta}}^{1,0}$$

with $\dim V_\eta^{1,0} = 1$; in particular, for such HS the image of $\mathbb{Q}(i)$ lies in $\mathrm{End}_{\mathrm{HS}}(V)$. The resulting MT domain is easily presented in the three forms from Section 12.1.B: as the circle conjugacy class $D_{\mathbf{t}}^+$ (taking $\mathbf{t} := \{\beta(i), Q\}$); as the noncompact Hermitian space $\mathrm{SU}(1,3)/S(U_1 \times U_3)$; and as a complex 3-ball. It is irreducible of Cartan type A_3.

EXERCISE 12.1.18   Show that the MT group of a generic HS in $D_{\mathbf{t}}^+$ has real points

$$M(\mathbb{R})/w_{\tilde{\varphi}}(\mathbb{R}^*) \cong U(2,1).$$

EXAMPLE 12.1.19 (Type (b))   [Mumford1969] constructs a quaternion algebra $\mathfrak{Q}$ over a totally real cubic field K, such that $\mathfrak{Q} \otimes_{\mathbb{Q}} \mathbb{R} \cong \mathbb{H} \oplus \mathbb{H} \oplus M_2(\mathbb{R})$, together with an embedding $\mathfrak{Q}^* \hookrightarrow \mathrm{GL}_8(\mathbb{Q})$. This yields a $\mathbb{Q}$-simple algebraic group

$$G := \mathrm{Res}_{K/\mathbb{Q}} U_{\mathfrak{Q}} \subset \mathrm{Sp}_8 \subset \mathrm{GL}(V)$$

with $G(\mathbb{R}) \cong \mathrm{SU}(2)^{\times 2} \times \mathrm{SL}_2(\mathbb{R})$, and the $G(\mathbb{R})$-orbit of

$$\varphi_0 : \mathbb{U} \to G,$$

$$a + ib \mapsto \mathrm{id}^{\times 2} \times \begin{pmatrix} a & b \\ -b & a \end{pmatrix}$$

yields an MT domain. Mumford shows that the generic HS it supports has trivial endomorphisms $\mathcal{E} = \mathbb{Q}$, so G is cut out by higher Hodge tensors.

REMARK 12.1.20   Though we limited ourselves to pure HS above to simplify the discussion, the Mumford–Tate business, and Proposition 12.1.13(a) in particular, still works in the more general setting of Definition 12.1.10.

12.2   LOCALLY SYMMETRIC VARIETIES

To construct quotients of Hermitian symmetric domains we'll need the following basic result:

PROPOSITION 12.2.1 *Let X be a topological space, with $x_0 \in X$, $\mathcal{G}$ be a locally compact group acting on X, and $\Gamma \leq \mathcal{G}$ be a discrete subgroup (i.e., one with no limit points). Assume*

(i) $K := \mathrm{stab}(x_0)$ is compact; and

(ii) $gK \mapsto gx_0 : \mathcal{G}/K \to X$ is a homeomorphism.
Then $\Gamma \backslash X$ is Hausdorff.

The proof is a nontrivial topology exercise. Writing $\pi : X \to \Gamma \backslash X$, the key points are

- π^{-1} of a compact set is compact; and

- the intersection of a discrete and compact set is finite.

COROLLARY 12.2.2 *Let $X = G(\mathbb{R})^+/K$ be a Hermitian symmetric domain, and $\Gamma \leq G(\mathbb{R})^+$ discrete and torsion-free. Then $\Gamma \backslash X$ has a unique complex-manifold structure for which π is a local isomorphism.*

REMARK 12.2.3 If Γ is not torsion-free then we get an orbifold.

EXAMPLE 12.2.4   (a) $X = \mathfrak{H}$, $G = \mathrm{SL}_2$ acting in the standard way, and

$$\Gamma = \Gamma(N) := \ker\{\mathrm{SL}_2(\mathbb{Z}) \to \mathrm{SL}_2(\mathbb{Z}/N\mathbb{Z})\}$$

with $N \geq 3$. The quotients $\Gamma \backslash X =: Y(N)$ are the **classical modular curves** classifying elliptic curves with *marked N-torsion*, an example of **level structure**.

(b) $X = \mathfrak{H}^n$, $G = \mathrm{Sp}_{2n}$, $\Gamma = \mathrm{Sp}_{2n}(\mathbb{Z})$. Then $\Gamma \backslash X$ is the **Siegel modular variety** classifying abelian n-folds with a fixed polarization.

(c) $X = \mathfrak{H} \times \cdots \times \mathfrak{H}$ (n times), $G = \mathrm{Res}_{F/\mathbb{Q}} \mathrm{SL}_2$ with F a totally real field of degree n over $\mathbb{Q}$, and $\Gamma = \mathrm{SL}_2(\mathcal{O}_F)$. The quotient $\Gamma \backslash X$ is a **Hilbert modular variety** classifying abelian n-folds with $\mathcal{E} \supset F$: the general member is of Albert type (I). We may view X as a proper MT subdomain of $\mathfrak{H}^n$. To get a more interesting level structure here, one could replace $\mathcal{O}_F$ by a proper ideal.

(d) $X = \{[v] \in \mathbb{P}^n(\mathbb{C}) \mid -|v_0|^2 + \sum_{i=1}^n |v_i|^2 < 0\}$, $G = \mathrm{Res}_{K/\mathbb{Q}} \mathrm{SU}(n,1)$ with K a quadratic imaginary field ($\mathrm{Hom}(K,\mathbb{C}) = \{\theta, \bar{\theta}\}$) and $\Gamma = \mathrm{SU}((n,1), \mathcal{O}_K)$. The quotient $\Gamma \backslash X$ is a **Picard modular variety**, classifying abelian $(n+1)$-folds with $\mathcal{E} \supset K$ in such a way that the dimension of the θ-eigenspace (resp. $\bar{\theta}$-eigenspace) in $T_0 A$ is 1 (resp. n). See the interesting treatment for $n = 2$ in [Holzapfel1995, especially Sec. 4.9].

All the discrete groups Γ arising here have been rather special.

DEFINITION 12.2.5   (a) *Let G be a $\mathbb{Q}$-algebraic group. Fix an embedding $G \overset{\iota}{\hookrightarrow} \mathrm{GL}_n$. A subgroup $\Gamma \leq G(\mathbb{Q})$ is*

arithmetic $\iff$ Γ *commensurable*[3] *with* $G(\mathbb{Q}) \cap GL_n(\mathbb{Z})$; *and*

congruence $\iff$ *for some* N, Γ *contains*
$$\Gamma(N) := G(\mathbb{Q}) \cap \{g \in GL_n(\mathbb{Z}) \mid g \underset{(N)}{\equiv} id\}$$
as a subgroup of finite index.

Congruence subgroups are arithmetic, and both notions are independent of the embedding ι.

(b) *Let* $\mathcal{G}$ *be a connected* real *Lie group. A subgroup* $\Gamma \leq \mathcal{G}$ *is* **arithmetic** *if there exist*

- *a* $\mathbb{Q}$-*algebraic group* G;
- *an arithmetic* $\Gamma_0 \leq G(\mathbb{Q})$; *and*
- *a homomorphism* $G(\mathbb{R})^+ \underset{\rho}{\twoheadrightarrow} \mathcal{G}$ *with compact kernel*

such that $\rho(\Gamma_0 \cap G(\mathbb{R})^+) = \Gamma$.

Part (b) is set up so that Γ will always contain a torsion-free subgroup of finite index.

THEOREM 12.2.6 ([BB1966]) *Let* $X = G(\mathbb{R})^+/K$ *be a Hermitian symmetric domain, and* $\Gamma < G(\mathbb{R})^+$ *a torsion-free arithmetic subgroup. Then* $X(\Gamma) := \Gamma \backslash X$ *is canonically a smooth quasi-projective algebraic variety, called a* **locally symmetric variety**.

REMARK 12.2.7 If we do not assume Γ torsion-free, we still get a quasi-projective algebraic variety, but it is an orbifold, hence not smooth and *not* called a locally symmetric variety.

IDEA OF PROOF. Construct a minimal, highly singular (Baily–Borel) compactification
$$X(\Gamma)^* := \Gamma \backslash \{X \amalg B\},$$
where B stands for "rational boundary components." Embed this in $\mathbb{P}^N$, using automorphic forms of sufficiently high weight, as a projective analytic—hence (by Chow/GAGA) projective algebraic—variety. The existence of enough automorphic forms to yield an embedding is a convergence question for certain Poincaré–Eisenstein series. $\qquad\square$

EXAMPLE 12.2.8 In the modular curve context (with $\Gamma = \Gamma(N)$), $B = \mathbb{P}^1(\mathbb{Q})$ and $X(\Gamma)^* \backslash X(\Gamma)$ is a finite set of points called **cusps**. We write $Y(N)$ (resp. $X(N)$) for $X(\Gamma)$ (resp. $X(\Gamma)^*$).

Recalling from Section 12.1.D that X is always an MT domain, we can give a Hodge-theoretic interpretation to the Baily–Borel compactification:

---

[3]That is, the intersection is of finite index in each.

PROPOSITION 12.2.9 *The boundary components B parametrize the possible*

$$\oplus_i \mathrm{Gr}_i^W H_{\mathrm{lim}}$$

for VHS into $X(\Gamma)$.

HEURISTIC IDEA OF PROOF. Assuming PGL_2 is not a quotient of G, the automorphic forms are Γ-invariant sections of $K_X^{\otimes N}$ for some $N \gg 0$. The canonical bundle K_X, which is pointwise isomorphic to $\bigwedge^d \mathfrak{g}^{(-1,1)}$, measures the change of the Hodge flag in every direction. So the boundary components parametrized by these sections must consist of *naive* limiting Hodge flags in $\partial \bar{X} \subset \check{X}$. In that limit, thinking projectively, the relation between periods that blow up at different rates (arising from different Gr_i^W) is fixed, which means *we cannot see extension data*. On the other hand, since $\exp(zN)$ does not change the $\mathrm{Gr}_i^W F^\bullet$, *this information is the same* for the naive limiting Hodge flag and the limiting mixed Hodge structure. $\qquad\square$

REMARK 12.2.10   (a) For the case of Siegel domains, the proposition is due to Carlson, Cattani, and Kaplan [CCK1980], but there seems to be no reference for the general statement.

   (b) There are other compactifications with different Hodge-theoretic interpretations:

   - the compactification of Borel–Serre [BS1973], which records Gr_i^W and adjacent extensions, at least in the Siegel case; and

   - the smooth toroidal compactifications of [AMRT1975], which capture the entire limit MHS.

   The latter is what the monograph of Kato and Usui [KU2009] generalizes (in a sense) to the non-Hermitian symmetric case, where $X(\Gamma)$ is not algebraic.

For the discussion of canonical models to come, we will need the following result:

THEOREM 12.2.11 [Borel1972] *Let $\mathcal{Y}$ be a quasi-projective algebraic variety over $\mathbb{C}$, and $X(\Gamma)$ a locally symmetric variety. Then any analytic map $\mathcal{Y} \to X(\Gamma)$ is algebraic.*

IDEA OF PROOF. Extend this to an analytic map $\bar{\mathcal{Y}} \to X(\Gamma)^*$, then use GAGA.

Suppose $X = \mathfrak{H}$ and $\mathcal{Y}$ is a curve; since Γ is torsion-free, $X(\Gamma) \cong \mathbb{C}\backslash\{\geq 2 \text{ points}\}$. Denote by $\mathbb{D}$ a small disk about the origin in $\mathbb{C}$. If a holomorphic $f : (\mathbb{D}\backslash\{0\}) \to X(\Gamma)$ does not extend to a holomorphic map from $\mathbb{D}$ to $\mathbb{P}^1$, then f has an essential singularity at 0. By the "big" Picard theorem, f takes all values of $\mathbb{C}$ except possibly one, a contradiction. Applying this argument to a neighborhood of each point of $\bar{\mathcal{Y}}\backslash\mathcal{Y}$ gives the desired extension.

The general proof uses the existence of a good compactification $\mathcal{Y} \subset \bar{\mathcal{Y}}$ (Hironaka) so that $\mathcal{Y}$ is locally $\mathbb{D}^{\times k} \times (\mathbb{D}^*)^{\times \ell}$. $\qquad\square$

12.3   COMPLEX MULTIPLICATION

A. CM-Abelian Varieties

A **CM-field** is a totally imaginary field E possessing an involution $\rho \in \mathrm{Gal}(E/\mathbb{Q}) =: \mathcal{G}_E$ such that $\phi \circ \rho = \bar{\phi}$ for each $\phi \in \mathrm{Hom}(E, \mathbb{C}) =: \mathcal{H}_E$.

EXERCISE 12.3.1   Show that then E^ρ is totally real, and $\rho \in Z(\mathcal{G}_E)$.

Denote by E^c a normal closure. For any decomposition

$$\mathcal{H}_E = \Phi \amalg \bar{\Phi},$$

(E, Φ) is a **CM-type**; this is equipped with a **reflex field**

$$
\begin{aligned}
E' &:= \mathbb{Q}\left(\left\{ \textstyle\sum_{\phi \in \Phi} \phi(e) \,\middle|\, e \in E \right\}\right) \subset E^c \\
&= \text{ fixed field of } \left\{ \sigma \in \mathcal{G}_{E^c} \mid \sigma\tilde{\Phi} = \tilde{\Phi} \right\},
\end{aligned}
$$

where $\tilde{\Phi} \subset \mathcal{H}_{E^c}$ consists of embeddings restricted on E to those in Φ. Composing Galois elements with a fixed choice of $\phi_1 \in \Phi$ gives an identification

$$\mathcal{H}_{E^c} \xleftarrow{\cong} \mathcal{G}_{E^c}$$

and a notion of inverse on $\mathcal{H}_{E^c}$. Define the **reflex type** by

$$\Phi' := \left\{ \tilde{\phi}^{-1}|_{E'} \,\middle|\, \tilde{\phi} \in \tilde{\Phi} \right\},$$

and **reflex norm** by

$$N_{\Phi'} : (E')^* \to E^*,$$

$$e' \mapsto \prod_{\phi' \in \Phi'} \phi'(e').$$

EXAMPLE 12.3.2   (a) All imaginary quadratic fields $\mathbb{Q}(\sqrt{-d})$ are CM; in this case, $N_{\Phi'}$ is the identity or complex conjugation.

(b) All cyclotomic fields $\mathbb{Q}(\zeta_n)$ are CM; and if $E/\mathbb{Q}$ is an abelian extension, then $E' = E^c = E$ and E is contained in some $\mathbb{Q}(\zeta_n)$. For cyclotomic fields we will write

$$\phi_j := \text{ embedding sending } \zeta_n \mapsto e^{2\pi ij/n}.$$

(c) The CM-type $(\mathbb{Q}(\zeta_5); \{\phi_1, \phi_2\})$ has reflex $(\mathbb{Q}(\zeta_5); \{\phi_1, \phi_3\})$.

The relationship of this to algebraic geometry is contained in the following proposition:

PROPOSITION 12.3.3   *(a) For a simple complex abelian g-fold A, the following are equivalent:*

(i) *The MT group of $H^1(A)$ is a torus.*

(ii) $\text{End}(A)_{\mathbb{Q}}$ *has (maximal) rank $2g$ over $\mathbb{Q}$.*

(iii) $\text{End}(A)_{\mathbb{Q}}$ *is a CM-field.*

(iv) $A \cong \mathbb{C}^g/\Phi(\mathfrak{a}) =: A_{\mathfrak{a}}^{(E,\Phi)}$ *for some CM-type (E,Φ) and ideal $\mathfrak{a} \subset \mathcal{O}_E$.*

(b) *Furthermore, any complex torus of the form $A_{\mathfrak{a}}^{(E,\Phi)}$ is algebraic.*

A **CM-abelian variety** is just a product of simple abelian varieties, each satisfying the conditions in (a). We will suppress the superscript (E,Φ) when the CM-type is understood. Note that in (i), the torus may be of dimension less than g, the so-called *degenerate* case.

EXAMPLE 12.3.4 $\Phi(\mathfrak{a})$ means the $2g$-lattice

$$\left\{ \left. \begin{pmatrix} \phi_1(a) \\ \vdots \\ \phi_g(a) \end{pmatrix} \right| a \in \mathfrak{a} \right\}.$$

For Example 12.3.2(c) above,

$$A_{\mathcal{O}_{\mathbb{Q}(\zeta_5)}} = \mathbb{C}^2 \Big/ \mathbb{Z}\left\langle \begin{pmatrix} 1 \\ 1 \end{pmatrix}, \begin{pmatrix} e^{2\pi i/5} \\ e^{4\pi i/5} \end{pmatrix}, \begin{pmatrix} e^{4\pi i/5} \\ e^{3\pi i/5} \end{pmatrix}, \begin{pmatrix} e^{6\pi i/5} \\ e^{2\pi i/5} \end{pmatrix} \right\rangle.$$

The interesting points in Proposition 12.3.3 are

- where does the CM-field come from?

- why is $A_{\mathfrak{a}}^{(E,\Phi)}$ polarized?

In lieu of a complete proof we address these issues.

PROOF OF 12.3.3(i) $\implies$ (iii). $V = H^1(A)$ is polarized by some Q. Set

$$\mathcal{E} := \text{End}_{\text{HS}}(H^1(A)) = \left(Z_{\text{GL}(V)}(M)\right)(\mathbb{Q}) \cup \{0\},$$

where the centralizer $Z_{\text{GL}(V)}(M)$ is a $\mathbb{Q}$-algebraic group and contains a maximal torus T. Since T commutes with M and is maximal, T contains M (possibly properly). One deduces that

- $T(\mathbb{Q}) \cup \{0\} =: E \underset{\eta}{(\hookrightarrow} \mathcal{E})$ is a field;

- V is a one-dimensional vector space over E; and

- E is actually all of $\mathcal{E}$.

M diagonalizes with respect to a Hodge basis

$$\omega_1, \ldots, \omega_g; \bar{\omega}_1, \ldots, \bar{\omega}_g$$

such that $\sqrt{-1}Q(\omega_i, \bar{\omega}_j) = \delta_{ij}$. The maximal torus in $\mathrm{GL}(V)$ that this basis defines, centralizes M, and hence must be T.

Now write $\mathcal{H}_E = \{\phi_1, \ldots, \phi_{2g}\}$, $E = \mathbb{Q}(\xi)$, and

$$m_\xi(\lambda) = \prod_{i=1}^{2g}(\lambda - \phi_i(\xi))$$

for the minimal polynomial of ξ, hence $\eta(\xi)$. Up to reordering, we therefore have

$$[\eta(\xi)]_\omega = \mathrm{diag}(\{\phi_i(\xi)\}_{i=1}^{2g})$$

for the matrix of "multiplication by ξ" with respect to the Hodge basis. Since $\eta(\xi) \in \mathrm{GL}(V)$ (a fortiori $\in \mathrm{GL}(V_\mathbb{R})$), and $\phi_j(\xi)$ determines ϕ_j,

$$\omega_{i+g} = \bar{\omega}_i \implies \phi_{i+g} = \bar{\phi}_i.$$

Define the **Rosati involution** $\dagger : \mathcal{E} \to \mathcal{E}$ by

$$Q(\varepsilon^\dagger v, w) = Q(v, \varepsilon w) \quad \forall v, w \in V.$$

This produces $\rho := \eta^{-1} \circ \dagger \circ \eta \in \mathcal{G}_E$, and we compute

$$\phi_{i+g}(e)Q(\omega_i, \omega_{i+g}) = Q(\omega_i, \eta(e)\omega_{i+g}) = Q(\eta(e)^\dagger \omega_i, \omega_{i+g})$$

$$= Q(\eta(\rho(e))\omega_i, \omega_{i+g}) = \phi_i(\rho(e))Q(\omega_i, \omega_{i+g}),$$

which yields $\phi_i \circ \rho = \bar{\phi}_i$. $\qquad\square$

PROOF OF 12.3.3(b). We have the following construction of $H^1(A)$: let V be a $2g$-dimensional $\mathbb{Q}$-vector space with identification

$$\beta : E \overset{\cong}{\to} V$$

inducing (via multiplication in E)

$$\eta : E \hookrightarrow \mathrm{End}(V).$$

Moreover, there is a basis $\omega = \{\omega_1, \ldots, \omega_g; \bar{\omega}_1, \ldots, \bar{\omega}_g\}$ of $V_\mathbb{C}$ with respect to which

$$[\eta_\mathbb{C}(e)]_\omega = \mathrm{diag}\{\phi_1(e), \ldots, \phi_g(e); \bar{\phi}_1(e), \ldots, \bar{\phi}_g(e)\},$$

and we set $V^{1,0} := \mathbb{C}\langle\omega_1, \ldots, \omega_g\rangle$. This gives $\eta(e) \in \mathrm{End}_{\mathrm{HS}}(V)$ and $V \underset{\mathrm{HS}}{\cong} H^1(A)$.

Now, there exists a $\xi \in E$ such that $\sqrt{-1}\phi_i(\xi) > 0$ for $i = 1, \ldots, g$, and we can put

$$Q(\beta(e), \beta(\tilde{e})) := \mathrm{Tr}_{E/\mathbb{Q}}(\xi \cdot e \cdot \rho(\tilde{e})) : V \times V \longrightarrow \mathbb{Q}.$$

Over $\mathbb{C}$, this becomes

$$[Q]_\omega = \begin{pmatrix} & & & \phi_1(\xi) & & \\ & 0 & & & \ddots & \\ \bar\phi_1(\xi) & & & & & \phi_g(\xi) \\ & \ddots & & & 0 & \\ & & \bar\phi_g(\xi) & & & \end{pmatrix}.$$

$\square$

REMARK 12.3.5 $N_{\Phi'}$ algebraizes to a homomorphism of algebraic groups

$$\mathcal{N}_{\Phi'} : \mathrm{Res}_{E'/\mathbb{Q}}\, \mathbb{G}_m \to \mathrm{Res}_{E/\mathbb{Q}}\, \mathbb{G}_m,$$

which gives $N_{\Phi'}$ on the $\mathbb{Q}$-points, and the MT group

$$M_{H^1(A)} \cong \mathrm{image}(\mathcal{N}_{\Phi'}).$$

Let E be a CM-field with $[E : \mathbb{Q}] = 2g$. In algebraic number theory we have

$\mathcal{I}(E)$ the monoid of nonzero ideals in $\mathcal{O}_E$;

$\mathcal{J}(E)$ the group of fractional ideals (of the form $e \cdot I$, $e \in E^*$ and $I \in \mathcal{I}(E)$); and

$\mathcal{P}(E)$ the subgroup of principal fractional ideals (of the form $(e) := e \cdot \mathcal{O}_E$, $e \in E^*$).

The (abelian!) **ideal class group**

$$\mathrm{Cl}(E) := \frac{\mathcal{J}(E)}{\mathcal{P}(E)},$$

or more precisely the **class number**

$$h_E := |\,\mathrm{Cl}(E)|,$$

expresses (if $\neq 1$) the failure of $\mathcal{O}_E$ to be a principal ideal domain (and to have unique factorization). Each class $\tau \in \mathrm{Cl}(E)$ has a representative $I \in \mathcal{I}(E)$ with norm bounded by the **Minkowski bound**, which implies h_E is finite.

Now let

$$\mathrm{Ab}(\mathcal{O}_E, \Phi) := \frac{\left\{ A_\mathfrak{a}^{(E,\Phi)} \,\middle|\, \mathfrak{a} \in \mathcal{J}(E) \right\}}{\mathrm{isomorphism}},$$

where the numerator denotes abelian g-folds with $\mathcal{O}_E \subset \mathrm{End}(A)$ acting on $T_0 A$ through Φ. They are all isogenous, and multiplication by any $e \in E^*$ gives an isomorphism

$$A_\mathfrak{a} \xrightarrow[\cong]{} A_{e\mathfrak{a}}.$$

Hence we get a bijection

$$\begin{array}{ccc} \mathrm{Cl}(E) & \xrightarrow[\cong]{} & \mathrm{Ab}(\mathcal{O}_E, \Phi), \\ [\mathfrak{a}] & \longmapsto & A_\mathfrak{a}. \end{array}$$

EXAMPLE 12.3.6   The class number of $\mathbb{Q}(\sqrt{-5})$ is 2. The elliptic curves $\mathbb{C}/\mathbb{Z}\langle 1, \sqrt{-5}\rangle$ and $\mathbb{C}/\mathbb{Z}\langle 2, 1+\sqrt{-5}\rangle$ are the representatives of $\mathrm{Ab}(\mathcal{O}_{\mathbb{Q}(\sqrt{-5})}, \{\phi\})$.

The key point now is to notice that for $A \in \mathrm{Ab}(\mathcal{O}_E, \Phi)$ and $\sigma \in \mathrm{Aut}(\mathbb{C})$,

$$^{\sigma}A \in \mathrm{Ab}(\mathcal{O}_E, \sigma\Phi).$$

This is because endomorphisms are given by algebraic cycles, so that the internal ring structure of $\mathcal{E}$ is left unchanged by Galois conjugation; what do change are the eigen-values of its action on $T_0 A$. From the definition of E' we see that

$$\mathrm{Gal}(\mathbb{C}/E') \text{ acts on } \mathrm{Ab}(\mathcal{O}_E, \Phi),$$

which suggests that the individual abelian varieties should be defined over an extension of E' of degree $h_{E'}$. This is not exactly true if $\mathrm{Aut}(A) \neq \{\mathrm{id}\}$, but the argument does establish that any CM-abelian variety is defined over $\bar{\mathbb{Q}}$.

B. Class Field Theory

In fact, it gets much better: not only is there a distinguished field extension H_L/L of degree h_L for any number field L; there is an isomorphism

$$\mathrm{Gal}(H_L/L) \xleftarrow{\cong} \mathrm{Cl}(L) . \tag{12.3.1}$$

To quote Chevalley, "L contains within itself the elements of its own transcendence."

IDEA OF THE CONSTRUCTION OF (12.3.1). Let $\tilde{L}/L$ be a degree-d extension which is

- **abelian**: Galois with $\mathrm{Gal}(\tilde{L}/L)$ abelian;

- **unramified**: for each prime $\mathfrak{p} \in \mathcal{I}(L)$, $\mathfrak{p} \cdot \mathcal{O}_{\tilde{L}} = \prod_{i=1}^{r} \mathfrak{P}_i$ for some $r | d$.

Writing $N(\mathfrak{p}) := |\mathcal{O}_L/\mathfrak{p}|$, the extension $(\mathcal{O}_{\tilde{L}}/\mathfrak{P}_i)/(\mathcal{O}_L/\mathfrak{p})$ of finite fields has degree $N(\mathfrak{p})^{d/r}$. The image of its Galois group in $\mathrm{Gal}(\tilde{L}/L)$ by

$$\mathrm{Gal}\left(\frac{(\mathcal{O}_{\tilde{L}}/\mathfrak{P}_i)}{(\mathcal{O}_L/\mathfrak{p})}\right) \xleftarrow{\cong} \left\{\sigma \in \mathrm{Gal}(\tilde{L}/L) \,\middle|\, \sigma\mathfrak{P}_i = \mathfrak{P}_i\right\} \hookrightarrow \mathrm{Gal}(\tilde{L}/L)$$

$$\underset{\text{by}}{\overset{\text{generated}}{\Big\{}} \quad \left\{\begin{array}{c} \alpha \to \alpha^{N(\mathfrak{p})} \\ (\mathrm{mod}\ \mathfrak{P}_i) \end{array}\right\} \longmapsto \hspace{4cm} =: \mathrm{Frob}_{\mathfrak{p}}$$

is (as the notation suggests) independent of i, yielding a map from

$$\left\{\begin{array}{c} \text{prime ideals} \\ \text{of } L \end{array}\right\} \longrightarrow \mathrm{Gal}(\tilde{L}/L).$$

Taking $\tilde{L}$ to be the Hilbert class field

$$H_L := \text{maximal unramified abelian extension of } L,$$

this leads (eventually) to (12.3.1).   □

More generally, given $I \in \mathcal{I}(L)$, we have the **ray class group mod** I,

$$\text{Cl}(I) = \frac{\text{fractional ideals prime to } I}{\text{principal fractional ideals with generator } \equiv 1 \ (\text{mod } I)}$$

and **ray class field mod** I,

$$L_I = \quad \begin{aligned} &\text{the maximal abelian extension of } L \text{ in which} \\ &\text{all primes } \equiv 1 \text{ mod } I \text{ split completely.} \end{aligned}$$

(Morally, L_I should be the maximal abelian extension in which primes dividing I are allowed to ramify, but this is not quite correct.) There is an isomorphism

$$\text{Gal}(L_I/L) \xleftarrow[\cong]{} \text{Cl}(I),$$

and $L_I \supseteq H_L$ with equality when $I = \mathcal{O}_L$.

EXAMPLE 12.3.7   For $L = \mathbb{Q}$, we have $L_{(n)} = \mathbb{Q}(\zeta_n)$.

To deal with the infinite extension

$$L^{\text{ab}} := \text{maximal abelian extension of } L \ (\subset \bar{\mathbb{Q}}),$$

we have to introduce the adéles.

In studying abelian varieties one considers, for $\ell \in \mathbb{Z}$ prime, the finite groups of ℓ-torsion points $A[\ell]$; multiplication by ℓ gives maps

$$\cdots \to A[\ell^{n+1}] \to A[\ell^n] \to \cdots.$$

If we do the same thing on the unit circle $S^1 \subset \mathbb{C}^*$, we get

$$\begin{array}{ccccccc} \cdots & \to & S^1[\ell^{n+1}] & \overset{\cdot \ell}{\to} & S^1[\ell^n] & \to & \cdots \\ & & \| & & \| & & \\ & \to & \mathbb{Z}/\ell^{n+1}\mathbb{Z} & \underset{\substack{\text{natural} \\ \text{map}}}{\to} & \mathbb{Z}/\ell^n\mathbb{Z} & \to & \end{array}$$

and one can define the ℓ-**adic integers** by the inverse limit

$$\mathbb{Z}_\ell := \varprojlim_n \ \mathbb{Z}/\ell^n\mathbb{Z}.$$

An element of this limit is by definition an infinite sequence of elements in $\mathbb{Z}/\ell^n\mathbb{Z}$ mapping to each other. There is the natural inclusion

$$\mathbb{Z} \hookrightarrow \mathbb{Z}_\ell,$$

and

$$\mathbb{Q}_\ell := \mathbb{Z}_\ell \otimes_\mathbb{Z} \mathbb{Q}.$$

Elements of $\mathbb{Q}_\ell$ can be written as power series $\sum_{i \geq n} a_i \ell^i$ for some $n \in \mathbb{Z}$ ($n \geq 0$ for elements of $\mathbb{Z}_\ell$). $\mathbb{Q}_\ell$ can also be thought of as the completion of $\mathbb{Q}$ with respect to the metric given by

$$d(x,y) = \frac{1}{\ell n} \text{ if } x - y = \ell^n \frac{a}{b} \text{ with } a, b \text{ relatively prime to } \ell.$$

The resulting topology on $\mathbb{Z}_\ell$ makes

$$U_n(\alpha) := \{\alpha + \lambda \ell^n \mid \lambda \in \mathbb{Z}_\ell\}$$

into "the open disk about $\alpha \in \mathbb{Z}_\ell$ of radius $1/\ell^n$". $\mathbb{Z}_\ell$ itself is compact and totally disconnected.

Now set

$$\hat{\mathbb{Z}} := \varprojlim_{n \in \mathbb{Z}} \mathbb{Z}/n\mathbb{Z},$$

which is isomorphic to $\prod_\ell \mathbb{Z}_\ell$ by the Chinese remainder theorem. The **finite adéles** appear naturally as

$$\mathbb{A}_f := \hat{\mathbb{Z}} \otimes_\mathbb{Z} \mathbb{Q} = \prod_\ell{}' \mathbb{Q}_\ell,$$

where $\prod'$ means the ∞-tuples with all but finitely many entries in $\mathbb{Z}_\ell$. The "full" **adéles** are constructed by writing

$$\begin{aligned} \mathbb{A}_\mathbb{Z} &:= \mathbb{R} \times \hat{\mathbb{Z}}, \\ \mathbb{A}_\mathbb{Q} &= \mathbb{Q} \otimes_\mathbb{Z} \mathbb{A}_\mathbb{Z} = \mathbb{R} \times \mathbb{A}_f, \end{aligned}$$

which generalizes for a number field L to

$$\mathbb{A}_L = L \otimes_\mathbb{Q} \mathbb{A}_\mathbb{Q} = \mathbb{R}^{[L:\mathbb{Q}]} \times \underbrace{\prod_{\mathfrak{P} \in \mathcal{I}(L)}{}' L_\mathfrak{P}}_{\mathbb{A}_{L,f}},$$

where $\prod'$ means all but finitely many entries in $(\mathcal{O}_L)_\mathfrak{P}$.

For a $\mathbb{Q}$-algebraic group G, we can define

$$G(\mathbb{A}_f) := \prod{}' G(\mathbb{Q}_\ell),$$

$$G(\mathbb{A}_\mathbb{Q}) := G(\mathbb{R}) \times G(\mathbb{A}_f),$$

with generalizations to $\mathbb{A}_L$ and $\mathbb{A}_{L,f}$. Here $\prod'$ simply means that for some (hence every) embedding $G \hookrightarrow GL_N$, all but finitely many entries lie in $G(\mathbb{Z}_\ell)$. The idéles

$$\mathbb{A}_{(f)}^\times = \mathbb{G}_m(\mathbb{A}_{(f)}),$$

$$\mathbb{A}^{\times}_{L(,f)} = (\mathrm{Res}_{L/\mathbb{Q}}\,\mathbb{G}_m)(\mathbb{A}_{(f)}) = \mathbb{G}_m(\mathbb{A}_{L(,f)})$$

were historically defined first; Weil introduced "adéle"—also (intentionally) a girl's name—as a contraction of "additive idéle". The usual norm $N_{\tilde{L}/L}$ and reflex norm $N_{\Phi'}$ extend to maps of idéles, using the formulation of these maps as morphisms of $\mathbb{Q}$-algebraic groups.

Returning to $S^1 \subset \mathbb{C}^*$, let ζ be an Nth root of unity and $a = (a_n) \in \hat{\mathbb{Z}}$; then

$$\zeta \mapsto \zeta^a := \zeta^{a_N} \text{ defines an action of } \hat{\mathbb{Z}}$$

on the torsion points of S^1 (which generate $\mathbb{Q}^{\mathrm{ab}}$).

The cyclotomic character

$$\chi : \mathrm{Gal}(\mathbb{Q}^{\mathrm{ab}}/\mathbb{Q}) \xrightarrow{\cong} \hat{\mathbb{Z}}^{\times} \cong \mathbb{Q}^{\times}\backslash\mathbb{A}^{\times}_{\mathbb{Q},f}$$

is defined by

$$\sigma(\zeta) =: \zeta^{\chi(\sigma)},$$

and we can think of it as providing a "continuous envelope"[4] for the action of a given σ on any finite order of torsion. The **Artin reciprocity map** is simply its inverse

$$\mathrm{art}_{\mathbb{Q}} := \chi^{-1}$$

for $\mathbb{Q}$. Assuming now that L is totally imaginary, this generalizes to

$$
\begin{array}{ccc}
L^{\times}\backslash\mathbb{A}^{\times}_{L,f} & \xrightarrow{\mathrm{art}_L} & \mathrm{Gal}(L^{\mathrm{ab}}/L) \\
\downarrow{\scriptstyle(\#)} & & \downarrow \\
L^{\times}\backslash\mathbb{A}^{\times}_{L,f}\Big/ N_{\tilde{L}/L}(\mathbb{A}^{\times}_{L,f}) & \xrightarrow{\cong} & \mathrm{Gal}(\tilde{L}/L).
\end{array}
$$

If $\tilde{L} = L_I$ then the double-coset turns out to be $\mathrm{Cl}(I)$, so this recovers the earlier maps for ray class fields. (For $\tilde{L} = H_L$, we can replace $N_{\tilde{L}/L}(\mathbb{A}^{\times}_{L,f})$ by $\widehat{\mathcal{O}_L}$.) The correspondence between

$$
\boxed{\begin{array}{c}\text{open subgroups}\\\text{of }\mathbb{A}^{\times}_{L,f}\end{array}} \longleftrightarrow \boxed{\begin{array}{c}\text{finite abelian}\\\text{extensions of }L\end{array}}
$$

is the essence of class field theory. Also note that $(\#)$ gives compatible maps to all the class groups of L, so that $\mathbb{A}^{\times}_{L,f}$ acts on them.

---

[4]I use this term because the automorphisms of $\mathbb{C}$ other than complex conjugation induce highly discontinuous (nonmeasurable!) maps on the complex points of a variety over $\mathbb{Q}$. But if one specifies a discrete set of points, it is sometimes possible to produce a continuous (even analytic/algebraic) automorphism acting in the same way on those points.

C. Main Theorem of CM

Now let us bring the adéles to bear upon abelian varieties. Taking the product of the **Tate modules**

$$T_\ell A := \varprojlim_n A[\ell^n] \ (= \text{rank-}2g \text{ free } \mathbb{Z}_\ell\text{-module})$$

of an abelian g-fold yields

$$T_f A := \prod_\ell T_\ell A,$$

$$V_f A := T_f A \otimes_{\mathbb{Z}} \mathbb{Q} \ (= \text{rank-}2g \text{ free } \mathbb{A}_f\text{-module})$$

with (for example)

$$\text{Aut}(V_f A) \cong \text{GL}_{2g}(\mathbb{A}_f).$$

The "main theorem," due to Shimura and Taniyama, is basically a detailed description of the action of $\text{Gal}(\mathbb{C}/E')$ on $\text{Ab}(\mathcal{O}_E, \Phi)$ and the torsion points of the (finitely many isomorphism classes of) abelian varieties it classifies.

THEOREM 12.3.8 *Let* $A_{[\mathfrak{a}]} \in \text{Ab}(\mathcal{O}_E, \Phi)$ *and* $\sigma \in \text{Gal}(\mathbb{C}/E')$ *be given. For any* $a \in \mathbb{A}^\times_{E',f}$ *with* $\text{art}_{E'}(a) = \sigma|_{(E')^{\text{ab}}}$, *we have*

(a) $^\sigma A_{[\mathfrak{a}]} \stackrel{\cong}{\to} A_{N_{\Phi'}(a) \cdot [\mathfrak{a}]}$ *(where* $N_{\Phi'}(a) \in \mathbb{A}^\times_{E,f}$, *and* $N_{\Phi'}(a) \cdot [\mathfrak{a}]$ *depends only on* $\sigma|_{H_{E'}}$*); and*

(b) *there exists a unique E-linear isogeny* $\alpha : A_{[\mathfrak{a}]} \to {}^\sigma A_{[\mathfrak{a}]}$ *such that*

$$\alpha\left(N_{\Phi'}(a) \cdot x\right) = {}^\sigma x \quad (\forall x \in V_f A).$$

SKETCH OF PROOF FOR (b).

- $^\sigma A \in \text{Ab}(\mathcal{O}_E, \Phi) \implies$ there exists (E-linear) isogeny $\alpha : A \to {}^\sigma A$.

- $V_f A$ is free of rank 1 over $\mathbb{A}_{E,f}$.

- The composition $V_f(A) \xrightarrow{\sigma} V_f({}^\sigma A) \xrightarrow{V_f(\alpha)^{-1}} V_f(A)$ is $\mathbb{A}_{E,f}$-linear, so is just multiplication by some $s \in \mathbb{A}^\times_{E,f}$.

- s is independent (up to $E^\times$) of the choice of α, defining the horizontal arrow of

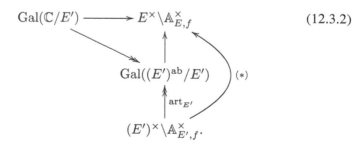

$$\text{(12.3.2)}$$

- A is defined over a number field k; the Shimura-Taniyama computation of the prime decompositions of the elements of $E \cong \operatorname{End}(A)_{\mathbb{Q}}$ reducing to various Frobenius maps (in residue fields of k) then shows that the vertical map $(*)$ in (12.3.2) is $N_{\Phi'}$. Hence

$$N_{\Phi'}(a) = s = V_f(\alpha)^{-1} \circ \sigma$$

which gives the formula in (b).

$\square$

So what does (b) mean? Like the cyclotomic character, we get a very nice interpretation when we restrict to the action on m-torsion points of A for any fixed $m \in \mathbb{N}$:

COROLLARY 12.3.9 *There exists a unique E-linear isogeny $\alpha_m : A \to {}^{\sigma}A$ such that*

$$\alpha_m(x) = {}^{\sigma}x \quad (\forall x \in A[m]).$$

That is, α_m provides a "continuous envelope" for the action of automorphisms of $\mathbb{C}$ on special points.

The reader who desires a more thorough treatment of the Main Theorem of CM may consult [Silverman1995] and [Milne2007].

12.4 SHIMURA VARIETIES

A. Three Key Adélic Lemmas

Besides the main theorem of CM, there is another (related) connection between the class field theory described in Section 12.3.B and abelian varieties. The tower of ray class groups associated to the ideals of a CM-field E can be expressed as

$$E^{\times} \backslash \mathbb{A}_{E,f}^{\times} \big/ \mathfrak{U}_I \cong T(\mathbb{Q}) \backslash T(\mathbb{A}_f) / \mathfrak{U}_I \tag{12.4.1}$$

where

$$\mathfrak{U}_I := \left\{ (a_{\mathfrak{p}})_{\mathfrak{p} \in \mathcal{I}(E) \text{ prime}} \in \mathbb{A}_{E,f}^{\times} \;\middle|\; \begin{array}{ll} a_{\mathfrak{p}} \in (\mathcal{O}_E)_{\mathfrak{p}} & \text{for all } \mathfrak{p}, \\ a_{\mathfrak{p}} \equiv 1 \bmod \mathfrak{p}^{\operatorname{ord}_{\mathfrak{p}} I} & \substack{\text{for the finitely} \\ \text{many } \mathfrak{p} \text{ dividing } I} \end{array} \right\}$$

is a compact open subgroup of $\mathbb{A}_{E,f}^{\times} = T(\mathbb{A}_f)$ and

$$T = \operatorname{Res}_{E/\mathbb{Q}} \mathbb{G}_m.$$

Expression (12.4.1) may be seen as parametrizing abelian varieties with CM by a type (E, Φ) and having fixed level structure—which "refines" the set parametrized by $\operatorname{Cl}(E)$. Shimura varieties give a way of extending this story to more general abelian varieties with other endomorphism and Hodge tensor structures, as well as the other families of Hodge structures parametrized by Hermitian symmetric domains.

The first fundamental result we will need is the following:

LEMMA 12.4.1 *For T any $\mathbb{Q}$-algebraic torus, and $K_f \subset T(\mathbb{A}_f)$ any open subgroup,*

$$T(\mathbb{Q})\backslash T(\mathbb{A}_f)/K_f$$

is finite.

SKETCH OF PROOF. This follows from the definition of compactness, assuming we can show that $T(\mathbb{Q})\backslash T(\mathbb{A}_f)$ is compact. For any number field F, the latter is closed in $T(F)\backslash T(\mathbb{A}_{F,f})$, and for some F over which T splits the latter is $(F^\times\backslash\mathbb{A}_{F,f}^\times)^{\dim(T)}$. Finally, by the Minkowski bound,

$$F^\times\backslash\mathbb{A}_{F,f}^\times/\widehat{\mathcal{O}_F} = \mathrm{Cl}(F) \text{ is finite,}$$

and $\widehat{\mathcal{O}_F}$ is compact (like $\hat{\mathbb{Z}}$), so $F^\times\backslash\mathbb{A}_{F,f}^\times$ is compact. $\square$

For a very different class of $\mathbb{Q}$-algebraic groups, we have the contrasting result:

LEMMA 12.4.2 *Suppose $G/\mathbb{Q}$ is semisimple and simply connected, of noncompact type;[5] then*

  (a) (***strong approximation***) $G(\mathbb{Q}) \subseteq G(\mathbb{A}_f)$ *is dense; and*

  (b) *for any open $K_f \subseteq G(\mathbb{A}_f)$, $G(\mathbb{A}_f) = G(\mathbb{Q}) \cdot K_f$.*

REMARK 12.4.3 Lemma 12.4.2(b) implies that the double coset $G(\mathbb{Q})\backslash G(\mathbb{A}_f)/K_f$ is trivial. Note that double cosets are now essential as we are in the nonabelian setting.

SKETCH OF (a) $\Longrightarrow$ (b). Given $(\gamma_\ell) \in G(\mathbb{A}_f)$, $U := (\gamma_\ell) \cdot K_f$ is an open subset of $G(\mathbb{A}_f)$, hence by (a) there exists a $g \in U \cap G(\mathbb{Q})$. Clearly $g = (\gamma_\ell) \cdot k$ for some $k \in K_f$, and so $(\gamma_\ell) = g \cdot k^{-1}$. $\square$

NONEXAMPLE 12.4.4 $\mathbb{G}_m$, which is of course reductive but not semisimple. If (b) held, then the ray class groups of $\mathbb{Q}$ (which are $\cong (\mathbb{Z}/\ell\mathbb{Z})^\times$) would be trivial. But even more directly, were $\mathbb{Q}^\times$ dense in $\mathbb{A}_f^\times$, there would be $q \in \mathbb{Q}^\times$ close to any $(a_\ell) \in \prod \mathbb{Z}_\ell^\times$. This forces [the image in $\mathbb{A}_f^\times$ of] q to lie in $\prod \mathbb{Z}_\ell^\times$, which means for each ℓ that (in lowest terms) the numerator and denominator of q are prime to ℓ. So $q = \pm 1$, a contradiction.

Finally, for a general $\mathbb{Q}$-algebraic group G, we have:

LEMMA 12.4.5 *The congruence subgroups of $G(\mathbb{Q})$ are precisely the $K_f \cap G(\mathbb{Q})$ (intersection in $G(\mathbb{A}_f)$) for compact open $K_f \subseteq G(\mathbb{A}_f)$.*

SKETCH OF PROOF. For $N \in \mathbb{N}$ the

$$K(N) := \left\{ (g_\ell)_{\ell \text{ prime}} \in G(\mathbb{A}_f) \; \middle| \; \begin{array}{ll} g_\ell \in G(\mathbb{Z}_\ell) & \text{for all } \ell, \\ g_\ell \equiv \mathbf{e} \bmod \ell^{\mathrm{ord}_\ell N} & \text{for each } \ell | N, \end{array} \right\}$$

---

[5]That is, none of its simple almost-direct $\mathbb{Q}$-factors G_i have $G_i(\mathbb{R})$ compact.

are compact open in $G(\mathbb{A}_f)$, and

$$
\begin{aligned}
K(N) \cap G(\mathbb{Q}) &= \left\{ g \in G(\mathbb{Z}) \;\middle|\; \begin{array}{c} g_\ell \equiv e \bmod \ell^{\operatorname{ord}_\ell N} \\ \text{for each } \ell \mid N \end{array} \right\} \\
&= \left\{ g \in G(\mathbb{Z}) \;\middle|\; g \equiv e \bmod N \right\} \quad = \quad \Gamma(N).
\end{aligned}
$$

In fact, the $K(N)$ are a *basis* of open subsets containing e. So any compact *open* K_f contains some $K(N)$, and

$$
(K_f \cap G(\mathbb{Q})) / (K(N) \cap G(\mathbb{Q})) \subseteq K_f / K(N)
$$

is a discrete subgroup of a compact set, and therefore finite.   □

In some sense, K_f is *itself* the congruence condition.

B. Shimura Data

A $\left\{ \begin{array}{l} \textbf{Shimura datum (SD)} \\ \text{(resp. \textbf{connected SD})} \end{array} \right.$ is a pair $\left(G, \left\{ \begin{array}{c} \tilde{X} \\ X \end{array} \right\} \right)$ consisting of

- $G := \left\{ \begin{array}{l} \text{reductive} \\ \text{semisimple} \end{array} \right.$ algebraic group defined over $\mathbb{Q}$

and

- $\left\{ \begin{array}{c} \tilde{X} \\ X \end{array} \right. := \text{a} \left\{ \begin{array}{c} G(\mathbb{R})\text{-} \\ G^{\mathrm{ad}}(\mathbb{R})^+\text{-} \end{array} \right.$ conjugacy class of homomorphisms

$$
\tilde{\varphi} : \mathbb{S} \to \left\{ \begin{array}{c} G_{\mathbb{R}} \\ G^{\mathrm{ad}}_{\mathbb{R}} \end{array} \right. ,
$$

satisfying the axioms

(SV1)   only $z/\bar{z}$, 1, $\bar{z}/z$ occur as eigenvalues of

$$
\mathrm{Ad} \circ \tilde{\varphi} : \mathbb{S} \to \mathrm{GL}\left(\mathrm{Lie}(G^{\mathrm{ad}})_{\mathbb{C}} \right) ;
$$

(SV2) $\Psi_{\mathrm{Ad}(\tilde{\varphi}(i))} \in \mathrm{Aut}(G^{\mathrm{ad}}_{\mathbb{R}})$ is Cartan;

(SV3) G^{ad} has no $\mathbb{Q}$-factor on which the projection

of every $(\mathrm{Ad} \circ) \tilde{\varphi} \in \left\{ \begin{array}{c} \tilde{X} \\ X \end{array} \right.$ is trivial

(SV4)   the weight homomorphism $\mathbb{G}_m \overset{w_{\tilde{\varphi}}}{\underset{\tilde{\varphi}}{\hookrightarrow \mathbb{S} \longrightarrow}} G$

(a priori defined over $\mathbb{R}$) is defined over $\mathbb{Q}$;

(SV5) $(Z^\circ/w_{\tilde\varphi}(\mathbb{G}_m))\,(\mathbb{R})$ is compact;[6] and

(SV6) Z° splits over a CM-field.

In the "connected" case, SV4–SV6 are trivial, while SV1–SV3 already imply

- X is a Hermitian symmetric domain (in the precise sense of in Section 12.1.B.III, with $\phi = \mu_{\tilde\varphi} \circ \jmath$); and

- G is of noncompact type (from SV3), but with $\ker\,(G(\mathbb{R})^+ \twoheadrightarrow \mathrm{Hol}(X)^+)$ compact (from SV2);

where the surjective arrow is defined by the action on the conjugacy class of $\tilde\varphi$.

In these definitions, axioms SV4–SV6 are sometimes omitted; for example, canonical models exist for Shimura varieties without them. We include them here from the beginning because they hold in the context of Hodge theory. Indeed, consider a full Mumford–Tate domain $\tilde X$ (cf. Section 12.1.D) for polarized $\mathbb{Q}$-Hodge structures with generic Mumford–Tate group G. Regarded as a pair, $(G, \tilde X)$ always satisfies

SV2:   by the second Hodge–Riemann bilinear relation (cf. Exercise 12.1.12);

SV3:   otherwise the generic MT group would be a proper subgroup of G;

SV4:   because the weight filtration is split over $\mathbb{Q}$;

SV5:   since G is a Mumford–Tate group, $G/w_{\tilde\varphi}(\mathbb{G}_m)$ contains a $G(\mathbb{R})^+$-conjugacy class of anisotropic maximal real tori, and these contain $Z^\circ_{\mathbb{R}}/w_{\tilde\varphi}(\mathbb{G}_m)$;

SV6:   since G is defined over $\mathbb{Q}$, the conjugacy class contains tori defined over $\mathbb{Q}$; and so $\tilde X$ contains a $\tilde\varphi$ factoring through some such rational T. This defines a polarized CM-Hodge structure, with MT group a $\mathbb{Q}$-torus $T_0 \subseteq T$ split over a CM-field (see Section 12.3). If $T_0 \not\supseteq Z^\circ$, then the projection of $\tilde\varphi$ to some $\mathbb{Q}$-factor of Z° is trivial and then it is trivial for all its conjugates, contradicting SV3. Hence $T_0 \supseteq Z^\circ$ and Z° splits over the CM-field.

Further, SV1 holds *if and only if* the IPR on $\tilde X$ is trivial.

What can one say about an arbitrary Shimura datum? First, any SD produces a connected SD by

- replacing G by G^{der} (which has the same G^{ad});

- replacing $\tilde X$ by a connected component X (which we may view as a $G^{\mathrm{ad}}(\mathbb{R})^+$-conjugacy class of homomorphisms $\mathrm{Ad} \circ \tilde\varphi$);

and so

$$\tilde X \text{ is a finite union of Hermitian symmetric domains.}$$

Moreover, by SV2 and SV5, one has that $\Psi_{\tilde\varphi(i)} \in \mathrm{Aut}(G_{\mathbb{R}}/w_{\tilde\varphi}(\mathbb{G}_m))$ is Cartan, so that there exist a symmetric bilinear form Q on a $\mathbb{Q}$-vector space V, and an embedding

---

[6]SV5 is sometimes weakened: cf. [Milne2005, Sec. 5].

$\rho : G/w_{\tilde{\varphi}}(\mathbb{G}_m) \hookrightarrow \text{Aut}(V, Q)$, such that $Q(\cdot, (\rho \circ \tilde{\varphi})(i)\bar{\cdot}) > 0$. As in the "Hodge domains" described in the lectures by Griffiths, for any such faithful representation $\tilde{\rho} : G \hookrightarrow \mathbb{G}O(V, Q)$ with Q polarizing $\tilde{\rho} \circ \tilde{\varphi}$, $\tilde{X}$ is realized as a finite union of MT domains with trivial IPR. Let M denote the MT group of $\tilde{X}$. If $G = M$, then $\tilde{X}$ is exactly a "full MT domain" as described in Section 12.1.D. However, in general, M is a normal subgroup of G with the same adjoint group; that the center can be smaller is seen by considering degenerate CM Hodge structures. Suffice it to say that (while the correspondence is slightly messy) all Shimura data, hence ultimately all Shimura varieties as defined below, have a Hodge-theoretic interpretation.

Now, given a *connected* SD (G, X), we add one more ingredient: let

$$\Gamma \leq G^{\text{ad}}(\mathbb{Q})^+$$

be a torsion-free arithmetic subgroup, with inverse image in $G(\mathbb{Q})^+$ a congruence subgroup. Its image $\bar{\Gamma}$ in $\text{Hol}(X)^+$ is

(i) (torsion-free and) arithmetic: since $\ker(G(\mathbb{R})^+ \twoheadrightarrow \text{Hol}(X)^+)$ compact;

(ii) isomorphic to Γ: $\Gamma \cap \ker(G(\mathbb{R})^+ \twoheadrightarrow \text{Hol}(X)^+) = \text{discrete} \cap \text{compact} = \text{finite}$, hence torsion (and there is no torsion).

We may write

$$X(\Gamma) := \Gamma \backslash X \underset{(ii)}{=} \bar{\Gamma} \backslash X \, ;$$

and $\bar{\Gamma} \backslash X$ is a locally symmetric variety by (i) and Baily–Borel. By Borel's theorem,

$$\Gamma \leq \Gamma' \implies X(\Gamma') \twoheadrightarrow X(\Gamma) \text{ is algebraic.} \tag{12.4.2}$$

DEFINITION 12.4.6 *The **connected Shimura variety** associated with* (G, X, Γ) *is*

$$\text{Sh}_{\Gamma}^{\circ}(G, X) := X(\Gamma).$$

REMARK 12.4.7 Every $X(\Gamma)$ is covered by an $X(\Gamma')$ with Γ' the image of a congruence subgroup of $G(\mathbb{Q})^+$. If one works with "sufficiently small" congruence subgroups of $G(\mathbb{Q})$, then

- they belong to $G(\mathbb{Q})^+$;

- they have a torsion-free image in $G^{\text{ad}}(\mathbb{Q})^+$;

- congruence $\implies$ arithmetic.

This will be tacit in what follows.

C. The Adélic Reformulation

Consider a connected SD (G, X) with G *simply connected*. By a result of Cartan, this means that $G(\mathbb{R})$ is connected, hence acts on X via Ad. Let $K_f \leq G(\mathbb{A}_f)$ be a ("sufficiently small") compact open subgroup and (referring to Lemma 12.4.5) $\Gamma = G(\mathbb{Q}) \cap K_f$ the corresponding subgroup of $G(\mathbb{Q})$. Replacing the earlier notation we write

$$\mathrm{Sh}^{\circ}_{K_f}(G, X)$$

for the associated locally symmetric variety.

PROPOSITION 12.4.8 *The connected Shimura variety* $\mathrm{Sh}^{\circ}_{K_f}(G, X)$ $(\cong \Gamma \backslash X)$ *is homeomorphic to*

$$G(\mathbb{Q}) \backslash X \times G(\mathbb{A}_f)/K_f,$$

where the action defining the double quotient is

$$g.(\tilde{\varphi}, a).k := (\underbrace{g.\tilde{\varphi}}_{g\tilde{\varphi}g^{-1}}, gak).$$

REMARK 12.4.9 If $G(\mathbb{R}) = G^{\mathrm{ad}}(\mathbb{R})^+$, then the quotient $X \times G(\mathbb{A}_f)/K_f$ can be written $G(\mathbb{A}_{\mathbb{Q}})/K$, where $K = K_{\mathbb{R}} \times K_f$ with $K_{\mathbb{R}} \subset G(\mathbb{R})$ maximal compact.

SKETCH OF PROOF OF PROPOSITION 12.4.8. First note that $[\tilde{\varphi}] \mapsto [(\tilde{\varphi}, 1)]$ gives a well-defined map from $\Gamma \backslash X$ to the double quotient, since

$$[\tilde{\varphi}] = [\tilde{\varphi}'] \implies \tilde{\varphi}' = \gamma.\tilde{\varphi} \ (\gamma \in \Gamma)$$

$$\implies (\tilde{\varphi}', 1) = \gamma.(\tilde{\varphi}, 1).\gamma^{-1}.$$

Now by assumption G is semisimple, simply connected, and of noncompact type. Lemma 12.4.2 implies that

$$G(\mathbb{A}_f) = G(\mathbb{Q}) \cdot K_f,$$

so that for any $(\tilde{\varphi}, a) \in X \times G(\mathbb{A}_f)$ we have

$$(\tilde{\varphi}, a) = (\tilde{\varphi}, gk) = g.(g^{-1}.\tilde{\varphi}, 1).k.$$

This implies that our map is surjective; for injectivity,

$$[(\tilde{\varphi}, 1)] = [(\tilde{\varphi}', 1)] \implies (\tilde{\varphi}', 1) = g.(\tilde{\varphi}, 1).k^{-1} = (g.\tilde{\varphi}, gk^{-1})$$

$$\implies g = k \in \Gamma, \quad [\tilde{\varphi}'] = [\tilde{\varphi}].$$

Finally, since K_f is open, $G(\mathbb{A}_f)/K_f$ is discrete and the map

$$\begin{aligned} X &\to X \times G(\mathbb{A}_f)/K_f, \\ \tilde{\varphi} &\mapsto (\tilde{\varphi}, [1]) \end{aligned}$$

is a homeomorphism, which continues to hold upon quotienting both sides by a discrete torsion-free subgroup.   □

More generally, given a Shimura datum $(G, \tilde{X})$ and compact open $K_f \leq G(\mathbb{A}_f)$, we simply *define* the **Shimura variety**

$$\mathrm{Sh}_{K_f}(G, \tilde{X}) := G(\mathbb{Q}) \backslash \tilde{X} \times G(\mathbb{A}_f) / K_f,$$

which will be a *finite disjoint union of locally symmetric varieties*. The disconnectedness, at first, would seem to arise from two sources:

(a) $\tilde{X}$ not connected (the effect of orbiting by $G(\mathbb{R})$);

(b) the failure of strong approximation for reductive groups (which gives multiple connected components even when $\tilde{X}$ is connected).

The first part of the theorem below says (a) does not contribute: the indexing of the components is "entirely arithmetic." First, some notation:

NOTATION 12.4.10

- X denotes a connected component of $\tilde{X}$;

- $G(\mathbb{R})_+$ is the preimage of $G^{\mathrm{ad}}(\mathbb{R})^+$ in $G(\mathbb{R})$, and $G(\mathbb{Q})_+ = G(\mathbb{Q}) \cap G(\mathbb{R})_+$;

- $\nu : G \to T$ denotes the maximal abelian quotient (from Section 12.1.A), and the composition $Z \hookrightarrow G \xrightarrow{\nu} T$ is an isogeny;

- $T(\mathbb{R})^\dagger := \mathrm{Im}(Z(\mathbb{R}) \to T(\mathbb{R}))$;

- $Y := T(\mathbb{Q})^\dagger \backslash T(\mathbb{Q})$.

THEOREM 12.4.11   *(i)* $G(\mathbb{Q})_+ \backslash X \times G(\mathbb{A}_f) / K_f \xrightarrow{\cong} G(\mathbb{Q}) \backslash \tilde{X} \times G(\mathbb{A}_f) / K_f$.

(ii) The map

$$G(\mathbb{Q})_+ \backslash X \times G(\mathbb{A}_f) / K_f \twoheadrightarrow G(\mathbb{Q})_+ \backslash G(\mathbb{A}_f) / K_f =: \mathcal{C}$$

"indexes" the connected components. If G^{der} is simply connected, we have

$$\mathcal{C} \xrightarrow[\nu]{\cong} T(\mathbb{Q})^\dagger \backslash T(\mathbb{A}_f) / \nu(K_f) \cong T(\mathbb{Q}) \backslash Y \times T(\mathbb{A}_f) / \nu(K_f).$$

Henceforth, "$\mathcal{C}$" denotes a set of representatives in $G(\mathbb{A}_f)$.

(iii) $\mathrm{Sh}_{K_f}(G, \tilde{X}) \cong \amalg_{g \in \mathcal{C}} \Gamma_g \backslash X$, *a finite union, where*

$$\Gamma_g := g K_f g^{-1} \cap G(\mathbb{Q})_+.$$

That $|\mathcal{C}| < \infty$ is by Lemma 12.4.1. The rest is in [Milne2005, Sec. 5] with two of the easier points (not requiring the hypothesis on G^{der}) described in the following exercise:

EXERCISE 12.4.12   (a) The preimage of $[1] \in \mathcal{C}$ is $\Gamma \backslash X \cong \mathrm{Sh}^\circ_{K_f^{\mathrm{der}}}(G^{\mathrm{der}}, X)$,
where $\Gamma = K_f \cap G(\mathbb{Q})_+$, and $K_f^{\mathrm{der}} \leq G^{\mathrm{der}}(\mathbb{A}_f)$ is some subgroup containing $K_f \cap G^{\mathrm{der}}(\mathbb{A}_f)$.

(b) For $\gamma \in \Gamma_g$, $(\gamma.\tilde{\varphi}, g) \equiv (\tilde{\varphi}, g)$ in $\mathrm{Sh}_{K_f}(G, \tilde{X})$.

The key point here is that (iii) is an *analytic description* of what will turn out to be $\mathrm{Gal}(\mathbb{C}/E)$-conjugates of $\Gamma \backslash X$, in analogy to the formula for $^\sigma A_{[\mathfrak{a}]}$ in the main theorem of CM. Here E is the reflex field of $(G, \tilde{X})$, which will turn out to be the minimal field of definition of this very clever disjoint union.

REMARK 12.4.13   (a) The definitions of Shimura varieties here are due to Deligne (in the late 1970s), who conjectured that they are fine moduli varieties for motives. The difficulty is in showing that the Hodge structures they parametrize are motivic.

(b) There is a natural construction of vector bundles $\mathcal{V}_{K_f}(\pi)$ on $\mathrm{Sh}_{K_f}(G, \tilde{X})$ (cf. [Milne1990], section III.2), holomorphic sections of which are (holomorphic) **automorphic forms** of level K_f and type π. (Here, π is a representation of the parabolic subgroup of $G(\mathbb{C})$ stabilizing a point of the compact dual $\check{X}$.) Their higher cohomology groups are called **automorphic cohomology**.

(c) The inverse system[7] $\mathrm{Sh}(G, \tilde{X})$ of all $\mathrm{Sh}_{K_f}(G, \tilde{X})$ is a reasonably nice scheme on which $G(\mathbb{A}_f)$ operates, and so for example one could consider

$$\varinjlim H^i(\mathrm{Sh}_{K_f}(G, \check{X}), \mathcal{O}(\mathcal{V}_{K_f}(\pi)))$$

as a representation of this group.

D. Examples

I. Zero-Dimensional Shimura Varieties

Let T be a $\mathbb{Q}$-algebraic torus satisfying SV5 and SV6, $\tilde{\varphi} : \mathbb{S} \to T$ a homomorphism of $\mathbb{R}$-algebraic groups satisfying SV4, and $K_f \leq T(\mathbb{A}_f)$ be compact open. Then

$$\mathrm{Sh}_{K_f}(T, \{\tilde{\varphi}\}) = T(\mathbb{Q}) \backslash T(\mathbb{A}_f) / K_f$$

is finite by Lemma 12.4.1. These varieties arise as $\mathcal{C}$ in the theorem, and also from CM-Hodge structures, which by definition have a torus as MT group.

II. Siegel Modular Variety

Begin with a $\mathbb{Q}$-**symplectic space** (V, ψ)—i.e., a vector space $V/\mathbb{Q}$ together with a nondegenerate alternating form $\psi : V \times V \to \mathbb{Q}$—and set $G =$

$$\mathbb{G}\mathrm{Sp}(V, \psi) := \left\{ g \in \mathrm{GL}(V) \;\middle|\; \begin{array}{c} \psi(gu, gv) = \chi(g)\psi(u, v) \; (\forall u, v \in V) \\ \text{for some } \chi(g) \in \mathbb{Q}^* \end{array} \right\}.$$

---

[7]It is suppressed in these notes since it is not needed for canonical models; it uses (12.4.2).

EXERCISE 12.4.14 Check that $\chi : G \to \mathbb{G}_m$ defines a character of G.

Now consider the spaces

$$X^{\pm} := \left\{ J \in \mathrm{Sp}(V, \psi)(\mathbb{R}) \;\middle|\; \begin{array}{c} J^2 = -e \text{ and} \\ \psi(u, Jv) \text{ is } \pm\text{-definite} \end{array} \right\}$$

of positive- and negative-definite symplectic complex structures on $V_{\mathbb{R}}$, and regard

$$\tilde{X} := X^+ \amalg X^-$$

as a set of homomorphisms via $\tilde{\varphi}(a + bi) := a + bJ$ (for $a + bi \in \mathbb{C}^* = \mathbb{S}(\mathbb{R})$). Then $G(\mathbb{R})$ acts transitively on $\tilde{X}$, and the datum $(G, \tilde{X})$ satisfies SV1–SV6. For any compact open K_f the attached Shimura variety is a **Siegel modular variety**.

Now consider the set

$$\mathcal{M}_{K_f} := \frac{\left\{ (A, Q, \eta) \;\middle|\; \begin{array}{c} A \text{ an abelian variety}/\mathbb{C}, \\ \pm Q \text{ a polarization of } H_1(A, \mathbb{Q}), \\ \eta : V_{\mathbb{A}_f} \to V_f(A) \,(\cong H_1(A, \mathbb{A}_f)) \text{ an} \\ \text{isomorphism sending } \psi \mapsto a \cdot Q \;(a \in \mathbb{A}_f) \end{array} \right\}}{\cong},$$

where an isomorphism of triples

$$(A, Q, \eta) \xrightarrow[\cong]{} (A', Q', \eta')$$

is an isogeny $f : A \to A'$ sending $Q' \mapsto q \cdot Q$ $(q \in \mathbb{Q}^*)$ such that for some $k \in K_f$ the diagram

$$
\begin{array}{ccc}
V_{\mathbb{A}_f} & \xrightarrow{\;\eta\;} & V_f(A) \\
{\scriptstyle \cdot k}\downarrow & & \downarrow{\scriptstyle f} \\
V_{\mathbb{A}_f} & \xrightarrow{\;\eta'\;} & V_f(A')
\end{array}
$$

commutes. $\mathcal{M}_{K_f}$ is a moduli space for polarized abelian varieties with K_f-level structure. Write $\tilde{\varphi}_A$ for the Hodge structure on $H_1(A)$, and choose an isomorphism $\alpha : H_1(A, \mathbb{Q}) \to V$ sending ψ to Q (up to $\mathbb{Q}^*$).

PROPOSITION 12.4.15 *The (well-defined) map*

$$\mathcal{M}_{K_f} \longrightarrow \mathrm{Sh}_{K_f}(G, \tilde{X})$$

induced by

$$(A, Q, \eta) \longmapsto (\alpha \circ \tilde{\varphi}_A \circ \alpha^{-1}, \alpha \circ \eta)$$

is a bijection.

III. Shimura Varieties of PEL Type

This time we take (V, ψ) to be a **symplectic** $(B, *)$**-module**, i.e.,

- (V, ψ) is a $\mathbb{Q}$-symplectic space;

- $(B, *)$ is a simple $\mathbb{Q}$-algebra with positive involution $*$ (that is, $\mathrm{tr}_{(B \otimes_{\mathbb{Q}} \mathbb{R})/\mathbb{R}}(b^* b) > 0$); and

- V is a B-module and $\psi(bu, v) = \psi(u, b^* v)$.

We put $G := \mathrm{Aut}_B(V) \cap \mathbb{G} \mathrm{Sp}(V, \psi)$, which is of generalized SL, Sp, or SO type (related to the Albert classification) according to the structure of $(B_{\bar{\mathbb{Q}}}, *)$. (Basically, G is cut out of $\mathbb{G} \mathrm{Sp}$ by fixing tensors in $T^{1,1} V$.) The (canonical) associated conjugacy class $\tilde{X}$ completes this to a Shimura datum, and the associated Shimura varieties parametrize **P**olarized abelian varieties with **E**ndomorphism and **L**evel structure (essentially a union of quotients of MT domains cut out by a subalgebra $\mathcal{E} \subseteq \mathrm{End}(V)$). They include the Hilbert and Picard modular varieties.

IV. Shimura Varieties of Hodge Type

This is a straightforward generalization of the example of Section 12.4.D.III, with G cut out of $\mathbb{G} \mathrm{Sp}(V, \psi)$ by fixing tensors of *all* degrees.

REMARK 12.4.16 In the examples of both Sections 12.4.D.III, IV, X is a subdomain of a Siegel domain, so "of Hodge type" excludes the type D and E Hermitian symmetric domains which still do yield Shimura varieties parametrizing equivalence classes of Hodge structures. So the last example is more general still:

V. Mumford–Tate Groups/Domains with Vanishing IPR

This was already partially dealt with in Section 12.4.B. In the notation from Section 12.1.D, let $G := M_{\mathbf{t}}$ and

$$\tilde{X} = M_{\mathbf{t}}(\mathbb{R}).\tilde{\varphi} =: D_{\mathbf{t}}$$

for some $\tilde{\varphi}$ with MT group $M_{\mathbf{t}}$. (Note that X will be $D_{\mathbf{t}}^+$.) Under the assumption that $D_{\mathbf{t}}$ has trivial infinitesimal period relation, each choice of level structure

$$K_f \leq M_{\mathbf{t}}(\mathbb{A}_f)$$

will produce Shimura varieties. These take the form

$$\amalg_{g \in \mathcal{C}} \Gamma_g \backslash M_{\mathbf{t}}^{\mathrm{ad}}(\mathbb{R})^+ / K_{\mathbb{R}} \quad (K_{\mathbb{R}} \text{ maximal compact})$$

with components parametrizing Γ_g-equivalence classes of higher weight Hodge structures.

    In addition to the examples in Section 12.1.D, one prototypical example is the MT group and domain for HS of weight 3 and type $(1, n, n, 1)$ with endomorphisms by an imaginary quadratic field, in such a way that the two eigenspaces are $V^{3,0} \oplus V^{2,1}$ and $V^{1,2} \oplus V^{3,0}$. (In particular, note that $M_{\mathfrak{t}}(\mathbb{R}) \cong \mathbb{G}U(1, n)$.) This has vanishing IPR and yields a Shimura variety.

REMARK 12.4.17   (i) Even in the not-necessarily-Hermitian-symmetric case, where

$$D_{\mathfrak{t}} = M(\mathbb{R})/H_{\tilde{\varphi}}$$

is a general MT domain, there is still a canonical homeomorphism

$$M(\mathbb{Q})\backslash M(\mathbb{A})/H_{\tilde{\varphi}} \times K_f \quad \cong \quad M(\mathbb{Q})\backslash D_{\mathfrak{t}} \times M(\mathbb{A}_f)/K_f$$

$$\overset{\text{homeo}}{\cong} \quad \amalg_{g \in \mathcal{C}} \Gamma_g \backslash D_{\mathfrak{t}}^{+} =: \mathrm{GS}_{K_f}(M, D_{\mathfrak{t}}).$$
$$(12.4.3)$$

The right-hand object is a **Griffiths–Schmid variety**, which is in general only complex analytic. The construction of vector bundles $\mathcal{V}_{K_f}(\pi)$ mentioned in Remark 12.4.13(b) extends to this setting. $\mathrm{GS}_{K_f}(M, D_{\mathfrak{t}})$ is algebraic if and only if $D_{\mathfrak{t}}^{+}$ fibers holomorphically or antiholomorphically over a Hermitian symmetric domain [GRT2013]. It is considered to be a Shimura variety under the slightly more stringent condition that the IPR be trivial.

(ii) Continuing in this more general setting, let $\alpha \in M(\mathbb{Q})$ and write $K_f \alpha K_f = \overset{\text{finite}}{\amalg_i} K_f a_i$ for some $a_i \in M(\mathbb{A}_f)$. This gives rise to an analytic correspondence

$$\sum_i \left\{ \left([m_\infty, m_f], [m_\infty, m_f a_i^{-1}]\right) \right\}_{[m_\infty, m_f] \in M(\mathbb{A})}$$

in the self-product of the left-hand side of (12.4.3). The endomorphism induced in automorphic cohomology groups

$$\mathfrak{A}_i(\pi) := H^i\left(\mathrm{GS}_{K_f}(M, D_{\mathfrak{t}}), \mathcal{O}(\mathcal{V}_{K_f}(\pi))\right)$$

is called the **Hecke operator** associated to α.

(iii) For simplicity, assume that the representations π are one-dimensional (so that the $\mathcal{V}_{K_f}(\pi)$ are line bundles) and that the Γ_g are cocompact.[8] In the Shimura variety case, the eigenvectors of Hecke operators in $\mathfrak{A}_0(\pi)$ are the arithmetically interesting automorphic forms. In the nonalgebraic Griffiths–Schmid case, provided π is "regular," we have $\mathfrak{A}_0(\pi) = \{0\}$; however, the automorphic cohomology $\mathfrak{A}_i(\pi)$ will typically be nonzero for some $i > 0$. It does not seem too far fetched to hope that the Hecke eigenvectors in *this* group might hold some yet-to-be-discovered arithmetic significance. Recent work of Carayol [Carayol2005] (in the *non*-co-compact case) seems particularly promising in this regard.

---

[8]Similar remarks apply in the non-co-compact case, except that one has to restrict to a subgroup of cuspidal classes in $\mathfrak{A}_i(\pi)$, whose behavior for $i > 0$ is not yet well understood.

12.5   FIELDS OF DEFINITION

Consider a period domain D for Hodge structures (on a fixed $\mathbb{Q}$-vector space V, polarized by a fixed bilinear form Q) with fixed Hodge numbers. The compact dual of a Mumford–Tate domain $D_M = M(\mathbb{R}).\tilde{\varphi} \subseteq D$ is the $M(\mathbb{C})$-orbit of the attached filtration $F_{\tilde{\varphi}}^\bullet$,

$$\check{D}_M = M(\mathbb{C}).F_{\tilde{\varphi}}^\bullet.$$

It is a connected component of the "MT Noether–Lefschetz locus" cut out of $\check{D}$ by the criterion of (a Hodge flag in $\check{D}$) having MT group contained in M,

$$\check{D}_M \subseteq \check{N}L_M \subseteq \check{D}.$$

Now, $\check{N}L_M$ is cut out by $\mathbb{Q}$ tensors hence defined over $\mathbb{Q}$, but its components[9] are permuted by the action of $\mathrm{Aut}(\mathbb{C})$. The fixed field of the subgroup of $\mathrm{Aut}(\mathbb{C})$ preserving $\check{D}_M$, is considered its field of definition; this is defined regardless of the vanishing of the IPR (or D_M being Hermitian symmetric). What is interesting in the Shimura variety case, is that this field has meaning "downstairs," for $\Gamma \backslash D_M$—even though the upstairs–downstairs correspondence is highly transcendental.

A. Reflex Field of a Shimura Datum

Let $(G, \tilde{X})$ be a Shimura datum; we start by repeating the definition just alluded to in this context. Recall that any $\tilde{\varphi} \in \tilde{X}$ determines a complex cocharacter of G by $z \mapsto \tilde{\varphi}_{\mathbb{C}}(z, 1) =: \mu_{\tilde{\varphi}}(z)$. (Complex cocharacters are themselves more general and essentially correspond to points of the compact dual.) Write, for any subfield $k \subset \mathbb{C}$,

$$\mathfrak{C}(k) := G(k) \backslash \mathrm{Hom}_k(\mathbb{G}_m, G_k)$$

for the *set* of $G(k)$-conjugacy classes of k-cocharacters. The Galois group $\mathrm{Gal}(k/\mathbb{Q})$ acts on $\mathfrak{C}(k)$, since $\mathbb{G}_m$ and G are $\mathbb{Q}$-algebraic groups.

The element

$$c(\tilde{X}) := [\mu_{\tilde{\varphi}}] \in \mathfrak{C}(\mathbb{C})$$

is independent of the choice of $\tilde{\varphi} \in \tilde{X}$.

DEFINITION 12.5.1 $E(G, \tilde{X})$ *is the fixed field of the subgroup of* $\mathrm{Aut}(\mathbb{C})$ *fixing* $c(\tilde{X})$ *as an element of* $\mathfrak{C}(\mathbb{C})$.

EXAMPLES 12.5.2   (a) Given A an abelian variety of CM-type (E, Φ), E' the associated reflex field (cf. Section 12.3.A), $\tilde{\varphi}$ the Hodge structure on $H^1(A)$, and

$$T = M_{\tilde{\varphi}} \subseteq \mathrm{Res}_{E/\mathbb{Q}} \mathbb{G}_m$$

the associated MT group. Then $\mu_{\tilde{\varphi}}(z)$ multiplies $H^{1,0}(A)$ (the Φ-eigenspaces for E) by z, and $H^{0,1}(A)$ (the $\bar{\Phi}$-eigenspaces for E) by 1. Clearly $\mathrm{Ad}(\sigma)\mu_{\tilde{\varphi}}(z)$

---

[9]which are $M(\mathbb{C})$-orbits, as M is (absolutely) connected; note that this does not mean $M(\mathbb{R})$ is connected.

(for $\sigma \in \text{Aut}(\mathbb{C})$) multiplies the $\sigma\Phi$-eigenspaces by z, while $T(\mathbb{C})$ acts trivially on $\text{Hom}_{\mathbb{C}}(\mathbb{G}_m, T_{\mathbb{C}})$. Consequently, σ fixes $c(\{\tilde{\varphi}\})$ if and only if σ fixes Φ, and so

$$E(T, \{\tilde{\varphi}\}) = E'.$$

(b) For an inclusion $(G', \tilde{X}') \hookrightarrow (G, \tilde{X})$, one has $E(G', \tilde{X}') \supseteq E(G, \tilde{X})$. Every $\tilde{X}$ has $\tilde{\varphi}$ factoring through rational tori, which are then CM-Hodge structures. Each torus arising in this manner splits over a CM-field, and so $E(G, \tilde{X})$ is always contained in a CM-field. (In fact, it is always either CM or totally real.)

(c) In the Siegel case, $E(G, \tilde{X}) = \mathbb{Q}$; while for a PEL Shimura datum, we have that $E(G, \tilde{X}) = \mathbb{Q}\left(\{\text{tr}(b|_{T_0 A})\}_{b \in B}\right)$.

Let T be a $\mathbb{Q}$-algebraic torus, μ a cocharacter defined over a finite extension $K/\mathbb{Q}$. Denote by

$$r(T, \mu) : \text{Res}_{K/\mathbb{Q}} \mathbb{G}_m \to T$$

the homomorphism given on rational points by

$$K^* \to T(\mathbb{Q}),$$

$$k \mapsto \prod_{\phi \in \text{Hom}(K, \bar{\mathbb{Q}})} \phi(\mu(k)).$$

As in Example 12.5.2(b), every $(G, \tilde{X})$ contains a CM-pair $(T, \{\tilde{\varphi}\})$. The field of definition of $\mu_{\tilde{\varphi}}$ and the reflex field $E(T, \{\tilde{\varphi}\})$ are the same; denote this by $E(\tilde{\varphi})$. The map

$$r(T, \mu_{\tilde{\varphi}}) : \text{Res}_{E(\tilde{\varphi})/\mathbb{Q}} \mathbb{G}_m \to T$$

yields on $\mathbb{A}_{\mathbb{Q}}$-points

$$\mathbb{A}_{E(\tilde{\varphi})}^{\times} \xrightarrow{r(T, \mu_{\tilde{\varphi}})} T(\mathbb{A}_{\mathbb{Q}}) \xrightarrow{\text{project}} T(\mathbb{A}_f).$$
$$\underbrace{\qquad\qquad\qquad\qquad}_{=: r_{\tilde{\varphi}}}$$

EXAMPLE 12.5.3 In Example 12.5.2(a) above, we have $E(\tilde{\varphi}) = E'$; assume for simplicity that $T = \text{Res}_{E/\mathbb{Q}} \mathbb{G}_m$. A nontrivial computation shows that the $r(T, \mu_{\tilde{\varphi}})$ part of this map is the adélicized reflex norm

$$\mathcal{N}_{\Phi'}(\mathbb{A}_{\mathbb{Q}}) : \mathbb{A}_{E'}^{\times} \to \mathbb{A}_E^{\times}.$$

B. Canonical Models

The Shimura varieties we have been discussing—i.e., $\text{Sh}_{K_f}(G, \tilde{X})$—are finite disjoint unions of locally symmetric varieties, and hence algebraic varieties defined a priori over $\mathbb{C}$. More generally, if $\mathcal{Y}$ is any complex algebraic variety, and $k \subset \mathbb{C}$ is a subfield, a **model of $\mathcal{Y}$ over k** is

- a variety $\mathcal{Y}_0$ over k; together with

- an isomorphism $\mathcal{Y}_{0,\mathbb{C}} \overset{\theta}{\underset{\cong}{\to}} \mathcal{Y}$.

For general algebraic varieties, it is *not* true that two models over the same field k are necessarily isomorphic *over that field*. But if we impose a condition on how $\mathrm{Gal}(\mathbb{C}/E_1')$ acts on a dense set of points on any model, then the composite isomorphism

$$\mathcal{Y}_{0,\mathbb{C}} \overset{\theta}{\underset{\cong}{\to}} \mathcal{Y} \overset{\tilde\theta}{\underset{\cong}{\to}} \widetilde{\mathcal{Y}}_{0,\mathbb{C}}$$

is forced to be $\mathrm{Gal}(\mathbb{C}/E_1')$-equivariant, making $\mathcal{Y}_0$ and $\widetilde{\mathcal{Y}}_0$ isomorphic over E_1'. Repeating this criterion for more point sets and number fields E_i', forces $\mathcal{Y}_0$ and $\widetilde{\mathcal{Y}}_0$ to be isomorphic over $\cap_i E_i'$.

To produce the dense sets of points we need the following definition:

DEFINITION 12.5.4   (a) *A point $\tilde\varphi \in \tilde X$ is a **CM-point**, if there exists a (minimal, $\mathbb{Q}$-algebraic) torus $\mathbb{T} \subset G$ such that $\tilde\varphi(\mathbb{S}(\mathbb{R})) \subset \mathbb{T}(\mathbb{R})$.*

(b) *$(\mathbb{T}, \{\tilde\varphi\})$ is then a **CM-pair** in $(G, \tilde X)$.*

REMARK 12.5.5   Such a $\tilde\varphi$ exists since in a $\mathbb{Q}$-algebraic group every conjugacy class of maximal real tori contains one defined over $\mathbb{Q}$. To get density in $\tilde X$, look at the orbit $G(\mathbb{Q}).\tilde\varphi$. To get density, more importantly, in $\mathrm{Sh}_{K_f}(G, \tilde X)$, look at the set $\{[(\tilde\varphi, a)]\}_{a \in G(\mathbb{A}_f)}$.

To produce the condition on Galois action, recall for any CM-field E' the Artin reciprocity map

$$\mathrm{art}_{E'} : \mathbb{A}_{E'}^\times \twoheadrightarrow \mathrm{Gal}\left((E')^{\mathrm{ab}}/E'\right).$$

DEFINITION 12.5.6   *A model $\left(M_{K_f}(G, \tilde X), \theta\right)$ of $\mathrm{Sh}_{K_f}(G, \tilde X)$ over $E(G, \tilde X)$ is **canonical**, if for every*

- *$CM\ (\mathbb{T}, \tilde\varphi) \subset (G, \tilde X)$;*

- *$a \in G(\mathbb{A}_f)$;*

- *$\sigma \in \mathrm{Gal}\left(E(\tilde\varphi)^{\mathrm{ab}}/E(\tilde\varphi)\right)$;*

- *$s \in \mathrm{art}_{E(\tilde\varphi)}^{-1}(\sigma) \subset \mathbb{A}_{E(\tilde\varphi)}^\times$;*

$\theta^{-1}[(\tilde\varphi, a)]$ is a point defined over $E(\tilde\varphi)^{\mathrm{ab}}$, and

$$\sigma.\theta^{-1}[(\tilde\varphi, a)] = \theta^{-1}.[(\tilde\varphi, r_{\tilde\varphi}(s)a)]. \qquad (12.5.1)$$

REMARK 12.5.7   In (12.5.1), the $r_{\tilde\varphi}$ is essentially a reflex norm.

The *uniqueness* of the canonical model is clear from the argument above—if one exists—since we can take the E'_i to be various $E(\tilde{\varphi})$ for CM $\tilde{\varphi}$, whose intersections are known to give $E(G, \tilde{X})$.

The *existence* of canonical models is known for all Shimura varieties by work of Deligne, Shih, Milne, and Borovoi. To see how it might come about for Shimura varieties of Hodge type, first note that by

- Baily–Borel

$\mathcal{Y} := \mathrm{Sh}_K(G, \tilde{X})$ is a variety over $\mathbb{C}$. Now we know that

- Sh_K is a moduli space for certain abelian varieties,

say $\mathcal{A} \to \mathcal{Y}$. Let $E = E(G, \tilde{X})$ and $\sigma \in \mathrm{Aut}(\mathbb{C}/E)$. Given $P \in \mathcal{Y}(\mathbb{C})$ we have an equivalence class $[A_P]$ of abelian varieties, and we define a map

$$({}^\sigma\mathcal{Y})(\mathbb{C}) \to \mathcal{Y}(\mathbb{C})$$

by

$$\sigma(P) \longmapsto [{}^\sigma A_P].$$

That ${}^\sigma A_P$ is still "in the family $\mathcal{A}$" follows from

- the definition of the reflex field

and

- **Deligne's theorem** (cf. [Deligne1982], or Chapter 11) that the Hodge tensors determining $\mathcal{A}$ are absolute.

That these maps produce regular (iso)morphisms

$$f_\sigma : {}^\sigma\mathcal{Y} \to \mathcal{Y}$$

boils down to

- Borel's theorem (Section 12.2).

Now $\mathcal{Y}$ has (for free) a model $\mathcal{Y}_0$ over some finitely generated extension L of E, and using

- $|\mathrm{Aut}(\mathcal{Y})| < \infty$ (cf. [Milne2005], Thm. 3.21)

we may deduce that for σ' fixing L

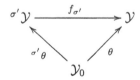

commutes. At this point it makes sense to spread $\mathcal{Y}_0$ out over E—i.e., take all $\mathrm{Gal}(\mathbb{C}/E)$-conjugates, viewed as a variety via

$$\mathcal{Y}_0 \longrightarrow \mathrm{Spec}\, L \longrightarrow \mathrm{Spec}\, E.$$

The diagram

$$
\begin{array}{ccc}
{}^\sigma\mathcal{Y} & \xrightarrow[\cong]{f_\sigma} & \mathcal{Y} \\
{}^\sigma\theta \big\uparrow \cong & & \cong \big\uparrow \theta \\
{}^\sigma\mathcal{Y}_0 & & \mathcal{Y}_0
\end{array}
$$

shows that the spread is constant; extending it over a quasi-projective base shows that $\mathcal{Y}$ has a model defined over a *finite* extension of E. (To get all the way down to E requires some serious descent theory.) Finally, that the action of $\mathrm{Aut}(\mathbb{C}/E)$ on the resulting model implied by the $\{f_\sigma\}$ satisfies (12.5.1) (hence yields a *canonical* model), is true by

- the main theorem of CM.

In fact, (12.5.1) is precisely encoding how Galois conjugation acts on various $\mathrm{Ab}(\mathcal{O}_E, \Phi)$ together with the level structure.

So the three key points are

1. the *entire* theory is used in the construction of canonical models;

2. $\mathrm{Sh}_{K_f}(G, \tilde{X})$ is defined over $E(G, \tilde{X})$ independently of K_f; and

3. the field of definition of a component $\mathrm{Sh}_{K_f}(G, \tilde{X})^+$ lies inside $E(G, \tilde{X})^{\mathrm{ab}}$ and gets larger as K_f shrinks (and the number of connected components increases).

C. Connected Components and VHS

Assume G^{der} is simply connected. The action on CM-points imposed by (12.5.1) turns out to force the following action on

$$\pi_0\left(\mathrm{Sh}_{K_f}(G, \tilde{X})\right) \cong T(\mathbb{Q})\backslash Y \times T(\mathbb{A}_f)/\nu(K_f),$$

where $G \xrightarrow{\nu} T$ is the maximal abelian quotient. For *any* $\tilde{\varphi} \in \tilde{X}$, put

$$r = r(T, \nu \circ \mu_{\tilde{\varphi}}) : \mathbb{A}^\times_{E(G,\tilde{X})} \to T(\mathbb{A}_\mathbb{Q}).$$

Then for $\sigma \in \mathrm{Gal}\left(E(G, \tilde{X})^{\mathrm{ab}}/E(G, \tilde{X})\right)$ and $s \in \mathrm{art}^{-1}_{E(G,\tilde{X})}(\sigma)$ one has

$$\sigma.[y, a] = [r(s)_\infty.y, \, r(s)_f \cdot a].$$

Assume for simplicity Y is trivial. Let $\mathbb{E}$ denote the field of definition of the component $\mathcal{S} := \mathrm{Sh}_{K_f}(G, \tilde{X})^+$ over $[(1,1)]$, which is a finite abelian extension of $E' := E(G, \tilde{X})$. From the above description of the Galois action, we get

$$\mathbb{E} = \text{fixed field of } \mathrm{art}_{E'} \left(r_f^{-1}\left(T(\mathbb{Q}) \cdot \nu(K_f) \right) \right).$$

That is, by virtue of the theory of canonical models we can essentially write down a minimal field of definition of the locally symmetric variety $\mathcal{S}$.

EXAMPLE 12.5.8 Take the Shimura datum $(G, \tilde{X}) = (\mathbb{T}, \{\tilde{\varphi}\})$ associated to an abelian variety with CM by E, so that E' is the reflex field (and $\mathbb{E}$ the field of definition of the point it lies over in a relevant Siegel modular variety). Let $K_f = \mathfrak{U}_I$ for $I \in \mathcal{I}(E)$ and consider the diagram

$$
\begin{array}{ccccc}
\mathbb{A}_{E',f}^{\times} & \xrightarrow{\;\;\mathcal{N}_{\Phi'}(=r_f)\;\;} & \mathbb{A}_{E,f}^{\times} & \longrightarrow & E^{\times}\backslash\mathbb{A}_{E,f}^{\times}/\mathfrak{U}_I \\
\downarrow{\scriptstyle\mathrm{art}_{E'}} & & \downarrow{\scriptstyle\mathrm{art}_E} & & \cong \;\big\downarrow{\scriptstyle\overline{\mathrm{art}_E}} \\
\mathrm{Gal}\left((E')^{\mathrm{ab}}/E'\right) & \xrightarrow{\;\;\mathfrak{N}_{\Phi'}\;\;} & \mathrm{Gal}\left(E^{\mathrm{ab}}/E\right) & \longrightarrow & \mathrm{Gal}\left(E_I/E\right),
\end{array}
$$

where $\mathfrak{N}_{\Phi'}$ exists (and is continuous) in such a way that the left-hand square commutes. Writing "FF" for fixed field, we get that

$$\mathbb{E} = \mathrm{FF}\left(\mathrm{art}_{E'}\left(\mathcal{N}_{\Phi'}^{-1}(E^{\times}\mathfrak{U}_I) \right) \right) = \mathrm{FF}\left(\mathfrak{N}_{\Phi'}^{-1}\left(\mathrm{Gal}(E^{\mathrm{ab}}/E_I) \right) \right).$$

In the case that the CM-abelian variety is an elliptic curve, $\mathcal{N}_{\Phi'}$ and $\mathfrak{N}_{\Phi'}$ are essentially the identity (and $E' = E$), so

$$\mathbb{E} = \mathrm{FF}\left(\mathrm{Gal}(E^{\mathrm{ab}}/E_I) \right) = E_I.$$

It is a well-known result that, for example, the j-invariant of a CM-elliptic curve generates (over the imaginary quadratic field E) its Hilbert class field $E_{(1)}$. We also see that the fields of definition of CM-points in the modular curve $X(N)$ are ray class fields modulo N.

We conclude by describing a possible application to variations of Hodge structure. Let $\mathcal{V} \to \mathcal{S}$ be a VHS with reference Hodge structure V_s over $s \in \mathcal{S}$. The underlying local system $\mathbb{V}$ produces a monodromy representation

$$\rho : \pi_1(\mathcal{S}) \to \mathrm{GL}(V_s),$$

and we denote $\rho(\pi_1(\mathcal{S})) =: \Gamma_0$ with **geometric monodromy group**

$$\Pi := \text{identity component of } \mathbb{Q}\text{-Zariski closure of } \Gamma_0.$$

Moreover, $\mathcal{V}$ has an MT group M, and we make the following two crucial assumptions:

- $\Pi = M^{\mathrm{der}}$; and

- D_M has vanishing IPR.

In particular, this means that the quotient of D_M by a congruence subgroup is a connected component of a Shimura variety, and that Π is as big as it can be.

For any compact open $K_f \subseteq M(\mathbb{A}_f)$ such that $\Gamma := K_f \cap M(\mathbb{Q}) \supseteq \Gamma_0$, $\mathcal{V}$ gives a period (analytic) mapping

$$\Psi^{an}_{K_f} : \mathcal{S}^{an}_{\mathbb{C}} \to \Gamma \backslash D_M \cong \left(\mathrm{Sh}_{K_f}(M, D_M)^+ \otimes_{\mathbb{E}} \mathbb{C} \right)^{an}.$$

This morphism is algebraic by Borel's theorem, and its minimal field of definition is (trivially) bounded below by the field of definition $\mathbb{E}$ of $\mathrm{Sh}_{K_f}(M, D_M)^+$, which henceforth we shall denote $\mathbb{E}(K_f)$.

The period mapping which gives the most information about $\mathcal{V}$, is the one attached to the *smallest* congruence subgroup $\Gamma \subset M(\mathbb{Q})$ containing Γ_0. Taking then the *largest* K_f with $K_f \cap M(\mathbb{Q})$ equal to this Γ, minimizes the resulting $\mathbb{E}(K_f)$. It is this last field which it seems natural to consider as the **reflex field of a VHS**—an "expected *lower* bound" for the field of definition of a period mapping of $\mathcal{V}$. Furthermore, if $\mathcal{V}$ arises (motivically) from $\mathcal{X} \xrightarrow{\pi} \mathcal{S}$, then assuming Deligne's absolute Hodge conjecture (cf. [Deligne1982] or Chapter 11), the $\bar{\mathbb{Q}}$-spread of π produces a period mapping into D_M modulo a *larger* Γ, and our "reflex field of $\mathcal{V}$" may be an *upper bound* for the minimal field of definition of *this* period map.

At any rate, the relations between fields of definition of

- varieties X_s;

- transcendental period points in D_M; and

- equivalence classes of period points in $\Gamma \backslash D_M$;

and hence between spreads of

- families of varieties;

- variations of Hodge structure; and

- period mappings

is very rich. Our suggested definition may be just one useful tool for investigating the case where the period map is into a Shimura subdomain of a period domain.

Bibliography

[AMRT1975]   A. Ash, D. Mumford, M. Rapoport, and Y. Tai, "Smooth compacti-
              fication of locally symmetric varieties", Math. Sci. Press, Brookline,
              MA, 1975.

[BB1966]   W. Baily and A. Borel, *Compactification of arithmetic quotients of
              bounded symmetric domains*, Ann. of Math (2) 84, 442–528.

[Borel1972]   A. Borel, *Some metric properties of arithmetic quotients of symmetric
              spaces and an extension theorem*, J. Diff. Geom. 6 (1972), 543–560.

[Borel1991]   ———, "Linear algebraic groups", Graduate Texts in Math. 126,
              Springer, New York, 1991.

[BS1973]   A. Borel and J.-P. Serre, *Corners and arithmetic groups*, Comm.
              Math. Helv. 4 (1973), 436–491.

[BP2009]   P. Brosnan and G. Pearlstein, *On the algebraicity of the zero locus of
              an admissible normal function*, arXiv:0910.0628, to appear in Com-
              pos. Math.

[Carayol2005] H. Carayol, *Cohomologie automorphe et compactifications par-
              tielles de certaines variétés de Griffiths–Schmid*, Compos. Math. 141
              (2005), 1081–1102.

[CCK1980]   J. Carlson, E. Cattani, and A. Kaplan, *Mixed Hodge structures and
              compactifications of Siegel's space*, in "Algebraic Geometry Angers,
              1979" (A. Beauville, Ed.), Sijthoff & Noordhoff, 1980, 77–105.

[CDK1995]   E. Cattani, P. Deligne, and A. Kaplan, *On the locus of Hodge classes*,
              J. Amer. Math. Soc. 8 (1995), 483–506.

[Deligne1979] P. Deligne, *Variétés de Shimura: interprétation modulaire, et tech-
              niques de construction de modéles canoniques*, in "Automorphic
              forms, representations and L-functions", Proc. Sympos. Pure Math.
              33, AMS, Providence, RI, 1979, 247–289.

[Deligne1982] ———, *Hodge cycles on abelian varieties (Notes by J. S. Milne)*,
              in "Hodge cycles, Motives, and Shimura varieties", Lect. Notes in
              Math. 900, Springer, New York, 1982, 9–100.

[GGK2010]   M. Green, P. Griffiths, and M. Kerr, "Mumford–Tate groups and do-
                mains: their geometry and arithmetic", Ann. Math. Stud., no. 183,
                Princeton University Press, 2012.

[GRT2013]   P. Griffiths, C. Robles, and D. Toledo, Quotients of non-classical flag
                domains are not algebraic, preprint, 2013.

[Holzapfel1995] R.-P. Holzapfel, "The ball and some Hilbert problems", Birkhäuser,
                Boston, 1995.

[KU2009]   K. Kato and S. Usui, "Classifying spaces of degenerating polarized
                Hodge structure", Ann. Math. Stud. 169, Princeton University Press,
                Princeton, NJ, 2009.

[Kodaira1954]   K. Kodaira, *On Kähler varieties of restricted type (an intrinsic char-
                acterization of algebraic varieties)*, Ann. of Math. 60 (1954), 28–48.

[Milne1990]   J. Milne, *Canonical models of (mixed) Shimura varieties and auto-
                morphic vector bundles*, in "Automorphic forms, Shimura varieties,
                and L-functions", Perspect. Math. 10, Acad. Press, 1990, 283–414.

[Milne2005]   ———, *Introduction to Shimura varieties*, in "Harmonic analysis,
                the trace formula and Shimura varieties" (Arthur and Kottwitz, Eds.),
                AMS, 2005.

[Milne2007]   ———, *The fundamental theorem of complex multiplication*,
                preprint, arXiv:0705.3446.

[Mumford1969]   D. Mumford, *A note of Shimura's paper "Discontinuous groups and
                abelian varieties"*, Math. Ann. 181 (1969), 345–351.

[Rotger2005]   V. Rotger, *Shimura varieties and their canonical models: notes on
                the course*, Centre de Recerca Mathemática, Bellatera, Spain, Sept.
                20–23, 2005.

[Serre1956]   J.-P. Serre, *Géométrie algébrique et géométrie analytique*, Annales
                de l'Institut Fourier 6 (1956), 1–42.

[Silverman1995] J. Silverman, "Advanced topics in the arithmetic of elliptic curves",
                Grad. Texts in Math. 151, Springer, New York, 1995.

Index

Milton Keynes UK
Ingram Content Group UK Ltd.
UKHW020959140124
435975UK00006BA/207